Cardiovascular Signaling in Health and Disease

Narasimham L. Parinandi • Thomas J. Hund
Editors

Cardiovascular Signaling in Health and Disease

Springer

Editors
Narasimham L. Parinandi
Division of Pulmonary, Critical Care, and Sleep Medicine
Department of Internal Medicine
Dorothy M. Davis Heart and Lung Research Institute, The Ohio State University Wexner Medical Center
Columbus, OH, USA

Thomas J. Hund
Departments of Biomedical Engineering and Internal Medicine
Dorothy M. Davis Heart and Lung Research Institute, The Ohio State University
Columbus, OH, USA

ISBN 978-3-031-08308-2 ISBN 978-3-031-08309-9 (eBook)
https://doi.org/10.1007/978-3-031-08309-9

This Springer imprint is published by the registered company Springer Nature Switzerland AG
The registered company address is: Gewerbestrasse 11, 6330 Cham, Switzerland

Preface on the Current Trends in Cardiovascular Signaling in Health and Disease

Cardiovascular diseases (CVDs) are major causes of morbidity, mortality, and heavy economic distress in the USA and worldwide. Prevention, treatment, and management of CVD remain daunting challenges for our healthcare system, which depends on advancements of scientific research in the field. It is becoming increasingly evident that signaling mechanisms at biochemical, molecular, and cellular levels in the blood vessels (vascular) and heart contribute to the underlying causes of development and progression of the CVDs. This book provides an overview of the state of the field regarding the cellular signaling mechanisms underlying the development and progression of life-threatening CVDs and is targeted at investigators and students interested in further discovery of efficient management and effective treatment of CVDs.

Calcium- and Stress-Dependent Signaling in Cardiac Myocytes

Calcium is arguably the most important second messenger in the heart, responsible for modulating a host of critical cell functions from contraction to cell death. Thus, the first part of our book begins with a chapter titled "Calcium-Dependent Signaling in Cardiac Myocytes" by Ko et al. addressing fundamental aspects of calcium signaling in cardiac myocytes with an overview of the key binding partners, molecular targets, and downstream effectors for calcium. The current understanding of how calcium dysregulation drives cardiac electrical and mechanical dysfunction in the setting of disease is addressed with a focus on novel, emerging concepts and potential therapeutic strategies. The discussion of calcium continues with a chapter titled "Organization of Ca2+ Signaling Microdomains in Cardiac Myocytes" addressing organization of microdomains for control of calcium signaling in cardiac myocytes. The importance of proper organization and structure for calcium microdomains is well established, but recent studies addressed in this chapter have delineated their biogenesis and maintenance, opening up new avenues for therapeutic intervention. Beyond calcium, cardiac myocytes have evolved an elaborate network of signaling nodes to coordinate their response to acute and chronic stress signals. In this regard, Ai and colleagues discuss the latest developments in our understanding of the pathophysiological roles for a very important class of stress-response serine/threonine kinases – the mitogen-activated protein kinase (MAPK) family. Protein tyrosine kinases (PTKs) constitute another class of important stress-dependent kinases with identified roles in insulin signaling, hypertrophy, and cell cycle regulation. PTKs also serve as targets for tyrosine kinase inhibitors (TKIs) that serve as common cancer drugs. In "Cardiotoxicity and Cardiac Cell Signaling", Scott and Smith summarize the growing body of literature linking TKIs to cardiac complications including electrical and mechanical dysfunction. Although much of the focus in the field tends to be on protein kinases with respect to posttranslational modification, protein phosphatases play an equally important role in modulating cardiac cell function. Abdullah et al. provide an overview on our current understanding of how protein phosphatases help balance protein phosphorylation to ensure proper regulation of critical cardiac functions, with a particular emphasis on their role in disease.

Reactive Oxygen Species and Lipid Signaling in Cardiac Myocytes

Reactive oxygen species (ROS) constitute a diverse family of oxidant molecules with molecular oxygen as a common precursor. While ROS support proper biological function throughout the body, they have been linked to a host of deleterious consequences through non-specific modification of proteins, lipids, and other molecules when produced in excess in cardiac myocytes. Mitochondria support energy production and are important sources for ROS in cardiac myocytes. In their chapter "Metabolic Regulation of Mitochondrial Dynamics and Cardiac Function," Akar and colleagues share insight into state of the field with respect to the dynamic nature of mitochondria and the implications for heart health and disease. Of course there are non-mitochondrial sources of ROS in myocytes, and among the most important are the plasma membrane NADPH oxidase system. Uppu and team contribute an update on a pathway linking NADPH oxidase, ROS production, and oxidized cholesterol species leading to altered mitochondrial function and apoptosis. There is growing appreciation for the physiological importance of cholesterol and other bioactive lipids beyond energy storage and organelle structure. Hernandez-Saavedra and Stanford delineate mounting data supporting a central function for bioactive lipids, in particular oxylipins, in regulating a host of cardiovascular functions in their chapter "Lipid Mediators in Cardiovascular Physiology and Disease."

Inflammatory Signaling, Fibrosis, and Cardiac Function

Dysregulation of the body's inflammatory response drives pathology across a wide range of cardiovascular diseases. Inflammasomes are multimeric protein complexes found in cells throughout the body, including in cardiac myocytes, which control the production of pro-inflammatory cytokines and amplify the inflammatory response to stress. In their chapter "Cardiac Inflammasome and Arrhythmia," Li and Dobrev lead the reader through recent findings about how the inflammasome contributes to development of arrhythmia, especially atrial fibrillation. There is growing appreciation in the field for bidirectional communication between inflammation and injury-induced cardiac remodeling including fibrosis. Graham and Sethu provide an update in "Myocardial Fibrosis: Cell Signaling and In Vitro Modeling" on the complex signaling involved in fibrosis with a focus on engineering and in vitro models for studying cell-cell communication.

Neural Regulation of Cardiac Rhythm

A treatment of cardiovascular signaling would not be complete without addressing the very important role that the autonomic nervous system plays in modulating cardiovascular function and disease. The author, Ripplinger, concludes the first part of our book with an evaluation of where the field stands in our understanding of signaling between the sympathetic and parasympathetic nervous systems and the heart. A detailed analysis of the key signaling pathways is provided along with an overview of state-of-the-art experimental models to facilitate dissection of the complicated feedback loops involved in neural regulation.

Reverse Cholesterol Transport in Atherosclerotic Cardiovascular Disease

In the chapter titled "Mechanisms of Lipoproteins and Reverse Cholesterol Transport in Atherosclerotic Cardiovascular Disease," Sucharski and Koenig comprehensively discuss reverse cholesterol transport in close association with lipoprotein mechanisms in atherosclerotic cardiovascular diseases (CVDs). Coronary heart disease (CHD) causes approximately 42% of all the cardiovascular disease-associated mortality in the USA. The *bad cholesterol* that is deposited in the arteries from low-density lipoprotein-associated cholesterol (LDL-C) has been known to increase the risk of CHD, atherosclerosis, myocardial infarction, and stroke. On the other hand, the elevated levels of high-density lipoprotein-associated cholesterol (HDL-C), the *good cholesterol*, have been recognized with the lower risk of CHD and known to play an important role in the reverse cholesterol transport (RCT) pathway. The RCT pathway is crucial and operates the cholesterol efflux from peripheral cells and tissues by HDL. SCARB1, ApoA-I, and ABCA1/ABCG1 variants are associated with atherosclerosis and coronary artery disease. Thus, understanding RCT mechanisms is of significant scientific interest. These aspects are discussed by Sucharski and Koenig in the chapter focusing on the lipoproteins, with a particular emphasis on the mechanisms of RCT, disease-associated variants, and current therapies.

Progression of the Atherosclerotic Plaque Regression

In the chapter titled "Atherosclerotic Plaque Regression: Future Perspective," Suseela et al. discuss in detail complex nature of the atherosclerotic plaque containing cholesterol, phospholipids, proteins, and their oxidatively modified species. The oxidized molecules in the plaque have been shown to undergo auto-oxidation generating lipid carbonyls and cyclized products, leading to the formation of complexes with proteins and plaque-destabilizing actions downstream. High-density lipoprotein cholesterol (HDL) is crucial for the reverse cholesterol transport (RCT). It has been shown that the HDL mimetics, drugs which elevate functional HDL, and dietary modifications can only cause 10–30% relief of the plaque burden. But, none of these strategies have been shown to scavenge or quench the oxidized lipids and/or oxidatively modified lipid species in the plaque. The authors emphasize that the molecules which scavenge and/or quench the lipid carbonyls can prevent formation of the carbonyl adducts leading to additional benefits. Furthermore, this chapter underscores the importance of improved plaque regression that could be possible with molecules capable of enhancing the functional HDL as well as scavenging the highly reactive lipid carbonyls.

Role of Bioactive Lipid, Phosphatidic Acid in Statin-Induced Myotoxicity

In the chapter titled "Role of Bioactive Lipid, Phosphatidic Acid in Hypercholesterolemia Drug-Induced Myotoxicity – Statin-Induced Phospholipase D (PLD) Lipid Signaling in Skeletal Muscle Cells," Tretter et al. present experimental findings and discuss the adverse actions of cholesterol-lowering drugs, statins, which lower cholesterol by inhibiting 3-hydroxy-3-methyl-glutaryl-CoA reductase (HMG-CoA reductase), the rate-limiting enzyme in the biosynthetic pathway of endogenous cholesterol. In spite of their efficacy for lowering endogenous cholesterol levels and offering cardiovascular protection in hypercholesterolemia patients, statins are known to cause skeletal muscle damage (myotoxicity and myalgia), but the mechanisms and treatment of statin-induced myotoxicity and myalgia are not established in detail. This chapter presents experimental findings on the involvement of phospholipase D (PLD) and PLD-mediated lipid signaling in the statin (mevastatin and simvastatin)-induced myotoxicity in C2C12 mouse skeletal muscle myoblast *in vitro*. This chapter emphasizes on the importance of endogenous cellular cholesterol depletion by statins, the PLD-mediated lipid signaling in statin-induced skeletal myocyte damage, and the importance of PLD inhibition in protecting against the statin-induced myotoxicity and myalgia in CVD patients who use statins to normalize high levels of endogenous cholesterol.

Cell-to-Cell Communication in the Vascular Endothelium

In the chapter titled "Cell-Cell Communication in the Vascular Endothelium," Ryan King et al. emphasize on the vasculature as a crucial organ responsible for the mass and energy transport throughout the body. The blood vessels consist of a layer of endothelial cells forming the interface with blood and pericytes modulating their function and the layers of smooth muscle cells which regulate the vascular tone. These operate and regulate the vascular function in response to biological and physiological cues in health and disease, which are critically dependent upon communication among the vascular cells that utilize a wide array of direct cell-to-cell, paracrine, and autocrine mechanisms. In this chapter, the authors review these mechanisms, focusing especially on the endothelial cells and the underlying structural and molecular mechanisms of regulation of the vascular cell-to-cell communication.

The Bioactive Phospholipid, Lysophosphatidic Acid Regulates Vascular Endothelial Barrier Integrity

In the chapter titled "Lysophosphatidic Acid Regulates Endothelial Barrier Integrity," Zhao et al. discuss the critical function of vasculature which operates as a vessel network for blood circulation between lungs and other organs. The authors emphasize on the endothelium as a major component of blood vessels, which forms the inner lining of the blood vessels, playing a central role in maintenance of the blood vessel integrity. The authors further highlight the important role of endothelial barrier that prevents vascular leak of the blood components into perivascular areas. Elevated vascular permeability and leak will cause tissue edema responsible for the acute inflammatory diseases. The authors identify the bioactive phospholipid, lysophosphatidic acid (LPA) that has several physiological and pathophysiological actions in a wide variety of cell types including the vascular endothelial cells (ECs). The authors in this chapter discuss on the LPA-mediated regulation of EC barrier property and present knowledge on LPA actions on the endothelial barrier function.

Role of Lipid Mediators in Regulation of Vascular Endothelial Barrier Integrity and Function

Fu et al., in the chapter titled "Regulation of Vascular Endothelial Barrier Integrity and Function by Lipid-Derived Mediators," provide a comprehensive and state-of-the-art discussion on the role of lipid-derived mediators in vascular endothelial barrier structural integrity and function. The authors have identified that the maintenance of endothelial cell (EC) integrity is critical for the vascular permeability and inflammation encountered among a plethora of pulmonary disorders and disease such as sepsis, ventilator-induced lung injury, and microbial infections. The authors present that the disruption of EC tight and adherens junctions lead to elevated vascular permeability, alveolar flooding, and pulmonary edema, whereas there is ample evidence which discloses that the vascular ECs are capable of annealing the junctions, leading to the restoration of the barrier function. In this regard, the authors have highlighted that the phenomenon of barrier restoration is assisted by naturally occurring barrier-enhancing lipids such as sphingosine-1-phosphate, prostaglandins, and oxidized phospholipids. The authors underscore the importance of an in-depth understanding of mechanisms of regulation of the vascular endothelial barrier restoration and stabilization which will open up novel therapies for vascular disorders.

Role of Iron in Diabetic Vascular Endothelial Dysfunction

In the chapter titled "Hyperglycemic Oxoaldehyde (Glyoxal)-Induced Vascular Endothelial Cell Damage Through Oxidative Stress Is Protected by Thiol Iron Chelator, Dimercaptosuccinic Acid – Role of Iron in Diabetic Vascular Endothelial Dysfunction," Gurney et al. discuss that in diabetic patients, the vascular endothelium is prone to damage mediated by the glucose-derived oxoaldehydes including methylglyoxal and glyoxal. Along these lines, the authors present evidence on glyoxal-induced the cytotoxicity, cytoskeletal alterations, and barrier dysfunction through the reactive oxygen species (ROS)-induced oxidative stress involving intracellular iron (Fe) and have shown protection of the thiol heavy metal chelator, dimercaptosuccinic acid (DMSA) against the glyoxal-induced damage of the vascular endothelial cells (ECs). Thus, the authors have discussed with experimental evidences on DMSA protection against the vascular EC damage caused by the hyperglycemic oxoaldehyde AGE precursor through the action of iron and oxidative stress that culminates into diabetic vascular endothelial dysfunction.

Columbus, OH, USA

Narasimham L. Parinandi
Thomas J. Hund

Acknowledgments

We are extremely grateful to the Hund family, Keila Rachael, Arthur Lev, and Gabriel Eitan and Parinandi family, Nagamani, Gurunadh, Srinivas Chidambaram, Carrie Ann, Grier Kiran, and Graley Karthik for their support in successful completion of this book.

We express our deepest gratitude to our contributors of chapters and the reviewers who have made this book possible.

We extend our special thanks to Mr. Patrick J. Oliver for his untiring support during the editorial process.

Narasimham L. Parinandi
Thomas J. Hund
Editors

Contents

Authors Biography

Narasimham L. Parinandi (pAri), PhD, is an associate professor in Department of Internal Medicine, The Ohio State University College of Medicine. Parinandi received his BSc (Hons) in botany with chemistry, zoology, and English and MSc in botany with environmental biology from Berhampur University, India, in 1975–77. From 1977 to 1980, he was a research fellow in environmental sciences at Andhra University, India. He earned his PhD (1986) from the University of Toledo, Toledo, OH, in biology and toxicology under the tutelage of Prof. Woon H. Jyung, an established zinc metabolism expert and aging biologist. During his graduate training at Toledo, he was exposed to the field of lipids by Prof. Max Funk, an expert lipoxygenase enzymologist from the lineage of Prof. Ned Porter. He did his postdoctoral fellowship (1986–90) at the Hormel Institute, University of Minnesota, the premier lipid institute in the USA, where he trained with Prof. Harald Schmid, a celebrity in the area of ether lipids and a pioneer in anandamide chemistry. At the Hormel Institute, University of Minnesota, Parinandi was associated with Prof. Ralph T. Holman (member of the National Academy of Sciences and pioneer in fatty acid and lipoxygenase biochemistry, who also coined the name *omega-3 fatty acid*) and conducted studies on omega-3 fatty acid dynamics in humans. He was also a research scientist/junior faculty at the Johns Hopkins University School of Medicine (1998–2002) under the mentorship of Prof. V. Natarajan, renowned lipid signaling expert, and Prof. Joe G.N. (Skip) Garcia, a celebrated lung vascular biologist. Parinandi has published nearly 125 peer-reviewed original scientific papers, reviews, and book chapters, and edited books on free radicals and antioxidant protocols with Prof. William Pryor, the legendary free radical and lipid peroxidation scientist; a book titled *Mitochondria in Lung Health and Disease* with Prof. Viswanathan Natarajan, a lipid signaling celebrity; and on measuring oxidants and oxidative stress with Prof. Lawrence J. Berliner, a celebrity in the field of free radical chemistry and biology. Parinandi has given more than 50 invited scientific lectures at the national level in the USA and at international institutions and conferences. He has also conducted and chaired several scientific conferences and symposia in the areas of oxidative stress and lipidology. He has teaching and mentoring experience of more than 30 years and mentored over 75 students, technicians, fellows, and junior faculty in his laboratory. He served as an editor of the Chemical Abstracts of the American Chemical Society. He has been a reviewer of over 70 peer-reviewed journals in the area of biochemistry, molecular biology, cell biology, and lipidomics. Parinandi has been on the editorial board of the *Molecular Biology Reports* (Springer), *Frontiers of Pharmacology*, *World Journal of GI Pharmacology*, *Current Chemical Research*, *Cell Biophysics and Biochemistry* (associate editor), and *The Protein Journal*. He has also received extramural funding from the National Institutes of Health (NIH), Department of Defense (DOD), American Thoracic Society (ATS), and International Academy of Oral Medicine and Toxicology (IAOMT) as a principal investigator (PI) and co-investigator (Co-I). Parinandi also serves as a reviewer of grant proposals of the NIH, AHA, DOD, US universities, Government of Israel, Government of Austria, and Government of South Africa. Parinandi has received awards including the gold medal for securing the highest GPA in the MS class of 1975–77 at Berhampur University, India; the Outstanding Teaching Assistant Award of the Biology Department at the University of Toledo in 1986; Distinguished Mentor Award of the Davis Heart & Lung Research Institute at The Ohio State University Wexner Medical Center in 2008, and the Distinguished Undergraduate Mentor Award of the Ohio State Undergraduate Research Program in 2009.

Thomas J. Hund, PhD, is Professor of Biomedical Engineering and Internal Medicine at The Ohio State University and a Fellow of the American Heart Association. He also serves as director and the William D. And Jacquelyn L. Wells Chair at the Dorothy M. Davis Heart and Lung Research Institute in the OSU Wexner Medical Center. Research in the Hund lab has defined novel molecular pathways for local control of cardiac ion channel activity with important implications for human cardiac arrhythmia and heart failure. His group has developed novel computational models and tools to study cardiac electrophysiology and arrhythmia that are routinely used by labs around the world. His approach is distinguished by a highly interdisciplinary style combining state-of-the-art computational and experimental techniques. He has published more than100 peer-reviewed articles, and although he has contributed several book chapters in the past, *Cardiovascular Signaling in Health and Disease* represents his first volume as co-editor. In his role as director, Dr. Hund oversees strategic planning and operations for one of the largest interdisciplinary institutes in the country with more than 700 faculty, staff, and trainees dedicated to the integrated study of heart and lung disease. In addition to his research and leadership achievements, Dr. Hund has been recognized for his dedication to promoting undergraduate and graduate education through curriculum development, didactic teaching, and mentoring. He has mentored dozens of students, postdocs, and fellows, the large majority of whom go on to successful scientific careers in industry or academia. Pre- and postdoctoral trainees in the Hund lab have been very successful in securing independent fellowship awards, including NIH K99/R00 "Pathway to Independence" Awards, NIH Ruth L. Kirchstein National Research Service Awards, and Pre- and Postdoctoral Fellowship Awards from the American Heart Association. Dr. Hund has also regularly taught several courses in the Department of Biomedical Engineering, including a popular graduate course on excitable cell engineering. Before joining Ohio State in July 2011, Dr. Hund was Assistant Professor of Internal Medicine and Biomedical Engineering at the University of Iowa. He received his BSE in biomedical engineering from Duke University and his MS and PhD in biomedical engineering from Case Western Reserve University, followed by postdoctoral training at Washington University School of Medicine in St. Louis.

Part I
Cardiac Signaling

Calcium-Dependent Signaling in Cardiac Myocytes

Christopher Y. Ko, Charlotte E. R. Smith, and Eleonora Grandi

Abstract Calcium (Ca) is a key regulator of cardiac function. Through interactions with various molecular binding partners, Ca controls both acute processes, such as ion channel gating and myofilament contraction, and long-term events such as transcriptional changes that regulate myocardial development, growth, and death. Cardiac myocyte Ca levels are modulated by complex networks of signaling mechanisms and precise subcellular structural organization that fine-tune the myocyte response to any given stimulus and allow for rhythmic contraction. On the other hand, disrupted Ca handling and Ca signaling abnormalities are well-established mediators of contractile dysfunction and transmembrane potential instabilities leading to arrhythmia. In this chapter, we discuss the most recent advances in understanding the complexities of Ca signaling in health and widespread cardiac disease, namely, heart failure and arrhythmia. We specifically focus on novel emerging aspects of Ca/calmodulin-dependent protein kinase II signaling and on ultrastructural changes that have been associated with these disease contexts. Unraveling these spatial and temporal aspects of Ca signaling is key to understanding the profound mechanistic consequences of Ca dysregulation for cardiac myocyte and organ function and imperative to inform future therapies that might improve disease outcomes.

Keywords Arrhythmia · Calcium · CaMKII · Heart failure · T-tubules

Christopher Y. Ko and Charlotte E. R. Smith contributed equally.

C. Y. Ko · C. E. R. Smith (✉) · E. Grandi (✉)
Department of Pharmacology, University of California at Davis, Davis, CA, USA
e-mail: cersmith@ucdavis.edu

N. L. Parinandi, T. J. Hund (eds.), *Cardiovascular Signaling in Health and Disease*, https://doi.org/10.1007/978-3-031-08309-9_1

Abbreviations

β-AR	β-Adrenergic receptor
μm	Micrometer
AA	Amino acid
AC	Adenylyl cyclase
AF	Atrial fibrillation
AP	Action potential
APD	Action potential duration
ATP	Adenosine triphosphate
Bin1	Amphiphysin II or Bridging Integrator 1
Ca	Calcium
Ca/CaM	Calcium-calmodulin
CaM	Calmodulin
CaMKII	Calcium/calmodulin-dependent protein kinase II
cAMP	Cyclic adenosine monophosphate
$[Ca]_i$	Intracellular calcium concentration
$Ca_v1.2$	L-type calcium channel
$Ca_v3.1$-3	T-type calcium channel
CICR	Calcium-induced calcium release
CPVT	Catecholaminergic polymorphic ventricular tachycardia
CRISPR	Clustered regularly interspaced short palindromic repeats
Cx43	Connexin-43
DAD	Delayed afterdepolarization
EAD	Early afterdepolarization
ECC	Excitation-contraction coupling
Epac	Exchange factor directly activated by cAMP
FKBP12.6	FK506 binding protein 12.6
GLUT	Glucose transporter
G_S	Stimulatory G protein
HCM	Hypertrophic cardiomyopathy
HDAC	Histone deacetylase
HF	Heart failure
HFpEF	Heart failure with preserved ejection fraction
HFrEF	Heart failure with reduced ejection fraction
I/R	Ischemia/reperfusion
I_{Ca}	Calcium current
$I_{Ca\text{-}L}$	L-type calcium current
$I_{Ca\text{-}T}$	T-type calcium current
I_K	Potassium current
I_{K1}	Inward rectifier potassium current
I_{Na}	Sodium current
$I_{Na,L}$	Late sodium current
$I_{Na,T}$	Transient sodium current

IP_3	Inositol trisphosphate
IP_3R	Inositol trisphosphate receptor
I_{ti}	Transient inward current
I_{to}	Transient outward potassium current
JPH2	Junctophilin-2
jSR	Junctional sarcoplasmic reticulum
K	Potassium
$K_{Ca}2.2$	Calcium-activated potassium channel
K_D	Dissociation constant
k_Da	Kilodalton
KI	Knock-in
$K_{ir}2.1$	Inward rectifier potassium channel
$K_{ir}6.2$	Inward rectifier potassium channel
$K_v1.4$	Voltage-gated potassium channel
$K_v4.2$	Voltage-gated potassium channel
$K_v4.3$	Voltage-gated potassium channel
$K_v7.1$	Voltage-gated potassium channel
LTCC	L-type calcium channel
MEF2	Myocyte enhancer factor 2
MI	Myocardial infarction
Na	Sodium
$[Na]_i$	Intracellular sodium concentration
Na_V	Voltage-dependent sodium channel
NCX	Sodium-calcium exchanger
NHE	Sodium-hydrogen exchanger
NKA	Sodium/potassium ATPase
nm	Nanometer
nM	Nanomolar
NOS	Nitric oxide synthase
O-GlcNAc	*O*-linked β-N-acetylglucosamine
PDE5	Phosphodiesterase 5
PKA	Protein kinase A
PLB	Phospholamban
PTM	Posttranslational modification
ROS	Reactive oxygen species
RyR	Ryanodine receptor
SCN5A	Sodium voltage-gated channel alpha subunit 5
SERCA	Sarco/endoplasmic reticulum calcium ATPase
SGLT	Sodium-glucose cotransporter
SK2	Small-conductance Calcium-activated potassium channel
SR	Sarcoplasmic reticulum
TA	Triggered action potential
TT	Transverse (t)-tubule
TTCC	T-type calcium channel
WT	Wild-type

Introduction

The heart is a complex and versatile organ capable of responding and adapting to a multitude of signals and stressors on every beat. The versatility of cardiac function, in turn, is highly contingent upon the many layers of functional integration in the heart, which comprise multiple spatial scales from the whole organ down to the molecular level. The multiple forms of signaling that exist among the cardiac system serve as the basis for communication and are an essential element that allows for the high degree of functional integration. The syncytium that forms at the tissue scale via gap junctional coupling, for example, allows for electrical communication between electrotonically coupled myocytes. The activation of adrenergic receptors upon binding circulating hormones serves to transmit information from the extracellular to the intracellular environment across the sarcolemma. Within the myocyte, cascades of molecular interactions are responsible for a multitude of functional and regulatory processes. Within these contexts, calcium (Ca) is a key, ubiquitous second messenger that is central to these processes.

Unfortunately, the qualities that are advantageous for normal healthy heart function can often add to the complications that emerge as the heart adapts and undergoes change in the disease state. As a consequence, heart disease still is the leading cause of death in the world, despite the significant advancements that have already been made in the field of cardiac research. Therefore, there remains a critical and continued need for unraveling the intricacies that govern heart function, in both physiology and disease, in order to better guide future therapeutic approaches and to improve clinical outcomes.

The goal of this chapter is to provide the most recent progress in the field on the ongoing efforts for understanding the complexities of cardiac function and disease, specifically as it pertains to Ca-dependent signaling in cardiac myocytes. The signaling mechanisms discussed will be largely relevant to cardiac myocyte pathophysiology, namely, as it relates to heart failure (HF) and arrhythmias, which are both complex and widely investigated clinical conditions resulting from perturbations in intracellular Ca-dependent signaling mechanisms. Among the many Ca-dependent proteins involved in these signaling processes, a special focus will be placed on Ca/calmodulin-dependent protein kinase II (CaMKII), a central regulator of excitation-contraction coupling (ECC) and Ca cycling, to highlight the recently identified importance of key posttranslational modifications (PTMs) functioning as specific molecular regulators in a variety of disease contexts. Finally, the importance of cellular remodeling and ultrastructural changes that occur in cardiac myocytes will be discussed in order to highlight the growing recognition of the tremendous, yet currently underappreciated, influence that spatiotemporal factors have on cardiac myocyte function.

Physiology

Contraction of the heart results from an increase in the cytosolic Ca concentration in cardiac myocytes subsequent to their electrical activation via ECC. Voltage-dependent opening of the sarcolemmal L-type Ca channels (LTCCs) allows for an initial influx of Ca that critically triggers Ca-dependent activation of ryanodine receptors (RyRs) located in the closely apposed sarcoplasmic reticulum (SR) membrane. RyR openings allow Ca to flow down its very large concentration gradient (three to four orders of magnitude) from the SR into the cytosol. This process initiates contraction of cardiac muscle and is termed Ca-induced Ca release (CICR) [1]. At the whole cell level, this synchronous activation causes the change in cytosolic Ca to exhibit a rapid rise (tens of milliseconds), followed by a prolonged decay over several hundred milliseconds.

In ventricular myocytes, projections of the surface membrane extend transversely into the cell center as an array of LTCC-containing transverse (t)-tubules (TTs) [2]. This facilitates the formation of dyads where the two membrane structures of the TT and juxtaposed "junctional" region of the SR (jSR) are separated by a 12–15 nm cleft [3, 4]. Here, LTCCs are coupled with intracellular RyR clusters that are areas of densely packed RyR tetramers (most often but not always within dyads and defined as regions with contiguous RyR antibody labeling in super-resolution light microscopy) [5]. The presence of dyads throughout the cell allows Ca release to occur simultaneously at the cell periphery and interior to ensure synchronous contraction [6].

Cardiac myocyte electrical activity, CICR, and contraction, are subject to modulation by several signaling pathways, which involve cascades of signaling molecules resulting in PTMs (e.g., phosphorylation) of target proteins [1]. The most widely studied signaling pathway in cardiac myocytes is that occurring in response to β-adrenergic stimulation, which leads to complex cardiac electrophysiological and Ca handling modulation resulting in increased heart rate (chronotropy), force of contraction (inotropy), speed of relaxation (lusitropy), and conduction (dromotropy) during the physiologic *fight or flight* response. Sympathetic activation involves changes in transmembrane potential homeostasis via both direct influences on sarcolemmal ion channels and transporters as well as indirect changes in Ca signaling that acutely regulate transmembrane fluxes and can lead to remodeling in the chronic (pathologic) setting [1].

It has been proposed that several of the downstream effects of β-adrenergic stimulation are mediated by CaMKII, which is activated in response to the resulting Ca elevation. CaMKII is required for increased chronotropy (sinoatrial node firing rate) [7] and inotropy (ventricular contractile force) [8] during the fight or flight response. As a central mediator of several essential processes in the heart, CaMKII regulates expression and function of ion channels, transporters, and myofilament proteins, thus modulating electrophysiology, Ca handling, contractile function, and metabolism. However, the role of CaMKII in mediating normal cardiovascular responses remains to be fully understood. On the other hand, many studies have highlighted the role of this kinase as a promoter of adverse cardiac remodeling, dysfunction, and arrhythmia. Not only is this Ca-dependent kinase activated by Ca elevation, but several other cAMP-dependent and cAmp-independent pathways have been involved, including oxidation [9], *O*-GlcNAcylation [10], *S*-nitrosylation [11], and the Epac/NOS-dependent pathway [12], which may contribute to CaMKII hyperactivation in various disease contexts. Indeed, CaMKII has emerged as a key molecule for transduction of myocardial stress response to various cardiac disease outcomes. We will review these avenues of CaMKII over-activation and their consequences for cardiac disease, with emphasis on HF and arrhythmia.

Pathophysiology

Heart Failure

HF is generally defined as the inability of the heart to meet the body's metabolic demands. Patients present with a number of debilitating symptoms including shortness of breath, fatigue, and edema, with the severity of disease increased when these symptoms occur at rest in addition to during attempted exercise [13, 14]. HF is the primary end point of many cardiac pathologies and a secondary comorbidity of diseases including hypertension and diabetes [15]. Around 6.2 million Americans have HF, with 2.97% of the population predicted to be affected by 2030 [16]. By this time, medical costs associated with HF are estimated to rise to almost $70 billion [15]. Given the additional impact of HF on patient morbidity and loss of labor, HF is a considerable burden on both healthcare and the economy.

Primary HF can be classified depending on whether failure is caused by impaired ventricular contraction or relaxation. Systolic HF, also known as HF with reduced ejection fraction (HFrEF), occurs when <40% of blood in the left ventricle is pumped out per beat [14, 17]. It is caused by impaired contractile function and is associated with dysregulation of intracellular Ca and ultrastructural remodeling [6]. By contrast, in HF with preserved ejection fraction (HFpEF), ejection fraction is normal at >50% [14]. Here pathology is the result of improper relaxation and compliance during diastole manifesting as reduced ventricular filling [18]. While ejection fraction is maintained above 50% in HFpEF, the total volume of blood in the ventricle is reduced due to impaired filling; thus, overall output is decreased [6, 14].

Gross structural remodeling characterizes both HFrEF and HFpEF, whereby ventricular chamber volume and wall thickness are altered, reflecting underlying changes in the size of individual myocytes (Fig. 1). HFrEF is typically accompanied by eccentric hypertrophy where myocytes become thinner and elongated resulting in chamber dilation and wall thinning [17]. Conversely, HFpEF more frequently presents with concentric hypertrophy in which chamber size is reduced and walls thickened, representative of increased cell width [6, 19, 20].

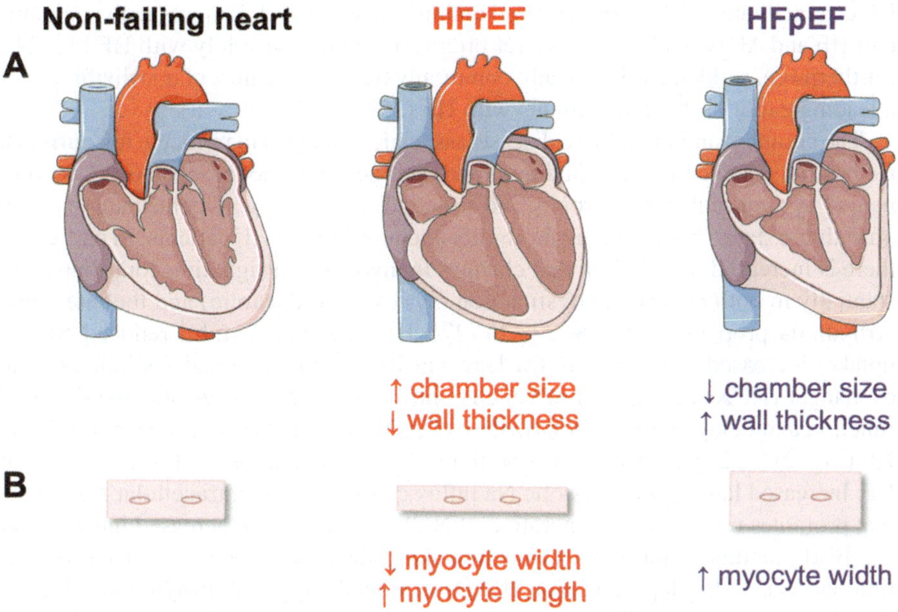

Fig. 1 Cardiac remodeling in HF. (**a**) Gross structure of the non-failing heart and differential remodeling observed in HFrEF and HFpEF. In HFrEF ventricular walls are thinned and chambers are enlarged, whereas in HFpEF walls are thickened and chamber size reduced. (**b**) Representative cellular remodeling underlying changes in gross structure in HF. Myocytes are typically thinned and elongated in HFrEF and conversely increased in width in HFpEF. (Created using Servier Medical Art)

Compared to HFrEF, less is known about HFpEF, which is also reflected in the lack of efficient and specific treatments for HFpEF. Recent studies have highlighted marked discrepancies in Ca handling and structural remodeling between etiologies that will be further discussed in section "Cardiac Myocyte Remodeling and Ultrastructural Change". Despite the differences between HFrEF and HFpEF, both have similar incidence and mortality rates [6, 21], highlighting the importance of examining both types of HF.

Arrhythmias

In addition to the morbidity associated with symptomatic HF, HF patients also exhibit an increased propensity to develop cardiac arrhythmias [22]. Atrial fibrillation (AF) often coexists with HF and can either be a cause or a consequence of failure [23]. The rapid atrial and often subsequently high ventricular rates in AF per se can cause hemodynamic changes that impair systolic and diastolic function leading to HF, while electrical, structural, neurohormonal, and metabolic alterations in

HF can facilitate AF development and maintenance [23, 24]. Patients with concomitant HF and AF typically have poorer prognosis than those solely with HF [22, 24]. Furthermore, sudden cardiac death, commonly related to ventricular arrhythmias, is a leading cause of death in patients with HF [25].

The complex and interactive disease-associated changes in myocyte ion currents, Ca handling, and contractile function (and their neurohormonal regulation), accompanied by ventricular hypertrophy and structural remodeling all contribute to a pro-arrhythmic state – exacerbating morbidity and mortality in HF patients. In particular, there is increased recognition that changes in myocyte Ca signaling contribute substantially to both contractile dysfunction (systolic and diastolic) and the integrated arrhythmia propensity [6, 18, 23, 26, 27]. HF myocytes exhibit reduced SR Ca uptake, increased diastolic SR Ca leak via RyR, and increased sodium-calcium exchanger (NCX) activity, all of which contribute to reduced systolic and diastolic function and delayed afterdepolarization (DAD)-mediated triggered arrhythmias in HF (Fig. 2) [6, 23]. As well as Ca, sodium (Na) dysregulation is also a hallmark of HF. Increased late I_{Na} and diastolic Na influx cause elevated intracellular Na ($[Na]_i$) that promotes reverse-mode operation of NCX and results in increased intracellular Ca ($[Ca]_i$)-mediated diastolic dysfunction alongside action potential (AP) prolongation and early afterdepolarization (EAD)-mediated triggered arrhythmias (Fig. 2) [28, 29]. Interestingly, defective Ca and Na homeostasis in HF have both been linked to altered modulation by CaMKII [29]. As such, the key role of CaMKII in both physiology and pathophysiology will be discussed in detail below.

Ca-Dependent CaMKII Signaling in the Cardiac Myocyte

Background

In the cardiac myocyte, the multifunctional serine/threonine kinase CaMKII is key in fine-tuning the intricate interplay among molecules responsible for many essential functions of the heart, such as AP generation, Ca cycling, and ECC. CaMKII is intimately linked to Ca signaling within the cardiac myocyte, whereby not only its activation and regulation are both dependent upon intracellular Ca levels, but also CaMKII itself is responsible for regulating many of the processes governing the levels of intracellular Ca on every beat. While the bidirectional feedback relationship between Ca and CaMKII is integral to the robust physiological function of the heart, the cross-talk between Ca and CaMKII-dependent regulatory pathways, both in the acute and chronic setting, adds complexity to the mechanistic understanding of cardiac disease. Upregulation in the expression and activity of CaMKIIδ, the predominant cardiac isoform, has been reported in human HF [30–33], and this was corroborated in animal studies in which cardiac-specific overexpression of the CaMKIIδ_C splice variant resulted in severe HF and arrhythmias in mice [34], while aortic banded mice lacking CaMKIIδ only developed ventricular hypertrophy without decompensating into HF [35]. Besides its involvement in HF [30–34, 36], chronic over-activation of CaMKII has been implicated in several other

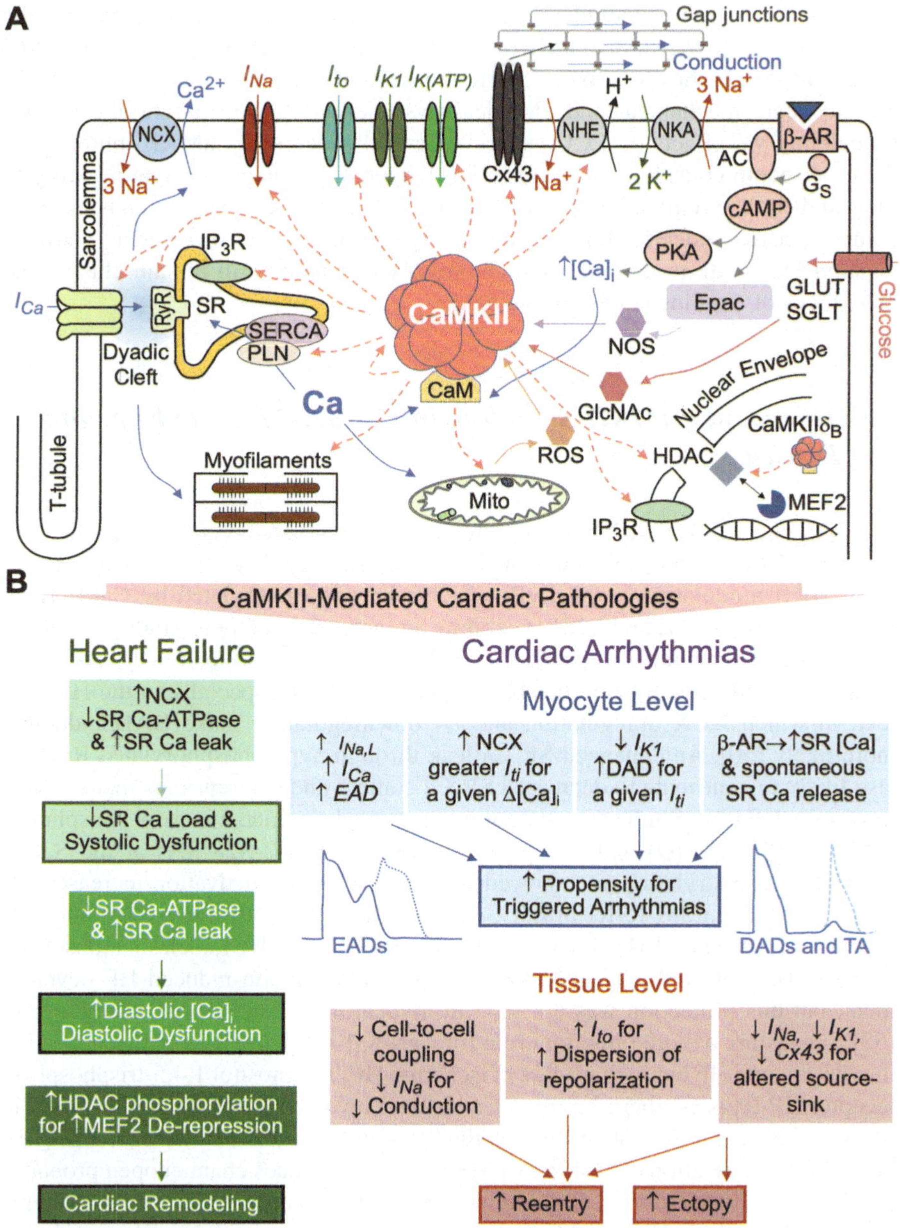

Fig. 2 Ca-dependent CaMKII signaling. (**a**) CaMKII regulates key proteins essential for myocyte function, such as the generation of the AP, Ca handling, contraction, and transcription. Chronic over-activation of CaMKII has been associated with several cardiac pathologies such as HF and arrhythmias at the cellular and tissue scales. Due to the ubiquitous nature of CaMKII regulation, the mechanisms underlying CaMKII-mediated cardiac pathologies are complex, integrative, and interconnected and can lead to long-term consequences like cellular and tissue remodeling. Recent studies have found that CaMKII may be susceptible to more disease stressors than once thought, increasing the threat that CaMKII poses on cardiac health. (**b**) Schematic of the main processes linking CaMKII activity to systolic and diastolic dysfunction and cell- and tissue-level pro-arrhythmic behavior

pathological conditions including cardiac hypertrophy [9, 37], diastolic and systolic dysfunction [38, 39], cardiac arrhythmias [40–43], and ischemia/reperfusion (I/R) injury [44, 45]. In these diseases, CaMKII regulation of cellular subsystems contributes to acute mechanical and electrical dysfunction as well as chronic cardiac remodeling via effects on ECC and cell survival processes. CaMKII inhibition is therefore being considered as a potential therapeutic strategy for treating arrhythmias and cardiac remodeling [46]. However, detailed molecular mechanisms and a comprehensive understanding of how the numerous factors involved in CaMKII signaling integrate to modulate cardiac myocyte function still remain elusive and are a focus of ongoing studies in the field.

Ca-Dependent CaMKII Signaling in Cardiac Myocyte Function and Disease

CaMKII regulates key Ca handling proteins that can have a direct influence on myocyte Ca transients and subcellular SR Ca release activity (Fig. 2). One major target of CaMKII modulation is the RyR, which when phosphorylated by CaMKII (at serines S2808, S2811, and S2814) exhibits an increased open probability, resulting in enhanced SR-mediated Ca leak, Ca sparks, and Ca waves [36, 47–49]. As a consequence, CaMKII activation has been linked to DADs, especially in the HF context, in which NCX upregulation and I_{K1} downregulation make DAD induction more likely [50]. An enhanced SR Ca leak through hyperphosphorylated RyR2 is also likely to contribute to decreased SR Ca content and Ca transients that are seen in systolic HF [36]. Studies in a pair of mutant knock-in mice that were phosphomimetic (RyR2-S2814D) and non-phosphorylatable (RyR2-S2814A) at the S2814 CaMKII phosphorylation site showed that acute CaMKII activation increased SR Ca leak, reduced CaM-RyR2 affinity, and caused an RyR2 shift into a pathological conformational state [51]. Interestingly, the RyR2-S2814A knock-in mouse was found to be protected against transverse aortic constriction-induced HF development, but this protection was not seen in myocardial infarction (MI)-induced HF [52], suggesting a more important role for CaMKII-mediated RyR phosphorylation in nonischemic HF but perhaps less in ischemic HF. The inositol 1,4,5-trisphosphate receptor (IP_3R) is another important Ca channel involved in the regulation of Ca in the myocyte, and it is localized at both the SR membrane and nuclear envelope [53]. While phosphorylation of IP_3R by CaMKII at S150 reduces channel open probability [53, 54], Ca release from IP_3R can potentiate RyR openings and arrhythmogenic effects, especially in HF and AF where IP_3 signaling is upregulated. The regulation of nuclear Ca dynamics by IP_3R is especially important in CaMKII-mediated excitation-transcription processes that contribute to cardiac remodeling and gene expression changes that occur in the HF phenotype [53, 55]. The direct involvement of CaMKII in regulating mitochondrial Ca handling proteins, such as the mitochondrial Ca uniporter and the mitochondrial permeability transition pore, has also been reported [56], though this role remains controversial [57] and the mechanistic details are continuing to be resolved.

CaMKII also regulates the main SR Ca reuptake mechanism via the SR Ca ATPase (SERCA) by phosphorylating T17 on phospholamban (PLB), which subsequently relieves its inhibitory effect on SERCA function [58]. The effect that PLB has on overall Ca handling, however, depends on the balance between SR Ca release, Ca load, and Ca reuptake. Phosphorylation of PLB can work to enhance Ca reuptake conditions in the myocyte and exert a protective mechanism or work to exacerbate spontaneous Ca leak activity and spontaneous activity, thereby enhancing arrhythmias. A decrease in SERCA, an increase in SERCA/PLB ratio, and a decrease in PLB phosphorylation have all been described in HF and are thought to contribute to slower SR Ca uptake in HF, leading to an increase in diastolic Ca [59]. This, along with enhanced NCX expression, also results in a decreased SR Ca load, diminished Ca transients, and weaker contraction. Just as the balance of Ca is maintained by SERCA and NCX, pH is maintained, in part, by the Na/H exchanger (NHE) on the sarcolemma membrane. NHE can be activated by CaMKII and can help to restore intracellular pH from acidosis following I/R injury [60].

In a model of CaMKIIδ_C transgenic mice, CaMKIIδ_C activation was implicated in the pathological progression of HF and the development of cardiac dysfunction [39]. In this study, CaMKIIδ activation during the early period of transaortic constriction promoted adaptive increases in Ca transients and nuclear transcriptional responses, while chronic progression of the nuclear Ca-CaMKIIδ_C axis contributed to eccentric hypertrophy and HF [39]. Notably, CaMKII regulates several key contractile proteins (Fig. 2), which include myosin binding protein C, troponin I, myosin light chain-2, and titin [61–65]. As such, CaMKII has also been implicated with systolic and diastolic dysfunction in the contexts of hypertrophic cardiomyopathy (HCM), late eccentric hypertrophy, and HF. Aberrant CaMKII-dependent titin phosphorylation occurs in end stage HF and may contribute to altered diastolic stress [64]. In mouse models of sarcomeric HCM (cardiac troponin T R92L and R92W) exhibiting disruptions in Ca homeostasis, CaMKII inhibition led to recovery of diastolic function coupled with improved SERCA activity and likely improvement in Ca handling in the R92W mutant, whereas R92L mutants showed worsened Ca handling, remodeling and function, highlighting a mutation-dependent role of activated CaMKII in HCM progression [38].

CaMKII also plays a critical role in regulating ion channels at the sarcolemmal membrane that affect Ca flux and handling in the cardiac myocyte (Fig. 2). Phosphorylation of the LTCC ($Ca_v1.2$) by CaMKII can increase LTCC current ($I_{Ca\text{-}L}$) amplitude (facilitation), slow inactivation [66, 67], and also accelerate recovery from inactivation [68]. The resulting enhancement in Ca influx, in turn, can initiate a positive feedback interaction between CaMKII activation and $I_{Ca\text{-}L}$ [69, 70] that promotes EAD development and altered RyR activity [71]. Furthermore, regulation of the T-type Ca channel (TTCC, $Ca_v3.1$–3) by CaMKII can result in increased TTCC current and open probability [72, 73].

Cardiac Na and K channels are also regulated by CaMKII (Fig. 2). CaMKII modulates Na channel kinetics similarly to a human mutation (SCN5A 1795InsD) associated with heritable arrhythmias [74], which involve both gain and loss of function in the cardiac I_{Na}. Notably, CaMKII significantly enhances the late Na current ($I_{Na,L}$), which has been found to contribute to AP prolongation and increased Na loading under pathological conditions [75], as well as increase the propensity for EADs in HF [76]. CaMKII-dependent phosphorylation also reduces Na channel steady-state availability and slows its recovery from inactivation, potentially predisposing to conduction defects, as discussed in the next paragraph. CaMKII regulates numerous cardiac K channels, including the voltage-gated ($K_v1.4$, $K_v4.2$, $K_v4.3$, $K_v7.1$), inward rectifier ($K_{ir}2.1$, $K_{ir}6.2$), and Ca-activated ($K_{Ca}2.2$, SK2) channels [65]. Among these, regulation of the transient outward current (I_{to}) has been the most extensively studied. CaMKII acutely regulates I_{to} subunit expression, trafficking, as well as I_{to} gating (leading to slower channel inactivation and more rapid recovery from inactivation) resulting in I_{to} increase [77]. Similarly, acute overexpression of CaMKII resulted in a significant increase in I_{K1} [77]. However, chronic CaMKII overexpression that leads to HF development was accompanied by reductions in I_{to} and I_{K1} densities [77], which may exacerbate repolarization abnormalities and lower the threshold for DAD-mediated spontaneous APs.

We have reviewed how CaMKII effects on RyR, SERCA/PLB, I_{Na}, I_{Ca}, and I_K can promote arrhythmia triggers such as EADs, DADs, and spontaneous APs at the cellular scale. Alterations in CaMKII-mediated Na and Ca handling are suspected to alter myocyte properties and also can manifest as changes in conduction velocity, transmural dispersion of repolarization, or vulnerability to reentry at the tissue scale, as implicated in Long QT and Brugada syndromes [9]. CaMKII activation was found to slow the inactivation of the fast, transient Na current ($I_{Na,T}$) and slow recovery from inactivation [29]. At the tissue scale, delaying the recovery of Na_Vs from inactivation was found to increase the slope of the APD restitution curve, which would enhance the likelihood of alternans and reentry [78]. Reduced Na_V conductance was found to increase the vulnerable window [79]. In structurally normal RyR2-P2328S CPVT mouse hearts which were more susceptible to atrial arrhythmia triggering, reduced upstroke velocity of monophasic APs, inter-atria conduction delays, and slowing of epicardial conduction velocity were observed [80]. These alterations suggested that Ca-dependent alterations in Na_Vs [81] could also promote functional reentry in other disease conditions like HF and also be associated with increased CaMKII activity. CaMKII-mediated loss of function on peak I_{Na} [74] could explain conduction slowing [82] independent of the structural and anatomical changes observed. Ventricular [83] and atrial simulations [84] also found that heterogeneous Na and Ca loading in cardiac tissue can predispose to reentrant arrhythmia. Furthermore, prerequisites for reentry (reduced conduction velocity, prolonged refractoriness, and increased susceptibility to conduction block) were demonstrated to be associated primarily to enhanced CaMKII effects on Na_Vs and increased oxidation in a multicellular mathematical model of the cardiac fiber [85]. Another important factor that affects conduction and the propensity for arrhythmias at the tissue scale is the degree of cell-to-cell coupling determined by connexin-43 (Cx43) expression and the formation of gap junctions between myocytes. In cardiac tissue, well-coupled repolarized myocytes act as an electrotonic "sink" that a "source" current from an AP must overcome in order to propagate in tissue [86]. Structural and electrophysiological remodeling processes that decrease the degree of coupling between myocytes, as occurs in HF [87], can alter this source-sink mismatch and lower the threshold for an arrhythmia trigger, such as an EAD, to propagate in tissue. Multiple sites of CaMKII phosphorylation of Cx43 that act to reduce gap junctional coupling have been reported [88], further implicating CaMKII regulation in the processes that govern susceptibility not only to arrhythmias but also to structural remodeling that occurs in pathophysiological contexts. A recent study demonstrated the possibility of a novel approach of harnessing the source-sink mismatch by implementing gene therapy approaches to create "stabilizer cells" overexpressing the inward rectifier K channel Kir2.1 in cardiac tissue vulnerable to arrhythmias to suppress arrhythmia triggers [89]. Progressive therapeutic approaches such as this may be key alternative avenues for addressing integrative mechanisms of cardiac pathologies.

Increasing evidence supports an important role of CaMKII in excitation-transcription coupling, especially with respect to long-term changes that occur in contexts such as cardiac remodeling in HF. CaMKII along with calcineurin, a CaM-dependent phosphatase, is directly involved in the Ca-mediated processes that could activate altered gene expression [90–92]. One key pathway is through CaMKII-dependent phosphorylation of histone deacetylases (HDACs) (e.g., HDAC4 and HDAC5). Upon phosphorylation by CaMKIIδ_B, HDAC4 can unbind and translocate out of the nucleus through chaperone proteins, thereby derepressing and allowing for hypertrophic transcription factors such as MEF2 (myocyte enhancer factor 2) to drive gene transcription [91, 92]. A parallel pro-hypertrophic Epac-mediated pathway involving PLC, IP$_3$R, and CaMKII activation as well as HDAC5 nuclear export and MEF2 activation has also been identified [12]. How CaMKII-mediated transcriptional changes regulate both the electrophysiological and structural remodeling is highly important in understanding the pathophysiology of HF and related Ca-mediated arrhythmias. Whether and how CaMKII regulates ultrastructural remodeling at the cellular level in parallel is also of great interest, but these mechanisms are not yet fully understood.

As is evident, CaMKII regulates many different targets and exerts a multitude of effects in the cardiac myocyte that involve multiple spatial and temporal scales. Unlike in controlled experimental contexts, these effects are not necessarily independent of each other but rather are involved in complex interdependent processes underlying cardiac function and disease. More recent studies have begun to investigate these interactions to better understand the complexities underlying mechanisms of disease. In the context of HF, a vicious cycle mechanism of positive feedback [93, 94] involving Na and Ca mishandling, upregulated CaMKII, and ROS – characteristic of cardiac diseases such as HF, long QT syndrome, and CPVT – was identified and demonstrated in CaMKIIδ mutant mouse and HF rabbit cardiac myocytes. These findings were consistent with a similar positive feedback mediated mechanism involving CaMKII activation and concurrent intracellular Na and Ca overload identified under conditions of hypokalemia-induced ventricular fibrillation [95].

Chronic over-activation of CaMKII is a major underlying contributor in the dysregulation of many of the described pathophysiological contexts in the myocyte. Since the regulation of Ca and CaMKII is mutually intertwined, elucidating the mechanisms that can explain how and why CaMKII becomes overactivated is an imperative goal that several research groups are striving to uncover at the molecular level. Many findings in this area of research have brought to light key insights that have helped to better understand the processes of CaMKII-mediated disease.

CaMKII Structure and Ca-Dependent Activation

The molecular mechanisms of CaMKII activation and deactivation and kinase activity are intricately linked to Ca. CaMKII self-assembles into a dodecameric complex (12 monomers) of 2 stacked hexameric rings. Each CaMKII monomer (56 kDa, 498-AA) consists of a catalytic domain (N-terminal), an autoinhibitory regulatory domain, and an association domain (C-terminal). Under baseline conditions, the regulatory domain is bound to and autoinhibits the catalytic domain such that CaMKII is in a closed conformational state (Fig. 3). As $[Ca]_i$ rises, Ca binds and

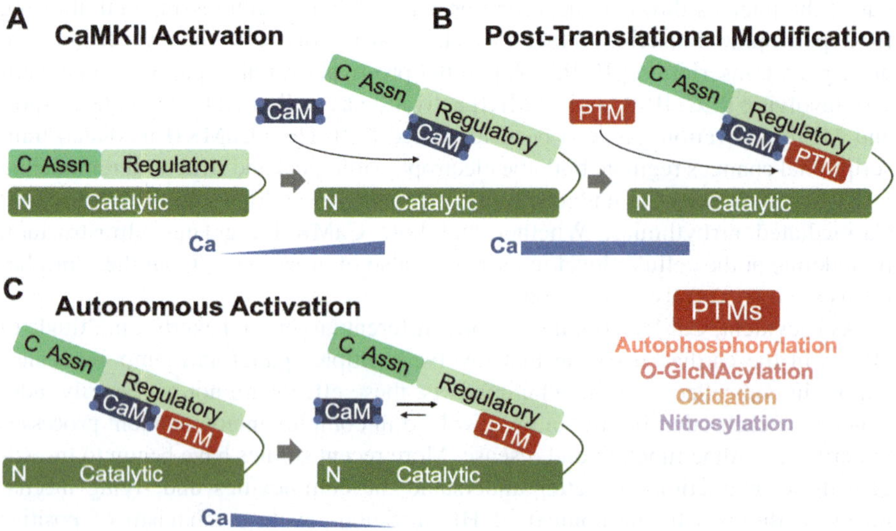

Fig. 3 Mechanisms of CaMKII activation and posttranslational modifications. (**a**) As $[Ca]_i$ increases within the cardiac myocyte, a monomer of CaMKII activates and adopts an open conformational state upon binding Ca/CaM. (**b**) General depiction of activated CaMKII monomer becoming posttranslationally modified. PTM represents any one of autophosphorylation at T287, oxidation at MM281/282, *O*-GlcNAcylation at S280, or *S*-nitrosylation at C273 or C290. (**c**) Posttranslationally modified CaMKII remains persistently active in open conformational state even after Ca/CaM dissociates when $[Ca]_i$ declines. CaM trapping may also occur as CaM affinity increases and dissociation rates are slowed. Chronic activation of CaMKII in autonomous activated state can promote HF and arrhythmias

forms a complex with calmodulin (Ca/CaM), which then can bind the regulatory domain of CaMKII, relieving the autoinhibition of the catalytic domain (Fig. 3).

CaMKII is then able to activate, adopting an open conformational state, and phosphorylate its molecular targets [96]. The low affinity that CaMKII has for Ca/CaM (K_D = 10–50 nM) is what allows for CaMKII to sensitively detect changes in $[Ca]_i$, which is especially important in the dyadic cleft in cardiac myocytes where high local $[Ca]_i$ changes occur during the cardiac AP. Typically, as $[Ca]_i$ declines in the myocyte, Ca/CaM unbinds and CaMKII deactivates. However, when $[Ca]_i$ elevation is prolonged, as occurs in many disease conditions like HF and arrhythmias, CaMKII monomers can autophosphorylate neighboring subunits at threonine 287 (T287) of the regulatory domain, prolonging the activated state of CaMKII [97, 98]. As a consequence, the affinity for Ca/CaM increases in a process called "CaM trapping" in which CaM release and CaMKII deactivation are slowed by ~100- to 1000-fold [99]. Even when CaM does dissociate, phosphorylated CaMKII can maintain a partially active "autonomous" open state (Fig. 3) [97, 98]. Together, "CaM trapping" and the autonomous activation of CaMKII contribute to a CaMKII "memory" effect responsible for the over-activation of CaMKII in cardiac pathologies.

There are four main isoforms of CaMKII – α, β, γ, and δ – and these are differentially expressed in various tissues with varying degrees of basal Ca/CaM affinity ($\gamma > \beta > \delta > \alpha$) [100, 101]. While neurons mostly express the α and β isoforms of CaMKII, cardiac myocytes mostly express CaMKIIδ, which is responsible for 85–90% of CaMKII activity, while CaMKIIγ is responsible for mostly the rest [101, 102]. Four of the at least 11 splice variants of CaMKIIδ are found differentially localized in the cardiac myocyte [103–105]. CaMKIIδ_A is localized primarily to t-tubule, sarcolemmal, and nuclear membranes. CaMKIIδ_B is concentrated mainly in the nucleus due to an 11-AA nuclear localization sequence. CaMKIIδ_C is the predominant splice variant in the cytoplasm and localizes largely at the z-lines. CaMKIIδ_9 is a lesser-studied splice variant expressed at similar levels to those of CaMKIIδ_B [106]. These splice variants all have the ability to heteromultimerize, thus diversifying the potential functional responses of CaMKII within the myocyte.

Posttranslational Modifications of CaMKII as Novel Mechanisms of Cardiac Disease

The recent discovery of several new PTMs of CaMKII has revealed that CaMKII may be susceptible to an even wider range of pathological stressors than once thought. Four of these PTMs include oxidation at MM281/282 [9], *O*-GlcNAcylation at S280 [10], and *S*-nitrosylation at C273 and C290 [11] (Fig. 3), and they now implicate CaMKII to stressors such as oxidative stress, diabetic hyperglycemia, and nitric oxide synthase activation, which often coexist in disease and are associated with morbidity, mortality, and healthcare costs involving cardiac disease. Biochemical in vitro studies have shown that these PTMs can prolong CaMKII activation similarly to how autophosphorylation of CaMKII at T287 promotes autonomous activation [10, 11, 104, 107, 108]. Additional studies have found that PTMs that promote autonomous activation of CaMKII are now directly implicated with cardiac disease. Oxidized CaMKII has been found to contribute to apoptosis post-MI [9] and AF [104], while *O*-GlcNAcylation was shown to contribute to hyperglycemia-induced SR Ca leak and arrhythmia [10]. Interestingly, *S*-nitrosylation was shown to have a dual effect: while *S*-nitrosylation at CaMKIIδ-C290 promoted autonomous activation, *S*-nitrosylation at C273 suppressed CaMKII activation by CaM [11]. The latter suppressive effect suggests that nitrosylation of CaMKII may confer a sex-dependent protective effect against damage from I/R in females, who tend to have higher basal levels of NOS and nitrosylation than males [109]. The mechanisms by which each of these PTMs contributes to their respectively associated pathologies are not fully understood and are now being clarified.

The mechanisms of PTM-dependent autonomous CaMKII activation and their pathophysiological consequences have begun to be investigated directly within the cardiac myocyte environment context in order to gain more physiologically relevant insights. In one recent study, the binding affinity and off-rate kinetics of CaM-CaMKIIδ interactions, an indicator of autonomous CaMKII activation, were directly measured in cardiac myocytes [110]. CaM was found to dissociate more slowly by a threefold factor from the phosphomimetic CaMKIIδ T287D mutant variant than from either WT or the phosphoresistant CaMKIIδ T287A mutant. CaM dissociated even more slowly from oxidized CaMKIIδ T287D, demonstrating the synergy of PTMs in their effects on the autonomous activation of CaMKII. Among the PTMs, studies uncovering the mechanistic role of CaMKII *O*-GlcNAcylation in diabetes and related arrhythmias have been especially prolific [10, 111–113]. In a study utilizing *O*-GlcNAcylation-resistant CRISPR S280A CaMKII-KI mice, hyperglycemia was shown to promote *O*-GlcNAcylation of CaMKII at S280 and induce arrhythmias via phosphorylation of RyRs and associated ROS increase in cardiac myocytes [111]. A subsequent study investigating Ca handling and electrophysiology demonstrated that high glucose-induced APD prolongation, APD alternans, Ca waves, and DADs were diminished in these CaMKIIδ-S280A-KI mice [10, 11, 104, 112]. Another study investigating the link between diabetic hyperglycemia and the increased risk of arrhythmias sought to determine whether hyperglycemia alone can be accountable for arrhythmias or whether it requires the presence of additional pathological factors. Even though hyperglycemia alone was sufficient to enhance cellular arrhythmias (i.e., APD prolongation, short-term APD variability, and alternans), a "second hit" greatly exacerbated cardiac arrhythmogenesis in diabetic hyperglycemia [114].

Given the recognized importance and significant clinical implications of chronic CaMKII activation and related Ca signaling alterations in cardiac disease, understanding the effects of these PTMs on CaMKII autonomy and their impact on cardiac myocyte function is especially important. Yet, many questions relating CaMKII to cardiac function and disease still remain. A recent study demonstrated the ability for activated CaMKIIδ to translocate from the TT/SR junction to its extra-dyadic targets within the myocyte [115], providing evidence contrary to the notion that CaMKIIδ is immobile and anchored within the myocyte. Another open question is how CaMKII activation dynamically changes from beat to beat in time with the changes in Ca within the myocyte. These spatiotemporal aspects of CaMKII activity and their effects on the myocyte remain largely unexplored and highlight the importance of understanding how the ultrastructure of the myocyte in the intracellular environment along with the spatial organization of proteins involved in ECC can impact cardiac myocyte function and disease.

Cardiac Myocyte Remodeling and Ultrastructural Change

Background

The gross anatomical and physiological changes that occur during the transition from healthy to failing myocardium are reflective of cellular and ultrastructural remodeling from within individual myocytes. In the healthy heart, subcellular architecture is optimally organized to ensure a uniform rise in Ca and thus contraction. This is primarily achieved through the presence of TT membrane invaginations that facilitate close apposition of LTCCs and RyRs to form dyads throughout the cell [2, 3]. This allows triggered Ca release to occur synchronously both at the surface and along TTs in the cell center. While Ca release from the SR is dependent on RyR properties [116–118], its spatial localization is associated with TT structure and organization due to the preferential expression of LTCCs on TTs [3, 119–121]. As a result of this, remodeling of either TTs or RyRs can dramatically affect the pattern of Ca release within cells.

While TTs were identified over 60 years ago [122], work in this area is resurgent due to the development of advanced imaging techniques that permit investigation of the interplay between TTs and RyRs and spatiotemporal patterns of subcellular Ca release. Recent work has shown there is extensive remodeling of the TT network in HF that is associated with dyadic disruption [123–128]. This is accompanied by contractile dysfunction through reduced trigger for Ca release [129, 130], RyR remodeling [131–135], and altered RyR modulation [94, 136–138]. While what instigates remodeling is yet to be elucidated, the subcellular changes that occur together culminate in the disease phenotypes we observe in vivo. As such, the mechanisms of impaired Ca release through ultrastructural remodeling in disease warrant continued investigation and will be the subject of this section.

Ultrastructural (T-Tubule) Remodeling

In the healthy ventricle, TT invaginations occur every ~1.8–2 μm at the z-line [2]. They project transversely with longitudinal elements to create a branching transverse-axial network within cells [2, 139–141]. To facilitate efficient ECC, all TTs are connected to the cellular SR network [141] with a recent study demonstrating that TTs are wrapped in voluminous SR that is often spaced less than 10 nm away [142]. The close apposition between TTs and the SR is imperative for effective CICR and is mediated by Junctophilin-2 (JPH2) interacting with LTCCs and anchoring the TT and SR membranes [143–145]. Interestingly, the density of the TT network varies between species [2, 123, 146, 147]. Narrow tubules with increased complexity are observed in animals with faster resting heart rates, likely due to the need for rapid AP propagation and Ca cycling [148, 149]. Though narrow TT lumens could restrict ion diffusion if static, TT luminal eccentricity changes with sarcomere shortening on a beat-by-beat basis to aid ion exchange [150]. As such, both the organization and plasticity of the TT network are key for cardiac function.

While the TT network is intricately arranged and ordered in the healthy heart, this is not the case in disease. Extensive TT remodeling has been reported in a number of HF pathologies induced by hypertrophy [124, 125], MI [124, 151–155], and tachypacing [123, 126, 129, 156], all of which result in impaired AP propagation [157] and dyadic uncoupling [123, 128, 158, 159]. Though it remains unclear whether changes to the TT network initiate the onset of disease or are an associated outcome, remodeling begins prior to function being impaired on the echocardiogram and progresses as the severity of disease pathology increases [125]. Patchy TT loss, with the TTs that remain being more longitudinally oriented, enlarged and branched are well-characterized features of HF [123–129, 151–156, 160] (Fig. 4). However, recent work has shown TT remodeling is dependent on whether there is systolic or diastolic impairment. In HFrEF, changes in TT structure reflect those previously described, yet in HFpEF TT density is conversely maintained or increased through proliferation and dilation [18, 27] (Fig. 4). As in physiological hypertrophy where TT density increases with exercise training [161], the enhancement of the TT network in HFpEF is thought to be compensatory, with a positive correlation between TT density and the level of diastolic dysfunction observed [6, 18]. Though impaired relaxation in HFpEF is partly thought to be associated with changes to the extracellular matrix [6, 162, 163], no difference in TT collagen deposition has been observed in HFpEF patients despite this being seen in HFrEF [18, 164]. This is thought to further support the notion of compensatory TT proliferation in HFpEF, highlighting the vast mechanistic differences in etiology in HFpEF versus the maladaptive TT remodeling and increased tubular collagen content seen in HFrEF [18]. Since TT collagen deposition could impair content exchange by making TTs stiffer and less able to distend, the impact of this, along with dyadic disorganization, should be considered when examining the impact of TT remodeling in HFrEF.

At present the cause of TT remodeling in HF is unclear. However, several TT regulatory proteins have been shown to be downregulated in the failing

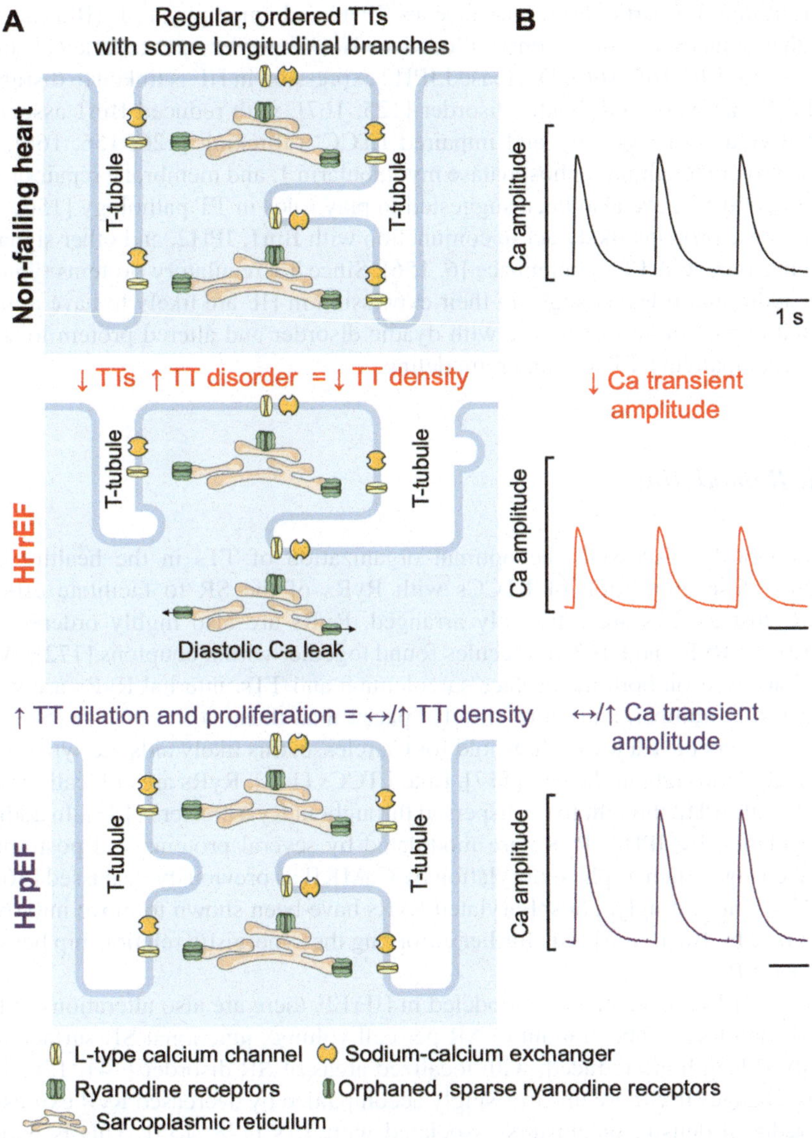

Fig. 4 Ultrastructural remodeling and consequent Ca release in HF. (**a**) Schematic diagrams of non-failing, HFrEF and HFpEF cardiac myocytes. In HFrEF there is patchy TT loss with remaining TTs increasingly longitudinally oriented, enlarged, and branched. Due to TT loss in HFrEF, there are orphaned RyRs that contribute to diastolic SR leak. In contrast, in HFpEF TT density is maintained or increased through dilation and proliferation. (**b**) Representative Ca transients for each group showing reduced transient amplitude in HFrEF and preserved/increased amplitude in HFpEF. (Diagrams created using Servier Medical Art with Ca transients recapitulating data from Kilfoil et al. [27])

myocardium. Of particular importance are JPH2 and amphiphysin II (Bin1), a protein that induces TT formation and localizes LTCCs to the sarcolemmal/TT membrane [126, 146, 165, 166]. Decreased JPH2 expression in HF is linked to disruption of the TT network and dyadic disorder [125, 167]; with reduced Bin1 associated with decreased TT density and impaired LTCC trafficking [126, 156, 168]. The z-disc protein telethonin, phosphatase myotubularin 1, and membrane repair protein Mitsugumin 53 have also been suggested to play roles in TT pathology [156, 169–171]. These proteins likely act in conjunction with Bin1, JPH2, and other signaling proteins to govern TT maintenance [6, 156]. Since TT regulatory proteins typically have additional roles, changes in their expression in HF are likely to have a multifaceted impact on ultrastructure with dyadic disorder and altered protein localization accompanying TT loss and remodeling.

RyR Remodeling

As previously discussed, the optimal organization of TTs in the healthy heart ensures close apposition of LTCCs with RyRs of the SR to facilitate efficient CICR. Just as TTs are intricately arranged, RyRs are also highly ordered with 80–85% of RyR and LTCC molecules found together within couplons [172]. While couplons exist on both the surface sarcolemma and TTs, internal RyRs are larger, increased in number, and more closely spaced than those at the surface [5]. Since larger clusters have a lower threshold for Ca release, this likely aids the synchronicity of ECC throughout the cell [117]. Like LTCCs [145], RyRs are colocalized with JPH2, with JPH2 thought to be dispersed throughout RyR clusters [173]. In addition to regulation by JPH2, RyRs are modulated by several proteins and posttranslational changes such as phosphorylation by CaMKII as previously discussed [36, 47, 48, 174]. Interestingly, phosphorylated RyRs have been shown to move into dyads organized by Bin1 [175] thus further enforcing the synergistic relationship between TTs and SR.

Just as TTs are adversely remodeled in HFrEF, there are also alterations to both the SR and RyRs. The amount of SR per cell volume, junctional SR surface area, and dyad length are reduced, with localized areas of SR disorder [141, 158, 159]. These changes in SR are unsurprisingly accompanied by decreased RyR expression and reduced density of clusters associated with TTs [128, 135]. This is perhaps compounded by decreased Bin1 expression reducing the localization of RyRs to Bin1-arranged dyads [175]. Indeed, due to TT remodeling in HFrEF, a greater number of RyRs are found outside of dyads, with orphaned RyRs remaining along the z-line but no longer coupled with TTs and thus lacking local control [131, 151] (Fig. 4). Interestingly, RyR clustering itself is also altered in the failing ventricle. Smaller, more dispersed, and more closely spaced RyR clusters predominate but are accompanied by large multi-cluster Ca release units [134, 137]. Remodeling of

RyRs in HFrEF is not exclusively caused by dyadic disruption through TT loss. Changes in RyRs are also associated with pathological hyperphosphorylation and oxidation, along with altered regulation by proteins such as FKBP12.6 [36, 137, 176–179]. While the causality of remodeling is multifaceted, it manifests as RyR hyperactivity resulting in Ca leak from the SR and diminished Ca transient amplitude (Fig. 4) [18, 27, 134, 138, 180].

Despite RyR remodeling being well characterized in HFrEF, very little is known about any changes in HFpEF. Contrary to systolic HF, TT preservation in HFpEF ensures dyads are maintained and thus stops RyR orphaning [18, 27]. Although structurally similar to the healthy heart, HFpEF RyRs are hyperphosphorylated as in HFrEF [27]. However, unlike in systolic HF, this is not associated with increased SR leak but rather diminished β-adrenergic drive [27]. As such, baseline Ca transient amplitude is unaltered or enhanced in HFpEF (Fig. 4), with impairment only presenting upon attempted inotropic stimulation [18, 27]. While further work is required to improve knowledge of HFpEF, the impact of RyR remodeling and dyadic disruption in HFrEF are highly detrimental to Ca release and will be discussed in more detail below.

Consequence of Structural Remodeling on Spatiotemporal Factors of Ca Release

The ultimate outcome of remodeling in HFrEF is reduced global Ca transient amplitude and thus contractile dysfunction. Though the mechanisms vary between models and disease state, this is typically a culmination of a number of factors. Less dyadic coupling between LTCCs and RyRs results in a dyssynchronous rise in Ca, with fewer LTCCs and reduced $I_{Ca\text{-}L}$ due to TT loss providing a smaller, less-effective trigger for CICR [126, 127, 129, 130, 151, 155, 160]. This is compounded by decreased SR Ca load caused by RyR leak and/or impaired SERCA activity generating a smaller Ca transient [176, 180, 181]. Taken together these alterations result in reduced contractility and thus impaired function.

Of great importance is the fact that remodeling is not uniform throughout failing ventricular cells. Spatial differences in TTs, dyad organization, and RyR clusters lead to heterogenous and inefficient triggered Ca release that is detrimental to contraction and has potential for arrhythmic activity. Indeed, intracellular Ca dysregulation clearly contributes to arrhythmia in a broad range of pathologies including HF, and HF myocytes and tissue are prone to triggered activity and alternans due to abnormal Ca handling [23, 182]. Patchy TT remodeling and heterogenous RyRs are associated with regions devoid of AP propagation, impaired triggered Ca release, and a dyssynchronous rise in systolic Ca [27, 127, 151, 155, 160, 183]. In addition to spatial differences in triggered Ca, diastolic release is also altered as a consequence of remodeling. Ca sparks are more frequently observed in hyperactive, orphaned RyRs outside of dyads [131, 138, 152], with slower to rise sparks and silent Ca leak associated with morphological changes in dyadic RyR clusters in modeling studies [134, 184]. Since RyR heterogeneity is likely to potentiate waves [117, 118] and SR leak and TT disruption promote alternans [185], this activity may underlie aberrant pro-arrhythmic Ca release observed in HF [131]. Ultimately since leak reduces SR Ca load, the main determinant of the systolic transient, this also impairs triggered release thus exacerbating the adverse effects of remodeling.

Though not covered in detail in this review, another important consideration of ultrastructural remodeling in HF is Ca removal by NCX. Since NCX-mediated removal of Ca from the cytosol is electrogenic, enhanced SR leak could cause after-depolarizations in the failing ventricle [50]. However, as NCX is preferentially expressed on TTs that are disrupted [3, 186], spatial differences in Ca handling are likely to occur depending on proximity of NCX to RyRs where the Ca release occurs. Though dyadic leak may result in an EAD or DAD due to close proximity to NCX, leak from orphaned RyRs is more likely to propagate resulting in waves [6]. Since both outcomes could occur simultaneously within cells, this is further evidence of the detrimental impact of ultrastructural remodeling in HF.

Potential Treatments

Given the adverse consequences of ultrastructural remodeling in HFrEF, it is unsurprising that TT and dyadic restoration are seen as ideal therapeutic targets. While a number of proteins play a role in TT regulation, Bin1 has emerged as a promising candidate for repair. It has been identified as a biomarker whereby circulating cardiac Bin1 correlates with cardiovascular risk and severity of remodeling in humans [187, 188]. Exogenous Bin1 delivered through adeno-associated virus has also been shown to both protect hearts prior to HF-induction and restore cardiac function in animals with preexisting failure [189, 190]. Gene therapy involving JPH2 has similarly improved outcomes in HF [167], with both Bin1 and JPH2 acting to limit TT remodeling and thus maintain Ca release [167, 189, 190]. While further work is needed to test the safety and efficacy of these treatments in patients, other works using preexisting therapies have also proved beneficial for reversal of remodeling. Both mechanical unloading and biventricular pacing have been shown to improve the homogeneity of TTs, RyRs, and Ca release in HF [132, 133, 152]. Similar restoration has been observed following treatment with tadalafil, a PDE5 inhibitor typically used to treat erectile dysfunction [156]. Interestingly in this study, reversal of TT remodeling was also associated with changes in Bin1 expression, further highlighting its importance [156]. Since unloading, resynchronization and tadalafil are already used in clinic, they appear to be ideal treatments for reversing ultrastructural remodeling and improving outcomes of HF.

Conclusions and Research Frontiers

Spatial and Temporal Heterogeneity of Ca-Dependent Signaling

Recent work has highlighted the spatial and temporal heterogeneity of Ca-dependent signaling in the heart. At the cellular level, CaMKIIδ has been shown to translocate from the dyad to reach other cellular targets [115], suggesting the timing and localization of its activation and deactivation could be important in regulating function. Since dyadic remodeling occurs in HF, it is likely that the localization of CaMKII is altered in addition to it being chronically activated [30–34, 36] as well as regulating hypertrophic gene transcription [12, 90–92]. This may contribute to the subcellular differences in Ca release associated with patchy, irregular TT patterns and variability in RyR coupling, cluster size, and phosphorylation that are also known to occur [36, 131, 134, 137, 151]. Direct regulation of ultrastructural remodeling by CaMKII is yet unknown, though CaMKII has been involved in remodeling of dendritic spines [191]. Taken together, it is clear that both CaMKII and TT remodeling in HF cause changes in discrete areas that lead to dyssynchronous Ca release within cells. This is ultimately compounded by unequal remodeling across the syncytium of the myocardium. In the healthy heart, transmural heterogeneity across the ventricular wall helps synchronize contraction and relaxation [192–194]. However, the pattern of heterogeneity is altered in HF leading to abnormal cardiac cycling [193, 194]. Theoretical studies suggest that TT degradation and consequent heterogenous SR Ca handling should promote both spontaneous Ca release, which drives DADs, and alternans in cardiac tissue [195]. While sophisticated models have been developed to describe detailed spatial and temporal characteristics of cardiac myocyte ECC, simulations of perturbed TT structure have been heuristic in their approach [185, 196–198], rather than tightly coupled to experiments. As remodeling in HF is non-uniform and dependent on causality, it is likely that some areas of the ventricle are more adversely affected than others. Indeed, regional variability in TT remodeling has been observed in HF patients [199]. However, it is unclear whether differences also exist between layers of the ventricular wall. At present the interaction between intercellular variability (e.g., transmural heterogeneity) and subcellular variability has never been investigated, and the mechanisms, extent, and importance of tissue homogenization are unknown. Since the dyssynchronous effects of subcellular heterogeneity would be amplified and further increased by transmural and regional differences at the tissue level, this area is a key topic for future work.

Considerations for Therapy

In addition to appreciating the impact of cellular remodeling at the whole-heart level, there are additional considerations for therapy. In order to reverse the aberrant changes in Ca handling associated with ultrastructural remodeling, much focus has been put on TT restoration. As previously discussed, this has already been shown to be beneficial in experimental studies and is now likely to progress to clinic. However, for this to be successful, EC coupling protein restoration is required in addition to TT repair to reverse any dyadic disruption not due to TT loss. Since impaired dyadic function has also been observed where TTs remain in the HF ventricle [155], this is a key concept for therapy. As the TT regulatory proteins Bin1 and JPH2 also regulate LTCCs and RyRs [145, 166, 173, 175], they should be able to target their trafficking back to the dyad, yet it is unclear whether the localization and activity of their modulators such as CaMKII can also be restored. A number of studies have already demonstrated the beneficial effects of CaMKII inhibition in improving function in animal models of maladaptive remodeling and human HF [30, 200, 201] as well as in mitigating susceptibility to arrhythmias [94, 95]. However, targeting CaMKII as a therapeutic approach presents a unique challenge given the need for preserving numerous essential CaMKII-dependent cellular processes and the wide range of CaMKII variants and PTMs. It is also unclear currently whether different modes of activation states underlie differences in physiological or pathological activity. Treatment for CaMKII overexpression and chronic activation in HF, thus, may need to be selectively targeted. Recent studies that have identified molecular mechanisms of chronic CaMKII activation may prove to be valuable in this regard. Alternatively, minimizing CaM affinity may serve to eradicate more widespread dysfunction associated with elevated diastolic Ca by limiting CaM trapping and preventing overactive CaMKII. Understanding how these numerous factors integrate to modulate Ca handling and affect cardiac myocyte physiology can yield valuable mechanistic insights that can lead to the development of therapeutic strategies for treating cardiac disease.

References

1. Bers DM. Cardiac excitation-contraction coupling. Nature. 2002;415(6868):198–205.
2. Soeller C, Cannell MB. Examination of the transverse tubular system in living cardiac rat myocytes by 2-photon microscopy and digital image–processing techniques. Circ Res. 1999;84(3):266–75.
3. Scriven DRL, Dan P, Moore EDW. Distribution of proteins implicated in excitation-contraction coupling in rat ventricular myocytes. Biophys J. 2000;79(5):2682–91.
4. Scriven DRL, Asghari P, Moore EDW. Microarchitecture of the dyad. Cardiovasc Res. 2013;98(2):169–76.
5. Shen X, van den Brink J, Hou Y, Colli D, Le C, Kolstad TR, MacQuaide N, Carlson CR, Kekenes-Huskey PM, Edwards AG, et al. 3D dSTORM imaging reveals novel detail of ryanodine receptor localization in rat cardiac myocytes. J Physiol. 2019;597(2):399–418.
6. Setterberg IE, Le C, Frisk M, Li J, Louch WE. The physiology and pathophysiology of T-tubules in the heart. Front Physiol. 2021;12:718404.
7. Wu Y, Gao Z, Chen B, Koval OM, Singh MV, Guan X, Hund TJ, Kutschke W, Sarma S, Grumbach IM, et al. Calmodulin kinase II is required for fight or flight sinoatrial node physiology. Proc Natl Acad Sci U S A. 2009;106(14):5972–7.

8. Wu Y, Luczak ED, Lee EJ, Hidalgo C, Yang J, Gao Z, Li J, Wehrens XH, Granzier H, Anderson ME. CaMKII effects on inotropic but not lusitropic force frequency responses require phospholamban. J Mol Cell Cardiol. 2012;53(3):429–36.
9. Anderson ME, Brown JH, Bers DM. CaMKII in myocardial hypertrophy and heart failure. J Mol Cell Cardiol. 2011;51(4):468–73.
10. Erickson JR, Pereira L, Wang L, Han G, Ferguson A, Dao K, Copeland RJ, Despa F, Hart GW, Ripplinger CM, et al. Diabetic hyperglycaemia activates CaMKII and arrhythmias by O-linked glycosylation. Nature. 2013;502(7471):372–6.
11. Erickson JR, Nichols CB, Uchinoumi H, Stein ML, Bossuyt J, Bers DM. S-Nitrosylation induces both autonomous activation and inhibition of calcium/calmodulin-dependent protein kinase II delta. J Biol Chem. 2015;290(42):25646–56.
12. Pereira L, Ruiz-Hurtado G, Morel E, Laurent AC, Metrich M, Dominguez-Rodriguez A, Lauton-Santos S, Lucas A, Benitah JP, Bers DM, et al. Epac enhances excitation-transcription coupling in cardiac myocytes. J Mol Cell Cardiol. 2012;52(1):283–91.
13. Inamdar AA, Inamdar AC. Heart failure: diagnosis, management and utilization. J Clin Med. 2016;5(7):62.
14. Del Buono MG, Buckley L, Abbate A. Primary and secondary diastolic dysfunction in heart failure with preserved ejection fraction. Am J Cardiol. 2018;122(9):1578–87.
15. Jackson SL, Tong X, King RJ, Loustalot F, Hong Y, Ritchey MD. National burden of heart failure events in the United States, 2006 to 2014. Circ Heart Fail. 2018;11(12):e004873.
16. Virani SS, Alonso A, Benjamin EJ, Bittencourt MS, Callaway CW, Carson AP, Chamberlain AM, Chang AR, Cheng S, Delling FN, et al. Heart disease and stroke statistics—2020 update: a report from the American Heart Association. Circulation. 2020;141(9):e139–596.
17. Nauta JF, Hummel YM, Tromp J, Ouwerkerk W, van der Meer P, Jin X, Lam CSP, Bax JJ, Metra M, Samani NJ, et al. Concentric vs. eccentric remodelling in heart failure with reduced ejection fraction: clinical characteristics, pathophysiology and response to treatment. Eur J Heart Fail. 2020;22(7):1147–55.
18. Frisk M, Le C, Shen X, Røe ÅT, Hou Y, Manfra O, Silva GJJ, van Hout I, Norden ES, Aronsen JM, et al. Etiology-dependent impairment of diastolic cardiomyocyte calcium homeostasis in heart failure with preserved ejection fraction. J Am Coll Cardiol. 2021;77(4):405–19.
19. Zile MR, Gottdiener JS, Hetzel SJ, McMurray JJ, Komajda M, McKelvie R, Baicu CF, Massie BM, Carson PE. Prevalence and significance of alterations in cardiac structure and function in patients with heart failure and a preserved ejection fraction. Circulation. 2011;124(23):2491–501.
20. Katz DH, Beussink L, Sauer AJ, Freed BH, Burke MA, Shah SJ. Prevalence, clinical characteristics, and outcomes associated with eccentric versus concentric left ventricular hypertrophy in heart failure with preserved ejection fraction. Am J Cardiol. 2013;112(8):1158–64.
21. Pfeffer MA, Shah AM, Borlaug BA. Heart failure with preserved ejection fraction in perspective. Circ Res. 2019;124(11):1598–617.
22. Maisel WH, Stevenson LW. Atrial fibrillation in heart failure: epidemiology, pathophysiology, and rationale for therapy. Am J Cardiol. 2003;91(6a):2d–8d.
23. Denham NC, Pearman CM, Caldwell JL, Madders GWP, Eisner DA, Trafford AW, Dibb KM. Calcium in the pathophysiology of atrial fibrillation and heart failure. Front Physiol. 2018;9:1380.
24. Anter E, Jessup M, Callans DJ. Atrial fibrillation and heart failure. Circulation. 2009;119(18):2516–25.
25. Tomaselli GF, Zipes DP. What causes sudden death in heart failure? Circ Res. 2004;95(8):754–63.
26. Eisner DA, Caldwell JL, Trafford AW, Hutchings DC. The control of diastolic calcium in the heart. Circ Res. 2020;126(3):395–412.
27. Kilfoil PJ, Lotteau S, Zhang R, Yue X, Aynaszyan S, Solymani RE, Cingolani E, Marbán E, Goldhaber JI. Distinct features of calcium handling and β-adrenergic sensitivity in heart failure with preserved versus reduced ejection fraction. J Physiol. 2020;598(22):5091–108.

28. Maltsev VA, Undrovinas A. Late sodium current in failing heart: friend or foe? Prog Biophys Mol Biol. 2008;96(1):421–51.
29. Grandi E, Herren A. CaMKII-dependent regulation of cardiac Na+ homeostasis. Front Pharmacol. 2014;5:41.
30. Sossalla S, Fluschnik N, Schotola H, Ort KR, Neef S, Schulte T, Wittkopper K, Renner A, Schmitto JD, Gummert J, et al. Inhibition of elevated Ca2+/calmodulin-dependent protein kinase II improves contractility in human failing myocardium. Circ Res. 2010;107(9):1150–61.
31. Hoch B, Meyer R, Hetzer R, Krause EG, Karczewski P. Identification and expression of delta-isoforms of the multifunctional Ca2+/calmodulin-dependent protein kinase in failing and nonfailing human myocardium. Circ Res. 1999;84(6):713–21.
32. Kirchhefer U, Schmitz W, Scholz H, Neumann J. Activity of cAMP-dependent protein kinase and Ca2+/calmodulin-dependent protein kinase in failing and nonfailing human hearts. Cardiovasc Res. 1999;42(1):254–61.
33. Bers DM. CaMKII inhibition in heart failure makes jump to human. Circ Res. 2010;107(9):1044–6.
34. Zhang T, Maier LS, Dalton ND, Miyamoto S, Ross J Jr, Bers DM, Brown JH. The deltaC isoform of CaMKII is activated in cardiac hypertrophy and induces dilated cardiomyopathy and heart failure. Circ Res. 2003;92(8):912–9.
35. Ling H, Zhang T, Pereira L, Means CK, Cheng H, Gu Y, Dalton ND, Peterson KL, Chen J, Bers D, et al. Requirement for Ca2+/calmodulin-dependent kinase II in the transition from pressure overload-induced cardiac hypertrophy to heart failure in mice. J Clin Invest. 2009;119(5):1230–40.
36. Ai X, Curran JW, Shannon TR, Bers DM, Pogwizd SM. Ca2+/calmodulin-dependent protein kinase modulates cardiac ryanodine receptor phosphorylation and sarcoplasmic reticulum Ca2+ leak in heart failure. Circ Res. 2005;97(12):1314–22.
37. Backs J, Backs T, Neef S, Kreusser MM, Lehmann LH, Patrick DM, Grueter CE, Qi X, Richardson JA, Hill JA, et al. The delta isoform of CaM kinase II is required for pathological cardiac hypertrophy and remodeling after pressure overload. Proc Natl Acad Sci U S A. 2009;106(7):2342–7.
38. Lehman SJ, Tal-Grinspan L, Lynn ML, Strom J, Benitez GE, Anderson ME, Tardiff JC. Chronic calmodulin-kinase II activation drives disease progression in mutation-specific hypertrophic cardiomyopathy. Circulation. 2019;139(12):1517–29.
39. Ljubojevic-Holzer S, Herren AW, Djalinac N, Voglhuber J, Morotti S, Holzer M, Wood BM, Abdellatif M, Matzer I, Sacherer M, et al. CaMKIIdeltaC drives early adaptive Ca(2+) change and late eccentric cardiac hypertrophy. Circ Res. 2020;127(9):1159–78.
40. Chelu MG, Sarma S, Sood S, Wang S, van Oort RJ, Skapura DG, Li N, Santonastasi M, Muller FU, Schmitz W, et al. Calmodulin kinase II-mediated sarcoplasmic reticulum Ca2+ leak promotes atrial fibrillation in mice. J Clin Invest. 2009;119(7):1940–51.
41. Voigt N, Li N, Wang Q, Wang W, Trafford AW, Abu-Taha I, Sun Q, Wieland T, Ravens U, Nattel S, et al. Enhanced sarcoplasmic reticulum Ca2+ leak and increased Na+-Ca2+ exchanger function underlie delayed afterdepolarizations in patients with chronic atrial fibrillation. Circulation. 2012;125(17):2059–70.
42. van Oort RJ, McCauley MD, Dixit SS, Pereira L, Yang Y, Respress JL, Wang Q, De Almeida AC, Skapura DG, Anderson ME, et al. Ryanodine receptor phosphorylation by calcium/calmodulin-dependent protein kinase II promotes life-threatening ventricular arrhythmias in mice with heart failure. Circulation. 2010;122(25):2669–79.
43. Wagner S, Maier LS, Bers DM. Role of sodium and calcium dysregulation in tachyarrhythmias in sudden cardiac death. Circ Res. 2015;116(12):1956–70.
44. Weinreuter M, Kreusser MM, Beckendorf J, Schreiter FC, Leuschner F, Lehmann LH, Hofmann KP, Rostosky JS, Diemert N, Xu C, et al. CaM Kinase II mediates maladaptive post-infarct remodeling and pro-inflammatory chemoattractant signaling but not acute myocardial ischemia/reperfusion injury. EMBO Mol Med. 2014;6(10):1231–45.

45. Ling H, Gray CB, Zambon AC, Grimm M, Gu Y, Dalton N, Purcell NH, Peterson K, Brown JH. Ca2+/calmodulin-dependent protein kinase II delta mediates myocardial ischemia/reperfusion injury through nuclear factor-kappaB. Circ Res. 2013;112(6):935–44.
46. Pellicena P, Schulman H. CaMKII inhibitors: from research tools to therapeutic agents. Front Pharmacol. 2014;5:21.
47. Yuchi Z, Lau K, Van Petegem F. Disease mutations in the ryanodine receptor central region: crystal structures of a phosphorylation hot spot domain. Structure. 2012;20(7):1201–11.
48. Guo T, Zhang T, Mestril R, Bers DM. Ca2+/calmodulin-dependent protein kinase II phosphorylation of ryanodine receptor does affect calcium sparks in mouse ventricular myocytes. Circ Res. 2006;99(4):398–406.
49. Purohit A, Rokita AG, Guan X, Chen B, Koval OM, Voigt N, Neef S, Sowa T, Gao Z, Luczak ED, et al. Oxidized Ca(2+)/calmodulin-dependent protein kinase II triggers atrial fibrillation. Circulation. 2013;128(16):1748–57.
50. Pogwizd SM, Schlotthauer K, Li L, Yuan W, Bers DM. Arrhythmogenesis and contractile dysfunction in heart failure: roles of sodium-calcium exchange, inward rectifier potassium current, and residual beta-adrenergic responsiveness. Circ Res. 2001;88(11):1159–67.
51. Uchinoumi H, Yang Y, Oda T, Li N, Alsina KM, Puglisi JL, Chen-Izu Y, Cornea RL, Wehrens XHT, Bers DM. CaMKII-dependent phosphorylation of RyR2 promotes targetable pathological RyR2 conformational shift. J Mol Cell Cardiol. 2016;98:62–72.
52. Respress JL, van Oort RJ, Li N, Rolim N, Dixit SS, deAlmeida A, Voigt N, Lawrence WS, Skapura DG, Skardal K, et al. Role of RyR2 phosphorylation at S2814 during heart failure progression. Circ Res. 2012;110(11):1474–83.
53. Bare DJ, Kettlun CS, Liang M, Bers DM, Mignery GA. Cardiac type 2 inositol 1,4,5-trisphosphate receptor: interaction and modulation by calcium/calmodulin-dependent protein kinase II. J Biol Chem. 2005;280(16):15912–20.
54. Maxwell JT, Natesan S, Mignery GA. Modulation of inositol 1,4,5-trisphosphate receptor type 2 channel activity by Ca2+/calmodulin-dependent protein kinase II (CaMKII)-mediated phosphorylation. J Biol Chem. 2012;287(47):39419–28.
55. Bers DM. Ca(2)(+)-calmodulin-dependent protein kinase II regulation of cardiac excitation-transcription coupling. Heart Rhythm. 2011;8(7):1101–4.
56. Joiner ML, Koval OM, Li J, He BJ, Allamargot C, Gao Z, Luczak ED, Hall DD, Fink BD, Chen B, et al. CaMKII determines mitochondrial stress responses in heart. Nature. 2012;491(7423):269–73.
57. Fieni F, Johnson DE, Hudmon A, Kirichok Y. Mitochondrial Ca2+ uniporter and CaMKII in heart. Nature. 2014;513(7519):E1–2.
58. Wegener AD, Simmerman HK, Liepnieks J, Jones LR. Proteolytic cleavage of phospholamban purified from canine cardiac sarcoplasmic reticulum vesicles. Generation of a low resolution model of phospholamban structure. J Biol Chem. 1986;261(11):5154–9.
59. Mattiazzi A, Kranias EG. The role of CaMKII regulation of phospholamban activity in heart disease. Front Pharmacol. 2014;5:5.
60. Vila-Petroff M, Mundina-Weilenmann C, Lezcano N, Snabaitis AK, Huergo MA, Valverde CA, Avkiran M, Mattiazzi A. Ca(2+)/calmodulin-dependent protein kinase II contributes to intracellular pH recovery from acidosis via Na(+)/H(+) exchanger activation. J Mol Cell Cardiol. 2010;49(1):106–12.
61. Tong CW, Wu X, Liu Y, Rosas PC, Sadayappan S, Hudmon A, Muthuchamy M, Powers PA, Valdivia HH, Moss RL. Phosphoregulation of cardiac inotropy via myosin binding protein-C during increased pacing frequency or beta1-adrenergic stimulation. Circ Heart Fail. 2015;8(3):595–604.
62. Cazorla O, Lucas A, Poirier F, Lacampagne A, Lezoualc'h F. The cAMP binding protein Epac regulates cardiac myofilament function. Proc Natl Acad Sci U S A. 2009;106(33):14144–9.
63. Eikemo H, Moltzau LR, Hussain RI, Nguyen CH, Qvigstad E, Levy FO, Skomedal T, Osnes JB. CaMKII in addition to MLCK contributes to phosphorylation of regulatory light chain in cardiomyocytes. Biochem Biophys Res Commun. 2016;471(1):219–25.

64. Hamdani N, Krysiak J, Kreusser MM, Neef S, Dos Remedios CG, Maier LS, Kruger M, Backs J, Linke WA. Crucial role for Ca2(+)/calmodulin-dependent protein kinase-II in regulating diastolic stress of normal and failing hearts via titin phosphorylation. Circ Res. 2013;112(4):664–74.
65. Hegyi B, Bers DM, Bossuyt J. CaMKII signaling in heart diseases: emerging role in diabetic cardiomyopathy. J Mol Cell Cardiol. 2019;127:246–59.
66. Anderson ME, Braun AP, Schulman H, Premack BA. Multifunctional Ca2+/calmodulin-dependent protein kinase mediates Ca(2+)-induced enhancement of the L-type Ca2+ current in rabbit ventricular myocytes. Circ Res. 1994;75(5):854–61.
67. Yuan W, Bers DM. Ca-dependent facilitation of cardiac Ca current is due to Ca-calmodulin-dependent protein kinase. Am J Phys. 1994;267(3 Pt 2):H982–93.
68. Guo J, Duff HJ. Calmodulin kinase II accelerates L-type Ca2+ current recovery from inactivation and compensates for the direct inhibitory effect of [Ca2+]i in rat ventricular myocytes. J Physiol. 2006;574(Pt 2):509–18.
69. Bers DM, Morotti S. Ca(2+) current facilitation is CaMKII-dependent and has arrhythmogenic consequences. Front Pharmacol. 2014;5:144.
70. Hashambhoy YL, Winslow RL, Greenstein JL. CaMKII-induced shift in modal gating explains L-type Ca(2+) current facilitation: a modeling study. Biophys J. 2009;96(5):1770–85.
71. Hashambhoy YL, Greenstein JL, Winslow RL. Role of CaMKII in RyR leak, EC coupling and action potential duration: a computational model. J Mol Cell Cardiol. 2010;49(4):617–24.
72. Welsby PJ, Wang H, Wolfe JT, Colbran RJ, Johnson ML, Barrett PQ. A mechanism for the direct regulation of T-type calcium channels by Ca2+/calmodulin-dependent kinase II. J Neurosci. 2003;23(31):10116–21.
73. Wolfe JT, Wang H, Perez-Reyes E, Barrett PQ. Stimulation of recombinant Ca(v)3.2, T-type, Ca(2+) channel currents by CaMKIIgamma(C). J Physiol. 2002;538(Pt 2):343–55.
74. Wagner S, Dybkova N, Rasenack EC, Jacobshagen C, Fabritz L, Kirchhof P, Maier SK, Zhang T, Hasenfuss G, Brown JH, et al. Ca2+/calmodulin-dependent protein kinase II regulates cardiac Na+ channels. J Clin Invest. 2006;116(12):3127–38.
75. Glynn P, Musa H, Wu X, Unudurthi SD, Little S, Qian L, Wright PJ, Radwanski PB, Gyorke S, Mohler PJ, et al. Voltage-gated sodium channel phosphorylation at Ser571 regulates late current, arrhythmia, and cardiac function in vivo. Circulation. 2015;132(7):567–77.
76. Moreno JD, Yang PC, Bankston JR, Grandi E, Bers DM, Kass RS, Clancy CE. Ranolazine for congenital and acquired late INa linked arrhythmias: in silico pharmacologic screening. Circ Res. 2013;113(7):e50–61.
77. Wagner M, Rudakova E, Schutz V, Frank M, Ehmke H, Volk T. Larger transient outward K(+) current and shorter action potential duration in Galpha(11) mutant mice. Pflugers Arch. 2010;459(4):607–18.
78. Qu Z, Karagueuzian HS, Garfinkel A, Weiss JN. Effects of Na(+) channel and cell coupling abnormalities on vulnerability to reentry: a simulation study. Am J Physiol Heart Circ Physiol. 2004;286(4):H1310–21.
79. Xie F, Qu Z, Garfinkel A, Weiss JN. Electrophysiological heterogeneity and stability of reentry in simulated cardiac tissue. Am J Physiol Heart Circ Physiol. 2001;280(2):H535–45.
80. King JH, Zhang Y, Lei M, Grace AA, Huang CL, Fraser JA. Atrial arrhythmia, triggering events and conduction abnormalities in isolated murine RyR2-P2328S hearts. Acta Physiol (Oxf). 2013;207(2):308–23.
81. King JH, Wickramarachchi C, Kua K, Du Y, Jeevaratnam K, Matthews HR, Grace AA, Huang CL, Fraser JA. Loss of Nav1.5 expression and function in murine atria containing the RyR2-P2328S gain-of-function mutation. Cardiovasc Res. 2013;99(4):751–9.
82. Grandi E, Puglisi JL, Wagner S, Maier LS, Severi S, Bers DM. Simulation of Ca-calmodulin-dependent protein kinase II on rabbit ventricular myocyte ion currents and action potentials. Biophys J. 2007;93(11):3835–47.

83. Xie Y, Liao Z, Grandi E, Shiferaw Y, Bers DM. Slow [Na]i changes and positive feedback between membrane potential and [Ca]i underlie intermittent early afterdepolarizations and arrhythmias. Circ Arrhythm Electrophysiol. 2015;8(6):1472–80.
84. Krogh-Madsen T, Christini DJ. Slow [Na(+)]i dynamics impacts arrhythmogenesis and spiral wave reentry in cardiac myocyte ionic model. Chaos. 2017;27(9):093907.
85. Christensen MD, Dun W, Boyden PA, Anderson ME, Mohler PJ, Hund TJ. Oxidized calmodulin kinase II regulates conduction following myocardial infarction: a computational analysis. PLoS Comput Biol. 2009;5(12):e1000583.
86. Xie Y, Sato D, Garfinkel A, Qu Z, Weiss JN. So little source, so much sink: requirements for afterdepolarizations to propagate in tissue. Biophys J. 2010;99(5):1408–15.
87. Glukhov AV, Fedorov VV, Kalish PW, Ravikumar VK, Lou Q, Janks D, Schuessler RB, Moazami N, Efimov IR. Conduction remodeling in human end-stage nonischemic left ventricular cardiomyopathy. Circulation. 2012;125(15):1835–47.
88. Huang RY, Laing JG, Kanter EM, Berthoud VM, Bao M, Rohrs HW, Townsend RR, Yamada KA. Identification of CaMKII phosphorylation sites in Connexin43 by high-resolution mass spectrometry. J Proteome Res. 2011;10(3):1098–109.
89. Liu MB, Priori SG, Qu Z, Weiss JN. Stabilizer cell gene therapy: a less-is-more strategy to prevent cardiac arrhythmias. Circ Arrhythm Electrophysiol. 2020;13(9):e008420.
90. Backs J, Olson EN. Control of cardiac growth by histone acetylation/deacetylation. Circ Res. 2006;98(1):15–24.
91. Wu X, Zhang T, Bossuyt J, Li X, McKinsey TA, Dedman JR, Olson EN, Chen J, Brown JH, Bers DM. Local InsP3-dependent perinuclear Ca2+ signaling in cardiac myocyte excitation-transcription coupling. J Clin Invest. 2006;116(3):675–82.
92. Wilkins BJ, Molkentin JD. Calcium-calcineurin signaling in the regulation of cardiac hypertrophy. Biochem Biophys Res Commun. 2004;322(4):1178–91.
93. Morotti S, Edwards AG, McCulloch AD, Bers DM, Grandi E. A novel computational model of mouse myocyte electrophysiology to assess the synergy between Na+ loading and CaMKII. J Physiol. 2014;592(6):1181–97.
94. Hegyi B, Pölönen R-P, Hellgren KT, Ko CY, Ginsburg KS, Bossuyt J, Mercola M, Bers DM. Cardiomyocyte Na(+) and Ca(2+) mishandling drives vicious cycle involving CaMKII, ROS, and ryanodine receptors. Basic Res Cardiol. 2021;116(1):58.
95. Pezhouman A, Singh N, Song Z, Nivala M, Eskandari A, Cao H, Bapat A, Ko CY, Nguyen T, Qu Z, et al. Molecular basis of hypokalemia-induced ventricular fibrillation. Circulation. 2015;132(16):1528–37.
96. Bayer KU, Harbers K, Schulman H. alphaKAP is an anchoring protein for a novel CaM kinase II isoform in skeletal muscle. EMBO J. 1998;17(19):5598–605.
97. Hudmon A, Schulman H, Kim J, Maltez JM, Tsien RW, Pitt GS. CaMKII tethers to L-type Ca2+ channels, establishing a local and dedicated integrator of Ca2+ signals for facilitation. J Cell Biol. 2005;171(3):537–47.
98. Singh P, Salih M, Tuana BS. Alpha-kinase anchoring protein alphaKAP interacts with SERCA2A to spatially position Ca2+/calmodulin-dependent protein kinase II and modulate phospholamban phosphorylation. J Biol Chem. 2009;284(41):28212–21.
99. Meyer T, Hanson PI, Stryer L, Schulman H. Calmodulin trapping by calcium-calmodulin-dependent protein kinase. Science. 1992;256(5060):1199–202.
100. Gaertner TR, Kolodziej SJ, Wang D, Kobayashi R, Koomen JM, Stoops JK, Waxham MN. Comparative analyses of the three-dimensional structures and enzymatic properties of alpha, beta, gamma and delta isoforms of Ca2+-calmodulin-dependent protein kinase II. J Biol Chem. 2004;279(13):12484–94.
101. Kreusser MM, Lehmann LH, Keranov S, Hoting MO, Kohlhaas M, Reil JC, Neumann K, Schneider MD, Hill JA, Dobrev D, et al. The cardiac CaMKII genes delta and gamma contribute redundantly to adverse remodeling but inhibit calcineurin-induced myocardial hypertrophy. Circulation. 2014;130(15):1262–73.

102. Grimm M, Ling H, Willeford A, Pereira L, Gray CB, Erickson JR, Sarma S, Respress JL, Wehrens XH, Bers DM, et al. CaMKIIdelta mediates beta-adrenergic effects on RyR2 phosphorylation and SR Ca(2+) leak and the pathophysiological response to chronic beta-adrenergic stimulation. J Mol Cell Cardiol. 2015;85:282–91.
103. Coultrap SJ, Zaegel V, Bayer KU. CaMKII isoforms differ in their specific requirements for regulation by nitric oxide. FEBS Lett. 2014;588(24):4672–6.
104. Erickson JR, Patel R, Ferguson A, Bossuyt J, Bers DM. Fluorescence resonance energy transfer-based sensor Camui provides new insight into mechanisms of calcium/calmodulin-dependent protein kinase II activation in intact cardiomyocytes. Circ Res. 2011;109(7):729–38.
105. Palomeque J, Rueda OV, Sapia L, Valverde CA, Salas M, Petroff MV, Mattiazzi A. Angiotensin II-induced oxidative stress resets the Ca2+ dependence of Ca2+-calmodulin protein kinase II and promotes a death pathway conserved across different species. Circ Res. 2009;105(12):1204–12.
106. Gray CB, Heller BJ. CaMKIIdelta subtypes: localization and function. Front Pharmacol. 2014;5:15.
107. Hudmon A, Schulman H. Neuronal CA2+/calmodulin-dependent protein kinase II: the role of structure and autoregulation in cellular function. Annu Rev Biochem. 2002;71:473–510.
108. Lai Y, Nairn AC, Gorelick F, Greengard P. Ca2+/calmodulin-dependent protein kinase II: identification of autophosphorylation sites responsible for generation of Ca2+/calmodulin-independence. Proc Natl Acad Sci U S A. 1987;84(16):5710–4.
109. Sun J, Steenbergen C, Murphy E. S-nitrosylation: NO-related redox signaling to protect against oxidative stress. Antioxid Redox Signal. 2006;8(9–10):1693–705.
110. Simon M, Ko CY, Rebbeck RT, Baidar S, Cornea RL, Bers DM. CaMKIIdelta post-translational modifications increase affinity for calmodulin inside cardiac ventricular myocytes. J Mol Cell Cardiol. 2021;161:53–61.
111. Lu S, Liao Z, Lu X, Katschinski DM, Mercola M, Chen J, Heller Brown J, Molkentin JD, Bossuyt J, Bers DM. Hyperglycemia acutely increases cytosolic reactive oxygen species via O-linked GlcNAcylation and CaMKII activation in mouse ventricular myocytes. Circ Res. 2020;126(10):e80–96.
112. Hegyi B, Fasoli A, Ko CY, Van BW, Alim CC, Shen EY, Ciccozzi MM, Tapa S, Ripplinger CM, Erickson JR, et al. CaMKII serine 280 O-GlcNAcylation links diabetic hyperglycemia to proarrhythmia. Circ Res. 2021;129(1):98–113.
113. Mesubi OO, Rokita AG, Abrol N, Wu Y, Chen B, Wang Q, Granger JM, Tucker-Bartley A, Luczak ED, Murphy KR, et al. Oxidized CaMKII and O-GlcNAcylation cause increased atrial fibrillation in diabetic mice by distinct mechanisms. J Clin Invest. 2021;131(2):e95747.
114. Hegyi B, Ko CY, Bossuyt J, Bers DM. Two-hit mechanism of cardiac arrhythmias in diabetic hyperglycemia: reduced repolarization reserve, neurohormonal stimulation and heart failure exacerbate susceptibility. Cardiovasc Res. 2021;117(14):2781–93.
115. Wood BM, Simon M, Galice S, Alim CC, Ferrero M, Pinna NN, Bers DM, Bossuyt J. Cardiac CaMKII activation promotes rapid translocation to its extra-dyadic targets. J Mol Cell Cardiol. 2018;125:18–28.
116. Wescott AP, Jafri MS, Lederer WJ, Williams GS. Ryanodine receptor sensitivity governs the stability and synchrony of local calcium release during cardiac excitation-contraction coupling. J Mol Cell Cardiol. 2016;92:82–92.
117. Galice S, Xie Y, Yang Y, Sato D, Bers DM. Size matters: ryanodine receptor cluster size affects arrhythmogenic sarcoplasmic reticulum calcium release. J Am Heart Assoc. 2018;7(13):e008724.
118. Xie Y, Yang Y, Galice S, Bers DM, Sato D. Size matters: ryanodine receptor cluster size heterogeneity potentiates calcium waves. Biophys J. 2019;116(3):530–9.
119. Brette F, Salle L, Orchard CH. Quantification of calcium entry at the T-tubules and surface membrane in rat ventricular myocytes. Biophys J. 2006;90(1):381–9.

120. Bhargava A, Lin X, Novak P, Mehta K, Korchev Y, Delmar M, Gorelik J. Super-resolution scanning patch clamp reveals clustering of functional ion channels in adult ventricular myocyte. Circ Res. 2013;112(8):1112–20.
121. Øyehaug L, Loose KØ, Jølle GF, Røe ÅT, Sjaastad I, Christensen G, Sejersted OM, Louch WE. Synchrony of cardiomyocyte Ca(2+) release is controlled by T-tubule organization, SR Ca(2+) content, and ryanodine receptor Ca(2+) sensitivity. Biophys J. 2013;104(8):1685–97.
122. Simpson FO, Oertelis SJ. The fine structure of sheep myocardial cells; sarcolemmal invaginations and the transverse tubular system. J Cell Biol. 1962;12(1):91–100.
123. Cannell MB, Crossman DJ, Soeller C. Effect of changes in action potential spike configuration, junctional sarcoplasmic reticulum micro-architecture and altered t-tubule structure in human heart failure. J Muscle Res Cell Motil. 2006;27(5–7):297–306.
124. Lyon AR, MacLeod KT, Zhang Y, Garcia E, Kanda GK, Lab MJ, Korchev YE, Harding SE, Gorelik J. Loss of T-tubules and other changes to surface topography in ventricular myocytes from failing human and rat heart. Proc Natl Acad Sci U S A. 2009;106(16):6854–9.
125. Wei S, Guo A, Chen B, Kutschke W, Xie YP, Zimmerman K, Weiss RM, Anderson ME, Cheng H, Song LS. T-tubule remodeling during transition from hypertrophy to heart failure. Circ Res. 2010;107(4):520–31.
126. Caldwell JL, Smith CE, Taylor RF, Kitmitto A, Eisner DA, Dibb KM, Trafford AW. Dependence of cardiac transverse tubules on the BAR domain protein amphiphysin II (BIN-1). Circ Res. 2014;115(12):986–96.
127. Seidel T, Navankasattusas S, Ahmad A, Diakos NA, Xu WD, Tristani-Firouzi M, Bonios MJ, Taleb I, Li DY, Selzman CH, et al. Sheet-like remodeling of the transverse tubular system in human heart failure impairs excitation-contraction coupling and functional recovery by mechanical unloading. Circulation. 2017;135(17):1632–45.
128. Crossman DJ, Ruygrok PN, Soeller C, Cannell MB. Changes in the organization of excitation-contraction coupling structures in failing human heart. PLoS One. 2011;6(3):e17901.
129. He J, Conklin MW, Foell JD, Wolff MR, Haworth RA, Coronado R, Kamp TJ. Reduction in density of transverse tubules and L-type Ca(2+) channels in canine tachycardia-induced heart failure. Cardiovasc Res. 2001;49(2):298–307.
130. Briston SJ, Caldwell JL, Horn MA, Clarke JD, Richards MA, Greensmith DJ, Graham HK, Hall MC, Eisner DA, Dibb KM, et al. Impaired beta-adrenergic responsiveness accentuates dysfunctional excitation-contraction coupling in an ovine model of tachypacing-induced heart failure. J Physiol. 2011;589(Pt 6):1367–82.
131. Song L-S, Sobie EA, McCulle S, Lederer WJ, Balke CW, Cheng H. Orphaned ryanodine receptors in the failing heart. Proc Natl Acad Sci U S A. 2006;103(11):4305–10.
132. Li H, Lichter JG, Seidel T, Tomaselli GF, Bridge JHB, Sachse FB. Cardiac resynchronization therapy reduces subcellular heterogeneity of ryanodine receptors, T-tubules, and Ca2+ Sparks produced by dyssynchronous heart failure. Circ Heart Fail. 2015;8(6):1105–14.
133. Sachse FB, Torres NS, Savio-Galimberti E, Aiba T, Kass DA, Tomaselli GF, Bridge JH. Subcellular structures and function of myocytes impaired during heart failure are restored by cardiac resynchronization therapy. Circ Res. 2012;110(4):588–97.
134. Kolstad TR, van den Brink J, MacQuaide N, Lunde PK, Frisk M, Aronsen JM, Norden ES, Cataliotti A, Sjaastad I, Sejersted OM, et al. Ryanodine receptor dispersion disrupts Ca(2+) release in failing cardiac myocytes. elife. 2018;7:e39427.
135. Hou Y, Bai J, Shen X, de Langen O, Li A, Lal S, Dos Remedios CG, Baddeley D, Ruygrok PN, Soeller C, et al. Nanoscale organisation of ryanodine receptors and Junctophilin-2 in the failing human heart. Front Physiol. 2021;12:724372.
136. Marx SO, Reiken S, Hisamatsu Y, Jayaraman T, Burkhoff D, Rosemblit N, Marks AR. PKA phosphorylation dissociates FKBP12.6 from the calcium release channel (ryanodine receptor): defective regulation in failing hearts. Cell. 2000;101(4):365–76.
137. Chen-Izu Y, Ward CW, Stark W Jr, Banyasz T, Sumandea MP, Balke CW, Izu LT, Wehrens XH. Phosphorylation of RyR2 and shortening of RyR2 cluster spacing in spontaneously hypertensive rat with heart failure. Am J Physiol Heart Circ Physiol. 2007;293(4):H2409–17.

138. Dries E, Santiago DJ, Gilbert G, Lenaerts I, Vandenberk B, Nagaraju CK, Johnson DM, Holemans P, Roderick HL, Macquaide N, et al. Hyperactive ryanodine receptors in human heart failure and ischaemic cardiomyopathy reside outside of couplons. Cardiovasc Res. 2018;114(11):1512–24.
139. Forbes MS, Hawkey LA, Sperelakis N. The transverse-axial tubular system (TATS) of mouse myocardium: its morphology in the developing and adult animal. Am J Anat. 1984;170(2):143–62.
140. Ogata T, Yamasaki Y. High-resolution scanning electron microscopic studies on the three-dimensional structure of the transverse-axial tubular system, sarcoplasmic reticulum and intercalated disc of the rat myocardium. Anat Rec. 1990;228(3):277–87.
141. Pinali C, Bennett H, Davenport JB, Trafford AW, Kitmitto A. Three-dimensional reconstruction of cardiac sarcoplasmic reticulum reveals a continuous network linking transverse-tubules: this organization is perturbed in heart failure. Circ Res. 2013;113(11):1219–30.
142. Rog-Zielinska EA, Moss R, Kaltenbacher W, Greiner J, Verkade P, Seemann G, Kohl P, Cannell MB. Nano-scale morphology of cardiomyocyte t-tubule/sarcoplasmic reticulum junctions revealed by ultra-rapid high-pressure freezing and electron tomography. J Mol Cell Cardiol. 2021;153:86–92.
143. Takeshima H, Komazaki S, Nishi M, Iino M, Kangawa K. Junctophilins: a novel family of junctional membrane complex proteins. Mol Cell. 2000;6(1):11–22.
144. van Oort RJ, Garbino A, Wang W, Dixit SS, Landstrom AP, Gaur N, De Almeida AC, Skapura DG, Rudy Y, Burns AR, et al. Disrupted junctional membrane complexes and hyperactive ryanodine receptors after acute junctophilin knockdown in mice. Circulation. 2011;123(9):979–88.
145. Gross P, Johnson J, Romero CM, Eaton DM, Poulet C, Sanchez-Alonso J, Lucarelli C, Ross J, Gibb AA, Garbincius JF, et al. Interaction of the joining region in Junctophilin-2 with the L-type Ca(2+) channel is pivotal for cardiac dyad assembly and intracellular Ca(2+) dynamics. Circ Res. 2021;128(1):92–114.
146. Hong T, Yang H, Zhang SS, Cho HC, Kalashnikova M, Sun B, Zhang H, Bhargava A, Grabe M, Olgin J, et al. Cardiac BIN1 folds T-tubule membrane, controlling ion flux and limiting arrhythmia. Nat Med. 2014;20(6):624–32.
147. Savio-Galimberti E, Frank J, Inoue M, Goldhaber JI, Cannell MB, Bridge JHB, Sachse FB. Novel features of the rabbit transverse tubular system revealed by quantitative analysis of three-dimensional reconstructions from confocal images. Biophys J. 2008;95(4):2053–62.
148. Brette F, Orchard C. T-tubule function in mammalian cardiac myocytes. Circ Res. 2003;92(11):1182–92.
149. Hong T, Shaw RM. Cardiac T-tubule microanatomy and function. Physiol Rev. 2017;97(1):227–52.
150. Rog-Zielinska EA, Scardigli M, Peyronnet R, Zgierski-Johnston CM, Greiner J, Madl J, O'Toole ET, Morphew M, Hoenger A, Sacconi L, et al. Beat-by-beat cardiomyocyte T-tubule deformation drives tubular content exchange. Circ Res. 2021;128(2):203–15.
151. Louch WE, Mork HK, Sexton J, Stromme TA, Laake P, Sjaastad I, Sejersted OM. T-tubule disorganization and reduced synchrony of Ca2+ release in murine cardiomyocytes following myocardial infarction. J Physiol. 2006;574(Pt 2):519–33.
152. Ibrahim M, Navaratnarajah M, Siedlecka U, Rao C, Dias P, Moshkov AV, Gorelik J, Yacoub MH, Terracciano CM. Mechanical unloading reverses transverse tubule remodelling and normalizes local Ca(2+)-induced Ca(2+)release in a rodent model of heart failure. Eur J Heart Fail. 2012;14(6):571–80.
153. Wagner E, Lauterbach MA, Kohl T, Westphal V, Williams GS, Steinbrecher JH, Streich JH, Korff B, Tuan HT, Hagen B, et al. Stimulated emission depletion live-cell super-resolution imaging shows proliferative remodeling of T-tubule membrane structures after myocardial infarction. Circ Res. 2012;111(4):402–14.
154. Pinali C, Malik N, Davenport JB, Allan LJ, Murfitt L, Iqbal MM, Boyett MR, Wright EJ, Walker R, Zhang Y, et al. Post-myocardial infarction T-tubules form enlarged branched

structures with dysregulation of Junctophilin-2 and bridging integrator 1 (BIN-1). J Am Heart Assoc. 2017;6(5):e004834.
155. Lipsett DB, Frisk M, Aronsen JM, Nordén ES, Buonarati OR, Cataliotti A, Hell JW, Sjaastad I, Christensen G, Louch WE. Cardiomyocyte substructure reverts to an immature phenotype during heart failure. J Physiol. 2019;597(7):1833–53.
156. Lawless M, Caldwell JL, Radcliffe EJ, Smith CER, Madders GWP, Hutchings DC, Woods LS, Church SJ, Unwin RD, Kirkwood GJ, et al. Phosphodiesterase 5 inhibition improves contractile function and restores transverse tubule loss and catecholamine responsiveness in heart failure. Sci Rep. 2019;9(1):6801.
157. Sacconi L, Ferrantini C, Lotti J, Coppini R, Yan P, Loew LM, Tesi C, Cerbai E, Poggesi C, Pavone FS. Action potential propagation in transverse-axial tubular system is impaired in heart failure. Proc Natl Acad Sci U S A. 2012;109(15):5815–9.
158. Wu HD, Xu M, Li RC, Guo L, Lai YS, Xu SM, Li SF, Lü QL, Li LL, Zhang HB, et al. Ultrastructural remodelling of Ca(2+) signalling apparatus in failing heart cells. Cardiovasc Res. 2012;95(4):430–8.
159. Zhang HB, Li RC, Xu M, Xu SM, Lai YS, Wu HD, Xie XJ, Gao W, Ye H, Zhang YY, et al. Ultrastructural uncoupling between T-tubules and sarcoplasmic reticulum in human heart failure. Cardiovasc Res. 2013;98(2):269–76.
160. Louch WE, Bito V, Heinzel FR, Macianskiene R, Vanhaecke J, Flameng W, Mubagwa K, Sipido KR. Reduced synchrony of Ca2+ release with loss of T-tubules-a comparison to Ca2+ release in human failing cardiomyocytes. Cardiovasc Res. 2004;62(1):63–73.
161. Kemi OJ, Hoydal MA, Macquaide N, Haram PM, Koch LG, Britton SL, Ellingsen O, Smith GL, Wisloff U. The effect of exercise training on transverse tubules in normal, remodeled, and reverse remodeled hearts. J Cell Physiol. 2011;226(9):2235–43.
162. Paulus WJ, Tschöpe C. A novel paradigm for heart failure with preserved ejection fraction: comorbidities drive myocardial dysfunction and remodeling through coronary microvascular endothelial inflammation. J Am Coll Cardiol. 2013;62(4):263–71.
163. Sweeney M, Corden B, Cook SA. Targeting cardiac fibrosis in heart failure with preserved ejection fraction: mirage or miracle? EMBO Mol Med. 2020;12(10):e10865.
164. Crossman DJ, Shen X, Jüllig M, Munro M, Hou Y, Middleditch M, Shrestha D, Li A, Lal S, Dos Remedios CG, et al. Increased collagen within the transverse tubules in human heart failure. Cardiovasc Res. 2017;113(8):879–91.
165. De La Mata A, Tajada S, O'Dwyer S, Matsumoto C, Dixon RE, Hariharan N, Moreno CM, Santana LF. BIN1 induces the formation of T-tubules and adult-like Ca(2+) release units in developing cardiomyocytes. Stem Cells. 2019;37(1):54–64.
166. Hong TT, Smyth JW, Gao D, Chu KY, Vogan JM, Fong TS, Jensen BC, Colecraft HM, Shaw RM. BIN1 localizes the L-type calcium channel to cardiac T-tubules. PLoS Biol. 2010;8(2):e1000312.
167. Reynolds JO, Quick AP, Wang Q, Beavers DL, Philippen LE, Showell J, Barreto-Torres G, Thuerauf DJ, Doroudgar S, Glembotski CC, et al. Junctophilin-2 gene therapy rescues heart failure by normalizing RyR2-mediated Ca2 + release. Int J Cardiol. 2016;225:371–80.
168. Hong TT, Smyth JW, Chu KY, Vogan JM, Fong TS, Jensen BC, Fang K, Halushka MK, Russell SD, Colecraft H, et al. BIN1 is reduced and Cav1.2 trafficking is impaired in human failing cardiomyocytes. Heart Rhythm. 2012;9(5):812–20.
169. Ibrahim M, Siedlecka U, Buyandelger B, Harada M, Rao C, Moshkov A, Bhargava A, Schneider M, Yacoub MH, Gorelik J, et al. A critical role for Telethonin in regulating t-tubule structure and function in the mammalian heart. Hum Mol Genet. 2013;22(2):372–83.
170. Lyon AR, Nikolaev VO, Miragoli M, Sikkel MB, Paur H, Benard L, Hulot JS, Kohlbrenner E, Hajjar RJ, Peters NS, et al. Plasticity of surface structures and beta(2)-adrenergic receptor localization in failing ventricular cardiomyocytes during recovery from heart failure. Circ Heart Fail. 2012;5(3):357–65.

171. Zhang C, Chen B, Wang Y, Guo A, Tang Y, Khataei T, Shi Y, Kutschke WJ, Zimmerman K, Weiss RM, et al. MG53 is dispensable for T-tubule maturation but critical for maintaining T-tubule integrity following cardiac stress. J Mol Cell Cardiol. 2017;112:123–30.
172. Scriven DRL, Asghari P, Schulson MN, Moore EDW. Analysis of Ca(v)1.2 and ryanodine receptor clusters in rat ventricular myocytes. Biophys J. 2010;99(12):3923–9.
173. Jayasinghe Isuru D, Baddeley D, Kong Cherrie H, Wehrens Xander H, Cannell Mark B, Soeller C. Nanoscale organization of junctophilin-2 and ryanodine receptors within peripheral couplings of rat ventricular cardiomyocytes. Biophys J. 2012;102(5):L19–21.
174. Asghari P, Scriven DR, Ng M, Panwar P, Chou KC, van Petegem F, Moore ED. Cardiac ryanodine receptor distribution is dynamic and changed by auxiliary proteins and post-translational modification. Elife. 2020;9:e51602.
175. Fu Y, Shaw SA, Naami R, Vuong CL, Basheer WA, Guo X, Hong T. Isoproterenol promotes rapid ryanodine receptor movement to bridging integrator 1 (BIN1)-organized dyads. Circulation. 2016;133(4):388–97.
176. Wehrens XH, Lehnart SE, Reiken S, Vest JA, Wronska A, Marks AR. Ryanodine receptor/calcium release channel PKA phosphorylation: a critical mediator of heart failure progression. Proc Natl Acad Sci U S A. 2006;103(3):511–8.
177. Sheard TMD, Hurley ME, Colyer J, White E, Norman R, Pervolaraki E, Narayanasamy KK, Hou Y, Kirton HM, Yang Z, et al. Three-dimensional and chemical mapping of intracellular signaling nanodomains in health and disease with enhanced expansion microscopy. ACS Nano. 2019;13(2):2143–57.
178. Terentyev D, Györke I, Belevych AE, Terentyeva R, Sridhar A, Nishijima Y, Carcache de Blanco E, Khanna S, Sen CK, Cardounel AJ, et al. Redox modification of ryanodine receptors contributes to sarcoplasmic reticulum Ca2+ leak in chronic heart failure. Circ Res. 2008;103(12):1466–72.
179. Benitah J-P, Perrier R, Mercadier J-J, Pereira L, Gómez AM. RyR2 and calcium release in heart failure. Front Physiol. 2021;12:734210.
180. Zima AV, Bovo E, Bers DM, Blatter LA. Ca^{2+} spark-dependent and -independent sarcoplasmic reticulum Ca^{2+} leak in normal and failing rabbit ventricular myocytes. J Physiol. 2010;588(Pt 23):4743–57.
181. Armoundas AA, Rose J, Aggarwal R, Stuyvers BD, O'Rourke B, Kass DA, Marbán E, Shorofsky SR, Tomaselli GF, William BC. Cellular and molecular determinants of altered Ca2+ handling in the failing rabbit heart: primary defects in SR Ca2+ uptake and release mechanisms. Am J Physiol Heart Circ Physiol. 2007;292(3):H1607–H18.
182. Wilson LD, Jeyaraj D, Wan X, Hoeker GS, Said TH, Gittinger M, Laurita KR, Rosenbaum DS. Heart failure enhances susceptibility to arrhythmogenic cardiac alternans. Heart Rhythm. 2009;6(2):251–9.
183. Crocini C, Coppini R, Ferrantini C, Yan P, Loew LM, Tesi C, Cerbai E, Poggesi C, Pavone FS, Sacconi L. Defects in T-tubular electrical activity underlie local alterations of calcium release in heart failure. Proc Natl Acad Sci U S A. 2014;111(42):15196–201.
184. Louch WE, Hake J, Mørk HK, Hougen K, Skrbic B, Ursu D, Tønnessen T, Sjaastad I, Sejersted OM. Slow Ca^{2+} sparks de-synchronize Ca^{2+} release in failing cardiomyocytes: evidence for altered configuration of Ca^{2+} release units? J Mol Cell Cardiol. 2013;58:41–52.
185. Nivala M, Song Z, Weiss JN, Qu Z. T-tubule disruption promotes calcium alternans in failing ventricular myocytes: mechanistic insights from computational modeling. J Mol Cell Cardiol. 2015;79:32–41.
186. Yang Z, Pascarel C, Steele DS, Komukai K, Brette F, Orchard CH. Na^+-Ca^{2+} exchange activity is localized in the T-tubules of rat ventricular myocytes. Circ Res. 2002;91(4):315–22.
187. Nikolova AP, Hitzeman TC, Baum R, Caldaruse AM, Agvanian S, Xie Y, Geft DR, Chang DH, Moriguchi JD, Hage A, et al. Association of a novel diagnostic biomarker, the plasma cardiac bridging integrator 1 score, with heart failure with preserved ejection fraction and cardiovascular hospitalization. JAMA Cardiol. 2018;3(12):1206–10.

188. Hitzeman TC, Xie Y, Zadikany RH, Nikolova AP, Baum R, Caldaruse AM, Agvanian S, Melmed GY, McGovern DPB, Geft DR, et al. cBIN1 score (CS) identifies ambulatory HFrEF patients and predicts cardiovascular events. Front Physiol. 2020;11:503.
189. Liu Y, Zhou K, Li J, Agvanian S, Caldaruse A-M, Shaw S, Hitzeman TC, Shaw RM, Hong T. In mice subjected to chronic stress, exogenous cBIN1 preserves calcium-handling machinery and cardiac function. JACC Basic Transl Sci. 2020;5(6):561–78.
190. Li J, Agvanian S, Zhou K, Shaw RM, Hong T. Exogenous cardiac bridging integrator 1 benefits mouse hearts with pre-existing pressure overload-induced heart failure. Front Physiol. 2020;11:708.
191. Zha XM, Dailey ME, Green SH. Role of Ca2+/calmodulin-dependent protein kinase II in dendritic spine remodeling during epileptiform activity in vitro. J Neurosci Res. 2009;87(9):1969–79.
192. Cordeiro JM, Greene L, Heilmann C, Antzelevitch D, Antzelevitch C. Transmural heterogeneity of calcium activity and mechanical function in the canine left ventricle. Am J Physiol Heart Circ Physiol. 2004;286(4):H1471–9.
193. Lou Q, Fedorov VV, Glukhov AV, Moazami N, Fast VG, Efimov IR. Transmural heterogeneity and remodeling of ventricular excitation-contraction coupling in human heart failure. Circulation. 2011;123(17):1881–90.
194. Glukhov AV, Fedorov VV, Lou Q, Ravikumar VK, Kalish PW, Schuessler RB, Moazami N, Efimov IR. Transmural dispersion of repolarization in failing and nonfailing human ventricle. Circ Res. 2010;106(5):981–91.
195. Gaeta SA, Bub G, Abbott GW, Christini DJ. Dynamical mechanism for subcellular alternans in cardiac myocytes. Circ Res. 2009;105(4):335–42.
196. Song Z, Liu MB, Qu Z. Transverse tubular network structures in the genesis of intracellular calcium alternans and triggered activity in cardiac cells. J Mol Cell Cardiol. 2018;114:288–99.
197. Shiferaw Y, Aistrup GL, Louch WE, Wasserstrom JA. Remodeling promotes proarrhythmic disruption of calcium homeostasis in failing atrial myocytes. Biophys J. 2020;118(2):476–91.
198. Sutanto H, van Sloun B, Schönleitner P, van Zandvoort M, Antoons G, Heijman J. The subcellular distribution of ryanodine receptors and L-type Ca(2+) channels modulates Ca(2+)-transient properties and spontaneous Ca(2+)-release events in atrial cardiomyocytes. Front Physiol. 2018;9:1108.
199. Crossman DJ, Young AA, Ruygrok PN, Nason GP, Baddelely D, Soeller C, Cannell MB. T-tubule disease: relationship between t-tubule organization and regional contractile performance in human dilated cardiomyopathy. J Mol Cell Cardiol. 2015;84:170–8.
200. Lebek S, Plossl A, Baier M, Mustroph J, Tarnowski D, Lucht CM, Schopka S, Florchinger B, Schmid C, Zausig Y, et al. The novel CaMKII inhibitor GS-680 reduces diastolic SR Ca leak and prevents CaMKII-dependent pro-arrhythmic activity. J Mol Cell Cardiol. 2018;118:159–68.
201. Duran J, Nickel L, Estrada M, Backs J, van den Hoogenhof MMG. CaMKIIdelta splice variants in the healthy and diseased heart. Front Cell Dev Biol. 2021;9:644630.

Organization of Ca^{2+} Signaling Microdomains in Cardiac Myocytes

Jing Li, Bradley Richmond, and TingTing Hong

Abstract Calcium signaling in cardiomyocytes regulates muscle contractile function and electrical signal propagation in the heart. While calcium influx in nodal cells controls the rhythmic heartbeat, calcium transients in working cardiomyocytes direct excitation-contraction coupling and energy-dependent diastolic relaxation to maintain beat-to-beat pump function. Calcium dysregulation is a hallmark of the pathophysiology of multiple heart diseases including hypertrophy, heart failure, and arrhythmia. Proper compartmentalization and regulation of intra-cardiomyocyte calcium signaling relies on the organization of key calcium-handling machinery to specific functional microdomains. These microdomains serve as signaling hubs to orchestrate cardiac activities whose disruption can contribute to disease progression. Thus, understanding the structure and function of the calcium signaling microdomains in cardiomyocytes is of great scientific interest and translational significance. In this chapter, we discuss the current knowledge concerning calcium signaling at major cardiomyocyte microdomains, microdomain remodeling in diseases, and targetable approaches for the development of new therapies.

Jing Li and Bradley Richmond contributed equally with all other contributors.

J. Li · B. Richmond
Department of Pharmacology and Toxicology, College of Pharmacy, University of Utah, Salt Lake City, UT, USA

T. Hong (✉)
Department of Pharmacology and Toxicology, College of Pharmacy, University of Utah, Salt Lake City, UT, USA

Nora Eccles Harrison Cardiovascular Research and Training Institute, University of Utah, Salt Lake City, UT, USA

Diabetes and Metabolism Research Center, University of Utah, Salt Lake City, UT, USA
e-mail: TingTing.Hong@pharm.utah.edu

N. L. Parinandi, T. J. Hund (eds.), *Cardiovascular Signaling in Health and Disease*, https://doi.org/10.1007/978-3-031-08309-9_2

Keywords Calcium signaling · Microdomains · Cardiomyocytes · Transverse tubule · Cardiac bridging integrator 1 · Ankyrin · Caveolae · Heart failure · Arrhythmia

Introduction

Human hearts consist of billions of cells activated in a coordinated sequence leading to synchronous contraction and relaxation during each heartbeat cycle. This coordination is responsible for the rhythmic and efficient pump function of the heart. For each heartbeat, a normal sequence of events starts with action potential initiation at the sinoatrial node followed by a rapid myocyte-to-myocyte spread of action potential through the conduction system and the working myocardium resulting in sequential activation of atria and ventricles. During diastolic relaxation, heart chambers dilate to allow blood refill and to prepare for next contraction. Effective contraction of heart chambers requires efficient intracellular calcium signaling events occurring at the individual cardiomyocyte level. As the contractile units building the chamber walls, working cardiomyocytes are large, rod-shaped cells with longitudinally aligned contractile myofibrils under the regulation of intracellular calcium concentrations. Effective beat-to-beat ventricular contraction and relaxation require coordinated calcium cycling in individual ventricular cardiomyocytes, which is often achieved by sequential calcium signaling events at compartmentalized intracellular microdomains.

Calcium Signaling in Cardiomyocytes

Normal beat-to-beat heart contraction requires proper excitation-contraction (EC) coupling between membrane action potential and cellular contraction. Efficient EC coupling is achieved by healthy intracellular calcium transients, which are a sequence of calcium signaling events starting from action potential-activated initial calcium entry, followed by a massive calcium release from the intracellular calcium-storing sarcoplasmic reticulum (SR), and completed with the subsequent diastolic calcium removal via SR reuptake and sarcolemmal exclusion.

The current accepted model of intracellular calcium transient development is described as calcium-induced-calcium release (CICR) [1]. Following each membrane action potential, opening of the sarcolemmal voltage-gated L-type calcium channels (LTCCs) mediates the initial calcium influx into the cytosol to then induce massive calcium release from the SR. After its original proposal in early 1980s, this model is supported by the subsequent finding of ryanodine receptors (RyRs) at the SR membrane which sense the calcium influx and cause the release of calcium from the SR [2, 3]. Later in 1993, the Lederer lab further identified calcium sparks from SR as the "elementary units" of calcium transients [4]. Optimal CICR requires a close physical association between the sarcolemmal LTCCs and the junctional SR (jSR) RyRs [1–4]. This is achieved by LTCC localization to the transverse tubules (t-tubules or TT) [5, 6]. The complexes of t-tubule LTCCs together with jSR membrane RyRs (approximately a 1:4 ratio) are known as dyads [7, 8]. During relaxation, the accumulated calcium needs to be removed from the cytosol by either calcium sequestration to SR lumen via the SR Ca^{2+}-ATPase 2a (SERCA2a) or calcium removal into the extracellular space by the Na^{+}/Ca^{2+} exchanger (NCX) and to a lesser extent by the sarcolemmal Ca^{2+}-ATPase [9]. Thus, a normal transient requires both optimal CICR initiated at dyads where LTCCs at TTs are coupled with RyRs at jSR and efficient energy-dependent diastolic calcium removal mainly through SERCA2a-mediated SR reuptake and NCX-mediated exclusion.

Impaired calcium transients are a pathophysiologic hallmark of failing myocytes [10–12]. Disturbed calcium transients in failing cardiomyocytes are attributed to TT remodeling [13–15] which uncouples LTCC-RyR dyads [11, 15–19] and diseased functional downregulation of calcium removal proteins including SERCA2a [20] and NCX, impairing cardiac inotropy and lusitropy. Given that calcium transients are generated by multiple channels and organelles, there are many disease-related alterations possible that can result in abnormal pump function. Historically, the focus has been on the altered calcium ion uptake into and release out of the SR. The typical failing calcium transient has a low peak amplitude and slowed decline [21–23]. The slow decline of the calcium transient is attributed to diseased regulation of SERCA2a and NCX [20, 24], whereas the reduction in calcium amplitude is due to hyperphosphorylation of RyRs [25, 26] that causes SR leakage and depletion of calcium storage in the SR. During acute or chronic stress such as sympathetic overdrive and pressure overload, altered β-adrenergic signaling and intracellular calcium homeostasis can contribute to pump failure by modulating the phosphorylation and functional states of the calcium-handling proteins.

In conclusion, a healthy calcium transient is needed for normal cardiac contraction and relaxation capacity in both resting states and under stressed conditions. In addition to supporting proper pump function, intracellular calcium homeostasis also helps maintain desired cardiomyocyte excitability. Disruption of normal excitability can cause life-threatening arrhythmias and sudden cardiac death. For instance, in dilated cardiomyopathy abnormal hyperphosphorylation-induced hyperactive and leaky RyR receptors cause SR calcium leak, decreasing EC coupling and promoting ventricular arrhythmias. Furthermore, outside of working cardiomyocytes, intracellular calcium signaling determinates a variety of different physiological functions in other types of cardiac myocytes. In pacemaker cells, for example, calcium influx through voltage-gated calcium channels controls automaticity and thus heart rate.

Taken together, intra-myocyte calcium signaling is critical to a range of key functions of the heart which ultimately sustain rhythmic and robust heartbeats.

Calcium Signaling Microdomains in Cardiomyocytes

The most well-studied microdomain-based regulation of calcium signaling is perhaps the modulation of intracellular distribution and compartmentalization of LTCC-RyR dyads, including regulation caused by β-adrenergic receptor (β-AR) activity. In this section, we will focus on the major intracellular microdomains involved in the regulation of LTCC-RyR dyads and diastolic calcium-handling proteins SERCA2a and NCX, including the cardiac bridging integrator 1 (cBIN1)-microdomains at the t-tubules, the ankyrin-spectrin microdomains, and the caveolae microdomains. Additionally, we will discuss how these microdomains remodel during disease progression.

TT/jSR Microdomains in Systolic and Diastolic Calcium Handling

As discussed earlier, EC coupling in cardiomyocytes is regulated at the LTCC-RyR dyads localized to microdomains within cardiac t-tubules. T-tubules are sarcolemmal invaginations that form and organize an interconnected and complex tubular membrane network. To achieve a variety of specific functional needs, t-tubules contain microdomains created by scaffolding proteins with distinct pools of ion channels, transporters, and signaling molecules clustered at compartmentalized subregions [27]. This is the case in dyad formation, as LTCC-RyR couplons are brought together by the membrane-molding and scaffolding protein cBIN1 [28, 29]. cBIN1 draws together and bridges t-tubule LTCCs to RyRs at the jSR membrane for synchronous CICR and efficient EC coupling. In this section, we will focus on the organization, functional roles, and abnormal remodeling of the t-tubule-localized cBIN1-microdomains. We will highlight their roles in the regulation of calcium signaling in normal and diseased cardiomyocytes. Possible cross talk between cBIN1-microdomains and other calcium signaling regulatory TT/jSR adaptor proteins like junctophilin-2 (JP2) will also be reviewed.

Organization of cBIN1 and JP2 Microdomains at TT/jSR Membrane

T-tubule cBIN1-microdomains are created by a membrane-scaffolding protein cBIN1, the cardiac isoform of BIN1 encoded by a splice variance derived from the *BIN1* gene. BIN1 is a phospholipid-binding and cytoskeleton-interacting protein belonging to the N-terminal Bin1-amphiphysin-Rvs (N-BAR) domain-containing protein superfamily [27, 30]. In the mammalian system, BIN1 (also known as amphiphysin 2) is encoded by a single gene of 20 exons and produces over ten protein isoforms via alternative splicing [28, 31–34]. Encoded by constitutive exons 1–10 in all variants, the N-terminal N-BAR domain of BIN1 forms intermolecular homodimers with a "banana"-shaped concave surface topology that binds to lipid bilayers and induces membrane invagination [35]. The C-terminal SH3 domain in BIN1, encoded by the last two constitutive exons [31, 33], is critical for BIN1 interaction with intracellular proteins including the microtubule and actin cytoskeleton [36–38]. Finally, the middle exons 11–18 encode a coiled-coil region in BIN1 which is heavily alternatively spliced with tissue and disease specificities [27]. For instance, the MYC-binding domain is formed by a constitutive exon 18 and a ubiquitously alternatively spliced exon 17 [39]. Other examples include the isoforms containing an exon 11-encoded phosphoinositide-binding motif which are highly expressed in skeletal muscle [34]. These isoforms provide a heightened binding affinity to phospholipids and thus have the membrane bending ability required for skeletal t-tubule biogenesis. In the neuronal system, a set of four continuous exons 13–16 are often co-spliced to encode an clathrin-associated protein (CLAP)-binding domain for endocytic neurotransmitter reuptake [32, 33].

In mammalian cardiomyocytes, alternate splicing of the *BIN1* gene produces four to six transcript variants and corresponding protein isoforms [28, 40]. Two BIN1 protein isoforms featuring the inclusion of exon 13, BIN1 + 13 and BIN1 + 13 + 17 (also known as cBIN1) [28], were detected in adult mouse ventricular cardiomyocytes. cBIN1 is the isoform that localizes to t-tubules, binds to the Z-disc structural protein α-actinin, and facilitates N-WASP-promoted cortical actin polymerization [28]. The unique domain encoded by co-splicing of exons 13 and 17 without exons 14–16 in cBIN1 allows the molecule to both interact with the lipid bilayer at t-tubules and the subsarcolemmal cortical actin. Both interactions are critical to the formation of microdomains anchored at the myofilament Z-discs [28]. Direct visualization by super-resolution fluorescent microscopy and transmission electron microscopy imaging revealed that cBIN1-microdomains are formed by contoured and multilayered membrane microfolds [29, 41]. In BIN1-deficient cardiomyocytes and the loss of domain-forming microfolds was rescued by exogenous cBIN1 but not by other BIN1 isoforms [28, 42]. Later studies identified that the exon 11-containing BIN1 isoform(s) highly expressed in skeletal muscle can also be detected in sheep cardiomyocytes [40]. Also, in human induced pluripotent stem cell (hiPSC)-derived cardiomyocytes, overexpression of the skeletal BIN1 isoform increased calcium transient amplitude and EC coupling [43]. The functional similarities and distinctions as well as impacts from intermolecular interactions among BIN1 isoforms in adult human ventricular cardiomyocytes await future detailed studies.

Another set of microdomains at the TT/jSR are the microdomains organized by a cardiac expressed JP2 whose unique binding affinity to both the TT and jSR membrane link the two membrane systems together. JP2 achieves the TT/jSR linkage function via a C-terminal hydrophobic domain which spans the jSR membrane and by its cytosolic membrane occupation and recognition nexus (MORN) domains which associate with the t-tubule membrane [44]. The membrane-tethering affinity of JP2 domains is modulated by posttranslational palmitoylation [45]. Theories concerning the roles of JP2 in organizing t-tubule structure and EC coupling are still evolving. Recently, it was reported that the joining region in JP2 interacts with and recruits LTCC and facilitates dyad assembly for normal cardiac CICR [46, 47]. Outside of its regulation of t-tubule structure and EC coupling, a separate role of JP2 has been reported as a gene transcriptional regulator when the N-terminal fragment of JP2 is liberated under stress and translocated to nucleus to modify gene transcription [48]. Future studies will be necessary to understand the full spectrum of JP2 functions and its contribution to healthy cardiomyocyte function.

Calcium Signaling at the cBIN1 and Other TT/jSR Microdomains

In mature ventricular cardiomyocytes, cBIN1-microdomains at t-tubules facilitate efficient EC coupling by organizing dyads formed by LTCCs and RyRs [27, 29]. T-tubule cBIN1-microdomains organize LTCC clusters by facilitating the microtubule-dependent forward trafficking of LTCCs [38] and by stabilizing LTCCs that are already delivered to the t-tubule surface. cBIN1-organized LTCC clusters are effectively coupled to RyRs via cBIN1-facilitated RyR recruitment to the jSR membrane [29]. This role of cBIN1 in recruiting RyRs is particularly important for proper dyad regulation during the flight-and-fight response driven under the acute activation of the sympathetic nervous system [29]. In normal cardiomyocytes with sufficient cBIN1 expression, the acute stress response causes a fast and responsive mobilization of cBIN1-microdomains in order to recruit active RyRs that are phosphorylated by protein kinase A (PKA). In the presence of intact cBIN1-microdomains, these stress-activated RyRs are effectively recruited to jSR to couple with LTCCs, forming larger and stronger dyads with elevated EC coupling gain [29] (Fig. 1a). Thus, cBIN1-microdomains assist cardiomyocytes in meeting functional systolic needs at resting states and particularly during acute stress responses. Recently, the cBIN1-microdomain has been identified as a critical regulator of diastolic calcium removal via its organization of the intracellular SERCA2a distribution along the SR membrane [42]. Subpopulations of SERCA2a near jSR are organized by cBIN1 to allow a quick decline of cytosolic calcium at the TT/jSR cleft during diastolic relaxation [42] (Fig. 1a). Such a role in diastolic calcium regulation indicates that cBIN1-microdomains may serve as a centralized calcium signaling hub in ventricular cardiomyocytes. This raises the possibility that cBIN1-microdomains can be targeted to induce simultaneous improvement in cardiac inotropy and lusitropy. Given the current lack of treatments for diastolic dysfunction, lusitropic protection offered by cBIN1-microdomains could be significant in the development of therapies for hearts with diastolic dysfunction.

In addition to regulating intracellular calcium cycling, cBIN1-microdomains also promote the electrical stability of the heart [28]. The tortuous membrane folds sculpted by cBIN1 dimers can restrict extracellular ion diffusion within t-tubule

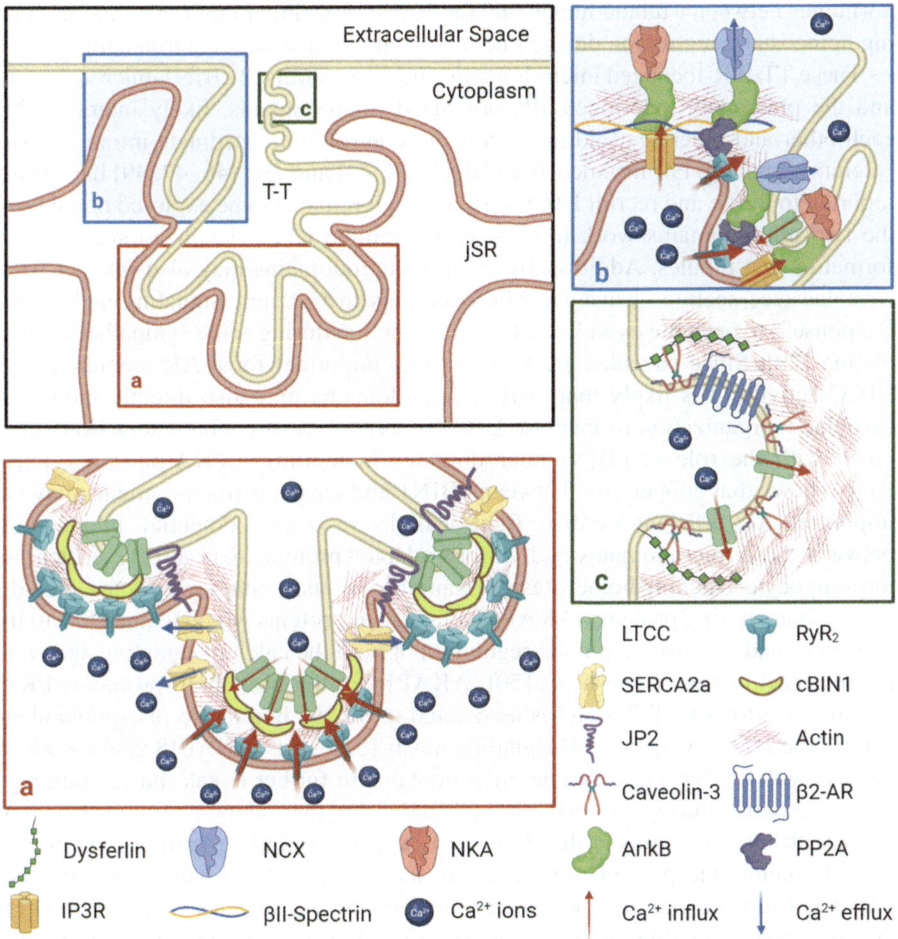

Fig. 1 Schematic illustration of calcium signaling microdomains in healthy cardiac transverse tubules. Box (**a**) cBIN1 and JP2 microdomains; Box (**b**) ankyrin-spectrin microdomains; Box (**c**) caveolae microdomains. Abbreviations are as follows: T-T transverse tubule, jSR junctional sarcoplasmic reticulum, LTCC L-type calcium channel, RyR_2 ryanodine receptor 2, SERCA2a sarcoendoplasmic reticulum Ca^{2+}-ATPase, cBIN1 cardiac bridging integrator 1, JP2 junctophilin-2, β2-AR β2 adrenergic receptor, NCX, a^+/Ca^{2+} exchanger, NKA sodium potassium ATPase, AnkB Ankyrin B, PP2A protein phosphatase 2A, IP3R inositol trisphosphate receptor. (Illustration created with BioRender.com)

lumen [28], maintaining cardiac homeostasis and membrane excitability [28]. During increasing rates of contraction, the slow diffusion zone causes outward-flowing ions like K^+ to accumulate while inward current ions like Na^+ and Ca^{2+} to deplete quickly, maintaining electrical stability of cardiomyocytes [28]. This slow diffusion zone may also restrict the exchange of other small molecules such as hormones between t-tubule lumen and the bulk extracellular environment, facilitating homeostatic regulation during fluctuations in extracellular environment.

These TT/jSR-localized microdomains, including both the cBIN1-microdomains and the previously mentioned JP2-organized microdomains, likely interact with each other and other scaffolding proteins to control and coordinate intra-myocyte calcium signaling. For instance, both BIN1 [38, 43] and JP2 [46, 47, 49] have been reported to traffic and recruit LTCCs. However, it remains to be explored if and how the two microdomains work in concert to promote LTCC localization and dyad formation at t-tubules. Additionally, cBIN1-microdomains may also interact with caveolae (see section "Caveolae Microdomains in Calcium Signaling and Stress Response") to promote dyad formation and function during acute sympathetic stimulation [29]. Since caveolae are known to be important for β-AR modulation of LTCC activity, it is likely that cBIN1-microdomains may also directly modulate caveolae reorganization to increase LTCC activity at dyads. This interaction combined with the role of cBIN1-microdomains in recruiting PKA-phosphorylated RyRs shows that cooperation between cBIN1 and caveolae microdomains may be important for efficient cardiac fight-or-flight response. Potential interactions between these microdomains and various adaptor proteins may also be critical to intra-myocyte calcium homeostasis. Examples of such adaptor proteins include A-kinase anchoring proteins (AKAPs), a family of proteins which tether PKA to its substrates and are involved in the regulation of both the calcium signaling and contractile machineries in the heart [50]. AKAP18 in the heart both promotes PKA phosphorylation of LTCCs and is associated to SR via binding to phospholamban (PLN), thereby serving as a SR signaling nexus [50]. Thus, AKAP18 could work in tandem with cBIN1 and caveolae microdomains to further tweak the calcium signaling response during sympathetic activation. Another adaptor protein that may work with microdomains is the BIN1 and caveolin-3-binding partner dysferlin, a 230-kD membrane protein with calcium and phospholipid-binding affinity [51] found at t-tubules and enriched at the subsarcolemmal vesicles [52, 53] (Fig. 1). Given its role in facilitating membrane repair [54, 55], dysferlin's interaction with cBIN1 and caveolae microdomains could be important for lipid and membrane trafficking at t-tubules. Disruption of these processes could result in de-tubulation which in turn impairs calcium transients and promotes cardiomyopathy. All these examples demonstrate the importance of identifying the complex molecular interactions and functional interplays among membrane microdomains and adaptor proteins at the sites near TT/jSR where the calcium transients initiate.

TT/jSR Microdomains, Calcium Signaling, and Heart Failure

Heart failure (HF) is a major cardiac syndrome with high mortality and morbidity. It is also a widespread challenge, with Americans over 45 years of age having overall lifetime risks for HF ranging from 20% to 45% [56]. Indeed, heart failure-related hospitalization in individuals 65 years of age and older is the single greatest cost to the Medicare budget [57]. A typical pathophysiology of failing cardiomyocytes includes weakened calcium transients [10–12] due to abnormal systolic calcium release from LTCC-RyR dyads [6, 11, 15–19] as well as impaired diastolic removal due to altered SERCA2a activity [20]. The central factor contributing to calcium transient weakening is the disorganization of t-tubules [13–15] leading to disruption of microdomains and calcium-handling machinery. Indeed, in the past decade, gross t-tubule network remodeling has gained substantial appreciation as a marker for the transition from hypertrophy to failure [14] and as a contributing mechanism of heart failure development and progression generally [15, 58–62]. Additionally, abnormal remodeling in calcium-regulating microdomains at TT/jSR regions such as the cBIN1, ankyrin B, JP2, and caveolae microdomains has recently drawn significant interest in the cardiovascular research field [27]. Today, new data are constantly accruing which reveal the importance of these t-tubule microdomains and their contribution to heart failure pathophysiology.

Since the original introduction of BIN1 into the cardiac field [38] and the cloning of the cardiac isoform in 2014 [28], a wealth of evidence has been generated connecting cBIN1-microdomain abnormalities to disturbed calcium transients in failing cardiomyocytes [29, 63–66]. cBIN1-microdomains are increasingly appreciated as the TT/jSR signaling hub central to the organization of CICR machinery including the LTCC-RyR dyad [29] and the SR calcium pump SERCA2a [42]. In failing cardiomyocytes, BIN1 proteins are transcriptionally decreased in human [63] and animal models of both systolic and diastolic heart failure [29, 42]. Moreover, BIN1 protein levels can recover during recovery of heart function following SERCA2a gene therapy [64]. Pathological outcomes resulting from reduced cBIN1 in failing cardiomyocytes include (a) diminished LTCC expression at t-tubule membrane due to reduced ability of cBIN1-deficient t-tubules to attract LTCC vesicles trafficking along microtubule highways [38], (b) disrupted LTCC clusters and reduced channel activity when cBIN1 scaffolds are missing, (c) loss of the restrictive diffusion zone and its protection against ventricular arrhythmia due to insufficient cBIN1 microfolds [28], (d) inefficient dyad recruitment of active RyRs during β-AR stress which both reduces the EC coupling gain needed for a proper stress response and causes the accumulation of leaky phosphorylated RyRs ("orphan" RyRs) that contribute to arrhythmia [29], and (e) diastolic calcium overload from reduced SERCA2a activity due to its improper redistribution along the SR [42]. Of great importance and excitement is that the converse is also true. Normalization of cBIN1-microdomains in failing cardiomyocytes by exogenous cBIN1 introduced by gene therapy can successfully reorganize the calcium-handling machinery in mouse hearts subjected to either chronic sympathetic overdrive [42] or pressure overload [42, 67]. Significantly, the restoration of this machinery results in the restoration of cardiac function. The functional importance of the cBIN1-microdomain in cardiomyocyte biology and heart failure pathophysiology, its involvement in both systolic and diastolic calcium regulation, and the recent successes of therapeutic interventions all indicate that this TT/jSR-localized membrane microdomain could be a master regulator of cardiomyocyte health. Because cBIN1-microdomains can modulate the entirety of the CICR calcium-handling machinery, cBIN1-microdomains offer a rich target in the development of new and effective therapies for heart failure.

cBIN1-microdomain deficiency also exacerbates β-AR-activated and PKA-promoted hyperphosphorylation of RyR_2 in cardiomyocytes [29], indicating microdomain-based modulation of the phosphorylation states and activities of calcium-handling proteins. As one of the key pathophysiologic hallmarks contributing to calcium mishandling in failing cardiomyocytes, altered phosphorylation states of calcium-handling proteins are caused by disrupted balance between protein kinases (PKA, PKC, Ca^{2+}/calmodulin-dependent protein kinase II (CAMKII), etc.) and protein phosphatases (PPs: PP1, PP2A, PP2B; calcineurin, etc.) [68–70]. Interestingly, as common targets of these kinases and phosphatases, the steady-state phosphorylation of calcium-handling proteins can be drastically different from each other [70, 71]. Under diseased conditions such as heart failure, RyR_2 is often hyper-phosphorylated and coexists with the opposite hypo-phosphorylated PLN [69, 70, 72, 73], indicating target-specific differential regulation. Although the underlying mechanisms are not completely elucidated, increasing evidence suggest that distinct scaffold proteins and their organized microdomains may differentially compartmentalize calcium-handling proteins with unique subsets of regulatory kinases and phosphatases. This intricate compartmentalization may contribute to the differences in the steady-state phosphorylation of protein targets such as RyR_2 and PLN. For instance, ankyrin B-microdomains (see section "Ankyrin-Spectrin Microdomains in Calcium Signaling") can localize PP2A to its primary target RyR_2 at the jSR to modulate RyR_2 phosphorylation [74, 75], while cBIN1-microdomains promote RyR_2 hyperphosphorylation by PKA but not CAMKII [29]. These data indicate that discrete microdomains may allocate divergent panels of kinases and phosphatases to specific protein targets, resulting in distinctive and microdomain-specific modulation of calcium-handling and overall intracellular calcium transients.

On the other hand, the roles of JP2-microdomains at the TT/jSR are complicated, and their functional importance in heart failure pathophysiology remains intriguing. Some studies suggest that knockdown of JP2 results in systolic heart failure [76] and that JP2 is downregulated [76, 77] or proteolytically cleaved in some models of heart failure [78]. However, other studies failed to detect JP2 changes in failing hearts [64, 65]. Interestingly, in a more recent study using the CASAAV (CRISPR/Cas9-AAV9-based somatic mutagenesis) to investigate cell-autonomous *JP2* function, it was found that loss of JP2 microdomains is more of a secondary effect rather than a primary cause of heart failure [79]. Nevertheless, the role of JP2 in linking TT/jSR where EC coupling dyads are located justifies further research into JP2-microdomain function, remodeling, and contribution to heart failure etiology and progression. Interactions between JP2 and other microdomains should also be explored to elucidate how functional interplays at TT/jSR regions could contribute to homeostatic control of beat-to-beat calcium cycling.

In summary, the junctional TT/jSR sites between t-tubules and jSR membrane house the calcium-handling machinery in cardiomyocytes critical to synchronous CICR and healthy calcium transients. Many scaffolding proteins at this specific TT/jSR region create a centralized calcium signaling hub by bringing calcium-handling proteins, receptors, enzymes (particularly kinases and phosphatases), plasma membrane lipids, adaptor proteins, and cytoskeleton components together into macromolecular complexes. This signaling hub not only maintains healthy calcium homeostasis at basal conditions, but it also controls effective calcium machinery mobilization and reorganization in response to stress. During disease progression, complex adaptive and maladaptive remodeling occurs at these microdomains, resulting in impaired calcium signaling and contributing to the worsening cycles of heart failure.

Ankyrin-Spectrin Microdomains in Calcium Signaling

In addition to the dyad-organizing cBIN1-microdomains at the TT/jSR regions, cardiac t-tubules are also enriched with microdomains organized by subsarcolemmal actin, ankyrins, and spectrins [80]. These well-studied ankyrin-spectrin microdomains are essential to the formation and correct localization of three key macromolecular complexes: NCX, sodium potassium ATPase (NKA), and the inositol trisphosphate receptor (IP3R) [81]. The NCX/NKA/IP3R microdomains organized by ankyrin B are critical to both cytosolic calcium decline during relaxation and the extrusion of excess calcium that might interfere with CICR efficiency. In addition, many other ankyrin-spectrin microdomains exist consisting of different spectrin and ankyrin protein isoforms and their unique set of binding partners. These microdomains are found at both the lateral sarcolemma and the intercalated discs of myocytes, and each is responsible for a distinct set of cellular functions critical to heart health [80]. In this section, we will discuss the organization of ankyrin-spectrin microdomains in cardiomyocytes, their roles in regulation of calcium signaling, and their relationships to cardiovascular diseases.

Organization of Ankyrin B-βII Spectrin Microdomains

The ankyrin-spectrin structural microdomains are formed by a subcortical actin/spectrin cytoskeleton network held together by the adaptor protein ankyrins. Spectrins are actin-binding polypeptides which provide structural support to the plasma membrane [82]. Two subunits of spectrin, α and β spectrin subunits, are encoded by two α and five β genes, respectively [83]. These spectrin subunits form antiparallel heterodimers which can assemble into a variety of hetero-tetramic spectrin molecules with differential localization and function [84]. For instance, while βIV spectrin binds to ankyrin G at intercalated disks [85], βII spectrin is expressed at t-tubules and catalyzes the formation of the larger ankyrin B macromolecular complex [86, 87]. Spectrins bind to ankyrins for interaction with integral membrane proteins. Of particular importance to calcium signaling, loss of βII spectrin in adult ventricular cardiomyocytes results in impaired expression of ankyrin B and its associated proteins [87, 88].

Ankyrins are adaptor proteins containing four domains including an N-terminal membrane-binding domain (MBD), a spectrin-binding domain (SBD), a death domain (DD), and a C-terminal domain (CTD) [89]. The MBD domain contains 24 ANK repeats with specific binding pockets for a number of membrane proteins [90–97] including ion channels, transporters, and pumps. These pockets are designed to facilitate simultaneous binding and multimodal regulation. The SBD in ankyrins is a highly conserved domain containing a tandem ZUF^N-ZUF^C-UPA structure with binding affinity to both spectrins [98, 99] and calcium-regulating proteins [74, 75, 87]. The DD and CTD together are recognized as the regulator domain (RD) with the capacity to modulate intermolecular interactions and localization of ankyrin-associated proteins [80, 100]. By bridging the actin/spectrin network to membrane proteins and their binding partners, ankyrins form large macromolecular complexes featuring compartmentalized functional regulation.

In vertebrates, there are three different ankyrin genes encoding ankyrin R (*ANK1*), ankyrin B (*ANK2*), and ankyrin G (*ANK3*). Alternative splicing of each of the three *ANK* genes generates multiple protein isoforms termed as small, canonical, and giant [80]. While ankyrin R is primarily expressed in the erythrocytes, smaller ankyrin R isoforms are expressed in striated muscle and bind to obscurins [101]. Ankyrin B and G are expressed ubiquitously across different types of cells, including cardiomyocytes [101]. Despite sequence similarity, ankyrin B and G contain distinct and non-compensatory functions that do not overlap. Ankyrin G and spectrin IV complex at the intercalated discs for interaction with Nav1.5, plakophilin, and CAMKII [85]. Ankyrin B is found at t-tubule, SR, and lateral membrane in myocytes comprising nodal cells, conduction system cells, atrial cardiomyocytes, and ventricular cardiomyocytes [86, 87, 101, 102]. In addition to the canonical 220-kD ankyrin B, two smaller isoforms of ankyrin B (188 and 212 kD) were also found in cardiomyocytes with distinct localization and functions [103]. While ankyrin B-212 kD expressed at M-line interacts exclusively with obscurins, the ankyrin B-188 kD isoform is expressed at the t-tubules of the ventricular cardiomyocytes and interacts with βII spectrin to regulate NCX expression at t-tubules. As ankyrin B is critical to regulation of cytosolic calcium removal, its binding partners and functional roles will be elucidated in the next section.

Ankyrin-Spectrin Microdomain-Regulated Calcium Signaling

The wide spectrum of ankyrin B-regulated myocardial signaling and activity is due to the large body of identified binding partners of ankyrin B in the heart. These partners include the MBD-bound ion channels and transporters (Cav1.3 in SAN only [97], Kir6.2 [96]), structural proteins (tubulin, β-catenin), and cell adhesion molecules (β-dystroglycan, dystrophin) [92, 104, 105]; the SBD-bound spectrins [98] (αII and βII spectrins) and protein phosphatase 2A (PP2A) [74]; and the HSP40 [106] and obscurin [107] that are bound to the RD in ankyrin B (see reviews in [100, 101]). This large pool of ankyrin B-binding partners enables the multimodal regulatory capacity of ankyrin B and allows the protein to modulate several cell biological processes in myocytes from calcium signaling to membrane excitability. In the SAN myocytes, ankyrin B regulates surface expression of the neuronal L-type calcium channels (Cav1.3) for normal pacemaker automaticity [97]. In ventricular cardiomyocytes, binding to the MBD domain of ankyrin B is required for the t-tubule localization of both NCX and NKA. These interactions form the well-established ankyrin B macromolecular complex containing NCX and NKA at the t-tubule membrane, as well as IP3R at the SR membrane [81]. Located next to the LTCC-RyR dyads, these ankyrin B microdomains help organize the functional NCX localization for efficient sarcolemmal calcium exclusion during diastole (Fig. 1b). This localization ensures a rapid calcium decline in the cleft between t-tubule and jSR membrane during relaxation. The $ZU5^c$ region of ankyrin B-SBD was later found to be associated with the B56α subunit of PP2A [74, 75], which localizes PP2A to its primary target RyR_2 at the jSR membrane. The interaction between ankyrin B-SBD and PP2A likely regulates the phosphorylation states of the RyR_2, a known determinator of RyR_2 activity and RyR_2-dependent SR calcium leak. By regulating both systolic SR calcium release via RyR_2 and diastolic calcium exclusion via t-tubule NCX, the t-tubule-localized ankyrin B-microdomains in ventricular cardiomyocytes play a critical role in maintaining intracellular calcium equilibrium for coordinated beat-to-beat heart contraction (Fig. 1b).

On the other hand, the ankyrin G and βIV spectrin complex at the intercalated discs is critical to Nav1.5 surface expression and channel activity modulated by CAMKII [85]. In addition to facilitating ankyrin G and CAMKII-dependent regulation of Nav1.5, βIV spectrin binding is critical to the localization and function of the mechano-sensitive K2p channel TREK-1 in ventricular cardiomyocytes [108]. In summary, the diverse roles of ankyrin-spectrin-organized ionophoric and signaling proteins are pivotal to the maintenance of myocyte contraction and membrane excitability for healthy and rhythmic heartbeats.

Ankyrin-Spectrin Microdomains in Cardiovascular Diseases

Abnormal remodeling in ankyrin-spectrin microdomains is associated with a variety of cardiovascular diseases. Due to its role in regulating the I_{Na} current at the intercalated discs, ankyrin G is associated with Brugada syndrome and arrhythmias in a *SCN5A* variant affecting the ankyrin G-binding motif of the $Na_v1.5$ channel itself (E1053K) [94]. *Ank3* conditional knockouts in mouse cardiomyocytes cause an ankyrin G deficiency as well as reduced heart rates, prolonged PR intervals, widened QRS complexes, and more frequent adrenergic stress-induced arrhythmias [109]. Ankyrin G is also a key nodal protein required for cardiac myofilament integration with the intercalated disks [110]. Deficiency of ankyrin G in mice and human is associated with the development and progression of dilated cardiomyopathy and heart failure [109].

Though ankyrin B loss most commonly manifests as arrhythmias, it has also been directly linked with a wide spectrum of cardiovascular diseases in human [101]. Loss of ankyrin B function in humans and mice is associated with a variety of changes in cardiac structure and function. At the myocyte level, ankyrin B loss disrupts ankyrin B-spectrin βII microdomains and thereby causes membrane ion channel mis-localization. This mis-localization can substantially disrupt intracellular calcium homeostasis and promote arrhythmias [111]. Of particular note is ankyrin B syndrome, an inheritable arrhythmogenic disease in humans originally classified as long QT syndrome 4 which develops from the loss of function of the *ANK2* gene [112]. In addition to arrhythmogenesis, ankyrin B syndrome is now associated with altered cardiac structure, excitability, and signaling. Thus, the list of clinical phenotypes of ankyrin B syndrome has expanded to include sick sinus node disease, heart rate variability, long QT syndrome, CPVT, arrhythmogenic cardiomyopathy, atrial fibrillation, acquired heart failure, and sudden cardiac death [101]. While the underlying mechanisms are yet to be elucidated, the observed phenotypic variability of ankyrin B syndrome is reflected in the fact that loss-of-function mutations can be found in all the four domains of ankyrin B. The first loss-of-function mutation discovered was an p.E1425G variant affecting the RD domain which was found in a French family over two decades ago [113]. Patients with this variant exhibit sinus node dysfunction, atrial fibrillation, prolonged QTc, CPVT, and sudden cardiac death. The more recently identified p.S6446F variant is found in the N-terminal MBD and causes improper NCX localization with subsequent cytosolic calcium overload [114]. Patients with the p.S6446F variant present congenital heart defect, Wolff-Parkinson-White syndrome, and cardiomyopathy. Variants in the SBD can disrupt ankyrin B's binding affinity to βII spectrin with or without altering NCX localization [87]. Interestingly, another variant identified as p.Q1283H in the SBD can cause impaired ankyrin B interaction with PP2A with a resultant RyR_2 hyperphosphorylation [115]. This connection links ankyrin B syndrome to altered EC coupling and dilated cardiomyopathy. In addition to loss-of-function mutations, *ANK2* haploinsufficiency due to reciprocal chromosomal translocation [116] also causes ankyrin B syndrome.

Most recently, ankyrin B syndrome has been associated with arrhythmogenic right ventricular cardiomyopathy (ARVC) [117]. Both ankyrin B variants p.E1458G and p.M1988T were detected in patients with ARVC. Intra-myocyte NCX localization was disrupted in ventricular tissue from patients housing the p.M1988T variant, yielding reduced plasma membrane expression and abnormal Z-line targeting. Meanwhile, mice with cardiomyocyte-specific knockout of the *Ank2* gene displayed ARVC-like phenotypes, including biventricular structural abnormalities, reduced ejection fraction, cardiac fibrosis, and exercise-induced death [117]. Cardiomyocytes from *Ank2* knockout mice showed mis-localization of β-catenin, a binding partner of the MBD in ankyrin B, though normal desmosomes at intercalated discs remain preserved. GSK-3β inhibitor, a pharmacological activator of β-catenin, was effective in reversing the ARVC phenotype in knockout mice, indicating its therapeutic potential for ankyrin B variant-driven ARVC in human.

Due to their essential roles in organizing ion channels and transporters to distinct functional compartments, the ankyrin-spectrin microdomains are emerging as key calcium signaling regulatory hubs in diverse cell types of cardiac myocytes. In addition to animal studies, the discovery of human diseases related to ankyrin B has further solidified the importance of the ankyrin-spectrin microdomains in human physiology and cardiac diseases. Elucidation of intricate molecular pathways for synergistic interactions with ankyrin B variants can help identify novel therapeutic targets to treat ankyrin B-associated diseases.

Caveolae Microdomains in Calcium Signaling and Stress Response

Another well-established class of membrane microdomain involved in cellular calcium regulation is the caveolae, which are cholesterol-rich sarcolemmal microdomains critical to β-adrenergic modulation of calcium handling. These microdomains facilitate effective regulation of calcium signaling during sympathetic stress by localizing key ion channels and catecholamine receptors to the caveolar membrane and near the t-tubules. Disruption of caveolae causes both β-adrenergic signaling breakdown and t-tubule remodeling. Caveolar dysfunction is associated with several diseases related to calcium signaling including cardiac hypertrophy, long QT syndrome, and arrhythmia.

Organization of Caveolae Microdomains in Cardiomyocytes

First observed in 1953 [118] and later found to be present in a wide variety of cell types [119], caveolae form flask-shaped invaginations in the sarcolemma of cardiomyocytes [120] that can increase the total sarcolemmal area by as much as 14–21% [121]. While these 50- to 90-nm-diameter [120] invaginations are located across the whole sarcolemma, they tend to concentrate strongly at the necks of t-tubules close to t-tubule opening [122–124]. The formation, shape, and organization of these invaginations are dependent primarily upon caveolin-3 (Cav3), a member of the membrane-scaffolding caveolin protein family which is unique to cardiomyocytes and skeletal muscle [125]. These proteins are composed of three domains including an N-terminal oligomerization domain, a caveolin scaffolding domain that binds membrane and other proteins, and a C-terminal intramembrane domain [126]. Cav3 oligomerizes into nonamers which further cluster and bind to the membrane to create the distinctive caveolar flask shape [127] (Fig. 1c). Caveolae formation also requires Cav3 interactions with a class of proteins known as cavins whose exact role has not yet been fully elucidated [119, 126]. Additionally, Cav3 has been shown to bind cholesterol and thereby concentrate cholesterol and sphingolipids to caveolae [119, 127]. Finally, Cav3 stabilizes caveolae by creating a strong network of cytoskeletal interaction with actin/microtubule scaffolding [119, 127] as well as with members of the dystrophin-glycoprotein complex [125, 128]. All these functions make Cav3 the central protein player in maintaining the shape, localization, stability, and lipid composition of caveolae.

Caveolae are involved in a wide range of cellular functions, including critical calcium signaling pathways. Some of these functions, including membrane tension regulation, endocytosis, lipid metabolism, and mechanosensory, depend primarily upon the shape and lipid chemistry of caveolae [119, 127]. However, many major caveolar functions come from the microdomain's concentration of key signaling proteins involved in a wide variety of cell-signaling pathways. The list of caveolae-associated signaling proteins is long, including Src kinases, G proteins, LTCCs,

HCN4, Kv1.5, Kv11.1, Kir6.2/Sur2a, Nav1.5, NCX1, β-AR2, adenylyl cyclase, insulin receptors, and various growth factors [119, 126, 127, 129–132]. While much remains unknown about the diverse roles of caveolae in these signaling pathways, it is evident that caveolae serve as multifaceted signaling hubs whose disruption could impact several crucial cell processes [133–138]. As detailed below, caveolae play a well-studied and important role in catecholamine regulation of calcium signaling in cardiomyocytes.

Caveolae-Related Organization of Calcium Signaling

Caveolae are critical regulators of cardiac calcium signaling particularly during sympathetic stimulation. Under acute sympathetic stress, catecholamine binding to β1 and β2 adrenergic receptors (β1-AR and β2-AR) at the cardiomyocyte surface alters intracellular calcium signaling to facilitate the faster and stronger contractions [139] of the fight-or-flight response (Fig. 1c). Not only do caveolae contain the entire β2-AR surface population in cardiomyocytes [133, 140, 141]; they also maintain populations of other proteins critical to stress signaling. Critically, localization to necks of t-tubules places caveolae near the CICR machinery [142]. Caveolae act in concert with a series of regulatory proteins to compartmentalize β2-AR signals to specific protein targets in this machinery. The net result of these interactions is increased calcium transient amplitude, tighter CICR synchronization, and increased TT/SR junctions [143]. Taken together, these may contribute to the increased cardiac output characteristic of sympathetic stress [139, 143].

Caveolae-clustered β2-AR stimulation influences calcium signaling through stimulatory and inhibitory G-protein cascades. On one hand, β2-AR couples with a stimulatory heterotrimeric G protein (Gs) whose alpha subunit (Gα) dissociates upon catecholamine activation of β2-AR. Gα then stimulates adenylyl cyclase (AC) to generate cAMP which in turn activates PKA to phosphorylate its targets including calcium-handling proteins. On the other hand, β2-AR stimulation can also spark an inhibitory G-protein cascade which is part of a negative feedback loop. In the inhibitory cascade, β2-AR stimulation causes the beta and gamma subunits (Gβγ) of an inhibitory G heterotetramer (Gi) to dissociate. Gβγ then inhibits the AC-cAMP-PKA axis, thus limiting PKA phosphorylation of calcium-handling proteins [133]. Interestingly, PKA phosphorylation of the β2-AR itself causes the receptor to favor association with Gi rather than Gs. This regulatory mechanism makes β2-AR activation biphasic. At first, β2-AR stimulation increases PKA activity via the Gs pathway, which is then followed by PKA phosphorylation of β2-ARs with a resultant negative feedback loop of upregulated Gi pathway. Given sufficient catecholamine stimulation, the Gi pathway becomes dominant [140, 144–146]. This biphasic pattern contrasts with signaling from the β1-ARs, which are spread out across sarcolemma and can associate only with Gs proteins without inhibitory self-regulation [133]. Such a β2-AR-mediated biphasic signaling suggests the functional significance of receptor compartmentalization in maintaining homeostasis within the local microenvironment, preventing the effects of potential overstimulation during stress.

By compartmentalizing β2-ARs, caveolae contribute to a fine-tuned stress signaling by restricting the effects of Gs and Gi to their own protein populations. In contrast to β1-AR-activated PKA phosphorylation of sarcolemmal and cytosolic targets in a much more diffused manner, localized β2-AR activation can more effectively modulate and control PKA-regulated calcium-handling proteins within subregions. One variety of PKA, known as PKA-RII, localizes to the membrane by binding AKAPs which are associated with PKA targets [142]. As discussed earlier, various AKAPs exist and associate with the calcium-handling machinery components LTCC, RyR_2, and PLN, making their phosphorylation by PKA very likely during sympathetic stimulation [147–150]. Furthermore, the target proteins themselves can also be clustered into caveolae. For instance, it has been established that caveolae maintain a subpopulation of Cav1.2 which is linked exclusively to β2-AR stimulation but are not involved in EC coupling, though their role in the stress response is unclear [151]. Increased channel conductance and calcium influx occur with PKA phosphorylation of the LTCC subunits Cav1.2 (α1c) and β2 and accessory protein Ahnak [152–155]. Due to caveolae localization to t-tubules, β2-AR stimulation also generates signals easily targeted to t-tubule proteins [133, 156]. In fact, disruption of t-tubule and the resultant caveolar β2-AR redistribution leads to catecholamine stimulation of cytosolic proteins typically targeted by β1-AR signaling [133, 141, 142]. This t-tubule compartmentalized β2-AR signaling is also achieved by concentrating phosphodiesterases (PDEs) at the desired protein targets. PDEs degrade cAMP and prevent cAMP spreading and activating off-target PKA [157–159]. Several kinds of PDEs are associated with t-tubules and calcium-handling proteins like RyR_2 and SERCA2a [141, 158, 160], localizing calcium signaling near t-tubules without spreading into the cytosol. In conclusion, caveolae causes β2-AR signaling to reach a specific set of calcium-handling protein targets without causing unwanted calcium signaling effects elsewhere in the cell.

Caveolae in cardiomyocytes may also work in concert with other calcium-handling microdomains such as the ones organized by cBIN1, JP2, and ankyrins. Accumulating evidence indicates that Cav3-organized caveolae interact with the calcium-handling cBIN1-microdomains at t-tubules. One original study in knockout mice indicates that BIN1 removal disrupts caveolar distribution in cardiomyocytes [161]. This codependency begins early with Cav3 association with BIN1 and t-tubules during t-tubule biogenesis, though a clear mechanistic role of Cav3 in the process is unknown [124, 162]. A recent study in zebrafish skeletal muscle indicates that cavin-dependent recycling of Cav3 between t-tubule and sarcolemmal membrane interplays with the BIN1 pathway to regulate t-tubule structure and biogenesis [163]. As discussed earlier, in cardiomyocytes, cBIN1-microdomains reorganize within minutes of β-AR activation, recruiting PKA-phosphorylated RyRs to dyads for effective LTCC-RyR couplon formation [29]. With the established role of caveolae in β2-AR signaling, it is logical to suspect that the two microdomains function in coordination to mobilize cardiomyocyte calcium signaling machinery in response to acute sympathetic stress. How the cBIN1-microdomain-modulated sympathetic response of EC coupling is connected to β2-AR-mediated sympathetic signaling remains as an interesting topic. Nevertheless, caveolae, caveolar proteins, and their binding partners are critical players in controlling proper calcium signaling and EC coupling at baseline and during stress. Breaking this tight local control of signaling disrupts proper calcium regulation, likely contributing to the caveolae-related heart dysfunction discussed in the next section.

Caveolae, Calcium Signaling, and Disease

Caveolae disruption is associated with a wide variety of illnesses throughout the body, including dystrophies, diabetes, cancer, osteoporosis, and lipodystrophy [119, 127, 164]. In the heart, caveolar dysfunction is linked to several cardiac diseases such as heart failure, hypertrophy, and arrhythmia.

Given the significance of caveolae in compartmentalizing β2-AR signaling, calcium-handling pathways in cardiomyocytes are usually vulnerable to caveolar dysfunction, although the relative contribution of caveolae in disease progression is difficult to determine. For instance, failing cardiomyocytes with weakened calcium transients often exhibit β2-AR dispersion out of caveolae [142] and t-tubules. Such uncoupling destroys proper cAMP compartmentalization, leading to off-target phosphorylation of cytosolic calcium-handling proteins following β2-AR activation. The disorganized β2-AR signaling and the resultant intrusion into the calcium signaling pathways of the β1-ARs could cause contractility and relaxation issues due to diffused and augmented catecholamine stimulation [133, 140–142]. This effect can be further complicated by β2-AR phosphorylation and biasing toward the Gi pathway, which is elevated in pressure overload-induced heart failure and causes detrimental maladaptive cardiac remodeling [165]. However, due to the coexisting of other complex calcium-disturbing problems including t-tubule remodeling [142], it is difficult to tease out the relative contribution of caveolar dysfunction to the pathophysiologic manifestation of failing cardiomyocytes.

Caveolar dysfunction has also been linked with the development of cardiac hypertrophy. Reduction in Cav3 levels and thus a decrease in the population of caveolae have been observed in models of pressure overload-induced hypertrophy [166]. Several Cav3 mutations are implicated in the development of these conditions, with some of them showing a reduction of caveolae at the cell surface [127, 167, 168]. Both the variety of Cav3 mutations and the complexity of the disorder make it difficult to determine the exact contributions of caveolar dysfunction to hypertrophic remodeling, which are partially allocated to β2-AR stimulation-activated ERKs and the MAPK pathways [134, 144]. Interestingly, blocking LTCCs localized to caveolae inhibits the development of hypertrophy while not causing changes in contractility, indicating a unique role of caveolar LTCC function tweaked by local β2-AR activity [169]. A recent study indicates that CAMKII in caveolae phosphorylates LTCCs subunits and poses a positive feedback loop further increasing LTCC activity in caveolae [170], aggravating cardiac hypertrophy. These data indicate that active local LTCCs with a resultant augmented calcium in caveolae, as well as activation of the downstream signaling pathways and maladaptive feedback loops, may serve as a key mechanism contributing to the development of cardiomyocyte hypertrophy.

Given that caveolae are pivotal to the subcellular localization of a variety of ion channels, caveolar dysfunction is often associated with abnormal ionic currents with increased arrhythmogenesis. *CAV3* mutations in human are associated with long QT syndromes and sudden infant death syndromes [130, 171]. Abnormalities associated with *CAV3* mutations range from increased depolarizing I_{Na} and I_{Ca} currents [171, 172] and reduced repolarizing inward rectifier I_{K1} current [171, 172] to decreased delayed rectifier I_{Ks} and I_{Kr} currents [173], together prolonging action potential and QT duration.

Altogether, caveolae are key pillars of healthy calcium signaling in cardiomyocytes, especially during sympathetic stress. Caveolae promote differential calcium signaling regulation in cardiomyocyte subregions by concentrating calcium channels and key stress-response receptors, facilitating highly localized and compartmentalized calcium signaling under both basal and stressed conditions. The wide range of pathways impacted by caveolae also makes these microdomains a potential target for new therapy development for several cardiac disorders, especially those where calcium signaling is compromised or altered by sympathetic stress.

Conclusions and Future Perspectives

In conclusion, the healthy rhythms, excitability, contractility, relaxation, and stress response of cardiomyocytes are all under the regulation of intra-myocyte calcium signaling. To achieve a diverse range of functional capacity, calcium signaling in cardiomyocytes is differentially regulated at various classes of membrane-based subcellular microdomains (summarized in Fig. 1). The complex organization, functions, and interactions of these calcium regulatory microdomains are critical to the maintenance of rhythmic and strong heartbeats. Disruption of these microdomains is connected to a wide range of cardiac diseases. In the case of chronic stress, microdomain remodeling occurs ahead of gross morphological changes in cardiomyocytes, which will then develop into maladaptive remodeling and functional decompensation. Future studies are needed to understand the formation, function, regulation, and interaction among these microdomains in healthy cardiomyocytes, as well as the molecular mechanisms underlying microdomain remodeling in cardiac diseases. Elucidation of these intra-cardiomyocyte calcium regulatory microdomains will help identify new druggable targets to exploit in the development of new and effective therapies to treat cardiovascular diseases like heart failure and sudden cardiac death.

References

1. Fabiato A. Calcium-induced release of calcium from the cardiac sarcoplasmic reticulum. Am J Phys. 1983;245(1):C1–14.
2. Pessah IN, Waterhouse AL, Casida JE. The calcium-ryanodine receptor complex of skeletal and cardiac muscle. Biochem Biophys Res Commun. 1985;128(1):449–56.
3. Inui M, Saito A, Fleischer S. Isolation of the ryanodine receptor from cardiac sarcoplasmic reticulum and identity with the feet structures. J Biol Chem. 1987;262(32):15637–42.
4. Cheng H, Lederer WJ, Cannell MB. Calcium sparks: elementary events underlying excitation-contraction coupling in heart muscle. Science. 1993;262(5134):740–4.
5. Bers DM. Cardiac excitation-contraction coupling. Nature. 2002;415(6868):198–205.
6. Scriven DR, Dan P, Moore ED. Distribution of proteins implicated in excitation-contraction coupling in rat ventricular myocytes. Biophys J. 2000;79(5):2682–91.
7. Bers DM, Stiffel VM. Ratio of ryanodine to dihydropyridine receptors in cardiac and skeletal muscle and implications for E-C coupling. Am J Phys. 1993;264(6 Pt 1):C1587–93.
8. Cannell MB, Cheng H, Lederer WJ. Spatial non-uniformities in [Ca2+]i during excitation-contraction coupling in cardiac myocytes. Biophys J. 1994;67(5):1942–56.
9. Bers DM. Calcium cycling and signaling in cardiac myocytes. Annu Rev Physiol. 2008;70:23–49.
10. Perreault CL, Shannon RP, Komamura K, Vatner SF, Morgan JP. Abnormalities in intracellular calcium regulation and contractile function in myocardium from dogs with pacing-induced heart failure. J Clin Invest. 1992;89(3):932–8.
11. Gomez AM, Valdivia HH, Cheng H, Lederer MR, Santana LF, Cannell MB, et al. Defective excitation-contraction coupling in experimental cardiac hypertrophy and heart failure. Science. 1997;276(5313):800–6.
12. Litwin SE, Zhang D, Bridge JH. Dyssynchronous Ca(2+) sparks in myocytes from infarcted hearts. Circ Res. 2000;87(11):1040–7.
13. Lyon AR, MacLeod KT, Zhang Y, Garcia E, Kanda GK, Lab MJ, et al. Loss of T-tubules and other changes to surface topography in ventricular myocytes from failing human and rat heart. Proc Natl Acad Sci U S A. 2009;106(16):6854–9.
14. Wei S, Guo A, Chen B, Kutschke W, Xie YP, Zimmerman K, et al. T-tubule remodeling during transition from hypertrophy to heart failure. Circ Res. 2010;107(4):520–31.
15. Louch WE, Sejersted OM, Swift F. There goes the neighborhood: pathological alterations in T-tubule morphology and consequences for cardiomyocyte Ca2+ handling. J Biomed Biotechnol. 2010;2010:503906.
16. Gomez AM, Guatimosim S, Dilly KW, Vassort G, Lederer WJ. Heart failure after myocardial infarction: altered excitation-contraction coupling. Circulation. 2001;104(6):688–93.
17. He J, Conklin MW, Foell JD, Wolff MR, Haworth RA, Coronado R, et al. Reduction in density of transverse tubules and L-type Ca(2+) channels in canine tachycardia-induced heart failure. Cardiovasc Res. 2001;49(2):298–307.
18. Louch WE, Mork HK, Sexton J, Stromme TA, Laake P, Sjaastad I, et al. T-tubule disorganization and reduced synchrony of Ca2+ release in murine cardiomyocytes following myocardial infarction. J Physiol. 2006;574(Pt 2):519–33.
19. Bito V, Heinzel FR, Biesmans L, Antoons G, Sipido KR. Crosstalk between L-type Ca2+ channels and the sarcoplasmic reticulum: alterations during cardiac remodelling. Cardiovasc Res. 2008;77(2):315–24.
20. Hasenfuss G. Alterations of calcium-regulatory proteins in heart failure. Cardiovasc Res. 1998;37(2):279–89.
21. Gwathmey JK, Copelas L, MacKinnon R, Schoen FJ, Feldman MD, Grossman W, et al. Abnormal intracellular calcium handling in myocardium from patients with end-stage heart failure. Circ Res. 1987;61(1):70–6.
22. Beuckelmann DJ, Nabauer M, Erdmann E. Intracellular calcium handling in isolated ventricular myocytes from patients with terminal heart failure. Circulation. 1992;85(3):1046–55.

23. Sipido KR, Stankovicova T, Flameng W, Vanhaecke J, Verdonck F. Frequency dependence of Ca2+ release from the sarcoplasmic reticulum in human ventricular myocytes from end-stage heart failure. Cardiovasc Res. 1998;37(2):478–88.
24. Hasenfuss G, Schillinger W, Lehnart SE, Preuss M, Pieske B, Maier LS, et al. Relationship between Na+-Ca2+-exchanger protein levels and diastolic function of failing human myocardium. Circulation. 1999;99(5):641–8.
25. Marx SO, Reiken S, Hisamatsu Y, Jayaraman T, Burkhoff D, Rosemblit N, et al. PKA phosphorylation dissociates FKBP12.6 from the calcium release channel (ryanodine receptor): defective regulation in failing hearts. Cell. 2000;101(4):365–76.
26. Wehrens XH, Lehnart SE, Reiken S, Vest JA, Wronska A, Marks AR. Ryanodine receptor/calcium release channel PKA phosphorylation: a critical mediator of heart failure progression. Proc Natl Acad Sci U S A. 2006;103(3):511–8.
27. Hong T, Shaw RM. Cardiac T-tubule microanatomy and function. Physiol Rev. 2017;97(1):227–52.
28. Hong T, Yang H, Zhang SS, Cho HC, Kalashnikova M, Sun B, et al. Cardiac BIN1 folds T-tubule membrane, controlling ion flux and limiting arrhythmia. Nat Med. 2014;20(6):624–32.
29. Fu Y, Shaw SA, Naami R, Vuong CL, Basheer WA, Guo X, et al. Isoproterenol promotes rapid ryanodine receptor movement to bridging integrator 1 (BIN1)-organized dyads. Circulation. 2016;133(4):388–97.
30. Prokic I, Cowling BS, Laporte J. Amphiphysin 2 (BIN1) in physiology and diseases. J Mol Med (Berl). 2014;92(5):453–63.
31. Wechsler-Reya R, Sakamuro D, Zhang J, Duhadaway J, Prendergast GC. Structural analysis of the human BIN1 gene. Evidence for tissue-specific transcriptional regulation and alternate RNA splicing. J Biol Chem. 1997;272(50):31453–8.
32. Butler MH, David C, Ochoa GC, Freyberg Z, Daniell L, Grabs D, et al. Amphiphysin II (SH3P9; BIN1), a member of the amphiphysin/Rvs family, is concentrated in the cortical cytomatrix of axon initial segments and nodes of ranvier in brain and around T tubules in skeletal muscle. J Cell Biol. 1997;137(6):1355–67.
33. Ge K, DuHadaway J, Du W, Herlyn M, Rodeck U, Prendergast GC. Mechanism for elimination of a tumor suppressor: aberrant splicing of a brain-specific exon causes loss of function of Bin1 in melanoma. Proc Natl Acad Sci U S A. 1999;96(17):9689–94.
34. Lee E, Marcucci M, Daniell L, Pypaert M, Weisz OA, Ochoa GC, et al. Amphiphysin 2 (Bin1) and T-tubule biogenesis in muscle. Science. 2002;297(5584):1193–6.
35. Peter BJ, Kent HM, Mills IG, Vallis Y, Butler PJ, Evans PR, et al. BAR domains as sensors of membrane curvature: the amphiphysin BAR structure. Science. 2004;303(5657):495–9.
36. Wigge P, Kohler K, Vallis Y, Doyle CA, Owen D, Hunt SP, et al. Amphiphysin heterodimers: potential role in clathrin-mediated endocytosis. Mol Biol Cell. 1997;8(10):2003–15.
37. Fernando P, Sandoz JS, Ding W, de Repentigny Y, Brunette S, Kelly JF, et al. Bin1 SRC homology 3 domain acts as a scaffold for myofiber sarcomere assembly. J Biol Chem. 2009;284(40):27674–86.
38. Hong TT, Smyth JW, Gao D, Chu KY, Vogan JM, Fong TS, et al. BIN1 localizes the L-type calcium channel to cardiac T-tubules. PLoS Biol. 2010;8(2):e1000312.
39. Sakamuro D, Elliott KJ, Wechsler-Reya R, Prendergast GC. BIN1 is a novel MYC-interacting protein with features of a tumour suppressor. Nat Genet. 1996;14(1):69–77.
40. Lawless M, Caldwell JL, Radcliffe EJ, Smith CER, Madders GWP, Hutchings DC, et al. Phosphodiesterase 5 inhibition improves contractile function and restores transverse tubule loss and catecholamine responsiveness in heart failure. Sci Rep. 2019;9(1):6801.
41. Fu Y, Hong T. BIN1 regulates dynamic t-tubule membrane. Biochim Biophys Acta. 2016;1863(7 Pt B):1839–47.
42. Liu Y, Zhou K, Li J, Agvanian S, Caldaruse AM, Shaw S, et al. In mice subjected to chronic stress, exogenous cBIN1 preserves calcium-handling machinery and cardiac function. JACC Basic Transl Sci. 2020;5(6):561–78.

43. De La Mata A, Tajada S, O'Dwyer S, Matsumoto C, Dixon RE, Hariharan N, et al. BIN1 induces the formation of T-tubules and adult-like Ca(2+) release units in developing cardiomyocytes. Stem Cells. 2019;37(1):54–64.
44. Landstrom AP, Beavers DL, Wehrens XH. The junctophilin family of proteins: from bench to bedside. Trends Mol Med. 2014;20(6):353–62.
45. Jiang M, Hu J, White FKH, Williamson J, Klymchenko AS, Murthy A, et al. S-Palmitoylation of junctophilin-2 is critical for its role in tethering the sarcoplasmic reticulum to the plasma membrane. J Biol Chem. 2019;294(36):13487–501.
46. Poulet C, Sanchez-Alonso J, Swiatlowska P, Mouy F, Lucarelli C, Alvarez-Laviada A, et al. Junctophilin-2 tethers T-tubules and recruits functional L-type calcium channels to lipid rafts in adult cardiomyocytes. Cardiovasc Res. 2021;117(1):149–61.
47. Gross P, Johnson J, Romero CM, Eaton DM, Poulet C, Sanchez-Alonso J, et al. Interaction of the joining region in Junctophilin-2 with the L-type Ca(2+) channel is pivotal for cardiac dyad assembly and intracellular Ca(2+) dynamics. Circ Res. 2021;128(1):92–114.
48. Guo A, Wang Y, Chen B, Wang Y, Yuan J, Zhang L, et al. E-C coupling structural protein junctophilin-2 encodes a stress-adaptive transcription regulator. Science. 2018;362(6421):eaan3303.
49. Zhang C, Chen B, Guo A, Zhu Y, Miller JD, Gao S, et al. Microtubule-mediated defects in junctophilin-2 trafficking contribute to myocyte transverse-tubule remodeling and Ca2+ handling dysfunction in heart failure. Circulation. 2014;129(17):1742–50.
50. Redden JM, Dodge-Kafka KL. AKAP phosphatase complexes in the heart. J Cardiovasc Pharmacol. 2011;58(4):354–62.
51. Davis DB, Doherty KR, Delmonte AJ, McNally EM. Calcium-sensitive phospholipid binding properties of normal and mutant ferlin C2 domains. J Biol Chem. 2002;277(25):22883–8.
52. Chase TH, Cox GA, Burzenski L, Foreman O, Shultz LD. Dysferlin deficiency and the development of cardiomyopathy in a mouse model of limb-girdle muscular dystrophy 2B. Am J Pathol. 2009;175(6):2299–308.
53. Demonbreun AR, Rossi AE, Alvarez MG, Swanson KE, Deveaux HK, Earley JU, et al. Dysferlin and myoferlin regulate transverse tubule formation and glycerol sensitivity. Am J Pathol. 2014;184(1):248–59.
54. Bansal D, Miyake K, Vogel SS, Groh S, Chen CC, Williamson R, et al. Defective membrane repair in dysferlin-deficient muscular dystrophy. Nature. 2003;423(6936):168–72.
55. Doherty KR, McNally EM. Repairing the tears: dysferlin in muscle membrane repair. Trends Mol Med. 2003;9(8):327–30.
56. Virani SS, Alonso A, Aparicio HJ, Benjamin EJ, Bittencourt MS, Callaway CW, et al. Heart disease and stroke statistics-2021 update: a report from the American Heart Association. Circulation. 2021;143(8):e254–743.
57. Virani SS, Alonso A, Benjamin EJ, Bittencourt MS, Callaway CW, Carson AP, et al. Heart disease and stroke statistics-2020 update: a report from the American Heart Association. Circulation. 2020;141(9):e139–596.
58. Brette F, Orchard C. T-tubule function in mammalian cardiac myocytes. Circ Res. 2003;92(11):1182–92.
59. Brette F, Orchard C. Resurgence of cardiac t-tubule research. Physiology (Bethesda). 2007;22:167–73.
60. Ibrahim M, Terracciano CM. Reversibility of T-tubule remodelling in heart failure: mechanical load as a dynamic regulator of the T-tubules. Cardiovasc Res. 2013;98(2):225–32.
61. Guo A, Zhang C, Wei S, Chen B, Song LS. Emerging mechanisms of T-tubule remodelling in heart failure. Cardiovasc Res. 2013;98(2):204–15.
62. Ferrantini C, Crocini C, Coppini R, Vanzi F, Tesi C, Cerbai E, et al. The transverse-axial tubular system of cardiomyocytes. Cell Mol Life Sci. 2013;70(24):4695–710.
63. Hong TT, Smyth JW, Chu KY, Vogan JM, Fong TS, Jensen BC, et al. BIN1 is reduced and Cav1.2 trafficking is impaired in human failing cardiomyocytes. Heart Rhythm. 2012;9(5):812–20.

64. Lyon AR, Nikolaev VO, Miragoli M, Sikkel MB, Paur H, Benard L, et al. Plasticity of surface structures and beta(2)-adrenergic receptor localization in failing ventricular cardiomyocytes during recovery from heart failure. Circ Heart Fail. 2012;5(3):357–65.
65. Caldwell JL, Smith CE, Taylor RF, Kitmitto A, Eisner DA, Dibb KM, et al. Dependence of cardiac transverse tubules on the BAR domain protein amphiphysin II (BIN-1). Circ Res. 2014;115(12):986–96.
66. Xu B, Fu Y, Liu Y, Agvanian S, Wirka RC, Baum R, et al. The ESCRT-III pathway facilitates cardiomyocyte release of cBIN1-containing microparticles. PLoS Biol. 2017;15(8):e2002354.
67. Li J, Agvanian S, Zhou K, Shaw RM, Hong T. Exogenous cardiac bridging integrator 1 benefits mouse hearts with pre-existing pressure overload-induced heart failure. Front Physiol. 2020;11:708.
68. Rapundalo ST. Cardiac protein phosphorylation: functional and pathophysiological correlates. Cardiovasc Res. 1998;38(3):559–88.
69. Luo M, Anderson ME. Mechanisms of altered Ca(2)(+) handling in heart failure. Circ Res. 2013;113(6):690–708.
70. Ikeda Y, Hoshijima M, Chien KR. Toward biologically targeted therapy of calcium cycling defects in heart failure. Physiology (Bethesda). 2008;23:6–16.
71. Reiken S, Gaburjakova M, Guatimosim S, Gomez AM, D'Armiento J, Burkhoff D, et al. Protein kinase A phosphorylation of the cardiac calcium release channel (ryanodine receptor) in normal and failing hearts. Role of phosphatases and response to isoproterenol. J Biol Chem. 2003;278(1):444–53.
72. Yano M, Ikeda Y, Matsuzaki M. Altered intracellular Ca2+ handling in heart failure. J Clin Invest. 2005;115(3):556–64.
73. Marks AR. Calcium cycling proteins and heart failure: mechanisms and therapeutics. J Clin Invest. 2013;123(1):46–52.
74. Bhasin N, Cunha SR, Mudannayake M, Gigena MS, Rogers TB, Mohler PJ. Molecular basis for PP2A regulatory subunit B56alpha targeting in cardiomyocytes. Am J Physiol Heart Circ Physiol. 2007;293(1):H109–19.
75. Little SC, Curran J, Makara MA, Kline CF, Ho HT, Xu Z, et al. Protein phosphatase 2A regulatory subunit B56alpha limits phosphatase activity in the heart. Sci Signal. 2015;8(386):ra72.
76. van Oort RJ, Garbino A, Wang W, Dixit SS, Landstrom AP, Gaur N, et al. Disrupted junctional membrane complexes and hyperactive ryanodine receptors after acute junctophilin knockdown in mice. Circulation. 2011;123(9):979–88.
77. Reynolds JO, Quick AP, Wang Q, Beavers DL, Philippen LE, Showell J, et al. Junctophilin-2 gene therapy rescues heart failure by normalizing RyR2-mediated Ca(2+) release. Int J Cardiol. 2016;225:371–80.
78. Wu CY, Chen B, Jiang YP, Jia Z, Martin DW, Liu S, et al. Calpain-dependent cleavage of junctophilin-2 and T-tubule remodeling in a mouse model of reversible heart failure. J Am Heart Assoc. 2014;3(3):e000527.
79. Guo Y, VanDusen NJ, Zhang L, Gu W, Sethi I, Guatimosim S, et al. Analysis of cardiac myocyte maturation using CASAAV, a platform for rapid dissection of cardiac myocyte gene function in vivo. Circ Res. 2017;120(12):1874–88.
80. El Refaey MM, Mohler PJ. Ankyrins and spectrins in cardiovascular biology and disease. Front Physiol. 2017;8:852.
81. Mohler PJ, Davis JQ, Bennett V. Ankyrin-B coordinates the Na/K ATPase, Na/Ca exchanger, and InsP3 receptor in a cardiac T-tubule/SR microdomain. PLoS Biol. 2005;3(12):e423.
82. Hund TJ, Mohler PJ. Cardiac spectrins: alternative splicing encodes functional diversity. J Mol Cell Cardiol. 2010;48(6):1031–2.
83. Bennett V, Baines AJ. Spectrin and ankyrin-based pathways: metazoan inventions for integrating cells into tissues. Physiol Rev. 2001;81(3):1353–92.
84. Bennett V, Healy J. Membrane domains based on ankyrin and spectrin associated with cell-cell interactions. Cold Spring Harb Perspect Biol. 2009;1(6):a003012.

85. Hund TJ, Koval OM, Li J, Wright PJ, Qian L, Snyder JS, et al. A beta(IV)-spectrin/CaMKII signaling complex is essential for membrane excitability in mice. J Clin Invest. 2010;120(10):3508–19.
86. Mohler PJ, Yoon W, Bennett V. Ankyrin-B targets beta2-spectrin to an intracellular compartment in neonatal cardiomyocytes. J Biol Chem. 2004;279(38):40185–93.
87. Smith SA, Sturm AC, Curran J, Kline CF, Little SC, Bonilla IM, et al. Dysfunction in the betaII spectrin-dependent cytoskeleton underlies human arrhythmia. Circulation. 2015;131(8):695–708.
88. Smith SA, Hughes LD, Kline CF, Kempton AN, Dorn LE, Curran J, et al. Dysfunction of the beta2-spectrin-based pathway in human heart failure. Am J Physiol Heart Circ Physiol. 2016;310(11):H1583–91.
89. Mohler PJ. Ankyrins and human disease: what the electrophysiologist should know. J Cardiovasc Electrophysiol. 2006;17(10):1153–9.
90. Koob R, Zimmermann M, Schoner W, Drenckhahn D. Colocalization and coprecipitation of ankyrin and Na+,K+-ATPase in kidney epithelial cells. Eur J Cell Biol. 1988;45(2):230–7.
91. Bourguignon LY, Jin H, Iida N, Brandt NR, Zhang SH. The involvement of ankyrin in the regulation of inositol 1,4,5-trisphosphate receptor-mediated internal Ca2+ release from Ca2+ storage vesicles in mouse T-lymphoma cells. J Biol Chem. 1993;268(10):7290–7.
92. Davis JQ, McLaughlin T, Bennett V. Ankyrin-binding proteins related to nervous system cell adhesion molecules: candidates to provide transmembrane and intercellular connections in adult brain. J Cell Biol. 1993;121(1):121–33.
93. Li ZP, Burke EP, Frank JS, Bennett V, Philipson KD. The cardiac Na+-Ca2+ exchanger binds to the cytoskeletal protein ankyrin. J Biol Chem. 1993;268(16):11489–91.
94. Mohler PJ, Rivolta I, Napolitano C, LeMaillet G, Lambert S, Priori SG, et al. Nav1.5 E1053K mutation causing Brugada syndrome blocks binding to ankyrin-G and expression of Nav1.5 on the surface of cardiomyocytes. Proc Natl Acad Sci U S A. 2004;101(50):17533–8.
95. Hortsch M, Nagaraj K, Godenschwege TA. The interaction between L1-type proteins and ankyrins--a master switch for L1-type CAM function. Cell Mol Biol Lett. 2009;14(1):57–69.
96. Li J, Kline CF, Hund TJ, Anderson ME, Mohler PJ. Ankyrin-B regulates Kir6.2 membrane expression and function in heart. J Biol Chem. 2010;285(37):28723–30.
97. Cunha SR, Hund TJ, Hashemi S, Voigt N, Li N, Wright P, et al. Defects in ankyrin-based membrane protein targeting pathways underlie atrial fibrillation. Circulation. 2011;124(11):1212–22.
98. Bennett V, Stenbuck PJ. The membrane attachment protein for spectrin is associated with band 3 in human erythrocyte membranes. Nature. 1979;280(5722):468–73.
99. Ipsaro JJ, Mondragon A. Structural basis for spectrin recognition by ankyrin. Blood. 2010;115(20):4093–101.
100. Nassal D, Yu J, Min D, Lane C, Shaheen R, Gratz D, et al. Regulation of cardiac conduction and arrhythmias by ankyrin/spectrin-based macromolecular complexes. J Cardiovasc Dev Dis. 2021;8(5):48.
101. Sucharski HC, Dudley EK, Keith CBR, El Refaey M, Koenig SN, Mohler PJ. Mechanisms and alterations of cardiac ion channels leading to disease: role of ankyrin-B in cardiac function. Biomol Ther. 2020;10(2):211.
102. Baines AJ, Pinder JC. The spectrin-associated cytoskeleton in mammalian heart. Front Biosci. 2005;10:3020–33.
103. Wu HC, Yamankurt G, Luo J, Subramaniam J, Hashmi SS, Hu H, et al. Identification and characterization of two ankyrin-B isoforms in mammalian heart. Cardiovasc Res. 2015;107(4):466–77.
104. Ayalon G, Hostettler JD, Hoffman J, Kizhatil K, Davis JQ, Bennett V. Ankyrin-B interactions with spectrin and dynactin-4 are required for dystrophin-based protection of skeletal muscle from exercise injury. J Biol Chem. 2011;286(9):7370–8.
105. Ayalon G, Davis JQ, Scotland PB, Bennett V. An ankyrin-based mechanism for functional organization of dystrophin and dystroglycan. Cell. 2008;135(7):1189–200.

106. Mohler PJ, Hoffman JA, Davis JQ, Abdi KM, Kim CR, Jones SK, et al. Isoform specificity among ankyrins. An amphipathic alpha-helix in the divergent regulatory domain of ankyrin-b interacts with the molecular co-chaperone Hdj1/Hsp40. J Biol Chem. 2004;279(24):25798–804.
107. Cunha SR, Mohler PJ. Obscurin targets ankyrin-B and protein phosphatase 2A to the cardiac M-line. J Biol Chem. 2008;283(46):31968–80.
108. Heurteaux C, Guy N, Laigle C, Blondeau N, Duprat F, Mazzuca M, et al. TREK-1, a K+ channel involved in neuroprotection and general anesthesia. EMBO J. 2004;23(13):2684–95.
109. Makara MA, Curran J, Little SC, Musa H, Polina I, Smith SA, et al. Ankyrin-G coordinates intercalated disc signaling platform to regulate cardiac excitability in vivo. Circ Res. 2014;115(11):929–38.
110. Makara MA, Curran J, Lubbers ER, Murphy NP, Little SC, Musa H, et al. Novel mechanistic roles for ankyrin-G in cardiac remodeling and heart failure. JACC Basic Transl Sci. 2018;3(5):675–89.
111. Mohler PJ, Bennett V. Ankyrin-based cardiac arrhythmias: a new class of channelopathies due to loss of cellular targeting. Curr Opin Cardiol. 2005;20(3):189–93.
112. Mohler PJ, Schott JJ, Gramolini AO, Dilly KW, Guatimosim S, duBell WH, et al. Ankyrin-B mutation causes type 4 long-QT cardiac arrhythmia and sudden cardiac death. Nature. 2003;421(6923):634–9.
113. Schott JJ, Charpentier F, Peltier S, Foley P, Drouin E, Bouhour JB, et al. Mapping of a gene for long QT syndrome to chromosome 4q25-27. Am J Hum Genet. 1995;57(5):1114–22.
114. Swayne LA, Murphy NP, Asuri S, Chen L, Xu X, McIntosh S, et al. Novel variant in the ANK2 membrane-binding domain is associated with ankyrin-B syndrome and structural heart disease in a first nations population with a high rate of long QT syndrome. Circ Cardiovasc Genet. 2017;10(1):e001537.
115. Zhu W, Wang C, Hu J, Wan R, Yu J, Xie J, et al. Ankyrin-B Q1283H variant linked to arrhythmias via loss of local protein phosphatase 2A activity causes ryanodine receptor hyperphosphorylation. Circulation. 2018;138(23):2682–97.
116. Huq AJ, Pertile MD, Davis AM, Landon H, James PA, Kline CF, et al. A novel mechanism for human cardiac ankyrin-B syndrome due to reciprocal chromosomal translocation. Heart Lung Circ. 2017;26(6):612–8.
117. Roberts JD, Murphy NP, Hamilton RM, Lubbers ER, James CA, Kline CF, et al. Ankyrin-B dysfunction predisposes to arrhythmogenic cardiomyopathy and is amenable to therapy. J Clin Invest. 2019;129(8):3171–84.
118. Palade GE. Fine structure of blood capillaries. J Appl Phys. 1953;24(11):1424.
119. Parton RG, del Pozo MA. Caveolae as plasma membrane sensors, protectors and organizers. Nat Rev Mol Cell Biol. 2013;14(2):98–112.
120. Levin KR, Page E. Quantitative studies on plasmalemmal folds and caveolae of rabbit ventricular myocardial-cells. Circ Res. 1980;46(2):244–55.
121. Page E. Quantitative ultrastructural analysis in cardiac membrane physiology. Am J Physiol. 1978;235(5):C147–C58.
122. Murphy RM, Mollica JP, Lamb GD. Plasma membrane removal in rat skeletal muscle fibers reveals caveolin-3 hot-spots at the necks of transverse tubules. Exp Cell Res. 2009;315(6):1015–28.
123. Verma V. High-voltage electron microscopical evidence on the connection between transverse tubules and invaginations of plasma-membrane in frog skeletal-muscle. Biol Cell. 1985;53(3):A22-A.
124. Lee E, Marcucci MJ, Daniell L, Pypaert M, Weisz OA, Ochoa G, et al. Amphiphysin 2 (BIN1) and T-tubule biogenesis in muscle. Mol Biol Cell. 2002;13:460a–1a.
125. Galbiati F, Engelman JA, Volonte D, Zhang XL, Minetti C, Li MM, et al. Caveolin-3 null mice show a loss of caveolae, changes in the microdomain distribution of the dystrophin-glycoprotein complex, and T-tubule abnormalities. J Biol Chem. 2001;276(24):21425–33.

126. Tanaka S, Fujio Y, Nakayama H. Caveolae-specific CaMKII signaling in the regulation of voltage-dependent calcium channel and cardiac hypertrophy. Front Physiol. 2018;9:1081.
127. Pradhan BS, Proszynski TJ. A role for caveolin-3 in the pathogenesis of muscular dystrophies. Int J Mol Sci. 2020;21(22):8736.
128. Song KS, Scherer PE, Tang ZL, Okamoto T, Li SW, Chafel M, et al. Expression of caveolin-3 in skeletal, cardiac, and smooth muscle cells - Caveolin-3 is a component of the sarcolemma and co-fractionates with dystrophin and dystrophin-associated glycoproteins. J Biol Chem. 1996;271(25):15160–5.
129. Wright PT, Nikolaev VO, O'Hara T, Diakonov I, Bhargava A, Tokar S, et al. Caveolin-3 regulates compartmentation of cardiomyocyte beta2-adrenergic receptor-mediated cAMP signaling. J Mol Cell Cardiol. 2014;67:38–48.
130. Balijepalli RC, Kamp TJ. Caveolae, ion channels and cardiac arrhythmias. Prog Biophys Mol Biol. 2008;98(2–3):149–60.
131. Balijepalli RC, Foell JD, Hall DD, Hell JW, Kamp TJ. Localization of cardiac L-type Ca2+ channels to a caveolar macromolecular signaling complex is required for beta(2)-adrenergic regulation. Proc Natl Acad Sci U S A. 2006;103(19):7500–5.
132. Scriven DRL, Klimek A, Asghari P, Bellve K, Moore EDW. Caveolin-3 is adjacent to a group of extradyadic ryanodine receptors. Biophys J. 2005;89(3):1893–901.
133. Xiao RP. Beta-adrenergic signaling in the heart: dual coupling of the beta2-adrenergic receptor to G(s) and G(i) proteins. Sci STKE. 2001;2001(104):re15.
134. Zou YZ, Komuro I, Yamazaki T, Kudoh S, Uozumi H, Kadowaki T, et al. Both G(s) and G(i) proteins are critically involved in isoproterenol-induced cardiomyocyte hypertrophy. J Biol Chem. 1999;274(14):9760–70.
135. Chesley A, Lundberg MS, Asai T, Xiao RP, Ohtani S, Lakatta EG, et al. The beta(2)-adrenergic receptor delivers an antiapoptotic signal to cardiac myocytes through G(i)-dependent coupling to phosphatidylinositol 3 '-kinase. Circ Res. 2000;87(12):1172–9.
136. Zhu WZ, Zheng M, Koch WJ, Lefkowitz RJ, Kobilka BK, Xiao RP. Dual modulation of cell survival and cell death by beta(2)-adrenergic signaling in adult mouse cardiac myocytes. Proc Natl Acad Sci U S A. 2001;98(4):1607–12.
137. Communal C, Singh K, Pimentel DR, Colucci WS. Norepinephrine stimulates apoptosis in adult rat ventricular myocytes by activation of the beta-adrenergic pathway. Circulation. 1998;98(13):1329–34.
138. Singh K, Xiao L, Remondino A, Sawyer DB, Colucci WS. Adrenergic regulation of cardiac myocyte apoptosis. J Cell Physiol. 2001;189(3):257–65.
139. Zhang DY, Anderson AS. The sympathetic nervous system and heart failure. Cardiol Clin. 2014;32(1):33–45, vii.
140. Xiang Y, Rybin VO, Steinberg SF, Kobilka B. Caveolar localization dictates physiologic signaling of beta 2-adrenoceptors in neonatal cardiac myocytes. J Biol Chem. 2002;277(37):34280–6.
141. Nikolaev VO, Moshkov A, Lyon AR, Miragoli M, Novak P, Paur H, et al. Beta2-adrenergic receptor redistribution in heart failure changes cAMP compartmentation. Science. 2010;327(5973):1653–7.
142. Gorelik J, Wright PT, Lyon AR, Harding SE. Spatial control of the betaAR system in heart failure: the transverse tubule and beyond. Cardiovasc Res. 2013;98(2):216–24.
143. Song LS, Wang SQ, Xiao RP, Spurgeon H, Lakatta EG, Cheng H. Beta-adrenergic stimulation synchronizes intracellular Ca(2+) release during excitation-contraction coupling in cardiac myocytes. Circ Res. 2001;88(8):794–801.
144. Daaka Y, Luttrell LM, Lefkowitz RJ. Switching of the coupling of the beta2-adrenergic receptor to different G proteins by protein kinase A. Nature. 1997;390(6655):88–91.
145. Devic E, Xiang Y, Gould D, Kobilka B. Beta-adrenergic receptor subtype-specific signaling in cardiac myocytes from beta(1) and beta(2) adrenoceptor knockout mice. Mol Pharmacol. 2001;60(3):577–83.

146. Zamah AM, Delahunty M, Luttrell LM, Lefkowitz RJ. Protein kinase A-mediated phosphorylation of the beta 2-adrenergic receptor regulates its coupling to Gs and Gi. Demonstration in a reconstituted system. J Biol Chem. 2002;277(34):31249–56.
147. Hulme JT, Lin TWC, Westenbroek RE, Scheuer T, Catterall WA. Beta-adrenergic regulation requires direct anchoring of PKA to cardiac Ca(V)1.2 channels via a leucine zipper interaction with A kinase-anchoring protein 15. Proc Natl Acad Sci U S A. 2003;100(22):13093–8.
148. Kapiloff MS, Jackson N, Airhart N. mAKAP and the ryanodine receptor are part of a multi-component signaling complex on the cardiomyocyte nuclear envelope. J Cell Sci. 2001;114(Pt 17):3167–76.
149. Lygren B, Carlson CR, Santamaria K, Lissandron V, McSorley T, Litzenberg J, et al. AKAP complex regulates Ca2+ re-uptake into heart sarcoplasmic reticulum. EMBO Rep. 2007;8(11):1061–7.
150. Yang JC, Drazba JA, Ferguson DG, Bond M. A-kinase anchoring protein 100 (AKAP100) is localized in multiple subcellular compartments in the adult rat heart. J Cell Biol. 1998;142(2):511–22.
151. Makarewich CA, Correll RN, Gao H, Zhang H, Yang B, Berretta RM, et al. A caveolae-targeted L-type Ca(2)+ channel antagonist inhibits hypertrophic signaling without reducing cardiac contractility. Circ Res. 2012;110(5):669–74.
152. Fu Y, Westenbroek RE, Scheuer T, Catterall WA. Basal and beta-adrenergic regulation of the cardiac calcium channel CaV1.2 requires phosphorylation of serine 1700. Proc Natl Acad Sci U S A. 2014;111(46):16598–603.
153. Fu Y, Westenbroek RE, Scheuer T, Catterall WA. Phosphorylation sites required for regulation of cardiac calcium channels in the fight-or-flight response. Proc Natl Acad Sci U S A. 2013;110(48):19621–6.
154. Haase H, Alvarez J, Petzhold D, Doller A, Behlke J, Erdmann J, et al. Ahnak is critical for cardiac Ca(v)1.2 calcium channel function and its beta-adrenergic regulation. FASEB J. 2005;19(14):1969–77.
155. Bunemann M, Gerhardstein BL, Gao T, Hosey MM. Functional regulation of L-type calcium channels via protein kinase A-mediated phosphorylation of the beta(2) subunit. J Biol Chem. 1999;274(48):33851–4.
156. Balijepalli RC, Foell JD, Hall DD, Hell JW, Kamp TJ. Localization of cardiac L-type Ca(2+) channels to a caveolar macromolecular signaling complex is required for beta(2)-adrenergic regulation. Proc Natl Acad Sci U S A. 2006;103(19):7500–5.
157. Karam S, Margaria JP, Bourcier A, Mika D, Varin A, Bedioune I, et al. Cardiac overexpression of PDE4B blunts beta-adrenergic response and maladaptive remodeling in heart failure. Circulation. 2020;142(2):161–74.
158. Lehnart SE, Wehrens XH, Reiken S, Warrier S, Belevych AE, Harvey RD, et al. Phosphodiesterase 4D deficiency in the ryanodine-receptor complex promotes heart failure and arrhythmias. Cell. 2005;123(1):25–35.
159. Leroy J, Richter W, Mika D, Castro LR, Abi-Gerges A, Xie M, et al. Phosphodiesterase 4B in the cardiac L-type Ca(2)(+) channel complex regulates Ca(2)(+) current and protects against ventricular arrhythmias in mice. J Clin Invest. 2011;121(7):2651–61.
160. Kerfant BG, Zhao D, Lorenzen-Schmidt I, Wilson LS, Cai S, Chen SRW, et al. PI3K gamma is required for PDE4, not PDE3, activity in subcellular microdomains containing the sarcoplasmic reticular calcium ATPase in cardiomyocytes. Circ Res. 2007;101(4):400–8.
161. Laury-Kleintop LD, Mulgrew JR, Heletz I, Nedelcoviciu RA, Chang MY, Harris DM, et al. Cardiac-specific disruption of Bin1 in mice enables a model of stress- and age-associated dilated cardiomyopathy. J Cell Biochem. 2015;116(11):2541–51.
162. Parton RG, Way M, Zorzi N, Stang E. Caveolin-3 associates with developing T-tubules during muscle differentiation. J Cell Biol. 1997;136(1):137–54.
163. Lo HP, Lim Y-W, Xiong Z, Martel N, Ferguson C, Ariotti NR, et al. Muscle-specific Cavin4 interacts with Bin1 to promote T-tubule formation and stability in developing skeletal muscle. bioRxiv. 2021:2021.01.13.426456.

164. Cronk LB, Ye B, Kaku T, Tester DJ, Vatta M, Makielski JC, et al. Novel mechanism for sudden infant death syndrome: persistent late sodium current secondary to mutations in caveolin-3. Heart Rhythm. 2007;4(2):161–6.
165. Zhu W, Petrashevskaya N, Ren S, Zhao A, Chakir K, Gao E, et al. Gi-biased beta2AR signaling links GRK2 upregulation to heart failure. Circ Res. 2012;110(2):265–74.
166. Bryant SM, Kong CHT, Watson JJ, Gadeberg HC, James AF, Cannell MB, et al. Caveolin 3-dependent loss of t-tubular ICa during hypertrophy and heart failure in mice. Exp Physiol. 2018;103(5):652–65.
167. Hayashi T, Arimura T, Ueda K, Shibata H, Hohda S, Takahashi M, et al. Identification and functional analysis of a caveolin-3 mutation associated with familial hypertrophic cardiomyopathy. Biochem Biophys Res Commun. 2004;313(1):178–84.
168. Traverso M, Gazzerro E, Assereto S, Sotgia F, Biancheri R, Stringara S, et al. Caveolin-3 T78M and T78K missense mutations lead to different phenotypes in vivo and in vitro. Lab Investig. 2008;88(3):275–83.
169. Makarewich CA, Correll RN, Gao H, Zhang HY, Yang BH, Berretta RM, et al. A caveolae-targeted L-type Ca2+ channel antagonist inhibits hypertrophic signaling without reducing cardiac contractility. Circ Res. 2012;110(5):669–U60.
170. Tonegawa K, Otsuka W, Kumagai S, Matsunami S, Hayamizu N, Tanaka S, et al. Caveolae-specific activation loop between CaMKII and L-type Ca2+ channel aggravates cardiac hypertrophy in alpha(1)-adrenergic stimulation. Am J Physiol Heart Circ Physiol. 2017;312(3):H501–H14.
171. Zaza A, Grandi E. Mechanisms of Cav3-associated arrhythmia: protein or microdomain dysfunction? Int J Cardiol. 2020;320:97–9.
172. Tyan L, Foell JD, Vincent KP, Woon MT, Mesquitta WT, Lang D, et al. Long QT syndrome caveolin-3 mutations differentially modulate K(v)4 and Ca(v)1.2 channels to contribute to action potential prolongation. J Physiol London. 2019;597(6):1531–51.
173. Guo J, Wang T, Li X, Shallow H, Yang T, Li W, et al. Cell surface expression of human ether-a-go-go-related gene (hERG) channels is regulated by caveolin-3 protein via the ubiquitin ligase Nedd4-2. J Biol Chem. 2012;287(40):33132–41.

Stress Kinase Signaling in Cardiac Myocytes

Xun Ai, Jiajie Yan, and Dan J. Bare

Abstract Stress-response kinases, the mitogen-activated protein kinases (MAPKs), are activated in response to the challenge of a myriad of stressors. c-jun N-terminal kinase (JNK), extracellular signal-regulated kinase (ERK), and MAPK p38 are the important members of the MAPK family in the heart. Extensive studies have revealed critical roles of activated MAPKs in the processes of cardiac injury, cardiac arrhythmias, heart failure, and other cardiovascular diseases. Advancing our understanding regarding the functional impacts of MAPKs in the development of heart diseases could shed new light on developing novel therapeutic approaches to improve cardiac function and prevent arrhythmia development in patients. This chapter summarizes relevant current knowledge on the pivotal roles of MAPKs in physiopathological and molecular remodeling in cardiac myocytes during the disease development and for the therapeutic potentials of developing MAPK inhibitors and/or activators.

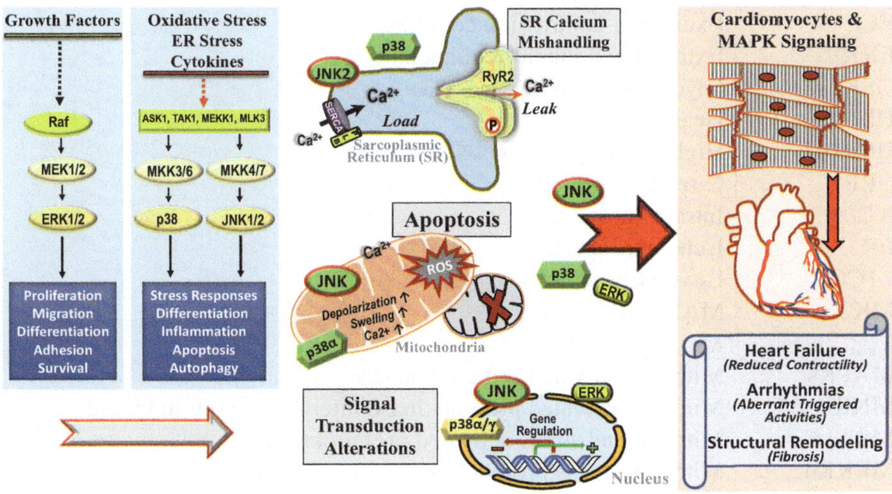

X. Ai (✉) · J. Yan · D. J. Bare
The Ohio State Univeristy, Physiology and Cell Biology, Columbus, OH, USA
e-mail: Xun.Ai@osumc.edu

N. L. Parinandi, T. J. Hund (eds.), *Cardiovascular Signaling in Health and Disease*, https://doi.org/10.1007/978-3-031-08309-9_3

Keywords Mitogen-activated protein kinases · Stress · Arrhythmia · Heart failure · Calcium homeostasis · Cardiovascular diseases

Abbreviations

ASK1-2	Apoptosis signal-regulating kinases-1/2
AF	Atrial fibrillation
BAMBI	BMP and activin membrane-bound inhibitor
Ca^{2+}-ATPase	Plasma membrane calcium/calmodulin-dependent ATPase or PMCA
CICR	Ca^{2+} induced Ca^{2+} release
I_{Ca}	Ca^{2+} influx
CaMKIIδ	Calcium calmodulin kinase IIδ
Ca^{2+}	Calcium
CVDs	Cardiovascular diseases
CHOP	C/EBP homologous protein
JNK	c-jun N-terminal kinase
Cx43	Connexin43
CSBP	Cytokinin-specific binding protein
DAD	Delayed afterdepolarization
DNMT1	DNA methyltransferase-1
DOC-1	Downstream-of-CHOP gene1
DWORF	Dwarf open reading frame
ECC	Excitation-contraction coupling
ERK	Extracellular signal-regulated kinase
GSK-3	Glycogen synthase kinase-3
HF	Heart failure
HMGB1	High mobility group box 1 protein
CRP	C-reactive protein
$[Ca^{2+}]_i$	Intracellular Ca^{2+}
I/R	Ischemia-reperfusion
LTCCs	L-type Ca^{2+} channels
MK2	MAPKAP kinase-2
V_{max}	Maximal rate
MAPKs	Mitogen-activated protein kinases
MNK	Mitogen-activated protein kinase interacting protein kinase
MSK1/2	Mitogen and stress-activated protein kinase1/2
MEKK4	Mitogen-kinase protein kinase kinase kinase-4
MI	Myocardial infarction
MyBP-C	Myosin-binding protein C
NCX	Na^{+} /Ca^{2+} exchanger
I_k	Outward potassium current
PKCε	Protein kinase Cε
PP1	Protein phosphatase 1

PLB	Phospholamban
PAF	Platelet-activating factor
$p70^{RSK}$	p70 ribosomal S6
$p90^{RSK}$	p90 ribosomal S6
I_{Na}	Rapid sodium influx
RISK	Reperfusion injury salvage kinase
RyR2	Ryanodine receptor channel-2
SR	Sarcoplasmic reticulum
SERCA2a	Sarcoplasmic reticulum Ca^{2+}-ATPase 2a
SPEG	Striated muscle enriched protein kinase
PTEN	Tensin homolog on chromosome 10
TNFα-R1	Tumor necrosis factor alpha receptor1
TnT	Troponin T
TGF-β	Transforming growth factor beta
TAC	Transverse aortic constriction
T-tubules	Transverse tubules
TPY	Threonine-proline-tyrosine phosphorylation motif

Introduction

The heart exhibits enhanced cellular stressors given its high level of energy consumption and output (i.e., oxidative stress, inflammatory stress, senescence stress) and with increasing age and cardiovascular diseases (CVDs), which leads to a higher susceptibility to additional extrinsic stress stimuli (i.e., ischemia, inflammation, excessive alcohol exposure, obesity) [1–9]. The mitogen-activated protein kinases (MAPKs) are stress kinases that are activated in response to both intrinsic and extrinsic stress challenges and critically regulate cell survival and growth. Over the years, MAPK activation has been found to be critical in the development of various diseases such as diabetes, cancer, Alzheimer's disease, as well as various cardiac diseases such as cardiac hypertrophy, heart failure, and atrial fibrillation [10–27]. Among ~500 recognized protein kinase genes in the human genome, the MAPK family is composed of three major members, c-jun NH2-terminal kinases (JNK), p38 mitogen-activated protein (MAP) kinase (p38), and extracellular signal-regulated kinases (ERK1/2). These three subfamilies have been the focus of extensive studies to uncover their contributions to pathological cardiac remodeling and disease development [10, 14–27].

Stress Kinase MAPK Signaling in the Heart

MAPKs are serine/threonine kinases that phosphorylate serine or threonine residues in a consensus sequence of Pro-X-Thr/Ser-Pro on the target protein [28]. The canonical pathways leading to MAPK activation require a kinase cascade in which MAP kinase kinases phosphorylate MAPKs that in turn activate MAPKs. The MAPK cascade controls the activity of downstream target proteins including numerous transcription factors, through the regulation of binding partners, protein conformational changes, protein stability, and subcellular localization [29].

JNK

JNK is an important member of the MAPK family that is activated in response to various stress challenges and regulates cell proliferation, differentiation, apoptosis, autophagy, cell survival, cell mobility, and cytokine production [10, 21, 28, 30–32]. The JNK kinase family was discovered in the early 1990s by Kyriakis and Avruch and reported as a novel protein named as pp54 MAP-2 kinase, which is activated by phosphorylation of the amino acid residues of threonine-183 and tyrosine-185 [33, 34]. Next, two isoforms were identified with molecular weights of 46 and 56 kDa and were named JNK1 and JNK2, respectively [35]. Later, it was revealed that these JNK kinases could be activated by various extracellular stimuli. Given that these JNKs contained a threonine-proline-tyrosine phosphorylation motif (TPY), they were characterized as a member of the MAPK family. Finally, JNK3 was discovered in 1995 as the last member of this MAPK subfamily and is mainly expressed in neurons [10, 22, 36, 37]. In the heart, JNK1 and JNK2 are the major isoforms, while JNK3 is expressed at a much lower level [24, 38].

JNKs function within a protein kinase cascade and are activated by dual phosphorylation of on the specific threonine-X-tyrosine motif by the upstream kinases MKK4 and MKK7 in response to stress challenges [21, 28, 31, 39]. JNK itself is a Ser/Thr kinase that phosphorylates its substrates on serine or threonine residues in a consensus sequence of Pro-X-Thr/Ser-Pro [28]. Numerous JNK substrates have been identified and include but are not limited to Ca^{2+}/calmodulin-dependent protein kinase II δ (CaMKIIδ), c-jun, Jun-B, ATF2, c-Myc, and p53 [28, 40]. JNK activation has been observed with aging, excessive binge alcohol-triggered "holiday heart syndrome," and with CVDs such as myocardial ischemic injury and heart failure. Accumulating evidence suggests that JNK signaling is critically involved in the development of diabetes, atrial fibrillation, and other CVDs such as heart failure (HF), ischemic myocardial infarction (MI), and atherosclerosis [9, 10, 20–22, 30, 41–44].

P38

p38 MAP kinase, also referred to as RK, p40, or CSBP2 (cytokinin-specific-binding protein 2), is also a member of the MAPK family and is ubiquitously expressed in all somatic cell types [45]. Although p38 participates in signaling cascades controlling cellular responses to cytokines and other stress stimuli, the function of the p38 kinases appears to be both protective and deleterious in the stressed heart [46]. The p38 was initially found as an unidentified 38-kDa protein, which exhibited increased phosphorylation of tyrosine residues by lipopolysaccharide [47]. Later, it was discovered that there are four identified genes of the p38 MAPK: p38α, p38β, p38γ, and p38δ. P38α shares sequence homology with p38β (~75%), p38γ (~62%), and p38δ (~61%), while p38γ and p38δ are ~70% identical [48, 49]. While the heart expresses all the sub-isoforms, p38α is the most abundant, followed by p38γ [50–52]. The canonical activation of p38 MAPK was found to be achieved through dual phosphorylation of the threonine (Thr)-glycine (Gly)-tyrosine(Tyr) motif in its activation loop [53]. The phosphorylation of this threonine-X-tyrosine motif of p38 in vivo is predominantly mediated by upstream MAPK kinases, MKK3 and MKK6, which are activated by MAPK kinase kinases such as TAK1, ASK1, DLK, and mitogen-kinase protein kinase kinase kinase 4 (MEKK4) [54–57]. This cascade can be instigated by one of multiple MKK-activating MAP3Ks in a stimulus-/stress-dependent manner relevant for specific cell type. Studies suggest that the different isoforms require differential activation of MAPK kinases for full activation, one such example being p38α which requires both MKK6 and MKK3 activation to be phosphorylated in response to cytokines, while p38δ is activated by MKK6, but negatively regulated by MKK3 [49]. In addition, low-molecular-weight GTP-binding proteins in the Rho family (i.e., Rac1, Cdc42, RhoA, and RhoF (Rif)) and heterotrimeric G-protein-coupled receptors could activate p38 [58–60]. Furthermore, the activity of p38 can also be modulated by dual-specificity phosphatases, such as PAC-1 and MKP-1 [53]. Finally, the autoactivation of p38, as a non-canonical mode of p38 activation, was observed through the activation of scaffolding proteins such as TAB1 along with the "priming" phosphorylation of Tyr-323 by a tyrosine kinase of the SYK family [61–63].

ERK

The ERK family, another MAPK member group, can be divided into five subfamilies named ERK1–ERK5. ERK1 and ERK2 share 90% homology and thus are usually referred to as ERK1/2 [64, 65]. And ERK1 and ERK2 (44kDa and 42kDa, respectively [66, 67]) are the most thoroughly investigated isoforms in the ERK family. The basic ERK signal transduction cascade has been shown to follow the typical MAPK cascade reaction (Ras-Raf-MEK-ERK pathway). ERK1/2 is activated by MEK1/2-mediated phosphorylation at Thr-183 and Tyr-185 [68]. This dual-site phosphorylation enhances the ERK1/2 activity by >1000 folds [68, 69]. ERK can also be activated by tyrosine kinase receptors and Gi/o-, Gq-, and Gs-coupled receptors via a range of different signaling pathways [70–75]. One of the best characterized MKKKs to activate ERK is Raf-1, a Ser/Thr protein kinase [76], which binds directly to activated GTP-bound Ras leading to a full activation of MKKK [77]. Once it is fully activated, Raf-1 phosphorylates and activates MKK1 or MKK2. MKK1/2, which in turn phosphorylates ERK1 or ERK2 (ERK1/2) on the Thr-X-Tyr motif in its activation loop, leads to ERK activation. Like JNK and p38, ERK is a Thr/serine (Ser) kinase that is normally located in the cytoplasm and translocates into the nucleus when it is activated [29, 78]. Fully activated ERKs phosphorylate a wide spectrum of substrates with a general amino acid consensus sequence of proline (Pro)-X-Ser/Thr-Pro that can be localized at the plasma membrane, in the cytosol and the nucleus that regulate important aspects of cell physiology including cell proliferation, differentiation, adhesion, migration, and survival [79–83].

MAPKs and Calcium Homeostasis in Myocytes

One of the hallmarks of a diseased heart is an altered protein phosphorylation state that critically contributes to ion transporter and channel dysfunctions leading to the disruption of Ca2+ homeostasis in response to both intrinsic and extrinsic stress stimuli in the heart [10, 20, 21, 22, 28, 31, 131, 132]. Calcium (Ca^{2+}) is an important cation in the conversion of an electrical signal to mechanical function (termed as excitation-contraction coupling) during each heartbeat to maintain normal cardiac function and is also important in the cellular signal transduction pathways that control myocyte survival and growth [84–91]. Excitation-contraction coupling (ECC) is an essential link between myocyte excitation (membrane depolarization of the action potential) and Ca^{2+} release from the sarcoplasmic reticulum (SR) for myocyte contraction, and this series of events is critical in the beat-to-beat cardiac muscle contraction and relaxation.

Normal and Abnormal Calcium (Ca^{2+}) Signaling in Myocytes

The cardiac action potential occurs when myocyte membrane potential is depolarized upon the initiation of a rapid sodium influx (I_{Na}) followed by Ca^{2+} influx (I_{Ca}), while repolarization occurs by the outward potassium current (I_k). During systole in normal cardiomyocytes, the Ca^{2+} entry via L-type Ca^{2+} channels (LTCCs) along with a much smaller amount of Ca^{2+} influx via the Na^+ /Ca^{2+} exchanger (NCX) activates ryanodine receptor channels (RyR2) on the SR membrane to release a large amount of SR stored Ca^{2+}. The LTCCs located in the plasma membrane are activated by the rapid sodium influx (I_{Na}) and depolarization of the myocyte membrane

[92–97]. A small amount of inward Ca^{2+} flux (I_{Ca}) through activated LTCCs triggers large quantities of Ca^{2+} to be released from the SR via cardiac ryanodine receptor type 2 (RyR2; also called Ca^{2+}-triggered SR Ca^{2+} release channels) to produce a large intracellular Ca^{2+} ($[Ca^{2+}]_i$) transient, driving myocyte contraction [89, 98–100]. This Ca^{2+}-induced Ca^{2+} release (CICR) event occurs locally within the clusters of RyR2 channels on the SR membrane that are in close proximity to LTCCs located on the plasma membrane [101, 102]. CICR is further facilitated by dyads, which are the structures consisting of terminal cisternae of SR, composed of clusters of RyR2 channels, paired with transverse tubules (T-tubules), and LTCCs [103]. Upon action potential arrival at the T-tubule, Ca^{2+} influx via LTCCs activates RyR2 channels on the cytosolic side of SR allowing for the occurrence of CICR, which activates neighboring RyR2 channels, resulting in a rapidly increased cytosolic Ca^{2+} [104, 105]. CICR is also the trigger for Ca^{2+}-troponin C binding, leading to myofilament activation and cardiac muscle contraction [106, 107]. During cardiac muscle relaxation, LTCCs close and terminate the influx of Ca^{2+}, and RyR2 channels usually are also closed. Meanwhile, the excess amount of cytosolic Ca^{2+} is removed mainly through cardiac sarcoplasmic reticulum Ca^{2+}-ATPase 2 (SERCA2) Ca^{2+} uptake back to the SR and Ca^{2+} extrusion from the myocyte to the extracellular space through NCX, while another small portion of Ca^{2+} is taken up by mitochondria via mitochondrial Ca^{2+} uniporters as well as a small Ca^{2+} efflux via the plasma membrane Ca^{2+}-ATPase (also known as plasma membrane calcium-/calmodulin-dependent ATPase or PMCA) [93, 98, 108–110]. Normal contraction of the heart requires high Ca^{2+} levels in systole and low levels in diastole [111, 112]. Therefore, SR Ca^{2+} release via RyR2 channels and reuptake via the predominating Ca^{2+} pump SERCA2a isoform (SERCA2a) and, to a much lesser extent SERCA2b isoform, critically mediate the cytoplasmic Ca^{2+} concentration, which is essential in cardiac contraction and relaxation of each heartbeat [87, 90].

Given the tightly regulated role of Ca^{2+} in ECC, even a small amount of aberrant Ca^{2+} release resulting from slowly developed pathological changes of the intracellular Ca^{2+} homeostasis can potentially have escalating negative consequences for the myocyte and ultimately the entire heart. With increasing age and abnormal stressed conditions (e.g., heart failure (HF), ischemia-reperfusion (I/R) injury, myocardial infarction (MI), post-MI, and excessive alcohol exposure), impaired Ca^{2+} homeostasis causes myocardial molecular remodeling, including aberrant gene expression, myocyte death, electrical and mechanical dysfunctions, contractile dysfunction, and triggered arrhythmic activities [24, 25, 113–122]. Abnormal Ca^{2+} dynamics such as reduced SR Ca^{2+} content via decreased uptake by SERCA2 and increased diastolic SR calcium leak via RyR2 channels are involved in the development of maladaptive cardiac remodeling. The abnormal diastolic SR Ca^{2+} leak via RyR2 openings may produce large/frequent Ca^{2+} sparks that may trigger propagating diastolic Ca^{2+} waves [113, 121, 123]. These aberrant Ca^{2+} waves result in an excess outward NCX current, which is electrogenic (3 Na^+ in, 1 Ca^{2+} out), and, thus, may produce triggered arrhythmic activities such as delayed afterdepolarization (DAD) that may initiate cardiac arrhythmias [98, 123]. Under certain pathological conditions such as HF, decreased SR Ca^{2+} refill during the SR Ca^{2+} cycling in myocytes due to reduced Ca^{2+} uptake by SERCA2a leads to a reduced Ca^{2+} transient amplitude and consequently decreased cardiac contractility as seen in the failing heart [124–127]. In the normal diastolic phase, CICR-mediated SR Ca^{2+} release shuts off almost completely (~99%). However, increased diastolic RyR2 channel activity leads to increased diastolic SR Ca^{2+} leak and further reduced SR Ca content, which results in a reduced systolic fractional Ca^{2+} release for a given I_{Ca} as the release trigger [123, 128, 129]. Meanwhile, this increased diastolic SR Ca^{2+} leakage along with an impaired SR Ca^{2+} uptake in HF slows down the intracellular Ca^{2+} decline and then elevates the amount of diastolic intracellular Ca^{2+} concentration, which leads to increased Na^{+} influx via NCX for removing the elevated intracellular Ca^{2+} outside of the cell membrane. As a result, increased diastolic SR Ca^{2+} leak promotes aberrant Ca^{2+} events (Ca^{2+} sparks and waves), and the inward NCX current produces abnormal triggered activities, DADs, to initiate atrial arrhythmias such as atrial fibrillation (AF) and ventricular arrhythmias including ventricular tachycardia and ventricular fibrillation; fatal types of cardiac arrhythmias [94, 98, 99, 123, 130].

MAPKs in Stress-Evoked Ca^{2+} Mishandling in Myocytes

JNK is a key member of the MAPK family, which plays a critical role in maladaptive cardiac remodeling [10, 20, 21, 30, 31, 43, 133–135]. Although the contributions of JNK1 in cellular apoptosis and proliferation as well as cardiac contractile function have been well studied, the function of JNK2, one of the two major cardiac isoforms, has received significantly less attention [10, 20, 133]. Recently, a causal role of JNK2 in abnormal Ca^{2+} handling was discovered for the first time in animal models and humans with both binge alcohol exposure and increasing age with preserved cardiac function and no history of cardiac arrhythmias or any major CVDs [20, 24–26]. JNK2, but not JNK1, drives a significant diastolic SR Ca^{2+} leak and a higher SR Ca^{2+} load at the same time in the stressed myocytes. This JNK2-enhanced SR Ca^{2+} uptake partially compensates for the greater diastolic SR Ca^{2+} leak and maintains a normal level of Ca^{2+} transients and normal cardiac function, while the greater JNK2-driven diastolic SR Ca^{2+} leak acts as a key contributor to enhanced atrial arrhythmic Ca^{2+} events and arrhythmia susceptibility [24, 25].

Very recently, JNK2 was identified as a previously unrecognized enhancer of SERCA2 function via the elevation of the maximal rate (V_{max}) of SERCA2 activity by phosphorylating SERCA2 protein [24]. The SERCA2 pump activity is known to be regulated by phospholamban (PLB), sarcolipin, myoregulin, striated muscle enriched protein kinase (SPEG) and DWORF (dwarf open reading frame) micropeptide [136–139]. All those regulators that are known to increase SERCA2 Ca^{2+} affinity (K_m) do not change the V_{max}, whereas, JNK2 significantly enhances the V_{max} of SERCA2-ATPase activity but not the K_m [24, 136–139]. Intriguingly, JNK2-enhanced SR Ca^{2+}($[Ca^{2+}]_{SR}$) load by itself (in the absence of CaMKII-dependent RyR2 channel sensitization) may not be sufficient to cause significant diastolic leak, a combined higher load and CaMKII-sensitized RyR2 channels may promote the diastolic leak as previously demonstrated [24, 113, 120, 122, 123]. Thus, JNK2 is an important stress-induced regulator driving to maintain a high SR Ca^{2+} uptake and load in order to preserve cardiac function, while JNK2 also drives a greater CaMKII-dependent SR Ca^{2+} leak to promote abnormal triggered activities (discussed in detail below), as seen in both humans and animal models.

The atrial action potential morphology differs from that of the ventricle, where the atrial action potential is generally shorter with a triangular shape due to a smaller Ca^{2+} influx and a more gradual repolarization period [140, 141]. Ca^{2+} handling in atrial myocytes while similar to that of ventricular myocytes has some important structural and molecular differences. Atrial myocytes are thinner and longer and exhibit a longer lag time between APs and Ca^{2+} transients at the center of the cell. This property of the atrial cell contributes to a higher instability of Ca^{2+} propagation, which is pro-arrhythmogenic [142]. Notably, atrial Ca^{2+} transient amplitude is smaller, and the rate of intracellular Ca^{2+} decay is higher than in ventricular myocytes due to a greater SERCA2 uptake and a stronger NCX removal of cytosolic Ca^{2+} during the diastolic phase [142, 143]. The increased SERCA2-dependent intracellular Ca^{2+} removal is attributed to a greater amount of SERCA2 protein and lower level of SERCA2 inhibitory protein PLB [142, 143]. This stronger cytosolic Ca^{2+} removal machinery in the atria leads to a higher SR Ca^{2+} content than that of ventricular myocytes, which makes atrial myocytes prone to diastolic SR Ca^{2+} leak when RyR2 is pathologically sensitized and SR Ca^{2+} overload is increased due to the dual functional impact of stress-activated JNK2 [24, 120, 123, 142, 144–146]. Thus, markedly increased JNK2 in atria exposed to stress stimuli (i.e., aging, binge alcohol) is a major contributor to continued pathological remodeling and enhanced AF risk [20, 24–26, 120]. T-tubules are a subcellular network involved in SR Ca^{2+} dynamics by the coupling of LTCCs to RyR2 channels on the SR membrane to allow a rapid intracellular Ca^{2+}-triggered SR Ca^{2+} release in response to electrical excitation [147–150]. Previously it was believed that atrial T-tubules were virtually absent in isolated myocytes from small rodents; [151, 152] however, accumulating evidence suggests that the T-tubule network is present in intact atrial myocytes and plays a functional role in atrial myocytes from both large mammalian species and small rodents (including humans, sheep, dogs, cows, and horses, rats, mice) [153–157]. These inconsistent research findings may likely be due to the nature of fast T-tubule structural deformation during myocyte isolation and preparation. However, atrial T-tubular networks are clearly less abundant and less well-organized compared to ventricular. In fact, rapid pacing-induced HF dogs showed reduced atrial T-tubular abundance that was linked to altered subcellular Ca^{2+} dynamics and AF development [122, 154, 155]. However, the involvement of the stress-response MAPKs in T-tubular remodeling remains unknown to date.

Like JNK, other MAPK members ERK and p38 are also involved in various types of stress-caused cardiac pathogenesis [10, 21, 22, 131, 132, 158]. Although enhanced activity of ERK or p38 alone may or may not be required or sufficient for facilitating cardiac hypertrophy, both ERK and p38 were found to be involved in pathological remodeling and AF development in the failing heart [159–165]. Hypertrophic stimuli lead to an increase in I_{Ca} and downregulation of SERCA2 expression via activated ERK [17–19]. While Ras, a GTPase, is able to activate

ERK through a Ras-Raf-MEK cascade [166], Ras signaling-activated ERK was found to contribute to downregulation of L-type Ca^{2+} channels and reduced channel activity along with reduced SERCA2 protein expression in cultured myocytes [15, 16]. Thus, Ras-ERK-modulated molecular remodeling led to decreased intracellular Ca^{2+} transients and impaired SR Ca^{2+} uptake, which could lead to enhanced arrhythmogenicity [167]. In both isolated cardiomyocytes and Langendorff-perfused intact hearts, activation of p38 MAPK signaling was found to induce SR Ca^{2+} overload through enhanced SERCA2 activity and increased SR Ca^{2+} uptake during cardiac I/R injury, which in turn prompted myocardial apoptosis [168]. Overall, the stress-response MAPK signaling cascades are critically involved in cardiac Ca^{2+} handling and stress-caused maladaptive cardiac remodeling. However, many questions remain unanswered such as the extent and detailed mechanisms of how the three MAPK members interact and functionally overlap with regards to Ca^{2+} handling in cardiac myocytes under physiological and stressed conditions.

MAPKs and Ca Handling Proteins in Myocytes

JNK2 was recently recognized as a critical activator and transcriptional regulator of CaMKIIδ, a highly validated pro-arrhythmic signal [25, 40, 120]. CaMKIIδ is a well-recognized regulator of Ca^{2+} dysregulation in cardiomyocytes through its critical contribution in phosphorylation of the Ca^{2+} handling proteins RyR2-Ser2815 (sensitizing RyR2 channels to increase diastolic SR Ca^{2+} leak) and PLB-Thr17 (elevating the SERCA activity to enhance the SR Ca^{2+} uptake), which results in triggered activities and arrhythmia pathogenesis in humans and animal models [25, 145, 146, 169–171]. In addition, CaMKIIδ also regulates other ion channels such as Ca^{2+}, Na^{+}, K^{+} channels as well as NCX and myofilament proteins including troponin T (TnT) and myosin-binding protein C (MyBP-C) via phosphorylation [172–191]. Thus, hyper-activated CaMKIIδ drives RyR2 channel-mediated diastolic Ca^{2+} dysfunction that causes and triggers arrhythmic activities but also contributes to cardiac contractile function. Intriguingly, recent studies revealed that JNK2 has specific actions in regulating both expression and activation of CaMKIIδ, which consequently drives CaMKIIδ-dependent SR Ca^{2+} mishandling in the stressed heart. Specifically, JNK2 and CaMKIIδ were found to be tethered with each other, and JNK2 increases phosphorylation of CaMKIIδ at the autophosphorylation site Thr286 to activate CaMKIIδ in a JNK2 dose-dependent manner [25]. Protein phosphatase 1 (PP1) is known to target this specific Thr286 site to dephosphorylate CaMKIIδ [192]. Thus, a possible interrelationship between PP1 and JNK2 might exist in regulating the CaMKIIδ activity, and this is worthy of further investigation.

Although CaMKIIδ is essential in regulating a large number of cellular substrates including ion channels, pumps, transporters, and transcription factors [25, 120, 172–191, 193], exactly how the CaMKIIδ gene and protein expression is controlled remains surprisingly understudied. A recent study reported for the first time that JNK2 downstream transcription factors c-jun and activating transcription factor 2 (ATF2) both bind to the CaMKIIδ gene promoter and regulate CaMKIIδ expression [40]. Surprisingly, c-jun was found to be a key transcription factor for the basal level expression of CaMKIIδ mRNAs and proteins. This was evidenced by the suppression of CaMKIIδ promoter baseline activity when c-jun was knocked out in the cells or the binding consensus sequence for c-jun was mutated to alter binding. Moreover, robustly activated JNK2, mimicking a stressed condition, significantly increases the binding of c-jun, but did not change the binding of ATF2, to the CaMKIIδ promoter, while JNK2 inhibition alleviated this enhanced c-jun binding activity. In addition, the JNK2 specific action in c-jun-regulated CaMKIIδ promoter activity was strongly supported by the suppressed CaMKIIδ promoter activity from JNK2 knockdown or suppressed activity [40]. These findings take on special translational importance given that the development of drugs that target CaMKIIδ activity has been considered as an appealing anti-arrhythmic intervention point; however, specificity issues driving off-target effects greatly hinder the clinical applications. Additionally, upstream or downstream components of the CaMKII signaling cascades are then considered as new potential therapeutic targets. Stress-driven JNK2 activation and the JNK2-CaMKII cross talk are likely critical mechanisms that couple stressors and maladaptive cardiac responses. Therefore, modulating JNK2 activity could be a likely therapeutic approach to prevent and treat cardiac arrhythmias.

An extensive number of studies have demonstrated that JNK1 activation is critically involved in preservation of cardiac function and promoting apoptosis after myocardial I/R, MI, and HF via the regulation of signaling pathways that modulate gene expression [10, 21, 165, 194–200]. Also, emerging evidence suggests that p38 regulates SERCA2 mRNA and protein expression via the transcription factors Egr-1 and SP1 [14]. However, the functional role of all three MAPK members and sub-isoforms of each kinase in Ca^{2+} handling proteins in cardiomyocytes under physiological conditions or stress challenges requires further investigation. Our advancement of understanding the functional role of MAPKs will likely aid the effort of developing novel preventive and therapeutic strategies for CVDs.

MAPKs and Molecular Remodeling in Myocytes

The MAPK signaling pathway transduces and integrates diverse stress stimuli into complex cytoplasmatic and nuclear processes and finally leads to altered cellular function including proliferation, gene expression, differentiation, and apoptosis. MAPKs regulate cellular processes via direct phosphorylation of downstream targets and/or indirectly regulate gene expression in maintaining normal cell function and cellular responses under stress stimuli challenges. Generally, the JNKs and p38 kinases regulate the stress or injury responses, while the ERKs are more specialized for mitogenic and growth factor stimulations [201, 202].

MAPKs and Gene Regulation in Myocytes

Gene regulation is one of the important functions of the MAPK family via the downstream transcription factors such as c-jun, ATF2, JunD, c-Fos, SRF, AP-1, c-Myc, MEF2, GATA, SMAD, STAT-1, and NFκB. The JNKs directly phosphorylate a number of their downstream transcription factors such as c-jun, ATF2, AP-1, JunD, Sp1, Elk1, and c-Myc [26, 40, 203–206]. The AP-1 complex is composed of homodimers of c-jun or heterodimers of c-jun/ATF2 or other combinations of transcription factors, which induce target gene expression by binding the AP-1 consensus site(s) in the promoter region of the gene or dissociating from the promoter region to upregulate or suppress the specific gene expression [40]. These activated transcription factors critically contribute to the transcriptional regulation of proteins that are involved in I/R caused myocardial injury and ATP depletion. JNK1 is also known to regulate several important genes (i.e., Notch1, SOD-3) in response to inflammation, oxidative stress, and heat stress through phosphorylation of several transcription factors (i.e., c-jun, Sp1, DAF-16) [207–209]. Under long-term stress challenges (i.e., aging, repeated binge alcohol exposure), JNK2, but not JNK1, is activated, which leads to reduced AP-1 activity but increased binding of c-jun and unchanged binding of ATF2 to the promoters of JNK2 target genes such as the "cell-to-cell communicator" gap junction protein connexin43(Cx43), "pro-arrhythmic kinase" CaMKIIδ, and "epigenetic regulatory molecule" DNA methyltransferase1 (DNMT1) to either suppress or enhance their transcriptional activities and gene expression. Consequently, this JNK2-mediated gene regulation leads to impaired cellular function including intercellular uncoupling between cardiac myocytes along with the slowing of electrical conduction or enhanced cardiac Ca^{2+}-mediated arrhythmic triggered activities or increases DNA methylation [20, 24–26, 40, 210].

Like the JNKs, p38 also directly phosphorylates transcription factors including ATF1/2/6, c-Myc, c-Fos, GATA4, MEF2A/C, SRF, STAT-1, and CHOP [53, 211–213]. Upon activation, p38 translocates into the nucleus and reenters the cytosol when inactivated which is essential for its function across all cell types [58]. The phosphorylation status and the interaction of p38 with other proteins determine its subcellular location and activities of its downstream targets [214]. For instance, activated p38/MAPKAP kinase-2 (MK2) can form a complex with activated p38 in the nucleus leading to the export of MK2 from the nucleus to phosphorylate its cytosolic substrates [215, 216]. The mitogen and stress-activated protein kinase 1/2 (MSK1/2) is a downstream target of p38 but also regulates the subcellular location of p38 and the p38/MSK complex in regulating the transcriptional activity of CREB, STAT3, and NFκB [217–221]. Moreover, p38-regulated transcription factors contribute to the upregulation of several important stress-response genes including phosphatase and tensin homolog on chromosome 10 (PTEN) that acts to limit the phosphorylation and activation of Akt as well as the transforming growth factor beta (TGF-β) signaling pathway in regulating cell growth/differentiation/apoptosis and other cellular functions [222, 223].

Activated ERK1/2, another MAPK family member, has been found to undergo nucleus-translocation and directly phosphorylate transcription factors and bind to chromatin inside the nuclei [224–226]. A recent discovery shows that autophosphorylation of ERK2 at the site of Thr188 (Thr208 in ERK1) promotes the nucleotide binding but attenuates ERK kinase activities, while inhibiting the upstream regulator MEK1/2 also abolishes the nucleotide binding and reduces the activity of ERK [227, 228]. ERK has diverse cytoplasmic targets such as the p90 ribosomal S6 ($p90^{RSK}$) family with isoforms 1–4, p70 ribosomal S6 ($p70^{RSK}$) [229], MNK [230], and glycogen synthase kinase-3 (GSK-3) [231], which consequently phosphorylate a wide range of substrates involved in gene transcription, translation, cell cycle regulation, and cell survival [232, 233]. Moreover, ERK can phosphorylate a complex family of transcription factors, the ternary complex factors (TCFs; SAP-1, Elk-1, Net, etc.), which are vertebrate ETS-domain proteins that link transcription to MAPK signaling in order to regulate the expression of c-Fos, c-Myc, and c-jun and in turn contribute to transcriptional regulation of various late-response genes that promote cell survival, cell division, and mobility, which are opposite to the regulatory role of JNK and p38 on those cellular responses as discussed above [158, 234–237]. In addition, several key transcription factors including FOXO, BETA2/NeurD1 (the basic helix-loop-helix protein partner), E4, and PDX-1 were found to be related to glucose regulation and insulin gene transcription, which also underlie the anti-apoptosis effect in myocytes and cardiac protective effect of ERK1/2 [238–241].

In summary, MAPK signaling pathways are key mediators of cell transcriptional responses to stress-induced extracellular signals. These pathways critically control gene expression in a number of ways including the phosphorylation of cytosolic proteins and regulation of transcription factors and co-regulatory proteins.

MAPKs and Apoptotic Signaling Pathways in Stress-Exposed Myocytes

Apoptosis is a highly regulated process in the cell composed of a balance between pro-death and pro-survival cell signals, and apoptotic cell death plays a pivotal role in myocyte survival in response to stress stimuli such as ischemic cardiac injury and heart failure. Apoptosis can be roughly divided into extrinsic apoptosis, meaning the apoptosis signaling coming from the environment, and intrinsic apoptosis, meaning the apoptosis signaling coming from the cell itself. The main route of apoptosis in the heart is intrinsic apoptosis, and the key initiator of intrinsic apoptosis is the mitochondrial release of cytochrome c [242]. Cytochrome c released from mitochondria forms a complex from pro-caspase 9 and its cofactor APAF-1 and eventually leads to the activation of caspase 9 and apoptosome formation in the cytosol followed by the activation of caspase 3 leading to apoptosis [242]. Various adverse conditions can lead to intrinsic apoptosis in cardiac myocytes including but not limited to redox stress, energy deprivation, and activation of $G\alpha_q$ signaling

[243–245]. On the other hand, activation of the extrinsic apoptotic signaling pathway is a common phenomenon in stressed hearts such as seen with myocardial ischemic I/R, postinfarction remodeling, end-stage heart failure, and diabetes [246–249]. Two families of independent receptors that mediate the extrinsic apoptosis signaling are mainly regulated by the Fas receptor (receptor for Fas ligand, FasL) and TNFα receptor 1 (TNFα-R1). Activation of Fas receptor and TNFα-R1 by their specific ligands leads to cleavage of pro-caspase 8 into caspase 8 which further activates caspase 3 to promote apoptosis. In both extrinsic and intrinsic apoptosis, the Bcl-2 family of proteins with Bcl-homology domains plays important roles. To date, this Bcl-2 family is known to contain two subgroups, anti-apoptotic proteins including Bcl-2, Bcl-X(L), Bcl-W, Bfl-1, and Mcf-1 and pro-apoptotic proteins including Bad, Bac, Bak, Bix, Box, Bid, Bim, Bnip3, and Nix [250, 251]. Phosphorylation of Bcl-2 family members plays a key role in regulating mitochondrial membrane integrity, and MAPK has been shown to regulate the Bcl-2 family by phosphorylation [250, 251]. While anti-apoptotic Bcl-2 proteins bind to the proteins forming mitochondrial pores, controlling their opening and closure, some pro-apoptotic Bcl-2 proteins including Bax and Bak can insert into mitochondrial outer membrane upon activation via phosphorylation and form pores into the mitochondria [252–255]. The Bcl-2 family achieve the anti-apoptotic effect through the inhibition of pro-apoptotic proteins (i.e., Bad, Bax, and Bim) [256–260].

JNK and Apoptotic Signaling Pathways in Myocytes

In the heart, JNK is activated in response to various stimuli signals including mechanical stretch [261, 262], pressure overload [263–265], I/R [266, 267], and catecholamine stimulation [263], which are known to activate apoptotic signaling pathways [262, 268, 269]. Apoptosis signal-regulating kinases (ASK, including ASK1 and ASK2) promote the activation of MKK4 and MKK7, which are the upstream activators of JNK [21]. However, mixed findings regarding the role of JNK in apoptosis have been reported, suggesting complicated stress responses of JNK signaling in different cell types and potentially different isoform activation at different time frames following various stress challenges. In ROS-challenged myocyte models (H_2O_2-treated or norepinephrine-treated adult rat ventricular myocytes), JNK activation is involved in ROS-induced apoptosis, evidenced by alleviated ROS production and reduced apoptosis through JNK-specific inhibiton [270, 271]. In addition, JNK has been localized to the mitochondria in cardiac myocytes and is known to promote the mitochondrial cytochrome c release in a capase-9-dependent but capase-8-independent manner, suggesting a direct functional role of JNK in mitochondria-associated appotosis [272]. This pro-apoptotic effect of JNK was also found in noncardiac mammalian cells via enhanced phosphorylation and degradation of anti-apoptotic proteins Bcl-2 and Bcl-X(L), which removes the inhibitory regulation of Bcl-2 and consequently activates the pro-apoptotic proteins Bad, Bax, and Bim [256–260]. Furthermore, suppressing JNK1 or the MKK4 pathway from overexpression of dominant negative proteins enhances apoptosis during NO treatment and I/R injuries, further supporting the anti-apoptotic effect of cardiac JNK in response to ischemic stress [273–275]. While JNK inhibition reduces apoptosis during numerous stress conditions including transient energy deprivation (glucose deprivation and mitochondria inhibition), I/R, and hyperglycemia in cultured H9c2 cells and myocytes, JNK-deficiency in fibroblasts activates mitochondrial apoptotic signaling [276–280]. Thus, the functional role of JNKs in cellular apoptotic pathways of the heart may vary depending on the cell type and context in response to different types of stress stimuli. While the findings discussed above are predominantly regarding the contributions of JNK1 in cellular apoptosis and proliferation that have been well studied [10, 20, 133], the functional roles of cardiac JNK2 and JNK3 (with less expression in the heart) in the apoptotic pathway under different stressed conditions remain largely unknown. Further investigations are clearly needed to enhance our understanding regarding the functional role of JNK in myocardial remodeling and maladaptive development.

p38 and Apoptotic Signaling Pathways in Myocytes

p38 is critically involved in cardiac apoptosis in various animal models of cardiac disease such as cardiac injury and HF [281–285]. p38 was found to activate p53 and promote apoptosis by enhancing the expression and translocation of Bax in mitochondria [286]. p38 can also phosphorylate Bcl-2 via translocation into mitochondria to suppress the anti-apoptotic effect of cytosolic Bcl-2 [287]. In platelet-activating factor (PAF)-treated H2c2 cardiac myocytes with elevated cytosolic Ca^{2+}, caspase 3 activity and mitochondria release of cytochrome c increase in a p38-dependent manner, further supporting the pivotal role of p38 in promoting myocyte apoptosis [288]. Also, a reduced level of Bcl-2 and enhanced apoptosis was found in cardiac myocytes with overexpression of wild-type p38 but not in the myocytes with overexpression of an inactive dominant negative form of p38α [283]. This pro-apoptotic function of p38α was further demonstrated in a mouse myocardial infarction model where overexpression of p38α alleviates the inhibition of apoptosis guard Bcl-X(L) and Bcl-2 reduces the Bcl-X(L) deamination consequently activating the pro-apoptotic signaling cascades [282]. The I/R injury response is crucially determined by mitochondrial function and activity and p38 inhibition during I/R decreases mitochondrial swelling and ultrastructural alterations and mitigates mitochondrial membrane depolarization [289]. There is also evidence that p38 activation during I/R contributes to cardiac damage by triggering intracellular Ca^{2+} overload [290]. While the pro-apoptotic role of p38α in myocytes has been well-recognized, the pro-apoptotic role of p38δ in deteriorating cardiac function was recently reported with chemotherapeutic doxorubicin-induced cardiomyopathy [291]. Moreover, p38 is a key player in α1-adrenoreceptor blocker doxazosin-induced apoptosis in cardiac myocytes that increases the risk of HF development [292, 293]. Although the overwhelming evidence suggests a pro-apoptotic role of p38, an anti-apoptosis role of p38 has also been described under certain stressed conditions. For instance, with specific β-adrenergic receptor signaling mediated through Gi-dependent receptors, p38 could exert anti-apoptotic roles, while osmotic stress also showed anti-apoptotic function of p38 via phosphorylation of small heat shock protein αB-crystallin [294–296]. Therefore, p38 may have a differential functional impact on myocyte apoptotic signaling pathways in response to different stress stimuli. In addition, the distinct functional roles of the different sub-isoforms of p38 under different stress stimuli require further investigation.

Other than a direct regulation of p38 in the apoptotic signaling cascade proteins, p38 can also promote apoptosis in cardiac myocytes by regulating apoptotic genes via a wide panel of transcription factors including but not limited to ATF2, AP-1, GATA, SMAD, STAT-1, and NFκB [297]. For instance, in I/R injury and hypoxia animal and cell models, p38-dependent phosphorylation of ATF2 leads to enhanced expression of phosphatase PTEN which limits the phosphorylation and activation of Akt, a powerful survival pathway regulator [222]. Suppressing p38 by overexpressing dominant negative p38 or PTEN both attenuated myocyte death, cardiac injury, and functional loss after the I/R injury [222]. Enhanced expression of FasL is a well-established route toward DNA fragmentation and apoptosis [298, 299]. p38 has been shown to upregulate Fas/FasL expression via increased phosphorylation of the downstream transcription factor STAT-1 at the Ser727 site to enhance its transcriptional activity in myocytes treated with angiotensin II, norepinephrine, and hypoxia [300, 301]. Moreover, p38-dependent AP-1 and GATA transcriptional regulation were also found to upregulate TGF-β expression, which triggers the activation of transcription factor SMAD-mediated apoptosis in angiotensin II-treated myocytes [223]. GADD153 (growth-arrest-and-DNA-damage-inducible protein 153 also known as C/EBP homologous protein (CHOP)) is one of the pro-apoptotic transcription factors engaged in response to enhanced ER stress. p38-dependent activation of GADD153 has been shown to promote the expression of downstream-of-CHOP gene 1(DOC-1), which encodes for a stress-induced form of carbonic anhydrase VI that catalyzes the formation of H_2CO_3 to increase cellular stress-induced apoptosis [302, 303]. In addition, p38-activated GADD153 enhances NFκB phosphorylation and nuclear translocation, which was found in doxorubicin-induced inflammation and apoptosis in cardiac myocytes [304]. Furthermore, inhibition of p38/NFκB signaling alleviates isoproterenol-induced cardiac dysfunction in a rat model [297]. Thus far, the evidence supports a key role for p38 in myocyte apoptosis and the development of cardiac maladaptive function.

ERK and Apoptotic Signaling Pathways in Myocytes

ERK activation is largely anti-apoptotic in the heart under various stressed conditions such as I/R injury, oxidative stress, hypoxia stimulation, and β-adrenergic stimulation [305, 306]. ERK inhibition enhanced the oxidative stress-induced cell injury and apoptosis, while also pharmacologically potentiated ERK activation showed a protective effect against apoptosis induced with a chemotherapy reagent doxorubicin in cardiac myocytes [307–309]. In contrast, activated ERK1/2 attenuates I/R injury in both cell and animal models [268, 310–313]. Another interesting finding is that the cell survival signaling through β2-adrenergic receptor activation has been shown to be regulated through activated ERK [314].

The cardiac protective roles of ERK1/2 during a stress challenge are multidimensional. ERK substrates include nuclear substrates (transcription factors) and cytosolic substrates promote apoptosis through activity and transcriptional regulation of certain key apoptosis-related proteins in the cytosol or nucleus. For example, activated cardiac ERK1/2 can phosphorylate cytosolic proteins such as phospholipase A2 and transcription-regulating kinases $p90^{SRK}$, GSK3, which are critically involved in cellular apoptosis under stressed conditions [315–322]. ERK1/2 also inhibits a key pro-apoptotic protein Bad by facilitating the protein kinase Cε (PKCε)-mediated phosphorylation of Bad in mitochondria and downregulates the expression of Bax, another pro-apoptotic protein in the Bcl-2 family, resulting in the inhibition of cytochrome c release from mitochondria in cardiac myocytes [312, 316–318]. Moreover, ERK-GATA4 signaling was shown to be involved in the anti-apoptosis function of ERK by upregulated expression of anti-apoptotic protein Bcl-X(L) via enhanced phosphorylation of the transcription factor GATA4 at the Ser105 site to enhance the promoter activity of Bcl-X(L) [312, 323]. It is known that activating ERK promotes the activation and nuclear translocation of transcription factor Nrf2 which promotes the expression of genes that potentiate the glutathione antioxidant response, while ERK also upregulates COX-2 via enhanced AP-1 and NFκB-2 transcriptional activity to sustain the cardiac myocyte survival and metabolism in response to an oxidative stress challenge [213, 324]. This convincing evidence suggests that ERK1/2 fulfills an anti-apoptotic role to promote cell survival and growth in response to stress stimuli to protect cardiac function. However, ERK was also found to be pro-apoptotic in certain disease models. For instance, in a diabetic rat model with upregulated HMGB1 (high mobility group box 1 protein), ERK was found to promote apoptosis via the activation of Ets-1, which eventually leads to enhanced Bax protein and caspase 3 activation [325].

Taken together, the three MAPK members fulfill unique but overlapping intracellular signaling mechanisms in responding to a myriad of mitogens and stressors mediating the signaling networks of cell survival and death as well as cardiac metabolism and pathogenesis in a cell type and context dependent manner.

Dynamic Relationships of MAPKs in Pathological Cardiac Remodeling in Stressed Hearts

Accumulating evidence suggests that the three MAPK family members play important roles in the development of pathological remodeling and maladaptive cardiac function during disease progression in a cellular context- and time-dependent manner [10, 21, 22, 131, 132, 326]. Cardiac remodeling encompasses the molecular and structural changes accompanying the electrical physiological and pathological changes in stress-exposed hearts. Those changes are manifested clinically in the progression of HF including increased heart size, deteriorated cardiac contractile function, and reduced cardiac output along with a series of clinical symptoms of

HF. Cardiomyocytes are the major cardiac cells involved in the remodeling process, although other components are also involved including the interstitium, fibroblasts, and coronary vasculature. Being terminally differentiated cells, cardiac myocytes respond to stress stimuli by adaptive growth, also known as hypertrophy, as the adaptive response in stress-exposed hearts [327]. Cardiac hypertrophy can be roughly divided into physiological (or adaptive) hypertrophy and pathological (or maladaptive) hypertrophy [328]. Physiological hypertrophy is reversible and occurs in response to continuously increased demand for cardiac function, which bears the main characteristics of increased cell size, increased fatty acid oxidation and protein synthesis, sarcomeric reorganization, and increased gene transcription related to cell growth [329, 330]. Pathological hypertrophy may also occur after myocardial ischemic injury, inflammatory heart muscle diseases (i.e., myocarditis), idiopathic dilated cardiomyopathy, or unfavorable cardiac volume loading such as hypertension, aortic stenosis, or valvular regurgitation. Pathological hypertrophy is usually accompanied by enhanced inflammation signaling, fetal gene expression, interstitial fibrosis, and risks of decompensation and progression toward to HF [329–332].

Regardless of cardiomyocyte MAPK activation under different stress conditions, a different temporal response to various stress stimuli clearly plays a role [161, 333–336]. Cardiac ERK1/2 plays an essential role in promoting myocyte survival and growth during the progression of adaptive hypertrophy and pathological maladaptive remodeling [337–339]. In response to stress stimuli, ERK1/2 can be activated by either G protein-coupled receptor (altered angiotensin II, endothelin-1, phenylephrine, catecholamines, etc.) [331, 332, 340–343], receptor tyrosine kinases (altered fibroblast growth factor, TGF-β1, growth differentiation factor 15, etc.) [344–346], or mechanical stretch [347]. Activated ERK drives a wide network of intricate pathways which activate secondary signaling to regulate cardiac myocyte cytosolic activity and gene regulation leading to cardiac hypertrophy without premature death or impaired cardiac function [84, 348–350]. While suppression of ERK1/2 signaling by inhibition of ERK1/2 or MEK1/2 with pharmacological agents and dominant negative protein overexpression abolishes hypertrophic response in myocytes during pathological insults (e.g., pressure overload, I/R injury, oxidative stress), this ERK inhibition however leads to aggravated cardiac function and exacerbated myocyte death [161, 324, 351–354]. Thus, ERK-mediated hypertrophic remodeling appears to be a beneficial early response to maintain a normal cardiac function under stress challenges.

A dynamic relationship between the three MAPKs was found in various cardiac disease models. While some studies observed only transient ERK activation in a pressure-overload mouse model induced with transverse aortic constriction (TAC), other reports suggest that ERK remained activated for 2- to 4-week post-TAC [333, 335, 336]. However, JNK and p38 were found to be constantly activated over the course of pressure overload [161, 333–336]. On the other hand, severity of the pressure overload and HF status likely plays a key role in the differential activation status of the three MAPK members [333, 335, 336]. For instance, JNK and p38 activation in the heart and in white blood cells were in positive correlation with the severity of pressure overload measured with trans-stenotic systolic pressure gradient and the severity of hypertrophy (left ventricular weight/body weight ratio) [355]. This was supported by another study in animals showing that JNK is significantly increased with a mild to severe pressure overload (35–85% aortic constriction), while p38 is activated in HF with a more severe pressure overload (85% aortic constriction), but ERK was only transiently activated [356]. In human hypertensive patients with uncontrolled blood pressure, p38 and JNK activation is significantly higher in white blood cells compared to patients with controlled blood pressure [355]. Myocardial I/R is another common and complicated pathophysiological stressor that can lead to myocardial stunning, arrhythmias, and eventually HF [357]. Different time scales of MAPK activation have been shown in the infarct or infarct border zone of post-MI mouse, and p38 phosphorylation is increased initially, while JNK phosphorylation increased approximately 2 weeks after MI; on the other hand, ERK phosphorylation increased after about 4 weeks after MI, likely contributing to the post-MI myocardial remodeling [358]. In a rat model of I/R injury, increased p38 and unchanged ERK occurred during the acute ischemia phase, while JNK was only increased during the reperfusion phase [359]. Similarly, in a long-term ischemia porcine model of coronary embolism, JNK, p38, and ERK activations were all enhanced, contributing to the postischemic remodeling from the microinfarction and apoptosis [360]. The enhanced phosphorylation of p38 and JNK after I/R injury further decreased the cell viability and promoted cardiac cell apoptosis [361] via decreased anti-apoptotic Bcl-2 and increased pro-apoptotic Bax [280]. Moreover, acute MI (6 hours after LAD ligation) could activate both JNK and p38, yet ERK was decreased, accompanied by increased activity of pro-apoptotic protein caspase-3 [362–364]. ROS has been shown to be critical in activating MAPKs in response to I/R injury. Suppression of ROS production alleviates MAPK activation and attenuates the loss of cardiac function and decreased infarct size [365]. This is supported by the findings of ROS-dependent JNK and p38 activation in post-MI ventricular tissues [366]. In animal models of diabetes and post-MI diabetes, increased ROS was found to be responsible for intensified myocardial injury and loss of cardiac function along with increased activation of JNK and p38, and the contribution of ROS-activated JNK and p38 was confirmed by the striking rescue effect of ROS suppression and stress kinase inhibiton [367, 368]. Moreover, antioxidant/anti-inflammatory agents also prove to enhance ERK1/2 activation and preserve the cardiac function along with suppressed activities of JNK and p38 [369, 370].

Overall, dynamic changes of the three MAPKs appear to occur at different time points during the pathological cardiac remodeling process in response to a myriad of mitogens and stressors. In general, the end result of this temporally dynamic activation of p38 and JNK leads to the aggravation of cardiac injury, promotes maladaptive cardiac remodeling, and impairs cardiac function, while the activation of ERK promotes cell survival and preserves cardiac function. To date, the MAPK family has received extensive interest due to the far-reaching implications its members manifest in signaling and cross talk with other signaling networks. Further understanding regarding the dynamic relationship between the three MAPK members under different stress conditions will be required for the development of effective therapeutic strategies and the discovery of novel therapeutic targets for early intervention and treatment of cardiac diseases.

MAPKs and Therapeutic Potentials

MAPKs are critically involved in cardiac pathological remodeling, and disease development (i.e., hypertrophy, ischemic injury, HF, and cardiac arrhythmias) positioning the modulation of MAPK activity as a novel point for therapeutic intervention. Pharmacological inhibition and genetic deletion of the three MAPK kinases has proven to change the course of stress-induced cardiac adaptive and maladaptive remodeling [333, 371]. Inhibition of JNK and p38 activities has been shown to reduce cardiac injury and preserve cardiac function in the stressed heart, and this was supported by manipulating upstream regulators of JNK and/or p38 to suppress their activities. The knockdown of Grb2 adaptor protein suppressed JNK and p38 activities (no effect on ERK), which effectively alleviated cardiac hypertrophy and apoptosis in pressure-overload hearts [334, 372]. Another example is that deletion/suppression of multiple upstream regulatory genes (i.e., BAMBI, NECTIN2, DKK3, and the 14-3-3 family) inhibits JNK and p38 activity along with exacerbated apoptosis, fibrosis, and cardiac function loss [373–376].

However, selective inhibition of each MAPK often offers different outcomes, which is in line with the evidence of differential involvement of the three MAPKs and their downstream signaling pathways at different stages during the progression of cardiac remodeling and disease development under different stress conditions. While some studies have shown a beneficial effect of competitive inhibition of the p38 signaling alone by overexpression of dominant negative p38 or pharmacological inhibition in promoting cardiac hypertrophy by preventing apoptosis and fibrosis and preserving cardiac function in response to pressure overload and I/R injury [281, 283, 372, 375, 377–379], other studies suggest that inhibiting p38 alone promotes hypertrophy but aggravates the deterioration of cardiac function in both pressure overload and I/R injury models and even abolishing the protective effects of ischemia preconditioning [380–383]. There are current clinical trials testing the safety and efficacy of p38 inhibition in myocardial infarction [384, 385]. The p38 inhibitor SB-681323 decreased the circulating inflammatory marker high-sensitivity C-reactive protein (hs-CRP) in statin-receiving patients undergoing elective percutaneous coronary intervention [386]. A phase 2 clinical trial suggests treating patients with non-ST-segment elevation myocardial infarction with the p38 inhibitor losmapimod (GW856553X) decreased circulating levels of inflammatory markers including IL-6 and hs-CRP short term (72 hour); however, no differences in inflammatory markers or risk of cardiovascular events were shown in a longer term trial (12 weeks) [384, 387, 388]. Similarly, JNK inhibition without the influence in ERK and p38 led to increased myocyte size; upregulated activation of pro-apoptotic proteins including p53, Bad, and Bax; and exacerbated deterioration of cardiac function in different disease models [389–391]. Thus, therapeutic potentials of selective inhibition of either JNK or p38 for cardiac function protection under certain pathological conditions need to be further evaluated. Recently, JNK2, an understudied JNK isoform, showed therapeutical potentials in anti-arrhythmogenesis [20, 24–26, 392]. While accumulating evidence suggests that pro-arrhythmic molecule CaMKII may represent a target for therapy and provoked the development of CAMKII inhibitors [113, 393, 394], CAMKII inhibitors have off-target effects that hinder their clinical application [395]. With the discovery of a novel form of pathogenic kinase-on-kinase cross talk, JNK2 directly regulates CAMKII activity, and the development of compounds directed at JNK2 may represent an alternative drug target. More research is warranted to explore JNK2 as an anti-arrhythmic drug target in patients.

Considering the protective role of ERK and detrimental role of JNK and p38 in cardiac injury and pathological remodeling, developing specific strategies for use of selective MAPK inhibitors has a high clinical potential in preventing and treating cardiac injury and pathological remodeling and ultimately preserving or improving cardiac function. In fact, activation of the ERK1/2 pathway has been identified as a central component of the so-called Reperfusion Injury Salvage Kinase (RISK) pathway [396]. ERK1/2-specific in vitro kinase activity in adult hearts subjected to 20 minutes of ischemia followed by 15 minutes of reperfusion was doubled [362]. In addition to this study, others support a protective role for the MEK1-ERK2-signaling pathway against I/R injury [268, 397, 398]. A thorough experimental dissection of the RISK pathway revealed a combination of two parallel cascades, PI3K-Akt and MEK1-ERK1/2, that produced a protective effect when blocked with the co-administration of both PI3K and ERK inhibitors at different time points [399]. Thus, broadly the RISK pathway refers to a group of pro-survival protein kinases, which confer cardioprotection when activated specifically at the time of reperfusion [399, 400]. The RISK pathway has recently been seen as a universal signaling cascade for cardioprotection and is likely recruited not only by ischemic conditioning but also by other pharmacological agents such insulin, bradykinin adenosine, or statins [401] shared by most cardioprotective therapies [402]. Therefore, the cardiac protective effect of activated ERK shed new light on drug development studies. Intriguingly, pharmacological reagents that are suppressive to JNK and p38 activities while augmenting the activation of ERK lead to attenuated scar formation and improved cardiac outcome, underscoring the functional cross talk between the MAPKs [282, 362–364]. As it has been discussed in this chapter, differential changes of the three MAPKs occur during the development of adaptive and maladaptive cardiac remodeling and cardiac disease progression in cellular context and time-dependent manners. Further investigation is urgently needed to enhance our understanding of the dynamic relationships between the three MAPK members under different stress conditions, which is essential for the development of effective therapeutic strategies and the discovery of novel therapeutic targets for early intervention and treatment of cardiac diseases.

Declarations This work was supported by National Institutes of Health [P01-HL06426, R01-AA024769, R01-HL113640, R01-HL146744 to XA].

References

1. Beckman KB, Ames BN. The free radical theory of aging matures. Physiol Rev. 1998;78:547–81.
2. Neuman RB, et al. Oxidative stress markers are associated with persistent atrial fibrillation. Clin Chem. 2007;53:1652–7. https://doi.org/10.1373/clinchem.2006.083923. clinchem.2006.083923 [pii]
3. Juhaszova M, Rabuel C, Zorov DB, Lakatta EG, Sollott SJ. Protection in the aged heart: preventing the heart-break of old age? Cardiovasc Res. 2005;66:233–44. https://doi.org/10.1016/j.cardiores.2004.12.020. S0008-6363(04)00592-9 [pii].
4. He BJ, et al. Oxidation of CaMKII determines the cardiotoxic effects of aldosterone. Nat Med. 2011;17:1610–8. https://doi.org/10.1038/nm.2506.
5. Belmin J, et al. Increased production of tumor necrosis factor and interleukin-6 by arterial wall of aged rats. Am J Physiol. 1995;268:H2288–93.
6. Ismahil MA, et al. Remodeling of the mononuclear phagocyte network underlies chronic inflammation and disease progression in heart failure: critical importance of the cardiosplenic axis. Circ Res. 2014;114:266–82. https://doi.org/10.1161/CIRCRESAHA.113.301720.
7. Judge S, Leeuwenburgh C. Cardiac mitochondrial bioenergetics, oxidative stress, and aging. Am J Physiol Cell Physiol. 2007;292:C1983–92. https://doi.org/10.1152/ajpcell.00285.2006. 00285.2006 [pii]
8. Yang Z, Shen W, Rottman JN, Wikswo JP, Murray KT. Rapid stimulation causes electrical remodeling in cultured atrial myocytes. J Mol Cell Cardiol. 2005;38:299–308. https://doi.org/10.1016/j.yjmcc.2004.11.015.
9. Li SY, et al. Aging induces cardiac diastolic dysfunction, oxidative stress, accumulation of advanced glycation endproducts and protein modification. Aging Cell. 2005;4:57–64. https://doi.org/10.1111/j.1474-9728.2005.00146.x. ACE146 [pii].
10. Rose BA, Force T, Wang Y. Mitogen-activated protein kinase signaling in the heart: angels versus demons in a heart-breaking tale. Physiol Rev. 2010;90:1507–46. https://doi.org/10.1152/physrev.00054.2009. 90/4/1507 [pii].
11. Yarza R, Vela S, Solas M, Ramirez MJ. c-Jun N-terminal Kinase (JNK) signaling as a therapeutic target for Alzheimer's disease. Front Pharmacol. 2015;6:321. https://doi.org/10.3389/fphar.2015.00321.
12. Xiao B, et al. Extracellular translationally controlled tumor protein promotes colorectal cancer invasion and metastasis through Cdc42/JNK/ MMP9 signaling. Oncotarget. 2016; https://doi.org/10.18632/oncotarget.10315. 10315 [pii].
13. Brozzi F, Eizirik DL. ER stress and the decline and fall of pancreatic beta cells in type 1 diabetes. Ups J Med Sci. 2016;121:133–9. https://doi.org/10.3109/03009734.2015.1135217.
14. Scharf M, et al. Mitogen-activated protein kinase-activated protein kinases 2 and 3 regulate SERCA2a expression and fiber type composition to modulate skeletal muscle and cardiomyocyte function. Mol Cell Biol. 2013;33:2586–602. https://doi.org/10.1128/MCB.01692-12.
15. Ho PD, et al. Ras reduces L-type calcium channel current in cardiac myocytes. Corrective effects of L-channels and SERCA2 on [Ca(2+)](i) regulation and cell morphology. Circ Res. 2001;88:63–9.
16. Ho PD, et al. The Raf-MEK-ERK cascade represents a common pathway for alteration of intracellular calcium by Ras and protein kinase C in cardiac myocytes. J Biol Chem. 1998;273:21730–5.

17. Huang H, Joseph LC, Gurin MI, Thorp EB, Morrow JP. Extracellular signal-regulated kinase activation during cardiac hypertrophy reduces sarcoplasmic/endoplasmic reticulum calcium ATPase 2 (SERCA2) transcription. J Mol Cell Cardiol. 2014;75:58–63. https://doi.org/10.1016/j.yjmcc.2014.06.018.
18. Hagiwara Y, et al. SHP2-mediated signaling cascade through gp130 is essential for LIF-dependent I CaL, [Ca2+]i transient, and APD increase in cardiomyocytes. J Mol Cell Cardiol. 2007;43:710–6. https://doi.org/10.1016/j.yjmcc.2007.09.004.
19. Takahashi E, et al. Leukemia inhibitory factor activates cardiac L-Type Ca2+ channels via phosphorylation of serine 1829 in the rabbit Cav1.2 subunit. Circ Res. 2004;94:1242–8. https://doi.org/10.1161/01.RES.0000126405.38858.BC.
20. Yan J, et al. c-Jun N-terminal kinase activation contributes to reduced connexin43 and development of atrial arrhythmias. Cardiovasc Res. 2013;97:589–97. https://doi.org/10.1093/cvr/cvs366.
21. Davis RJ. Signal transduction by the JNK group of MAP kinases. Cell. 2000;103:239–52.
22. Karin M, Gallagher E. From JNK to pay dirt: jun kinases, their biochemistry, physiology and clinical importance. IUBMB Life. 2005;57:283–95. https://doi.org/10.1080/15216540500097111. H527470040088211 [pii]
23. Ramos JW. The regulation of extracellular signal-regulated kinase (ERK) in mammalian cells. Int J Biochem Cell Biol. 2008;40:2707–19. https://doi.org/10.1016/j.biocel.2008.04.009.
24. Yan J, et al. JNK2, a newly-identified SERCA2 enhancer, augments an arrhythmic [Ca(2+)]SR leak-load relationship. Circ Res. 2021;128:455–70. https://doi.org/10.1161/CIRCRESAHA.120.318409.
25. Yan J, et al. Role of stress Kinase JNK in binge alcohol-evoked atrial arrhythmia. J Am Coll Cardiol. 2018;71:1459–70. https://doi.org/10.1016/j.jacc.2018.01.060. S0735-1097(18)30435-2 [pii].
26. Yan J, et al. The stress kinase JNK regulates gap junction Cx43 gene expression and promotes atrial fibrillation in the aged heart. J Mol Cell Cardiol. 2017;114:105–15. https://doi.org/10.1016/j.yjmcc.2017.11.006.
27. Ai L, et al. Inhibition of Abeta Proteotoxicity by Paeoniflorin in Caenorhabditis elegans through regulation of oxidative and heat shock stress responses. Rejuvenation Res. 2018;21:304–12. https://doi.org/10.1089/rej.2017.1966.
28. Bogoyevitch MA, Kobe B. Uses for JNK: the many and varied substrates of the c-Jun N-terminal kinases. Microbiol Mol Biol Rev. 2006;70:1061–95. https://doi.org/10.1128/MMBR.00025-06. 70/4/1061 [pii]
29. Widmann C, Gibson S, Jarpe MB, Johnson GL. Mitogen-activated protein kinase: conservation of a three-kinase module from yeast to human. Physiol Rev. 1999;79:143–80.
30. Karin M. Inflammation-activated protein kinases as targets for drug development. Proc Am Thorac Soc. 2005;2:386–390; discussion 394-385. https://doi.org/10.1513/pats.200504-034SR. 2/4/386 [pii]
31. Raman M, Chen W, Cobb MH. Differential regulation and properties of MAPKs. Oncogene. 2007;26:3100–12. https://doi.org/10.1038/sj.onc.1210392.
32. Shimizu S, et al. Involvement of JNK in the regulation of autophagic cell death. Oncogene. 2010;29:2070–82. https://doi.org/10.1038/onc.2009.487. onc2009487 [pii].
33. Kyriakis JM, Avruch J. pp54 microtubule-associated protein 2 kinase. A novel serine/threonine protein kinase regulated by phosphorylation and stimulated by poly-L-lysine. J Biol Chem. 1990;265:17355–63.
34. Kyriakis JM, Brautigan DL, Ingebritsen TS, Avruch J. pp54 microtubule-associated protein-2 kinase requires both tyrosine and serine/threonine phosphorylation for activity. J Biol Chem. 1991;266:10043–6.
35. Hu MC, Qiu WR, Wang YP. JNK1, JNK2 and JNK3 are p53 N-terminal serine 34 kinases. Oncogene. 1997;15:2277–87. https://doi.org/10.1038/sj.onc.1201401.

36. Mohit AA, Martin JH, Miller CA. p493F12 kinase: a novel MAP kinase expressed in a subset of neurons in the human nervous system. Neuron. 1995;14:67–78. https://doi.org/10.1016/0896-6273(95)90241-4.
37. Nakano R, Nakayama T, Sugiya H. Biological properties of JNK3 and its function in neurons, astrocytes, pancreatic β-cells and cardiovascular cells. Cells. 2020;9 https://doi.org/10.3390/cells9081802.
38. Kuan CY, et al. The Jnk1 and Jnk2 protein kinases are required for regional specific apoptosis during early brain development. Neuron. 1999;22:667–76., S0896-6273(00)80727-8 [pii].
39. Moriguchi T, et al. A novel SAPK/JNK kinase, MKK7, stimulated by TNFalpha and cellular stresses. EMBO J. 1997;16:7045–53. https://doi.org/10.1093/emboj/16.23.7045.
40. Gao X, et al. Transcriptional regulation of stress kinase JNK2 in pro-arrhythmic CaMKIIdelta expression in the aged atrium. Cardiovasc Res. 2018;114:737–46. https://doi.org/10.1093/cvr/cvy011. 4817461 [pii].
41. Izumi Y, Kim S, Murakami T, Yamanaka S, Iwao H. Cardiac mitogen-activated protein kinase activities are chronically increased in stroke-prone hypertensive rats. Hypertension. 1998;31:50–6.
42. Peart JN, Gross ER, Headrick JP, Gross GJ. Impaired p38 MAPK/HSP27 signaling underlies aging-related failure in opioid-mediated cardioprotection. J Mol Cell Cardiol. 2007;42:972–80. https://doi.org/10.1016/j.yjmcc.2007.02.011. S0022-2828(07)00050-8 [pii].
43. Rich MW. Epidemiology of atrial fibrillation. J Interv Card Electrophysiol. 2009;25:3–8. https://doi.org/10.1007/s10840-008-9337-8.
44. Wang MC, Bohmann D, Jasper H. JNK signaling confers tolerance to oxidative stress and extends lifespan in Drosophila. Dev Cell. 2003;5:811–6., S153458070300323X [pii].
45. Ono K, Han J. The p38 signal transduction pathway: activation and function. Cell Signal. 2000;12:1–13. https://doi.org/10.1016/s0898-6568(99)00071-6. S0898-6568(99)00071-6 [pii].
46. Han J, Lee JD, Bibbs L, Ulevitch RJ. A MAP kinase targeted by endotoxin and hyperosmolarity in mammalian cells. Science. 1994;265:808–11. https://doi.org/10.1126/science.7914033.
47. Han J, Lee JD, Tobias PS, Ulevitch RJ. Endotoxin induces rapid protein tyrosine phosphorylation in 70Z/3 cells expressing CD14. J Biol Chem. 1993;268:25009–14.
48. Cuenda A, Rousseau S. p38 MAP-kinases pathway regulation, function and role in human diseases. Biochim Biophys Acta. 2007;1773:1358–75. https://doi.org/10.1016/j.bbamcr.2007.03.010.
49. Remy G, et al. Differential activation of p38MAPK isoforms by MKK6 and MKK3. Cell Signal. 2010;22:660–7. https://doi.org/10.1016/j.cellsig.2009.11.020.
50. Seta K, Sadoshima J. What is the unique function of SAPK3/p38gamma in cardiac myocytes? J Mol Cell Cardiol. 2002;34:597–600. https://doi.org/10.1006/jmcc.2002.2000.
51. Court NW, dos Remedios CG, Cordell J, Bogoyevitch MA. Cardiac expression and subcellular localization of the p38 mitogen-activated protein kinase member, stress-activated protein kinase-3 (SAPK3). J Mol Cell Cardiol. 2002;34:413–26. https://doi.org/10.1006/jmcc.2001.1523.
52. Dingar D, et al. Effect of pressure overload-induced hypertrophy on the expression and localization of p38 MAP kinase isoforms in the mouse heart. Cell Signal. 2010;22:1634–44. https://doi.org/10.1016/j.cellsig.2010.06.002. S0898-6568(10)00167-1 [pii].
53. Raingeaud J, et al. Pro-inflammatory cytokines and environmental stress cause p38 mitogen-activated protein kinase activation by dual phosphorylation on tyrosine and threonine. J Biol Chem. 1995;270:7420–6. https://doi.org/10.1074/jbc.270.13.7420. S0021-9258(18)71780-8 [pii].
54. Moriguchi T, et al. A novel kinase cascade mediated by mitogen-activated protein kinase kinase 6 and MKK3. J Biol Chem. 1996;271:13675–9. https://doi.org/10.1074/jbc.271.23.13675.
55. Raingeaud J, Whitmarsh AJ, Barrett T, Derijard B, Davis RJ. MKK3- and MKK6-regulated gene expression is mediated by the p38 mitogen-activated protein kinase signal transduction pathway. Mol Cell Biol. 1996;16:1247–55. https://doi.org/10.1128/MCB.16.3.1247.

56. Chang L, Karin M. Mammalian MAP kinase signalling cascades. Nature. 2001;410:37–40. https://doi.org/10.1038/35065000. 35065000 [pii].
57. Cuadrado A, Nebreda AR. Mechanisms and functions of p38 MAPK signalling. Biochem J. 2010;429:403–17. https://doi.org/10.1042/BJ20100323. BJ20100323 [pii].
58. Zarubin T, Han J. Activation and signaling of the p38 MAP kinase pathway. Cell Res. 2005;15:11–8. https://doi.org/10.1038/sj.cr.7290257.
59. Marinissen MJ, Chiariello M, Pallante M, Gutkind JS. A network of mitogen-activated protein kinases links G protein-coupled receptors to the c-jun promoter: a role for c-Jun NH2-terminal kinase, p38s, and extracellular signal-regulated kinase 5. Mol Cell Biol. 1999;19:4289–301.
60. Zhang S, et al. Rho family GTPases regulate p38 mitogen-activated protein kinase through the downstream mediator Pak1. J Biol Chem. 1995;270:23934–6.
61. Ge B, et al. MAPKK-independent activation of p38alpha mediated by TAB1-dependent autophosphorylation of p38alpha. Science. 2002;295:1291–4. https://doi.org/10.1126/science.1067289.
62. Salvador JM, et al. Alternative p38 activation pathway mediated by T cell receptor-proximal tyrosine kinases. Nat Immunol. 2005;6:390–5. https://doi.org/10.1038/ni1177.
63. Mocsai A, Ruland J, Tybulewicz VL. The SYK tyrosine kinase: a crucial player in diverse biological functions. Nat Rev Immunol. 2010;10:387–402. https://doi.org/10.1038/nri2765.
64. Boulton TG, Cobb MH. Identification of multiple extracellular signal-regulated kinases (ERKs) with antipeptide antibodies. Cell Regul. 1991;2:357–71. https://doi.org/10.1091/mbc.2.5.357.
65. Boulton TG, et al. ERKs: a family of protein-serine/threonine kinases that are activated and tyrosine phosphorylated in response to insulin and NGF. Cell. 1991;65:663–75. https://doi.org/10.1016/0092-8674(91)90098-j. 0092-8674(91)90098-J [pii].
66. Sturgill TW, Ray LB, Erikson E, Maller JL. Insulin-stimulated MAP-2 kinase phosphorylates and activates ribosomal protein S6 kinase II. Nature. 1988;334:715–8. https://doi.org/10.1038/334715a0.
67. Kohno M, Pouyssegur J. Alpha-thrombin-induced tyrosine phosphorylation of 43,000- and 41,000-Mr proteins is independent of cytoplasmic alkalinization in quiescent fibroblasts. Biochem J. 1986;238:451–7.
68. Prowse CN, Deal MS, Lew J. The complete pathway for catalytic activation of the mitogen-activated protein kinase, ERK2. J Biol Chem. 2001;276:40817–23. https://doi.org/10.1074/jbc.M105860200. S0021-9258(20)77927-5 [pii].
69. Ahn NG, et al. Multiple components in an epidermal growth factor-stimulated protein kinase cascade. In vitro activation of a myelin basic protein/microtubule-associated protein 2 kinase. J Biol Chem. 1991;266:4220–7., S0021-9258(20)64310-1 [pii].
70. Lowes VL, Ip NY, Wong YH. Integration of signals from receptor tyrosine kinases and g protein-coupled receptors. Neurosignals. 2002;11:5–19., 57317 [pii] 57317.
71. Wang LP, et al. Erythropoietin decreases the occurrence of myocardial fibrosis by inhibiting the NADPH-ERK-NF-x03BA;B pathway. Cardiology. 2016;133:97–108. https://doi.org/10.1159/000440995. 000440995 [pii].
72. Chen LJ, et al. Angiotensin-converting enzyme 2 ameliorates renal fibrosis by blocking the activation of mTOR/ERK signaling in apolipoprotein E-deficient mice. Peptides. 2016;79:49–57. https://doi.org/10.1016/j.peptides.2016.03.008. S0196-9781(16)30043-2 [pii].
73. Hao G, et al. Ets-1 upregulation mediates angiotensin II-related cardiac fibrosis. Int J Clin Exp Pathol. 2015;8:10216–27.
74. Peng K, et al. Novel EGFR inhibitors attenuate cardiac hypertrophy induced by angiotensin II. J Cell Mol Med. 2016;20:482–94. https://doi.org/10.1111/jcmm.12763.
75. Li Q, Liu X, Wei J. Ageing related periostin expression increase from cardiac fibroblasts promotes cardiomyocytes senescent. Biochem Biophys Res Commun. 2014;452:497–502. https://doi.org/10.1016/j.bbrc.2014.08.109. S0006-291X(14)01541-1 [pii].

76. Wellbrock C, Karasarides M, Marais R. The RAF proteins take centre stage. Nat Rev Mol Cell Biol. 2004;5:875–85. https://doi.org/10.1038/nrm1498. nrm1498 [pii].
77. Muslin AJ. Role of raf proteins in cardiac hypertrophy and cardiomyocyte survival. Trends Cardiovasc Med. 2005;15:225–9. https://doi.org/10.1016/j.tcm.2005.06.008. S1050-1738(05)00089-7 [pii].
78. Chen Z, et al. MAP kinases. Chem Rev. 2001;101:2449–76., cr000241p [pii].
79. Hatano N, et al. Essential role for ERK2 mitogen-activated protein kinase in placental development. Genes Cells. 2003;8:847–56., 680 [pii].
80. Saba-El-Leil MK, et al. An essential function of the mitogen-activated protein kinase Erk2 in mouse trophoblast development. EMBO Rep. 2003;4:964–8. https://doi.org/10.1038/sj.embor.embor939. embor939 [pii].
81. Blasco RB, et al. c-Raf, but not B-Raf, is essential for development of K-Ras oncogene-driven non-small cell lung carcinoma. Cancer Cell. 2011;19:652–63. https://doi.org/10.1016/j.ccr.2011.04.002. S1535-6108(11)00147-4 [pii].
82. Yao Y, et al. Extracellular signal-regulated kinase 2 is necessary for mesoderm differentiation. Proc Natl Acad Sci U S A. 2003;100:12759–64. https://doi.org/10.1073/pnas.2134254100. 2134254100 [pii].
83. Alvarez E, et al. Pro-Leu-Ser/Thr-Pro is a consensus primary sequence for substrate protein phosphorylation. Characterization of the phosphorylation of c-myc and c-jun proteins by an epidermal growth factor receptor threonine 669 protein kinase. J Biol Chem. 1991;266:15277–85., S0021-9258(18)98613-8 [pii].
84. Bueno OF, et al. The MEK1-ERK1/2 signaling pathway promotes compensated cardiac hypertrophy in transgenic mice. EMBO J. 2000;19:6341–50. https://doi.org/10.1093/emboj/19.23.6341.
85. Fabiato A. Time and calcium dependence of activation and inactivation of calcium-induced release of calcium from the sarcoplasmic reticulum of a skinned canine cardiac Purkinje cell. J Gen Physiol. 1985;85:247–89.
86. Rougier O, Vassort G, Garnier D, Gargouil YM, Coraboeuf E. Existence and role of a slow inward current during the frog atrial action potential. Pflugers Arch. 1969;308:91–110.
87. Afzal N, Dhalla NS. Differential changes in left and right ventricular SR calcium transport in congestive heart failure. Am J Physiol. 1992;262:H868–74. https://doi.org/10.1152/ajpheart.1992.262.3.H868.
88. Makarewich CA, et al. The DWORF micropeptide enhances contractility and prevents heart failure in a mouse model of dilated cardiomyopathy. Elife. 2018;7 https://doi.org/10.7554/eLife.38319.
89. Minamisawa S, Sato Y, Cho MC. Calcium cycling proteins in heart failure, cardiomyopathy and arrhythmias. Exp Mol Med. 2004;36:193–203. https://doi.org/10.1038/emm.2004.27.
90. Santulli G, Xie W, Reiken SR, Marks AR. Mitochondrial calcium overload is a key determinant in heart failure. Proc Natl Acad Sci U S A. 2015;112:11389–94. https://doi.org/10.1073/pnas.1513047112.
91. Smith GL, Eisner DA. Calcium Buffering in the Heart in Health and Disease. Circulation. 2019;139:2358–71. https://doi.org/10.1161/circulationaha.118.039329.
92. Fernández-Miranda G, Romero-Garcia T, Barrera-Lechuga TP, Mercado-Morales M, Rueda A. Impaired activity of Ryanodine receptors contributes to calcium mishandling in Cardiomyocytes of metabolic syndrome rats. Front Physiol. 2019;10:520. https://doi.org/10.3389/fphys.2019.00520.
93. Gorski PA, Ceholski DK, Hajjar RJ. Altered myocardial calcium cycling and energetics in heart failure--a rational approach for disease treatment. Cell Metab. 2015;21:183–94. https://doi.org/10.1016/j.cmet.2015.01.005.
94. Lindner M, Böhle T, Beuckelmann DJ. Ca2+-handling in heart failure--a review focusing on Ca2+ sparks. Basic Res Cardiol. 2002;97(Suppl 1):I79–82. https://doi.org/10.1007/s003950200034.

95. Santulli G, Nakashima R, Yuan Q, Marks AR. Intracellular calcium release channels: an update. J Physiol. 2017;595:3041–51. https://doi.org/10.1113/jp272781.
96. Shiferaw Y, Watanabe MA, Garfinkel A, Weiss JN, Karma A. Model of intracellular calcium cycling in ventricular myocytes. Biophys J. 2003;85:3666–86. https://doi.org/10.1016/s0006-3495(03)74784-5.
97. Tamayo M, et al. Intracellular calcium mishandling leads to cardiac dysfunction and ventricular arrhythmias in a mouse model of propionic acidemia. Biochim Biophys Acta Mol Basis Dis. 2020;1866:165586. https://doi.org/10.1016/j.bbadis.2019.165586.
98. Bers DM. Calcium fluxes involved in control of cardiac myocyte contraction. Circ Res. 2000;87:275–81.
99. Ather S, Respress JL, Li N, Wehrens XH. Alterations in ryanodine receptors and related proteins in heart failure. Biochim Biophys Acta. 1832;2425-2431:2013. https://doi.org/10.1016/j.bbadis.2013.06.008.
100. Lugenbiel P, et al. Atrial fibrillation complicated by heart failure induces distinct remodeling of calcium cycling proteins. Plos One. 2015;10:e0116395. https://doi.org/10.1371/journal.pone.0116395.
101. Hu ST, et al. Altered intracellular Ca2+ regulation in chronic rat heart failure. J Physiol Sci. 2010;60:85–94. https://doi.org/10.1007/s12576-009-0070-6.
102. Lehnart SE, et al. Sarcoplasmic reticulum proteins in heart failure. Ann N Y Acad Sci. 1998;853:220–30. https://doi.org/10.1111/j.1749-6632.1998.tb08270.x.
103. Wescott AP, Jafri MS, Lederer WJ, Williams GS. Ryanodine receptor sensitivity governs the stability and synchrony of local calcium release during cardiac excitation-contraction coupling. J Mol Cell Cardiol. 2016;92:82–92. https://doi.org/10.1016/j.yjmcc.2016.01.024.
104. Gambardella J, Trimarco B, Iaccarino G, Santulli G. New insights in cardiac calcium handling and excitation-contraction coupling. Adv Exp Med Biol. 2018;1067:373–85. https://doi.org/10.1007/5584_2017_106.
105. Landstrom AP, Dobrev D, Wehrens XHT. Calcium signaling and cardiac arrhythmias. Circ Res. 2017;120:1969–93. https://doi.org/10.1161/circresaha.117.310083.
106. Maier LS. CaMKIIdelta overexpression in hypertrophy and heart failure: cellular consequences for excitation-contraction coupling. Brazil J Med Biol Res = Revista brasileira de pesquisas medicas e biologicas / Sociedade Brasileira de Biofisica [et al]. 2005;38:1293–302. https://doi.org/10.1590/s0100-879x2005000900002.
107. Marks AR. Calcium cycling proteins and heart failure: mechanisms and therapeutics. J Clin Invest. 2013;123:46–52. https://doi.org/10.1172/jci62834.
108. Balke CW, Shorofsky SR. Alterations in calcium handling in cardiac hypertrophy and heart failure. Cardiovasc Res. 1998;37:290–9. https://doi.org/10.1016/s0008-6363(97)00272-1.
109. Dobrev D, Wehrens XHT. Calcium-mediated cellular triggered activity in atrial fibrillation. J Physiol. 2017;595:4001–8. https://doi.org/10.1113/jp273048.
110. Ottolia M, Torres N, Bridge JH, Philipson KD, Goldhaber JI. Na/Ca exchange and contraction of the heart. J Mol Cell Cardiol. 2013;61:28–33. https://doi.org/10.1016/j.yjmcc.2013.06.001.
111. Eisner DA, Caldwell JL, Kistamás K, Trafford AW. Calcium and excitation-contraction coupling in the heart. Circ Res. 2017;121:181–95. https://doi.org/10.1161/circresaha.117.310230.
112. Kistamás K, et al. Calcium handling defects and cardiac arrhythmia syndromes. Front Pharmacol. 2020;11:72. https://doi.org/10.3389/fphar.2020.00072.
113. Ai X, Curran JW, Shannon TR, Bers DM, Pogwizd SM. Ca2+/calmodulin-dependent protein kinase modulates cardiac ryanodine receptor phosphorylation and sarcoplasmic reticulum Ca2+ leak in heart failure. Circ Res. 2005;97:1314–22. https://doi.org/10.1161/01.RES.0000194329.41863.89.
114. Bers DM. Calcium cycling and signaling in cardiac myocytes. Annu Rev Physiol. 2008;70:23–49. https://doi.org/10.1146/annurev.physiol.70.113006.100455.
115. Bovo E, Mazurek SR, Zima AV. Oxidation of ryanodine receptor after ischemia-reperfusion increases propensity of Ca(2+) waves during beta-adrenergic receptor stimu-

lation. Am J Physiol Heart Circ Physiol. 2018;315:H1032–40. https://doi.org/10.1152/ajpheart.00334.2018.
116. Breckenridge R. Heart failure and mouse models. Dis Model Mech. 2010;3:138–43. https://doi.org/10.1242/dmm.005017.
117. Denham NC, et al. Calcium in the pathophysiology of atrial fibrillation and heart failure. Front Physiol. 2018;9:1380. https://doi.org/10.3389/fphys.2018.01380.
118. Kashimura T, et al. In the RyR2(R4496C) mouse model of CPVT, beta-adrenergic stimulation induces Ca waves by increasing SR Ca content and not by decreasing the threshold for Ca waves. Circ Res. 2010;107:1483–9. https://doi.org/10.1161/CIRCRESAHA.110.227744.
119. Liu B, et al. Ablation of HRC alleviates cardiac arrhythmia and improves abnormal Ca handling in CASQ2 knockout mice prone to CPVT. Cardiovasc Res. 2015;108:299–311. https://doi.org/10.1093/cvr/cvv222. cvv222 [pii].
120. Yan J, et al. Stress signaling JNK2 crosstalk with CaMKII underlies enhanced atrial Arrhythmogenesis. Circ Res. 2018; https://doi.org/10.1161/CIRCRESAHA.117.312536.
121. Respress JL, et al. Role of RyR2 phosphorylation at S2814 during heart failure progression. Circ Res. 2012;110:1474–83. https://doi.org/10.1161/circresaha.112.268094.
122. Yeh YH, et al. Calcium-handling abnormalities underlying atrial arrhythmogenesis and contractile dysfunction in dogs with congestive heart failure. Circ Arrhythm Electrophysiol. 2008;1:93–102. https://doi.org/10.1161/CIRCEP.107.754788.
123. Bers DM. Cardiac sarcoplasmic reticulum calcium leak: basis and roles in cardiac dysfunction. Annu Rev Physiol. 2014;76:107–27. https://doi.org/10.1146/annurev-physiol-020911-153308.
124. Houser SR, Piacentino V 3rd, Weisser J. Abnormalities of calcium cycling in the hypertrophied and failing heart. J Mol Cell Cardiol. 2000;32:1595–607. https://doi.org/10.1006/jmcc.2000.1206.
125. Louch WE, et al. Reduced synchrony of Ca2+ release with loss of T-tubules-a comparison to Ca2+ release in human failing cardiomyocytes. Cardiovasc Res. 2004;62:63–73. https://doi.org/10.1016/j.cardiores.2003.12.031.
126. Sjaastad I, Wasserstrom JA, Sejersted OM. Heart failure -- a challenge to our current concepts of excitation-contraction coupling. J Physiol. 2003;546:33–47. https://doi.org/10.1113/jphysiol.2002.034728.
127. Xie LH, Sato D, Garfinkel A, Qu Z, Weiss JN. Intracellular Ca alternans: coordinated regulation by sarcoplasmic reticulum release, uptake, and leak. Biophys J. 2008;95:3100–10. https://doi.org/10.1529/biophysj.108.130955.
128. Bassani JW, Yuan W, Bers DM. Fractional SR Ca release is regulated by trigger Ca and SR Ca content in cardiac myocytes. Am J Physiol. 1995;268:C1313–9.
129. Shannon TR, Ginsburg KS, Bers DM. Potentiation of fractional sarcoplasmic reticulum calcium release by total and free intra-sarcoplasmic reticulum calcium concentration. Biophys J. 2000;78:334–43. https://doi.org/10.1016/S0006-3495(00)76596-9.
130. Desantiago J, et al. Arrhythmogenic effects of beta2-adrenergic stimulation in the failing heart are attributable to enhanced sarcoplasmic reticulum Ca load. Circ Res. 2008;102:1389–97. https://doi.org/10.1161/CIRCRESAHA.107.169011. CIRCRESAHA.107.169011 [pii].
131. Kyriakis JM, Avruch J. Mammalian mitogen-activated protein kinase signal transduction pathways activated by stress and inflammation. Physiol Rev. 2001;81:807–69.
132. Kyoi S, et al. Opposing effect of p38 MAP kinase and JNK inhibitors on the development of heart failure in the cardiomyopathic hamster. Cardiovasc Res. 2006;69:888–98. https://doi.org/10.1016/j.cardiores.2005.11.015. S0008-6363(05)00527-4 [pii].
133. Bogoyevitch MA. The isoform-specific functions of the c-Jun N-terminal Kinases (JNKs): differences revealed by gene targeting. Bioessays. 2006;28:923–34. https://doi.org/10.1002/bies.20458.
134. Go AS, et al. Prevalence of diagnosed atrial fibrillation in adults: national implications for rhythm management and stroke prevention: the AnTicoagulation and Risk Factors in Atrial Fibrillation (ATRIA) Study. JAMA. 2001;285:2370–5., jcc10004 [pii].

135. Benjamin EJ, et al. Independent risk factors for atrial fibrillation in a population-based cohort. The Framingham Heart Study. JAMA. 1994;271:840–4.
136. Chen Z, Stokes DL, Rice WJ, Jones LR. Spatial and dynamic interactions between phospholamban and the canine cardiac Ca2+ pump revealed with use of heterobifunctional cross-linking agents. J Biol Chem. 2003;278:48348–56. https://doi.org/10.1074/jbc.M309545200.
137. Nelson BR, et al. A peptide encoded by a transcript annotated as long noncoding RNA enhances SERCA activity in muscle. Science. 2016;351:271–5. https://doi.org/10.1126/science.aad4076.
138. Bluhm WF, Kranias EG, Dillmann WH, Meyer M. Phospholamban: a major determinant of the cardiac force-frequency relationship. Am J Physiol Heart Circ Physiol. 2000;278:H249–55.
139. Quan C, et al. SPEG controls calcium reuptake into the sarcoplasmic reticulum through regulating SERCA2a by its second Kinase-Domain. Circ Res. 2019;124:712–26. https://doi.org/10.1161/CIRCRESAHA.118.313916.
140. Hume JR, Uehara A. Ionic basis of the different action potential configurations of single guinea-pig atrial and ventricular myocytes. J Physiol. 1985;368:525–44. https://doi.org/10.1113/jphysiol.1985.sp015874.
141. Amos GJ, et al. Differences between outward currents of human atrial and subepicardial ventricular myocytes. J Physiol. 1996;491(Pt 1):31–50. https://doi.org/10.1113/jphysiol.1996.sp021194.
142. Walden AP, Dibb KM, Trafford AW. Differences in intracellular calcium homeostasis between atrial and ventricular myocytes. J Mol Cell Cardiol. 2009;46:463–73. https://doi.org/10.1016/j.yjmcc.2008.11.003.
143. Freestone NS, et al. Differential lusitropic responsiveness to beta-adrenergic stimulation in rat atrial and ventricular cardiac myocytes. Pflugers Arch. 2000;441:78–87.
144. Venetucci LA, Trafford AW, O'Neill SC, Eisner DA. The sarcoplasmic reticulum and arrhythmogenic calcium release. Cardiovasc Res. 2008;77:285–92. https://doi.org/10.1093/cvr/cvm009.
145. Chelu MG, et al. Calmodulin kinase II-mediated sarcoplasmic reticulum Ca2+ leak promotes atrial fibrillation in mice. J Clin Invest. 2009;119:1940–51.
146. Neef S, et al. CaMKII-dependent diastolic SR Ca2+ leak and elevated diastolic Ca2+ levels in right atrial myocardium of patients with atrial fibrillation. Circ Res. 2010;106:1134–44. https://doi.org/10.1161/CIRCRESAHA.109.203836. CIRCRESAHA.109.203836 [pii].
147. Franzini-Armstrong C, Protasi F, Tijskens P. The assembly of calcium release units in cardiac muscle. Ann N Y Acad Sci. 2005;1047:76–85. https://doi.org/10.1196/annals.1341.007.
148. Wang SQ, Song LS, Lakatta EG, Cheng H. Ca2+ signalling between single L-type Ca2+ channels and ryanodine receptors in heart cells. Nature. 2001;410:592–6. https://doi.org/10.1038/35069083.
149. Ibrahim M, et al. Prolonged mechanical unloading affects cardiomyocyte excitation-contraction coupling, transverse-tubule structure, and the cell surface. FASEB J. 2010;24:3321–9. https://doi.org/10.1096/fj.10-156638.
150. Brette F, Orchard C. T-tubule function in mammalian cardiac myocytes. Circ Res. 2003;92:1182–92. https://doi.org/10.1161/01.RES.0000074908.17214.FD.
151. Berlin JR. Spatiotemporal changes of Ca2+ during electrically evoked contractions in atrial and ventricular cells. Am J Physiol. 1995;269:H1165–70.
152. Forbes MS, Van Niel EE, Purdy-Ramos SI. The atrial myocardial cells of mouse heart: a structural and stereological study. J Struct Biol. 1990;103:266–79.
153. Richards MA, et al. Transverse tubules are a common feature in large mammalian atrial myocytes including human. Am J Physiol Heart Circ Physiol. 2011;301:H1996–2005. https://doi.org/10.1152/ajpheart.00284.2011.
154. Lenaerts I, et al. Ultrastructural and functional remodeling of the coupling between Ca2+ influx and sarcoplasmic reticulum Ca2+ release in right atrial myocytes from experimental persistent atrial fibrillation. Circ Res. 2009;105:876–85. https://doi.org/10.1161/CIRCRESAHA.109.206276.

155. Dibb KM, et al. Characterization of an extensive transverse tubular network in sheep atrial myocytes and its depletion in heart failure. Circ Heart Fail. 2009;2:482–9. https://doi.org/10.1161/CIRCHEARTFAILURE.109.852228.
156. Wakili R, et al. Multiple potential molecular contributors to atrial hypocontractility caused by atrial tachycardia remodeling in dogs. Circ Arrhythm Electrophysiol. 2010;3:530–41. https://doi.org/10.1161/CIRCEP.109.933036.
157. Frisk M, et al. Variable t-tubule organization and Ca2+ homeostasis across the atria. Am J Physiol Heart Circ Physiol. 2014;307:H609–20. https://doi.org/10.1152/ajpheart.00295.2014.
158. Yoon S, Seger R. The extracellular signal-regulated kinase: multiple substrates regulate diverse cellular functions. Growth Factors. 2006;24:21–44. https://doi.org/10.1080/02699050500284218.
159. Li D, et al. Effects of angiotensin-converting enzyme inhibition on the development of the atrial fibrillation substrate in dogs with ventricular tachypacing-induced congestive heart failure. Circulation. 2001;104:2608–14.
160. Nishida K, et al. p38alpha mitogen-activated protein kinase plays a critical role in cardiomyocyte survival but not in cardiac hypertrophic growth in response to pressure overload. Mol Cell Biol. 2004;24:10611–20. https://doi.org/10.1128/MCB.24.24.10611-10620.2004.
161. Purcell NH, et al. Genetic inhibition of cardiac ERK1/2 promotes stress-induced apoptosis and heart failure but has no effect on hypertrophy in vivo. Proc Natl Acad Sci U S A. 2007;104:14074–9. https://doi.org/10.1073/pnas.0610906104.
162. Wang Y, et al. Cardiac muscle cell hypertrophy and apoptosis induced by distinct members of the p38 mitogen-activated protein kinase family. J Biol Chem. 1998;273:2161–8.
163. Zechner D, Thuerauf DJ, Hanford DS, McDonough PM, Glembotski CC. A role for the p38 mitogen-activated protein kinase pathway in myocardial cell growth, sarcomeric organization, and cardiac-specific gene expression. J Cell Biol. 1997;139:115–27.
164. Li M, et al. p38 MAP kinase mediates inflammatory cytokine induction in cardiomyocytes and extracellular matrix remodeling in heart. Circulation. 2005;111:2494–502. https://doi.org/10.1161/01.CIR.0000165117.71483.0C.
165. Cardin S, et al. Evolution of the atrial fibrillation substrate in experimental congestive heart failure: angiotensin-dependent and -independent pathways. Cardiovasc Res. 2003;60:315–25.
166. Avruch J, et al. Ras activation of the Raf kinase: tyrosine kinase recruitment of the MAP kinase cascade. Recent Prog Horm Res. 2001;56:127–55.
167. Zheng M, et al. Sarcoplasmic reticulum calcium defect in Ras-induced hypertrophic cardiomyopathy heart. Am J Physiol Heart Circ Physiol. 2004;286:H424–33. https://doi.org/10.1152/ajpheart.00110.2003.
168. Zhu S, et al. Luteolin enhances sarcoplasmic reticulum Ca2+-ATPase activity through p38 MAPK signaling thus improving Rat cardiac function after ischemia/reperfusion. Cell Physiol Biochem. 2017;41:999–1010. https://doi.org/10.1159/000460837. 000460837 [pii].
169. Greiser M, et al. Distinct contractile and molecular differences between two goat models of atrial dysfunction: AV block-induced atrial dilatation and atrial fibrillation. J Mol Cell Cardiol. 2009;46:385–94. https://doi.org/10.1016/j.yjmcc.2008.11.012.
170. Chiang DY, et al. Loss of microRNA-106b-25 cluster promotes atrial fibrillation by enhancing ryanodine receptor type-2 expression and calcium release. Circ Arrhythm Electrophysiol. 2014;7:1214–22. https://doi.org/10.1161/CIRCEP.114.001973.
171. Voigt N, et al. Cellular and molecular mechanisms of atrial arrhythmogenesis in patients with paroxysmal atrial fibrillation. Circulation. 2014;129:145–56. https://doi.org/10.1161/CIRCULATIONAHA.113.006641.
172. Mustroph J, Maier LS, Wagner S. CaMKII regulation of cardiac K channels. Front Pharmacol. 2014;5:20. https://doi.org/10.3389/fphar.2014.00020.
173. Nerbonne JM. Repolarizing cardiac potassium channels: multiple sites and mechanisms for CaMKII-mediated regulation. Heart Rhythm. 2011;8:938–41. https://doi.org/10.1016/j.hrthm.2011.01.018.

174. Hegyi B, et al. Enhanced depolarization drive in failing Rabbit ventricular myocytes: calcium-dependent and β-adrenergic effects on late sodium, L-type calcium, and sodium-calcium exchange currents. Circ Arrhythm Electrophysiol. 2019;12:e007061. https://doi.org/10.1161/circep.118.007061.
175. Wagner S, et al. Ca2+/calmodulin-dependent protein kinase II regulates cardiac Na+ channels. J Clin Invest. 2006;116:3127–38. https://doi.org/10.1172/JCI26620.
176. Yoon JY, Ho WK, Kim ST, Cho H. Constitutive CaMKII activity regulates Na+ channel in rat ventricular myocytes. J Mol Cell Cardiol. 2009;47:475–84. https://doi.org/10.1016/j.yjmcc.2009.06.020.
177. Xu L, et al. Analysis of Na(+)/Ca (2+) exchanger (NCX) function and current in murine cardiac myocytes during heart failure. Mol Biol Rep. 2012;39:3847–52. https://doi.org/10.1007/s11033-011-1163-x.
178. Yang Y, et al. Xanthine oxidase inhibitor allopurinol improves atrial electrical remodeling in diabetic rats by inhibiting CaMKII/NCX signaling. Life Sci. 2020;259:118290. https://doi.org/10.1016/j.lfs.2020.118290.
179. Ronkainen JJ, et al. Ca2+-calmodulin-dependent protein kinase II represses cardiac transcription of the L-type calcium channel alpha(1C)-subunit gene (Cacna1c) by DREAM translocation. J Physiol. 2011;589:2669–86. https://doi.org/10.1113/jphysiol.2010.201400.
180. Wang Y, et al. Ca2+/calmodulin-dependent protein kinase II-dependent remodeling of Ca2+ current in pressure overload heart failure. J Biol Chem. 2008;283:25524–32. https://doi.org/10.1074/jbc.M803043200.
181. Van Wagoner DR, et al. Atrial L-type Ca2+ currents and human atrial fibrillation. Circ Res. 1999;85:428–36.
182. Christ T, et al. L-type Ca2+ current downregulation in chronic human atrial fibrillation is associated with increased activity of protein phosphatases. Circulation. 2004;110:2651–7. https://doi.org/10.1161/01.CIR.0000145659.80212.6A. 01.CIR.0000145659.80212.6A [pii].
183. Zhang R, et al. A dynamic alpha-beta inter-subunit agonist signaling complex is a novel feedback mechanism for regulating L-type Ca2+ channel opening. FASEB J. 2005;19:1573–5. https://doi.org/10.1096/fj.04-3283fje.
184. Greer-Short A, et al. Calmodulin kinase II regulates atrial myocyte late sodium current, calcium handling, and atrial arrhythmia. Heart Rhythm. 2020;17:503–11. https://doi.org/10.1016/j.hrthm.2019.10.016.
185. McCauley MD, et al. Ion channel and structural remodeling in obesity-mediated atrial fibrillation. Circ Arrhythm Electrophysiol. 2020;13:e008296. https://doi.org/10.1161/CIRCEP.120.008296.
186. Yoo S, et al. Oxidative stress creates a unique, CaMKII-mediated substrate for atrial fibrillation in heart failure. JCI Insight. 2018;3 https://doi.org/10.1172/jci.insight.120728.
187. Syeda F, et al. PITX2 modulates atrial membrane potential and the antiarrhythmic effects of sodium-channel blockers. J Am Coll Cardiol. 2016;68:1881–94. https://doi.org/10.1016/j.jacc.2016.07.766.
188. Justo F, et al. Inhibition of the cardiac late sodium current with eleclazine protects against ischemia-induced vulnerability to atrial fibrillation and reduces atrial and ventricular repolarization abnormalities in the absence and presence of concurrent adrenergic stimulation. Heart Rhythm. 2016;13:1860–7. https://doi.org/10.1016/j.hrthm.2016.06.020.
189. Henry BL, et al. Relaxin suppresses atrial fibrillation in aged rats by reversing fibrosis and upregulating Na+ channels. Heart Rhythm. 2016;13:983–91. https://doi.org/10.1016/j.hrthm.2015.12.030.
190. Jaquet K, Fukunaga K, Miyamoto E, Meyer HE. A site phosphorylated in bovine cardiac troponin T by cardiac CaM kinase II. Biochim Biophys Acta. 1995;1248:193–5. https://doi.org/10.1016/0167-4838(95)00028-s.
191. Hartzell HC, Glass DB. Phosphorylation of purified cardiac muscle C-protein by purified cAMP-dependent and endogenous Ca2+-calmodulin-dependent protein kinases. J Biol Chem. 1984;259:15587–96.

192. Hidalgo CG, et al. The multifunctional Ca(2+)/calmodulin-dependent protein kinase II delta (CaMKIIdelta) phosphorylates cardiac titin's spring elements. J Mol Cell Cardiol. 2013;54:90–7. https://doi.org/10.1016/j.yjmcc.2012.11.012.
193. Rokita AG, Anderson ME. New therapeutic targets in cardiology: arrhythmias and Ca2+/calmodulin-dependent kinase II (CaMKII). Circulation. 2012;126:2125–39. https://doi.org/10.1161/CIRCULATIONAHA.112.124990.
194. Dhanasekaran DN, Reddy EP. JNK-signaling: a multiplexing hub in programmed cell death. Genes Cancer. 2017;8:682–94. https://doi.org/10.18632/genesandcancer.155.
195. Knight RJ, Buxton DB. Stimulation of c-Jun kinase and mitogen-activated protein kinase by ischemia and reperfusion in the perfused rat heart. Biochem Biophys Res Commun. 1996;218:83–8. https://doi.org/10.1006/bbrc.1996.0016.
196. Mohammad J, et al. JNK inhibition blocks piperlongumine-induced cell death and transcriptional activation of heme oxygenase-1 in pancreatic cancer cells. Apoptosis. 2019;24:730–44. https://doi.org/10.1007/s10495-019-01553-9.
197. Omura T, et al. Activation of mitogen-activated protein kinases in in vivo ischemia/reperfused myocardium in rats. J Mol Cell Cardiol. 1999;31:1269–79. https://doi.org/10.1006/jmcc.1999.0959.
198. Pombo CM, et al. The stress-activated protein kinases are major c-Jun amino-terminal kinases activated by ischemia and reperfusion. J Biol Chem. 1994;269:26546–51.
199. Schaeffer HJ, Weber MJ. Mitogen-activated protein kinases: specific messages from ubiquitous messengers. Mol Cell Biol. 1999;19:2435–44. https://doi.org/10.1128/mcb.19.4.2435.
200. Zhang W, et al. USP49 inhibits ischemia-reperfusion-induced cell viability suppression and apoptosis in human AC16 cardiomyocytes through DUSP1-JNK1/2 signaling. J Cell Physiol. 2019;234:6529–38. https://doi.org/10.1002/jcp.27390.
201. Garrington TP, Johnson GL. Organization and regulation of mitogen-activated protein kinase signaling pathways. Curr Opin Cell Biol. 1999;11:211–8. https://doi.org/10.1016/s0955-0674(99)80028-3.
202. Kehat I, Molkentin JD. Extracellular signal-regulated kinase 1/2 (ERK1/2) signaling in cardiac hypertrophy. Ann N Y Acad Sci. 2010;1188:96–102. https://doi.org/10.1111/j.1749-6632.2009.05088.x. NYAS5088 [pii].
203. Kaoud TS, et al. From in Silico discovery to intra-cellular activity: targeting JNK-protein interactions with small molecules. ACS Med Chem Lett. 2012;3:721–5. https://doi.org/10.1021/ml300129b.
204. Morooka H, Bonventre JV, Pombo CM, Kyriakis JM, Force T. Ischemia and reperfusion enhance ATF-2 and c-Jun binding to cAMP response elements and to an AP-1 binding site from the c-jun promoter. J Biol Chem. 1995;270:30084–92. https://doi.org/10.1074/jbc.270.50.30084.
205. Sugden PH, Clerk A. "Stress-responsive" mitogen-activated protein kinases (c-Jun N-terminal kinases and p38 mitogen-activated protein kinases) in the myocardium. Circ Res. 1998;83:345–52. https://doi.org/10.1161/01.res.83.4.345.
206. Zeke A, Misheva M, Reményi A, Bogoyevitch MA. JNK signaling: regulation and functions based on complex protein-protein partnerships. Microbiol Mol Biol Rev. 2016;80:793–835. https://doi.org/10.1128/mmbr.00043-14.
207. Oh SW, et al. JNK regulates lifespan in Caenorhabditis elegans by modulating nuclear translocation of forkhead transcription factor/DAF-16. Proc Natl Acad Sci U S A. 2005;102:4494–9. https://doi.org/10.1073/pnas.0500749102. 0500749102 [pii].
208. Zhao L, et al. The transcription factor DAF-16 is essential for increased longevity in C. elegans exposed to Bifidobacterium longum BB68. Sci Rep. 2017;7:7408. https://doi.org/10.1038/s41598-017-07974-3. 10.1038/s41598-017-07974-3 [pii].
209. Pan M, et al. JNK1 induces Notch1 expression to regulate genes governing photoreceptor production. Cell. 2019;8 https://doi.org/10.3390/cells8090970. E970 [pii]. cells8090970 [pii].

210. Yang M, et al. Causal roles of stress kinase JNK2 in DNA methylation and binge alcohol withdrawal-evoked behavioral deficits. Pharmacol Res. 2021;164:105375. https://doi.org/10.1016/j.phrs.2020.105375.
211. Vlahopoulos SA, et al. The role of ATF-2 in oncogenesis. Bioessays. 2008;30:314–27. https://doi.org/10.1002/bies.20734.
212. Braz JC, et al. Targeted inhibition of p38 MAPK promotes hypertrophic cardiomyopathy through upregulation of calcineurin-NFAT signaling. J Clin Invest. 2003;111:1475–86. https://doi.org/10.1172/JCI17295.
213. Chiu PY, Chen N, Leong PK, Leung HY, Ko KM. Schisandrin B elicits a glutathione antioxidant response and protects against apoptosis via the redox-sensitive ERK/Nrf2 pathway in H9c2 cells. Mol Cell Biochem. 2011;350:237–50. https://doi.org/10.1007/s11010-010-0703-3.
214. Wood CD, Thornton TM, Sabio G, Davis RA, Rincon M. Nuclear localization of p38 MAPK in response to DNA damage. Int J Biol Sci. 2009;5:428–37. https://doi.org/10.7150/ijbs.5.428.
215. Ben-Levy R, Hooper S, Wilson R, Paterson HF, Marshall CJ. Nuclear export of the stress-activated protein kinase p38 mediated by its substrate MAPKAP kinase-2. Curr Biol. 1998;8:1049–57. https://doi.org/10.1016/s0960-9822(98)70442-7. S0960-9822(98)70442-7 [pii]
216. White A, Pargellis CA, Studts JM, Werneburg BG, Farmer BT 2nd. Molecular basis of MAPK-activated protein kinase 2:p38 assembly. Proc Natl Acad Sci U S A. 2007;104:6353–8. https://doi.org/10.1073/pnas.0701679104 0701679104 [pii].
217. Drobic B, Perez-Cadahia B, Yu J, Kung SK, Davie JR. Promoter chromatin remodeling of immediate-early genes is mediated through H3 phosphorylation at either serine 28 or 10 by the MSK1 multi-protein complex. Nucleic Acids Res. 2010;38:3196–208. https://doi.org/10.1093/nar/gkq030. gkq030 [pii].
218. Vermeulen L, De Wilde G, Van Damme P, Vanden Berghe W, Haegeman G. Transcriptional activation of the NF-kappaB p65 subunit by mitogen- and stress-activated protein kinase-1 (MSK1). EMBO J. 2003;22:1313–24. https://doi.org/10.1093/emboj/cdg139.
219. Wierenga AT, Vogelzang I, Eggen BJ, Vellenga E. Erythropoietin-induced serine 727 phosphorylation of STAT3 in erythroid cells is mediated by a MEK-, ERK-, and MSK1-dependent pathway. Exp Hematol. 2003;31:398–405. https://doi.org/10.1016/s0301-472x(03)00045-6. S0301472X03000456 [pii].
220. Pierrat B, Correia JS, Mary JL, Tomas-Zuber M, Lesslauer W. RSK-B, a novel ribosomal S6 kinase family member, is a CREB kinase under dominant control of p38alpha mitogen-activated protein kinase (p38alphaMAPK). J Biol Chem. 1998;273:29661–71. https://doi.org/10.1074/jbc.273.45.29661. S0021-9258(19)59366-8 [pii].
221. Cargnello M, Roux PP. Activation and function of the MAPKs and their substrates, the MAPK-activated protein kinases. Microbiol Mol Biol Rev. 2011;75:50–83. https://doi.org/10.1128/MMBR.00031-10. 75/1/50 [pii].
222. Qian J, et al. Regulation of phosphatase and tensin homolog on chromosome 10 in response to hypoxia. Am J Physiol Heart Circ Physiol. 2012;302:H1806–17. https://doi.org/10.1152/ajpheart.00929.2011. ajpheart.00929.2011 [pii].
223. Schroder D, Heger J, Piper HM, Euler G. Angiotensin II stimulates apoptosis via TGF-beta1 signaling in ventricular cardiomyocytes of rat. J Mol Med (Berl). 2006;84:975–83. https://doi.org/10.1007/s00109-006-0090-0.
224. Lenormand P, et al. Growth factors induce nuclear translocation of MAP kinases (p42mapk and p44mapk) but not of their activator MAP kinase kinase (p45mapkk) in fibroblasts. J Cell Biol. 1993;122:1079–88. https://doi.org/10.1083/jcb.122.5.1079.
225. Whitmarsh AJ, Davis RJ. Transcription factor AP-1 regulation by mitogen-activated protein kinase signal transduction pathways. J Mol Med (Berl). 1996;74:589–607. https://doi.org/10.1007/s001090050063.

226. Winston LA, Hunter T. Intracellular signalling: putting JAKs on the kinase MAP. Curr Biol. 1996;6:668–71. https://doi.org/10.1016/s0960-9822(09)00445-x. S0960-9822(09)00445-X [pii].
227. Lawrence MC, et al. Chromatin-bound mitogen-activated protein kinases transmit dynamic signals in transcription complexes in beta-cells. Proc Natl Acad Sci U S A. 2008;105:13315–20. https://doi.org/10.1073/pnas.0806465105. 0806465105 [pii].
228. McReynolds AC, et al. Phosphorylation or mutation of the ERK2 activation loop alters oligonucleotide binding. Biochemistry. 2016;55:1909–17. https://doi.org/10.1021/acs.biochem.6b00096.
229. Toda T, Shimanuki M, Yanagida M. Fission yeast genes that confer resistance to staurosporine encode an AP-1-like transcription factor and a protein kinase related to the mammalian ERK1/MAP2 and budding yeast FUS3 and KSS1 kinases. Genes Dev. 1991;5:60–73. https://doi.org/10.1101/gad.5.1.60.
230. Joshi S, Platanias LC. Mnk kinase pathway: cellular functions and biological outcomes. World J Biol Chem. 2014;5:321–33. https://doi.org/10.4331/wjbc.v5.i3.321.
231. McCubrey JA, et al. GSK-3 as potential target for therapeutic intervention in cancer. Oncotarget. 2014;5:2881–911. https://doi.org/10.18632/oncotarget.2037. 2037 [pii].
232. Casalvieri KA, Matheson CJ, Backos DS, Reigan P. Selective targeting of RSK isoforms in cancer. Trends Cancer. 2017;3:302–12. https://doi.org/10.1016/j.trecan.2017.03.004. S2405-8033(17)30059-6 [pii]
233. Anjum R, Blenis J. The RSK family of kinases: emerging roles in cellular signalling. Nat Rev Mol Cell Biol. 2008;9:747–58. https://doi.org/10.1038/nrm2509. nrm2509 [pii].
234. Dhillon AS, Hagan S, Rath O, Kolch W. MAP kinase signalling pathways in cancer. Oncogene. 2007;26:3279–90. https://doi.org/10.1038/sj.onc.1210421. 1210421 [pii].
235. Hollenhorst PC, McIntosh LP, Graves BJ. Genomic and biochemical insights into the specificity of ETS transcription factors. Annu Rev Biochem. 2011;80:437–71. https://doi.org/10.1146/annurev.biochem.79.081507.103945.
236. Xu Z, et al. CCL19 suppresses angiogenesis through promoting miR-206 and inhibiting Met/ERK/Elk-1/HIF-1alpha/VEGF-A pathway in colorectal cancer. Cell Death Dis. 2018;9:974. https://doi.org/10.1038/s41419-018-1010-2. 10.1038/s41419-018-1010-2 [pii].
237. Yang R, et al. EGFR activates GDH1 transcription to promote glutamine metabolism through MEK/ERK/ELK1 pathway in glioblastoma. Oncogene. 2020;39:2975–86. https://doi.org/10.1038/s41388-020-1199-2. 10.1038/s41388-020-1199-2 [pii].
238. Khoo S, et al. Regulation of insulin gene transcription by ERK1 and ERK2 in pancreatic beta cells. J Biol Chem. 2003;278:32969–77. https://doi.org/10.1074/jbc.M301198200. S0021-9258(20)83856-3 [pii].
239. Greer EL, Brunet A. FOXO transcription factors at the interface between longevity and tumor suppression. Oncogene. 2005;24:7410–25. https://doi.org/10.1038/sj.onc.1209086. 1209086 [pii].
240. Farhan M, et al. FOXO signaling pathways as therapeutic targets in cancer. Int J Biol Sci. 2017;13:815–27. https://doi.org/10.7150/ijbs.20052. ijbsv13p0815 [pii].
241. Wang X, Chen WR, Xing D. A pathway from JNK through decreased ERK and Akt activities for FOXO3a nuclear translocation in response to UV irradiation. J Cell Physiol. 2012;227:1168–78. https://doi.org/10.1002/jcp.22839.
242. Bishopric NH, Andreka P, Slepak T, Webster KA. Molecular mechanisms of apoptosis in the cardiac myocyte. Curr Opin Pharmacol. 2001;1:141–50. https://doi.org/10.1016/s1471-4892(01)00032-7S1471-4892(01)00032-7 [pii].
243. Stefanelli C, et al. Polyamines directly induce release of cytochrome c from heart mitochondria. Biochem J. 2000;347(Pt 3):875–80.
244. Bialik S, et al. The mitochondrial apoptotic pathway is activated by serum and glucose deprivation in cardiac myocytes. Circ Res. 1999;85:403–14. https://doi.org/10.1161/01.res.85.5.403.

245. Del Re DP, Amgalan D, Linkermann A, Liu Q, Kitsis RN. Fundamental mechanisms of regulated cell death and implications for heart disease. Physiol Rev. 2019;99:1765–817. https://doi.org/10.1152/physrev.00022.2018.
246. Freude B, et al. Apoptosis is initiated by myocardial ischemia and executed during reperfusion. J Mol Cell Cardiol. 2000;32:197–208. https://doi.org/10.1006/jmcc.1999.1066. S0022-2828(99)91066-0 [pii].
247. Guerra S, et al. Myocyte death in the failing human heart is gender dependent. Circ Res. 1999;85:856–66. https://doi.org/10.1161/01.res.85.9.856.
248. Sam F, et al. Progressive left ventricular remodeling and apoptosis late after myocardial infarction in mouse heart. Am J Physiol Heart Circ Physiol. 2000;279:H422–8. https://doi.org/10.1152/ajpheart.2000.279.1.H422.
249. Zhou YT, et al. Lipotoxic heart disease in obese rats: implications for human obesity. Proc Natl Acad Sci U S A. 2000;97:1784–9. https://doi.org/10.1073/pnas.97.4.178497/4/1784 [pii].
250. Singh V, et al. Phosphorylation: implications in cancer. Protein J. 2017;36:1–6. https://doi.org/10.1007/s10930-017-9696-z. 10.1007/s10930-017-9696-z [pii].
251. Wang XX, Zhang B, Xia R, Jia QY. Inflammation, apoptosis and autophagy as critical players in vascular dementia. Eur Rev Med Pharmacol Sci. 2020;24:9601–14. https://doi.org/10.26355/eurrev_202009_23048. 23048 [pii].
252. Mattson MP, Kroemer G. Mitochondria in cell death: novel targets for neuroprotection and cardioprotection. Trends Mol Med. 2003;9:196–205. https://doi.org/10.1016/s1471-4914(03)00046-7. S1471491403000467 [pii]
253. Gross A, McDonnell JM, Korsmeyer SJ. BCL-2 family members and the mitochondria in apoptosis. Genes Dev. 1999;13:1899–911. https://doi.org/10.1101/gad.13.15.1899.
254. Cory S, Huang DC, Adams JM. The Bcl-2 family: roles in cell survival and oncogenesis. Oncogene. 2003;22:8590–607. https://doi.org/10.1038/sj.onc.1207102. 1207102 [pii].
255. Chittenden T. BH3 domains: intracellular death-ligands critical for initiating apoptosis. Cancer Cell. 2002;2:165–6. https://doi.org/10.1016/s1535-6108(02)00128-9. S1535610802001289 [pii].
256. Fan M, et al. Vinblastine-induced phosphorylation of Bcl-2 and Bcl-XL is mediated by JNK and occurs in parallel with inactivation of the Raf-1/MEK/ERK cascade. J Biol Chem. 2000;275:29980–5. https://doi.org/10.1074/jbc.M003776200. S0021-9258(18)44324-4 [pii].
257. Yamamoto K, Ichijo H, Korsmeyer SJ. BCL-2 is phosphorylated and inactivated by an ASK1/Jun N-terminal protein kinase pathway normally activated at G(2)/M. Mol Cell Biol. 1999;19:8469–78. https://doi.org/10.1128/MCB.19.12.8469.
258. Donovan N, Becker EB, Konishi Y, Bonni A. JNK phosphorylation and activation of BAD couples the stress-activated signaling pathway to the cell death machinery. J Biol Chem. 2002;277:40944–9. https://doi.org/10.1074/jbc.M206113200. S0021-9258(19)72207-8 [pii].
259. Deng Y, Ren X, Yang L, Lin Y, Wu X. A JNK-dependent pathway is required for TNFalpha-induced apoptosis. Cell. 2003;115:61–70. https://doi.org/10.1016/s0092-8674(03)00757-8. S0092867403007578 [pii].
260. Lei K, Davis RJ. JNK phosphorylation of Bim-related members of the Bcl2 family induces Bax-dependent apoptosis. Proc Natl Acad Sci U S A. 2003;100:2432–7. https://doi.org/10.1073/pnas.0438011100. 0438011100 [pii].
261. Yano M, Kim S, Izumi Y, Yamanaka S, Iwao H. Differential activation of cardiac c-jun amino-terminal kinase and extracellular signal-regulated kinase in angiotensin II-mediated hypertension. Circ Res. 1998;83:752–60. https://doi.org/10.1161/01.res.83.7.752.
262. Seko Y, Takahashi N, Tobe K, Kadowaki T, Yazaki Y. Pulsatile stretch activates mitogen-activated protein kinase (MAPK) family members and focal adhesion kinase (p125(FAK)) in cultured rat cardiac myocytes. Biochem Biophys Res Commun. 1999;259:8–14. https://doi.org/10.1006/bbrc.1999.0720. S0006-291X(99)90720-9 [pii].
263. Ramirez MT, et al. The MEKK-JNK pathway is stimulated by alpha1-adrenergic receptor and ras activation and is associated with in vitro and in vivo cardiac hyper-

trophy. J Biol Chem. 1997;272:14057–61. https://doi.org/10.1074/jbc.272.22.14057. S0021-9258(19)62512-3 [pii].
264. He H, Li HL, Lin A, Gottlieb RA. Activation of the JNK pathway is important for cardiomyocyte death in response to simulated ischemia. Cell Death Differ. 1999;6:987–91. https://doi.org/10.1038/sj.cdd.4400572.
265. Choukroun G, et al. Regulation of cardiac hypertrophy in vivo by the stress-activated protein kinases/c-Jun NH(2)-terminal kinases. J Clin Invest. 1999;104:391–8. https://doi.org/10.1172/JCI6350.
266. Shvedova M, Anfinogenova Y, Atochina-Vasserman EN, Schepetkin IA, Atochin DN. c-Jun N-Terminal Kinases (JNKs) in Myocardial and Cerebral Ischemia/Reperfusion Injury. Front Pharmacol. 2018;9:715. https://doi.org/10.3389/fphar.2018.00715.
267. Duplain H. Salvage of ischemic myocardium: a focus on JNK. Adv Exp Med Biol. 2006;588:157–64. https://doi.org/10.1007/978-0-387-34817-9_14.
268. Yue TL, et al. Inhibition of extracellular signal-regulated kinase enhances Ischemia/Reoxygenation-induced apoptosis in cultured cardiac myocytes and exaggerates reperfusion injury in isolated perfused heart. Circ Res. 2000;86:692–9. https://doi.org/10.1161/01.res.86.6.692.
269. Laderoute KR, Webster KA. Hypoxia/reoxygenation stimulates Jun kinase activity through redox signaling in cardiac myocytes. Circ Res. 1997;80:336–44. https://doi.org/10.1161/01.res.80.3.336.
270. Kwon SH, Pimentel DR, Remondino A, Sawyer DB, Colucci WS. H(2)O(2) regulates cardiac myocyte phenotype via concentration-dependent activation of distinct kinase pathways. J Mol Cell Cardiol. 2003;35:615–21. https://doi.org/10.1016/s0022-2828(03)00084-1. S0022282803000841 [pii].
271. Remondino A, et al. Beta-adrenergic receptor-stimulated apoptosis in cardiac myocytes is mediated by reactive oxygen species/c-Jun NH2-terminal kinase-dependent activation of the mitochondrial pathway. Circ Res. 2003;92:136–8. https://doi.org/10.1161/01.res.0000054624.03539.b4.
272. Aoki H, et al. Direct activation of mitochondrial apoptosis machinery by c-Jun N-terminal kinase in adult cardiac myocytes. J Biol Chem. 2002;277:10244–50. https://doi.org/10.1074/jbc.M112355200. S0021-9258(19)36158-7 [pii].
273. Andreka P, et al. Cytoprotection by Jun kinase during nitric oxide-induced cardiac myocyte apoptosis. Circ Res. 2001;88:305–12.
274. Dougherty CJ, et al. Activation of c-Jun N-terminal kinase promotes survival of cardiac myocytes after oxidative stress. Biochem J. 2002;362:561–71.
275. Hreniuk D, et al. Inhibition of c-Jun N-terminal kinase 1, but not c-Jun N-terminal kinase 2, suppresses apoptosis induced by ischemia/reoxygenation in rat cardiac myocytes. Mol Pharmacol. 2001;59:867–74.
276. Tournier C, et al. Requirement of JNK for stress-induced activation of the cytochrome c-mediated death pathway. Science. 2000;288:870–4. https://doi.org/10.1126/science.288.5467.870. 8440 [pii].
277. Gabai VL, et al. Suppression of stress kinase JNK is involved in HSP72-mediated protection of myogenic cells from transient energy deprivation. HSP72 alleviates the stewss-induced inhibition of JNK dephosphorylation. J Biol Chem. 2000;275:38088–94. https://doi.org/10.1074/jbc.M006632200. S0021-9258(20)88528-7 [pii].
278. Ma L, et al. Ginsenoside Rb3 protects cardiomyocytes against ischemia-reperfusion injury via the inhibition of JNK-mediated NF-kappaB pathway: a mouse cardiomyocyte model. Plos One. 2014;9:e103628. https://doi.org/10.1371/journal.pone.0103628. PONE-D-14-00008 [pii].
279. Pan Y, et al. Inhibition of JNK phosphorylation by a novel curcumin analog prevents high glucose-induced inflammation and apoptosis in cardiomyocytes and the development of diabetic cardiomyopathy. Diabetes. 2014;63:3497–511. https://doi.org/10.2337/db13-1577. db13-1577 [pii].

280. Li C, et al. Quercetin attenuates cardiomyocyte apoptosis via inhibition of JNK and p38 mitogen-activated protein kinase signaling pathways. Gene. 2016;577:275–80. https://doi.org/10.1016/j.gene.2015.12.012. S0378-1119(15)01483-3 [pii].
281. Ma XL, et al. Inhibition of p38 mitogen-activated protein kinase decreases cardiomyocyte apoptosis and improves cardiac function after myocardial ischemia and reperfusion. Circulation. 1999;99:1685–91. https://doi.org/10.1161/01.cir.99.13.1685.
282. Ren J, Zhang S, Kovacs A, Wang Y, Muslin AJ. Role of p38alpha MAPK in cardiac apoptosis and remodeling after myocardial infarction. J Mol Cell Cardiol. 2005;38:617–23. https://doi.org/10.1016/j.yjmcc.2005.01.012. S0022-2828(05)00033-7 [pii].
283. Kaiser RA, et al. Targeted inhibition of p38 mitogen-activated protein kinase antagonizes cardiac injury and cell death following ischemia-reperfusion in vivo. J Biol Chem. 2004;279:15524–30. https://doi.org/10.1074/jbc.M313717200. S0021-9258(19)63956-6 [pii].
284. Zhang L, Guo Z, Wang Y, Geng J, Han S. The protective effect of kaempferol on heart via the regulation of Nrf2, NF-kappabeta, and PI3K/Akt/GSK-3beta signaling pathways in isoproterenol-induced heart failure in diabetic rats. Drug Dev Res. 2019;80:294–309. https://doi.org/10.1002/ddr.21495.
285. Clark JE, Sarafraz N, Marber MS. Potential of p38-MAPK inhibitors in the treatment of ischaemic heart disease. Pharmacol Ther. 2007;116:192–206. https://doi.org/10.1016/j.pharmthera.2007.06.013. S0163-7258(07)00150-7 [pii].
286. Kim SJ, Hwang SG, Shin DY, Kang SS, Chun JS. p38 kinase regulates nitric oxide-induced apoptosis of articular chondrocytes by accumulating p53 via NFkappa B-dependent transcription and stabilization by serine 15 phosphorylation. J Biol Chem. 2002;277:33501–8. https://doi.org/10.1074/jbc.M202862200. S0021-9258(20)74511-4 [pii].
287. Torcia M, et al. Nerve growth factor inhibits apoptosis in memory B lymphocytes via inactivation of p38 MAPK, prevention of Bcl-2 phosphorylation, and cytochrome c release. J Biol Chem. 2001;276:39027–36. https://doi.org/10.1074/jbc.M102970200. S0021-9258(20)74167-0 [pii].
288. Zhao D, et al. PAF exerts a direct apoptotic effect on the rat H9c2 cardiomyocytes in Ca2+-dependent manner. Int J Cardiol. 2010;143:86–93. https://doi.org/10.1016/j.ijcard.2009.01.068. S0167-5273(09)00128-4 [pii].
289. Kumphune S, Surinkaew S, Chattipakorn SC, Chattipakorn N. Inhibition of p38 MAPK activation protects cardiac mitochondria from ischemia/reperfusion injury. Pharm Biol. 2015;53:1831–41. https://doi.org/10.3109/13880209.2015.1014569.
290. D'Oria R, et al. The role of oxidative stress in cardiac disease: from physiological response to injury factor. Oxid Med Cell Longev. 2020;2020:5732956. https://doi.org/10.1155/2020/5732956.
291. George SA, et al. p38delta genetic ablation protects female mice from anthracycline cardiotoxicity. Am J Physiol Heart Circ Physiol. 2020;319:H775–86. https://doi.org/10.1152/ajpheart.00415.2020.
292. Oyadomari S, Mori M. Roles of CHOP/GADD153 in endoplasmic reticulum stress. Cell Death Differ. 2004;11:381–9. https://doi.org/10.1038/sj.cdd.4401373. 4401373 [pii].
293. Eiras S, et al. Doxazosin induces activation of GADD153 and cleavage of focal adhesion kinase in cardiomyocytes en route to apoptosis. Cardiovasc Res. 2006;71:118–28. https://doi.org/10.1016/j.cardiores.2006.03.014. S0008-6363(06)00127-1 [pii]
294. Hoover HE, Thuerauf DJ, Martindale JJ, Glembotski C. C. alpha B-crystallin gene induction and phosphorylation by MKK6-activated p38. A potential role for alpha B-crystallin as a target of the p38 branch of the cardiac stress response. J Biol Chem. 2000;275:23825–33. https://doi.org/10.1074/jbc.M003864200. S0021-9258(19)66047-3 [pii].
295. Communal C, Colucci WS, Singh K. p38 mitogen-activated protein kinase pathway protects adult rat ventricular myocytes against beta -adrenergic receptor-stimulated apoptosis. Evidence for Gi-dependent activation. J Biol Chem. 2000;275:19395–400. https://doi.org/10.1074/jbc.M910471199.

296. Communal C, Colucci WS. The control of cardiomyocyte apoptosis via the beta-adrenergic signaling pathways. Arch Mal Coeur Vaiss. 2005;98:236–41.
297. Li L, Hao J, Jiang X, Li P, Sen H. Cardioprotective effects of ulinastatin against isoproterenol-induced chronic heart failure through the PI3KAkt, p38 MAPK and NF-kappaB pathways. Mol Med Rep. 2018;17:1354–60. https://doi.org/10.3892/mmr.2017.7934.
298. Hsu SC, Gavrilin MA, Tsai MH, Han J, Lai MZ. p38 mitogen-activated protein kinase is involved in Fas ligand expression. J Biol Chem. 1999;274:25769–76. https://doi.org/10.1074/jbc.274.36.25769. S0021-9258(19)55339-X [pii].
299. Kawahara A, Enari M, Talanian RV, Wong WW, Nagata S. Fas-induced DNA fragmentation and proteolysis of nuclear proteins. Genes Cells. 1998;3:297–306. https://doi.org/10.1046/j.1365-2443.1998.00189.x.
300. Stephanou A, et al. Induction of apoptosis and Fas receptor/Fas ligand expression by ischemia/reperfusion in cardiac myocytes requires serine 727 of the STAT-1 transcription factor but not tyrosine 701. J Biol Chem. 2001;276:28340–7. https://doi.org/10.1074/jbc.M101177200. S0021-9258(19)31644-8 [pii].
301. Sharov VG, et al. Hypoxia, angiotensin-II, and norepinephrine mediated apoptosis is stimulus specific in canine failed cardiomyocytes: a role for p38 MAPK, Fas-L and cyclin D1. Eur J Heart Fail. 2003;5:121–9. https://doi.org/10.1016/s1388-9842(02)00254-4. S1388984202002544 [pii]
302. Wang XZ, et al. Identification of novel stress-induced genes downstream of chop. EMBO J. 1998;17:3619–30. https://doi.org/10.1093/emboj/17.13.3619.
303. Sok J, et al. CHOP-Dependent stress-inducible expression of a novel form of carbonic anhydrase VI. Mol Cell Biol. 1999;19:495–504. https://doi.org/10.1128/MCB.19.1.495.
304. Guo RM, et al. Activation of the p38 MAPK/NF-kappaB pathway contributes to doxorubicin-induced inflammation and cytotoxicity in H9c2 cardiac cells. Mol Med Rep. 2013;8:603–8. https://doi.org/10.3892/mmr.2013.1554.
305. Sheng Z, et al. Cardiotrophin 1 (CT-1) inhibition of cardiac myocyte apoptosis via a mitogen-activated protein kinase-dependent pathway. Divergence from downstream CT-1 signals for myocardial cell hypertrophy. J Biol Chem. 1997;272:5783–91. https://doi.org/10.1074/jbc.272.9.5783. S0021-9258(18)41273-2 [pii].
306. Parrizas M, LeRoith D. Insulin-like growth factor-1 inhibition of apoptosis is associated with increased expression of the bcl-xL gene product. Endocrinology. 1997;138:1355–8. https://doi.org/10.1210/endo.138.3.5103.
307. Chen YL, Loh SH, Chen JJ, Tsai CS. Urotensin II prevents cardiomyocyte apoptosis induced by doxorubicin via Akt and ERK. Eur J Pharmacol. 2012;680:88–94. https://doi.org/10.1016/j.ejphar.2012.01.034. S0014-2999(12)00104-5 [pii].
308. Su HF, et al. Oleylethanolamide activates Ras-Erk pathway and improves myocardial function in doxorubicin-induced heart failure. Endocrinology. 2006;147:827–34. https://doi.org/10.1210/en.2005-1098. en.2005-1098 [pii].
309. Montgomery MD, et al. An Alpha-1A Adrenergic receptor agonist prevents acute doxorubicin cardiomyopathy in male mice. Plos One. 2017;12:e0168409. https://doi.org/10.1371/journal.pone.0168409. PONE-D-16-31909 [pii].
310. Lips DJ, et al. MEK1-ERK2 signaling pathway protects myocardium from ischemic injury in vivo. Circulation. 2004;109:1938–41. https://doi.org/10.1161/01.CIR.0000127126.73759.23. 01.CIR.0000127126.73759.23 [pii].
311. Zhu W, et al. MAPK superfamily plays an important role in daunomycin-induced apoptosis of cardiac myocytes. Circulation. 1999;100:2100–7. https://doi.org/10.1161/01.cir.100.20.2100.
312. Nebigil CG, Etienne N, Messaddeq N, Maroteaux L. Serotonin is a novel survival factor of cardiomyocytes: mitochondria as a target of 5-HT2B receptor signaling. FASEB J. 2003;17:1373–5. https://doi.org/10.1096/fj.02-1122fje. 02-1122fje [pii].
313. Iwai-Kanai E, et al. Basic fibroblast growth factor protects cardiac myocytes from iNOS-mediated apoptosis. J Cell Physiol. 2002;190:54–62. https://doi.org/10.1002/jcp.10036. 10.1002/jcp.10036 [pii].

314. Shizukuda Y, Buttrick PM. Subtype specific roles of beta-adrenergic receptors in apoptosis of adult rat ventricular myocytes. J Mol Cell Cardiol. 2002;34:823–31. https://doi.org/10.1006/jmcc.2002.2020. S0022282802920201 [pii].
315. Erikson RL. Structure, expression, and regulation of protein kinases involved in the phosphorylation of ribosomal protein S6. J Biol Chem. 1991;266:6007–10., S0021-9258(18)38072-4 [pii].
316. Baines CP, et al. Mitochondrial PKCepsilon and MAPK form signaling modules in the murine heart: enhanced mitochondrial PKCepsilon-MAPK interactions and differential MAPK activation in PKCepsilon-induced cardioprotection. Circ Res. 2002;90:390–7. https://doi.org/10.1161/01.res.0000012702.90501.8d.
317. Smith JA, Poteet-Smith CE, Malarkey K, Sturgill TW. Identification of an extracellular signal-regulated kinase (ERK) docking site in ribosomal S6 kinase, a sequence critical for activation by ERK in vivo. J Biol Chem. 1999;274:2893–8. https://doi.org/10.1074/jbc.274.5.2893. S0021-9258(19)88070-5 [pii].
318. Valks DM, et al. Phenylephrine promotes phosphorylation of Bad in cardiac myocytes through the extracellular signal-regulated kinases 1/2 and protein kinase A. J Mol Cell Cardiol. 2002;34:749–63. https://doi.org/10.1006/jmcc.2002.2014. S0022282802920146 [pii].
319. Cho J, Rameshwar P, Sadoshima J. Distinct roles of glycogen synthase kinase (GSK)-3alpha and GSK-3beta in mediating cardiomyocyte differentiation in murine bone marrow-derived mesenchymal stem cells. J Biol Chem. 2009;284:36647–58. https://doi.org/10.1074/jbc.M109.019109. S0021-9258(20)37474-3 [pii].
320. Bopassa JC, Eghbali M, Toro L, Stefani E. A novel estrogen receptor GPER inhibits mitochondria permeability transition pore opening and protects the heart against ischemia-reperfusion injury. Am J Physiol Heart Circ Physiol. 2010;298:H16–23. https://doi.org/10.1152/ajpheart.00588.2009. 00588.2009 [pii].
321. Pabbidi MR, et al. Inhibition of cAMP-dependent PKA activates beta2-Adrenergic receptor stimulation of cytosolic phospholipase A2 via Raf-1/MEK/ERK and IP3-dependent Ca2+ signaling in atrial myocytes. PLoS One. 2016;11:e0168505. https://doi.org/10.1371/journal.pone.0168505. PONE-D-16-28898 [pii].
322. Micova P, et al. Chronic intermittent hypoxia affects the cytosolic phospholipase A2alpha/cyclooxygenase 2 pathway via beta2-adrenoceptor-mediated ERK/p38 stimulation. Mol Cell Biochem. 2016;423:151–63. https://doi.org/10.1007/s11010-016-2833-8. 10.1007/s11010-016-2833-8 [pii].
323. Kitta K, et al. Hepatocyte growth factor induces GATA-4 phosphorylation and cell survival in cardiac muscle cells. J Biol Chem. 2003;278:4705–12. https://doi.org/10.1074/jbc.M211616200. S0021-9258(19)32767-X [pii].
324. Adderley SR, Fitzgerald DJ. Oxidative damage of cardiomyocytes is limited by extracellular regulated kinases 1/2-mediated induction of cyclooxygenase-2. J Biol Chem. 1999;274:5038–46. https://doi.org/10.1074/jbc.274.8.5038. S0021-9258(19)87829-8 [pii].
325. Wang WK, et al. HMGB1 mediates hyperglycaemia-induced cardiomyocyte apoptosis via ERK/Ets-1 signalling pathway. J Cell Mol Med. 2014;18:2311–20. https://doi.org/10.1111/jcmm.12399.
326. Maruyama J, Naguro I, Takeda K, Ichijo H. Stress-activated MAP kinase cascades in cellular senescence. Curr Med Chem. 2009;16:1229–35.
327. Chien KR, Knowlton KU, Zhu H, Chien S. Regulation of cardiac gene expression during myocardial growth and hypertrophy: molecular studies of an adaptive physiologic response. FASEB J. 1991;5:3037–46. https://doi.org/10.1096/fasebj.5.15.1835945.
328. Rohini A, Agrawal N, Koyani CN, Singh R. Molecular targets and regulators of cardiac hypertrophy. Pharmacol Res. 2010;61:269–80. https://doi.org/10.1016/j.phrs.2009.11.012. S1043-6618(09)00281-3 [pii].
329. Nadal-Ginard B, Kajstura J, Leri A, Anversa P. Myocyte death, growth, and regeneration in cardiac hypertrophy and failure. Circ Res. 2003;92:139–50. https://doi.org/10.1161/01.res.0000053618.86362.df.

330. Bernardo BC, Weeks KL, Pretorius L, McMullen JR. Molecular distinction between physiological and pathological cardiac hypertrophy: experimental findings and therapeutic strategies. Pharmacol Ther. 2010;128:191–227. https://doi.org/10.1016/j.pharmthera.2010.04.005. S0163-7258(10)00079-3 [pii].
331. Simpson P. Norepinephrine-stimulated hypertrophy of cultured rat myocardial cells is an alpha 1 adrenergic response. J Clin Invest. 1983;72:732–8. https://doi.org/10.1172/JCI111023.
332. Knowlton KU, et al. The alpha 1A-adrenergic receptor subtype mediates biochemical, molecular, and morphologic features of cultured myocardial cell hypertrophy. J Biol Chem. 1993;268:15374–80., S0021-9258(18)82267-0 [pii].
333. Ai F, et al. Schisandrin B attenuates pressure overload-induced cardiac remodeling in mice by inhibiting the MAPK signaling pathway. Exp Ther Med. 2019;18:4645–52. https://doi.org/10.3892/etm.2019.8154. ETM-0-0-8154 [pii].
334. Esposito G, et al. Cardiac overexpression of a G(q) inhibitor blocks induction of extracellular signal-regulated kinase and c-Jun NH(2)-terminal kinase activity in in vivo pressure overload. Circulation. 2001;103:1453–8.
335. You J, et al. Differential cardiac hypertrophy and signaling pathways in pressure versus volume overload. Am J Physiol Heart Circ Physiol. 2018;314:H552–62. https://doi.org/10.1152/ajpheart.00212.2017ajpheart.00212.2017 [pii].
336. Liu Y, et al. Cardiac-specific PID1 overexpression enhances pressure overload-induced cardiac hypertrophy in mice. Cell Physiol Biochem. 2015;35:1975–85. https://doi.org/10.1159/000374005. 000374005 [pii].
337. Lavoie H, Gagnon J, Therrien M. ERK signalling: a master regulator of cell behaviour, life and fate. Nat Rev Mol Cell Biol. 2020;21:607–32. https://doi.org/10.1038/s41580-020-0255-7. 10.1038/s41580-020-0255-7 [pii].
338. Bodart JF, Chopra A, Liang X, Duesbery N. Anthrax, MEK and cancer. Cell Cycle. 2002;1:10–5., 11020115 [pii].
339. McCubrey JA, et al. Roles of the Raf/MEK/ERK pathway in cell growth, malignant transformation and drug resistance. Biochim Biophys Acta. 2007;1773:1263–84. https://doi.org/10.1016/j.bbamcr.2006.10.001. S0167-4889(06)00315-6 [pii].
340. Clerk A, Bogoyevitch MA, Anderson MB, Sugden PH. Differential activation of protein kinase C isoforms by endothelin-1 and phenylephrine and subsequent stimulation of p42 and p44 mitogen-activated protein kinases in ventricular myocytes cultured from neonatal rat hearts. J Biol Chem. 1994;269:32848–57., S0021-9258(20)30069-7 [pii].
341. Xiao L, et al. MEK1/2-ERK1/2 mediates alpha1-adrenergic receptor-stimulated hypertrophy in adult rat ventricular myocytes. J Mol Cell Cardiol. 2001;33:779–87. https://doi.org/10.1006/jmcc.2001.1348. S0022-2828(01)91348-3 [pii].
342. Yue TL, et al. Extracellular signal-regulated kinase plays an essential role in hypertrophic agonists, endothelin-1 and phenylephrine-induced cardiomyocyte hypertrophy. J Biol Chem. 2000;275:37895–901. https://doi.org/10.1074/jbc.M007037200. S0021-9258(20)88503-2 [pii].
343. Cipolletta E, et al. Targeting the CaMKII/ERK Interaction in the Heart Prevents Cardiac Hypertrophy. PLoS One. 2015;10:e0130477. https://doi.org/10.1371/journal.pone.0130477. PONE-D-14-54390 [pii].
344. Heger J, et al. Growth differentiation factor 15 acts anti-apoptotic and pro-hypertrophic in adult cardiomyocytes. J Cell Physiol. 2010;224:120–6. https://doi.org/10.1002/jcp.22102.
345. House SL, et al. Fibroblast growth factor 2 mediates isoproterenol-induced cardiac hypertrophy through activation of the extracellular regulated Kinase. Mol Cell Pharmacol. 2010;2:143–54. https://doi.org/10.4255/mcpharmacol.10.20.
346. Dobaczewski M, Chen W, Frangogiannis NG. Transforming growth factor (TGF)-beta signaling in cardiac remodeling. J Mol Cell Cardiol. 2011;51:600–6. https://doi.org/10.1016/j.yjmcc.2010.10.033S0022-2828(10)00431-1 [pii].

347. Ruwhof C, van der Laarse A. Mechanical stress-induced cardiac hypertrophy: mechanisms and signal transduction pathways. Cardiovasc Res. 2000;47:23–37. https://doi.org/10.1016/s0008-6363(00)00076-6. S0008-6363(00)00076-6 [pii].
348. Molkentin JD, Robbins J. With great power comes great responsibility: using mouse genetics to study cardiac hypertrophy and failure. J Mol Cell Cardiol. 2009;46:130–6. https://doi.org/10.1016/j.yjmcc.2008.09.002. S0022-2828(08)00584-1 [pii].
349. Ruppert C, et al. Interference with ERK(Thr188) phosphorylation impairs pathological but not physiological cardiac hypertrophy. Proc Natl Acad Sci U S A. 2013;110:7440–5. https://doi.org/10.1073/pnas.1221999110 1221999110 [pii].
350. Lorenz K, Schmitt JP, Schmitteckert EM, Lohse MJ. A new type of ERK1/2 autophosphorylation causes cardiac hypertrophy. Nat Med. 2009;15:75–83. https://doi.org/10.1038/nm.1893. nm.1893 [pii].
351. Lorenz K, Schmitt JP, Vidal M, Lohse MJ. Cardiac hypertrophy: targeting Raf/MEK/ERK1/2-signaling. Int J Biochem Cell Biol. 2009;41:2351–5. https://doi.org/10.1016/j.biocel.2009.08.002. S1357-2725(09)00225-8 [pii].
352. van Berlo JH, Elrod JW, Aronow BJ, Pu WT, Molkentin JD. Serine 105 phosphorylation of transcription factor GATA4 is necessary for stress-induced cardiac hypertrophy in vivo. Proc Natl Acad Sci U S A. 2011;108:12331–6. https://doi.org/10.1073/pnas.1104499108. 1104499108 [pii].
353. Harris IS, et al. Raf-1 kinase is required for cardiac hypertrophy and cardiomyocyte survival in response to pressure overload. Circulation. 2004;110:718–23. https://doi.org/10.1161/01.CIR.0000138190.50127.6A. 01.CIR.0000138190.50127.6A [pii].
354. Punn A, Mockridge JW, Farooqui S, Marber MS, Heads RJ. Sustained activation of p42/p44 mitogen-activated protein kinase during recovery from simulated ischaemia mediates adaptive cytoprotection in cardiomyocytes. Biochem J. 2000;350(Pt 3):891–9.
355. Esposito G, et al. Induction of mitogen-activated protein kinases is proportional to the amount of pressure overload. Hypertension. 2010;55:137–43. https://doi.org/10.1161/HYPERTENSIONAHA.109.135467. HYPERTENSIONAHA.109.135467 [pii].
356. Sopontammarak S, et al. Mitogen-activated protein kinases (p38 and c-Jun NH2-terminal kinase) are differentially regulated during cardiac volume and pressure overload hypertrophy. Cell Biochem Biophys. 2005;43:61–76. https://doi.org/10.1385/CBB:43:1:061. CBB:43:1:061 [pii].
357. Rentrop KP, Feit F. Reperfusion therapy for acute myocardial infarction: concepts and controversies from inception to acceptance. Am Heart J. 2015;170:971–80. https://doi.org/10.1016/j.ahj.2015.08.005. S0002-8703(15)00515-3 [pii].
358. Yeh CC, et al. Distinctive ERK and p38 signaling in remote and infarcted myocardium during post-MI remodeling in the mouse. J Cell Biochem. 2010;109:1185–91. https://doi.org/10.1002/jcb.22498.
359. Bogoyevitch MA, et al. Stimulation of the stress-activated mitogen-activated protein kinase subfamilies in perfused heart. p38/RK mitogen-activated protein kinases and c-Jun N-terminal kinases are activated by ischemia/reperfusion. Circ Res. 1996;79:162–73.
360. Liu T, et al. Coronary Microembolization induces Cardiomyocyte apoptosis in swine by activating the LOX-1-dependent mitochondrial pathway and Caspase-8-dependent pathway. J Cardiovasc Pharmacol Ther. 2016;21:209–18. https://doi.org/10.1177/1074248415599265. 1074248415599265 [pii].
361. Song CL, et al. Down-regulation of microRNA-320 suppresses cardiomyocyte apoptosis and protects against myocardial ischemia and reperfusion injury by targeting IGF-1. Oncotarget. 2016; https://doi.org/10.18632/oncotarget.9240. 9240 [pii].
362. Milano G, et al. A peptide inhibitor of c-Jun NH2-terminal kinase reduces myocardial ischemia-reperfusion injury and infarct size in vivo. Am J Physiol Heart Circ Physiol. 2007;292:H1828–35. https://doi.org/10.1152/ajpheart.01117.2006. 01117.2006 [pii].

363. Jeong CW, et al. Curcumin protects against regional myocardial ischemia/reperfusion injury through activation of RISK/GSK-3beta and inhibition of p38 MAPK and JNK. J Cardiovasc Pharmacol Ther. 2012;17:387–94. https://doi.org/10.1177/1074248412438102. 1074248412438102 [pii].
364. Liu X, Gu J, Fan Y, Shi H, Jiang M. Baicalin attenuates acute myocardial infarction of rats via mediating the mitogen-activated protein kinase pathway. Biol Pharm Bull. 2013;36:988–94., DN/JST.JSTAGE/bpb/b13-00021 [pii].
365. Zhu Z, et al. All-trans retinoic acid ameliorates myocardial ischemia/reperfusion injury by reducing cardiomyocyte apoptosis. Plos One. 2015;10:e0133414. https://doi.org/10.1371/journal.pone.0133414PONE-D-15-03759 [pii].
366. Yang R, et al. Sodium tanshinone IIA sulfonate protects cardiomyocytes against oxidative stress-mediated apoptosis through inhibiting JNK activation. J Cardiovasc Pharmacol. 2008;51:396–401. https://doi.org/10.1097/FJC.0b013e3181671439. 00005344-200804000-00008 [pii].
367. Samuel SM, et al. Thioredoxin-1 gene therapy enhances angiogenic signaling and reduces ventricular remodeling in infarcted myocardium of diabetic rats. Circulation. 2010;121:1244–55. https://doi.org/10.1161/CIRCULATIONAHA.109.872481. CIRCULATIONAHA.109.872481 [pii].
368. Liu J, et al. Palmitate promotes autophagy and apoptosis through ROS-dependent JNK and p38 MAPK. Biochem Biophys Res Commun. 2015;463:262–7. https://doi.org/10.1016/j.bbrc.2015.05.042. S0006-291X(15)00957-2 [pii].
369. Suchal K, et al. Kaempferol attenuates myocardial ischemic injury via inhibition of MAPK signaling pathway in experimental model of myocardial ischemia-reperfusion injury. Oxid Med Cell Longev. 2016;2016:7580731. https://doi.org/10.1155/2016/7580731.
370. Zhang N, et al. Nobiletin attenuates cardiac dysfunction, oxidative stress, and inflammatory in streptozotocin: induced diabetic cardiomyopathy. Mol Cell Biochem. 2016;417:87–96. https://doi.org/10.1007/s11010-016-2716-z. 10.1007/s11010-016-2716-z [pii].
371. Rapacciuolo A, et al. Important role of endogenous norepinephrine and epinephrine in the development of in vivo pressure-overload cardiac hypertrophy. J Am Coll Cardiol. 2001;38:876–82. https://doi.org/10.1016/s0735-1097(01)01433-4. S0735-1097(01)01433-4 [pii].
372. Zhang S, et al. The role of the Grb2-p38 MAPK signaling pathway in cardiac hypertrophy and fibrosis. J Clin Invest. 2003;111:833–41. https://doi.org/10.1172/JCI16290.
373. Villar AV, et al. BAMBI (BMP and activin membrane-bound inhibitor) protects the murine heart from pressure-overload biomechanical stress by restraining TGF-beta signaling. Biochim Biophys Acta. 2013;1832:323–35. https://doi.org/10.1016/j.bbadis.2012.11.007. S0925-4439(12)00258-X [pii].
374. Satomi-Kobayashi S, et al. Deficiency of nectin-2 leads to cardiac fibrosis and dysfunction under chronic pressure overload. Hypertension. 2009;54:825–31. https://doi.org/10.1161/HYPERTENSIONAHA.109.130443. HYPERTENSIONAHA.109.130443 [pii].
375. Zhang S, et al. Role of 14-3-3-mediated p38 mitogen-activated protein kinase inhibition in cardiac myocyte survival. Circ Res. 2003;93:1026–8. https://doi.org/10.1161/01.RES.0000104084.88317.91. 01.RES.0000104084.88317.91 [pii].
376. Zhang Y, et al. Dickkopf-3 attenuates pressure overload-induced cardiac remodelling. Cardiovasc Res. 2014;102:35–45. https://doi.org/10.1093/cvr/cvu004. cvu004 [pii].
377. Sari FR, et al. Attenuation of CHOP-mediated myocardial apoptosis in pressure-overloaded dominant negative p38alpha mitogen-activated protein kinase mice. Cell Physiol Biochem. 2011;27:487–96. https://doi.org/10.1159/000329970. 000329970 [pii].
378. van Eickels M, et al. 17beta-estradiol attenuates the development of pressure-overload hypertrophy. Circulation. 2001;104:1419–23. https://doi.org/10.1161/hc3601.095577.
379. Wu L, et al. Loss of TRADD attenuates pressure overload-induced cardiac hypertrophy through regulating TAK1/P38 MAPK signalling in mice. Biochem Biophys Res Commun. 2017;483:810–5. https://doi.org/10.1016/j.bbrc.2016.12.104. S0006-291X(16)32164-7 [pii].

380. Liu J, et al. Pressure overload induces greater hypertrophy and mortality in female mice with p38alpha MAPK inhibition. J Mol Cell Cardiol. 2006;41:680–8. https://doi.org/10.1016/j.yjmcc.2006.07.007. S0022-2828(06)00720-6 [pii].
381. Gorog DA, et al. Inhibition of p38 MAPK activity fails to attenuate contractile dysfunction in a mouse model of low-flow ischemia. Cardiovasc Res. 2004;61:123–31. https://doi.org/10.1016/j.cardiores.2003.09.034. S0008636303006655 [pii].
382. Mocanu MM, Baxter GF, Yue Y, Critz SD, Yellon DM. The p38 MAPK inhibitor, SB203580, abrogates ischaemic preconditioning in rat heart but timing of administration is critical. Basic Res Cardiol. 2000;95:472–8. https://doi.org/10.1007/s003950070023.
383. Nakano A, Cohen MV, Critz S, Downey JM. SB 203580, an inhibitor of p38 MAPK, abolishes infarct-limiting effect of ischemic preconditioning in isolated rabbit hearts. Basic Res Cardiol. 2000;95:466–71. https://doi.org/10.1007/s003950070022.
384. Melloni C, et al. The study of LoSmapimod treatment on inflammation and InfarCtSizE (SOLSTICE): design and rationale. Am Heart J. 2012;164:646–653 e643. https://doi.org/10.1016/j.ahj.2012.07.030. S0002-8703(12)00581-9 [pii].
385. O'Donoghue ML, et al. Rationale and design of the LosmApimod To Inhibit p38 MAP kinase as a TherapeUtic target and moDify outcomes after an acute coronary syndromE trial. Am Heart J. 2015;169:622–630 e626. https://doi.org/10.1016/j.ahj.2015.02.012. S0002-8703(15)00115-5 [pii].
386. Sarov-Blat L, et al. Inhibition of p38 mitogen-activated protein kinase reduces inflammation after coronary vascular injury in humans. Arterioscler Thromb Vasc Biol. 2010;30:2256–63. https://doi.org/10.1161/ATVBAHA.110.209205. ATVBAHA.110.209205 [pii].
387. Newby LK, et al. Losmapimod, a novel p38 mitogen-activated protein kinase inhibitor, in non-ST-segment elevation myocardial infarction: a randomised phase 2 trial. Lancet. 2014;384:1187–95. https://doi.org/10.1016/S0140-6736(14)60417-7. S0140-6736(14)60417-7 [pii].
388. O'Donoghue ML, et al. Effect of Losmapimod on cardiovascular outcomes in patients hospitalized with acute myocardial infarction: a randomized clinical trial. JAMA. 2016;315:1591–9. https://doi.org/10.1001/jama.2016.3609. 2511223 [pii].
389. Liu W, et al. Deprivation of MKK7 in cardiomyocytes provokes heart failure in mice when exposed to pressure overload. J Mol Cell Cardiol. 2011;50:702–11. https://doi.org/10.1016/j.yjmcc.2011.01.013. S0022-2828(11)00053-8 [pii].
390. Liu W, et al. Cardiac-specific deletion of mkk4 reveals its role in pathological hypertrophic remodeling but not in physiological cardiac growth. Circ Res. 2009;104:905–14. https://doi.org/10.1161/CIRCRESAHA.108.188292. CIRCRESAHA.108.188292 [pii].
391. Calamaras TD, et al. Mixed lineage kinase-3 prevents cardiac dysfunction and structural remodeling with pressure overload. Am J Physiol Heart Circ Physiol. 2019;316:H145–59. https://doi.org/10.1152/ajpheart.00029.2018.
392. Yan J, et al. Stress signaling JNK2 crosstalk with CaMKII underlies enhanced atrial Arrhythmogenesis. Circ Res. 2018;122:821–35. https://doi.org/10.1161/CIRCRESAHA.117.312536. CIRCRESAHA.117.312536 [pii].
393. Gao X, et al. Transcriptional regulation of stress Kinase JNK2 in pro-arrhythmic CaMKIIdelta expression in the aged atrium. Cardiovasc Res. 2018; https://doi.org/10.1093/cvr/cvy011.
394. Zhang R, et al. Calmodulin kinase II inhibition protects against structural heart disease. Nat Med. 2005;11:409–17.
395. Anderson ME. Calmodulin kinase and L-type calcium channels; a recipe for arrhythmias? Trends Cardiovasc Med. 2004;14:152–61. https://doi.org/10.1016/j.tcm.2004.02.005. S1050173804000283 [pii].
396. Hausenloy DJ, Yellon DM. New directions for protecting the heart against ischaemia-reperfusion injury: targeting the Reperfusion Injury Salvage Kinase (RISK)-pathway. Cardiovasc Res. 2004;61:448–60. https://doi.org/10.1016/j.cardiores.2003.09.024. S0008636303006540 [pii].

397. Das A, Salloum FN, Xi L, Rao YJ, Kukreja RC. ERK phosphorylation mediates sildenafil-induced myocardial protection against ischemia-reperfusion injury in mice. Am J Physiol Heart Circ Physiol. 2009;296:H1236–43. https://doi.org/10.1152/ajpheart.00100.2009. 00100.2009 [pii].
398. Yang X, et al. Cardioprotection by mild hypothermia during ischemia involves preservation of ERK activity. Basic Res Cardiol. 2011;106:421–30. https://doi.org/10.1007/s00395-011-0165-0.
399. Hausenloy DJ, Tsang A, Mocanu MM, Yellon DM. Ischemic preconditioning protects by activating prosurvival kinases at reperfusion. Am J Physiol Heart Circ Physiol. 2005;288:H971–6. https://doi.org/10.1152/ajpheart.00374.2004. 00374.2004 [pii].
400. Schulman D, Latchman DS, Yellon DM. Urocortin protects the heart from reperfusion injury via upregulation of p42/p44 MAPK signaling pathway. Am J Physiol Heart Circ Physiol. 2002;283:H1481–8. https://doi.org/10.1152/ajpheart.01089.2001. 01089.2001 [pii].
401. Rossello X, Yellon DM. A critical review on the translational journey of cardioprotective therapies! Int J Cardiol. 2016;220:176–84. https://doi.org/10.1016/j.ijcard.2016.06.131. S0167-5273(16)31141-X [pii].
402. Hausenloy DJ, Tsang A, Yellon DM. The reperfusion injury salvage kinase pathway: a common target for both ischemic preconditioning and postconditioning. Trends Cardiovasc Med. 2005;15:69–75, S1050-1738(05)00025-3 [pii]. https://doi.org/10.1016/j.tcm.2005.03.001.

Intracellular Cardiac Signaling Pathways Altered by Cancer Therapies

Shane S. Scott, Ashley N. Greenlee, Ethan J. Schwendeman, Somayya J. Mohammad, Michael T. Naughton, Anna Matzko, Mamadou Diallo, Matthew Stein, Rohith Revan, Taborah Z. Zaramo, Gabriel Shimmin, Shwetabh Tarun, Joel Ferrall, Thai H. Ho, and Sakima A. Smith

Abstract Recent advancements in targeted and immune-specific cancer therapies have improved patient survivorship. Specifically, tyrosine kinase inhibitors (TKIs), immune checkpoint inhibitors (ICIs), and chimeric antigen receptor (CAR) T-cell therapies have emerged as efficacious treatments for several cancers, including metastatic renal cell carcinoma (mRCC), hematologic malignancies, and solid cancers, respectively. These therapies are associated with complete and durable responses, but as the population of cancer survivors increases, a myriad of unwanted cardiac adverse effects (AEs) have been recognized. Cardiac AEs may be further com-

---

Authors Shane S. Scott and Ashley N. Greenlee contributed equally to the conceptional framework, proof outline and completion of this work.

S. S. Scott · A. N. Greenlee · E. J. Schwendeman · S. J. Mohammad · M. T. Naughton · A. Matzko · M. Diallo · M. Stein · R. Revan · T. Z. Zaramo · G. Shimmin · S. Tarun · J. Ferrall
Davis Heart and Lung Research Institute, The Ohio State University, Columbus, OH, USA
e-mail: Shane.Scott@osumc.edu; Ashley.Greenlee@osumc.edu; Ethan.Schwendeman@osumc.edu; Somayya.Mohammad@osumc.edu; Naughton.38@osu.edu; Anna.Matzko@osumc.edu; Matthew.Stein@osumc.edu; Taborah.Zaramo@osumc.edu; shimmin.2@buckeyemail.osu.edu; tarun.shwetabh@medstudent.pitt.edu; ferrall.18@osu.edu

T. H. Ho
Division of Medical Oncology, Department of Internal Medicine, Mayo Clinic Arizona, Phoenix, AZ, USA
e-mail: Ho.Thai@mayo.edu

S. A. Smith (✉)
Davis Heart and Lung Research Institute, The Ohio State University, Columbus, OH, USA

Division of Cardiology, Department of Internal Medicine, The Ohio State University College of Medicine, Columbus, OH, USA

Dorothy M. Davis Heart and Lung Research Institute Wexner Medical Center, The Ohio State University, Columbus, OH, USA
e-mail: Sakima.Smith@osumc.edu

N. L. Parinandi, T. J. Hund (eds.), *Cardiovascular Signaling in Health and Disease*, https://doi.org/10.1007/978-3-031-08309-9_4

pounded by the approval of dual combination therapy including TKIs and ICIs used in kidney and liver cancers which may precipitate additive insults to the myocardium. The most common AEs include hypertension, de novo arrhythmias, myocarditis, and heart failure (HF), which confer significant risk of morbidity and mortality. Therefore, it is important to understand the mechanism by which these drugs lead to cardiac dysfunction. In this chapter, we provide a comprehensive review of the most relevant studies demonstrating the impact of TKIs, ICIs, and CAR T-cell therapies on cardiomyocyte signaling and function. Strategies used to prevent, reduce, and treat cardiac dysfunction associated with these cancer therapies will also be explored.

Keywords Tyrosine kinase inhibitors (TKIs) · Immune checkpoint inhibitors (ICIs) · CAR T therapy · Cardiotoxicity · Arrhythmias · Cardiac dysfunction · Cardio-oncology · VEGFR · PDGFR · cKIT

Part I: Introduction

In the United States, the cancer death rate has decreased in large part due to improved treatments of several cancers [1]. Three treatments essential to this rate reduction are tyrosine kinase inhibitors (TKIs), immune checkpoint inhibitors (ICIs), and chimeric antigen receptor T-cell (CAR T) therapies [1–3]. TKIs have emerged as first-line therapies for several malignancies including metastatic renal cell carcinoma (RCC), thyroid cancer, sarcomas, colorectal cancer (CC), and chronic myeloid leukemia (CML) [1, 4–6]. Over the last decade, ICIs have become the primary treatment method for many solid tumors by activating the patient's own immune system to target cancer [1, 7]. CAR T-cell therapy has revolutionized the treatment of hematological malignancies, notably, acute lymphoblastic leukemia (ALL), myeloma, and lymphoma [2, 8]. While these drugs have remained popular options, frequent cardiac adverse effects (AEs) have been recognized for all three (Fig. 1). Cardiovascular complications and the potential cardiotoxic effects of TKIs, ICIs, and CAR T-cell therapies are especially relevant among cancer patients who are older, have received previous cycles of cardiotoxic chemotherapy, or have preexisting cardiovascular risk factors [2, 9, 10]. Refining these therapies requires practitioners to have familiarity with the cardiotoxicity profiles and mechanisms that mediate the cardiotoxic effects of these anti-cancer agents.

Tyrosine Kinase Inhibitor Therapy

Small-molecule TKIs target the kinase domains of receptors, which are evolutionarily conserved leading to "off"-target kinase inhibition and/or lack of specificity. While the inhibition of these receptors on cancer cells is responsible for the observed

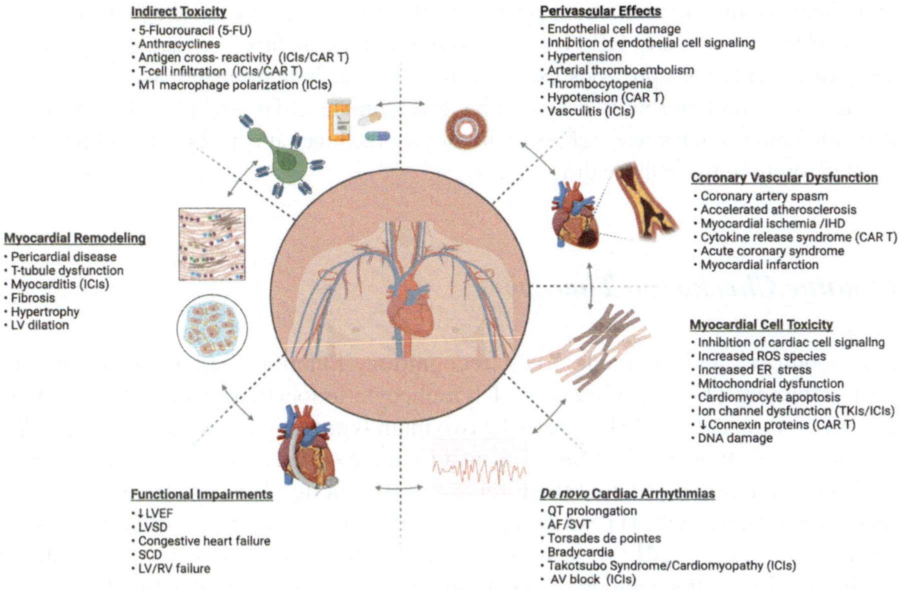

Fig. 1 Direct and indirect cardiovascular complications of emerging cancer therapies. *Abbreviations: AF* atrial fibrillation; *AV* atrioventricular; *CAR T* chimeric antigen receptor T-cell therapy; *DNA* deoxyribonucleic acid; *ER* endoplasmic reticulum; *ICIs* immune checkpoint inhibitors; *IHD* ischemic heart disease; *LV/RV* left/right ventricle; *LVEF* left ventricular ejection fraction; *LVSD* left ventricular systolic dysfunction; *ROS* reactive oxygen species; *SVT* supraventricular tachycardia; *SCD* sudden cardiac death. (Created with Biorender.com)

decrease in cancer growth and progression, these receptors are also present on other cell types, including vascular endothelial cells (ECs), cardiomyocytes, and fibroblasts. TKI-induced cardiotoxicity is mediated by TKIs that preferentially target the vascular endothelial growth factor receptor (VEGFR), platelet-derived growth factor receptor (PDGFR), fibroblast growth factor receptor (FGFR), and stem cell factor (SCF) receptor (c-KIT) [5, 11]. Hypertension (HTN), arrhythmias, QT prolongation, atrial fibrillation (AF), left ventricular systolic dysfunction (LVSD), and heart failure (HF) are among the most common TKI-mediated cardiac effects reported [5, 9, 11]. It has been hypothesized that VEGFR inhibition and subsequent changes in the balance of endothelin-1 (ET-1) and nitric oxide (NO) are responsible for HTN, the most common side effect associated with TKI use [5, 11, 12]. There is evidence to suggest that the "off"-target inhibition of cardiac VEGF and PDGF receptors is responsible for cardiac dysfunction, through reduced phosphatidylinositol 3-kinase (PI3K)-protein kinase B (Akt), Ras/MEK/ERK, and phospholipase C (PLC-γ)-inositol phosphate (IP3)/calcium (Ca^{2+}) signaling pathways and is responsible for cardiac dysfunction. Further, the "off"-target inhibition of cardiac enzymes, in addition to activated CaMKII, may lead to dysregulation in ion channels and ion homeostasis, which can contribute to increased incidence of arrhythmias in this population [5, 9, 13, 14]. Decreases in energy metabolism following inhibition of adenosine 5'-monophosphate–activated protein kinase–mammalian target of rapamycin (AMPK-mTOR)-signaling promote mitochondrial dysfunction, reduce cardiac cell survival, and increase apoptosis, leading to HTN, LVSD, and HF [15, 16]. Additionally, concurrent or sequential use of TKIs in patients with previous exposure to other cardiotoxic drugs may exacerbate cardiac injury and morbidity accelerating development of HF and mortality. As such, it may be important for researchers to investigate the mechanisms by which TKIs augment cardiotoxicity of these drugs.

Immune Checkpoint Therapy

ICIs modulate and augment T-cell recognition, function, and immune activity against tumors [17–19]. Cytotoxic T-lymphocyte-associated protein 4 (CTLA-4) and programmed death 1 (PD1) are the two main lymphocytic immune checkpoints, and their inhibition by ICIs enable the host immune system to recognize cancer cells as foreign antigens and target them for T-cell destruction [20]. However, ICI treatment is associated with HTN, myocarditis, arrhythmias, and pericardial and vascular disease [2, 21, 22]. It is hypothesized that the mechanism of ICI-related cardiotoxicity results from increased T-cell infiltration into vascular endothelial cell walls and T-cell activity against cardiomyocytes leading to CTLA-4-induced HTN and myocarditis [17, 23]. Alternatively, inhibition of cardiac PD-1 receptors along with myocardial and vascular T-cell infiltration is thought to contribute to PD-1 inhibitor-induced cardiac dysfunction [24, 25]. Interestingly, in comparison to CTLA-4 inhibitors, epidemiological evidence shows an increased prevalence of fatal cardiac toxicity with PD1 inhibitors [17].

CAR T-Cell Therapy

The most target for CAR T-cell therapy is the CD19 protein, which is widely expressed on B-cell malignancies. The cardiac AEs associated with CAR T-cell therapies are both age- and time-dependent (20). Patients taking CAR T-cell therapies experience cardiac inflammation, hypotension, de novo AF, and LVSD [2]. While the pathophysiology of CAR T-mediated cardiac AEs is not well understood, two mechanisms have been proposed [2, 3, 27, 28]. One mechanism is thought to be analogous to proinflammatory cytokine interleukin-6 (IL-6)-mediated cardiomyopathy due to cytokine release syndrome (CRS) and sepsis [3, 28]. CRS occurs when supraphysiologic levels of inflammatory cytokines are released by activated CAR T-cells and other immune cells and has been reported in 70–90% of patients in CAR T-cell therapy with increasing severity based on CRS grade [29]. IL-6 is a primary contributor to the pathogenesis of CRS, inducing capillary leak, hypotension, complement activation, and myocardial dysfunction. Specifically, during infectious and inflammatory disease states, IL-6 is thought to mediate myocardial depression [30–32]. Another mechanism of CAR T-cell therapy-induced cardiac AEs involves potential cross-reactivity affinity-enhanced T-cells with unrelated peptides expressed by normal myocardial tissue and cells [3, 27]. Notably, Linette et al. showed that after treatment with T-cells targeting melanoma-associated antigen 3 (MAGE-A3), patients developed fever, progressive cardiogenic shock, and death, due to "off"-target cross-reactivity of T-cell receptors against the striated muscle-specific protein titin [3, 27].

Taken together, the emerging chemotherapy-related cardiac dysfunction associated with newer mechanisms of action such as TKIs, ICIs, and CAR T-cell therapies warrants increased patient risk factor surveillance and further research toward understanding the mechanisms of their oncologic cardiovascular insults. With immune-based therapies, durable complete responses are observed further highlighting the need for long-term monitoring of cardiovascular events in cancer survivorship clinics. Clinically available cardioprotective drugs such as β-blockers, angiotensin-converting enzyme inhibitors (ACEIs), angiotensin receptor blockers (ARBs), statins, and sodium–glucose cotransporter-2 (SGLT2) inhibitors have been shown to reduce cardiovascular-related morbidity [16, 33, 34]. Support for their concurrent use with cancer chemotherapy drugs is evidence by their ability to reduce cardiovascular toxicity in vitro along cardiac dysfunction and HF in vivo [16, 35, 36]. In a large cohort of CML patients with known HTN, Mulas et al. found that patients receiving second- and third-generation TKIs with ACEIs and ARBs experienced a reduction in arterial occlusive events compared to other cardioprotective drug classes [36]. Further, drugs such as statins, which possess pleiotropic effects including their ability to reduce systemic inflammation, may reduce the inflammation and cardiac morbidity observed with ICIs, and CAR T-cell therapies and should be investigated further. More importantly, the reversibility of the side effect and difference in half lives of small molecules (days) versus immunotherapies (weeks) is a key differentiator. Cessation of small-molecule therapy resolves most side effects, whereas immunotherapy/CAR T side effects require additional steroid and/or IL-6 therapy to reverse the side effects. In this review, we will discuss the most recent understanding of the clinical cardiovascular sequelae associated with TKIs, ICIs, and CAR T-cell therapies and the underlying mechanisms by which these drugs alter intracellular cardiac signaling and cardiovascular function, with an emphasis on disrupted tyrosine kinase signaling with TKIs. We will also review current challenges in diagnosis of cardiac events and the available cardioprotective strategies being used to combat and alleviate chemotherapy-related cardiac dysfunction associated with these cancer therapies.

Part II: Tyrosine Kinases Inhibitor Therapy

Tyrosine Kinase Signaling

The amplification of signals into larger biological responses allows cells to respond to extracellular and intracellular stimuli by a mechanism called signal transduction [37]. Tyrosine kinases and their receptors play central roles in transducing extracellular signals in response to growth factors and cytokines to amplify intracellular adaptation [16, 38]. Ligand binding of tyrosine kinase receptors leads to the activation of their intracellular kinase domain, which catalyzes the transfer of phosphate residues from adenosine triphosphate (ATP) to tyrosine residues on their target substrates [39]. These activated substrates translocate to different subcellular domains to regulate cell growth, differentiation, metabolism, migration, motility, and cell death. Tyrosine kinases are grouped into receptor (RTKs) and nonreceptor tyrosine kinases (NRTKs). RTKs are embedded in the cell membrane with an extracellular ligand-binding domain and an intracellular kinase domain that signals to the interior of the cell [40]. Conversely, NRTKs are located at various subcellular domains within the cell. Together, RTKs and NRTKs mediate transduction of both extracellular and intracellular signals, respectively.

Abnormal Tyrosine Kinase Activity and Signaling in Cancer

Tyrosine kinases are normally quiescent until activated by extracellular ligands such as growth factors (e.g., VEGF, PDGF, and SCF) or intracellular stimuli such as oxidant stress, which can lead to the activation of NRTKs. Tight regulation of tyrosine kinase activity, specifically a balance between phosphorylation and dephosphorylation of tyrosine residues by tyrosine kinases and phosphatases, is critical for controlling signal timing and duration [37, 38]. The dysregulation of tyrosine kinase signaling due to constitutive activation of several tyrosine kinases is responsible for tumorigenesis, tumor angiogenesis, and metastasis in cancers including breakpoint cluster region-Abelson 1 (Bcr-Abl1) in CML and c-KIT and PDGFR-α in gastrointestinal stromal tumor (GIST) [37, 38]. It has been estimated that tyrosine kinases are aberrantly activated in a variety of human cancers that overexpress VEGF resulting in tumor progression and metastasis [37]. With increased tumor burden, existing vasculature becomes insufficient and hypoxic conditions stimulate the tumor cells to produce proangiogenic cytokines such as VEGF. Some cancers also express the VEGF homolog placental growth factor (PLGF), which acts as a ligand for VEGFR-1, while others express FGF and PDGF—all of which act, in part, to upregulate the expression of VEGF. The importance of the VEGF signaling pathway in cancer pathobiology makes it an ideal therapeutic target for reducing cancer growth and metastases. Furthermore, almost all the proangiogenic growth factors that were identified in tumor angiogenesis are ligands of RTKs targeted by TKIs [37].

Tyrosine Kinase Inhibitors

Tyrosine kinases can be inhibited by monoclonal antibodies or small molecules. Small-molecule inhibitors used in cancer therapy are predominantly directed at tyrosine kinases, with the serine/threonine kinase superfamily being another target. These small-molecule inhibitors target and block the ATP-binding site of both subclasses of tyrosine kinases [11, 37, 38]. Monoclonal antibodies competitively bind to RTKs and prevent activation of kinases [39]. TKIs can directly inhibit the trans-autophosphorylation of the intracellular kinase domains as well as the phosphorylation and recruitment of downstream signaling proteins, terminating the signaling cascade [37, 38]. Similarly, following cellular entry, TKIs block NRTK signal transduction by targeting intracellular kinases. The evolutionarily conserved receptor kinase domains of these small molecules allow them to be promiscuous in binding multiple receptors, allowing for a broader range of use but also a greater potential for "off"-target effects. Examples of these multi-targeted TKIs include axitinib, cabozantinib, pazopanib, sunitinib, and sorafenib (Table 1) [5, 6]. These drugs have been shown to inhibit many receptors including VEGFR-1, VEGFR-2, and VEGFR-3; PDGFR-α and PDGFR-β; FGFR-1, FGFR-2, and FGFR-3; c-KIT; interleukin-2 receptor-inducible T-cell kinase; and transmembrane glycoprotein receptor tyrosine kinase [11, 41]. Malignant cells can co-opt signaling cascades responsible for increased proliferation and angiogenesis, along with inhibition of apoptosis through mutations, overexpression, or constitutive activation of tyrosine kinase proteins integral to these pathways [42]. Broad targeted inhibition of RTKs confers efficacy of TKI's as cancer drugs, however, this lack of specificity is culpable for cardiovascular toxicity and injury observed in patients. Below we discuss some of the popular receptor targets for TKIs, their roles in normal cardiovascular physiology, and how their inhibition can promote cardiotoxicity.

Tyrosine Kinase Receptors and Cardiomyopathy

VEGF/VEGFR Signaling

Regardless of the mechanism, there are strong data suggesting that VEGF/VEGFR signaling is important for maintaining normal function of ECs, vasculature, and cardiomyocytes [37]. VEGFR-1, -2, and -3 are all expressed on ECs, where their activation mediate angiogenesis in tissues [37]. In the vasculature endothelium, VEGFR-1 and VEGFR-2 activation by VEGF-A and VEGF-B binding increases Ras/Raf-1/MEK/ERK, PI3K/Akt, and PLC-γ /IP3/Ca^{2+} signaling [5]. Increased endothelial intracellular signaling triggers a series of phosphorylation cascades, which culminates into increased vascular permeability, as well as cell migration, proliferation, and survival [5]. Cardiomyocytes express VEGFR-1 and VEGFR-2, and their activation plays an important role in the development, maintenance, and survival of myocardial endothelial cells and cardiomyocytes [5]. In cardiovascular diseases, increased VEGF secretion and cardiomyocyte VEGF/VEGFR signaling are essential for responding to myocardial stress and injury [5, 37]. Supporting this, cardiomyocytes upregulate the expression of VEGFR-1 and VEGFR-2 in response to hypoxia in vitro and in vivo [43]. In a transgenic murine model with VEGFR-1

Table 1 TKI therapies, their receptor targets, incidence of cardiotoxicity, and proposed mechanisms of action

Agents	Receptor target(s)	Cardiotoxicity type (rate)	Preclinical model(s)	Proposed cardiotoxic signaling mechanisms	Indication(s)	Reference
Axitinib[a]	c-KIT PDGFR-α/β VEGFR-1, VEGFR-2, VEGFR-3	AT (1%) DVT (1%) HF (2%) HTN (40–57.7%) QT (0.66%) TE (3%)	NA	Targets multiple tyrosine kinase receptors, leading to the downregulation of pro-survival signaling, oxidative stress, and reduced angiogenesis; dysregulation in NO and calcineurin/NFAT signaling precipitates HTN/increased afterload	RCC	[48, 73, 86, 90, 110, 145, 170]
Cabozantinib[a]	CDK RET VEGFR-2	HTN (20–80.8%) PE (2.3%) TP (11–39.7%) TE (7%)	NA	Downregulation of endothelial NO synthase due to inhibition of VEGFR-2 resulting in disruption of the balance NO to ET-1, precipitating increased peripheral resistance, increased blood pressure, and vasoconstriction[a]	MTC RCC	[73, 76, 77, 145, 158, 177]
Cediranib[a]	c-KIT FGFR1 PDGFR-α/β VEGFR-1, VEGFR-2, VEGFR-3	HTN (41%) PE (6%) QT (0.33–0.7%) TP (35%)	NA	NA	CC GBM NSCLC OC	[5, 11, 112, 145, 176]
Dasatinib[b]	Bcr-Abl B-RAF c-KIT EphA2 PDGFR-α/β Src kinase	Atrial ischemic events (5%) HF (2–4%) CV ischemia (2–4%) LVSD (2–4%) QT (0.4–3%) SVT	Rat primary cardiomyocytes	Cellular apoptosis caused by increases in cardiomyocyte ER stress/ROS signaling	CML Ph-ALL	[42, 73, 90, 93, 110, 113, 114, 116, 145]
Erlotinib	EGFR (ERBB1)	TE (3.9–11%) QT (14.3%)	NA	NA	NSCLC PC	[90, 111, 179]

Agents	Receptor target(s)	Cardiotoxicity type (rate)	Preclinical model(s)	Proposed cardiotoxic signaling mechanisms	Indication(s)	Reference
Gefitinib	EGFR (ERBB1)	MI (1.2%) QT (20.0%)	In vitro H9c2 cells and in vitro rat cardiomyocytes	Cardiac oxidative stress and cardiac apoptosis pathways promote hypertrophy subsequent to increased expression of β-MHC and BNP mRNA and protein in in vivo and in vitro rat models	NSCLC	[71, 111, 144, 173]
Ibrutinib	BTK	AF (3.3–13%) HF (3.7%) HTN (17–78.3%) MI (1.4%) SCD (1.1%)	NA	The mechanism is not well understood but is associated with BTK-regulated inhibition of PI3K/Akt signaling	CML GVHD MCL MZL WM	[90, 95, 96]
Imatinib	Bcr-Abl c-KIT PDGFR-α/β	HF (0.7–1.7%) HTN (0.1–4.3%) IHD (1.8%) LVSD (<0.1–2.7%) QT (<0.5–19.5 %)	Rat primary cardiomyocytes	Cellular apoptosis due to ER stress, mitochondrial dysfunction, and increased ROS was observed in cultured cells and murine hearts treated with imatinib	CML CMML DFSP GIST GM HES Ph-ALL	[42, 73, 84, 92, 93, 111–116, 145, 160, 181]
Intedanib[a]	VEGFR-1, VEGFR-2, VEGFR-3PDGFR FGFR	NA	NA	NA	NSCLC OC	[5]
Lapatinib	EGFR (ERBB-1, ERBB-2)	HF (0.2–2%) ↓ LVEF (0.2%) LVSD (0.2–1.5%) QT (1.7–16%)	NA	The ratio of pro-apoptotic BCL-XS to BCL-XL proteins is increased, leading to several side effects, including reduced cardiac contractility, ATP deletion, and cardiac cell death via mitochondrial induced apoptosis	HER2+ BC	[71, 73, 84, 90, 93, 110, 112, 145]

(continued)

Table 1 (continued)

Agents	Receptor target(s)	Cardiotoxicity type (rate)	Preclinical model(s)	Proposed cardiotoxic signaling mechanisms	Indication(s)	Reference
Lenvatinib[a]	c-KIT FGFR PDGFR-α/β RET VEGFR-1, VEGFR-2, VEGFR-3	HTN (40–90%) PE (2.7–3.4%) QT (1.5–11%)	NA	Targets multiple tyrosine kinase receptors, leading to downregulation of prosurvival signaling and angiogenesis; reduces NO and calcineurin/NFAT signaling, which contributes to cardiotoxicities	MTC RCC	[48, 73, 110, 145, 165, 172]
Nilotinib	Bcr-Abl c-KIT DDR-1, DDR-2 PDGFR-α/β	HTN (5.9–24%), IHD (3.7–15.2%) LVSD (1%) MI (7%) QT (0.68–10%) SCD (0.6%) TE (1–10%)	Culture rat primary cardiomyocytes	Activated ER stress response triggers cardiac cell injury and death; inhibition of hERG potassium channels, reducing I_{Kr} currents leading to QT prolongation	CML GM	[73, 86, 90, 92, 93, 110, 114, 116, 145, 180]
Nintedanib[a]	FGFR FLT3 PDGFR RET VEGFR-1, -2, -3	AT (2.5%) HTN (5.2%) MI (1.5–2.7%) QT (3.3%) SCD (0.5%)	NA	NA	IPF NSCLC	[5, 73, 112, 171]
Pazopanib[a, b]	B-RAF c-KIT FGFR-1, FGFR-3 MCSFR-1 PDGFR-α/β VEGFR-1, VEGFR-2, VEGFR-3	HF (0.3–13%) HTN (37–52%) LVSD (7–11%) MI (1–3%) QT (1.04–13.0%) SCD TdP (0.3%) TP (41%) TE (1–5%)	HL-1 cardiomyocytes C57BL/6 Mice	Inhibition of cardiac FGFR1/2, FLT3 and VEGF receptors resulting in impaired cardiac stress response, reduced contractility and reduced P13K/Akt signaling; activation of cardiomyocyte apoptotic pathways have also been proposed	MTC STS RCC	[5, 11, 48, 93, 110, 111, 147, 165, 168]

Agents	Receptor target(s)	Cardiotoxicity type (rate)	Preclinical model(s)	Proposed cardiotoxic signaling mechanisms	Indication(s)	Reference
Ponatinib[a]	Bcr-Abl c-KIT FGFR1-3 FLT3 PDGFR Src kinase VEGFR-2	AF (4%) AT and VT (23%) HF (7–9%) HTN (25–68) LVSD (7%) QT (3.7%) TP (37–46%) MI (2%)	hiPSC-induced cardiomyocytes; zebrafish; NRVMs	Increased accumulation of ROS and mitochondrial dysfunction Inhibition of cardiac Akt and ERK pro-survival signaling pathways, leading to cardiomyocyte apoptosis	ALL CML GBM GIST Ph-ALL NSCLC	[73, 78–80, 90, 112, 113, 145]
Regorafenib[a, b]	B-RAF c-KIT FGFR1 PDGFR-β RET VEGFR-1, VEGFR-2, VEGFR-3	HF (1–3%) HTN (1–76%) MI (0.9%) Thrombosis (1%)	Rat H9c2 cardiomyocytes	Reduced cell proliferation and altered ATP content; altered mitochondrial membrane potential, structure and mtDNA content	CC GIST HCC	[73, 115, 142, 145, 175]
Sorafenib[a, b]	B-RAF/C-RAF c-KIT FLT3 PDGFR-β VEGFR-1, VEGFR-2, VEGFR-3	AF (5.1%)[c] AT (1.7%) HF (0.2–0.3%) HTN (1.7–43%) ↓ LVEF (6.5–13%) LVSD (1–8%) MI (2–3%) QT (21.2%)	Zebrafish NRVMs	Mitochondrial dysfunction and apoptosis are promoted by the inhibition of the Ras/Raf-1/Mek/Erk, caused by ROS expression and increased CaMKII, which reduces cardiac survival	HCC Melanoma RCC	[5, 48, 71, 73, 84, 90, 93, 111, 145, 166]

(continued)

Table 1 (continued)

Agents	Receptor target(s)	Cardiotoxicity type (rate)	Preclinical model(s)	Proposed cardiotoxic signaling mechanisms	Indication(s)	Reference
Sunitinib[a, b]	c-KIT CSF-1R FLT3 PDGFR-α/β RET VEGFR-1, VEGFR-2, VEGFR-3	AT (1.4%) HF (2–16%) HTN (5–68.1%) ↓ LVEF (0.4–28%) LVSD (2.7–19%) QT (0.52–20.0%) TdP (<0.1%) TP (62.5–78%) TE (3%) MI (4%)	NRVMs; Swiss-webster mice; Rat H9c2 cardiomyocytes; C57BL/6J mice	Impaired AMPK-mTOR signaling and energy homeostasis promotes death and cardiomyocyte autophagy Mitochondrial damage, cytochrome C release, and caspase 9 activation precedes the initiation of the mitochondrial apoptotic pathway in vivo and in vitro Contributes to cardiomyocyte apoptosis in the presence of HTN or other underlying cardiac pathology	GIST PNET RCC	[5, 11, 15, 73, 84, 90, 93, 110, 111, 113, 133, 143, 158, 169, 177, 178]
Telatenib[a]	c-KIT PDGFR-β VEGFR-2, VEGFR-3	HTN (23%)	NA	Reduced formation of NO by endothelial cells, reduced responsiveness of vascular smooth muscles to NO, increased vasoconstricting stimuli, reduced stability of the vascular wall, and microvascular rarefaction	HCC RCC	[85, 145, 167]
Tivozanib[a]	VEGFR-1, VEGFR-2, VEGFR-3	HTN (43.1–44%)	NA	NA	BC GIM RCC	[145, 174, 177]
Vandetanib[a]	EGFR PDGFR-β RET VEGFR-1, VEGFR-2, VEGFR-3	HF (0.4%) HTN (24-33) QT (4.3–66.7%) SCD TdP	Postmortem human cardiac tissue	Myocyte degeneration is induced in the papillary muscles of the myocardium as well as the subendocardial zones of the myocardium	MTC NSCLC	[5, 86, 90, 93, 110, 112, 113, 132, 145]
Vatalanib[a]	c-KIT PDGFR- α VEGFR-1, VEGFR-2, VEGFR-3	HTN (23%) PE (5.7%)	NA	Reduces tyrosine kinase activity of VEGFR, reduces NO production, and increases capillary permeability and endothelial cell proliferation	CC MESO NSCLC PC	[145]

Agents	Receptor target(s)	Cardiotoxicity type (rate)	Preclinical model(s)	Proposed cardiotoxic signaling mechanisms	Indication(s)	Reference
Vemurafenib[b]	B-RAF	QT (1.5–34.3%)	HEK293T; isolated canine Purkinje fibers	cAMP activity is increased with inhibition of BRAF, as well as subsequent increases in PKA. This results in hERG channel phosphorylation, causing a reduction in their ability to open in response during repolarization, leading to QT interval prolongation	ECD MM	[11, 110–113, 115]

Abbreviations: AF atrial fibrillation; *ALL* acute lymphocytic leukemia; *AMPK* AMP-activated protein kinase; *AT* arterial thromboembolism; *ATP* adenosine triphosphate; *BC* breast cancer; *Bcr-Abl* breakpoint cluster region-Abelson; *BNP* brain natriuretic peptide; *BTK* Bruton's tyrosine kinase; *CaMKII* calcium/calmodulin-dependent protein kinase; *cAMP* cyclic adenosine monophosphate; *CC* colorectal cancer; *CDK* cyclin-dependent kinase; c-*KIT* stem cell factor receptor; *CML* chronic myeloid leukemia; *CMML* chronic myelomonocytic leukemia; *CSF-1R* colony-stimulating factor 1 receptor; *DFSP* dermatofibrosarcoma protuberans; *DVT* deep vein thrombosis; *ECD* Erdheim-Chester disease; *EGFR* epidermal growth factor receptor; *DDR-1, DDR-2* discoidin domain receptor 1,2; *EPHA2* ephrin type-A receptor 2; *ER* endoplasmic reticulum; *ERK* extracellular-signal-regulated kinase; *ET-1* endothelin-1; *EGFR* epidermal growth factor receptor; *FGFR-1/ FGFR-2* fibroblast growth factor receptor; *FLT3* FMS-related tyrosine kinase 3; *GBM* glioblastoma; *GIM* gastrointestinal malignancies; *GIST* gastrointestinal stromal tumor; *GM* gynecologic melanoma; *GVHD* graft-versus-host disease; *HCC* hepatocellular carcinoma; *HEK293T* human embryonic kidney cells 293 T; *hERG* human ether-a-go-go-related gene; *HES* hypereosinophilic syndrome; *hiPSC* human induced pluripotent stem cells; *HF* heart failure; *HTN* hypertension; *HO1* heme oxygenase 1; *IHD* ischemic heart disease; *IPF* idiopathic pulmonary fibrosis; *LVEF* left ventricular ejection fraction; *LVSD* left ventricular systolic dysfunction; *MCL* mantle cell lymphoma; *MCSFR-1* macrophage colony-stimulating factor-1 receptor; *MESO* mesothelioma; *MHC* myosin heavy chain; *MI* myocardial ischemia/infarction; *MM* malignant melanoma; *MTC* medullary thyroid cancer; *mtDNA* mitochondrial DNA; *MZL* marginal zone lymphoma; *NO* nitric oxide; *NRVMs* neonatal rat ventricular myocytes; *NSCLC* non-small-cell lung cancer; *OC* ovarian cancer; *PC* pancreatic cancer; *PDGFR* platelet-derived growth factor receptors; *PE* pulmonary embolism; *Ph-ALL* Philadelphia-chromosome-positive acute lymphoblastic leukemia; *PI3K* phosphoinositide 3-kinase; *PKA* protein kinase A; *PNET* pancreatic neuroendocrine tumors; *QT* QT prolongation; *RC* renal cell carcinoma; *RET* rearranged during transfection; *ROS* reactive oxygen species; *SCD* sudden cardiac death; *Src* short for sarcoma-proto-oncogene; *ST* soft tissue sarcoma; *SVT* supraventricular tachycardia; *TdP* torsades de pointes; *TE* thromboembolism; *TKI* tyrosine kinase inhibitors; *TP* thrombocytopenia; *VEGFR* vascular endothelial growth factor receptors; *VT* venous thrombosis; *WM* Waldenstrom's macroglobulinemia

Note (s):

[a]All VEGFR-TKIs have the potential to cause hypertension via this molecular mechanism. Further, the mechanisms leading to VEGFR-TKIs is multi-factorial and might be related to microvascular dysfunction, ATP depletion in the mitochondria, myocardial proapoptotic kinases, microvascular dysfunction, and profound vasoconstriction

[b]All B-RAF inhibitors have the potential to promote QT prolongation by this mechanism. (NA) Indicates that to the authors knowledge there are no preclinical studies, which directly evaluated these drugs on cardiomyocyte tissue

[c]Administered with fluorouracil

deleted from ECs, increased angiogenesis and cardiomyocyte hypertrophy were observed suggesting a role for paracrine signaling or cross between ECs and cardiomyocytes. Further, VEGF signaling in myocardial remodeling requires a balance of VEGFR-1 and VEGFR-2 activation, which respectively block and promote cardiomyocyte hypertrophy [43, 44].

PDGF/ PDGFR Signaling

To date, 11 TKIs have been approved by the US Food and Drug Administration (FDA) to treat cancers such as chronic myelomonocytic leukemia (CML), GIST, and glioblastoma—all of which overexpress PDGFR-α/β [18]. However, PDGFRs are ubiquitously expressed on human and mouse cardiomyocytes as well as ECs, which suggests that TKI inhibition of cardiomyocyte PDGFR signaling may be one of the key pathways responsible for the increased incidence of cardiovascular dysfunction in patients treated with these drugs. While the role of cardiovascular PDGFR signaling is still largely unknown, PDGFR has been implicated in regulation of angiogenesis, tissue fibrosis [45–47], and cardiac response to mechanically induced pressure overload. In cardiomyocytes, downstream signal transduction cascades affected by PDGFR-β include PLC-γ /IP3/Ca^{2+} and PI3K/p-Akt pathways [45, 48]. Yue et al. found that PDGFR-β signaling is depressed in aging myocardium, while myocyte-specific activation of PDGFR-β-mediated PI3K/p-Akt signaling promotes cardiomyocyte proliferation, cardiac regeneration, and systolic function[45]. Mouse hearts with increased left ventricular (LV) pressure overload due to transverse aortic constriction (TAC) show upregulated PDGFR-β [5]. In cardiac-specific PDGFR-β murine knockout models, reduced LV function and reduced angiogenesis were observed post TAC in comparison with wild-type mice after TAC [49]. Further, the administration of PDGF in rat models improved cardiac function, protected cardiomyocytes, and reduced post-myocardial infarction (MI) size [5, 50, 51]. Inactivation of PDGFR-β signaling, however, leads to cardiac abnormalities, including ventricular septal defects, stage-dependent embryonic ventricular dilation, absence of coronary vascular smooth muscle cells, and hypertrabeculation of the myocardium[45, 52–56].

c-KIT Signaling

Dysregulation of c-KIT receptor signaling has been observed in a wide variety of cancers including, acute myeloid leukemia, GIST, and small-cell lung cancer [37, 57]. Small-molecule TKIs that potently inhibit c-KIT in vitro [58] and in vivo [59–62] include dasatinib, pazopanib, and quizartinib. These drugs markedly disrupt hematopoietic progenitor cells, while imatinib, crenolanib, sunitinib, and sorafenib have shown moderate and negligible activity against c-KIT in vivo and in vitro [42, 58]. Insights about how c-KIT inhibition could lead to cardiotoxicity were provided by mouse treatment with a variant of imatinib that does not target ABL kinase but retains activity against c-KIT. In the study, the imatinib variant was not associated with cardiac dysfunction, suggesting that imatinib-induced cardiotoxicity is mediated by inhibition of the ABL kinase [37, 63], casting some doubt on the role of c-KIT inhibition in the development of TKI-induced cardiotoxicity. Notwithstanding, it is still possible for drugs like dasatinib and nilotinib, which also target c-KIT and are far more potent than imatinib in vitro [42], to mediate cardiac disease via c-KIT inhibition.

In cardiovascular biology, c-KIT was initially used as a marker to identify and enrich adult cardiac stem/progenitor cells that can differentiate into cardiomyocytes, ECs, and smooth muscle cells in vitro as well as in vivo after myocardial injury [18]. However, the expression of c-KIT in cardiac cells declines rapidly from embryonic stages and is almost absent in the adult myocardium [64]. A role for c-KIT in maintaining the normal cardiac function was demonstrated in a murine model with constitutive activation of c-KIT receptor [65]. Here, prolonged c-KIT activation was associated with increased cardiac myogenic and vasculogenic reparative potential after injury and significant improvement in survival [65]. In a heterozygous c-KIT mouse model, with one c-KIT allele deleted and the other encoding c-KIT with reduced kinase activity, it was demonstrated that reduced c-KIT expression impaired the homing of bone marrow–derived proangiogenic stem/progenitor cells to regions of infarction, which lead to impaired cardiac recovery following MI and declined cardiac function with aging [5, 66–69]. Remodeling of the LV architecture (e.g., LV chamber dilatation and hypertrophy) and LV function (e.g., reduced left ventricular ejection fraction (LVEF)) was demonstrated in aging c-KIT mutant mice [69]. Notably, in the phase I study, Cardiosphere-Derived Autologous Stem Cells to Reverse Ventricular Dysfunction (CADEUCES) [70], the transplantation of ex vivo expanded human c-KIT+ cardiac cells into MI patients resulted in mild cardiac functional improvement [64]. Nonetheless, a larger patient population is required to further validate the benefit of transplanting cardiosphere-derived cells that contain c-KIT+ cells (CADEUCES) [64]. Although the cardiomyogenic potential of adult c-KIT+ cells is controversial, it is conceivable that in response to cardiac injury and aging, inhibition of c-KIT activity is linked to cardiac remodeling. Moreover, reduced cardiac c-KIT activity may impair normal cardiac response to stress in TKI-treated patients. Furthermore, a substantial proportion of patients who receive these therapies are at an advanced age with underlying cardiovascular comorbidities and may be especially susceptible to an abnormal stress response due to c-KIT inhibition.

TKI-Induced Cardiovascular Dysfunction

Hypertension

HTN has emerged as is the most common cardiovascular abnormality observed in patients receiving TKIs that target the VEGF/VEGFR signaling pathway with incidence ranging from 1.7 to 80.8% in clinical trials [71–73] (Table 1). Among the TKIs that promote de novo HTN in patients are axitinib, ponatinib, pazopanib, carbozantinib, and sunitinib, many of which require dose reductions or treatment discontinuation because of this [11, 74–80]. Interestingly, HTN has been shown to serve as a predictive measure for better outcomes in RCC patients treated with pazopanib, axitinib, and sunitinib [11, 74, 75]. The cell signaling pathways mediating HTN include reduction in PI3K/Akt [5, 11] and AMPK signaling pathways [16].

Basal VEGFR-2 signaling in ECs increases PI3K/Akt kinase activity and downstream vasodilatory nitric oxide (NO), while decreasing the vasoconstrictor, endothelin-1 (ET-1). As such, TKI inhibition of the VEGFR-2 receptor leads to increased production and concentration of plasma ET-1 [81], which can cause sustained vasoconstriction leading to hypertension [5]. HTN induced by the TKI sunitinib is associated with increased circulating ET-1 levels [12]. In addition, the PI3K/Akt pathway plays an important role in cell survival and vasodilation resulting from downstream NO. Inhibition of VEGFR-2 on ECs by TKIs further diminishes production of NO leading to impairment in cardiomyocyte function [82]. Reduced PI3K/Akt signaling or PLC-γ /IP3/Ca^{2+} signaling due to cardiomyocyte VEGFR-2 inhibition may result in impaired calcium cycling which alters cardiomyocyte contractility.

Hassinoff et al. first demonstrated the role of AMPK signaling in TKI-induced cardiotoxicity (57,58). In the study, sunitinib potently inhibited the enzyme activity of AMPK at therapeutically relevant concentrations. However, metformin, an AMPK-activating antidiabetic drug did not protect cardiomyocytes from sunitinib treatment [83]. Preclinical studies by Kerkela and colleagues also showed that biologically relevant concentrations of sunitinib alters cardiomyocyte metabolism and ATP depletion via sunitinib inhibition of activated AMPK in neonatal rat ventricular cardiomyocytes and in the heart [15]. Although this group did not report on whether sunitinib inhibition resulted in HTN, a recent study by Ren et al. showed that sunitinib-induced HTN is also mediated via regulation of AMPK-mTOR signaling and warrants further investigation [16]. Mice treated with sunitinib exhibited dramatically elevated blood pressures in comparison with controls [16]. In the same study, treatment with the SGLT2 inhibitor empagliflozin ameliorated sunitinib-induced HTN possibly through blocking of sunitinib-mediated inhibition of AMPK-mTOR signaling [16]. While the findings support that SGLT2 inhibitor therapy could possibly be used as cardioprotective approach for cardiovascular complications among patients receiving sunitinib, these favorable effects require validation in clinical trials.

Another hypothesized mechanism for TKI-induced HTN is functional capillary rarefaction (i.e., a decrease in perfused microvessels) or anatomic rarefaction (i.e., a reduction in vascular density). Moreover, tissue hypoxia resulting from capillary rarefaction contributes to increased total peripheral resistance and afterload, precipitating the development of HTN [85]. Reduction in global and organ-specific capillary density [86] is coupled with impaired angiogenesis, which are directly involved in the development of HTN. As previously discussed, VEGF/VEGFR-2 activation on ECs increases PI3K/Akt signaling that is responsible for EC proliferation, survival, and angiogenesis. Therefore, prolonged VEGF/VEGFR-2 inhibition by TKIs may lead to endothelial dysfunction and regression of capillary networks [11]. To date, capillary rarefaction due to VEGF signaling inhibition has been demonstrated in telatinib-treated patients [85]. In a phase I trial using telatinib, a potent inhibitor of VEGFR-2, VEGFR-3, PDGFR, and c-KIT, measurements of skin blood flux with laser Doppler flow, and buccal mucosa capillary density with side-stream dark field imaging were significantly decreased [85]. These findings support that TKI-induced HTN occurs via ECs dysfunction due to inhibition of PI3K/Akt signaling promoting regression of capillary networks. Further, it is conceivable that HTN observed in patients treated with TKIs is mediated by renal impairment after capillary rarefaction in renal glomeruli leading to the activation of the renin-angiotensin-aldosterone system (RAAS) and increased secretion of renin from the juxtaglomerular apparatus [11]. Increased renin promotes synthesis and release of angiotensin II (Ang-II), a potent vasoconstrictor from the adrenal glands leading to sustained systemic vasoconstriction. The activation of Ang-II-type-I receptor signaling induces NO synthesis, and subsequent generation of reactive oxygen species (ROS) which can precipitate EC damage. To date, however, plasma levels of renin are significantly decreased or unchanged in rats and patients treated with sunitinib, and coronary microvasculature in these same rodents became less responsive to exogenous Ang-II [12]. In human studies, sorafenib did not lead to significant changes in serum levels of aldosterone, plasma renin, ET-1, or urotensin II [87]. Despite lack of evidence supporting significant increases in RAAS mediators, RAAS inhibitors (ACEIs and ARBs) are used to successfully treat hypertensive patients with CML undergoing therapy with second- or third-generation TKIs [36] and improved survival in patents with mRCC [88, 89]. This confounding evidence suggests that the mechanism of TKI-induced HTN may occur through increased expression of Ang-II receptors or increased ACE productions and is therefore responsible for the cardiovascular protection conferred by RAAS inhibitors. Nonetheless, preclinical and clinical studies are required to support this conjecture.

Arrhythmias

De novo arrhythmias are common among the receptor TKIs that target VEGF/VEGFR signaling and include atrial fibrillation (AF), supraventricular tachycardia (SVT), bradycardia, and QT prolongation that can lead to life-threatening *torsade de pointes* (Table 1) [5, 11, 86, 90–92]. The pathophysiology of AF is complex, making it difficult to confidently ascribe causation which can be multi-factorial in patients with cancer. At baseline, arrhythmias can be present in 16–36% of cancer patients supporting a "multiple hit" hypotheses mechanism [93, 94]. Notwithstanding, ibrutinib (3.3–13%) [95, 96], nilotinib (0.68–10%), and sorafenib (5.1%) are the most common TKIs associated with AF, and sorafenib-induced AF is exacerbated when used concurrently with other chemotherapies (Table 1). In a clinical trial, the incidence of AF in patients treated with sorafenib was 5.1% when used in conjunction with 5-fluorouracil (5-FU) [5, 63].

Reduction in PI3K/Akt signaling is also a potential mechanism for AF in patients taking TKIs or other VEGF inhibitors [97, 98]. In mouse models, altered PI3K/Akt/Ca^{2+} signaling pathway underlies the pathogenesis of AF [5, 99]. The PI3K protein kinases are critical for cardiac response to stress [100, 101]. Class 1A (PI3Kα) and class 1B (PI3Kγ) notably elicit opposing effects on cardiomyocyte health [102, 103]. Both prolonged engagement and depletion of the PI3Kγ signaling pathway result in multiple AEs [104, 105], which are partially attributed to increased Ca^{2+} influx via LTCCs and Akt activity. Differential densities of calcium currents $I_{Ca,L}$ were observed in mouse ventricular myocytes lacking PTEN ($PTEN^{-/-}$), $PI3K\gamma^{-/-}$, and PI3Kγ compared to controls in response to stress/stimuli [103, 105]. Pharmacological stimulation of β_2-adernergic receptor in the $PI3K\gamma^{-/-}$ model led to increased $I_{Ca,L}$ densities as well as Ca^{2+} spark resulting in fatal arrhythmias [105]. In contrast, upregulation of PI3Kγ was observed in wild-type models of TAC-induced HF [103, 106]. Recently, Garnier et al. published atrial genetic mapping of patients with AF showing upregulation and increased expression of PI3Kγ. Thus, in excitable cells, dysregulation of PI3Kγ and PI3Kγ-mediated Ca^{2+} channels trafficking is associated with arrhythmia and HF [107]. Furthermore, developmental or adult loss of PI3Kα/β results in severe Ca^{2+} handling abnormalities, HF, and death [108], which is due to the role of PI3K α/β signaling in maintaining T-tubule architecture. Appropriate LTCC localization and organization at the T-tubules are essential for the rapid and synchronous Ca^{2+} release in response to APs, promoting coordinated contraction [108]. Of note, PI3Kα/β signaling mediates proper localization of JPH-2 (junctophilin-2). Supporting this, double PI3K knockout myocytes had less JPH-2 in T-tubule networks [108]. Genetic studies also demonstrated that constitutive activation of PI3Kα in the heart resulted in increased number of myofilaments and thicker myofibrils, compared to dominant-negative PI3Kα hearts, which display fewer myofilaments and thinner myofibrils [109]. In another study, mice expressing a dominant-negative PI3K mutant with reduced activity developed AF, while increased PI3K activity reduced atrial fibrosis and improved conduction [5, 99].

Studies in human atrial tissue supports a link between reduced PI3K activation and AF susceptibility [99]. For example, Pretorius et al. observed reduced PI3K activity in human atrial appendages isolated from AF patients compared to appendages from patients in sinus rhythm [99]. In addition to cardiac signaling, TKI-induced AF may be potentiated by other myocardial toxic effects and dysfunction. For example, LV contractile dysfunction or hypertension may create a substrate for arrhythmia.

QT interval prolongation occurs more often with vandetanib (7–18%), lapatinib (6.1–16%), nilotinib (~1–10%), and vemurafenib (1.5–34.3%) TKIs [13, 91, 92, 110–112]. Less incidence of QT prolongation has been observed in patients treated with pazopanib and axitinib [110]. Cancer patients are at an increased risk of developing QT interval prolongation due to older age, underlying disease, and concomitant medications. Incidence of QT interval prolongation in patients taking vandetanib alone was between 8% and 11% versus 1.2% in controls, and increased to 22% in patients treated with vandetanib plus other chemotherapy [6]. Analogous to other kinds of drug-induced arrhythmias, TKI-induced arrhythmia is predominantly associated with ion channel inhibition, electrolyte imbalances/derangements, underlying comorbidities and concomitant medications (e.g., antiemetics, cardiac medications, antibiotics) [93].

TKIs and Ion Homeostasis

A growing volume of published literature supports that TKIs induce changes in the electrical activity of the heart. TKIs can lead to arrhythmias due to dysregulation in ion channels and ion homeostasis [5, 11, 14]. In this section, we will systematically discuss recent reports describing the effects of TKIs on cardiomyocyte metabolic and signaling pathways involved in ion homeostasis and ion channels' regulation.

TKI-Mediated Potassium Ion Channel Dysfunction

TKI-mediated changes in cardiomyocyte K^+ centers on multiple reports that point to TKI inhibition of the KCNH2 (Kv11.1) or the human ether-a-go-go (hERG) channel, which regulates K^+ efflux out of the cardiomyocyte during repolarization and comprises the "rapid" delayed rectifier current (I_{Kr}) in intact heart myofibers [11, 113]. The integrity of I_{Kr} is necessary for repolarization of heart tissue in preparation for the next beat, and alterations in I_{Kr} reduce K^+ efflux from the cell leading to QT prolongation. Potent hERG channel inhibitors such as vandetanib and lapatinib are commonly associated with QT prolongation [90], while dasatinib, sunitinib, and nilotinib [113–116] have been demonstrated to be potent inhibitors of hERG channels in vitro. To date, TKI-mediated inhibition of hERG channels occurs through inhibition of cardioprotective enzymes B-RAF-1 and C-RAF-1 leading to the posttranslational modification of hERG channels. B-RAF inhibition with increased cyclic adenosine monophosphate (cAMP) signaling promotes protein kinase A (PKA)-induced phosphorylation and inhibition hERG channels [11, 117–119]. Hyperphosphorylation of hERG channels reduces their ability to open during action potentials (AP) and decreases I_{Kr}, which promotes the development of QT prolongation and arrhythmias. These inhibitors are also PDGFR inhibitors, which results in reduced PI3K/Akt and Ras/Raf-1/MEK/ERK pathway signaling [113, 120]. Lu et al. demonstrated that TKI-drug-induced long QT interval prolongation can occur through reduction in PI3K/Akt signaling and alterations in multiple ion currents [120]. Suppression of PI3K signaling in canine cardiac myocytes by TKIs (e.g., nilotinib) and mouse hearts lacking the

PI3K p110α catalytic subunit resulted in prolonged action potentials and QT intervals [120].

TKI Dysregulation of Calcium-Mediated Signaling

Calcium (Ca^{2+}) channels are integral to cardiomyocyte depolarization, APs, and excitation-contraction coupling. Dysregulation in Ca^{2+} homeostasis can lead to prolonged QT intervals, which predisposes the development of early afterdepolarizations (EADs). Delayed afterdepolarizations are also caused by intracellular Ca^{2+} overload and occurs after repolarization. Demonstrated mediators of TKI-induced Ca^{2+} dysregulation involve increased ROS and Ca^{2+}/calmodulin (CaM)-dependent protein kinase II (CaMKII). ROS regulate physiological functions and are by-products of metabolism, but at supraphysiologic concentrations, ROS have deleterious effects including endoplasmic reticulum (ER) and mitochondrial oxidative damage leading to impaired Ca^{2+} handling and cardiovascular injury. CaMKII is a multi-functional protein that also plays a role in Ca^{2+} cycling via the phosphorylation of multiple proteins such as L-type Ca^{2+} channels (LTCC), phospholamban, and ryanodine receptor 2 (RyR2) [121, 122]. CaMKII can be posttranslationally modified and activated via ROS-independent phosphorylation at threonine position 287 by Ca^{2+}/CaM or ROS-dependent oxidation at the methionine 207/208 site [123, 124].

Evidence supporting the link between ROS, CaMKII, and calcium dysregulation has been observed in cardiomyocytes treated with sorafenib. In rat ventricular cardiomyocytes, Ma and colleagues showed that sorafenib treatment increased incidence of preventricular contractions and LVSD secondary to significant increases in phosphorylated and oxidized CaMKII, as well as cytosolic and mitochondrial ROS [14]. Additionally, they showed that sorafenib dysregulated Ca^{2+} homeostasis that was ablated by the CaMKII inhibitor KN-93 and ROS scavenger, MPG [14]. In another study, McMullen and colleagues demonstrated the role of ROS and CaMKII in sunitinib- and imatinib-mediated toxicity in cardiac fibroblasts [125]. Oxidized CaMKII as well as cardiac fibroblast mitochondrial superoxide production was significantly increased in response to sunitinib and imatinib [125], while sunitinib also increased phosphorylated CaMKII [125]. Though channel function was assessed in the preceding studies, it is conceivable that increased activation of CaMKII and subsequent phosphorylation of cardiac LTCC or RyR2 facilitates influx of Ca^{2+} into cardiomyocytes. The consequence of this increased cytosolic Ca^{2+} may contribute to de novo arrhythmias observed in patients undergoing TKI therapy.

TKIs and Sodium Channels: An Unexplored Territory

Literature reports on TKI-mediated ion channel dysfunction are predominantly centered on potassium and calcium channels. However, TKI-induced arrhythmias via changes in voltage-gated sodium channels, especially Nav1.5 encoded by SCN5A gene, are also conceivable. Nav1.5 is essential for cardiac membrane excitability, where it is responsible for the initiation of APs and is regulated through phosphorylation [126, 127]. Improper inactivation of Nav1.5, however, can lead to an increase in late sodium current ($I_{Na,L}$). While late sodium currents are observed under normal conditions, pathological increases predispose the development of arrhythmias and HF [128]. Notably, perturbations in $I_{Na,L}$ during the initial phase of cardiac APs can impact other ion channels involved in generating APs.

As previously mentioned, TKIs can lead to significant increases PKA and CaMKII kinases. PKA and CaMKII activity are known regulators of Nav1.5, and their hyperphosphorylation of Nav1.5 increases $I_{Na,L}$. For example, Hegyi and colleagues demonstrated that both activated PKA and CaMKII in response to β-adrenergic stimulation can alter $I_{Na,L}$ in rabbit ventricular cardiomyocytes [129]. Koval et al. first identified increased CaMKII-mediated phosphorylation of Nav1.5 at serine position 571 (S571) in vitro and from failing mouse, canine, and human hearts [126]. Later, Glynn et al. provided the first in vivo evidence that phosphorylation at S571 increases in $I_{Na,L}$, which promoted abnormal repolarization, impaired intracellular Ca^{2+} handling, and increased susceptibility to arrhythmia [127]. In the same study, a change in this serine to a glutamic acid E, S571E, mimicked hyperphosphorylation of Nav1.5, and an increase in $I_{Na,L}$ [127]. Electrophysiology and echocardiographic studies revealed that S571E mutation led to prolonged action potentials, preventricular contractions, and a significant decrease in ejection fraction (EF) and fractional shortening [127]. However, these electrical and functional abnormalities were ablated in the presence of $I_{Na,L}$ blockers such as flecainide and ranolazine [127]. Notably, S571 is required for maladaptive remodeling and arrhythmias in response to TAC [127]. PKA and CaMKII can also posttranslationally modify other positions in the Nav1.5 channel protein. Iqbal et al. extensively reviewed these PKA- and CaMKII-sensitive positions in Nav1.5 and the reported consequences, as well as other kinases that can target Nav1.5 [130]. Overall, these data suggest that Nav1.5 may be a potential mediator of TKI-induced arrhythmias and that $I_{Na,L}$ blockers could serve as a possible treatment.

TKIs and Heart Failure

Functional changes in the heart after TKI treatment range from asymptomatic LVSD to cardiogenic shock, HF, MI, and sudden cardiac death (SCD) [91, 93, 131]. In clinical trials, a greater than 10% drop in LVEF or an LVEF below 50% in the presence of heart failure (HF) is among the most important heart-related functional changes that occur in response to TKI therapy, leading to treatment discontinuation [93]. Sunitinib, axitinib, sorafenib, and vandetanib are most associated with reduced LVEF and symptomatic HF [93, 131–133], while dasatinib and bosutinib demonstrated the highest incidence of HF [91]. In a meta-analysis composed of more than 10,000 patients from 5 phase II and 16 randomized phase III trials, HF risks with all FDA-approved TKIs were evaluated [134]. Ghatalia et al. reported a significant risk (relative risk (RR) = 2.69, $p < 0.001$) for all grades of HF was observed with TKIs that target VEGFR compared to no TKI use [134]. Notwithstanding, this number may underestimate the true prevalence of HF with TKI use because patients with severe cardiac comorbidities are often excluded from therapeutic trials [11]. When comparing more selective TKIs (e.g., axitinib) to nonspecific multi-targeted TKIs (e.g., sunitinib, sorafenib, and pazopanib), a similar RR for HF was observed [93, 134]. In a different study, it was shown that the right ventricle, perhaps due to its thinner wall, is more vulnerable to TKI therapy in the absence of pulmonary HTN than the left ventricle [135]. These data are inconclusive however, because right-sided HF usually goes underreported in clinical trials [136].

TKI-induced HF is hypothesized to occur through dysregulation of several pathways involved in pathological survival and abnormal proliferation of cancer cells that also regulate the survival of normal cells, including cardiomyocytes. Direct inhibition of the Raf/MEK/ERK pathway was shown to mediate sorafenib-induced cardiomyocyte apoptosis in rat and zebrafish cardiomyocytes (Fig. 2) [137]. Direct inhibition of VEGFR-2/PDGFR-β signaling blockade and downstream inhibition of the PI3K/Akt pathway are also implicated in cardiomyocyte apoptosis [5, 11, 138, 139]. In a preclinical rodent model, constitutive action of Akt in the setting of cardiac ischemia-reperfusion injury prevented cardiomyocyte apoptosis and improved cardiac function [140]. Similarly, in another study, exogenous administration of VEGF prevented cardiomyocyte apoptosis and preserved cardiac function [139] and contractility [43] possibly through improved VEGF/VEGF-2/PI3K/Akt-mediated inhibition of proapoptotic proteins [94]. In addition, animals treated with VEGF exhibited diminished fibrosis and increased contractile myocardium after infarction [43]. At baseline, Akt activation prevents cardiomyocyte cell death due to inhibition of BCL-2-antagonist of cell death (BAD), a proapoptotic protein (Fig. 2) [11, 138, 141]. Inhibition of BAD due to reduced PI3K/Akt signaling leads to accumulation of BAD at the surface of the mitochondria, leading to mitochondrial dysfunction and increased cytoplasmic cytochrome C (Cyt C) and caspases 3 and 9 that may lead to cardiomyocyte apoptosis (Table 1) [5]. Supporting this, sunitinib was shown to also inhibit the ribosomal S6 kinase (RSK1), an integral regulator of cell survival via BAD phosphorylation [83]. H9c2 ventricular myoblast cells treated with regorafenib resulted in disrupted mitochondrial function, altered mitochondrial membrane potential and structure, along with reduced mitochondrial DNA content [142]. It is unknown whether inhibition of the PI3K/Akt pathway by TKI-induced VEGFR-2 blockade contributes to cardiomyocyte apoptosis and HF in patients treated with TKIs and is a potential pathway for investigation [11, 37]. PI3K/Akt signaling in the heart is also modulated by several other receptor pathways not affected by TKIs; as such, further research should be conducted to determine the exact mechanism by which TKI-induced cardiomyocyte apoptosis is required. Additionally, the diversified functions of VEGF/VEGFR signaling indicate the complex regulation of their downstream signaling cascades.

Another signaling pathway postulated to mediate cardiomyocyte death is the AMPK-mTOR pathway [15, 83, 84, 143]. Hasinoff et al. first demonstrated that sunitinib potently inhibits AMPK, leading to cardiotoxic effects [83]. Under normal circumstances, AMPK signaling maintains ATP concentrations for cardiomyocyte function. Due to limited energy reservoirs in the cardiomyocyte, ATP depletion can rapidly lead to impaired cardiac contractility and cardiac injury. Recently, it was demonstrated that sunitinib inhibited the phosphorylation and activation of AMPK, which led to cardiomyocyte cytotoxicity in vivo and in vitro along with impaired LVEF and LVSD in vivo [16]. Interestingly, empagliflozin, a selective SGLT2 inhibitor, ameliorated sunitinib-induced cardiac dysfunction and cardiomyocyte death in mice and cell viability in H9c2 cells as evidenced by TUNEL assays [16]. Sunitinib also inhibits autophagy through the inhibition of AMPK-mTOR signaling, an evolutionarily conserved cellular degradation process with vital roles in both cell survival and cell death [16]. Activation of the ER stress response [114] and oxidative stress pathways (i.e., increased caspase-3, p53) [144] due to direct TKI toxicity may contribute to cell death and HF. ER stress can occur due to TKI-reduced PLC-γ/IP3/Ca^{2+} signaling. Subsequent impaired Ca^{2+} cycling can lead to mitochondrial swelling and dysfunction, increased ROS, and an increased ratio of pro-apoptotic BCL-XS to BCL-XL proteins (Fig. 2) [11, 83, 145]. Similarly, mitochondrial damage may lead to ATP depletion, reduced cardiac contractility, and cardiac cell death (Fig. 2) [15, 71, 84].

Cardiomyocyte hypertrophy and remodeling is another possible mechanism of TKI-mediated HF. As stated above, a balance of VEGFR-1 and VEGFR-2 activation on cardiomyocytes is essential for maintaining cardiomyocyte homeostasis [43]. Kivela and colleagues demonstrated that bidirectional ECs and cardiomyocyte crosstalk is essential for this balance [146]. In the study, deletion of EC VEGFR-1 in adult mice increased the coronary vasculature and induced cardiomyocyte hypertrophy [146]. Notably, both angiogenesis and cardiomyocyte growth were reversed by blocking VEGFR-2 or deleting endothelial VEGFR-2 [146]. These data suggest that TKI inhibition of endothelial VEGFR-1 may promote increased signaling through VEGFR-2 in the setting of prolonged VEGF exposure, which results in cardiomyocyte hypertrophy and contributes to cardiomyopathy and HF. On the other hand, VEGFR-1 activation on cardiomyocytes markedly improves rat cardiac function after MI by directly protecting cardiomyocytes from apoptosis and upregulation of genes driving compensatory hypertrophic response [43]. As such, TKI-induced inhibition of VEGFR-1 may prevent cardiac response to injury and promote the development of HF.

Given the lack of kinase selectivity that TKIs exhibit, it is likely that inhibition of a wide array of kinases are responsible for the cardiotoxic effects of these drugs. Inhibition of FGFR-1 and FGFR-2 signaling in addition to VEGF-signaling inhibition by TKIs may reduce systolic function. At baseline, FGFR signaling mediates cell proliferation, differentiation, survival, and angiogenesis [11, 147–153]. FGF1 stimulation protects the heart from ischemic disease by inhibiting cardiomyocyte apoptosis [150, 154]. Four weeks post injury, this treatment reduces heart scarring and wall thinning and markedly rescues cardiac function [151]. Further, transient overexpression of FGFR-1 increased cardiac contractility, while chronic expression promoted hypertrophy and preserved systolic function [155]. FGF2 regulates autophagy and ubiquitinated protein accumulation induced by myocardial

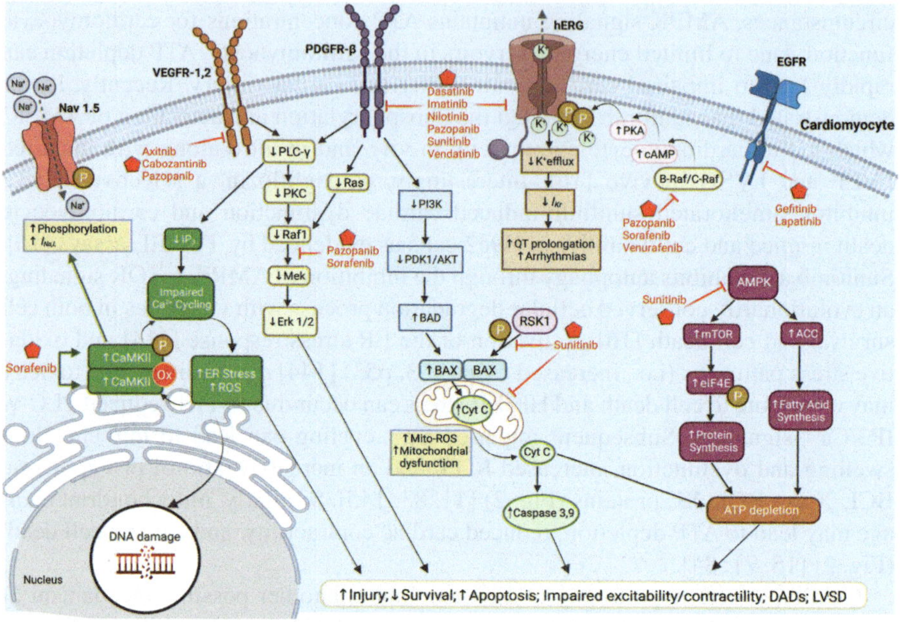

Fig. 2 Intracellular signaling pathways mediating tyrosine kinase inhibitor cardiotoxicity. TKI-mediated inhibition of VEGFR-2 results in the downregulation of PLC-γ which alters IP3 and Ras/Raf/Mek/Erk signal transduction cascades. Reduced IP_3 contributes LVSD and reduced myocardial contractility, which results from impaired calcium cycling and increased ER stress and ROS. Increased Ca^{2+} and ROS is proposed to activate and increase CaMKII phosphorylation and oxidation, respectively, which further promotes dysregulation in Ca^{2+} homeostasis. Increased CAMKII activity was also proposed mediated hyperphosphorylation and activation of Nav 1.5 sodium channel and increasing late Na^+ current ($I_{Na}{}^+$), which is a harbinger of arrhythmias. Increased ER ROS and Mito-ROS may also lead to DNA damage. Reduced Ras/Raf/Mek/Erk signaling is also directly affected by PDGFR-β inhibition resulting in decreased cardiomyocyte survival, increased apoptosis, and LVSD. Inhibition of PDGFR-β reduces PI3K/Akt signaling and upregulates the proapoptotic proteins, BAD and BAX leading to mitochondrial dysfunction, release of Cyt C, activation of caspases 3 and 9, ATP depletion, and cell death. AMPK inhibition by sunitinib also depletes ATP due to increased energy sink from mTOR and ACC-mediated protein and fatty acid synthesis, respectively. Loss of ATP contributes to cardiomyocyte injury and death. TKIs also directly inhibit myocyte Kv11.1/hERG channels disrupting K^+ currents. Inhibition of cytoplasmic C-RAF/B-RAF enzymes increases cAMP promoting PKA phosphorylation and inhibition of hERG channels, leading to QT prolongation and the development of arrhythmias. Other receptors inhibited by TKIs include EGFR. (Created with Biorender.com). *Abbreviations: ACC* acetyl-coenzyme A carboxylase; *Akt* protein kinase B; *AMPK* adenosine 5'-monophosphate-activated protein kinase; *ATP* adenosine triphosphate; *BAD* BCL2-antagonist of cell death; *BAX* BCL2-associated X protein; *CaMKII* Ca^{2+}/calmodulin-dependent protein kinase II; *cAMP* cyclic adenosine monophosphate; *Cyt C* cytochrome c; *DADs* delayed after depolarizations; *EGFR* epidermal growth factor receptor; *elF4E* eukaryotic translation initiation factor 4E; *Erk* extracellular signal–regulated kinase 1/2; *hERG* human ether-à-go-go; *IP_3* inositol-trisphosphate-3-kinase; *LVSD* left ventricular systolic dysfunction; *Mek* mitogen–activated protein kinase; *mTOR* mammalian target of rapamycin; *Nav 1.5* voltage gated sodium channel isoform 1.5; *PDK1* phosphoinositide-dependent kinase 1; *PI3K* phosphatidyl inositol-3-kinase; *PLC-γ* phospholipase C gamma; *PKA/C* protein kinase A/C; *PDGFR*-β platelet-derived growth factor receptor; *Raf-1* rapidly accelerated fibrosarcoma-1; *Ras* rat sarcoma virus protein; *RSK1* ribosomal s6 kinase 1; *ROS* reactive oxygen species; *TKI* tyrosine kinase inhibitor

ischemia/reperfusion via the activation of the PI3K/Akt/mTOR pathway [153]. In the absence of FGFR-2, researchers observed increased thrombocytosis, poor vascular function, and impaired cardiac response to ischemia [147]. In another study, FGFR-2 loss ablated physiological hypertrophic response to pressure overload [156]. Finally, it is also possible that direct or current inhibition of PDGFR-β and c-KIT may disrupt coronary microvasculature by blocking stress-induced coronary angiogenesis contributing to TKI-induced HF phenotype [49].

TKIs and Thromboembolism

Arterial thrombosis (AT) is a less frequent, yet life-threatening TKI-induced complication that results in MI, stroke, and critical limb ischemia. Vascular events, presumably from prothrombotic complications are most observed with TKIs targeted to the VEGFR and Bcr-Abl1 translocation [157]. Of the drugs that target Bcr-Abl1, ponatinib and nilotinib have higher rates of acute arterial thrombotic events compared to bosutinib, imatinib, and dasatinib [157]. The wide spectrum of kinases inhibited by ponatinib along with its strong activity against all members of the VEGF family have raised questions regarding whether these properties contribute to the high rate for arterial thrombotic AEs. Although overall rates of AT are lower, a meta-analysis of trials with VEGF-targeted agents suggests that VEGF inhibitors confers a threefold risk of AT events compared with control subjects [157, 158]. Among anti-VEGFRs, sunitinib and sorafenib, which also possess cross-reactivity to Bcr-Abl1, PDGF, and FGFR, have highest rates of AT [157]. As such, these drugs should be used with caution when treating patients with a history of coronary artery disease, previous embolic stroke, or at a high risk of thrombotic events.

While endothelial injury and dysfunction is unlikely the sole cause of TKI-related AT events, it is the focus of most research. Potential mechanisms responsible for arterial thrombotic complications are suggested to be multi-factorial, involving the status of the endothelium, coagulation pathway, fibrinolysis enzymes, and platelet adhesion [157]. Loss of EC integrity, injury-coupled hemostasis, and hypercoagulability (i.e., Virchow's triad) contribute to the development of thrombotic events. VEGF signaling is important in EC survival and function. Hypoxia induced by VEGFR inhibition could also exacerbate the risk for thrombosis via overproduction of erythropoietin leading to increased blood viscosity and hematocrit [90, 159]. VEGF/VEGFR-2/PI3K/Akt/Ca^{2+} and VEGF/VEGFR-2/PLC-γ/PKC signaling pathways are also involved in maintaining balance of prothrombotic or proinflammatory effects. Reduced VEGF receptor-mediated signaling leads to reduced activation of endothelial nitric oxide synthase (eNOS), NO production, and prostacyclin (PGI_2 a potent platelet inhibitor) via loss of PI3K/Akt and PLC-γ/PKC activity, respectively. Together, loss of NO and PGI_2 may promote thrombosis in the setting of endothelial injury with subsequent coagulation activation [90, 157, 160].

In support, ponatinib was shown to increase platelet adhesion in the presence or absence of platelet activation [157] and hyperactive platelets [161]. In situ immunofluorescent imaging of large and small vessels from hyperlipidemic mice with atherosclerosis following 24h exposure to ponatinib revealed markedly increased endothelial-associated von Willebrand factor (VWF) multimers, resulting in platelet adhesion [162]. TKIs are also implicated in atherosclerotic progression, which contributes to thrombosis. Indeed, nilotinib has been shown to exacerbate dyslipidemia and hypertriglyceridemia [163]. Together, this evidence suggests that VWF, hyperactive platelets, and dyslipidemia may play a role in TKI-mediated atherosclerotic progression [164] and could explain mechanisms for worsening atherosclerosis reported in patients [157].

Monitoring and Treatment of TKI-Induced Cardiotoxicity

Management paradigms for TKI-related cardiotoxicity have centered on using antihypertensives to reduce HTN, dose reduction, and/or drug discontinuation [182]. ACEIs, ARBs, and non-dihydropyridine calcium channel blockers are usually used to treat HTN, while the β-blockers nebivolol and carvedilol have been used to prevent progression of TKI-cardiomyopathy to LV dysfunction [11, 93, 131]. Co-treatment of pazopanib plus metoprolol or diltiazem prevented pazopanib-related QT interval prolongation [11, 183]. Since there is evidence for reduced VEGFR-mediated NO production and disrupted ET-1 and NO balance in TKI-induced HTN, NO-producing drugs including isosorbide dinitrate or mononitrate and ET-1 receptor blockers may have potential uses [182, 184]. Drug interactions between antihypertensives and TKIs should also be taken into consideration. It has been shown that cabozantinib and pazopanib are metabolized by the CYP3A4 enzyme. Consequently, to maintain the therapeutic doses and plasma clearance of these drugs, antihypertensive drugs that inhibit CYP3A4 should be avoided [11, 182].

Statins and SGLT2 inhibitors also exert cardioprotective effects by thwarting the development and/or progression of several cardiovascular diseases [16, 185, 186] and can protect the heart from cancer therapy-induced cardiac injury [131]. In atorvastatin- and dasatinib-treated H9c2 cardiomyocytes, reduced cardiomyocyte cell death and restoration in homeostasis were observed [35]. Hung et al. also showed that statins improved overall patient survival [187]. However, a specific role for statins in reducing TKI-induced cardiovascular AEs remains to be further investigated [187]. Recently, the SGLT2 inhibitor empagliflozin was shown to ablate sunitinib-induced cardiac changes in vivo and in vitro [16]. Further, empagliflozin prevented sunitinib-induced HTN and reduced LVEF in vivo and cardiomyocyte death and cell viability in vitro [16]. These findings indicate that SGLT2 inhibitor therapy serves as a potential cardioprotective approach for combating sunitinib-dependent cardiac AEs but requires validation in clinical trials [16].

Arrhythmias in patients on TKIs may be mediated by hERG channel inhibition, impaired Ca^{2+} and sodium homeostasis, and diarrhea. As such, reducing the risk of TKI-related arrhythmias requires optimization and monitoring of patient electrolytes in conjunction with monthly ECGs prior to and during treatment [11]. Further, diuretics and other electrolyte-depleting drugs should be avoided [93, 182, 188].

Echocardiographic monitoring has demonstrated a role in detecting early signs of HF in patients taking TKIs [11]. In a retrospective study following TKI-treated patients, early changes in left ventricular strain were observed in echocardiograms, which may serve as a precursory marker of TKI-induced systolic dysfunction [189, 190]. Using velocity vector imaging, Moustafa and colleagues also identified early subclinical cardiac chamber dysfunction secondary to TKI treatment in patients with mRCC [135]. Therefore, close echocardiographic surveillance should be used in all patients at interval durations before and during TKI therapy. Furthermore, a standardized LVEF cut off for TKI dose adjustment or discontinuation should be considered.

Conclusion and Future Directions

The prevalence of TKI-mediated cardiovascular complications remains high and can lead to increased comorbidity with HTN as well as life-threatening cardiac effects including arrhythmias and HF. An understanding of the intracellular signaling cascades associated with TKI-related cardiotoxicity is critical to understanding the clinical cardiovascular sequelae observed with this drug class. Therefore, evaluating the VEGF/VEGFR, PDGF/PGDR, and SCF/c-KIT signaling axes will provide important clues toward uncovering the cause of TKI-induced HTN, arrhythmias, and HF observed with TKI therapy. Unfortunately, the nature of TKIs and their inhibition of numerous signaling pathways make the investigation of the pathophysiological mechanisms underlying their cardiotoxicity challenging. Further, there are no proven strategies or biomarkers that predict TKI-induced cardiac dysfunction. A single approach is unlikely to address this issue. Through interdisciplinary partnerships between cardiologists and oncologists, a standardized algorithm for cardiac surveillance should be established for patients on TKI therapy. Clinical trials should be designed to explore the utility of concurrent cardioprotective drug use with TKIs. Inclusion of patients with cardiovascular comorbidities that better reflect the real-world population may also provide insight into the true prevalence of TKI-associated cardiac AEs. In addition, further investigation of TKI-mediated arrhythmias via disruption of cardiac ion channels (especially Nav 1.5) may provide insight into how TKIs promote cardiac arrhythmogenesis. The findings summarized in this review demonstrates that further research into the general role of tyrosine kinases in cardiac biology is critical for combating chemotherapy-related cardiotoxicity observed with TKI, toward improving patient safety.

Part III: Immune Checkpoint Inhibitors

T-cell activation in response to antigen presentation is a critical step in the adaptive immune response [191, 192]. Recently, it was discovered that T cells could recognize and mount an immune response against tumor cells [193]. However, tumors cells can adapt ways to evade T-cell recognition and destruction by increasing expression of programmed cell death receptor 1 (PD-1), programmed cell death ligand 1 (PD-L1), and cytotoxic T-lymphocyte-associated protein 4 (CTLA-4), preventing T-cell activation [193]. To elicit a robust immune response against these cancer cells, immune checkpoint inhibitors (ICIs) were developed to sequester PD-1, PD-L1, and CTLA-4, thereby preventing tumor-mediated T-cell suppression [191, 192]. Though initially designed and approved for the treatment of melanoma in 2014, today, ICIs are one of the most efficacious therapeutic agents for an expanding number of advanced cancers including melanoma, RCC, colorectal cancer, non-small-cell lung cancer, and urothelial carcinoma (Table 2) [19]. Despite their efficacy with durable complete responses, prolonged activation of T cells can result in a wide range of systemic adverse events including short- and long-term cardiotoxicities [194, 195]. While the cardiotoxic profile of ICIs is low (<1%), ICI-related cardiotoxicities are often described as fulminant and progress to cardiac death with a high mortality rate (>20%) [192, 195–197].

According to the American Society of Clinical Oncology, the incidence of ICI-driven myocarditis, pericardial disorders, de novo arrhythmias, and HF was less than 0.1%, yet the mortality due to these immune-related AEs (IRAEs) was 27–46% [192, 193]. Of note, the observed clinical incidences of the IRAEs are nine to 100 times higher (0.9–10%) [196, 198]. ICI pharmacovigilance data analysis on cardiotoxicities reported myocarditis (0.31–1.2%), pericardial disease (between (0.12% and 0.46%), arrhythmia (0.14–0.15%), and HF (0.72%). However, D'Souza et al. reported an ICI-associated increased relative risk of myocarditis and pericarditis at 1.8% as well as arrhythmia at 2.9% within 6 months of first dose, while the relative risk for HF peaked at 2.5% 1 year post first dose [198, 199]. Additionally, due to a highly variable and mutagenic tumor microenvironment, clinicians are turning to multi-model and multi-targeted therapies, enabling them to inhibit CTLA-4 and PD-1/PD-L1 simultaneously. ICI combination therapy is especially efficacious in advanced cancers with low PD-L1 expression, as it induces increased tumor cell expression of PD-L1 and T-cell infiltration (Table 2) [200]. This treatment approach carries a higher risk of IRAEs and mortality (76%) [189, 191]. While AE rates seem low, the growing popularity of these drugs suggests that thousands of patients could be impacted. Furthermore, considering the early onset and high mortality rate of cardiotoxicities, it is vital to understand the mechanisms via which ICI mediate reversible and irreversible cardiac AEs.

T-Cell Activation and Regulation

In addition to T-cell receptor (TCR) binding of antigens on an antigen-presenting cell (APC), T-cell activation requires binding of B7 molecule (CD80/CD86) on APC to CD28 receptors on T cells [201, 202]. CD28 binding initiates recruitment and activation of T-cell PI3K/Akt intracellular signaling [202], which promotes cell cycle progression and proliferation [203]. To regulate immune response under normal circumstances, CTLA-4 translocates to T-cell surface and competes with CD28 to bind B7 proteins on the APC in order to block co-stimulation [201]. CTLA-4 is

Table 2 FDA-approved ICI therapies, their receptor targets, indications, and incidence of cardiotoxicity

ICI Name	Target (s)	Indications	Cardiotoxicity type (rate, %)	Reference
Ipilimumab (YERVOY)	CTLA-4	Advanced melanoma Metastatic melanoma Advanced RCC Metastatic CRC HCC NSCLC Pleural mesothelioma	Myocarditis[a] (<0.1–14.60%) Fatal myocarditis[a] (<0.1%) Pericardial disease[a] (<1–8.92%) Arrhythmia (8.28%) TS (0.85%) HF (N/A) ACS (N/A) Vasculitis[a]	[17, 227, 254, 257]
Pembrolizumab (KEYTRUDA)	PD-1	Advanced RCC Cervical Cancer cHL Esophageal cancer Gastric cancer HCC Melanoma Metastatic SCLC MCC NSCLC Refractory PMBCL SCCHN TNBC UC	Myocarditis[a] (0.5–16.10%) Pericardial disease[a] (2–8.0%) Arrhythmia[a] (4–6.84%) TS (0.85 %) HLM (0.5%) HF (0.4%) ACS (N/A) MI[a] (2%) Cardiac tamponade[a] (2%) Vasculitis[a] Cardiac arrest[a]	[17, 227, 254, 257]
Nivolumab (OPDIVO)	PD-1	Advanced RCC cHL CRC Esophageal cancer HCC HNSCC Melanoma Metastatic NSCLC/ SCLC SCCHN UC	Myocarditis[a] (<1–13.40%) Pericardial disease[a] (14.10%) Arrhythmia (6.44%) Ventricular arrhythmia[a] (1-10%) TS (0.54%) HF[a] (N/A) ACS (N/A) AF[a] (N/A)	[17, 227, 254, 257]
Avelumab (BAVENCIO)	PD-L1	Advanced UC Metastatic UC Advanced RCC Metastatic MCC	Myocarditis[a] (25 %) Pericardial disease (12.5%) Arrhythmia (12.50%) TS (N/A) HF (N/A) ACS (N/A) MI[a] (N/A) SCD[b] (1.2%) CHF[b] (1.8%)	[17, 227]

(continued)

Table 2 (continued)

ICI Name	Target (s)	Indications	Cardiotoxicity type (rate, %)	Reference
Atezolizumab (TECENTRIQ)	PD-L1	Advanced SCLC HCC Melanoma Metastatic NSCLC TNBC UC	Myocarditis[a] (≤1.0–10.6%) Pericardial disease[a] (17%) Arrhythmia[a] (6.38%) TS (1.06%) HF[a] (N/A) ACS (N/A) MI[a] (0.6%) Cardiac arrest[a] (0.4%) Prolonged QT[a] (N/A) Decreased LVEF[a] (N/A) VT[a] (N/A)	[17, 227, 254, 257]
Cemiplimab	PD-1	Advanced SCC	Myocarditis (N/A)	[254]
Durvalumab (IMFINZI)	PD-L1	UC Stage III NSCLC SCLC	Myocarditis[a] (<1–11.8%) Pericardial disease (11.8%) Arrhythmia (N/A) TS (N/A) HF (N/A) ACS (N/A)	[17, 227, 254, 257]
Ipilimumab + nivolumab	CTLA-4 PD-1	Metastatic melanoma RCC	Myocarditis (0.27–2.4%) TS (N/A)	[17, 237, 257]
Durvalumab + tremelimumab	PD-L1 CTLA-4	Endometrial cancer	TS (N/A) Arrhythmia (N/A)	[17, 237]

Abbreviations: ACS acute coronary syndrome; *AF* atrial fibrillation; *cHL* classical Hodgkin lymphoma; *CHF* congestive heart failure; *CRC* colorectal cancer; *HF* heart failure; *CTLA-4* cytotoxic T-lymphocyte-associated protein 4; *HCC* hepatocellular carcinoma; *HLM* Hodgkin lymphoma myocarditis; *LVEF* left ventricular ejection fraction; *MCC* Merkel cell carcinoma; *MI* myocardial infarction; *NSCLC* non-small-cell lung cancer; *PMBCL* primary mediastinal large B-cell lymphoma; *PD-1* programmed cell death receptor 1; *PDL-1* programmed cell death-ligand 1; *RCC* renal cell carcinoma; *SCC* squamous cell carcinoma; *SCCHN* squamous cell carcinoma of the head and neck; *SCD* sudden cardiac death; *SCLC* small-cell lung cancer; *TNBC* triple-negative breast cancer; *TS* Takotsubo syndrome; *UC* urothelial carcinoma; *VT* ventricular tachycardia

[a]Included in the Food and Drug Administration label (up to date as of March 2022)

[b]Adverse effects in combination with the tyrosine kinase inhibitor, axitinib

expressed on all regulatory T cells (T_{reg}) as well as differentiated helper ($CD4^{+}$) and cytotoxic ($CD8^{+}$) T cells, with differential expression rates stimulated by the host environment [22, 204]. CTLA-4 effectiveness in cancer management [200] is credited to CTLA-4's ability to compete with T-cell co-stimulators at immune checkpoints [200]. CTLA-4 competition results in reduced T-cell proliferation, angiogenesis, and cytokine secretion via inhibition of PI3K/Akt signaling and nuclear transcription factor (NF-κB) pathways [202, 205, 206]. Ipilimumab is the first and only FDA-approved anti-CTLA-4 ICI therapy on the market, with indications for advanced melanoma, renal cell carcinoma, and metastatic colorectal cancer.

PD-1 is expressed on activated T cells as well as B cells, and thus has a broader inhibitory capacity compared to CTLA-4 [202]. After T-/B-cell activation, PD-1 expression is upregulated. Conversely, PD-L1 is expressed on multiple cell types including T/B cell, T_{reg}, APC, tumor cells, and many more [207]. PD-1 activation reduces PI3K/Akt signaling and inhibits cellular proliferation, angiogenesis, and cytokine secretion [200, 202, 208]. There are three FDA-approved anti-PD-1 pharmaceuticals (pembrolizumab, nivolumab, and cemiplimab) and three approved anti-PD-L1 pharmaceuticals (atezolizumab, avelumab, and durvalumab), which are prescribed as agents for advanced cancer management [200]. In one study, anti-PD-L1 drugs demonstrated better efficacy compared to anti-PD1, due to increased ligand blocking versus receptor blocking [199]. Life-threatening consequences of releasing the brakes off the immune regulators (i.e., CTLA-4, PD-1, and PD-L1) include unregulated systemic AEs and organ-specific toxicities, especially short- and long-term cardiotoxicities [194, 195, 209].

Mechanisms of ICI-Mediated Cardiac Disorders

ICIs enhance activation and maintenance of T cells by launching a robust immune response against advanced cancers. However, increased T-cell PI3k/Akt signaling, maturation, proliferation, and angiogenesis are not limited to the cancer microenvironment. Specifically, activation of myocardial T-cell populations or increased infiltration of activated T cells in the myocardium can promote cardiac inflammation, dysfunction, and injury (Fig. 3). Moreover, both mouse and human cardiomyocytes express CTLA-4 and PD-1 receptors and PD-L1 ligands, which can be directly inhibited by ICI exposure [22, 207]. However, despite increases in ICI-induced myocardial diseases, there is limited knowledge of pathological mechanisms and consequences of blocking cardiac CTLA-4, PD-1, and PD-L1. Furthermore, preclinical studies designed to recapitulate clinical manifestations of ICI-induced AEs have produced mixed results. Despite observations of overt cardiomyopathy in PD-1 knockout mice [210–212], researchers have had difficulty instigating cardiac inflammation with only PD-1/PD-L1 antibodies. Direct ICI-mediated inhibition of cardiomyocyte CTLA-4 and PD-1 receptors can result in increased localized activation of PI3K/Akt signaling pathways. A consequence of increased PI3K/Akt signaling is impaired calcium handling and cardiac remodeling, which contribute to the development of arrhythmias and HF (Fig. 3) [103, 216]. Lymphocytic infiltration in the myocardium and pericardium is associated with myocarditis and pericarditis [217], while increased exposure to systemic and localized cytokines is associated with multiple cardiovascular disorders resulting in HF [218, 219].

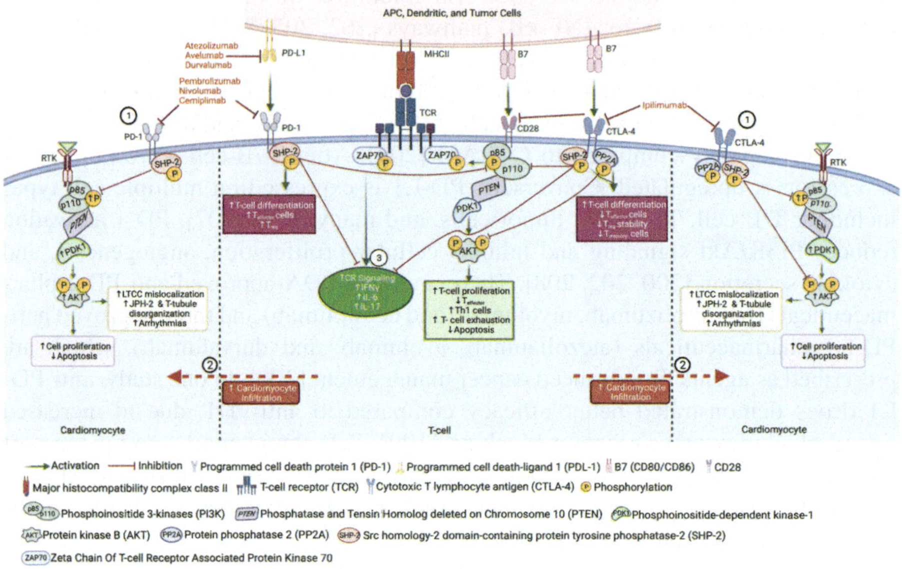

Fig. 3 Proposed mechanisms of ICI-mediated cardiotoxicity. Immune checkpoint inhibitors (ICIs) act to prevent T-cell exhaustion by blocking the binding of co-inhibitor receptors (B7 or PD-L1 ligand) on antigen-presenting cells (APCs), dendritic cells, and tumor cells to activated T cells (via CTLA-4 and PD-1 receptors, respectively). ICIs are monoclonal antibodies that bind to CTLA-4, PD-1, and/or PD-L1. While blocking these interactions between tumor cell and T cell promotes T-cell-mediated cytotoxic killing of tumor cells, several hypotheses have been proposed to explain ICI-associated cardiotoxicity including inhibition of myocardial PD-1 and CTLA-4 receptors. First (1), direct binding of ICIs to CTLA-4, and PD-1 receptors expressed on normal cardiomyocytes lead to dysregulated PI3k/Akt/Ca^{2+} signaling, reduced SHIP-2 phosphorylation, and subsequent increases in Akt. Increased cardiac Akt activity has been linked to increased remodeling of T-tubules including increased LTCC channel mislocalization, JPH-2 and T-tubule disorganization and development of arrhythmias. Increased Akt activity can also promote apoptosis. Second, increased T-cell infiltration of cardiac tissue due to reduced activation threshold for self-reactive T cells (2). Inhibition of immunologic checkpoints may dampen normal regulation of autoimmune process leading to nonspecific T-cell recognition and cytolysis of normal body cells, where damage of endothelial cells results in vasculitis or atherosclerosis and myocardial and/or pericardial attack leads to myocarditis and pericarditis. In final, there is evidence to suggest that increases in systemic proinflammatory cytokines promote increased T-cell infiltration and increase likelihood of T-cell interaction with cardiac antigens precipitating pericardial and myocardial disease. Finally, antigen cross-reactivity is also another proposed mechanism. Ordinary cardiomyocytes and tumor cells may also share or possess homologous antigens, which can be recognized and activate T cells. (Created with Biorender.com). *Abbreviations: APC* antigen-presenting cell; *AKT* protein kinase B; *B7* integral membrane protein B7; *CD28* cluster of differentiation 28; *CTLA-4* cytotoxic T-lymphocyte-associated protein 4; *INFγ* interferon gamma; *IL-6* interleukin 6; *IL-17* interleukin 17; *JPH-2* junctophilin 2; *LTCC* L-type calcium channels; *mTORC1* mechanistic target of rapamycin complex 2; *p* phosphate; *PD-1* programmed cell death protein 1; *PDK1* phosphoinositide-dependent kinase1; *PD-L1* programmed cell death protein ligand 1; *PP2A* protein phosphatase 2A; *PTEN* phosphatase and tensin homolog; *MHCII* major histocompatibility complex class II; *p85* phosphoinositide 3-kinase subunit 85; *p110* phosphoinositide 3-kinase subunit 110; *RTK* receptor tyrosine kinase; *SHP-2* Src homology region 2; *TCR* T-cell receptor; $T_{effector}$ effector T cells; *Treg* regulatory T cell; *ZAP70* zeta chain of T-cell receptor associated protein kinase 70

Arrhythmias

Arrhythmias are rarely reported in patients with ICI therapy (Table 2). Despite low incidence in clinical trials, real-world data suggest the incidence of these AE is much higher than expected and are likely to increase as use of these drugs continues to dominate the landscape. Pharmacovigilance analysis reported that 0.97% of the ICI-mediated AEs are arrhythmogenic in nature including supraventricular arrhythmias (0.71%), conductive disorders (0.12%), QT prolongation (0.07%), and ventricular arrhythmias (0.07%) [196]. In a multi-center study investigating the cardiotoxicities associated with ICI mono/combination therapy, AF was reported in three of the eight case studies [220]. ICI-associated arrhythmias in a 30-patient study reported incidences of AF, ventricular arrhythmia, conduction disorders, and Takotsubo syndrome at 30%, 27%, 17%, and 14% of patients, respectively [221]. In this cohort, 80% of the cardiovascular mortalities associated with conduction abnormalities were in patients with exposure to combination therapy [221]. Indeed, several case studies assessing incidences of major adverse cardiac events or myocarditis provide support for increased risk of arrhythmias in patients with combination therapy [198, 213, 214, 222].

While it is unknown whether arrhythmias are a primary IRAE of ICI exposure or secondary to myocarditis, the mechanisms for ICI-mediated arrhythmias includes altered PI3K/Akt/Ca^{2+} signaling, as previously discussed in TKI section, and direct inhibition of CTLA-4 receptors expressed on cardiomyocytes [223]. It is hypothesized that arrhythmias are triggered due to inhibition of cardiac CTLA-4, which interacts with the tyrosine phosphatase SHP-2 and serine/threonine phosphatase (PP2A) proteins responsible for the dephosphorylation and inactivation of Akt [203, 224, 225]. Therefore, direct inhibition of cardiomyocyte CTLA-4 by ICIs may also increase activation of Akt without changes to PI3K activation, contributing to cardiomyocyte dysfunction. In line with this, CTLA-$4^{+/-}$ plus PD-$1^{-/-}$ mouse model electrophysiological studies revealed arrhythmias [226].

Myocarditis and Pericarditis

Myocarditis and pericardial disease are the two most common cardiotoxicities associated with ICI therapy [193, 196, 227, 228], and patients with previous history of diabetes, cardiac disease, and autoimmune disease also have increased risk of autoimmune myocarditis [214, 229]. In clinical trials, the prevalence of myocarditis is highest with combination therapy (1.33%) followed by anti-PD-1/PDL1 monotherapy (0.32–41%) and lowest in anti-CTLA-4 (0.07–0.27%) [193, 196]. However, these numbers are underestimated due to lack of routine diagnostic imaging and cardiac monitoring, as well as exclusion of patients with underlying diseases [212]. ICI-associated myocarditis also has the highest reported mortality rate ranging from 39.7% [17] to 42.1% [230] within a median time of ~30 days from symptom onset to death (Table 2) [196]. Pericardial disease prevalence is lower than myocarditis (0.36% in combination therapy to 0.16% with anti-CTLA-4 alone) [196]. Despite differences in incidence rates, there are strong mechanistic overlaps centered on reduced self-tolerance, infiltration of activated T cells, and subsequent cytokine release [217, 227, 231, 232].

In ICI-mediated autoimmune myocarditis, several mechanisms have been demonstrated in preclinical and clinical studies. In preclinical studies, knockout of PD-1 in the autoimmune susceptible MRL mouse strain resulted in fatal myocarditis associated with increased auto antibodies against cardiac myosin, cardiac dilation, T-cell infiltration, and increased PD-L1 expression on cardiomyocytes [210, 212]. This finding is supported by poor tolerance of ICI therapy in patients with preexisting autoimmune diseases [212, 213]. While these patients were absent from immunotherapy clinical trials, ongoing trials (NCT03816345) will aid in further elucidating the safety of ICIs such as nivolumab in patients with a variety of autoimmune diseases [212]. In a model of $CD8^+$ T-cell-mediated myocarditis, Grabie and colleagues showed that myocardial PD-L1 is predominantly localized on the myocardial endothelium and is responsible for immune-mediated cardiac injury and leukocyte inflammation [233]. Furthermore, inhibition of PD-L1 in this model converted transient myocarditis to lethal disease.

Another study revealed that pressure overload disrupts cardiac immune cell tolerance allowing for cardiac self-antigens on MHCI receptors to induce T-cell cytotoxic immune responses [220, 227] and provides support for a mechanism in which massive cytokine production by activated T cells negatively impacts LV function [234]. Histopathology of myocardial biopsies and autopsies in ICI-exposed patients have confirmed increased cytotoxic $CD8^+$ T-cell infiltration [222] and macrophages [215]. Similarly, anti-mouse PD-1 inhibitor promoted M1 macrophage polarization and cardiac injury by modulating the microRNA-34a/Kruppel-like factor 4-signaling pathway and inducing myocardial inflammation [235]. The increased myocardial and pericardial infiltration of cytotoxic T cells was observed in multiple CTLA-4 and PD-L1 knockout mice models, which developed severe or fatal myocarditis [17, 217, 232], and is reinforced by similar findings in cynomolgus monkeys undergoing combination therapy with ipilimumab and nivolumab [236]. Furthermore, a shared antigen homolog between tumor cells and skeletal muscle tissue was found to be present in T-cell-infiltrated cardiomyocytes of patients with myocarditis diagnosis [213]. Laubli and colleagues supported this finding, demonstrating that the lineage of the cardiac infiltrating T cells are the same as tumor infiltrating cells [215]. Thus, there is a breakdown in cardiac and tumor immune tolerance due to CTLA-4 and PD-1 inhibition as well expansion of T cells targeting a mutually expressed antigen [227].

Additionally, immature but antigen-specific T cells play a role in the pathogenesis of myocarditis [237]. For instance, cardiac α-myosin heavy chain peptides can induce and activate T-cell clones that escape selection to destroy cardiac tissue (Fig. 3) [238]. Studies in retinoic acid receptor-related orphan nuclear receptor (RORγT)-deficient mice (*Rorc*$^{-/-}$) demonstrated resistance to myocarditis, implying that T helper 17 (Th17) cells are pathogenic in this model [239]. Notably, Th17 cell-induced myocarditis was mediated by IL-6 [240, 241]. A significant association between circulating IL-6 levels and development of HF was described in a study with 961 patients [241]. A study examining the tissue and serum of patients with myocarditis reported a critical role for IL-17, a cytokine produced by Th17 cells [240]. Here, Th17 cells production of IL-17A induced cardiac fibrosis, which was required for myocarditis to progress to dilated cardiomyopathy [240]. Furthermore, increased serum concentrations of IL-17 and Th17 have been reported with ICI therapy [242].

Systemic inflammation along with tissue damage was found at autopsy in a small sample of melanoma patients who were exposed to ICIs [243, 244]. Significantly elevated inflammatory markers like NF-kB (11.1% vs. 5%, $p < 0.005$) have been reported in cardiac tissue biopsies from patients with myocarditis [245]. NF-kB activation is an important regulator of proinflammatory cytokines IL-1 and IL-6 [246], and prolonged exposure of cardiomyocytes to NF-kB results in chronic inflammation and cardiac tissue damage.

Heart Failure

The incidence of HF with ICI therapy ranges from 0.72% [196] to 83% [221]. Mechanistically, ICI-related HF is likely multi-factorial, including increased circulation cardiac troponin (cTnI), along with increased myocardial levels of activated PI3K/Akt, NF-kB, IL-6 and T-cell infiltration, all of which are associated with cardiac remodeling and HF development [211, 216, 247]. Recent approval of TKI and ICI combination therapies (cabozantinib plus nivolumab, axitinib plus pembrolizumab, axitinib plus avelumab, lenvatinib plus pembrolizumab) also raises concern for synergistic cardiotoxic side effects. HF and sudden cardiac death were observed in PD-$1^{-/-}$ mice, suggesting that the immune inhibitors play a critical role in maintaining normal cardiac function [248]. In preclinical studies, BALB/c mice lacking PD-1 developed overt dilated cardiomyopathy, reduced cardiac function, and antibody deposition on cardiomyocytes [248]. Later, it was demonstrated that autoantibodies against cTnI are responsible for dilated cardiomyopathy in PD-1-deficient mice [211]. Furthermore, Okazaki et al. showed that administration of monoclonal cTnI antibodies promoted cardiac dilatation and dysfunction in wild-type mice and augmented the voltage-dependent L-type Ca^{2+} current. PI3Kγ deficiency has demonstrated cardio protection against isoproterenol-induced HF [249], while overactivation of PI3Kγ activates PI3Kα and induces cardiac hypertrophy and MI [125]. In addition, overexpression of Akt and mTOR in transgenic mice results in increased cardiomyocyte size, reduced cardiac output, hypertrophy, and ischemia [250]. In another study, moderate Akt and mTOR activity was cardioprotective, via reduction in apoptosis.

T-cell infiltration can also mediate ICI-related HF. In LV tissue isolated from nonischemic end-stage HF patients, a significant number of T cells can be found adhered to vascular ECs [218]. T-cell infiltration in the LV is also associated with

LV myocyte hypertrophy and fibrosis [218]. In T-cell depletion models (TCRα$^{-/-}$ and pharmacologically depleted T-cell mice), induction of LV pressure overload resulted in significantly reduced LV fibrosis, improved LV function, and reduced HF [218]. The authors concluded that LV recruitment of activated T cells in the setting of pressure overload contributes to myocardial dysfunction, remodeling, and ultimately HF [251]. Furthermore, higher numbers of circulating T cells and effector T-cell subsets have been observed in human and mice models with LV pressure overload [251]. There is evidence to suggest regulatory/effector T-cell imbalance as a marker of myocardial dysfunction and remodeling in HF. CD3$^+$ T cells isolated from patients with severe HF demonstrated increased adhesion to cardiac ECs, suggesting the mechanism of increased T-cell recruitment to the myocardium in HF [218]. Additionally, direct inhibition of CTLA-4 may also cause HF. In a case-control study with ~2000 patients, a CTLA-4 polymorphism resulted in reduced CTLA-4 function and significantly increased risk of HF [23]. Finally, one prevailing hypothesis for the incidence of HF with ICI exposure is exacerbations of preexisting cardiovascular dysfunction, genetic predisposition, chemotherapy, or radiation [252]. The above evidence suggests that ICI-associated HF is a result of multifactorial dysregulation of cardiac tissue environment.

Monitoring and Treatment of ICI-Associated Cardiotoxicity

Currently, elevated cardiac troponin has shown to be the most reliable marker for ICI-related myocarditis with a sensitivity of 94–97% and a high specificity. However, management challenges can occur in cases of subclinical myocarditis, where patients can have asymptomatic increases in cardiac troponin. As such, Bonaca et al. developed a new approach for diagnosing ICI-associated myocarditis, which incorporates cardiac MRIs and biopsies [253]. Recently, Stein-Merlob and colleagues also documented a proposed algorithm for treating ICI-myocarditis [254]. While echocardiograms are routinely employed to assess LV function in suspected cases of cardiotoxicity, decreased LVEF was reported in only 40–49% of the patients with myocarditis diagnosis [246]. Thus, a cardiac MRI along with close monitoring of cardiac troponin would better detect myocarditis [191]. Corticosteroids, including glucocorticoids and prednisone, are the conventional pharmacological agents prescribed to manage myocarditis in patients with elevated troponin and/or decreased LVEF [227, 234, 254–256]. While the use of immunosuppressive agents has shown limited success in managing either ICI-mediated myocarditis or pericardial disease [227, 256, 257], recently, the use of infliximab, anti-thymocyte globulin ATG, or intravenous immunoglobulin has shown some success with ICI-mediated HF management [227, 256, 257].

Conclusion and Future Directions

Immune checkpoint inhibitors have only been in clinical use for about a decade, and all long-term complications and organ toxicities remain unknown. However, life-threatening cardiac AEs must be addressed. ICI use promotes increased circulation and infiltration of effector T cells that are not limited to the tumors they treat. The evolving and resistant tumor microenvironment further increases the use of combination ICI therapy, which can be more cardiotoxic than ICI alone. As such, further examination of the mechanisms involved in provoking an ICI-associated and heightened autoimmune response toward the myocardium is necessary to combat these effects. Furthermore, clinical management strategies, such as preventive prophylaxis as well as cardiovascular monitoring in patients with ICI therapy, must be developed.

Part IV: CAR T-Cell Immunotherapy

Chimeric antigen receptor (CAR) T-cell therapy has revolutionized cancer treatment for highly clonal neoplasms such as lymphoma, leukemia, and myeloma (Table 3). Currently, there are four FDA-approved CAR T-cell therapies (tisagenlecleucel, idecabtagenevicleucel, axicabtagene ciloleucel, and brexucabtagene autoleucel) with over than a hundred more being developed in clinical trials [258, 259]. Tisagenlecleucel, axicabtagene ciloleucel, and brexucabtagene autoleucel target the CD19 antigen (Table 3), which is frequently overexpressed in a variety of B-cell malignancies. These three drugs have improved clinical outcomes for pediatric and adult patients with relapsed or refractory (r/r) B-cell acute lymphoblastic leukemias (B-ALL), subtypes of non-Hodgkin lymphoma, and mantle cell lymphoma [260]. Antigen recognition and activation by CAR T-cells promote the release of proinflammatory cytokines (Fig. 4) (interferon gamma (IFNγ); interleukin-1, interleukin-2, interleukin-6 (IL-1, IL-2, IL-6); and tumor necrosis factor alpha (TNFα)) to induce a cytotoxic response against cancer cells. These cytokines promote an immune response via T-cell proliferation and differentiation, along with the recruitment and activation of macrophages and monocytes [29, 261]. The robust immune response is responsible for targeted destruction of cancer cells by CAR T cell with unparalleled response rates ranging from 50% to 93% in the r/r setting [29, 262–264]. Despite their ability to reduce mortality, adverse cardiovascular events and cytokine release syndrome (CRS), the most prevalent side effect reported in 70% to 90% of patients with increasing severity [29, 260–264], are common. Continued and sustained release of these cytokines due to lack of regulation by costimulatory receptors leads to supraphysiologic levels of inflammatory cytokines and the development of CRS (Fig. 4) [261]. CRS may present in patients as mild to moderate (grades 1–2) or develop life-threatening symptoms (grades 3–4), and severe symptoms have been reported between 14% and 27% of patients [265–267].

While there is limited information on cardiotoxicity associated with CAR T-cell therapy, it is hypothesized that CAR T-cell-mediated cardiac AEs are associated

Table 3 Currently FDA-approved CAR T-cell therapies, their receptor target, and incidence of cardiotoxicity

Agent(s)	Target	CRS (%)[d]	Drug-specific cardiovascular toxicity	Reported incidence (%)	Reference(s)
Tisagenlecleucel (KYMRIAH)	CD19	45–79%	Hypotension[a] (RIS)[c] Hypertension[a] Tachycardia[a] Arrhythmia[a,b] LVSD Cardiac arrest[a] HF[a]	26–33% (9–17%) 2–19% 4–26% 6–30% 4.0% 4.0% 2.7–7%	[30, 262, 263]
Axicabtagene ciloleucel (YESCARTA)	CD19	84–96%	Hypotension (RIS)[c] Hypertension[a] Tachycardia[a] Arrhythmia[a,] Cardiomyopathy Cardiac arrest[a] HF[a]	38–59% (14%) 15% 39–57% 21–23% 2% 1–4% 2–6%	[29, 30, 267]
Brexucabtagene autoleucel (TECARTUS)	CD19	91%	Hypotension (RIS)[c] Hypertension[a] Tachycardia[a] Bradycardia[a] Arrhythmia[a] Cardiac arrest[a] HF[a] Palpitations[a] Prolonged QT[a]	51–57% (22%) 18% 31–63% 10% 10–15% N/A 4% 3% N/A	[30, 264, 267]
Idecabtagene vicleucel (ABECMA)	BCMA	76– 85%	Hypotension[a] Hypertension[a] Tachycardia[a] Atrial fibrillation[a] Cardiomyopathy[a]	17% 11% 19% 4.7% 1.6%	[267]
Lisocabtagene maraleucel (BREYANZI)	CD19	46%	Hypotension[a] Hypertension[a] Tachycardia[a] Arrhythmia[a] Cardiomyopathy[a] HF[a] Cardiac arrest[a]	22–26% 14% 16–25% <1–6% <1–1.5% N/A N/A	[267]

Abbreviations: BCMA B-cell maturation antigen; *CD19* B-lymphocyte antigen CD19; *CRS* cytokine release syndrome; *HF* heart failure; *LVSD* left ventricular systolic dysfunction; *N/A* not applicable; *RIS* required inotropic support

[a]Included in the Food and Drug Administration label (up to date as of March 2022)

[b]Arrhythmia includes atrial fibrillation, supraventricular tachycardia, ventricular extrasystoles

[c]Values in parenthesis represents the percentage of total patients in requiring inotropic support (RIS) due to diagnosis of hypotension or shock in clinical trials

[d]There is heterogeneity in the CRS grading system used in clinical trials. Tisagenlecleucel trials follow Penn Criteria of CRS grading, while the Lee Criteria for CRS grading was used for axicabtagene ciloleucel, brexucabtagene autoleucel, idecabtagene vicleucel, and lisocabtagene maraleucel

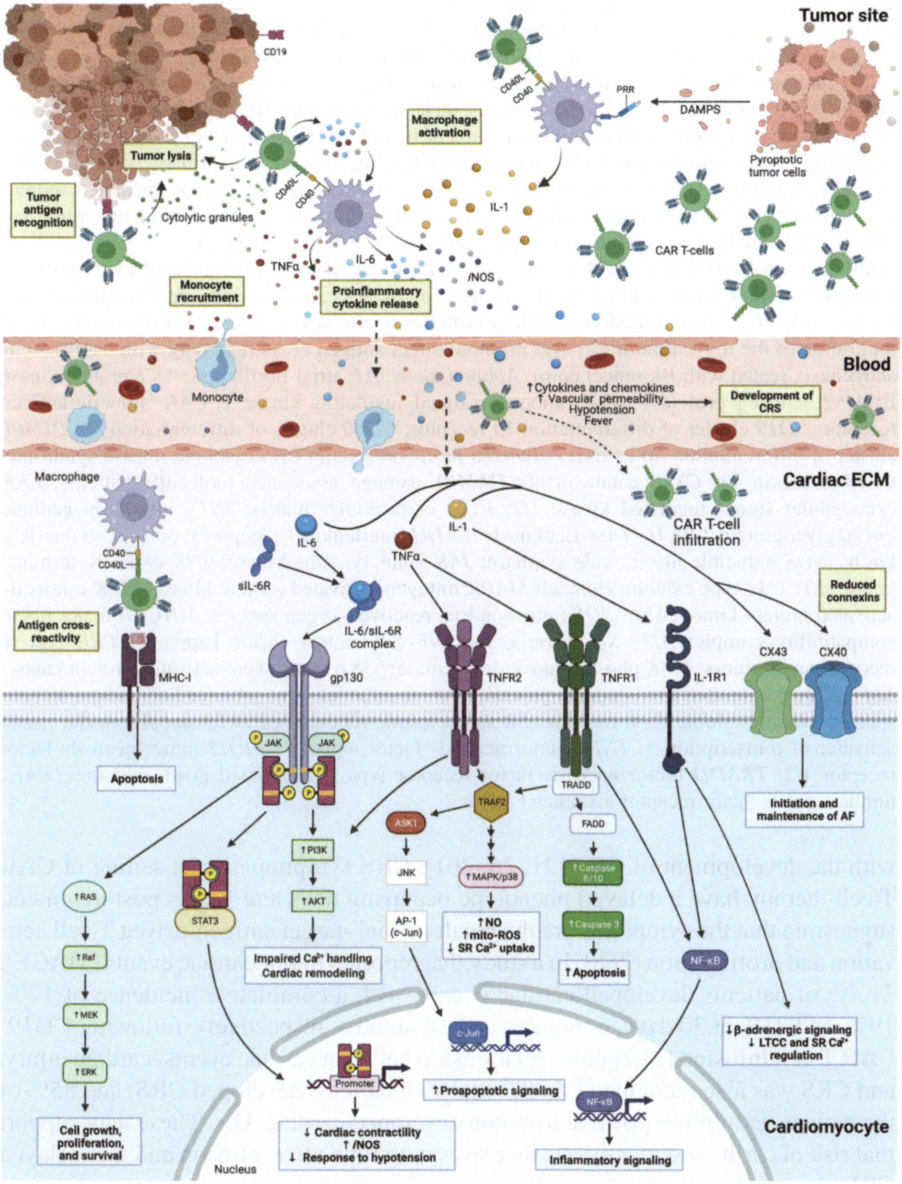

Fig. 4 Mechanisms and signaling pathways mediating the development of systemic CRS and CAR T-cell cardiotoxicity. At lymph node tumor sites, autologous CART cells encounter and recognize tumor antigens on CD19 positive cancer cells, leading activated T-cell release of cytolytic granules (i.e., granzymes and perforins) that promote tumor cell lysis and death. At the tumor site, macrophages can also interact with activated T cells via CD40L to CD40 binding. This binding promotes the release of INFγ, TNFα, GM-CSF, and activation of macrophages. Activated macrophages release proinflammatory cytokines (IL-1, IL-6, TNFα, INFγ, *i*NOS), which mediate increases monocyte recruitment from blood and into myocardial tissue. Unchecked

(continued)

Fig. 4. (continued) overstimulation of immune cells also precipitates supra-physiologic increases of cytokines and development of CRS. Elevated blood cytokines levels increase membrane permeability and vascular leak causing fever, hypotension, hypoxia, multi-organ toxicity and cardiac dysfunction including tachycardia, myocarditis, de novo arrythmias, HF, and sudden cardiac death. There is evidence to suggest that direct activation of IL-1, IL-6, and TNFα signaling in cardiomyocytes may also promote perturb Ca^{2+} homeostasis leading to impaired cardiac contractility and cardiac remodeling. Activation of TNFR1 and TNFR2 may precipitate caspase-mediated apoptosis, increased NO, increased mitoROS, and reduced SR Ca^{2+} uptake. IL-1 activation of cardiomyocyte IL-1R1signaling reduces ß-adrenergic signaling, LTCC $I_{Ca}{}^{2+}$, and SR Ca^{2+} cycling. Both cardiomyocyte IL-1R1 and TNFR2 activation can also promote NFkB-mediated increases transcription of inflammation genes. Finally, increased cytokines can also elicit a decrease in connexin protein (CX43/CX40) expression, which contributes to the development of arrhythmias. T-cell infiltration of the myocardium can also promote direct antigen cross-reactivity with cardiac self-antigens. (Created with Biorender.com). *Abbreviations: AF* atrial fibrillation; *AKT* protein kinase B; *AP-1* activator protein 1; *ASK1* apoptosis signal–regulating kinase 1; *CAR* chimeric antigen receptor; *CD19* cluster of differentiation 19 receptor; *CD40* cluster of differentiation 40; *CD40L* cluster of differentiation 40 ligand; *c–Jun* Jun proto–oncogene; *CRS* cytokine release syndrome; *CX43* connexin 43; *CX40* connexin 40; *DAMPS* damage–associated molecular patterns; *ERK* extracellular signal–regulated kinase 1/2; *ECM* extracellular matrix; *INFγ* interferon gamma; *gp130* glycoprotein 130; *IL–1* interleukin–1; *IL–1R1* interleukin 1 receptor type 1; *IL–6* interleukin 6; *iNOS* inducible nitric oxide synthase; *JAK* janus tyrosine kinase; *JNK* c–jun N–terminal kinase; *LTCC* L–type calcium channel; *MAPK* mitogen–activated protein kinase; *MEK* mitogen–activated protein kinase; *Mito-ROS* mitochondrial reactive oxygen species; *MHCI/II* major histocompatibility complex 1/2; *NO* nitric oxide; *NF–kB* nuclear factor kappa B; *PRR* pattern recognition receptors; *PI3K* phosphoinositide 3–kinase; *p38* p38 mitogen–activated protein kinase; *Raf* rapidly accelerated fibrosarcoma protein; *RAS* rat sarcoma virus protein; *ROS* radical oxygen species; *sIL–6R* soluble IL–6 receptor; *SR* sarcoplasmic reticulum; *STAT3* signal transducer and activator of transcription 3; *TNFa* tumor necrosis factor alpha; *TNFR1/2* tumor necrosis factor receptor 1/2; *TRADD* tumor necrosis factor receptor type 1–associated death domain; *TRAF2* tumor necrosis factor receptor associated factor 2

with the development of CRS [21, 26, 261]. CRS symptoms in the setting of CAR T-cell therapy have a delayed phenotype occurring days and weeks post treatment, suggesting that the symptoms are the result of "on"-target antigen-driven T-cell activation and proliferation [268]. In a study that reported major cardiac events (MACE), 21.4% of patients developed cardiac events, with a cumulative incidence of 17%, 19%, and 21% at 30 days, 6 months, and 12 months, respectively, following CD19-CAR T-cell infusion [2]. A close relationship between cardiac events, cardiac injury, and CRS was also recognized. In the study, all patients developed CRS, and 55% of the patients exhibiting positive troponin developed cardiac AEs. These data support that risk of cardiovascular events increases with time after infusion and that delayed CRS phenotypes are due to a culmination of immune system response and activation. In clinical trials, CRS-mediated tachycardia and fever with associated fever, hypotension, hypoxia, and cardiac events have been reported in both pediatric and adult patients [29, 262–264]. Troponin elevation, hypotension, arrhythmia, ST segment alteration, reduced LVEF, decompensated heart failure, cardiogenic shock requiring vasopressor inotropic support, and sudden cardiac death (SCD) have also been recognized in patients from CAR T-cell infusion [29, 262–265].

Along with time-dependent symptoms, reactions to CAR T-cell therapy vary with age. In adult patients, hypotension is present at higher rates in comparison to pediatric and young patients. In a meta-analysis of three clinical trials, hypotension ranged from 26% to 59%, and hypotension or shock requiring vasopressor support occurred in 9–29% of patients [29, 262–264]. Interestingly, cardiotoxic complications associated with CAR T-cell therapy in pediatric and young adult patients appear to be self-limited [269, 270], as most patients return to pretreatment baseline, even after cardiac arrest. In contrast, in adult patients that developed cardiotoxicity while receiving CAR T-cell therapy, resolution did not always follow cardiac events, and in some cases cardiotoxicities proved fatal. In one study, where 83% of patients developed CRS (grade ≥2), serial echocardiograms revealed that 10.3% of patients developed de novo or progressive cardiomyopathy within 30 days following CAR T-cell infusion [271]. In this cohort, LVEF improved in 75%, while 25% succumbed [271]. Other less frequent cardiovascular events reported in younger patients include left ventricular dysfunction (4%), cardiac failure (2.7%), and cardiac arrest (4%) [263]. In a retrospective study assessing cardiovascular outcomes in patients treated with axicabtagene ciloleucel, tisagenlecleucel, and an investigational CD19 CAR T-cell therapy, 12% of the patients developed cardiovascular events [21]. Six SCDs, six accounts of acute HF, and five occurrences of *new-onset* arrhythmias were also reported [21].

Overall, the major risk factor for cardiovascular dysfunction in CAR T-cell therapy appears to be hemodynamic stress due to CRS. As such, most clinical management strategies aimed at ameliorating the cardiac toxicities induced by CAR T-cell therapy are focused on reducing CRS [26, 30, 271]. However, CRS alone cannot explain all the cardiac AEs observed in patients taking CAR T-cell therapy. Further, there is a knowledge gap in defining the direct effects of CAR T-cell therapy on individual components of the cardiovascular system (CVS). Previous treatment with anthracycline-containing treatment regimens, allogeneic stem cell transplantation, and irradiation therapy also increase the risk of cardiac disease, despite exclusion criteria for preexisting or recent cardiovascular events, in clinical trials. Therefore, it is vital to understand the mechanisms by which CAR T-cell therapy mediates reversible and irreversible immune-related cardiac AEs.

CAR T-Cell Mechanism of Action

Under normal physiologic conditions, the presentation of antigens to the T cells via cell major histocompatibility complex (MHC)/human leukocyte antigen (HLA) proteins promotes the recognition of foreign or cancer cells. However, presentation and recognition of the tumor antigen is not enough to activate T cells. Full activation of T cells requires co-stimulation with B7 molecules on the antigen-presenting cell to CD28 receptors on T cells. Cancer cells evade this essential checkpoint by activating immune checkpoint receptors such as PD-1 and CTLA-4, which prevents the co-stimulation and full activation of T cells. As such, CAR T-cell therapy focuses on increased recognition of malignant cells by T cells. CAR T-cell therapy overcomes cancer-mediated T-cell inhibition by generating non-MHC/HLA-dependent receptors bypassing the requirement of co-stimulation for activation, thereby avoiding immune checkpoint receptors highly expressed by cancer cells.

The chimeric antigen receptor utilized in CAR T-cell therapy consists of extracellular, transmembrane, and intracellular domains. The extracellular domain contains a single-chain variable fragment created from variable heavy and light chains of an antibody that targets a non-HLA-dependent specific tumor antigen. The intracellular domain fuses with T-cell activation signaling domains, leading to T-cell activation. Uptake of the CAR by extracted patient T cells occurs through lentiviral or retroviral vectors. After expansion, the CAR T cells are infused back into the patient. In the bloodstream, engagement of CAR with its cognate antigen (e.g., CD19) initiates the activation of CAR T cells via the intracellular domain, followed by cytolytic/apoptotic destruction of malignant tumors cells expressing the antigen. Below we will describe the current understanding of both direct and indirect "off"-target effects of CAR T-cell therapy in the CVS.

Proposed Mechanisms of Cardiotoxicity in CAR T-Cell Therapy

While leveraging the immune system to target previously unrecognized cancer cells is an exciting new paradigm, widespread inflammation can promote multi-organ insults such as neurotoxicity and CVS complications. The clinical constellation of CAR T-cell therapy-associated cardiotoxicity can mirror sepsis, with hypotension, tachycardia, and decreased LVEF due to CRS. Most patients who respond to CAR T-cell therapy develop some degree of CRS. Therefore, it is important to address the signaling pathways of CRS-associated cytokines to elucidate potential mechanisms of cardiotoxicity, even though CRS-mediated cardiac AEs are unlikely mediated by an individual cytokine.

JAK/STAT Signaling Pathway

IFNγ and IL-6 propagate signaling via the Janus kinase (JAK)/signal transducers and activators of transcription (STAT) pathway. In this pathway, cytokine receptor binding and activation of JAK lead to receptor phosphorylation and recruitment of STAT transcription factors. STAT translocation to the nucleus upregulates the transcription of cytokine-responsive genes. STAT5α, STAT3, and STAT6 have a cardioprotective role in animal models. STAT 6 and STAT5a were selectively activated in rat hearts in response to ischemia/reperfusion [272], while cardiomyocyte-specific STAT3 knockout mice spontaneously developed heart dysfunction with advancing age. In addition, JAK/STATs mediate angiotensin II signaling, which promotes pathological cardiovascular conditions (Fig. 4). Based on the role of these studies, it is believed that uncontrolled JAK/STAT signaling due to activation by cytokine signaling may mediate CAR T-cell-CRS-induced cardiotoxicity.

IL-6

Levels of interleukin-6, a pleiotropic proinflammatory cytokine, rapidly increase following initiation of CAR T-cell therapy and are involved in the development of CRS via T-cell recruitment, chemokine release, and inflammation [26, 273]. These chemokines, coupled with inflammation, can trigger widespread increases in vascular permeability and capillary leaks, which contribute to tissue hypoxia, hypotension, and shock (Fig. 4). IL-6 signals through the interleukin-6 receptor (IL-6R) and the IL-6 family common receptor gp130 [274]. Though evidence supports that membrane-bound IL-6R is not expressed on cardiomyocytes, IL-6 signaling pathways are activated in cardiomyocytes when IL-6 to soluble IL-6R (sIL-6R) complex binds and/or interacts with gp130 (Fig. 4) [275]. Glycoprotein 130 (gP130) signaling regulates cell survival, apoptosis, growth, proliferation, and differentiation. Downstream IL-6/IL-6R interacts with gp130 and signals through either JAK/STAT (specifically STAT3), Ras/Raf/MEK/ERK, or PI3K/Akt pathways.

IL-6 pleiotropy underlies its function as a cardioprotector and a cardiotoxic cytokine. Acute elevation of IL-6 has cardioprotective effects, however, when chronically elevated, IL-6 induces maladaptive hypertrophy, fibrosis, and decreased contractile function, leading to MI and HF [28, 271, 274]. Cardiomyocytes generate IL-6 in response to ischemia-reperfusion injury and myocardial infarction. Here, IL-6 acts to depress cardiac function by decreasing basal contractility by decreasing β-adrenergic responsiveness [274, 276]. This IL-6-induced reduction in cardiac contractility is driven by IL-6/JAK/STAT3 signaling [274, 277]. Furthermore, an IL-6-mediated decrease in β-adrenergic responsiveness dampens the physiological response to combat hypotension, further promoting progression to shock (Fig. 4). Yu et al. demonstrated that activation of IL-6/JAK/STAT3 signaling promotes de novo synthesis and activation of Ca^{2+}-independent inducible nitric oxide synthase (*i*NOS) proteins (Fig. 4) [277]. β-adrenergic signaling is essential for cardiac contractility, and Ca^{2+} handling and IL-6/JAK/STAT3-induced reduction in cardiac contractility, which along with the production of vasodilatory NO may lead to the systemic hypotension observed in CAR T-cell therapy [274, 277]. In pediatric and young adult patients, hypotension is the primary cardiovascular event reported. In the pivotal phase 2 trial with tisagenlecleucel, 29% of pediatric patients developed hypotension [266]. Other studies with pediatric and young adult patients found that 18–38% of patients will develop grade 3–4 hypotension, of which 13–27% of required vasopressor support and ~33% of patients required admission to the intensive care unit [263, 266, 267, 269, 270, 278]. In a retrospective study, Fitzgerald et al. revealed that patients who developed cardiovascular toxicity developed fluid-refractory shock that necessitated the use of alpha agonists [279].

In CAR T-cell-driven CRS, chronically elevated and long-term exposure of myocardium to IL-6 signaling could contribute to pathology, remodeling, and loss of cardiac function. Supporting this, continuous IL-6 signaling and increased expression of IL-6 have been observed in the left ventricles of patients with dilated cardiomyopathy (DCM) [280–282], cardiogenic shock [283], and end-stage HF [284]. In rodent models, increased and continuous expression of STAT3, the downstream target of IL-6, resulted in cardiac hypertrophy independent of other cardiac insults [274, 285]. In addition, chronic IL-6 expression can cause myocardial stunning and mechanical dysfunction leading to myocarditis [26, 274]. To date, myocarditis provides the most striking pathologic example of IL-6 in the heart. In myocarditis, IL-6 mediates autoimmune myocarditis which may lead to DCM and HF [286]. When presented concomitantly with other well-known cardiac risk factors, IL-6 may also be a contributing risk factor for QT prolongation and the development of *torsade de pointes* [287]. Together these data suggest that continuous or long-term myocardial exposure to IL-6 signaling contributes to pathology, loss of cardiac function, and remodeling in the setting of CAR T-driven CRS.

TNFα

TNFα is a proinflammatory cytokine with a broad range of homeostatic and pathophysiological processes. Once thought to be mostly produced by immune cells, TNFα is also expressed on endothelial, epithelial, smooth muscle cells, and cardiomyocytes. TNFα pathology may contribute to vascular dysfunction, atherosclerosis development and progression, and adverse cardiovascular remodeling following myocardial infarction and HF [288]. Secreted TNFα protein activates one of two receptors, tumor necrosis factor receptor 1 (TNFR1), which is ubiquitously expressed, and tumor necrosis factor receptor 2 (TNFR2), found mainly on immune and cardiac cells [255, 289]. TNFR1 activation leads to NF-kB signaling, MAPK/C--Jun-signaling, and the caspase apoptotic pathway. TNFR2 activation results in activation of the PI3k/Akt pathway and proangiogenic pathways in addition to slower NF-kB signaling in comparison to TNFR1 [255]. TNFα impairs NO formation and bioavailability through TNFR1 signaling and stimulates ROS production via superoxide generation (Fig. 4) [288].

In healthy patients, cardiac TNFα levels are very low, with the presence of protein and transcripts restricted to the microvessels of the heart [290]. In vivo administration of exogenous TNFα in rat and dog hearts leads to cardio-depressant effects. TNFα also led to a dose-dependent reduction of Ca^{2+} uptake by the sarcoplasmic reticulum (SR) mediated by p38 MAPK signaling (Fig. 4) [291–294]. Suppression of intracellular Ca^{2+} handling and Ca^{2+} leak from the SR increases caspase 8 activity, which increases myocardial NO and mitochondrial ROS production culminating in cardiac injury [295]. Cardiac myocytes have also been shown to undergo apoptosis after TNFα stimulation in vitro [296]. As such, cardiac exposure to TNFα signaling may contribute to pathology and loss of cardiac function observed in patients taking CAR T-cell therapy.

Interleukin 1

Interleukin-1 (IL-1) is a multi-functional proinflammatory cytokine with a well-defined role in inflammatory and reparative responses, as well as cardiovascular diseases such as atherosclerosis, acute myocardial infarction, and HF [297]. IL-1 signal transduction leads to activation of the NF-kB system, initiating a wide variety of inflammatory cytokines, chemokines, adhesion molecules, colony-stimulating factors, and mesenchymal growth factor genes [298]. In addition, type II cyclooxygenase, *i*NOS, and phospholipase A2 are highly sensitive to IL-1 and mediate some of IL-1's signaling through prostaglandins and nitric oxide. Cumulatively, these inflammatory mediators increase vascular smooth muscle relaxation and can lead to a widespread increase in vascular permeability and decreased vascular resistance, thereby causing tissue hypoxia and shock [298, 299]. There is mitigation of IL-1 signaling in normal physiology via interleukin-1 receptor agonist type II, which acts as a decoy receptor binding excess IL-1 and preventing further signal transduction [300, 301]. However, in a situation where supraphysiologic levels of IL-1 are present and sustained, regulatory control cannot compensate leading to pathology.

Experimental investigations have also demonstrated that IL-1 may enhance cardiomyocyte apoptosis in the ischemic myocardium, enhance the post-infarction inflammatory response, and mediate adverse cardiovascular remodeling through altering matrix metalloproteinase expression post-MI [298]. IL-1 impairs downstream β-adrenergic receptor signaling and reduces both L-type calcium channels (LTCC) and SR cytoplasmic Ca^{2+} handling (Fig. 4) [297]. This dysregulation of Ca^{2+} homeostasis has been proposed in the development of AF in response to CAR T-cell treatments [265]. Downregulation of connexins 43 and 40 by IL-6, in conjunction with inflammatory signaling from ILs and TNFα, has been associated with the initiation and maintenance of AF (Fig. 4). Changes in function and expression of Ca^{2+} and K^+ channels due to cytokines can lead to prolonged action potential duration, predisposing patients to arrhythmias [265, 302, 303].

Off-Target Affects

Though CAR T-cells are modified with enhanced affinity against a tumor-specific antigen, safety concerns arise due to potential cross-reactivity with unrelated peptides expressed by normal tissue. To date, the only documentation of direct interaction of CAR T-cells in cardiac pathophysiology was demonstrated by Linette et al. who showed CAR T-cell cross-reactivity with titin, a striated muscle protein in the heart [27]. In the study, two patients developed fever and progressive cardiogenic shock with death after receiving CAR T-cells targeting MAGE-A3 [27]. The cardiogenic shock was attributed to autoimmunity due to "off"-target cross-reactivity with titin presented on HLA-A*01 protein [27]. This tragic event calls for more accurate antigen targeting of CAR T-cell therapies to minimize future occurrences. During stages of drug development, strategies to assess and mitigate the risk of "off"-target

CAR T-cell-mediated cardiotoxicity should be identified a priori using peptide scanning and complex cardiomyocyte cell culture.

Monitoring of CAR T-Cell Associated Cardiotoxicity

With the recognition of CRS-associated cardiotoxicity in CAR T-cell therapy, a standardized pretreatment screening protocol and cardiotoxicity treatment guideline are required. Currently, there are no formally accepted guidelines for the management of cardiotoxicity induced by CAR T-cell therapy, and cardiac disease is not an absolute contraindication to proceed with treatment. For now, only one site-specific institutional guideline has been established/proposed [304]. As such, current treatments center on supportive care for hemodynamic instability and CRS [26, 30, 271]. Pretreatment cardiovascular considerations in patients receiving CAR T-cell therapy should include assessment of CV risk factors before treatment initiation [30], with special considerations for patients who have had prior exposure to cardiotoxic regimens with recognized cardiotoxic profile comorbidities (i.e., anthracyclines, allogeneic stem cell transplantation, irradiation) [26]. Patient inclusion criteria for multiple CAR T-cell clinical trials have required normal LVEF as well as the absence of cardiac arrhythmias and myocardial infarction [305]. However, the incidence of underlying CVS disease is prevalent in real-world populations, and baseline cardiac evaluation for approved CAR T-cell therapies varies between institutions.

Clinical surveillance during and after CAR T-cell Infusion should center on monitoring for CRS. Due to the strong relationship between CRS and CAR T-cell therapy-related cardiotoxicity, clinical recognition and mitigation of CRS are essential to minimizing adverse cardiovascular outcomes. In the absence of a standardized CRS grading schema, however, this can differ from center to center. High-level CRS clinical surveillance during and after CAR T-cell infusion is necessary to evaluate hypotension. Cardiac monitoring recommendations should include routine blood pressures, 12-lead ECGs, and cardiac biomarker assessments (e.g., troponin and brain natriuretic peptide) in patients with clinical evidence of CRS. Though there are limited data detailing cardiac biomarker assessments for CAR T-cell therapy, elevated troponin levels have been associated with an increased risk of developing cardiovascular AEs [279]. As such, cardiac troponin elevation should be considered and assessed particularly in those that have known preexisting cardiovascular disease. Finally, when hypotension or tachycardia is present, volume resuscitation and monitoring for vascular leak and pulmonary edema should be started. For hemodynamically unstable patients, transfer to the intensive care unit is recommended.

Current and Proposed Treatments for CAR T-Cell-Induced Cardiotoxicity

In response to elevated IL-6 levels resulting from severe CRS, tocilizumab, a first-line treatment of CRS-associated toxicity has been used to prevent IL-6 signal transduction and inflammation. Tocilizumab is a monoclonal antibody that competitively inhibits the binding of IL-6 to its receptor IL-6R, blocking both classical and trans-IL-6 signaling via direct binding to both membrane bound and soluble forms of IL-6R. However, tocilizumab may reduce the efficacy of CAR T-cell therapy by blocking cell signaling transduction [31]. While current research suggests that tocilizumab is an effective means of preventing CRS progression, more investigation is needed to determine the precise severity of CRS that warrants the use of tocilizumab [266, 278, 306]. Siltuximab is a monoclonal antibody that directly binds to IL-6, preventing the activation of immune effector cells without disabling the overall function of IL-6R [31, 307] and may serve as a potential alternative to tocilizumab. Though not FDA-approved for use in CRS induced by CAR T-cell therapy, siltuximab should be considered due to its higher affinity to IL-6 and as an alternative to patients not responding to tocilizumab. For patients with severe CRS and refractory to other interventions, corticosteroid therapy is recommended as a second-line agent [308, 309]. Corticosteroid therapy reduces systemic inflammation and can be administered in conjunction with tocilizumab for severe CRS [31, 307, 310]. While administration of systemic corticosteroids does not impart a detrimental impact on the efficacy of the CAR T-cell therapy, its use to treat or prevent adverse cardiac events is poorly understood. Further, the administration of steroids in a study of pediatric patients did not reduce the risk of developing cardiac dysfunction [269]. As such, more investigation is required to justify the use of corticosteroids in CAR T-cell-CRS-mediated cardiotoxicity.

Anti-IL-1 therapy represents another avenue for treating CRS-mediated CAR T-cell cardiac AEs. In phase II and III clinical trials, IL-1 blockade prevented recurrent atherothrombotic cardiovascular events, improved exercise capacity in HF patients, and prevention of HF following MI [297]. Anakinra, an FDA-approved anti-IL-1 treatment for rheumatoid arthritis, has indicated potential therapeutic use in preclinical models of CRS [311]. In two animal model studies, IL-1 blockade prevented CRS, thereby reducing mortality [312, 313]. An ongoing phase II clinical trial (NCT04359784) continues to explore the role of anakinra in patients. In a case report, it was demonstrated that tocilizumab and anakinra use in a patient treated with anti-B-cell maturation antigen CAR T-cell therapy improved patient symptoms [311]. In the report, one dose of tocilizumab was given followed by a 10-day taper of anakinra [311]. Significant improvement of the patient's symptoms shortly occurred after initiation of anakinra. These data support the possibility that IL-1 blockage can decrease the duration and severity of the CRS and potentially prevent cardiac dysfunction [311].

Another possible therapy of CAR T-cell-induced CRS is anti-TNFα monoclonal antibodies (e.g., infliximab) or etanercept, a soluble TNFα receptor sequester. While there are limited reports of the TNFα inhibitor therapy for CAR T-cell-induced CRS, a few case studies provide hints for its use [267, 314–316]. A case of grade 3 CRS in a patient treated with CAR T-cell therapy for Epstein-Barr virus-associated lymphoma was mitigated within hours after receiving treatment with etanercept in conjunction with methylprednisolone, dopamine, and norepinephrine [267]. In addition, three patients treated with etanercept showed excellent clinical response after being treated for multiple myeloma [314]. In patients being treated for pre-B-cell ALL treatment consisting of etanercept, tocilizumab, and corticosteroids ablated CRS [315]. Therefore, as clinical trials for CAR T-cell therapy in multiple myeloma and pre-B-cell ALL increase, the role of TNFα should be further explored.

Conclusion and Future Directions

As therapeutic applications grow for CD19 CAR T-cell therapy, our understanding of the cardiotoxic events associated with treatment needs to be expanded. Toxicity is a major barrier to new CAR T-cell therapy use, and as FDA-approved therapies include more patients, risk stratification, recognition of CRS, and management of cardiotoxic outcomes are needed to reduce patient morbidity and mortality associated with their use. There are minimal data available on the molecular mechanisms, identification, and management of cardiotoxicity with CAR T-cell therapy. Therefore, cardio-oncology researchers must seek to address these gaps. For now, selective assessment of cardiovascular clinical symptoms, cardiac biomarkers, and imaging-based indices of cardiac function can enhance our understanding of cardiotoxicity during and after CAR T-cell therapy. Though, Stein-Merlob and colleagues recently documented a proposed algorithm for treating CAR T-cell therapy-associated cardiotoxicity [254], multi-institutional collaboration is needed for data collection in this patient group toward implementing evidence-based practice guidelines that optimize patient outcomes and reduce CAR T-cell therapy-associated cardiotoxicity.

Disclosure Statement The authors have no relevant affiliations or financial involvement with any organization or entity with a financial interest in or financial conflict with the subject matter or materials discussed in the manuscript. This includes employment, consultancies, honoraria, stock ownership or options, expert testimony, grants, or patents received or pending, or royalties.

References

1. Siegel RL, Miller KD, Fuchs HE, Jemal A. Cancer statistics, 2021. CA A Cancer J Clin. 2021;71:7–33.
2. Lefebvre B, Kang Y, Smith AM, Frey NV, Carver JR, Scherrer-Crosbie M. Cardiovascular effects of CAR T cell therapy. JACC: CardioOncology. 2020;2:193–203.
3. Lui JE. CV complications of CAR T-cell therapy - American College of Cardiology. Cell Therapy. 2020;6
4. Sasaki K, Strom SS, O'Brien S, et al. Relative survival in patients with chronic-phase chronic myeloid leukaemia in the tyrosine-kinase inhibitor era: analysis of patient data from six prospective clinical trials. Lancet Haematol. 2015;2:e186–93.

5. Dobbin SJH, Petrie MC, Myles RC, Touyz RM, Lang NN. Cardiotoxic effects of angiogenesis inhibitors. Clin Sci. 2021;135:71–100.
6. Sternberg CN, Davis ID, Mardiak J, et al. Pazopanib in locally advanced or metastatic renal cell carcinoma: results of a randomized phase III trial. JCO. 2010;28:1061–8.
7. Robert C. A decade of immune-checkpoint inhibitors in cancer therapy. Nat Commun. 2020;11:3801.
8. June CH, O'Connor RS, Kawalekar OU, Ghassemi S, Milone MC. CAR T cell immunotherapy for human cancer. Science. 2018;359:1361–5.
9. Pinkhas D, Ho T, Smith S. Assessment of pazopanib-related hypertension, cardiac dysfunction and identification of clinical risk factors for their development. Cardio-Oncology. 2017;3:5.
10. Assuncao BMBL, Handschumacher MD, Brunner AM, Yucel E, Bartko PE, Cheng K-H, Campos O, Fathi AT, Tan TC, Scherrer-Crosbie M. Acute leukemia is associated with cardiac alterations before chemotherapy. J Am Soc Echocardiogr. 2017;30:1111–8.
11. Justice CN, Derbala MH, Baich TM, Kempton AN, Guo AS, Ho TH, Smith SA. The impact of Pazopanib on the cardiovascular system. J Cardiovasc Pharmacol Ther. 2018;23:387–98.
12. Kappers MHW, van Esch JHM, Sluiter W, Sleijfer S, Danser AHJ, van den Meiracker AH. Hypertension induced by the tyrosine kinase inhibitor sunitinib is associated with increased circulating Endothelin-1 levels. Hypertension. 2010;56:675–81.
13. Shah DR, Shah RR, Morganroth J. Tyrosine kinase inhibitors: their on-target toxicities as potential indicators of efficacy. Drug Saf. 2013;36:413–26.
14. Ma W, Liu M, Liang F, Zhao L, Gao C, Jiang X, Zhang X, Zhan H, Hu H, Zhao Z. Cardiotoxicity of sorafenib is mediated through elevation of ROS level and CaMKII activity and dysregulation of calcium homoeostasis. Basic Clin Pharmacol Toxicol. 2020;126:166–80.
15. Kerkela R, Woulfe KC, Durand J-B, Vagnozzi R, Kramer D, Chu TF, Beahm C, Chen MH, Force T. Sunitinib-induced cardiotoxicity is mediated by off-target inhibition of AMP-activated protein kinase. Clin Transl Sci. 2009;2:15–25.
16. Ren C, Sun K, Zhang Y, Hu Y, Hu B, Zhao J, He Z, Ding R, Wang W, Liang C. Sodium–Glucose CoTransporter-2 inhibitor Empagliflozin Ameliorates sunitinib-induced cardiac dysfunction via regulation of AMPK–mTOR signaling pathway–mediated autophagy. Front Pharmacol. 2021;12:664181.
17. Wang DY, Salem J-E, Cohen JV, et al. Fatal toxic effects associated with immune checkpoint inhibitors: a systematic review and meta-analysis. JAMA Oncol. 2018;4:1721.
18. Marino F, Scalise M, Cianflone E, et al. Role of c-Kit in myocardial regeneration and aging. Front Endocrinol. 2019;10:371.
19. Palaskas N, Lopez-Mattei J, Durand JB, Iliescu C, Deswal A. Immune checkpoint inhibitor myocarditis: pathophysiological characteristics, diagnosis, and treatment. JAHA. 2020; https://doi.org/10.1161/JAHA.119.013757.
20. Buchbinder EI, Desai A. CTLA-4 and PD-1 pathways: similarities, differences, and implications of their inhibition. Am J Clin Oncol. 2016;39:98–106.
21. Alvi RM, Frigault MJ, Fradley MG, et al. Cardiovascular events among adults treated with chimeric antigen receptor T-cells (CAR-T). J Am Coll Cardiol. 2019;74:3099–108.
22. Zhou Y-W, Zhu Y-J, Wang M-N, Xie Y, Chen C-Y, Zhang T, Xia F, Ding Z-Y, Liu J-Y. Immune checkpoint inhibitor-associated cardiotoxicity: current understanding on its mechanism, diagnosis and management. Front Pharmacol. 2019;10:1350.
23. Zhao Y, Gu Z, Zhao X, Wang L, Gao M. Association of *CTLA4* and *NFKB1* polymorphisms with chronic heart failure: a case-control study in the Chinese population. Eur J Inflamm. 2019;17:205873921985285.
24. Sharpe AH, Pauken KE. The diverse functions of the PD1 inhibitory pathway. Nat Rev Immunol. 2018;18:153–67.
25. Tay WT, Fang Y-H, Beh ST, Liu Y-W, Hsu L-W, Yen C-J, Liu P-Y. Programmed cell death-1: programmed cell death-ligand 1 interaction protects human cardiomyocytes against T-cell mediated inflammation and apoptosis response in vitro. IJMS. 2020;21:2399.

26. Stein-Merlob AF, Rothberg MV, Holman P, Yang EH. Immunotherapy-associated cardiotoxicity of immune checkpoint inhibitors and chimeric antigen receptor T cell therapy: diagnostic and management challenges and strategies. Curr Cardiol Rep. 2021;23:11.
27. Linette GP, Stadtmauer EA, Maus MV, et al. Cardiovascular toxicity and titin cross-reactivity of affinity-enhanced T cells in myeloma and melanoma. Blood. 2013;122:863–71.
28. Pathan N, Hemingway CA, Alizadeh AA, et al. Role of interleukin 6 in myocardial dysfunction of meningococcal septic shock. The Lancet. 2004;363:203–9.
29. Neelapu SS, Locke FL, Bartlett NL, et al. Axicabtagene ciloleucel CAR T-cell therapy in refractory large B-cell lymphoma. N Engl J Med. 2017;377:2531–44.
30. Burns EA, Gentille C, Trachtenberg B, Pingali SR, Anand K. Cardiotoxicity associated with anti-CD19 chimeric antigen receptor T-cell (CAR-T) therapy: recognition, risk factors, and management. Diseases. 2021;9:20.
31. Murthy H, Iqbal M, Chavez JC, Kharfan-Dabaja MA. Cytokine release syndrome: current perspectives. ITT. 2019;8:43–52.
32. Riegler LL, Jones GP, Lee DW. Current approaches in the grading and management of cytokine release syndrome after chimeric antigen receptor T-cell therapy. TCRM. 2019;15:323–35.
33. Kalay N, Basar E, Ozdogru I, et al. Protective effects of carvedilol against Anthracycline-induced cardiomyopathy. J Am Coll Cardiol. 2006;48:2258–62.
34. Cardinale D, Colombo A, Lamantia G, Colombo N, Civelli M, De Giacomi G, Rubino M, Veglia F, Fiorentini C, Cipolla CM. Anthracycline-induced cardiomyopathy. J Am Coll Cardiol. 2010;55:213–20.
35. Enoma E, Wei L, Chen H. The impact of statin on chemotherapy-induced cardiotoxicity. FASEB J. 2019; https://doi.org/10.1096/fasebj.2019.33.1_supplement.833.12.
36. Mulas O, Caocci G, Stagno F, et al. Renin angiotensin system inhibitors reduce the incidence of arterial thrombotic events in patients with hypertension and chronic myeloid leukemia treated with second- or third-generation tyrosine kinase inhibitors. Ann Hematol. 2020;99:1525–30.
37. Chen MH, Kerkelä R, Force T. Mechanisms of cardiac dysfunction associated with tyrosine kinase inhibitor CANCER THERAPEUTICS. Circulation. 2008;118:84–95.
38. Du Z, Lovly CM. Mechanisms of receptor tyrosine kinase activation in cancer. Mol Cancer. 2018;17:58.
39. Metibemu DS, Akinloye OA, Akamo AJ, Ojo DA, Okeowo OT, Omotuyi IO. Exploring receptor tyrosine kinases-inhibitors in cancer treatments. Egypt J Med Hum Genet. 2019;20:35.
40. Qin S, Li A, Yi M, Yu S, Zhang M, Wu K. Recent advances on anti-angiogenesis receptor tyrosine kinase inhibitors in cancer therapy. J Hematol Oncol. 2019;12:27.
41. Welsh SJ, Fife K. Pazopanib for the treatment of renal cell carcinoma. Future Oncol. 2015;11:1169–79.
42. Orphanos GS, Ioannidis GN, Ardavanis AG. Cardiotoxicity induced by tyrosine kinase inhibitors. Acta Oncol. 2009;48:964–70.
43. Zentilin L, Puligadda U, Lionetti V, et al. Cardiomyocyte VEGFR-1 activation by VEGF-B induces compensatory hypertrophy and preserves cardiac function after myocardial infarction. FASEB J. 2010;24:1467–78.
44. Hiratsuka S, Minowa O, Kuno J, Noda T, Shibuya M. Flt-1 lacking the tyrosine kinase domain is sufficient for normal development and angiogenesis in mice. PNAS. 1998;95:9349–54.
45. Yue Z, Chen J, Lian H, et al. PDGFR-β signaling regulates cardiomyocyte proliferation and myocardial regeneration. Cell Rep. 2019;28:966–978.e4.
46. Andrae J, Gallini R, Betsholtz C. Role of platelet-derived growth factors in physiology and medicine. Genes Dev. 2008;22:1276–312.
47. Pontén A, Bergsten Folestad E, Pietras K, Eriksson U. Platelet-derived growth factor D induces cardiac fibrosis and proliferation of vascular smooth muscle cells in heart-specific transgenic mice. Circ Res. 2005;97:1036–45.
48. Hahn VS, Zhang KW, Sun L, Narayan V, Lenihan DJ, Ky B. Heart failure with targeted cancer therapies: mechanisms and cardioprotection. Circ Res. 2021;128:1576–93.

49. Chintalgattu V, Ai D, Langley RR, et al. Cardiomyocyte PDGFR-β signaling is an essential component of the mouse cardiac response to load-induced stress. J Clin Invest. 2010;120:472–84.
50. Hsieh PCH. Controlled delivery of PDGF-BB for myocardial protection using injectable self-assembling peptide nanofibers. J Clin Investig. 2005;116:237–48.
51. Hsieh PCH, MacGillivray C, Gannon J, Cruz FU, Lee RT. Local controlled intramyocardial delivery of platelet-derived growth factor improves postinfarction ventricular function without pulmonary toxicity. Circulation. 2006;114:637–44.
52. Bjarnegård M, Enge M, Norlin J, Gustafsdottir S, Fredriksson S, Abramsson A, Takemoto M, Gustafsson E, Fässler R, Betsholtz C. Endothelium-specific ablation of PDGFB leads to pericyte loss and glomerular, cardiac and placental abnormalities. Development. 2004;131:1847–57.
53. Hellström M, Kalén M, Lindahl P, Abramsson A, Betsholtz C. Role of PDGF-B and PDGFR-beta in recruitment of vascular smooth muscle cells and pericytes during embryonic blood vessel formation in the mouse. Development. 1999;126:3047–55.
54. Mellgren AM, Smith CL, Olsen GS, Eskiocak B, Zhou B, Kazi MN, Ruiz FR, Pu WT, Tallquist MD. Platelet-derived growth factor receptor β signaling is required for efficient epicardial cell migration and development of two distinct coronary vascular smooth muscle cell populations. Circ Res. 2008;103:1393–401.
55. Tallquist MD, Soriano P. Cell autonomous requirement for PDGFRα in populations of cranial and cardiac neural crest cells. Development. 2003;130:507–18.
56. Zhang H, Jia X, Han F, Zhao J, Zhao Y, Fan Y, Yuan X. Dual-delivery of VEGF and PDGF by double-layered electrospun membranes for blood vessel regeneration. Biomaterials. 2013;34:2202–12.
57. Abbaspour Babaei M, Kamalidehghan B, Saleem M, Zaman Huri H, Ahmadipour F. Receptor tyrosine kinase (c-Kit) inhibitors: a potential therapeutic target in cancer cells. DDDT. 2016;10:2443–59.
58. Galanis A, Mark L. Inhibition of c-Kit by tyrosine kinase inhibitors. Haematologica. 2015;100:e77–9.
59. Talpaz M, Paquette R, Blackwood-Chirchir MA, Decillis AP. Dasatinib in Imatinib-resistant philadelphia chromosome–positive leukemias. N Engl J Med. 2006;354:2531–41.
60. Demetri GD, Lo Russo P, MacPherson IRJ, Wang D, Morgan JA, Brunton VG, Paliwal P, Agrawal S, Voi M, Evans TRJ. Phase I dose-escalation and pharmacokinetic study of Dasatinib in patients with advanced solid tumors. Clin Cancer Res. 2009;15:6232–40.
61. Brazzelli V, Grasso V, Barbaccia V, Manna G, Rivetti N, Zecca M, Giorgiani G, Vassallo C, Borroni G. Hair depigmentation and Vitiligo-like lesions in a leukaemic paediatric patient during chemotherapy with Dasatinib. Acta Derm Venerol. 2012;92:218–9.
62. Hurwitz HI, Dowlati A, Saini S, Savage S, Suttle AB, Gibson DM, Hodge JP, Merkle EM, Pandite L. Phase I trial of Pazopanib in patients with advanced cancer. Clin Cancer Res. 2009;15:4220–7.
63. Fernández A, Sanguino A, Peng Z, et al. An anticancer C-Kit kinase inhibitor is reengineered to make it more active and less cardiotoxic. J Clin Invest. 2007;117:4044–54.
64. Zhou B, Wu SM. Reassessment of c-Kit in cardiac cells: a complex interplay between expression, fate, and function. Circ Res. 2018;123:9–11.
65. Di Siena S, Gimmelli R, Nori SL, et al. Activated c-Kit receptor in the heart promotes cardiac repair and regeneration after injury. Cell Death Dis. 2016;7:e2317.
66. Ayach BB, Yoshimitsu M, Dawood F, et al. Stem cell factor receptor induces progenitor and natural killer cell-mediated cardiac survival and repair after myocardial infarction. PNAS. 2006;103:2304–9.
67. Fazel S. Cardioprotective c-kit+ cells are from the bone marrow and regulate the myocardial balance of angiogenic cytokines. J Clin Investig. 2006;116:1865–77.
68. Cimini M, Fazel S, Zhuo S, Xaymardan M, Fujii H, Weisel RD, Li R-K. c-Kit Dysfunction impairs myocardial healing after infarction. Circulation. 2007; https://doi.org/10.1161/CIRCULATIONAHA.107.708107.

69. Ye L, Zhang EY, Xiong Q, Astle CM, Zhang P, Li Q, From AHL, Harrison DE, Zhang JJ. Aging Kit Mutant mice develop cardiomyopathy. PLoS One. 2012;7:e33407.
70. Makkar RR, Smith RR, Cheng K, et al. Intracoronary cardiosphere-derived cells for heart regeneration after myocardial infarction (CADUCEUS): a prospective, randomised phase 1 trial. The Lancet. 2012;379:895–904.
71. Chaar M, Kamta J, Ait-Oudhia S. Mechanisms, monitoring, and management of tyrosine kinase inhibitors-associated cardiovascular toxicities. OTT. 2018;11:6227–37.
72. Choueiri TK, Halabi S, Sanford BL, et al. Cabozantinib versus sunitinib as initial targeted therapy for patients with metastatic renal cell carcinoma of poor or intermediate risk: the Alliance A031203 CABOSUN trial. JCO. 2017;35:591–7.
73. Jin Y, Xu Z, Yan H, He Q, Yang X, Luo P. A comprehensive review of clinical cardiotoxicity incidence of FDA-approved small-molecule kinase inhibitors. Front Pharmacol. 2020;11:891.
74. Rini BI, Cohen DP, Lu DR, Chen I, Hariharan S, Gore ME, Figlin RA, Baum MS, Motzer RJ. Hypertension as a biomarker of efficacy in patients with metastatic renal cell carcinoma treated with sunitinib. J Natl Cancer Inst. 2011;103:763–73.
75. Rini BI, Schiller JH, Fruehauf JP, et al. Diastolic blood pressure as a biomarker of Axitinib efficacy in solid tumors. Clin Cancer Res. 2011;17:3841–9.
76. Zhang X, Shao Y, Wang K. Incidence and risk of hypertension associated with cabozantinib in cancer patients: a systematic review and meta-analysis. Expert Rev Clin Pharmacol. 2016;9:1109–15.
77. Abou-Alfa GK, Meyer T, Cheng A-L, et al. Cabozantinib in patients with advanced and progressing hepatocellular carcinoma. N Engl J Med. 2018;379:54–63.
78. Singh AP, Umbarkar P, Tousif S, Lal H. Cardiotoxicity of the BCR-ABL1 tyrosine kinase inhibitors: emphasis on ponatinib. Int J Cardiol. 2020;316:214–21.
79. Tan FH, Putoczki TL, Stylli SS, Luwor RB. Ponatinib: a novel multi-tyrosine kinase inhibitor against human malignancies. OTT. 2019;12:635–45.
80. Singh AP, Glennon MS, Umbarkar P, Gupte M, Galindo CL, Zhang Q, Force T, Becker JR, Lal H. Ponatinib-induced cardiotoxicity: delineating the signalling mechanisms and potential rescue strategies. Cardiovasc Res. 2019;115:966–77.
81. Li W, Croce K, Steensma DP, McDermott DF, Ben-Yehuda O, Moslehi J. Vascular and metabolic implications of novel targeted cancer therapies. J Am Coll Cardiol. 2015;66:1160–78.
82. Colliva A, Braga L, Giacca M, Zacchigna S. Endothelial cell–cardiomyocyte crosstalk in heart development and disease. J Physiol. 2020;598:2923–39.
83. Hasinoff BB, Patel D, O'Hara KA. Mechanisms of myocyte cytotoxicity induced by the multiple receptor tyrosine kinase inhibitor sunitinib. Mol Pharmacol. 2008;74:1722–8.
84. Force T, Krause DS, Van Etten RA. Molecular mechanisms of cardiotoxicity of tyrosine kinase inhibition. Nat Rev Cancer. 2007;7:332–44.
85. Steeghs N, Gelderblom H, Roodt J, op 't, Christensen O, Rajagopalan P, Hovens M, Putter H, Rabelink TJ, de Koning E. Hypertension and rarefaction during treatment with Telatinib, a small molecule angiogenesis inhibitor. Clin Cancer Res. 2008;14:3470–6.
86. Chung R, Tyebally S, Chen D, Kapil V, Walker JM, Addison D, Ismail-Khan R, Guha A, Ghosh AK. Hypertensive cardiotoxicity in cancer treatment—systematic analysis of adjunct, conventional chemotherapy, and novel therapies—epidemiology, incidence, and pathophysiology. JCM. 2020;9:3346.
87. Veronese ML, Mosenkis A, Flaherty KT, Gallagher M, Stevenson JP, Townsend RR, O'Dwyer PJ. Mechanisms of hypertension associated With BAY 43-9006. JCO. 2006;24:1363–9.
88. Pinter M, Jain RK. Targeting the renin-angiotensin system to improve cancer treatment: implications for immunotherapy. Sci Transl Med. 2017;9:eaan5616.
89. McKay RR, Rodriguez GE, Lin X, Kaymakcalan MD, Hamnvik O-PR, Sabbisetti VS, Bhatt RS, Simantov R, Choueiri TK. Angiotensin system inhibitors and survival outcomes in patients with metastatic renal cell carcinoma. Clin Cancer Res. 2015;21:2471–9.
90. Chang H-M, Okwuosa TM, Scarabelli T, Moudgil R, Yeh ETH. Cardiovascular complications of cancer therapy. J Am Coll Cardiol. 2017;70:2552–65.

91. Cirmi S, El Abd A, Letinier L, Navarra M, Salvo F. Cardiovascular toxicity of tyrosine kinase inhibitors used in chronic myeloid leukemia: an analysis of the FDA adverse event reporting system database (FAERS). Cancer. 2020;12:826.
92. Hochhaus A, Saglio G, Hughes TP, et al. Long-term benefits and risks of frontline nilotinib vs imatinib for chronic myeloid leukemia in chronic phase: 5-year update of the randomized ENESTnd trial. Leukemia. 2016;30:1044–54.
93. Zamorano JL, Lancellotti P, Rodriguez Muñoz D, et al. 2016 ESC Position Paper on cancer treatments and cardiovascular toxicity developed under the auspices of the ESC Committee for Practice Guidelines: The Task Force for cancer treatments and cardiovascular toxicity of the European Society of Cardiology (ESC). Eur Heart J. 2016;37:2768–801.
94. Pituskin E, Haykowsky M, Mackey JR, Thompson RB, Ezekowitz J, Koshman S, Oudit G, Chow K, Pagano JJ, Paterson I. Rationale and design of the Multidisciplinary Approach to Novel Therapies in Cardiology Oncology Research Trial (MANTICORE 101 - Breast): a randomized, placebo-controlled trial to determine if conventional heart failure pharmacotherapy can prevent trastuzumab-mediated left ventricular remodeling among patients with HER2+ early breast cancer using cardiac MRI. BMC Cancer. 2011;11:318.
95. Dickerson T, Wiczer T, Waller A, et al. Hypertension and incident cardiovascular events following ibrutinib initiation. Blood. 2019;134:1919–28.
96. Leong DP, Caron F, Hillis C, Duan A, Healey JS, Fraser G, Siegal D. The risk of atrial fibrillation with ibrutinib use: a systematic review and meta-analysis. Blood. 2016;128:138–40.
97. Morabito A, Piccirillo MC, Falasconi F, De Feo G, Del Giudice A, Bryce J, Di Maio M, De Maio E, Normanno N, Perrone F. Vandetanib (ZD6474), a dual inhibitor of Vascular Endothelial Growth Factor Receptor (VEGFR) and Epidermal Growth Factor Receptor (EGFR) Tyrosine Kinases: current status and future directions. Oncology. 2009;14:378–90.
98. Holden SN, Eckhardt SG, Basser R, de Boer R, Rischin D, Green M, Rosenthal MA, Wheeler C, Barge A, Hurwitz HI. Clinical evaluation of ZD6474, an orally active inhibitor of VEGF and EGF receptor signaling, in patients with solid, malignant tumors. Ann Oncol. 2005;16:1391–7.
99. Pretorius L, Du X-J, Woodcock EA, et al. Reduced Phosphoinositide 3-Kinase (p110α) activation increases the susceptibility to atrial fibrillation. Am J Pathol. 2009;175:998–1009.
100. Mao S, Luo X, Li Y, He C, Huang F, Su C. Role of PI3K/AKT/mTOR pathway associated oxidative stress and cardiac dysfunction in Takotsubo syndrome. CNR. 2020;17:35–43.
101. McMullen JR, Boey EJH, Ooi JYY, Seymour JF, Keating MJ, Tam CS. Ibrutinib increases the risk of atrial fibrillation, potentially through inhibition of cardiac PI3K-Akt signaling. Blood. 2014;124:3829–30.
102. Vanhaesebroeck B, Guillermet-Guibert J, Graupera M, Bilanges B. The emerging mechanisms of isoform-specific PI3K signalling. Nat Rev Mol Cell Biol. 2010;11:329–41.
103. Ghigo A, Laffargue M, Li M, Hirsch E. PI3K and calcium signaling in cardiovascular disease. Circ Res. 2017;121:282–92.
104. Liang W, Oudit GY, Patel MM, Shah AM, Woodgett JR, Tsushima RG, Ward ME, Backx PH. Role of phosphoinositide 3-Kinase α, protein kinase C, and L-Type Ca $^{2+}$ channels in mediating the complex actions of Angiotensin II on mouse cardiac contractility. Hypertension. 2010;56:422–9.
105. Ghigo A, Perino A, Mehel H, et al. Phosphoinositide 3-Kinase γ protects against catecholamine-induced ventricular arrhythmia through protein Kinase A–mediated regulation of distinct phosphodiesteraseS. Circulation. 2012;126:2073–83.
106. Patrucco E, Notte A, Barberis L, et al. PI3K% modulates the cardiac response to chronic pressure overload by distinct Kinase-dependent and -independent effects. Cell. 2004;118:375.
107. Garnier A, Bork NI, Jacquet E, et al. Mapping genetic changes in the cAMP-signaling cascade in human atria. J Mol Cell Cardiol. 2021;155:10–20.
108. Wu C-YC, Jia Z, Wang W, et al. PI3Ks maintain the structural integrity of T-tubules in cardiac myocytes. PLoS One. 2011;6:e24404.
109. Waardenberg AJ, Bernardo BC, Ng DCH, et al. Phosphoinositide 3-Kinase (PI3K(p110α)) directly regulates key components of the Z-disc and cardiac structure. J Biol Chem. 2011;286:30837–46.

110. Buza V, Rajagopalan B, Curtis AB. Cancer treatment–induced arrhythmias: focus on chemotherapy and targeted therapies. Circ Arrhythm Electrophysiol. 2017; https://doi.org/10.1161/CIRCEP.117.005443.
111. Kloth JSL, Pagani A, Verboom MC, Malovini A, Napolitano C, Kruit WHJ, Sleijfer S, Steeghs N, Zambelli A, Mathijssen RHJ. Incidence and relevance of QTc-interval prolongation caused by tyrosine kinase inhibitors. Br J Cancer. 2015;112:1011–6.
112. Porta-Sánchez A, Gilbert C, Spears D, Amir E, Chan J, Nanthakumar K, Thavendiranathan P. Incidence, diagnosis, and management of QT prolongation induced by cancer therapies: a systematic review. JAHA. 2017; https://doi.org/10.1161/JAHA.117.007724.
113. Lamore SD, Kohnken RA, Peters MF, Kolaja KL. Cardiovascular toxicity induced by kinase inhibitors: mechanisms and preclinical approaches. Chem Res Toxicol. 2020;33:125–36.
114. Freebern W, Fang H, Slade M, Wells S, Canale J, Megill J, Grubor B, Shi H, Fletcher A. In vitro cardiotoxicity potential comparative assessments of chronic myelogenous leukemia tyrosine kinase inhibitor therapies: dasatinib, imatinib and nilotinib. Blood. 2007;110:4582.
115. Bronte E, Bronte G, Novo G, et al. What links BRAF to the heart function? new insights from the cardiotoxicity of BRAF inhibitors in cancer treatment. Oncotarget. 2015;6:35589–601.
116. Xu Z, Cang S, Yang T, Liu D. Cardiotoxicity of tyrosine kinase inhibitors in chronic myelogenous leukemia therapy. Hematol Rep. 2009;1:4.
117. Mellor HR, Bell AR, Valentin J-P, Roberts RRA. Cardiotoxicity associated with targeting Kinase pathways in cancer. Toxicol Sci. 2011;120:14–32.
118. Hudmon A, Schulman H. Structure–function of the multifunctional Ca2+/calmodulin-dependent protein kinase II. Biochem J. 2002;364:593–611.
119. van Oort RJ, McCauley MD, Dixit SS, et al. Ryanodine Receptor phosphorylation by Calcium/Calmodulin-dependent protein Kinase II promotes life-threatening ventricular arrhythmias in mice with heart failure. Circulation. 2010;122:2669–79.
120. Lu Z, Wu C-YC, Jiang Y-P, Ballou LM, Clausen C, Cohen IS, Lin RZ. Suppression of Phosphoinositide 3-Kinase signaling and alteration of multiple ion currents in drug-induced long QT syndrome. Sci Transl Med. 2012;4:131ra50–131ra50.
121. Ling H, Gray CBB, Zambon AC, Grimm M, Gu Y, Dalton N, Purcell NH, Peterson K, Brown JH. Ca $^{2+}$ /Calmodulin-dependent protein Kinase II δ mediates myocardial ischemia/reperfusion injury through nuclear factor-κB. Circ Res. 2013;112:935–44.
122. Siri-Angkul N, Song Z, Fefelova N, Gwathmey JK, Chattipakorn SC, Qu Z, Chattipakorn N, Xie L-H. Activation of TRPC (Transient Receptor Potential Canonical) channel currents in iron overloaded cardiac myocytes. Circ Arrhythm Electrophysiol. 2021; https://doi.org/10.1161/CIRCEP.120.009291.
123. Erickson JR, Pereira L, Wang L, et al. Diabetic hyperglycaemia activates CaMKII and arrhythmias by O-linked glycosylation. Nature. 2013;502:372–6.
124. Swaminathan PD, Purohit A, Soni S, et al. Oxidized CaMKII causes cardiac sinus node dysfunction in mice. J Clin Invest. 2011;121:3277–88.
125. McMullen CJ, Chalmers S, Wood R, Cunningham MR, Currie S. Sunitinib and Imatinib display differential cardiotoxicity in adult rat cardiac fibroblasts that involves a role for calcium/calmodulin dependent protein Kinase II. Front Cardiovasc Med. 2021;7:630480.
126. Koval OM, Snyder JS, Wolf RM, et al. Ca $^{2+}$ /Calmodulin-dependent protein kinase II–based regulation of voltage-gated Na $^{+}$ channel in cardiac disease. Circulation. 2012;126:2084–94.
127. Glynn P, Musa H, Wu X, et al. Voltage-gated sodium channel phosphorylation at Ser571 regulates late current, arrhythmia, and cardiac function in vivo. Circulation. 2015;132:567–77.
128. Horváth B, Hézső T, Kiss D, Kistamás K, Magyar J, Nánási PP, Bányász T. Late sodium current inhibitors as potential antiarrhythmic agents. Front Pharmacol. 2020;11:413.
129. Hegyi B, Pölönen R-P, Hellgren KT, Ko CY, Ginsburg KS, Bossuyt J, Mercola M, Bers DM. Cardiomyocyte Na+ and Ca2+ mishandling drives vicious cycle involving CaMKII, ROS, and ryanodine receptors. Basic Res Cardiol. 2021;116:58.
130. Iqbal SM, Lemmens-Gruber R. Phosphorylation of cardiac voltage-gated sodium channel: potential players with multiple dimensions. Acta Physiol. 2019; https://doi.org/10.1111/apha.13210.

131. Curigliano G, Lenihan D, Fradley M, et al. Management of cardiac disease in cancer patients throughout oncological treatment: ESMO consensus recommendations. Ann Oncol. 2020;31:171–90.
132. Scheffel RS, Dora JM, Siqueira DR, Burttet LM, Cerski MR, Maia AL. Toxic cardiomyopathy leading to fatal acute cardiac failure related to vandetanib: a case report with histopathological analysis. Eur J Endocrinol. 2013;168:K51–4.
133. Chu TF, Rupnick MA, Kerkela R, et al. Cardiotoxicity associated with tyrosine kinase inhibitor sunitinib. Lancet. 2007;370:2011–9.
134. Ghatalia P, Morgan CJ, Je Y, Nguyen PL, Trinh Q-D, Choueiri TK, Sonpavde G. Congestive heart failure with vascular endothelial growth factor receptor tyrosine kinase inhibitors. Crit Rev Oncol Hematol. 2015;94:228–37.
135. Moustafa S, Ho TH, Shah P, Murphy K, Nelluri BK, Lee H, Wilansky S, Mookadam F. Predictors of incipient dysfunction of all cardiac chambers after treatment of metastatic renal cell carcinoma by tyrosine kinase inhibitors: Cardiac Chambers' Dysfunction with TKIs. J Clin Ultrasound. 2016;44:221–30.
136. Groarke JD, Choueiri TK, Slosky D, Cheng S, Moslehi J. Recognizing and managing left ventricular dysfunction associated with therapeutic inhibition of the vascular endothelial growth factor signaling pathway. Curr Treat Options Cardiovasc Med. 2014;16:335.
137. Cheng H, Kari G, Dicker AP, Rodeck U, Koch WJ, Force T. A novel preclinical strategy for identifying cardiotoxic kinase inhibitors and mechanisms of cardiotoxicity. Circ Res. 2011;109:1401–9.
138. Datta SR, Dudek H, Tao X, Masters S, Fu H, Gotoh Y, Greenberg ME. Akt phosphorylation of BAD couples survival signals to the cell-intrinsic death machinery. Cell. 1997;91:231–41.
139. Friehs I, Barillas R, Vasilyev NV. Vascular endothelial growth factor prevents apoptosis and preserves contractile function in hypertrophied infant heart. Circulation. 2006;114:I-290-I–295.
140. Matsui T, Tao J, del Monte F, Lee K-H, Li L, Picard M, Force TL, Franke TF, Hajjar RJ, Rosenzweig A. Akt activation preserves cardiac function and prevents injury after transient cardiac ischemia in vivo. Circulation. 2001;104:330–5.
141. del Peso L, Gonzalez-Garcia M, Page C, Herrera R, Nunez G. Interleukin-3-induced phosphorylation of BAD through the protein kinase Akt. Science. 1997;278:687–9.
142. Boran T, Akyildiz AG, Jannuzzi AT, Alpertunga B. Extended regorafenib treatment can be linked with mitochondrial damage leading to cardiotoxicity. Toxicol Lett. 2021;336:39–49.
143. Greineder CF, Kohnstamm S, Ky B. Heart failure associated with Sunitinib: lessons learned from animal models. Curr Hypertens Rep. 2011;13:436–41.
144. Korashy HM, Attafi IM, Ansari MA, Assiri MA, Belali OM, Ahmad SF, IA AL-A, Anazi FEA, Alhaider AA. Molecular mechanisms of cardiotoxicity of gefitinib in vivo and in vitro rat cardiomyocyte: role of apoptosis and oxidative stress. Toxicol Lett. 2016;252:50–61.
145. Bronte E, Galvano A, Novo G, Russo A. Cardiotoxic effects of anti-VEGFR tyrosine kinase inhibitors. In: Cardio-Oncology; 2017. p. 69–89.
146. Kivelä R, Hemanthakumar KA, Vaparanta K, et al. Endothelial cells regulate physiological cardiomyocyte growth via VEGFR2-mediated paracrine signaling. Circulation. 2019;139:2570–84.
147. Virag JAI, Rolle ML, Reece J, Hardouin S, Feigl EO, Murry CE. Fibroblast growth factor-2 regulates myocardial infarct repair. Am J Pathol. 2007;171:1431–40.
148. Lieu C, Heymach J, Overman M, Tran H, Kopetz S. Beyond VEGF: inhibition of the fibroblast growth factor pathway and antiangiogenesis. Clin Cancer Res. 2011;17:6130–9.
149. Baird A, Esch F, Mormede P, Ueno N, Ling N, Bohlen P. Molecular characterization of fibroblast growth factor: distribution and biological activities in various tissues. Recent Prog Horm Res. 1986;42:143–205.
150. Xie Y, Su N, Yang J, et al. FGF/FGFR signaling in health and disease. Sig Transduct Target Ther. 2020;5:181.
151. Engel FB, Hsieh PCH, Lee RT, Keating MT. FGF1/p38 MAP kinase inhibitor therapy induces cardiomyocyte mitosis, reduces scarring, and rescues function after myocardial infarction. Proc Natl Acad Sci. 2006;103:15546–51.

152. Baines C, Molkentin J. STRESS signaling pathways that modulate cardiac myocyte apoptosis. J Mol Cell Cardiol. 2005;38:47–62.
153. Wang Z-G, Wang Y, Huang Y, et al. bFGF regulates autophagy and ubiquitinated protein accumulation induced by myocardial ischemia/reperfusion via the activation of the PI3K/Akt/mTOR pathway. Sci Rep. 2015;5:9287.
154. Cuevas P, Reimers D, Carceller F, Martinez-Coso V, Redondo-Horcajo M, Saenz de Tejada I, Giménez-Gallego G. Fibroblast growth factor-1 prevents myocardial apoptosis triggered by ischemia reperfusion injury. Eur J Med Res. 1997;2:465–8.
155. Cilvik SN, Wang JI, Lavine KJ, et al. Fibroblast growth factor receptor 1 signaling in adult cardiomyocytes increases contractility and results in a hypertrophic cardiomyopathy. PLoS One. 2013;8:e82979.
156. Schultz JEJ, Witt SA, Nieman ML, Reiser PJ, Engle SJ, Zhou M, Pawlowski SA, Lorenz JN, Kimball TR, Doetschman T. Fibroblast growth factor-2 mediates pressure-induced hypertrophic response. J Clin Invest. 1999;104:709–19.
157. Wu MD, Moslehi JJ, Lindner JR. Arterial thrombotic complications of tyrosine kinase inhibitors. ATVB. 2020; https://doi.org/10.1161/ATVBAHA.120.314694.
158. Choueiri TK, Schutz FAB, Je Y, Rosenberg JE, Bellmunt J. Risk of arterial thromboembolic events with Sunitinib and Sorafenib: a systematic review and meta-analysis of clinical trials. JCO. 2010;28:2280–5.
159. Kamba T, McDonald DM. Mechanisms of adverse effects of anti-VEGF therapy for cancer. Br J Cancer. 2007;96:1788–95.
160. Pandey AK, Singhi EK, Arroyo JP, Ikizler TA, Gould ER, Brown J, Beckman JA, Harrison DG, Moslehi J. Mechanisms of VEGF (Vascular Endothelial Growth Factor) inhibitor–associated hypertension and vascular disease. Hypertension. 2018; https://doi.org/10.1161/HYPERTENSIONAHA.117.10271.
161. Merkulova A, Mitchell SC, Stavrou EX, Forbes GL, Schmaier AH. Ponatinib treatment promotes arterial thrombosis and hyperactive platelets. Blood Adv. 2019;3:2312–6.
162. Latifi Y, Moccetti F, Wu M, et al. Thrombotic microangiopathy as a cause of cardiovascular toxicity from the BCR-ABL1 tyrosine kinase inhibitor ponatinib. Blood. 2019;133:1597–606.
163. Rea D, Mirault T, Cluzeau T, Gautier J-F, Guilhot F, Dombret H, Messas E. Early onset hypercholesterolemia induced by the 2nd-generation tyrosine kinase inhibitor nilotinib in patients with chronic phase-chronic myeloid leukemia. Haematologica. 2014;99:1197–203.
164. Wu Y-L, Soo RA, Locatelli G, Stammberger U, Scagliotti G, Park K. Does c-Met remain a rational target for therapy in patients with EGFR TKI-resistant non-small cell lung cancer? Cancer Treat Rev. 2017;61:70–81.
165. Takahashi S, Kiyota N, Yamazaki T, et al. A Phase II study of the safety and efficacy of lenvatinib in patients with advanced thyroid cancer. Future Oncol. 2019;15:717–26.
166. Strumberg D, Clark JW, Awada A, Moore MJ, Richly H, Hendlisz A, Hirte HW, Eder JP, Lenz H, Schwartz B. Safety, pharmacokinetics, and preliminary antitumor activity of Sorafenib: a review of four phase I trials in patients with advanced refractory solid tumors. Oncologist. 2007;12:426–37.
167. Strumberg D, Schultheis B, Adamietz IA, et al. Phase I dose escalation study of telatinib (BAY 57-9352) in patients with advanced solid tumours. Br J Cancer. 2008;99:1579–85.
168. Motzer RJ, Hutson TE, Cella D, et al. Pazopanib versus Sunitinib in metastatic renal-cell carcinoma. N Engl J Med. 2013;369:722–31.
169. Motzer RJ, Nosov D, Eisen T, et al. Tivozanib versus Sorafenib as initial targeted therapy for patients with metastatic renal cell carcinoma: results from a phase III trial. JCO. 2013;31:3791–9.
170. Gunnarsson O, Pfanzelter N, Cohen R, Keefe S. Evaluating the safety and efficacy of axitinib in the treatment of advanced renal cell carcinoma. CMAR. 2015;7:65.
171. Corte T, Bonella F, Crestani B, Demedts MG, Richeldi L, Coeck C, Pelling K, Quaresma M, Lasky JA. Safety, tolerability and appropriate use of nintedanib in idiopathic pulmonary fibrosis. Respir Res. 2015;16:116.
172. Boss DS, Glen H, Beijnen JH, et al. A phase I study of E7080, a multitargeted tyrosine kinase inhibitor, in patients with advanced solid tumours. Br J Cancer. 2012;106:1598–604.

173. Alhoshani A, Alanazi FE, Alotaibi MR, et al. EGFR inhibitor Gefitinib induces cardiotoxicity through the modulation of cardiac PTEN/Akt/FoxO3a pathway and reactive metabolites formation: *In Vivo* and *in Vitro* Rat studies. Chem Res Toxicol. 2020;33:1719–28.
174. Agulnik M, Costa RLB, Milhem M, et al. A phase II study of tivozanib in patients with metastatic and nonresectable soft-tissue sarcomas. Ann Oncol. 2017;28:121–7.
175. Chen J, Wang J. Risk of regorafenib-induced cardiovascular events in patients with solid tumors: a systematic review and meta-analysis. Medicine. 2018;97:e12705.
176. Zhu AX, Ancukiewicz M, Supko JG, et al. Efficacy, safety, pharmacokinetics, and biomarkers of Cediranib monotherapy in advanced hepatocellular carcinoma: a phase II study. Clin Cancer Res. 2013;19:1557–66.
177. Bæk Møller N, Budolfsen C, Grimm D, Krüger M, Infanger M, Wehland ME, Magnusson N. Drug-induced hypertension caused by Multikinase inhibitors (Sorafenib, Sunitinib, Lenvatinib and Axitinib) in renal cell carcinoma treatment. IJMS. 2019;20:4712.
178. Khakoo AY, Kassiotis CM, Tannir N, Plana JC, Halushka M, Bickford C, Trent J, Champion JC, Durand J-B, Lenihan DJ. Heart failure associated with sunitinib malate: a multitargeted receptor tyrosine kinase inhibitor. Cancer. 2008;112:2500–8.
179. Kus T, Sevinc A, Aktas G, Kalender M, Camci C. Could erlotinib treatment lead to acute cardiovascular events in patients with lung adenocarcinoma after chemotherapy failure? OTT. 2015;8:1341.
180. Jain P, Kantarjian H, Boddu PC, et al. Analysis of cardiovascular and arteriothrombotic adverse events in chronic-phase CML patients after frontline TKIs. Blood Adv. 2019;3:851–61.
181. Kerkelä R, Grazette L, Yacobi R, et al. Cardiotoxicity of the cancer therapeutic agent imatinib mesylate. Nat Med. 2006;12:908–16.
182. Milling RV, Grimm D, Krüger M, Grosse J, Kopp S, Bauer J, Infanger M, Wehland M. Pazopanib, Cabozantinib, and Vandetanib in the treatment of progressive medullary thyroid cancer with a special focus on the adverse effects on hypertension. Int J Mol Sci. 2018;19:3258.
183. Akman T, Erbas O, Akman L, Yilmaz AU. Prevention of Pazopanib-induced prolonged cardiac repolarization and proarrhythmic effects. Arq Bras Cardiol. 2014; https://doi.org/10.5935/abc.20140138.
184. Kappers MHW, de Beer VJ, Zhou Z, Danser AHJ, Sleijfer S, Duncker DJ, van den Meiracker AH, Merkus D. Sunitinib-induced systemic vasoconstriction in Swine is endothelin mediated and does not involve nitric oxide or oxidative stress. Hypertension. 2012;59:151–7.
185. Cholesterol Treatment Trialists' (CTT) Collaborators. The effects of lowering LDL cholesterol with statin therapy in people at low risk of vascular disease: meta-analysis of individual data from 27 randomised trials. Lancet. 2012;380:581–90.
186. Mills EJ, Rachlis B, Wu P, Devereaux PJ, Arora P, Perri D. Primary prevention of cardiovascular mortality and events with statin treatments. J Am Coll Cardiol. 2008;52:1769–81.
187. Hung M-S, Chen I-C, Lee C-P, Huang R-J, Chen P-C, Tsai Y-H, Yang Y-H. Statin improves survival in patients with EGFR-TKI lung cancer: a nationwide population-based study. PLoS One. 2017;12:e0171137.
188. Medeiros BC, Possick J, Fradley M. Cardiovascular, pulmonary, and metabolic toxicities complicating tyrosine kinase inhibitor therapy in chronic myeloid leukemia: strategies for monitoring, detecting, and managing. Blood Rev. 2018;32:289–99.
189. Geyer H, Caracciolo G, Abe H, Wilansky S, Carerj S, Gentile F, Nesser H-J, Khandheria B, Narula J, Sengupta PP. Assessment of myocardial mechanics using speckle tracking echocardiography: fundamentals and clinical applications. J Am Soc Echocardiogr. 2010;23:351–69.
190. Thavendiranathan P, Poulin F, Lim K-D, Plana JC, Woo A, Marwick TH. Use of myocardial strain imaging by echocardiography for the early detection of cardiotoxicity in patients during and after cancer chemotherapy. J Am Coll Cardiol. 2014;63:2751–68.
191. Michel L, Rassaf T, Totzeck M. Cardiotoxicity from immune checkpoint inhibitors. IJC Heart Vasc. 2019;25:100420.
192. Michel L, Totzeck M, Lehmann L, Finke D. Emerging role of immune checkpoint inhibitors and their relevance for the cardiovascular system. Herz. 2020;45:645–51.

193. Jagielska B, Ozdowska P, Gepner K, Kubala S, Siedlecki JA, Sarnowski TJ, Sarnowska E. Cardiotoxicity danger in immunotherapy. IUBMB Life. 2020;72:1160–7.
194. Spiers L, Coupe N, Payne M. Toxicities associated with checkpoint inhibitors—an overview. Rheumatology. 2019;58:vii7–vii16.
195. Kichloo A, Albosta M, Dahiya D, Guidi JC, Aljadah M, Singh J, Shaka H, Wani F, Kumar A, Lekkala M. Systemic adverse effects and toxicities associated with immunotherapy: a review. WJCO. 2021;12:150–63.
196. Salem J-E, Manouchehri A, Moey M, et al. Cardiovascular toxicities associated with immune checkpoint inhibitors: an observational, retrospective, pharmacovigilance study. Lancet Oncol. 2018;19:1579–89.
197. Zarifa A, Lopez-Mattei J, Palaskas N, Iliescu C, Durand J-B, Kim PY. Immune checkpoint inhibitors (ICIs)-related cardiotoxicity. In: Naing A, Hajjar J, editors. Immunotherapy. Cham: Springer International Publishing; 2020. p. 277–85.
198. D'Souza M, Nielsen D, Svane IM, et al. The risk of cardiac events in patients receiving immune checkpoint inhibitors: a nationwide Danish study. Eur Heart J. 2021;42:1621–31.
199. De Sousa LA, Battin C, Jutz S, et al. Therapeutic PD-L1 antibodies are more effective than PD-1 antibodies in blocking PD-1/PD-L1 signaling. Sci Rep. 2019;9:11472.
200. Jiang Y, Chen M, Nie H, Yuan Y. PD-1 and PD-L1 in cancer immunotherapy: clinical implications and future considerations. Hum Vaccin Immunother. 2019;15:1111–22.
201. Grosso JF, Jure-Kunkel MN. CTLA-4 blockade in tumor models: an overview of preclinical and translational research. Cancer Immun. 2013;13:14.
202. Parry RV, Chemnitz JM, Frauwirth KA, Lanfranco AR, Braunstein I, Kobayashi SV, Linsley PS, Thompson CB, Riley JL. CTLA-4 and PD-1 receptors inhibit T-cell activation by distinct mechanisms. Mol Cell Biol. 2005;25:11.
203. Guo Q, Lu T, Chen Y, Su Y, Zheng Y, Chen Z, Chen C, Lin S, Pan J, Yuan X. Genetic variations in the PI3K-PTEN-AKT-mTOR pathway are associated with distant metastasis in nasopharyngeal carcinoma patients treated with intensity-modulated radiation therapy. Sci Rep. 2016;6:37576.
204. Chan DV, Gibson HM, Aufiero BM, Wilson AJ, Hafner MS, Mi Q-S, Wong HK. Differential CTLA-4 expression in human CD4+ versus CD8+ T cells is associated with increased NFAT1 and inhibition of CD4+ proliferation. Genes Immun. 2014;15:25–32.
205. Kubsch S, Graulich E, Knop J, Steinbrink K. Suppressor activity of anergic T cells induced by IL-10-treated human dendritic cells: association with IL-2- and CTLA-4-dependent G1 arrest of the cell cycle regulated by p27Kip1. Eur J Immunol. 2003;33:1988–97.
206. Harlin H, Hwang KW, Palucki DA, Kim O, Thompson CB, Boothby M, Alegre M-L. CTLA-4 engagement regulates NF-κB activation in vivo. Eur J Immunol. 2002;32:2095.
207. Liang SC, Latchman YE, Buhlmann JE, Tomczak MF, Horwitz BH, Freeman GJ, Sharpe AH. Regulation of PD-1, PD-L1, and PD-L2 expression during normal and autoimmune responses. Eur J Immunol. 2003;33:2706–16.
208. Banna GL, Cantale O, Bersanelli M, Del Re M, Friedlaender A, Cortellini A, Addeo A. Are anti-PD1 and anti-PD-L1 alike? The non-small-cell lung cancer paradigm. Oncol Rev. 2020; https://doi.org/10.4081/oncol.2020.490.
209. Abdelrahim M, Mamlouk O, Lin H, et al. Incidence, predictors, and survival impact of acute kidney injury in patients with melanoma treated with immune checkpoint inhibitors: a 10-year single-institution analysis. OncoImmunology. 2021;10:1927313.
210. Wang J, Okazaki I, Yoshida T, Chikuma S, Kato Y, Nakaki F, Hiai H, Honjo T, Okazaki T. PD-1 deficiency results in the development of fatal myocarditis in MRL mice. Int Immunol. 2010;22:443–52.
211. Okazaki T, Tanaka Y, Nishio R, et al. Autoantibodies against cardiac troponin I are responsible for dilated cardiomyopathy in PD-1-deficient mice. Nat Med. 2003;9:1477–83.
212. Xu S, Sharma UC, Tuttle C, Pokharel S. Immune checkpoint inhibitors: cardiotoxicity in preclinical models and clinical studies. Front Cardiovasc Med. 2021;8:619650.
213. Johnson DB, Balko JM, Compton ML, et al. Fulminant myocarditis with combination immune checkpoint blockade. N Engl J Med. 2016;375:1749–55.

214. Ganatra S, Neilan TG. Immune checkpoint inhibitor-associated myocarditis. Oncology. 2018;23:879–86.
215. Läubli H, Balmelli C, Bossard M, Pfister O, Glatz K, Zippelius A. Acute heart failure due to autoimmune myocarditis under pembrolizumab treatment for metastatic melanoma. J Immunother Cancer. 2015;3:11.
216. Ghigo A, Li M. Phosphoinositide 3-kinase: friend and foe in cardiovascular disease. Front Pharmacol. 2015; https://doi.org/10.3389/fphar.2015.00169.
217. Tarrio ML, Grabie N, Bu D, Sharpe AH, Lichtman AH. PD-1 Protects against Inflammation and Myocyte Damage in T Cell-Mediated Myocarditis. JI. 2012;188:4876–84.
218. Nevers T, Salvador AM, Grodecki-Pena A, Knapp A, Velázquez F, Aronovitz M, Kapur NK, Karas RH, Blanton RM, Alcaide P. Left ventricular T-cell recruitment contributes to the pathogenesis of heart failure. Circ Heart Fail. 2015;8:776–87.
219. Komarowska I, Coe D, Wang G, et al. Hepatocyte growth factor receptor c-Met instructs T cell cardiotropism and promotes T cell migration to the heart via Autocrine Chemokine Release. Immunity. 2015;42:1087–99.
220. Heinzerling L, Ott PA, Hodi FS, Husain AN, Tajmir-Riahi A, Tawbi H, Pauschinger M, Gajewski TF, Lipson EJ, Luke JJ. Cardiotoxicity associated with CTLA4 and PD1 blocking immunotherapy. J Immunother Cancer. 2016;4:50.
221. Escudier M, Cautela J, Malissen N, et al. Clinical features, management, and outcomes of immune checkpoint inhibitor–related cardiotoxicity. Circulation. 2017;136:2085–7.
222. Mahmood SS, Fradley MG, Cohen JV, et al. Myocarditis in patients treated with immune checkpoint inhibitors. J Am Coll Cardiol. 2018;71:1755–64.
223. Schneider H, Prasad KV, Shoelson SE, Rudd CE. CTLA-4 binding to the lipid kinase phosphatidylinositol 3-kinase in T cells. J Exp Med. 1995;181:351–5.
224. Schneider H, Valk E, Leung R, Rudd CE. CTLA-4 activation of Phosphatidylinositol 3-Kinase (PI 3-K) and Protein Kinase B (PKB/AKT) sustains T-cell anergy without cell death. PLoS One. 2008;3:e3842.
225. Guo H, German P, Bai S, et al. The PI3K/AKT pathway and renal cell carcinoma. J Genet Genomics. 2015;42:343–53.
226. Wei SC, Meijers WC, Axelrod ML, et al. A Genetic Mouse Model recapitulates immune checkpoint inhibitor–associated myocarditis and supports a mechanism-based therapeutic interventioN. Cancer Discov. 2021;11:614–25.
227. Upadhrasta S, Elias H, Patel K, Zheng L. Managing cardiotoxicity associated with immune checkpoint inhibitors. Chronic Dis Transl Med. 2019;5:6–14.
228. Mir H, Alhussein M, Alrashidi S, Alzayer H, Alshatti A, Valettas N, Mukherjee SD, Nair V, Leong DP. Cardiac complications associated with checkpoint inhibition: a systematic review of the literature in an important emerging area. Can J Cardiol. 2018;34:1059–68.
229. Lipe DN, Rajha E, Wechsler AH, Gaeta S, Palaskas NL, Alhajji Z, Viets-Upchurch J, Chaftari P. Cardiotoxicity associated with immune checkpoint inhibitors and CAR T-cell therapy. Am J Emerg Med. 2021;50:51–8.
230. Fan Q, Hu Y, Yang C, Zhao B. Myocarditis following the use of different immune checkpoint inhibitor regimens: a real-world analysis of post-marketing surveillance data. Int Immunopharmacol. 2019;76:105866.
231. de Almeida DVP, Gomes JR, Haddad FJ, Buzaid AC. Immune-mediated Pericarditis with Pericardial Tamponade during Nivolumab therapy. J Immunother. 2018;41:329–31.
232. Tivol EA, Borriello F, Schweitzer AN, Lynch WP, Bluestone JA, Sharpe AH. Loss of CTLA-4 leads to massive lymphoproliferation and fatal multiorgan tissue destruction, revealing a critical negative regulatory role of CTLA-4. Immunity. 1995;3:541–7.
233. Grabie N, Gotsman I, DaCosta R, Pang H, Stavrakis G, Butte MJ, Keir ME, Freeman GJ, Sharpe AH, Lichtman AH. Endothelial programmed death-1 Ligand 1 (PD-L1) regulates $CD8^+$ T-cell–mediated injury in the heart. Circulation. 2007;116:2062–71.
234. Yang S, Asnani A. Cardiotoxicities associated with immune checkpoint inhibitors. Curr Probl Cancer. 2018;42:422–32.

235. Xia W, Zou C, Chen H, Xie C, Hou M. Immune checkpoint inhibitor induces cardiac injury through polarizing macrophages via modulating microRNA-34a/Kruppel-like factor 4 signaling. Cell Death Dis. 2020;11:575.
236. Ji C, Roy MD, Golas J, et al. Myocarditis in cynomolgus monkeys following treatment with immune checkpoint inhibitors. Clin Cancer Res. 2019;25:4735–48.
237. Kishimoto C, Abelmann WH. In vivo significance of T cells in the development of Coxsackievirus B3 myocarditis in mice. Immature but antigen-specific T cells aggravate cardiac injury. Circ Res. 1990;67:589–98.
238. Lv H, Havari E, Pinto S, et al. Impaired thymic tolerance to α-myosin directs autoimmunity to the heart in mice and humans. J Clin Invest. 2011;121:1561–73.
239. Yamashita T, Iwakura T, Matsui K, et al. IL-6-mediated Th17 differentiation through RORγt is essential for the initiation of experimental autoimmune myocarditis. Cardiovasc Res. 2011;91:640–8.
240. Myers JM, Cooper LT, Kem DC, Stavrakis S, Kosanke SD, Shevach EM, Fairweather D, Stoner JA, Cox CJ, Cunningham MW. Cardiac myosin-Th17 responses promote heart failure in human myocarditis. JCI Insight. 2016; https://doi.org/10.1172/jci.insight.85851.
241. Blanton RM, Carrillo-Salinas FJ, Alcaide P. T-cell recruitment to the heart: friendly guests or unwelcome visitors? Am J Physiol Heart Circul Physiol. 2019;317:H124–40.
242. von Euw E, Chodon T, Attar N, Jalil J, Koya RC, Comin-Anduix B, Ribas A. CTLA4 blockade increases Th17 cells in patients with metastatic melanoma. J Transl Med. 2009;7:35.
243. McDowall LM, Fernando SL, Ange N, Yun J, Chia KKM. Immune checkpoint inhibitor–mediated myocarditis and ventricular tachycardia storm. HeartRhythm Case Rep. 2019;5:497–500.
244. Giglio D, Berntsson H, Fred Å, Ny L. Immune checkpoint inhibitor-induced polymyositis and myasthenia gravis with fatal outcome. Case Rep Oncol. 2020;13:1252–7.
245. Alter P, Rupp H, Maisch B. Activated nuclear transcription factor κB in patients with myocarditis and dilated cardiomyopathy—relation to inflammation and cardiac function. Biochem Biophys Res Commun. 2006;339:180–7.
246. Gordon JW, Shaw JA, Kirshenbaum LA. Multiple facets of NF-κB in the heart: to be or not to NF-κB. Circ Res. 2011;108:1122–32.
247. Woodall MC, Ciccarelli M, Woodall BP, Koch WJ. G protein–coupled receptor kinase 2: a link between myocardial contractile function and cardiac metabolism. Circ Res. 2014;114:1661–70.
248. Nishimura H, Okazaki T, Tanaka Y, et al. Autoimmune dilated cardiomyopathy in PD-1 receptor-deficient mice. Science. 2001;291:319–22.
249. Oudit GY, Crackower MA, Eriksson U, et al. Phosphoinositide 3-kinase γ–deficient mice are protected from isoproterenol-induced heart failure. Circulation. 2003;108:2147–52.
250. Song X, Kusakari Y, Xiao C-Y, Kinsella SD, Rosenberg MA, Scherrer-Crosbie M, Hara K, Rosenzweig A, Matsui T. mTOR attenuates the inflammatory response in cardiomyocytes and prevents cardiac dysfunction in pathological hypertrophy. Am J Physiol Cell Physiol. 2010;299:C1256–66.
251. Kallikourdis M, Martini E, Carullo P, et al. T cell costimulation blockade blunts pressure overload-induced heart failure. Nat Commun. 2017;8:14680.
252. De Ruysscher D, Dingemans A-M, Vooijs M, Heymans S. PD-(L)1 inhibition and cardiac damage: a relevant toxicity? J Thorac Oncol. 2018;13:478–9.
253. Bonaca MP, Olenchock BA, Salem J-E, et al. Myocarditis in the setting of cancer therapeutics: proposed case definitions for emerging clinical syndromes in cardio-oncology. Circulation. 2019;140:80–91.
254. Stein-Merlob AF, Rothberg MV, Ribas A, Yang EH. Cardiotoxicities of novel cancer immunotherapies. Heart. 2021;107:1694–703.
255. Yang S, Wang J, Brand DD, Zheng SG. Role of TNF–TNF receptor 2 signal in regulatory T cells and its therapeutic implications. Front Immunol. 2018;9:784.
256. Jain V, Mohebtash M, Rodrigo ME, Ruiz G, Atkins MB, Barac A. Autoimmune myocarditis caused by immune checkpoint inhibitors treated With antithymocyte globulin. J Immunother. 2018;41:332–5.

257. Chen D-Y, Huang W-K, Chien-Chia WV, Chang W-C, Chen J-S, Chuang C-K, Chu P-H. Cardiovascular toxicity of immune checkpoint inhibitors in cancer patients: a review when cardiology meets immuno-oncology. J Formos Med Assoc. 2020;119:1461–75.
258. Tang J, Shalabi A, Hubbard-Lucey VM. Comprehensive analysis of the clinical immuno-oncology landscape. Ann Oncol. 2018;29:84–91.
259. Nair R, Westin J. CAR T-Cells. In: Naing A, Hajjar J, editors. Immunotherapy. Cham: Springer International Publishing; 2020. p. 215–33.
260. Liu L, Bi E, Ma X, et al. Enhanced CAR-T activity against established tumors by polarizing human T cells to secrete interleukin-9. Nat Commun. 2020;11:5902.
261. Neelapu SS, Tummala S, Kebriaei P, et al. Chimeric antigen receptor T-cell therapy — assessment and management of toxicities. Nat Rev Clin Oncol. 2018;15:47–62.
262. Schuster SJ, Bishop MR, Tam CS, et al. Tisagenlecleucel in adult relapsed or refractory diffuse large B-cell lymphoma. N Engl J Med. 2019;380:45–56.
263. Maude SL, Laetsch TW, Buechner J, et al. Tisagenlecleucel in children and young adults with B-cell lymphoblastic leukemia. N Engl J Med. 2018;378:439–48.
264. Wang M, Munoz J, Goy A, et al. KTE-X19 CAR T-cell therapy in relapsed or refractory mantle-cell lymphoma. N Engl J Med. 2020;382:1331–42.
265. Simbaqueba CC, Aponte MP, Kim P, Deswal A, Palaskas NL, Iliescu C, Jahangir E, Yang EH, Steiner RE, Lopez-Mattei J. Cardiovascular complications of Chimeric antigen receptor T-cell therapy: the Cytokine release syndrome and associated arrhythmias. J Immunother Precis Oncol. 2020;3:113–20.
266. Maude SL, Frey N, Shaw PA, et al. Chimeric antigen receptor T cells for sustained remissions in leukemia. N Engl J Med. 2014;371:1507–17.
267. Lee DW, Kochenderfer JN, Stetler-Stevenson M, et al. T cells expressing CD19 chimeric antigen receptors for acute lymphoblastic leukaemia in children and young adults: a phase 1 dose-escalation trial. The Lancet. 2015;385:517–28.
268. Porter D, Frey N, Wood PA, Weng Y, Grupp SA. Grading of cytokine release syndrome associated with the CAR T cell therapy tisagenlecleucel. J Hematol Oncol. 2018;11:35.
269. Shalabi H, Sachdev V, Kulshreshtha A, et al. Impact of cytokine release syndrome on cardiac function following CD19 CAR-T cell therapy in children and young adults with hematological malignancies. J Immunother Cancer. 2020;8:e001159.
270. Burstein DS, Maude S, Grupp S, Griffis H, Rossano J, Lin K. Cardiac profile of chimeric antigen receptor T cell therapy in children: a single-institution experience. Biol Blood Marrow Transplant. 2018;24:1590–5.
271. Ganatra S, Redd R, Hayek SS, Parikh R, Azam T, Yanik GA, Spendley L, Nikiforow S, Jacobson C, Nohria A. Chimeric antigen receptor T-cell therapy–associated cardiomyopathy in patients with refractory or relapsed non-Hodgkin lymphoma. Circulation. 2020;142:1687–90.
272. Hwang J-R, Byeon Y, Kim D, Park S-G. Recent insights of T cell receptor-mediated signaling pathways for T cell activation and development. Exp Mol Med. 2020;52:750–61.
273. Scheller J, Chalaris A, Schmidt-Arras D, Rose-John S. The pro- and anti-inflammatory properties of the cytokine interleukin-6. Biochimica et Biophysica Acta (BBA) - Mol Cell Res. 2011;1813:878–88.
274. Fontes JA, Rose NR, Čiháková D. The varying faces of IL-6: from cardiac protection to cardiac failure. Cytokine. 2015;74:62–8.
275. Matsushita K, Iwanaga S, Oda T, Kimura K, Shimada M, Sano M, Umezawa A, Hata J, Ogawa S. Interleukin-6/soluble interleukin-6 receptor complex reduces infarct size via inhibiting myocardial apoptosis. Lab Invest. 2005;85:1210–23.
276. Prabhu SD. Cytokine-induced modulation of cardiac function. Circ Res. 2004;95:1140–53.
277. Yu X, Kennedy RH, Liu SJ. JAK2/STAT3, not ERK1/2, mediates interleukin-6-induced activation of inducible nitric-oxide synthase and decrease in contractility of adult ventricular myocytes. J Biol Chem. 2003;278:16304–9.
278. Davila ML, Riviere I, Wang X, et al. Efficacy and toxicity management of 19-28z CAR T cell therapy in B cell acute lymphoblastic leukemia. Sci Transl Med. 2014; https://doi.org/10.1126/scitranslmed.3008226.

279. Fitzgerald JC, Weiss SL, Maude SL, et al. Cytokine release syndrome after chimeric antigen receptor T cell therapy for acute lymphoblastic leukemia. Crit Care Med. 2017;45:e124–31.
280. Podewski EK, Hilfiker-Kleiner D, Hilfiker A, Morawietz H, Lichtenberg A, Wollert KC, Drexler H. Alterations in Janus Kinase (JAK)-Signal Transducers and Activators of Transcription (STAT) signaling in patients with end-stage dilated cardiomyopathy. Circulation. 2003;107:798–802.
281. Plenz G, Song ZF, Reichenberg S, Tjan TD, Robenek H, Deng MC. Left-ventricular expression of interleukin-6 messenger-RNA higher in idiopathic dilated than in ischemic cardiomyopathy. Thorac Cardiovasc Surg. 1998;46:213–6.
282. Buzás K, Megyeri K, Hõgye M, Csanády M, Bogáts G, Mándi Y. Comparative study of the roles of cytokines and apoptosis in dilated and hypertrophic cardiomyopathies. Eur Cytokine Netw. 2004;15:53–9.
283. Kataja A, Tarvasmäki T, Lassus J, et al. Kinetics of procalcitonin, C-reactive protein and interleukin-6 in cardiogenic shock – insights from the CardShock study. Int J Cardiol. 2021;322:191–6.
284. Kubota T, Miyagishima M, Alvarez RJ, et al. Expression of proinflammatory cytokines in the failing human heart: comparison of recent-onset and end-stage congestive heart failure. J Heart Lung Transplant. 2000;19:819–24.
285. Kunisada K, Negoro S, Tone E, Funamoto M, Osugi T, Yamada S, Okabe M, Kishimoto T, Yamauchi-Takihara K. Signal transducer and activator of transcription 3 in the heart transduces not only a hypertrophic signal but a protective signal against doxorubicin-induced cardiomyopathy. Proc Natl Acad Sci U S A. 2000;97:315–9.
286. Cihakova D, Rose NR. Chapter 4 pathogenesis of myocarditis and dilated cardiomyopathy. In: Advances in immunology. Elsevier; 2008. p. 95–114.
287. Lazzerini PE, Laghi-Pasini F, Bertolozzi I, et al. Systemic inflammation as a novel QT-prolonging risk factor in patients with torsades de pointes. Heart. 2017;103:1821–9.
288. Cicha I, Urschel K. TNF-α; in the cardiovascular system: from physiology to therapy. IJICMR. 2015;7:9.
289. Defer N, Azroyan A, Pecker F, Pavoine C. TNFR1 and TNFR2 signaling interplay in cardiac myocytes. J Biol Chem. 2007;282:35564–73.
290. Al-Lamki RS, Brookes AP, Wang J, et al. TNF receptors differentially signal and are differentially expressed and regulated in the human heart. Am J Transplant. 2009;9:2679–96.
291. Thielmann M, Dörge H, Martin C, et al. Myocardial dysfunction with Coronary microembolization: signal transduction through a sequence of nitric oxide, tumor necrosis factor-α, and Sphingosine. Circ Res. 2002;90:807–13.
292. Skyschally A, Gres P, Caster P, Sand A, Boengler K, Schulz R, Heusch G. Reduced calcium responsiveness characterizes contractile dysfunction following coronary microembolization. Basic Res Cardiol. 2008;103:552–9.
293. Tsai C-T, Wu C-K, Lee J-K, Chang S-N, Kuo Y-M, Wang Y-C, Lai L-P, Chiang F-T, Hwang J-J, Lin J-L. TNF- down-regulates sarcoplasmic reticulum Ca2+ ATPase expression and leads to left ventricular diastolic dysfunction through binding of NF- B to promoter response element. Cardiovasc Res. 2015;105:318–29.
294. Friedrichs GS, Swillo RE, Jow B, Bridal T, Numann R, Warner LM, Killar LM, Sidek K. Sphingosine modulates myocyte electrophysiology, induces negative inotropy, and decreases survival after myocardial ischemia. J Cardiovasc Pharmacol. 2002;39:18–28.
295. Fauconnier J, Meli AC, Thireau J, et al. Ryanodine receptor leak mediated by caspase-8 activation leads to left ventricular injury after myocardial ischemia-reperfusion. Proc Natl Acad Sci U S A. 2011;108:13258–63.
296. Krown KA, Page MT, Nguyen C, Zechner D, Gutierrez V, Comstock KL, Glembotski CC, Quintana PJ, Sabbadini RA. Tumor necrosis factor alpha-induced apoptosis in cardiac myocytes. Involvement of the sphingolipid signaling cascade in cardiac cell death. J Clin Invest. 1996;98:2854–65.
297. Buckley LF, Abbate A. Interleukin-1 blockade in cardiovascular diseases: a clinical update. Eur Heart J. 2018;39:2063–9.

298. Bujak M, Frangogiannis NG. The role of IL-1 in the pathogenesis of heart disease. Arch Immunol Ther Exp (Warsz). 2009;57:165–76.
299. Pruitt JH, Copeland EM, Moldawer LL. Interleukin-1 and Interleukin-1 Antagonism in sepsis, systemic inflammatory response syndrome, and septic shock. Shock. 1995;3:235–51.
300. Colotta F, Dower SK, Sims JE, Mantovani A. The type II 'decoy' receptor: a novel regulatory pathway for interleukin 1. Immunol Today. 1994;15:562–6.
301. Colotta F, Re F, Muzio M, Bertini R, Polentarutti N, Sironi M, Giri JG, Dower SK, Sims JE, Mantovani A. Interleukin-1 type II receptor: a Decoy target for IL-1 that is regulated by IL-4. Science. 1993;261:472–5.
302. Lazzerini PE, Laghi-Pasini F, Acampa M, et al. Systemic inflammation rapidly induces reversible atrial electrical remodeling: the role of Interleukin-6–mediated changes in Connexin expression. JAHA. 2019; https://doi.org/10.1161/JAHA.118.011006.
303. Hu Y-F, Chen Y-J, Lin Y-J, Chen S-A. Inflammation and the pathogenesis of atrial fibrillation. Nat Rev Cardiol. 2015;12:230–43.
304. Ghosh AK, Chen DH, Guha A, Mackenzie S, Walker JM, Roddie C. CAR T cell therapy–related cardiovascular outcomes and management. JACC: CardioOncology. 2020;2:97–109.
305. Brudno JN, Kochenderfer JN. Toxicities of chimeric antigen receptor T cells: recognition and management. Blood. 2016;127:3321–30.
306. Brammer JE, Braunstein Z, Katapadi A, et al. Early toxicity and clinical outcomes after chimeric antigen receptor T-cell (CAR-T) therapy for lymphoma. J Immunother Cancer. 2021;9:e002303.
307. Dal'bo N, Patel R, Parikh R, Shah SP, Guha A, Dani SS, Ganatra S. Cardiotoxicity of contemporary anticancer immunotherapy. Curr Treat Options Cardiovasc Med. 2020;22:62.
308. Jamal FA, Khaled SK. The cardiovascular complications of chimeric antigen receptor T cell therapy. Curr Hematol Malig Rep. 2020;15:130–2.
309. Lobenwein D, Kocher F, Dobner S, Gollmann-Tepeköylü C, Holfeld J. Cardiotoxic mechanisms of cancer immunotherapy – a systematic review. Int J Cardiol. 2021;323:179–87.
310. Asnani A. Cardiotoxicity of immunotherapy: incidence, diagnosis, and management. Curr Oncol Rep. 2018;20:44.
311. Jatiani SS, Aleman A, Madduri D, Chari A, Cho HJ, Richard S, Richter J, Brody J, Jagannath S, Parekh S. Myeloma CAR-T CRS management with IL-1R Antagonist Anakinra. Clin Lymphoma Myeloma Leuk. 2020;20:632–636.e1.
312. Norelli M, Camisa B, Barbiera G, et al. Monocyte-derived IL-1 and IL-6 are differentially required for cytokine-release syndrome and neurotoxicity due to CAR T cells. Nat Med. 2018;24:739–48.
313. Giavridis T, van der Stegen SJC, Eyquem J, Hamieh M, Piersigilli A, Sadelain M. CAR T cell–induced cytokine release syndrome is mediated by macrophages and abated by IL-1 blockade. Nat Med. 2018;24:731–8.
314. Zhang L, Wang S, Xu J, Zhang R, Zhu H, Wu Y, Zhu L, Li J, Chen L. Etanercept as a new therapeutic option for cytokine release syndrome following chimeric antigen receptor T cell therapy. Exp Hematol Oncol. 2021;10:16.
315. Grupp SA, Kalos M, Barrett D, et al. Chimeric antigen receptor–modified T cells for acute lymphoid leukemia. N Engl J Med. 2013;368:1509–18.
316. Rodríguez-Otero P, Prósper F, Alfonso A, Paiva B, San Miguel JFS. CAR T-cells in multiple myeloma are ready for prime time. JCM. 2020;9:3577.

Protein Phosphatase Signaling in Cardiac Myocytes

Danielle Abdallah, Nipun Malhotra, and Mona El Refaey

Abstract A delicate balance of protein phosphorylation, performed by protein kinases, and dephosphorylation, performed by protein phosphatases, is required for the proper regulation of various cardiac functions such as Ca^{2+} signaling and excitation-contraction coupling. Abnormalities in this regulatory mechanism may contribute to the initiation and progression of a host of cardiovascular phenotypes, a major cause of mortality in the United States. Serine-threonine phosphatases, including protein phosphatase 1 (PP1), protein phosphatase 2A (PP2A), and protein phosphatase 2B (PP2B), perform the vast majority of dephosphorylation events in the heart. Their role in cardiovascular disease, specifically heart failure and atrial fibrillation, along with their distinctive structure, genetics, localization, and binding partners is reviewed. Specific emphasis is given to PP2A and mouse models that have explored its activity in the heart.

Keywords Kinase-phosphatase balance · PP1 · PP2A · PP2B · Heart failure · Arrhythmia

---

D. Abdallah · N. Malhotra
The Frick Center for Heart Failure and Arrhythmia, Dorothy M. Davis Heart and Lung Research Institute, Columbus, OH, USA

Department of Physiology and Cell Biology, The Ohio State University, Columbus, OH, USA

M. El Refaey (✉)
The Frick Center for Heart Failure and Arrhythmia, Dorothy M. Davis Heart and Lung Research Institute, Columbus, OH, USA

Department of Surgery/Division of Cardiac Surgery, The Ohio State University, Columbus, OH, USA
e-mail: Mona.elrefaey@osumc.edu

N. L. Parinandi, T. J. Hund (eds.), *Cardiovascular Signaling in Health and Disease*, https://doi.org/10.1007/978-3-031-08309-9_5

Introduction

Cardiovascular disease (CVD) is associated with one death every 36 seconds and is the leading cause of mortality in the United States [1]. CVD comprises a variety of conditions including heart failure (HF), stroke, congenital heart disease, coronary artery disease, and arrhythmias. On average, CVD causes 1 out of every 4 deaths, accumulating to about 655,000 deaths per year [2]. CVD results in enormous economic costs for the United States. Direct costs of these cardiovascular conditions totaled $213.8 billion in 2014–2015, which included health care services, medications, and lost productivity due to death [2]. Research on the pathogenic mechanisms underlying CVD is crucial and urgent due to its increasing prevalence, severity, and economic burden.

Management of patients with CVD is mainly symptomatic, aiming at relieving the symptoms rather than eliminating the cause of the illness. Unfortunately, current medications may cause side effects or fail to completely combat the underlying mechanism of the illness. The pathogenesis of CVD is often mediated by dysfunction in various signaling pathways; therefore, treatments attempt to control these pathways [3, 4]. Many processes, such as excitation-contraction coupling, myofilament regulation, and intercellular signaling, are widely regulated by the phosphorylation and dephosphorylation of proteins [5]. Kinases are enzymes that mediate protein phosphorylation and have been heavily studied in connection to CVD. Beta-blockers suppress kinase activity and have become a standard for the treatment of CVD [6]. Kinase inhibitors, traditionally used to treat cancers, have also been studied as a therapy for CVD. However, due to the highly conserved amino acid sequence among different kinases, kinase inhibitors suffer from poor specificity and toxicity [7]. Therefore, novel targets for the treatment of CVD are being explored.

Emerging evidence highlights the important role of protein phosphatases, which work antagonistically with kinases to remove phosphate groups. In the heart, protein phosphatase 2A (PP2A), alongside other protein phosphatases, is involved in regulating the healthy cardiac state as well as cardiac pathophysiology. Further understanding of the role of protein phosphatases in cardiovascular health and disease can lead to the development of novel targets for treatments. This chapter focuses on the role of serine-threonine phosphatases in the heart: protein phosphatase 1 (PP1), PP2A, and protein phosphatase 2B (PP2B), with a specific focus on PP2A. Their distinctive structure, genetics, functions, regulation, and role in CVD, specifically HF and atrial fibrillation (AF), will be reviewed.

The Kinase-Phosphatase Axis

Cellular activity, in general, depends on the regulation of protein function. This can occur in a variety of ways, with control at the level of protein transcription, translation, and/or degradation. Notably, one of the most versatile regulatory mechanisms is the reversible process of protein phosphorylation. For normal cardiac function, it is critical to have a balance of phosphorylation, performed by kinase enzymes, and dephosphorylation, performed by protein phosphatases. The importance of this level of control has been previously reviewed [8–10].

Kinases act on a specific set of target proteins, which can be defined by the local environment, chemical interactions between the kinase and target, and/or regulation by scaffolding or adaptor proteins [11]. Removal of phosphate groups by phosphatase enzymes is regulated differently, mostly through interaction with phosphatase-associated proteins [12]. Kinases are generally more abundant and substrate targeting is more specific than that of protein phosphatases; therefore, kinases have historically been heavily studied. The human genome contains more than 500 kinases but only about 40 serine-threonine phosphatases [13, 14]. However, enormous diversity results from the protein phosphatases due to their holoenzyme composition. By forming complexes with a wide range of regulatory subunits, serine-threonine phosphatases can act to control many different cellular processes.

Proper control of reversible protein phosphorylation is key for healthy cardiac function. Beta-blockers, a staple for the treatment of CVD, suffer from poor specificity and may result in undesired side effects. There are more than 30 protein kinase inhibitors approved for the treatment of cancer and other diseases, but none to date is indicated as therapy for CVD [7, 15]. The pathogenic kinase activity is inherently different in cancer and CVD, making the overlap of treatments complicated. In cancer, abnormal kinase activity is due to genetic variants, whereas in CVD, aberrant kinase activity results from enhanced stimulation by the activated neurohormonal system [7]. Nonetheless, many kinase enzymes, such as $Ca^{2+/}$calmodulin-dependent protein kinase II delta (CaMKII δ) and protein kinase C (PKC), have been investigated for the treatment of HF and arrhythmia. These kinases are known to control various cardiac processes, such as vasoconstriction and hypertrophy, and therefore inhibitors are currently being explored to treat heart disease [7].

With the increasing prevalence of CVD, the search for novel targets is warranted. Protein phosphatases control many of the same processes that kinases do, and therefore provide a fascinating option for further research. Traditionally, PP2A has been studied as a target in cancer therapy, as it is well established as a cell cycle regulator. Evidence has shown that PP2A inhibitors promote tumor growth, highlighting the role of PP2A as a tumor suppressor [16]. In many cancers, PP2A is genetically altered or inactivated, providing rationale for therapeutic studies that have explored the reactivation of PP2A in cancer [16]. The role of protein phosphatases in cardiac pathophysiology is still not clear and requires more research.

Protein Phosphatase Families and Genetics

There are three main protein phosphatase (PP) families, categorized by the known amino acid sites for protein phosphorylation: tyrosine, serine-threonine, and dual-specificity phosphatases [17]. Although not the focus of this chapter, other groups have reviewed the role of tyrosine and dual-specificity phosphatases in cardiovascular health and disease [18–21]. In the heart, the serine-threonine phosphatases are responsible for 90% of the dephosphorylation events [17]. This family includes PP1, PP2A, and PP2B, all of which have been linked to the pathology of heart failure and arrhythmias [22–24]. Each of these enzymes follows a holoenzyme composition, making for combinatorial complexity [5]. Table 1 summarizes the genetics discussed in this section.

The PP1 holoenzyme is ubiquitously expressed in most cell types, including atrial myocytes [24]. It consists of a catalytic subunit that is encoded by three separate genes. PPP1CA, PPP1CB, and PPP1CC encode PP1α, PP1β, and PP1γ, respectively [25]. Further diversification is achieved by alternative splicing of PPP1CC into PP1γ_1 and PP1γ_2 [25]. At the amino acid level, each isoform is ~90% identical, with differences present mostly at their terminal domains [26]. While most isoforms are widely expressed in the human body, PP1γ is expressed in a tissue-specific manner in the brain and testis [22]. The activity of PP1 is also controlled through the interaction with one or more regulatory subunits or interactors. There are currently around 200 known genes that encode regulatory subunits/interactors that combine with the catalytic subunit of PP1, with estimated hundreds more to be discovered [25].

Table 1 A summary of the genes encoding protein subunits that compose the three major serine-threonine phosphatases in the heart. For PP2A regulatory subunits, only those expressed in the heart are listed

Serine-threonine protein phosphatases						
	PP1		PP2A		PP2B	
	Gene/protein	Refs.	Gene/protein	Refs.	Gene/protein	Refs.
Catalytic subunit	PPP1CA/ PP1α PPP1CB/ PP1β PPP1CC/ PP1γ	[25]	PPP2CA/ PP2ACα PPP2CB/ PP2ACβ	[28, 29]	PPP3CA/ CNAα PPP3CB/ CNAβ PPP3CC/ CNAγ	[23]
Scaffolding subunit	–	–	PPP2R1A/ PP2AAα PPP2R1B/ PP2AAβ	[28, 29]	–	–
Regulatory subunits	>200 genes/ subunits	[25]	*Class B*: PPP2R2A/B55α PPP2R2B/ B55β *Class B′*: PPP2R5A/ B56α PPP2R5B/ B56β PPP2R5C/ B56γ PPP2R5D/ B56δ PPP2R5E/ B56ε *Class B″*: PPP2R3A/ PR72/ PR130 *Class B‴*: PPP2R4/ PTPA	[28, 29]	PPP3R1/ CNBα PPP3R2/ CNBβ	[33]

PP1 does not exist freely in cardiac cells. Instead, the regulatory subunits compete to form a holoenzyme complex with the PP1 catalytic subunits and determine the subcellular localization and substrate specificity [22]. Most regulatory subunits are known to associate with all isoforms of PP1, making for a diverse protein interactome [27]. This is due to PP1 containing multiple surface grooves that bind regulatory interactors at short linear motifs of four to eight residues [26]. Additionally, most regulatory subunits have multiple points of binding with PP1, and different subunits can share PP1 interaction sites [27]. However, each regulatory subunit binds to the docking motifs in a specific combination, resulting in diversity of possible PP1 dimers [27]. Most PP1-interacting proteins are structurally unrelated, but contain similar functional domains that can influence PP1-anchoring, substrate targeting, and/or PP1 inhibition [26]. Interestingly, PP1 is known to share common substrates with another key protein phosphatase in the heart, PP2A [22].

The holoenzyme composition of PP2A is even more complex. Although most phosphatases consist of two interacting subunits, PP2A can function not only as a dimer but as a trimer as well [17]. The dimer contains a catalytic (C) and scaffolding (A) subunit, each with two isoforms encoded by distinct genes [28, 29]. The two scaffolding isoforms, PP2AAα and PP2AAβ are encoded by PPP2R1A and PPP2R1B, respectively [28, 29]. Both isoforms are composed of 15 HEAT repeats, which help them to bind the C and B-type regulatory subunits [4]. Similarly, PP2ACα is encoded by PPP2CA and PP2ACβ is encoded by PPP2CB, both of which share 97% sequence identity [30, 31]. The Aα and Cα subunits are more highly expressed and critical for proper PP2A activity, proven by the lethality of PPP2CA and PPP2R1A knockout mouse models [31]. The C subunit is originally synthesized as an inactive, unstable protein, and interacts with biogenesis regulators that transition it to a catalytically active state [31]. For example, PPP2R4 encodes PTPA, which promotes folding of the C subunit into an active conformation [31]. Additionally, post-translational modifications of the catalytic subunit may interfere with PP2A function [28].

Approximately one-third of PP2A activity occurs as a dimer complex, but the great majority of PP2A holoenzymes exist as hetero-trimers, interacting with an interchangeable regulatory subunit [14]. There are numerous regulatory subunits, grouped into four families: PP2AB, PP2AB', PP2AB", and PP2AB"'. Each family is encoded by separate genes, which produce multiple isoforms and splice variants, resulting in proteins with different roles in controlling PP2A function [28]. For example, the B56α class of regulators, encoded by PPP2R5A, has been proven necessary for the regulation of $Na_v1.5$ activity in the heart [17]. Of the 13 known regulatory subunits, all but four are present in cardiac tissue [28]. The resulting product of all three subunits determines substrate specificity, cell- and tissue-specific expression, and subcellular targeting [16].

PP2B, also referred to as calcineurin (CN), is a Ca^{2+}-dependent phosphatase with significant roles in gene transcription. Interestingly, PP2B is the only protein phosphatase to be regulated by a second messenger and links Ca^{2+} and phosphorylation-dependent signaling pathways [22, 32]. PP2B exists as a holoenzyme composed of a catalytic subunit (CNA) and a Ca^{2+}-binding regulatory subunit (CNB), with other interacting subunits sometimes present [24]. There are three mammalian genes for the CNA subunit: PPP3CA, PPP3CB, and PPP3CC, giving rise to three isoforms, with CNAα and CNAβ being widely expressed, most highly in the brain, and CNAγ being testis-specific [23]. Alternative splicing of the CNAα gene results in two transcripts, CNAα_1 and CNAα_2, and that of CNAβ results in three transcripts, CNAβ_1, CNAβ_2, and CNAβ_3 [32]. The regulatory CNB subunit is expressed by two genes, CNBα/PPP3R1 and CNBβ/PPP3R2 [33]. CNBα brings about two isoforms: CNBα_1, which is ubiquitously expressed, and CNBα_2 which shows testis-specific expression [32].

Localization and Binding Partners of Protein Phosphatases

PP1, PP2A, and PP2B make up the great majority of protein phosphatase activity in the heart. Each protein has distinct subcellular localization and binding partners within cardiomyocytes. Most studies on PP1 focus on localization of the catalytic subunit, however much of the localization is dependent on the interaction with the various regulatory subunits [25]. The PP1-interacting subunits can determine localization at the nucleus, plasma membrane, or near actin in the cardiomyocyte. Studies utilizing immunohistochemistry have shown that PP2B/calcineurin is mostly localized to the T-tubules in adult cardiomyocytes, where it plays a role in depolarization of the plasma membrane [33]. Its localization and activity can also change throughout development. In neonatal mice, calcineurin activity seemed to distribute homogenously throughout cardiomyocytes and respond to individual Ca^{2+} pulses [33]. Overall, calcineurin has been found to be tethered near sites of Ca^{2+} release and interacting with binding partners involved in a wide variety of cellular processes [33]. The remainder of this section will focus on the localization and binding partners of PP2A.

The PP2A holoenzyme shows distinct expression and localization in different chambers of the human heart. The catalytic and scaffolding subunits appear to have higher expression in the right atria and ventricle when compared to the left atria and ventricle [28]. At the cellular level, the catalytic and scaffolding subunits show broad distribution throughout ventricular myocytes, both in the cytosol and nucleus [24]. Nine PP2A regulatory subunits and their splice variants have been observed to be expressed in the heart, including B55α, B55β, B56α, B56β, B56γ, B56δ, B56ε, PR72/PR130, and PR53/PPP2R4 [28]. The regulatory subunits show no difference in protein expression in different heart chambers, but do have specific subcellular localization [28]. For example, some regulatory subunits can exist near the Z-line/T--tubule region or within the myocyte nucleus [28]. Additionally, several regulatory subunits appear to show different expression levels in the heart across various species. Rat and mouse samples, compared to human and dog samples, display significantly higher expression levels of PPP2R5A and PPP2R5E and significantly lower expression levels of PPP2R2A and PPP2R5C [28].

PP2A is involved in the regulation and localization of cardiac ion channels, transporters, and regulatory molecules for cardiac contractility and relaxation. Binding partners of PP2A can interact with the catalytic, scaffolding, or regulatory subunits. For example, B56α, a PP2A regulatory subunit expressed in the heart, relies on an interaction with ankyrin-B, an adapter protein, for proper functioning [34]. Ankryin-B and B56α have been found to co-localize, specifically at the M-lines in neonatal cardiomyocytes, and co-immunoprecipitate [34]. In a mouse model of cardiac ankyrin-B deficiency, B56α expression was increased compared to wild-type mice; moreover, the PP2A holoenzyme did not function properly in ankyrin-B deficient myocytes, due to less efficient localization. This resulted in increased phosphorylation of ryanodine receptor 2 (RyR2) and arrhythmia burden [34, 35]. The PP2A-B56α complex has also been shown to co-localize with the Nav1.5 channel in the heart at the intercalated disc [17]. This complex plays a role in the regulation of Nav1.5 phosphorylation and the arrhythmogenic "late" component of the Nav channel current, $I_{Na, L}$ [17]. B56α also targets PP2A to dephosphorylate contractile machinery, including cardiac myosin binding protein-C (cMyBP-C) and the inhibitory subunit of troponin (TnI) [36].

Through dephosphorylation, PP2A plays a critical role in the regulation of other membrane targets including the L-type calcium-channel (Cav1.2), RyR2, Na^+/Ca^+ exchanger (NCX), and Na^+/K^+ ATPase [34, 35, 37, 38]. PP2A also forms a macromolecular complex with p21-activated kinase 1 (PAK1), a serine-threonine protein kinase, and connexin-43 (Cx-43), one of the connexins that compromises membrane gap junctions in myocytes [39]. In a rabbit model of HF, increased activity of PAK1, which has been shown to activate PP2A, contributes to significantly increased dephosphorylation of Cx-43 and impaired intercellular coupling [39]. Lastly, stimuli such as cAMP, Ca^{2+} ions, and ceramides can act as second messengers that directly or indirectly regulate PP2A activity [4]. Of relevance for future work, particularly for defining new therapeutic avenues, is to better understand how local subunits are modulated both through transcriptional and post-translational mechanisms in vivo. For example, the literature suggests that cell and animal phenotypes associated with loss of PP2A scaffolding are inconsistent with phenotypes associated with reduction or even loss of PP2A regulatory subunits. In fact, loss of PP2A subunits may paradoxically increase PP2A activity. As each holoenzyme may be comprised of a multitude of splice forms and subunit composition, we expect that there will be significant complexity in unraveling these signaling pathways, particularly in animal models where subunit composition is likely to be remodeled immediately.

Mouse Models

Mouse models have long been an important tool for in vivo studies in biomedical research. Numerous mouse models have been developed to better understand the role of serine-threonine phosphatases in the diseased and healthy heart. This section will focus on different types of mouse models, including total body or cardiac-specific knockout models, knock-in models of mutated protein phosphatase subunits, and more. Figure 1 summarizes the mouse models discussed in this section.

Most mouse models looking at PP1 activity in the heart show that increased PP1 activity contributes to the progression of cardiac pathophysiology. One study created mouse models with a cardiac-specific overexpression of the catalytic PP1α subunit by inserting the cDNA fragment of the coding sequence for PPP1CA downstream of the αMHC promoter [40]. These transgenic (TG) mice developed functional abnormalities at 3 months of age, such as a significant reduction in fractional shortening [40]. At 6 months of age, echocardiographic and molecular studies showed cardiac enlargement, extensive interstitial fibrosis, and a re-expression of fetal gene isoforms in TG mice, characterized as dilated cardiac hypertrophy and heart failure [40]. Notably, PP1 activity is regulated by two proteins that become active upon phosphorylation by protein kinase A (PKA): inhibitor 1 (I-1) and I-2.

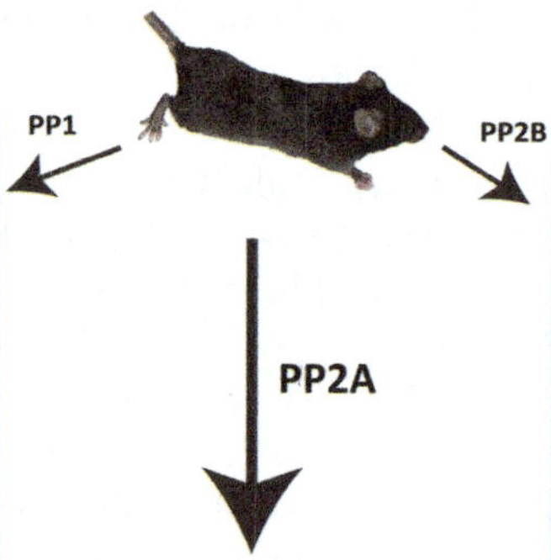

Model	Phenotype(s)
Overexpression of PP1Cα [40]	• Reduced fractional shortening at 3 mo. • Cardiac hypertrophy leading to heart failure • Extensive interstitial fibrosis
KO of I-1, an inhibitor of PP1 [40]	• Increase in PP1 activity • Impaired contractility in knockout hearts
KO of PPP1R3A [42]	• Impaired PP1 targeting • Abnormal SR-Ca^{2+} release • Increased AF susceptibility

Model	Phenotype(s)
Cardiac-specific over-activated calcineurin [61]	• Cardiac hypertrophy • Dilated cardiomyopathy • Heart failure
Cardiac-specific KO of Carabin/TBC1D10C [64]	• Enhanced pressure overload-induced cardiac hypertrophy
Cardiac-specific overexpression of Carabin/TBC1D10C [64]	• Preserved cardiac contractile function • Decreased levels of hypertrophic and fibrotic markers

Catalytic Subunit		Scaffolding Subunit		Regulatory Subunit	
Model	Phenotype(s)	Model	Phenotype(s)	Model	Phenotype(s)
Total body KO of PP2ACα [43,44]	• Embryonically lethal	Total body KO of PP2AAα [51,52]	• Embryonically lethal	Cardiac-specific overexpression of B56α [54]	• Increased PP2A activity localized to the cytoplasm and myofilament fraction • Increased myofilament Ca2+ sensitivity • Impaired contractile response to β-adrenergic stimulation
Cardiac-specific deletion of exons 3-5 of PP2ACα [45]	• Cardiac hypertrophy • Decrease in glucose utilization in the heart	Expression of dominant negative mutant AΔ5 [53]	• Dilated cardiomyopathy • Enlarged hearts • Weakness, slow movement	Knockdown of B56α [35]	• Increased PP2A activity leading to reduced phosphorylation of RyR2 • Decreased heart rates • Increased heart rate variability • Conduction defects • Altered PP2A localization
Cardiac-specific deletion of exon 2 of PP2ACα [46]	• Dilated cardiac hypertrophy • Increase in cardiac tissue fibrosis			Total body KO of B56α [17]	• Reduced phosphorylation of Nav1.5 • Decreased sensitivity to isoproterenol-induced induction of $I_{Na,L}$
Overexpression of PP2ACα [47]	• Dilated cardiomyopathy • Increased adverse remodeling following MI • No impact on overall survival			Total body KO of B56γ [57]	• Improper heart development • Reduced number of ventricular myocytes

Fig. 1 Overview of mouse models modulating PP1, PP2A, and PP2B, as well as their regulatory subunits

Therefore, this same study generated I-1-deficient mouse models that also showed impaired cardiac function compared to wide type mice after a resulting increase in PP1 activity [40]. When examining samples of non-failing and failing human hearts with dilated cardiomyopathy, they found that although protein levels of I-1 were similar in each, I-1 was predominantly in its inactive form in failing hearts [40]. Another study confirmed the idea that decreased I-1 activity along with increased PP1 activity contributes to the progression of HF [41]. The contribution of PP1 to AF pathogenesis has recently been studied in knockout mice lacking a novel PP1 regulatory subunit within the RyR2 channel complex, PPP1R3A [42]. It was found that PPP1R3A-deficient mice had impaired PP1 targeting associated with increased phosphorylation of RyR2 and phospholamban (PLB), abnormal sarcoplasmic reticulum-Ca^{2+} release, and increased AF susceptibility [42].

PP2A has been shown to play a role in cancer, neurodegenerative disorders, diabetes, and renal failure [23]. Importantly, through mouse models, the function and regulation of PP2A and its role in CVD have been explored. Genetically modified mouse models have been produced to alter the catalytic, scaffolding, and/or regulatory subunits of PP2A. The PP2ACα and PP2AAα subunits, as previously noted, are critical for proper PP2A function. Total body knockout models of PP2ACα were proven to be embryonically lethal [43, 44], therefore conditional knockout models have been developed. One study created a cardiac-specific deletion of exons 3–5 of PP2ACα at the neonatal stage (using the *Cre-loxP* strategy) and found cardiac hypertrophy during early development, confirmed by increases in the expression of hypertrophic markers atrial and brain natriuretic peptide (ANP and BNP) [45]. This study also explored metabolic remodeling, finding that PP2ACα-knockout hearts exhibited a decrease in glucose utilization and a compensatory increase in the expression of genes that control fatty acid utilization [45]. Another study generated a cardiac-specific deletion of exon 2 of PP2ACα using a tamoxifen-inducible *Cre* recombinase activated at 3 months of age [46]. Echocardiography was used to assess cardiac function before and after tamoxifen injection and showed markers indicating dilated cardiac hypertrophy [46]. Additionally, Masson trichrome staining revealed a significant increase in cardiac fibrosis in PP2ACα-knockout mice [46]. Thus, heart dysfunction in these knockout models was found to be caused by reduced PP2A activity due to the lack of expression of the PP2ACα subunit. Similar conclusions were made in a study of changes in post-translational regulation of the PP2A catalytic subunit in samples of human HF compared to non-failing heart samples. The results revealed a favoring inactivation of the PP2A holoenzyme in human HF, mediated by an increase in phosphorylation at Tyr-307 and a decrease in methylation at Leu-309 [28].

Interestingly, excess PP2A activity has also been shown to contribute to heart failure. Expression levels of the catalytic subunit of PP2A were seen to have a two-fold increase in both ischemic and non-ischemic failing human heart samples when compared to non-diseased hearts [28]. One study created a transgenic mouse model overexpressing the PP2ACα subunit [47]. The adult mice underwent chronic left anterior descending coronary artery (LAD) ligation to induce myocardial infarction and were analyzed before and 28 days after surgery. Although the overall survival rate of the TG mice was not impacted, the PP2A overexpression led to the development of dilated cardiomyopathy and seemed to increase the extent of adverse remodeling. This remodeling was characterized by a significant increase in myocyte hypertrophy and interstitial fibrosis [47]. Overall, this study reinforced the negative effects of increased phosphatase activity in HF, but also revealed an unexpectedly good survival rate of the TG mice, highlighting restoration mechanisms that are so far unknown.

Numerous studies have generated models modifying the PPP2R1A gene (encoding the Aα subunit), highlighting its role in female fertility, liver fibrosis, spleen atrophy, and mitotic progression [48–50]. Mouse models altering the PP2A scaffolding subunit have also been developed, although cardiac-specific studies exist to a much lesser extent than for the catalytic and regulatory subunits. The Aα subunit is more highly expressed and critical for PP2A function when compared to the Aβ subunit, therefore models surrounding the Aα subunits will be the focus of this section. Similar to models of PP2ACα, total body knockout models of the Aα subunit have been found to be embryonically lethal [51, 52]. Therefore, one study created TG mice expressing high levels of a dominant negative mutant of the A subunit, AΔ5, that reduced the PP2A holoenzyme to core enzyme ratio [53]. Starting at day one after birth, the AΔ5-TG mice displayed a dilated cardiomyopathy phenotype. Additionally, at 7–12 months of age, 25% of the AΔ5-TG mice showed enlarged hearts and symptoms of weakness, slow movement, and an increased respiratory rate that lead to death a few weeks later [53].

Due to the plethora of regulatory B-subunits, PP2A strains have been developed to better understand the role of each. Although relatively few cardiac-specific mouse models have been created when compared to the number of regulatory subunits that exist, these models are important to explore the physiologic functions of PP2A hetero-trimers. Multiple studies have manipulated B56α. One generated mice with cardiomyocyte-directed overexpression of this subunit, which resulted in enhanced PP2A activity [54]. The increased PP2A-B56α activity was localized to the cytoplasm and myofilament fraction, leading to decreased basal phosphorylation of several contractile proteins. This resulted in increased myofilament Ca^{2+} sensitivity, along with increased basal contractility but an impaired contractile response to β-adrenergic stimulation [54]. Prior in vitro work has also suggested that B56α can promote arrhythmia susceptibility by dissociating PP2A activity from the RyR2, leading to hyper-phosphorylation of RyR2 and abnormal Ca^{2+} cycling [55, 56]. Another study classified B56α as an auto-inhibitor of cardiac PP2A-dependent activity in vivo [35]. Mice deficient in B56α did not display compensatory changes in the amount of other PP2A subunits, but did show increased PP2A activity that resulted in decreased heart rates, heart rate variability, conduction defects, and increased sensitivity to parasympathetic agonist [35]. The $B56\alpha^{+/-}$ mice also displayed significant reductions in RyR2 phosphorylation, yet no change in the phosphorylation status of myofilament contractile proteins [35]. Lastly, the $B56\alpha^{+/-}$ and $B56\alpha^{-/-}$ myocytes displayed increased perinuclear and nuclear localization of PP2A/A compared to wild-type myocytes, revealing the influence of regulatory subunits on subcellular targeting of the core enzyme subunits [35]. B56α has been further identified as a novel target for the treatment of arrhythmia due to the role of PP2A-B56α in regulating the primary cardiac Na_V channel, $Na_V1.5$ [17]. B56α KO myocytes were found to display decreased sensitivity to isoproterenol-induced induction of the arrhythmogenic $I_{Na,L}$ and reduced phosphorylation of Nav1.5 [17]. These studies highlight the significant role that B56α has in regulating PP2A activity.

Additional PP2A models manipulating B-type subunits have been developed. One study concluded that B56γ is required for proper heart development after creating B56γ knockout mice from an embryonic cell line. The lack of PP2A-B56γ activity resulted in insufficient cardiomyocyte maturation, including the formation of an incomplete ventricular septum and a reduction in the number of ventricular myocytes [57]. Additionally, knockout mice had deficits in neuromuscular control and developed obesity at 3–6 months of age, suggesting additional problems caused by a lack of B56γ [57]. Embryonic lethality in total-body KO models of other PP2A regulators, including PPP2R5D (encodes B56δ) [58], PPP2R6A (encodes STRN) [59], and PPP2R4 (encodes PTPA) [60], suggests the importance of select PP2A trimers for survival. To increase understanding of the role of these regulators in the heart, cardiac-specific knockout models should be explored.

PP2B has also been well characterized in relation to hypertrophic heart disease. In 1998, a study by Molkentin et al. created a mouse model using α-MHC to overexpress PP2B. It showed that the sustained activity of PP2B/calcineurin led to cardiac hypertrophy that progresses to decompensated HF and fatal cardiac arrhythmias [61]. Subsequent research has confirmed this finding and studied ways to negatively regulate PP2B for cardiac protection, with a large focus on the role of endogenous inhibitors of PP2B [62, 63]. For example, one study highlighted the cardioprotective role of Carabin/TBC1D10C (TBC1 domain family, member 10C), a GTPase-activating protein and known inhibitor of PP2B. In animal models of HF and in patients with cardiomyopathy, levels of TBC1D10C were found to be reduced. Mouse models were also created utilizing cardiac-specific knockouts of TBC1D10C, and they exhibited enhanced pressure overload-induced cardiomyopathy [64]. This study is one of many that has validated the causal role of PP2B in cardiac pathology.

Protein Phosphatases in Cardiovascular Disease

In a normal heart, cyclic systolic and diastolic events rest on accurate regulation of a variety of cardiac processes such as ion channel functioning, Ca^{2+} handling, and the transmission of Ca^{2+} sensitivity to downstream contractile machinery and neighboring myocytes. This control is widely achieved by the dephosphorylation of key proteins by PP1, PP2A, and PP2B [65]. These serine-threonine protein phosphatases present an attractive drug target in CVD; however, this requires a better understanding of their substrate specificity, target localization, and role in cardiac pathology. This section will focus on the known role of serine-threonine phosphatases in two types of CVD, HF and AF (Fig. 2).

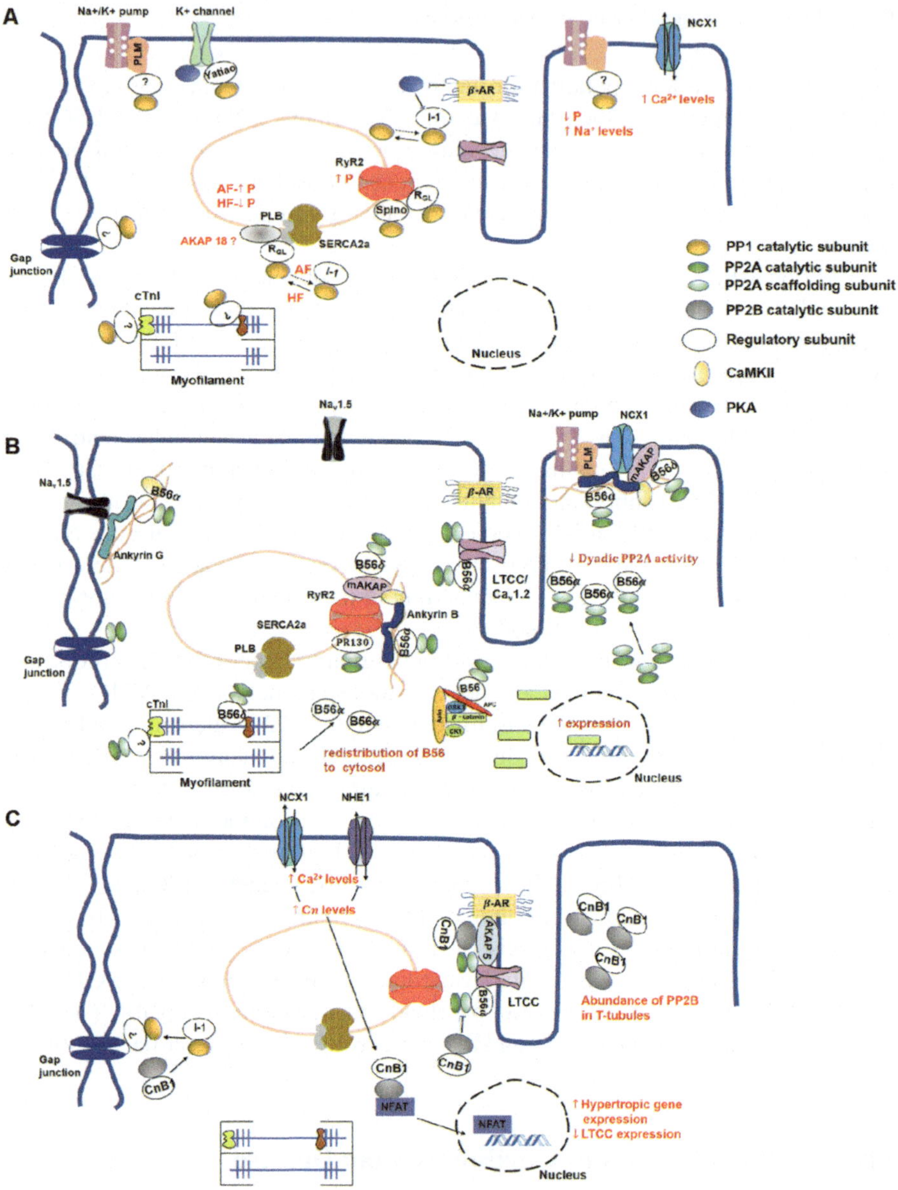

Fig. 2 Schematic overview of protein phosphatases interacting with regulatory subunits, ion channels, and other key binding partners in cardiomyocytes. Known molecular events that occur in heart failure (HF) and atrial fibrillation (AF) are depicted. Figures are specific for (**a**) PP1, (**b**) PP2A and (**c**) PP2B. Abbreviations: sodium/potassium pump (Na+/K+ pump), potassium channel (K+ channel), phospholemman (PLM), β-adrenergic receptor (β-AR), ryanodine receptor 2 (RyR2), inhibitor 1 (I-1), inhibitor 2 (I-2), phosphorylation (P), A kinase anchoring protein (AKAP), sarcoplasmic reticulum Ca^{2+}-adenosine triphosphatase (SERCA2a), Na^+/Ca^+ exchanger (NCX1), inhibitory troponin subunit (cTnI), L-type calcium-channel (LTCC/Cav1.2), Ca^{2+}-/calmodulin-dependent protein kinase II delta (CaMKII δ), protein kinase C (PKC), glycogen synthase kinase 3 (GSK3), casein kinase 1 (CK1), phospholamban (PLB), muscle-specific A kinase anchoring protein (mAKAP), *adenomatous polyposis coli* gene product (APC), Na^+/H^+ exchanger (NHE1), Calcineurin regulatory subunit (CnB1), nuclear factor of activated T-cells (NFAT), and protein kinase A (PKA)

PP1 in Cardiovascular Disease

AF is the most prevalent and progressive arrhythmic disorder and is characterized by an impaired electrophysiological function that can cause reduced atrial excitability, disrupted calcium flux/homeostasis, and structural changes in the heart [66]. The abnormal Ca^{2+} flux in AF is partially caused by a kinase-phosphatase imbalance, which impairs the regulation of key elements of the sarcoplasmic reticulum (SR) and contractile myofilaments [67]. This imbalance has been studied and validated using protein tyrosine kinase K (PTK) and PP-specific inhibitors [68, 69]. One study found that another AF attribute is reduced targeting of the PP1-R_{GL} holoenzyme to the sarcoplasmic reticulum Ca^{2+}-adenosine triphosphatase (SERCA2A)/PLB macromolecular complex, making SERCA2A more active and increasing Ca^{2+} loading of the SR. The increased load may cause larger spontaneous Ca^{2+} release and late after-depolarizations that promote AF [67].

HF is characterized by desensitization of the β-adrenoceptor (β-AR) and its downstream signaling partners, including I1. This broadly increases PP1 activity during HF, yet phosphorylation varies across different subcellular locations of the failing heart [70]. For example, the PP1-I1 complex dephosphorylates phospholemman (PLM) at an increased rate in HF. This causes diminished Na^{+}/K^{+} ATPase activity, resulting in increased intracellular Na^{+} levels and thus an increase in intracellular Ca^{2+} load [71, 72]. Notably, the events outlining the regulation of I-1 provide a point of crosstalk among phosphatases and also between PKA and PP1 [73, 74]. Additionally, heightened activity of the PP1-R_{GL} holoenzyme activates PLB (a SERCA2A inhibitory subunit) thereby inactivating SERCA2A, resulting in reduced Ca^{2+} uptake and SR load (impaired diastolic Ca^{2+} sequestration) in HF [42, 75].

The overall intracellular Ca^{2+} level and phosphorylation state of contractile machinery contribute to a normal or abnormal heartbeat. In diastolic HF, there is reduced Ca^{2+} decay together with myofilament impairment, causing increased myofilament Ca^{2+} sensitivity. The reduced phosphorylation of the inhibitory troponin subunit (cTnI) of the thin filament and the cardiac myosin binding protein C (cMyBP-C) of the thick filament, both of which are targeted by PP1, contributes to this phenotype [68, 76, 77].

Compared to the known 200+ PP1 interactors, a small number has been identified and validated to date. This includes the most studied: PPP1R1A (I-1), PPP1R2 (I-2), PPP1R9B (spinophilin), PPP1R3A (R_{GL}), PPP1R7 (SDS22), PPP1R8, and PPP1R18/phostensin. Recent advances in bioinformatics have led to the identification of some putative regulatory subunits such as cold-shock domain protein A (CSDA), found in the sarcolemma of cardiomyocytes, and phosphodiesterase type 5A (PDE5A), a cyclic guanosine monophosphate (cGMP)-dependent-specific phosphodiesterase that plays a crucial role in oxidative stress and cardiac hypertrophy [78].

PP2A in Cardiovascular Disease

PP2A undertakes a major load of phospho-regulation in cardiac cells as evident by its heterogeneous expression across different cardiac chambers and disease phenotypes, which has previously been reviewed [28]. AF is marked by increased global PP2A activity from increased PP2A-C expression, whereas both increased and decreased PP2A-C expressions have been reported in HF [28, 29, 79].

Under normal cardiac conditions, Na^+ channels open transiently and are quickly inactivated. However, some channels remain active or close and reopen, resulting in the generation of a persistent or late sodium current ($I_{Na,L}$) which is pro-arrhythmic [17]. In a pressure-overload mouse model, hyperphosphorylation of $Na_v1.5$ by CAMKII at Ser571 was shown to increase $I_{Na,L}$ and generate arrhythmogenic triggers [80]. Interestingly, the PP2A-B56α signaling complex is localized at the intercalated disc of the myocyte with $Na_v1.5$, ankyrin-G, β_{IV} spectrin, and CaMKII δ. PP2A-B56 α, within this complex, is critical for action potential (AP) duration by modulating the phosphorylation of $Na_v1.5$ and the pathogenic $I_{Na,L}$. [17, 81]. Alterations in the $Ca_v1.2$ channel current (denoted $I_{Ca,L}$) have also been reported in heart arrhythmias. In particular, the dephosphorylation of $Ca_v1.2$ by PP2A results in a reduction of $I_{Ca,L}$ and subsequent AP shortening, which is a hallmark of AF [69, 79]. This dephosphorylation event can occur either by the PP2A-C dimer or a trimeric holoenzyme [37, 82].

PP2A, like PP1, also plays a major role in one of the core mechanisms for Ca^{2+} release, which is phospho-regulation of RyR2. Hyperphosphorylation of RyR2 and subsequent Ca^{2+} sparks and after-depolarization events have been noted in both HF and AF, resulting from a decrease in PP2A activity [55, 67]. PP2A targeting of RyR2 is achieved through interaction within a large macromolecular complex including key proteins such as PKA, CaMKII, and PP1. Anchoring of PP2A to RyR2 in this complex can be achieved through direct attachment via PR130, a PP2A regulatory subunit, or through indirect coupling with other proteins in the complex [83]. These include muscle-specific A kinase anchoring protein (mAKAP) and ankyrin-B via B56δ and B56α, respectively [56, 84]. These interactions are critical for proper control of RyR2 phosphorylation events and thus play a role in calcium release. For example, local PP2A activity is regulated through the PP2A-B56α holoenzyme either by targeting PP2A to the cardiac dyad to modulate CaMKII-dependent RyR2 phosphorylation or by sequestering free, more active PP2A-C dimers [35].

PP1 and PP2A play a key role in the regulation of cardiac contractility. Both PPs play a role in the dephosphorylation of cMyBP-C and TnI. cMyBP-C phosphorylation modulates contractility by altering the cross-bridge kinetics, while TnI phosphorylation modulates Ca^{2+} sensitivity to regulate contractility. Notably, both PPs dephosphorylate TnI but at different target residues. Interestingly, the targeting of PP2A to the contractile machinery is dependent on the B56α regulatory subunit [54, 76]. B56α was also identified as a binding partner to ankyrin-B, which regulates Na^+/K^+ ATPase and NCX1. NCX1 plays a significant role in Ca^{2+} homeostasis in the normal and failing heart [85]. Notably, PKA, PKC, and other regulatory enzymes such as PP1, PP2A, and mAKAP all form a macromolecular complex with NCX1. The presence of two kinases and two phosphatases within the complex suggests their role in precise and specific regulation of NCX1 function [86].

Cardiac hypertrophy and myocardial fibrosis have been reported in CVD and were linked to reduced PP2A activity [46]. This pathologic feature underlies a crucial epithelial-mesenchymal transition, which results from the reactivation of key developmental genes in the adult heart [87]. Interestingly, cardiomyocyte-specific deletion of PP2A resulted in alteration of the Akt/GSK3β/β-catenin pathway [46]. Notably, the levels of β-catenin play a significant role in this gene reactivation and is regulated by an intricate β-catenin destruction complex composed of PP2A along with a scaffolding protein: Axin, *adenomatous polyposis coli* gene product (APC), casein kinase 1 (CK1), and glycogen synthase kinase 3 (GSK3) [88]. Importantly, a reduction in the expression of the PP2A modulator casein kinase-2 interacting protein-1 (CKIP-1) and the pro-hypertrophic genes have all been reported in HF [89].

Finally, gap junctions, composed of connexins, are essential for myocyte-myocyte communication. PP1 and PP2A were suggested to regulate Cx-43 based on the localization of the PPs, phosphorylation sites on Cx-43, and reduced gap junction linked with increased phosphatase activity. In HF and AF models, conduction abnormalities resulting from modulations of gap junctions (i.e. dephosphorylation, downregulation, and lateralization of Cx-43) were closely linked to increased PP1/PP2A co-localization with Cx-43 and increased phosphatase activity [39, 90–92].

PP2B in Cardiovascular Disease

The PP2B/calcineurin holoenzyme has only one regulatory subunit in the heart, CnB1, as opposed to the plethora that exist for PP1 and PP2A [93]. Additionally, in contrast to PP1 and PP2A, whose regulatory subunits aid in localization and substrate targeting, the catalytic subunits of PP2B play this role [94, 95]. In AF and HF, Ca^{2+} sparks promote abnormal PP2B activation and affect overall excitability by increasing the nuclear factor of activated T-cells (NFAT), which reduces L-type Ca^{2+} channel (LTCC) expression [96]. This increased activation of PP2B, along with its localization at the T-tubules and its proximity to AKAP5 and LTCC, indirectly increases $I_{Ca,L}$ by modulating the PP2A-LTCC interaction, although the exact mechanism is unclear [97–99]. The complex consisting of PP2B, LTCC, AKAP5, and caveolin 3 (CAV3) may provide a link to β-AR trafficking, which is abnormal in heart failure [33, 100].

Abnormal activation of NCX1 and Na^+/K^+ exchanger (NHE1) provides a platform to stimulate the PP2B/NFAT cascade, leading to cardiac hypertrophy [101, 102]. Notably, the role of PP2B in cardiac hypertrophy is its main contribution to the development of HF and AF.

Conclusions and Future Directions

In summary, serine-threonine phosphatases are key regulatory proteins that control a wide array of processes in cardiac myocytes. Their holoenzyme composition allows for complex regulation of cellular processes, as well as specificity in cellular localization. This chapter has explored previous research revealing that serine-threonine phosphatases contribute to the pathogenesis of cardiovascular diseases, specifically HF and AF. In order to consider these protein phosphatases as potential therapeutic targets, more knowledge is needed on their specific function, localization, and regulation in the heart. This gap in knowledge can be addressed with more cardiac-specific mouse models of PP1, PP2A, PP2B, and their abundant regulatory subunits.

Although specific protein phosphatase targeting may be essential for future HF and AF therapy, the role of cross-talk between protein phosphatases and kinases must also be considered when designing novel interventions. Through the examples we have highlighted in this chapter, several feedback loops together regulate the protein phosphatase activity in various subcellular microdomains (Fig. 3). In particular, I-1 plays a predominant role in facilitating crosstalk between different types of phosphatases and kinases. Notably, β-AR signals negatively regulate PP1 activity by PKA-dependent I-1 activation, whereas PP2A and PP2B positively regulate PP1 activity via I-1 repression. Moreover, I-2 is either autoregulated by its PP1

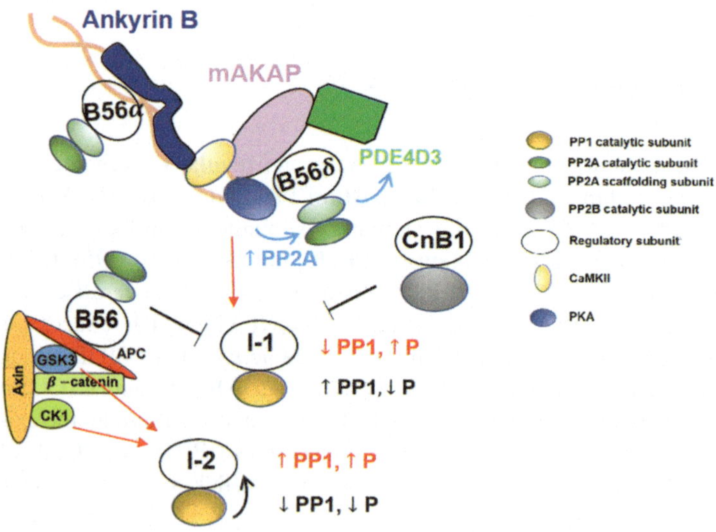

Fig. 3 Schematic of crosstalk between phosphatases, kinases, and other crucial proteins. Abbreviations: muscle-specific A kinase anchoring protein (mAKAP), phosphorylation (P), inhibitor 1 (I-1), inhibitor 2 (I-2), glycogen synthase kinase 3 (GSK3), casein kinase 1 (CK1), *adenomatous polyposis coli* gene product (APC), protein kinase A (PKA), $Ca^{2+-/}$calmodulin-dependent protein kinase II delta (CaMKII δ), and cAMP-specific phosphodiesterase (PDE4D3)

holoenzyme or by kinases like GSK3 and CK1. Also, GSK3 and CK1 are part of the β-catenin destruction complex, which is involved in PP2A-mediated structural remodeling. In addition, PKA facilitates crosstalk between the PP2A-B56δ holoenzyme and cAMP-specific phosphodiesterase (PDE4D3). These elaborate crosstalk events highlight the difficulty in identifying isolated targets for cardiovascular therapy. Deeper investigation into these complex phosphatase-phosphatase and phosphatase-kinase feedback loops would contribute significantly to the advancement of treatments for cardiovascular diseases. Given the speed of remodeling observed in the heart, it is critical that care be taken when interpreting and comparing in vitro and in vivo models.

References

1. Prevention CfDCa. Underlying cause of death, 1999–2018. 2018.
2. Virani SSAA, Benjamin EJ, Bittencourt MS, Callaway CW, Carson AP, et al. Heart disease and stroke statistics—2020 update: a report from the American Heart Association. Circulation. 2020;141:e139–596.
3. Lubbers ER, Mohler PJ. Roles and regulation of protein phosphatase 2A (PP2A) in the heart. J Mol Cell Cardiol. 2016;101:127–33.
4. Lambrecht C, Haesen D, Sents W, Ivanova E, Janssens V. Structure, regulation, and pharmacological modulation of PP2A phosphatases. Methods Mol Biol. 2013;1053:283–305.
5. Baskaran R, Velmurugan BK. Protein phosphatase 2A as therapeutic targets in various disease models. Life Sci. 2018;210:40–6.
6. Reiken S, Wehrens XH, Vest JA, Barbone A, Klotz S, Mancini D, et al. Beta-blockers restore calcium release channel function and improve cardiac muscle performance in human heart failure. Circulation. 2003;107(19):2459–66.
7. Shahin R, Shaheen O, El-Dahiyat F, Habash M, Saffour S. Research advances in kinase enzymes and inhibitors for cardiovascular disease treatment. Future Sci OA. 2017;3(4):FSO204.
8. Ruvolo PP. The broken "Off" switch in cancer signaling: PP2A as a regulator of tumorigenesis, drug resistance, and immune surveillance. BBA Clin. 2016;6:87–99.
9. Sato PY, Chuprun JK, Schwartz M, Koch WJ. The evolving impact of g protein-coupled receptor kinases in cardiac health and disease. Physiol Rev. 2015;95(2):377–404.
10. Cohen P. The origins of protein phosphorylation. Nat Cell Biol. 2002;4(5):E127–30.
11. Ubersax JA, Ferrell JE Jr. Mechanisms of specificity in protein phosphorylation. Nat Rev Mol Cell Biol. 2007;8(7):530–41.
12. Chatterjee J, Kohn M. Targeting the untargetable: recent advances in the selective chemical modulation of protein phosphatase-1 activity. Curr Opin Chem Biol. 2013;17(3):361–8.
13. Kumar R, Singh VP, Baker KM. Kinase inhibitors for cardiovascular disease. J Mol Cell Cardiol. 2007;42(1):1–11.
14. Sents W, Ivanova E, Lambrecht C, Haesen D, Janssens V. The biogenesis of active protein phosphatase 2A holoenzymes: a tightly regulated process creating phosphatase specificity. FEBS J. 2013;280(2):644–61.
15. Panicker RC, Chattopadhaya S, Coyne AG, Srinivasan R. Allosteric small-molecule serine/threonine kinase inhibitors. Adv Exp Med Biol. 2019;1163:253–78.
16. Sangodkar J, Farrington CC, McClinch K, Galsky MD, Kastrinsky DB, Narla G. All roads lead to PP2A: exploiting the therapeutic potential of this phosphatase. FEBS J. 2016;283(6):1004–24.
17. El Refaey M, Musa H, Murphy NP, Lubbers ER, Skaf M, Han M, et al. Protein phosphatase 2A regulates cardiac Na(+) channels. Circ Res. 2019;124(5):737–46.

18. Jiang L, Ren L, Guo X, Zhao J, Zhang H, Chen S, et al. Dual-specificity phosphatase 9 protects against cardiac hypertrophy by targeting ASK1. Int J Biol Sci. 2021;17(9):2193–204.
19. Gao PP, Qi XW, Sun N, Sun YY, Zhang Y, Tan XN, et al. The emerging roles of dual-specificity phosphatases and their specific characteristics in human cancer. Biochim Biophys Acta Rev Cancer. 1876;2021(1):188562.
20. Wade F, Belhaj K, Poizat C. Protein tyrosine phosphatases in cardiac physiology and pathophysiology. Heart Fail Rev. 2018;23(2):261–72.
21. Abdelsalam SS, Korashy HM, Zeidan A, Agouni A. The role of protein tyrosine phosphatase (PTP)-1B in cardiovascular disease and its interplay with insulin resistance. Biomolecules. 2019;9(7):286.
22. Weber S, Meyer-Roxlau S, Wagner M, Dobrev D, El-Armouche A. Counteracting protein kinase activity in the heart: the multiple roles of protein phosphatases. Front Pharmacol. 2015;6:270.
23. Heijman J, Dewenter M, El-Armouche A, Dobrev D. Function and regulation of serine/threonine phosphatases in the healthy and diseased heart. J Mol Cell Cardiol. 2013;64:90–8.
24. Heijman J, Ghezelbash S, Wehrens XH, Dobrev D. Serine/threonine phosphatases in atrial fibrillation. J Mol Cell Cardiol. 2017;103:110–20.
25. Chiang DY, Heck AJ, Dobrev D, Wehrens XH. Regulating the regulator: insights into the cardiac protein phosphatase 1 interactome. J Mol Cell Cardiol. 2016;101:165–72.
26. Ferreira M, Beullens M, Bollen M, Van Eynde A. Functions and therapeutic potential of protein phosphatase 1: insights from mouse genetics. Biochim Biophys Acta, Mol Cell Res. 2019;1866(1):16–30.
27. Heroes E, Lesage B, Gornemann J, Beullens M, Van Meervelt L, Bollen M. The PP1 binding code: a molecular-lego strategy that governs specificity. FEBS J. 2013;280(2):584–95.
28. DeGrande ST, Little SC, Nixon DJ, Wright P, Snyder J, Dun W, et al. Molecular mechanisms underlying cardiac protein phosphatase 2A regulation in heart. J Biol Chem. 2013;288(2):1032–46.
29. Wijnker PJ, Boknik P, Gergs U, Muller FU, Neumann J, dos Remedios C, et al. Protein phosphatase 2A affects myofilament contractility in non-failing but not in failing human myocardium. J Muscle Res Cell Motil. 2011;32(3):221–33.
30. Eichhorn PJ, Creyghton MP, Bernards R. Protein phosphatase 2A regulatory subunits and cancer. Biochim Biophys Acta. 2009;1795(1):1–15.
31. Reynhout S, Janssens V. Physiologic functions of PP2A: lessons from genetically modified mice. Biochim Biophys Acta, Mol Cell Res. 2019;1866(1):31–50.
32. Herzig S, Neumann J. Effects of serine/threonine protein phosphatases on ion channels in excitable membranes. Physiol Rev. 2000;80(1):173–210.
33. Chaklader M, Rothermel BA. Calcineurin in the heart: new horizons for an old friend. Cell Signal. 2021;87:110134.
34. Bhasin N, Cunha SR, Mudannayake M, Gigena MS, Rogers TB, Mohler PJ. Molecular basis for PP2A regulatory subunit B56alpha targeting in cardiomyocytes. Am J Physiol Heart Circ Physiol. 2007;293(1):H109–19.
35. Little SC, Curran J, Makara MA, Kline CF, Ho HT, Xu Z, et al. Protein phosphatase 2A regulatory subunit B56alpha limits phosphatase activity in the heart. Sci Signal. 2015;8(386):ra72.
36. Yin X, Cuello F, Mayr U, Hao Z, Hornshaw M, Ehler E, et al. Proteomics analysis of the cardiac myofilament subproteome reveals dynamic alterations in phosphatase subunit distribution. Mol Cell Proteomics. 2010;9(3):497–509.
37. Hall DD, Feekes JA, Arachchige Don AS, Shi M, Hamid J, Chen L, et al. Binding of protein phosphatase 2A to the L-type calcium channel Cav1.2 next to Ser1928, its main PKA site, is critical for Ser1928 dephosphorylation. Biochemistry. 2006;45(10):3448–59.
38. Kimura T, Han W, Pagel P, Nairn AC, Caplan MJ. Protein phosphatase 2A interacts with the Na,K-ATPase and modulates its trafficking by inhibition of its association with arrestin. PLoS One. 2011;6(12):e29269.

39. Ai X, Jiang A, Ke Y, Solaro RJ, Pogwizd SM. Enhanced activation of p21-activated kinase 1 in heart failure contributes to dephosphorylation of connexin 43. Cardiovasc Res. 2011;92(1):106–14.
40. Carr AN, Schmidt AG, Suzuki Y, del Monte F, Sato Y, Lanner C, et al. Type 1 phosphatase, a negative regulator of cardiac function. Mol Cell Biol. 2002;22(12):4124–35.
41. Ishikawa K, Fish KM, Tilemann L, Rapti K, Aguero J, Santos-Gallego CG, et al. Cardiac I-1c overexpression with reengineered AAV improves cardiac function in swine ischemic heart failure. Mol Ther. 2014;22(12):2038–45.
42. Alsina KM, Hulsurkar M, Brandenburg S, Kownatzki-Danger D, Lenz C, Urlaub H, et al. Loss of protein phosphatase 1 regulatory subunit PPP1R3A promotes atrial fibrillation. Circulation. 2019;140(8):681–93.
43. Gotz J, Probst A, Ehler E, Hemmings B, Kues W. Delayed embryonic lethality in mice lacking protein phosphatase 2A catalytic subunit Calpha. Proc Natl Acad Sci U S A. 1998;95(21):12370–5.
44. Gu P, Qi X, Zhou Y, Wang Y, Gao X. Generation of Ppp2Ca and Ppp2Cb conditional null alleles in mouse. Genesis. 2012;50(5):429–36.
45. Dong D, Li L, Gu P, Jin T, Wen M, Yuan C, et al. Profiling metabolic remodeling in PP2Acalpha deficiency and chronic pressure overload mouse hearts. FEBS Lett. 2015;589(23):3631–9.
46. Li L, Fang C, Xu D, Xu Y, Fu H, Li J. Cardiomyocyte specific deletion of PP2A causes cardiac hypertrophy. Am J Transl Res. 2016;8(4):1769–79.
47. Hoehn M, Zhang Y, Xu J, Gergs U, Boknik P, Werdan K, et al. Overexpression of protein phosphatase 2A in a murine model of chronic myocardial infarction leads to increased adverse remodeling but restores the regulation of beta-catenin by glycogen synthase kinase 3beta. Int J Cardiol. 2015;183:39–46.
48. Chen L, Guo P, Li W, Fang F, Zhu W, Fan J, et al. Perturbation of specific signaling pathways is involved in initiation of mouse liver fibrosis. Hepatology. 2021;73(4):1551–69.
49. Fan JL, Wang FP, Wang S, Liu XL, Wu XN, Chen W, et al. Phenotype and mechanism of inducible ppp2r1a knockout mouse model. Zhonghua Yu Fang Yi Xue Za Zhi. 2018;52(5):530–7.
50. Hu MW, Wang ZB, Jiang ZZ, Qi ST, Huang L, Liang QX, et al. Scaffold subunit Aalpha of PP2A is essential for female meiosis and fertility in mice. Biol Reprod. 2014;91(1):19.
51. Lange L, Marks M, Liu J, Wittler L, Bauer H, Piehl S, et al. Patterning and gastrulation defects caused by the t(w18) lethal are due to loss of Ppp2r1a. Biol Open. 2017;6(6):752–64.
52. Ruediger R, Ruiz J, Walter G. Human cancer-associated mutations in the Aalpha subunit of protein phosphatase 2A increase lung cancer incidence in Aalpha knock-in and knockout mice. Mol Cell Biol. 2011;31(18):3832–44.
53. Brewis N, Ohst K, Fields K, Rapacciuolo A, Chou D, Bloor C, et al. Dilated cardiomyopathy in transgenic mice expressing a mutant A subunit of protein phosphatase 2A. Am J Physiol Heart Circ Physiol. 2000;279(3):H1307–18.
54. Kirchhefer U, Brekle C, Eskandar J, Isensee G, Kucerova D, Muller FU, et al. Cardiac function is regulated by B56alpha-mediated targeting of protein phosphatase 2A (PP2A) to contractile relevant substrates. J Biol Chem. 2014;289(49):33862–73.
55. Belevych AE, Sansom SE, Terentyeva R, Ho HT, Nishijima Y, Martin MM, et al. MicroRNA-1 and -133 increase arrhythmogenesis in heart failure by dissociating phosphatase activity from RyR2 complex. PLoS One. 2011;6(12):e28324.
56. Terentyev D, Belevych AE, Terentyeva R, Martin MM, Malana GE, Kuhn DE, et al. miR-1 overexpression enhances Ca(2+) release and promotes cardiac arrhythmogenesis by targeting PP2A regulatory subunit B56alpha and causing CaMKII-dependent hyperphosphorylation of RyR2. Circ Res. 2009;104(4):514–21.
57. Varadkar P, Despres D, Kraman M, Lozier J, Phadke A, Nagaraju K, et al. The protein phosphatase 2A B56gamma regulatory subunit is required for heart development. Dev Dyn. 2014;243(6):778–90.
58. Kapfhamer D, Berger KH, Hopf FW, Seif T, Kharazia V, Bonci A, et al. Protein Phosphatase 2a and glycogen synthase kinase 3 signaling modulate prepulse inhibition of the acous-

tic startle response by altering cortical M-Type potassium channel activity. J Neurosci. 2010;30(26):8830–40.
59. Garza AE, Rariy CM, Sun B, Williams J, Lasky-Su J, Baudrand R, et al. Variants in striatin gene are associated with salt-sensitive blood pressure in mice and humans. Hypertension. 2015;65(1):211–7.
60. Sents W, Meeusen B, Kalev P, Radaelli E, Sagaert X, Miermans E, et al. PP2A inactivation mediated by PPP2R4 haploinsufficiency promotes cancer development. Cancer Res. 2017;77(24):6825–37.
61. Molkentin JD, Lu JR, Antos CL, Markham B, Richardson J, Robbins J, et al. A calcineurin-dependent transcriptional pathway for cardiac hypertrophy. Cell. 1998;93(2):215–28.
62. Bisserier M, Berthouze-Duquesnes M, Breckler M, Tortosa F, Fazal L, de Regibus A, et al. Carabin protects against cardiac hypertrophy by blocking calcineurin, Ras, and Ca2+/calmodulin-dependent protein kinase II signaling. Circulation. 2015;131(4):390–400; discussion.
63. Lemmens K, Segers VF, Demolder M, Michiels M, Van Cauwelaert P, De Keulenaer GW. Endogenous inhibitors of hypertrophy in concentric versus eccentric hypertrophy. Eur J Heart Fail. 2007;9(4):352–6.
64. Zhu X, Fang J, Gong J, Guo JH, Zhao GN, Ji YX, et al. Cardiac-specific EPI64C blunts pressure overload-induced cardiac hypertrophy. Hypertension. 2016;67(5):866–77.
65. MacDougall LK, Jones LR, Cohen P. Identification of the major protein phosphatases in mammalian cardiac muscle which dephosphorylate phospholamban. Eur J Biochem. 1991;196(3):725–34.
66. Van Wagoner DR, Nerbonne JM. Molecular basis of electrical remodeling in atrial fibrillation. J Mol Cell Cardiol. 2000;32(6):1101–17.
67. El-Armouche A, Boknik P, Eschenhagen T, Carrier L, Knaut M, Ravens U, et al. Molecular determinants of altered Ca2+ handling in human chronic atrial fibrillation. Circulation. 2006;114(7):670–80.
68. Neumann J, Boknik P, Herzig S, Schmitz W, Scholz H, Gupta RC, et al. Evidence for physiological functions of protein phosphatases in the heart: evaluation with okadaic acid. Am J Phys. 1993;265(1 Pt 2):H257–66.
69. Greiser M, Halaszovich CR, Frechen D, Boknik P, Ravens U, Dobrev D, et al. Pharmacological evidence for altered src kinase regulation of I (Ca,L) in patients with chronic atrial fibrillation. Naunyn Schmiedeberg's Arch Pharmacol. 2007;375(6):383–92.
70. Wittkopper K, Dobrev D, Eschenhagen T, El-Armouche A. Phosphatase-1 inhibitor-1 in physiological and pathological beta-adrenoceptor signalling. Cardiovasc Res. 2011;91(3):392–401.
71. Fischer TH, Eiringhaus J, Dybkova N, Saadatmand A, Pabel S, Weber S, et al. Activation of protein phosphatase 1 by a selective phosphatase disrupting peptide reduces sarcoplasmic reticulum Ca(2+) leak in human heart failure. Eur J Heart Fail. 2018;20(12):1673–85.
72. El-Armouche A, Wittkopper K, Fuller W, Howie J, Shattock MJ, Pavlovic D. Phospholemman-dependent regulation of the cardiac Na/K-ATPase activity is modulated by inhibitor-1 sensitive type-1 phosphatase. FASEB J. 2011;25(12):4467–75.
73. El-Armouche A, Rau T, Zolk O, Ditz D, Pamminger T, Zimmermann WH, et al. Evidence for protein phosphatase inhibitor-1 playing an amplifier role in beta-adrenergic signaling in cardiac myocytes. FASEB J. 2003;17(3):437–9.
74. El-Armouche A, Bednorz A, Pamminger T, Ditz D, Didie M, Dobrev D, et al. Role of calcineurin and protein phosphatase-2A in the regulation of phosphatase inhibitor-1 in cardiac myocytes. Biochem Biophys Res Commun. 2006;346(3):700–6.
75. El-Armouche A, Eschenhagen T. Beta-adrenergic stimulation and myocardial function in the failing heart. Heart Fail Rev. 2009;14(4):225–41.
76. Kuster DW, Bawazeer AC, Zaremba R, Goebel M, Boontje NM, van der Velden J. Cardiac myosin binding protein C phosphorylation in cardiac disease. J Muscle Res Cell Motil. 2012;33(1):43–52.

77. van der Velden J. Diastolic myofilament dysfunction in the failing human heart. Pflugers Arch. 2011;462(1):155–63.
78. Barallobre-Barreiro J, Mayr M. Affinity proteomics for phosphatase interactions in atrial fibrillation. J Am Coll Cardiol. 2015;65(2):174–6.
79. Christ T, Boknik P, Wohrl S, Wettwer E, Graf EM, Bosch RF, et al. L-type Ca2+ current downregulation in chronic human atrial fibrillation is associated with increased activity of protein phosphatases. Circulation. 2004;110(17):2651–7.
80. Toischer K, Hartmann N, Wagner S, Fischer TH, Herting J, Danner BC, et al. Role of late sodium current as a potential arrhythmogenic mechanism in the progression of pressure-induced heart disease. J Mol Cell Cardiol. 2013;61:111–22.
81. Chen TC, Law B, Kondratyuk T, Rossie S. Identification of soluble-protein phosphatases that dephosphorylate voltage-sensitive sodium-channels in rat-brain. J Biol Chem. 1995;270(13):7750–6.
82. Xu H, Ginsburg KS, Hall DD, Zimmermann M, Stein IS, Zhang M, et al. Targeting of protein phosphatases PP2A and PP2B to the C-terminus of the L-type calcium channel Ca v1.2. Biochemistry. 2010;49(48):10298–307.
83. Bers DM. Macromolecular complexes regulating cardiac ryanodine receptor function. J Mol Cell Cardiol. 2004;37(2):417–29.
84. Dodge-Kafka KL, Bauman A, Mayer N, Henson E, Heredia L, Ahn J, et al. cAMP-stimulated protein phosphatase 2A activity associated with muscle A kinase-anchoring protein (mAKAP) signaling complexes inhibits the phosphorylation and activity of the cAMP-specific phosphodiesterase PDE4D3. J Biol Chem. 2010;285(15):11078–86.
85. Morad M, Cleemann L, Menick DR. NCX1 phosphorylation dilemma: a little closer to resolution. Focus on "Full-length cardiac Na+/Ca2+ exchanger 1 protein is not phosphorylated by protein kinase A". Am J Physiol Cell Physiol. 2011;300(5):C970–C3.
86. Schulze DH, Muqhal M, Lederer WJ, Ruknudin AM. Sodium/calcium exchanger (NCX1) macromolecular complex. J Biol Chem. 2003;278(31):28849–55.
87. Foulquier S, Daskalopoulos EP, Lluri G, Hermans KCM, Deb A, Blankesteijn WM. WNT signaling in cardiac and vascular disease. Pharmacol Rev. 2018;70(1):68–141.
88. Kimelman D, Xu W. beta-catenin destruction complex: insights and questions from a structural perspective. Oncogene. 2006;25(57):7482–91.
89. Ling SK, Sun Q, Li YH, Zhang L, Zhang PF, Wang XG, et al. CKIP-1 inhibits cardiac hypertrophy by regulating class II histone deacetylase phosphorylation through recruiting PP2A. Circulation. 2012;126(25):3028–U467.
90. Rucker-Martin C, Milliez P, Tan S, Decrouy X, Recouvreur M, Vranckx R, et al. Chronic hemodynamic overload of the atria is an important factor for gap junction remodeling in human and rat hearts. Cardiovasc Res. 2006;72(1):69–79.
91. Ai X, Pogwizd SM. Connexin 43 downregulation and dephosphorylation in nonischemic heart failure is associated with enhanced colocalized protein phosphatase type 2A. Circ Res. 2005;96(1):54–63.
92. Igarashi T, Finet JE, Takeuchi A, Fujino Y, Strom M, Greener ID, et al. Connexin gene transfer preserves conduction velocity and prevents atrial fibrillation. Circulation. 2012;125(2):216–25.
93. Maillet M, Davis J, Auger-Messier M, York A, Osinska H, Piquereau J, et al. Heart-specific deletion of CnB1 reveals multiple mechanisms whereby calcineurin regulates cardiac growth and function. J Biol Chem. 2010;285(9):6716–24.
94. Li X, Li J, Martinez EC, Froese A, Passariello CL, Henshaw K, et al. Calcineurin Abeta-specific anchoring confers isoform-specific compartmentation and function in pathological cardiac myocyte hypertrophy. Circulation. 2020;142(10):948–62.
95. Chen J, Balakrishnan-Renuka A, Hagemann N, Theiss C, Chankiewitz V, Chen J, et al. A novel interaction between ATOH8 and PPP3CB. Histochem Cell Biol. 2016;145(1):5–16.
96. Qi XY, Yeh YH, Xiao L, Burstein B, Maguy A, Chartier D, et al. Cellular signaling underlying atrial tachycardia remodeling of L-type calcium current. Circ Res. 2008;103(8):845–54.

97. Dittmer PJ, Dell'Acqua ML, Sather WA. Ca2+/calcineurin-dependent inactivation of neuronal L-type Ca2+ channels requires priming by AKAP-anchored protein kinase A. Cell Rep. 2014;7(5):1410–6.
98. Wang Y, Tandan S, Hill JA. Calcineurin-dependent ion channel regulation in heart. Trends Cardiovasc Med. 2014;24(1):14–22.
99. Oliveria SF, Dittmer PJ, Youn DH, Dell'Acqua ML, Sather WA. Localized calcineurin confers Ca2+-dependent inactivation on neuronal L-type Ca2+ channels. J Neurosci. 2012;32(44):15328–37.
100. Fraser ID, Cong M, Kim J, Rollins EN, Daaka Y, Lefkowitz RJ, et al. Assembly of an A kinase-anchoring protein-beta(2)-adrenergic receptor complex facilitates receptor phosphorylation and signaling. Curr Biol. 2000;10(7):409–12.
101. Nakamura TY, Iwata Y, Arai Y, Komamura K, Wakabayashi S. Activation of Na+/H+ exchanger 1 is sufficient to generate Ca2+ signals that induce cardiac hypertrophy and heart failure. Circ Res. 2008;103(8):891–9.
102. Katanosaka Y, Iwata Y, Kobayashi Y, Shibasaki F, Wakabayashi S, Shigekawa M. Calcineurin inhibits Na+/Ca2+ exchange in phenylephrine-treated hypertrophic cardiomyocytes. J Biol Chem. 2005;280(7):5764–72.

Metabolic Regulation of Mitochondrial Dynamics and Cardiac Function

Michael W. Rudokas, Marine Cacheux, and Fadi G. Akar

Abstract Increasing lines of evidence highlight cardiac mitochondria as key regulators of myocardial function and arrhythmias in response to acute injury and chronic diseases. In recent years, mitochondria have been shown to form highly *dynamic* organelles that continuously *fuse* and *divide*. These morphological changes, caused by complex fusion and fission events, are essential for normal embryonic development as well as other fundamental processes. Disruption of the balance between mitochondrial fusion and fission results in either highly interconnected mitochondrial networks (favoring fusion) or abnormally fragmented mitochondria (favoring fission). Our understanding of these opposing processes has been markedly advanced by the identification of key proteins that regulate mitochondrial dynamics. Until recently, however, the functional importance of these proteins to the pathogenesis of cardiovascular disorders has received little attention. In this chapter, we discuss recent advances in our understanding of how altered regulation and expression of pro-fusion and pro-fission mitochondrial dynamics proteins impact cardiac function in common cardiovascular diseases.

Keywords Mitochondria · Fusion · Fission · Mitochondrial dynamics proteins · AMPK

M. W. Rudokas · M. Cacheux
Electro-biology & Arrhythmia Therapeutics Laboratory, Cardiovascular Research Center (Y-CVRC), Yale University, New Haven, CT, USA

F. G. Akar (✉)
Electro-biology & Arrhythmia Therapeutics Laboratory, Cardiovascular Research Center (Y-CVRC), Yale University, New Haven, CT, USA

Section of Cardiology, Yale University School of Medicine & School of Engineering and Applied Sciences, New Haven, CT, USA
e-mail: fadi.akar@yale.edu; https://medicine.yale.edu/profile/fadi_akar; https://medicine.yale.edu/lab/akar

N. L. Parinandi, T. J. Hund (eds.), *Cardiovascular Signaling in Health and Disease*, https://doi.org/10.1007/978-3-031-08309-9_6

Cardiac mitochondria fulfill an integral role in cellular energy production via the synthesis of ATP through oxidative phosphorylation [1, 2]. As a metabolic by-product of this process, reactive oxygen species (ROS) are also generated eliciting diverse cell signaling functions, and under certain conditions, promoting cell death [3–6]. Although mitochondria were traditionally viewed as static arbiters of cell death and survival pathways, they have more recently become recognized for their highly dynamic nature, which allows them to form adaptive and interactive networks across cardiomyocytes [7, 8]. Indeed, these networks of mitochondria regulate cardiac function through finely tuned inter-organelle communication. Naturally, this recognition has reinvigorated our interest in understanding, at a fundamental level, the exact role that cardiac mitochondria in general and mitochondrial network dynamics in particular play in both health and disease [9]. In this chapter, we discuss recent advances in our understanding of how changes in the architecture of the intricate mitochondrial network, dictated by fusion and fission events, impact cardiac function in the context of metabolic disease.

Mitochondrial Dynamics: Fusion and Fission Events

Mitochondrial ultrastructure and function are regulated by a multitude of mechanisms that impact the mitochondrial dynamics processes of fusion and fission. Mitochondrial fusion involves the merging of the outer (OMM) and inner (IMM) membranes of neighboring mitochondria to form elaborate interactive mitochondrial networks with synchronized mitochondrial membrane potential and therefore function. Additionally, fusion serves to mix and unify contents of the mitochondrial compartment, including mitochondrial DNA [10]. Mitochondrial fusion is a highly choreographed process that requires key membrane-bound GTPases, namely, mitofusin 1 (Mfn1), mitofusin 2 (Mfn2), and optic atrophy protein 1 (Opa1) [11–14]. The elaborate process of fusion commences with the tethering of mitofusins on the OMM of two neighboring mitochondria. Propelled by a GTP hydrolysis-driven conformational change, the tethered mitofusins merge the two OMMs. Next, a similar process occurs by the tethering of Opa1 and cardiolipin to cause fusion of the IMM [15].

On the opposite end of the spectrum, mitochondrial fission serves to divide rather than unify mitochondria. This conserved process evolved to achieve mitochondrial inheritance, recycle damaged organelles, and release pro-apoptotic factors [10]. The major regulator of mitochondrial fission is the GTPase dynamin-related protein 1 (Drp1). Drp1 is recruited from the cytoplasm to OMM sites, preconstricted by the endoplasmic reticulum [16, 17], via interaction with Drp1 adapter proteins: mitochondrial dynamics proteins of 49 and 51 kDa (Mid49 and Mid51), mitochondrial fission factor (Mff), and possibly mitochondrial fission 1 (Fis1) in mammals [18–20]. Drp1 binding forms multimeric spirals which trigger GTP hydrolysis causing a Drp1 conformational change that prompts the tightening of the spiral [21]. The GTPase dynamin 2 (Dnm2) has been suggested to form a collar-like assembly around this constriction point ultimately resulting in the mitochondrial division by severing the inner and outer mitochondrial membranes [22]. However, more recent research has demonstrated that Dnm2 is dispensable in mitochondrial fission, while Drp1 is not, indicating that Drp1 is sufficient for constricting *and* severing the mitochondrial membranes [23].

Mitochondrial fusion and fission events arise continuously in the native myocardium to maintain a healthy metabolic state. Disruption of these highly regulated processes represents an early adaptation to stress. This can be seen in various cardiac maladies in which dysregulated mitochondrial dynamics lead to either the onset or progression of disease [24]. The following sections will examine the role of cardiac mitochondrial dynamics in common metabolic diseases that lead to cardiac dysfunction and sudden cardiac death, including acute ischemia/reperfusion injury, cardiac hypertrophy, and heart failure (HF).

Diabetes-Related Cardiac Ischemia and Reperfusion Injury

Type-2 diabetes mellitus (T2DM) is a global public health epidemic that continues to expand in both its incidence and prevalence. Recent analyses by the CDC estimate that over 30 million adults in the USA alone are affected [25]. The leading cause of morbidity and mortality in T2DM patients is cardiovascular disease [26]. Indeed, diabetic patients are prone to multiple cardiovascular disorders, including a very high burden of coronary artery disease that culminates in acute ischemia/reperfusion (I/R) events and chronic myocardial infarction (MI).

Central to the cardiovascular pathophysiology of T2DM is oxidative stress driven by increased production and impaired scavenging of reactive oxygen species (ROS). While the sources of ROS in T2DM are multi-factorial, many of the pathophysiological complications of T2DM have now been linked to hyperglycemia-mediated mitochondrial ROS overproduction [27, 28]. Increased generation of ROS in hyperglycemic conditions has been recently shown to require dynamic changes in mitochondrial morphology. Specifically, Yu et al. [29] elegantly demonstrated that mitochondrial fragmentation mediated by the fission process is a necessary component of high glucose-induced ROS overproduction and associated changes in mitochondrial respiration in H9c2 cells. This strongly suggests that mitochondrial fission may be a previously unrecognized hub in the control of ROS production in hyperglycemia-associated disorders [30].

As noted above, cardiovascular complications in diabetes are causally associated with hyperglycemia-induced ROS overproduction. Hyperglycemic conditions increase the input of metabolic substrates into mitochondria, resulting in T2DM-related lipotoxicity and impaired mitochondrial respiratory capacity. Previous studies identified defective mitochondrial respiration with substrates of complexes I, II, and IV and impaired state 4 ➔ 3 energetic transition in isolated mitochondria from leptin receptor-deficient (*db/db*) mice with T2DM [31]. This, in turn, caused a major overflow of mitochondrial ROS that was triggered by high glucose conditions.

Hyperglycemia promotes mitochondrial fragmentation, which in turn, contributes to ROS overproduction [29, 32]. Mitochondrial dynamics, specifically fission, have been implicated in excess ROS generation in cardiac cell lines [30]. In addition, lipotoxicity generated-ROS has been shown to modify the posttranslational state of Drp1 and Opa1 leading to mitochondrial dysfunction, likely due to altered mitochondrial ultrastructure [33]. Supporting evidence for these in vitro observations has come from animal models of diabetes. A streptozotocin mouse model of type 1 diabetes was found to exhibit smaller cardiac interfibrillar mitochondria (IFM) with an enhanced apoptotic propensity compared to nondiabetic mice [34]. Notably, the subsarcolemmal mitochondrial (SSM) population in these experiments did not undergo similar ultrastructural remodeling indicating inherent differences in the response of various mitochondrial subpopulations [34]. Interestingly, the same group reported opposite changes in the mitochondrial subpopulations in *db/db* mice, in which morphological and functional dysfunctions were detected in the SSM with minimal effects on the IFM [35]. The notion that various mitochondrial subpopulations may be subject to differential remodeling in various metabolic diseases was also reaffirmed in a model of combined high-fat diet with streptozotocin challenge, in which the size of IFM but not SSM mitochondria was decreased. Paradoxically, however, SSM but not IFM mitochondria appeared to generate more ROS highlighting an apparent discrepancy between gross abnormalities in mitochondrial ultrastructure and function, at least in this model [36]. Finally, in human atrial myocardium, T2DM was associated with IFM network fragmentation without a change in mitochondrial density [37].

Because ischemic events are typically more frequent and severe in diabetic patients compared to their nondiabetic counterparts, a detailed understanding of the direct effects of I/R per se on mitochondrial dynamics is important. Simulated ischemia in an atrial lineage myogenic cell line (HL-1) revealed extensive mitochondrial fragmentation involving p38 MAPK signaling [38]. Further studies in another cardiac cell line (H9c2) corroborated this observation of ischemia-induced mitochondrial fragmentation [39, 40]. Strikingly, Liu et al. found that hypoxia and reoxygenation of H9c2 cells led to the emergence of toroidal donut-shaped mitochondria [40]. These uniquely shaped organelles were believed to represent a beneficial adaptation for functional mitochondrial recovery possibly due to the opening of potassium channels or altered activity of the mitochondrial permeability transition pore. Whether similar mitochondrial adaptations and morphologies arise in native cardiac myocytes is yet to be determined.

Mitochondria in adult cardiac myocytes undergo Drp1-mediated fission in response to I/R. Indeed, ischemic injury in adult mouse hearts was found to cause mitochondrial fragmentation akin to what was observed in heterologous expression systems [41]. Similar studies supported this observation of mitochondrial fragmentation in adult rodent hearts exposed to I/R [42–44]. I/R-related oxidative phosphorylation inhibition, ROS production, mitochondrial membrane potential depolarization, Ca^{2+} overloading, and other mitochondrial signaling pathways can mediate mitochondrial fragmentation in response to acute injury [45, 46]. Nonetheless, this idea of mitochondrial fragmentation arising as a response to I/R is not universally held. For one, adult rat cardiomyocytes, following I/R, have also been shown to exhibit mitochondrial swelling in the absence of an increase in total mitochondrial number [47]. Therefore, additional studies are needed to further clarify the exact effects of I/R on mitochondrial ultrastructure, function, and dynamics.

While the exact changes in mitochondrial morphology due to I/R may not be fully agreed upon, many studies have clearly demonstrated that the inhibition of Drp1 in the setting of I/R is cardioprotective. In HL-1 cells, pharmacologic Drp1 inhibition with the mitochondrial division inhibitor-1 (MDIVI-1) or expression of a dominant negative Drp1 mutant both led to an increase in elongated mitochondria, a decrease in mitochondrial permeability transition pore sensitivity, and a reduction in the rate of cell death in response to simulated I/R [41]. Although MDIVI-1 may exert nonspecific effects independent of its Drp1 inhibitory action [48], a more specific pharmacologic Drp1 inhibitor, Drpitor1a, was also found to inhibit mitochondrial fission and preserve cardiac function in I/R, reaffirming this strategy as a potentially viable therapeutic option [49]. Similar to direct Drp1 inhibition, its indirect suppression with the antiapoptotic and pro-proliferative kinase Pim-1 also preserved mitochondrial phenotype in neonatal rat cardiomyocytes exposed to simulated ischemia [50]. Another study which expressed the dominant negative mutant of Drp1 in rat hearts in vivo demonstrated a decrease in infarct size and cell death in response to I/R with improved cardiac function [44]. A heterozygous Drp1 knockout mouse, which exhibited a 60% decrease in myocardial Drp1 protein expression, also developed a smaller infarct size compared to control mice after I/R [51]. Taken together, these studies indicate that the prevention of mitochondrial fission through Drp1 downregulation is indeed cardioprotective against I/R injury. Yet, seemingly contradictory evidence has also been presented in the literature highlighting a more nuanced role for Drp1-mediated fission in the regulation of I/R injury. Cardiac-specific Drp1 knockout mice exhibited more elongated and damaged mitochondria and a decrease in autophagic flux, which led to mitochondrial dysfunction and consequently left ventricular dysfunction followed by premature death at 13 weeks [42]. Heterozygous versions of these mice exhibited normal left ventricular function at 12 weeks of age but were more susceptible to I/R events, as seen by an increase in infarct size compared to control mice [42]. This suggests that a basal level of Drp1 expression is required to maintain normal mitochondrial and cardiac homeostasis, but that acute Drp1 inhibition pharmacologically or through gene silencing may be a viable therapeutic strategy for I/R. Importantly, Drp1 inhibition should be a temporary measure, because chronic inhibition of this critical mitochondrial division protein most likely interferes with adaptative mechanisms needed to maintain mitochondrial homeostasis in the long term.

At the opposite end of the mitochondrial dynamics spectrum, several studies have found that promotion of mitochondrial fusion, as opposed to inhibition of fission, is equally advantageous for preventing the deleterious effects of I/R. Overexpression of the pro-fusion regulatory protein Opa1 elicited clear cardioprotective effects against I/R injury in a genetic mouse model [52]. Pharmacological stimulation of fusion improved mitochondrial function, reduced infarct size, and attenuated cardiac apoptosis in a rat model of I/R [53]. Simulated ischemia in H9c2 cells reduced Opa1 protein levels. The functional significance of this remodeling was borne out of studies in which short hairpin RNA mediated reduction of Opa1 expression caused an increase in apoptosis and mitochondrial fragmentation [39]. In a mouse model with decreased cardiac Opa1 expression, infarct size was significantly greater following I/R [54]. Overexpression of Mfn2 in neonatal rat ventricular myocytes prevented cell death following simulated I/R [55]. Mice with cardiac-specific Mfn2 depletion exhibited an increase in cardiac cell death following I/R [56]. Lastly, activation of Akt signaling, whose mitochondrial elongating effects are associated with increased Mfn1 activity, also reduced myocardial infarct size likely by decreasing the extent of mitochondrial fragmentation following I/R [57].

Despite the aforementioned body of evidence, the role of mitochondrial fusion in I/R injury is also nuanced with some seemingly contradictory findings. For example, adult cardiac myocytes with Mfn2 deletion were found to be protected from (not prone to) cell death due to hypoxia/reoxygenation [58]. Furthermore, Mfn2 deletion mitigated infarct size likely by suppressing apoptosis after I/R [58]. Other studies, although not directly employing I/R methodologies, showed that Mfn2 silencing in H9c2 cells or Mfn1 in cardiomyocytes led to protection against ROS-mediated apoptosis [59, 60]. Finally, overexpression of Opa1 decreased mitochondrial fragmentation but did not confer additional protection against simulated ischemia-induced apoptosis in H9c2 cells [39].

Clearly, I/R injury elicits direct effects on cardiac mitochondrial morphology, but which mitochondrial dynamics processes and proteins are responsible for these morphological changes still requires further clarification. It appears that inhibition of the fission regulatory protein Drp1 on a short time scale is cardioprotective against I/R injury while its chronic and complete inhibition may be detrimental, likely by disrupting the delicate balance between fusion and fission processes in the long term. Similarly, the impact of altered expression of fusion regulatory proteins in I/R requires further investigation to address fundamental discrepancies. Indeed, it is likely that cardioprotection against I/R requires a finely tuned activation of a host of mitochondrial dynamics proteins rather than a sledgehammer approach aimed at artificially increasing or decreasing the expression of only one of those interacting players. Overall, while significant progress has been made, more studies are still needed to address the conflicting evidence and to create a more comprehensive understanding of mitochondrial dynamics in all cell types of the heart before modulation of these processes can be leveraged for the treatment of I/R-related cardiac dysfunction, especially in the context of metabolic disease in which classically cardioprotective pathways that converge on mitochondria are generally impaired [61].

Diabetes-Related Cardiac Chronic Remodeling, Hypertrophy, and Failure

In addition to acute ischemic injury, T2DM is also associated with adverse chronic cardiac remodeling that can culminate in left ventricular hypertrophy (LVH) and heart failure (HF) typically with a preserved ejection fraction [62, 63]. LVH in response to increased systemic pressure, a common manifestation of metabolic disease, is an adaptive mechanism that helps maintain cardiac output in the face of external stress. However, because LVH is associated with adverse remodeling, it ultimately becomes a leading risk factor for HF and ventricular arrhythmias [64]. Changes in mitochondrial energetics are a hallmark of hypertrophied myocytes. Numerous studies have shown that cardiac mitochondria are indeed central to the pathophysiology of LVH. For example, a marked shift in substrate utilization (free fatty acid to glucose) for energy production has been documented in various animal models and humans with hypertrophy [65, 66].

A clear association between hypertrophy and altered mitochondrial dynamics/ultrastructure has also been observed. Exposure of neonatal rat cardiomyocytes to norepinephrine, a hypertrophy-inducing catecholamine, caused substantial mitochondrial fragmentation and decreased metabolic function [67]. In this study by Pennanen et al. [67], mitochondrial fragmentation was likely due to an increase in cytoplasmic Ca^{2+}, which prompted the recruitment of Drp1 to mitochondria. Finally, treatment with a dominant negative form of Drp1 not only inhibited mitochondrial fission but also prevented norepinephrine-mediated hypertrophic growth [67]. Additional studies in multiple experimental models further confirmed the role of Drp1 in the pathogenesis of adverse remodeling and highlighted its inhibition as a viable anti-hypertrophic target [68, 69]. However, Drp1 is also known to regulate essential processes, such as mitophagy, which may counterbalance some of the benefits that stem from its inhibition. In that regard, the challenge of a cardiac-specific heterozygous Drp1 knockout mouse model with pressure overload was associated with worse cardiac remodeling likely reflecting the impact of Drp1 on mitophagy [70].

In addition to Drp1, other regulatory proteins that promote fusion rather than fission also appear to play a major role in hypertrophic disorders. Mfn2 was found to be downregulated in a variety of in vitro and in vivo models of hypertrophy with a few exceptions [71]. In neonatal rat cardiomyocytes, sole inhibition of Mfn2 led to mitochondrial fragmentation and cellular hypertrophy [67]. Conversely, Mfn2 overexpression prevented cellular and organ level hypertrophy and mitochondrial dysfunction in angiotensin II-induced models of LVH [72, 73]. It is believed that Mfn2 provides these protective effects through the promotion of mitochondrial fusion and/or autophagy [72]. Opa1 may also have a role in hypertrophy, as Opa1 haploinsufficient mice subjected to transversal aortic constriction were reported to develop almost double the cardiac hypertrophy phenotype compared to control mice [74]. More studies are required to fully understand the function of fusion regulatory proteins in LVH, both with and without the added influence of metabolic disease.

Complicating matters further, hypertrophy is a general term that can be divided into sub-classifications with differential effects on mitochondrial dynamics processes. Typically, an early phase of concentric ("compensated") hypertrophy is followed by eccentric volume overload-induced hypertrophy ("decompensated"), which then devolves into heart failure [75, 76]. An initial study demonstrated that "compensated" hypertrophy was accompanied by decreased Drp1, increased Opa1, and no changes to Mfn1/2 expression [77]. This appears to contradict the aforementioned studies which clearly documented decreases in Mfn2 as well as the importance of Drp1 in the development of hypertrophy. Therefore, studies investigating the specific hypertrophic phases that lead to eventual HF need to be continued to capture the range of pathophysiological changes in mitochondrial dynamics during disease progression. Such studies will help inform which changes may be deleterious and which may be adaptive during the course of remodeling.

Once a heart has progressed from hypertrophy to failure, mitochondrial fragmentation is clearly evident, indicating major dysregulation of mitochondrial dynamics processes [39, 78, 79]. Furthermore, the level of mitochondrial fragmentation appears to correlate with the severity of HF [78]. Again, specific mitochondrial dynamics processes and regulatory proteins may vary depending on the stage and etiology of HF. For example, Opa1 was reported to be decreased in a rat model of ischemic HF, but another study using a similar model showed that Opa1 expression was preserved until much later in the course of HF [39, 80]. Similarly, Opa1 was shown to be decreased in human ischemic HF but not in nonischemic HF [39].

Beyond changes in overall expression, altered posttranslational modifications of mitochondrial dynamics proteins may be a major cause of mitochondrial dysfunction in HF. As an example, the expression of forkhead box O3a (FOXO3a) transcription factor and its effectors were found to be increased in rodent and human HF and to correlate with disease severity [81, 82]. FOXO3a upregulates BCL2 interacting protein 3 (BNIP3), which leads to mitochondrial membrane depolarization, mitochondrial fragmentation, and apoptosis [83]. This is believed to be mediated either via activation of the Ca^{2+}-dependent phosphatase calcineurin, which dephosphorylates the serine-637 site on Drp1 increasing its translocation to mitochondria and promoting fission or via BNIP3-induced inhibition of the fission-suppressing Drp1 phosphorylation by protein kinase A (PKA) [84, 85].

Many studies have focused on the rescue of mitochondrial dysfunction in metabolic disease by targeting mitochondrial dynamics. A previously mentioned study used overexpression of a dominant negative form of Drp1 in cell lines to prevent high glucose-induced mitochondrial fragmentation and ROS overproduction [29]. Overexpression of Mfn2, which stimulated mitochondrial elongation and prevented an increase in ROS due to high glucose exposure, was also used [29, 30]. Promotion of fusion by overexpression of Opa1 rescued mitochondrial fragmentation and mitochondrial function in neonatal cardiac myocytes under high glucose conditions [86]. Furthermore, suppression of Drp1 by small interfering RNAs in H9c2 cells exposed to oxidative stress reduced the extent of mitochondrial fragmentation and restored insulin signal transduction [87]. The implication that insulin signaling is tied to mitochondrial dynamics in cardiac cells was further investigated in another study, which demonstrated that Opa1 levels and consequently mitochondrial dynamics are regulated by insulin-mediated Akt-mTOR-NFκB signaling [88].

Pharmacological studies using the mitochondrial division inhibitors (MDIVI-1 & P110), which have distinct mechanisms of action on Drp1 [89], have also been used to reduce myocyte apoptosis, in vitro and infarct size, in vivo [90]. The role of Drp1 in the pathogenesis of diabetes-related cardiovascular disorders has received limited attention [29, 87, 91, 92]. A notable recent study provided compelling evidence of increased Drp1 translocation to mitochondria and decreased mitochondrial size consistent with increased fission in response to I/R injury in a mouse model of T2DM [93]. In vivo delivery of MDIVI-1 to these mice inhibited Drp1 translocation to mitochondria and decreased mitochondrial fission. Furthermore, pharmacological inhibition of Drp1 reduced MI size and decreased serum cardiac troponin and lactate dehydrogenase levels [93].

In summary, cardiac mitochondrial dynamics appear to play a significant and underappreciated role in common metabolic diseases that lead to cardiac dysfunction and sudden cardiac death. Hence, a better understanding of upstream mechanisms that regulate mitochondrial dynamics proteins may offer new therapeutic avenues. In what follows, we focus on one axis of this regulation that is mediated by signaling via the master metabolic sensor AMP-activated protein kinase (AMPK).

Metabolic Control of Mitochondrial Dynamics: The Role of AMPK

Recent studies have shown robust regulation of mitochondrial fission in the heart by a signaling cascade involving AMPK [91, 94, 95], a serine/threonine kinase that is ubiquitously expressed in various cell types and tissues in the body, including the heart. As a master metabolic sensor, AMPK plays a fundamental role in the regulation of the cardiac response to acute and chronic stressors. In a pivotal study, AMPK was shown to regulate the fission portion of mitochondrial dynamics [96]. Specifically, Toyama et al. demonstrated that direct pharmacological activation of AMPK promoted mitochondrial fission likely via direct phosphorylation at serine-155 and serine-173 on Mff, the mitochondrial outer-membrane protein [97]. Mff is an essential adapter protein involved in the recruitment of Drp1 to the mitochondrial membrane [98]. This AMPK-driven mechanism for fission was experimentally supported by studies documenting an increase in Drp1 mitochondrial localization only when the Mff phosphorylation sites were intact. The necessity for AMPK-mediated phosphorylation of Mff to control mitochondrial fission was further confirmed in skeletal muscle cells [99]. The serine-45 site of Armadillo repeat-containing protein 10 (ARMC10) was also found to be yet another direct AMPK phosphorylation substrate involved in the control of mitochondrial dynamics [100]. ARMC10 is localized at the OMM where it associates with Mff and Drp1 likely participating in the regulation of mitochondrial fission. The role of AMPK in the control of mitochondrial fission was also determined from studies in which CRISPR gene editing was used to disrupt AMPKα1 and/or AMPKα2 in vitro. In these studies, AMPK depletion was found to abrogate the mitochondrial fission response induced by rotenone [11].

In the context of diabetes, however, mitochondrial fission can be inhibited by activation of AMPK pathways that appear to alter Drp1 phosphorylation [91]. Accumulating (but largely circumstantial) evidence is pointing toward a critical role of this so-called AMPK-Drp1 axis in the cardioprotective effects of various interventions in T2DM. For example, Zhou et al. demonstrated that the sodium/glucose cotransporter 2 (SGLT2) inhibitor empagliflozin improved myocardial function, preserved the integrity of the microvascular barrier, sustained eNOS phosphorylation, and improved endothelium-dependent relaxation [91]. Importantly, these beneficial effects were mediated by AMPK-dependent inhibition of mitochondrial fission [91]. The importance of this axis in modulating long-term cardiac remodeling is further exemplified by a recent study in which overexpression of Sirtuin-3 reduced cardiac fibrosis, improved myocardial function, inhibited the inflammatory response, and reduced cardiomyocyte death by attenuating mitochondrial fission [94]. Once again, AMPK inhibition was sufficient to reactivate Drp1 and abrogate the protective effects of Sirtuin-3 on mitochondrial fission, reaffirming the role of the AMPK-Drp1 axis in this process [94].

Concluding Remarks

Mitochondria are highly dynamic organelles that undergo fusion and fission events. These morphological changes allow them to form interactive functional networks that span the entire cardiomyocyte. Disruption to the balance between mitochondrial fusion and fission is tied to the development of cardiac myocyte dysfunction in metabolic diseases. Furthermore, mitochondrial dynamics have been implicated in the initiation and progression of associated cardiac disorders that culminate in heart failure and likely sudden cardiac death. Investigations into the control of mitochondrial dynamics through upstream metabolic signaling cascades, including regulation by the master metabolic sensor AMPK, are ongoing in our laboratory and many others. These studies, we hope, will provide new avenues for the treatment of cardiac dysfunction in metabolic diseases.

References

1. Gustafsson AB, Gottlieb RA. Heart mitochondria: gates of life and death. Cardiovasc Res. 2008;77(2):334–43.
2. O'Rourke B. Apoptosis: rekindling the mitochondrial fire. Circ Res. 1999;85(10):880–3.
3. Zorov DB, Juhaszova M, Sollott SJ. Mitochondrial ROS-induced ROS release: an update and review. Biochim Biophys Acta. 2006;1757(5–6):509–17.
4. Droge W. Free radicals in the physiological control of cell function. Physiol Rev. 2002;82(1):47–95.
5. Biary N, Akar FG. A brighter side of ROS revealed by selective activation of beta-adrenergic receptor subtypes. J Physiol. 2010;588(Pt 16):2973–4.
6. Becker LB. New concepts in reactive oxygen species and cardiovascular reperfusion physiology. Cardiovasc Res. 2004;61(3):461–70.
7. Aon MA. From isolated to networked: a paradigmatic shift in mitochondrial physiology. Front Physiol. 2010;1(1):20.
8. Aon MA, Cortassa S, Akar FG, Brown DA, Zhou L, O'Rourke B. From mitochondrial dynamics to arrhythmias. Int J Biochem Cell Biol. 2009;41(10):1940–8.

9. Michelakis ED. Mitochondrial medicine: a new era in medicine opens new windows and brings new challenges. Circulation. 2008;117(19):2431–4.
10. Westermann B. Mitochondrial fusion and fission in cell life and death. Nat Rev Mol Cell Biol. 2010;11(12):872–84.
11. Chen H, Detmer SA, Ewald AJ, Griffin EE, Fraser SE, Chan DC. Mitofusins Mfn1 and Mfn2 coordinately regulate mitochondrial fusion and are essential for embryonic development. J Cell Biol. 2003;160(2):189–200.
12. Cipolat S, Martins de Brito O, Dal Zilio B, Scorrano L. OPA1 requires mitofusin 1 to promote mitochondrial fusion. Proc Natl Acad Sci U S A. 2004;101(45):15927–32.
13. Meeusen S, McCaffery JM, Nunnari J. Mitochondrial fusion intermediates revealed in vitro. Science. 2004;305(5691):1747–52.
14. Song Z, Ghochani M, McCaffery JM, Frey TG, Chan DC. Mitofusins and OPA1 mediate sequential steps in mitochondrial membrane fusion. Mol Biol Cell. 2009;20(15):3525–32.
15. Tilokani L, Nagashima S, Paupe V, Prudent J. Mitochondrial dynamics: overview of molecular mechanisms. Essays Biochem. 2018;62(3):341–60.
16. Friedman JR, Lackner LL, West M, DiBenedetto JR, Nunnari J, Voeltz GK. ER tubules mark sites of mitochondrial division. Science. 2011;334(6054):358–62.
17. Lewis SC, Uchiyama LF, Nunnari J. ER-mitochondria contacts couple mtDNA synthesis with mitochondrial division in human cells. Science. 2016;353(6296):aaf5549.
18. Gandre-Babbe S, van der Bliek AM. The novel tail-anchored membrane protein Mff controls mitochondrial and peroxisomal fission in mammalian cells. Mol Biol Cell. 2008;19(6):2402–12.
19. James DI, Parone PA, Mattenberger Y, Martinou JC. hFis1, a novel component of the mammalian mitochondrial fission machinery. J Biol Chem. 2003;278(38):36373–9.
20. Palmer CS, Osellame LD, Laine D, Koutsopoulos OS, Frazier AE, Ryan MT. MiD49 and MiD51, new components of the mitochondrial fission machinery. EMBO Rep. 2011;12(6):565–73.
21. van der Bliek AM, Shen Q, Kawajiri S. Mechanisms of mitochondrial fission and fusion. Cold Spring Harb Perspect Biol. 2013;5(6):a011072.
22. Lee JE, Westrate LM, Wu H, Page C, Voeltz GK. Multiple dynamin family members collaborate to drive mitochondrial division. Nature. 2016;540(7631):139–43.
23. Kamerkar SC, Kraus F, Sharpe AJ, Pucadyil TJ, Ryan MT. Dynamin-related protein 1 has membrane constricting and severing abilities sufficient for mitochondrial and peroxisomal fission. Nat Commun. 2018;9(1):5239.
24. Forte M, Schirone L, Ameri P, Basso C, Catalucci D, Modica J, et al. The role of mitochondrial dynamics in cardiovascular diseases. Br J Pharmacol. 2021;178(10):2060–76.
25. Prevention CfDCa. National Diabetes Statistics Report, 2020. Atlanta: Centers for Disease Control and Prevention, U.S. Dept of Health and Human Services; 2020.
26. Go AS, Mozaffarian D, Roger VL, Benjamin EJ, Berry JD, Borden WB, et al. Heart disease and stroke statistics--2013 update: a report from the American Heart Association. Circulation. 2013;127(1):e6–e245.
27. Brownlee M. Biochemistry and molecular cell biology of diabetic complications. Nature. 2001;414(6865):813–20.
28. Shah MS, Brownlee M. Molecular and cellular mechanisms of cardiovascular disorders in diabetes. Circ Res. 2016;118(11):1808–29.
29. Yu T, Robotham JL, Yoon Y. Increased production of reactive oxygen species in hyperglycemic conditions requires dynamic change of mitochondrial morphology. Proc Natl Acad Sci U S A. 2006;103(8):2653–8.
30. Yu T, Sheu SS, Robotham JL, Yoon Y. Mitochondrial fission mediates high glucose-induced cell death through elevated production of reactive oxygen species. Cardiovasc Res. 2008;79(2):341–51.
31. Tocchetti CG, Caceres V, Stanley BA, Xie C, Shi S, Watson WH, et al. GSH or palmitate preserves mitochondrial energetic/redox balance, preventing mechanical dysfunc-

tion in metabolically challenged myocytes/hearts from type 2 diabetic mice. Diabetes. 2012;61(12):3094–105.
32. Vanhorebeek I, De Vos R, Mesotten D, Wouters PJ, De Wolf-Peeters C, Van den Berghe G. Protection of hepatocyte mitochondrial ultrastructure and function by strict blood glucose control with insulin in critically ill patients. Lancet. 2005;365(9453):53–9.
33. Tsushima K, Bugger H, Wende AR, Soto J, Jenson GA, Tor AR, et al. Mitochondrial reactive oxygen species in lipotoxic hearts induce post-translational modifications of AKAP121, DRP1, and OPA1 that promote mitochondrial fission. Circ Res. 2018;122(1):58–73.
34. Williamson CL, Dabkowski ER, Baseler WA, Croston TL, Alway SE, Hollander JM. Enhanced apoptotic propensity in diabetic cardiac mitochondria: influence of subcellular spatial location. Am J Physiol Heart Circ Physiol. 2010;298(2):H633–42.
35. Dabkowski ER, Baseler WA, Williamson CL, Powell M, Razunguzwa TT, Frisbee JC, et al. Mitochondrial dysfunction in the type 2 diabetic heart is associated with alterations in spatially distinct mitochondrial proteomes. Am J Physiol Heart Circ Physiol. 2010;299(2):H529–40.
36. Koncsos G, Varga ZV, Baranyai T, Boengler K, Rohrbach S, Li L, et al. Diastolic dysfunction in prediabetic male rats: role of mitochondrial oxidative stress. Am J Physiol Heart Circ Physiol. 2016;311(4):H927–H43.
37. Montaigne D, Marechal X, Coisne A, Debry N, Modine T, Fayad G, et al. Myocardial contractile dysfunction is associated with impaired mitochondrial function and dynamics in type 2 diabetic but not in obese patients. Circulation. 2014;130(7):554–64.
38. Brady NR, Hamacher-Brady A, Gottlieb RA. Proapoptotic BCL-2 family members and mitochondrial dysfunction during ischemia/reperfusion injury, a study employing cardiac HL-1 cells and GFP biosensors. Biochim Biophys Acta. 2006;1757(5–6):667–78.
39. Chen L, Gong Q, Stice JP, Knowlton AA. Mitochondrial OPA1, apoptosis, and heart failure. Cardiovasc Res. 2009;84(1):91–9.
40. Liu X, Hajnoczky G. Altered fusion dynamics underlie unique morphological changes in mitochondria during hypoxia-reoxygenation stress. Cell Death Differ. 2011;18(10):1561–72.
41. Ong SB, Subrayan S, Lim SY, Yellon DM, Davidson SM, Hausenloy DJ. Inhibiting mitochondrial fission protects the heart against ischemia/reperfusion injury. Circulation. 2010;121(18):2012–22.
42. Ikeda Y, Shirakabe A, Maejima Y, Zhai P, Sciarretta S, Toli J, et al. Endogenous Drp1 mediates mitochondrial autophagy and protects the heart against energy stress. Circ Res. 2015;116(2):264–78.
43. Wang JX, Jiao JQ, Li Q, Long B, Wang K, Liu JP, et al. miR-499 regulates mitochondrial dynamics by targeting calcineurin and dynamin-related protein-1. Nat Med. 2011;17(1):71–8.
44. Zepeda R, Kuzmicic J, Parra V, Troncoso R, Pennanen C, Riquelme JA, et al. Drp1 loss-of-function reduces cardiomyocyte oxygen dependence protecting the heart from ischemia-reperfusion injury. J Cardiovasc Pharmacol. 2014;63(6):477–87.
45. Hernandez-Resendiz S, Prunier F, Girao H, Dorn G, Hausenloy DJ, Action E-CC. Targeting mitochondrial fusion and fission proteins for cardioprotection. J Cell Mol Med. 2020;24(12):6571–85.
46. Ong SB, Hausenloy DJ. Mitochondrial dynamics as a therapeutic target for treating cardiac diseases. Handb Exp Pharmacol. 2017;240:251–79.
47. Sharp WW, Fang YH, Han M, Zhang HJ, Hong Z, Banathy A, et al. Dynamin-related protein 1 (Drp1)-mediated diastolic dysfunction in myocardial ischemia-reperfusion injury: therapeutic benefits of Drp1 inhibition to reduce mitochondrial fission. FASEB J. 2014;28(1):316–26.
48. Bordt EA, Clerc P, Roelofs BA, Saladino AJ, Tretter L, Adam-Vizi V, et al. The putative Drp1 inhibitor mdivi-1 is a reversible mitochondrial complex I inhibitor that modulates reactive oxygen species. Dev Cell. 2017;40(6):583–94 e6.
49. Wu D, Dasgupta A, Chen KH, Neuber-Hess M, Patel J, Hurst TE, et al. Identification of novel dynamin-related protein 1 (Drp1) GTPase inhibitors: therapeutic potential of Drpitor1 and Drpitor1a in cancer and cardiac ischemia-reperfusion injury. FASEB J. 2020;34(1):1447–64.
50. Din S, Mason M, Volkers M, Johnson B, Cottage CT, Wang Z, et al. Pim-1 preserves mitochondrial morphology by inhibiting dynamin-related protein 1 translocation. Proc Natl Acad Sci U S A. 2013;110(15):5969–74.

51. Bouche L, Kamel R, Tamareille S, Garcia G, Villedieu C, Pillot B, et al. DRP1 haploinsufficiency attenuates cardiac ischemia/reperfusion injuries. PLoS One. 2021;16(3):e0248554.
52. Varanita T, Soriano ME, Romanello V, Zaglia T, Quintana-Cabrera R, Semenzato M, et al. The OPA1-dependent mitochondrial cristae remodeling pathway controls atrophic, apoptotic, and ischemic tissue damage. Cell Metab. 2015;21(6):834–44.
53. Maneechote C, Palee S, Kerdphoo S, Jaiwongkam T, Chattipakorn SC, Chattipakorn N. Balancing mitochondrial dynamics via increasing mitochondrial fusion attenuates infarct size and left ventricular dysfunction in rats with cardiac ischemia/reperfusion injury. Clin Sci (Lond). 2019;133(3):497–513.
54. Le Page S, Niro M, Fauconnier J, Cellier L, Tamareille S, Gharib A, et al. Increase in cardiac ischemia-reperfusion injuries in Opa1+/– mouse model. PLoS One. 2016;11(10):e0164066.
55. Olmedo I, Pino G, Riquelme JA, Aranguiz P, Diaz MC, Lopez-Crisosto C, et al. Inhibition of the proteasome preserves Mitofusin-2 and mitochondrial integrity, protecting cardiomyocytes during ischemia-reperfusion injury. Biochim Biophys Acta Mol basis Dis. 2020;1866(5):165659.
56. Zhao T, Huang X, Han L, Wang X, Cheng H, Zhao Y, et al. Central role of mitofusin 2 in autophagosome-lysosome fusion in cardiomyocytes. J Biol Chem. 2012;287(28):23615–25.
57. Ong SB, Hall AR, Dongworth RK, Kalkhoran S, Pyakurel A, Scorrano L, et al. Akt protects the heart against ischaemia-reperfusion injury by modulating mitochondrial morphology. Thromb Haemost. 2015;113(3):513–21.
58. Papanicolaou KN, Khairallah RJ, Ngoh GA, Chikando A, Luptak I, O'Shea KM, et al. Mitofusin-2 maintains mitochondrial structure and contributes to stress-induced permeability transition in cardiac myocytes. Mol Cell Biol. 2011;31(6):1309–28.
59. Papanicolaou KN, Ngoh GA, Dabkowski ER, O'Connell KA, Ribeiro RF Jr, Stanley WC, et al. Cardiomyocyte deletion of mitofusin-1 leads to mitochondrial fragmentation and improves tolerance to ROS-induced mitochondrial dysfunction and cell death. Am J Physiol Heart Circ Physiol. 2012;302(1):H167–79.
60. Shen T, Zheng M, Cao C, Chen C, Tang J, Zhang W, et al. Mitofusin-2 is a major determinant of oxidative stress-mediated heart muscle cell apoptosis. J Biol Chem. 2007;282(32):23354–61.
61. Xie C, Hu J, Motloch LJ, Karam BS, Akar FG. The classically cardioprotective agent diazoxide elicits arrhythmias in type 2 diabetes mellitus. J Am Coll Cardiol. 2015;66(10):1144–56.
62. Eguchi K, Boden-Albala B, Jin Z, Rundek T, Sacco RL, Homma S, et al. Association between diabetes mellitus and left ventricular hypertrophy in a multiethnic population. Am J Cardiol. 2008;101(12):1787–91.
63. Kenny HC, Abel ED. Heart failure in type 2 diabetes mellitus. Circ Res. 2019;124(1):121–41.
64. Jin H, Nass RD, Joudrey PJ, Lyon AR, Chemaly ER, Rapti K, et al. Altered spatiotemporal dynamics of the mitochondrial membrane potential in the hypertrophied heart. Biophys J. 2010;98(10):2063–71.
65. Leong HS, Brownsey RW, Kulpa JE, Allard MF. Glycolysis and pyruvate oxidation in cardiac hypertrophy--why so unbalanced? Comp Biochem Physiol A Mol Integr Physiol. 2003;135(4):499–513.
66. Aung LHH, Jumbo JCC, Wang Y, Li P. Therapeutic potential and recent advances on targeting mitochondrial dynamics in cardiac hypertrophy: a concise review. Mol Ther Nucleic Acids. 2021;25:416–43.
67. Pennanen C, Parra V, Lopez-Crisosto C, Morales PE, Del Campo A, Gutierrez T, et al. Mitochondrial fission is required for cardiomyocyte hypertrophy mediated by a Ca2+-calcineurin signaling pathway. J Cell Sci. 2014;127(Pt 12):2659–71.
68. Chang YW, Chang YT, Wang Q, Lin JJ, Chen YJ, Chen CC. Quantitative phosphoproteomic study of pressure-overloaded mouse heart reveals dynamin-related protein 1 as a modulator of cardiac hypertrophy. Mol Cell Proteomics. 2013;12(11):3094–107.
69. Hasan P, Saotome M, Ikoma T, Iguchi K, Kawasaki H, Iwashita T, et al. Mitochondrial fission protein, dynamin-related protein 1, contributes to the promotion of hypertensive cardiac hypertrophy and fibrosis in Dahl-salt sensitive rats. J Mol Cell Cardiol. 2018;121:103–6.

70. Shirakabe A, Zhai P, Ikeda Y, Saito T, Maejima Y, Hsu CP, et al. Drp1-dependent mitochondrial autophagy plays a protective role against pressure overload-induced mitochondrial dysfunction and heart failure. Circulation. 2016;133(13):1249–63.
71. Fang L, Moore XL, Gao XM, Dart AM, Lim YL, Du XJ. Down-regulation of mitofusin-2 expression in cardiac hypertrophy in vitro and in vivo. Life Sci. 2007;80(23):2154–60.
72. Xiong W, Ma Z, An D, Liu Z, Cai W, Bai Y, et al. Mitofusin 2 participates in mitophagy and mitochondrial fusion against angiotensin II-induced cardiomyocyte injury. Front Physiol. 2019;10:411.
73. Yu H, Guo Y, Mi L, Wang X, Li L, Gao W. Mitofusin 2 inhibits angiotensin II-induced myocardial hypertrophy. J Cardiovasc Pharmacol Ther. 2011;16(2):205–11.
74. Piquereau J, Caffin F, Novotova M, Prola A, Garnier A, Mateo P, et al. Down-regulation of OPA1 alters mouse mitochondrial morphology, PTP function, and cardiac adaptation to pressure overload. Cardiovasc Res. 2012;94(3):408–17.
75. Frey N, Katus HA, Olson EN, Hill JA. Hypertrophy of the heart: a new therapeutic target? Circulation. 2004;109(13):1580–9.
76. Dorn GW 2nd, Robbins J, Sugden PH. Phenotyping hypertrophy: eschew obfuscation. Circ Res. 2003;92(11):1171–5.
77. Tang Y, Mi C, Liu J, Gao F, Long J. Compromised mitochondrial remodeling in compensatory hypertrophied myocardium of spontaneously hypertensive rat. Cardiovasc Pathol. 2014;23(2):101–6.
78. Sabbah HN, Sharov V, Riddle JM, Kono T, Lesch M, Goldstein S. Mitochondrial abnormalities in myocardium of dogs with chronic heart failure. J Mol Cell Cardiol. 1992;24(11):1333–47.
79. Schaper J, Froede R, Hein S, Buck A, Hashizume H, Speiser B, et al. Impairment of the myocardial ultrastructure and changes of the cytoskeleton in dilated cardiomyopathy. Circulation. 1991;83(2):504–14.
80. Javadov S, Rajapurohitam V, Kilic A, Hunter JC, Zeidan A, Said Faruq N, et al. Expression of mitochondrial fusion-fission proteins during post-infarction remodeling: the effect of NHE-1 inhibition. Basic Res Cardiol. 2011;106(1):99–109.
81. Chaanine AH, Jeong D, Liang L, Chemaly ER, Fish K, Gordon RE, et al. JNK modulates FOXO3a for the expression of the mitochondrial death and mitophagy marker BNIP3 in pathological hypertrophy and in heart failure. Cell Death Dis. 2012;3:265.
82. Galasso G, De Rosa R, Piscione F, Iaccarino G, Vosa C, Sorriento D, et al. Myocardial expression of FOXO3a-Atrogin-1 pathway in human heart failure. Eur J Heart Fail. 2010;12(12):1290–6.
83. Chaanine AH, Kohlbrenner E, Gamb SI, Guenzel AJ, Klaus K, Fayyaz AU, et al. FOXO3a regulates BNIP3 and modulates mitochondrial calcium, dynamics, and function in cardiac stress. Am J Physiol Heart Circ Physiol. 2016;311(6):H1540–H59.
84. Cereghetti GM, Stangherlin A, Martins de Brito O, Chang CR, Blackstone C, Bernardi P, et al. Dephosphorylation by calcineurin regulates translocation of Drp1 to mitochondria. Proc Natl Acad Sci U S A. 2008;105(41):15803–8.
85. Cribbs JT, Strack S. Reversible phosphorylation of Drp1 by cyclic AMP-dependent protein kinase and calcineurin regulates mitochondrial fission and cell death. EMBO Rep. 2007;8(10):939–44.
86. Makino A, Suarez J, Gawlowski T, Han W, Wang H, Scott BT, et al. Regulation of mitochondrial morphology and function by O-GlcNAcylation in neonatal cardiac myocytes. Am J Physiol Regul Integr Comp Physiol. 2011;300(6):R1296–302.
87. Watanabe T, Saotome M, Nobuhara M, Sakamoto A, Urushida T, Katoh H, et al. Roles of mitochondrial fragmentation and reactive oxygen species in mitochondrial dysfunction and myocardial insulin resistance. Exp Cell Res. 2014;323(2):314–25.
88. Parra V, Verdejo HE, Iglewski M, Del Campo A, Troncoso R, Jones D, et al. Insulin stimulates mitochondrial fusion and function in cardiomyocytes via the Akt-mTOR-NFkappaB-Opa-1 signaling pathway. Diabetes. 2014;63(1):75–88.

89. Cassidy-Stone A, Chipuk JE, Ingerman E, Song C, Yoo C, Kuwana T, et al. Chemical inhibition of the mitochondrial division dynamin reveals its role in Bax/Bak-dependent mitochondrial outer membrane permeabilization. Dev Cell. 2008;14(2):193–204.
90. Maneechote C, Palee S, Kerdphoo S, Jaiwongkam T, Chattipakorn SC, Chattipakorn N. Differential temporal inhibition of mitochondrial fission by Mdivi-1 exerts effective cardioprotection in cardiac ischemia/reperfusion injury. Clin Sci (Lond). 2018;132(15):1669–83.
91. Zhou H, Wang S, Zhu P, Hu S, Chen Y, Ren J. Empagliflozin rescues diabetic myocardial microvascular injury via AMPK-mediated inhibition of mitochondrial fission. Redox Biol. 2018;15:335–46.
92. Yu J, Maimaitili Y, Xie P, Wu JJ, Wang J, Yang YN, et al. High glucose concentration abrogates sevoflurane post-conditioning cardioprotection by advancing mitochondrial fission but dynamin-related protein 1 inhibitor restores these effects. Acta Physiol (Oxf). 2017;220(1):83–98.
93. Ding M, Dong Q, Liu Z, Liu Z, Qu Y, Li X, et al. Inhibition of dynamin-related protein 1 protects against myocardial ischemia-reperfusion injury in diabetic mice. Cardiovasc Diabetol. 2017;16(1):19.
94. Liu J, Yan W, Zhao X, Jia Q, Wang J, Zhang H, et al. Sirt3 attenuates post-infarction cardiac injury via inhibiting mitochondrial fission and normalization of AMPK-Drp1 pathways. Cell Signal. 2019;53:1–13.
95. Wang Q, Zhang M, Torres G, Wu S, Ouyang C, Xie Z, et al. Metformin suppresses diabetes-accelerated atherosclerosis via the inhibition of Drp1-mediated mitochondrial fission. Diabetes. 2017;66(1):193–205.
96. Toyama EQ, Herzig S, Courchet J, Lewis TL Jr, Loson OC, Hellberg K, et al. Metabolism. AMP-activated protein kinase mediates mitochondrial fission in response to energy stress. Science. 2016;351(6270):275–81.
97. Ducommun S, Deak M, Sumpton D, Ford RJ, Nunez Galindo A, Kussmann M, et al. Motif affinity and mass spectrometry proteomic approach for the discovery of cellular AMPK targets: identification of mitochondrial fission factor as a new AMPK substrate. Cell Signal. 2015;27(5):978–88.
98. Otera H, Wang C, Cleland MM, Setoguchi K, Yokota S, Youle RJ, et al. Mff is an essential factor for mitochondrial recruitment of Drp1 during mitochondrial fission in mammalian cells. J Cell Biol. 2010;191(6):1141–58.
99. Seabright AP, Fine NHF, Barlow JP, Lord SO, Musa I, Gray A, et al. AMPK activation induces mitophagy and promotes mitochondrial fission while activating TBK1 in a PINK1-Parkin independent manner. FASEB J. 2020;34(5):6284–301.
100. Chen Z, Lei C, Wang C, Li N, Srivastava M, Tang M, et al. Global phosphoproteomic analysis reveals ARMC10 as an AMPK substrate that regulates mitochondrial dynamics. Nat Commun. 2019;10(1):104.

[illegible]

NADPH Oxidase System Mediates Cholesterol Secoaldehyde-Induced Oxidative Stress and Cytotoxicity in H9c2 Cardiomyocytes

Laura Laynes, Achuthan C. Raghavamenon, Deidra S. Atkins-Ball, and Rao M. Uppu

Abstract An oxidized cholesterol species, 3β-hydroxy-5-oxo-5,6-secocholestan-6-al (cholesterol secoaldehyde, ChSeco, or atheronal-A), likely to be formed in vivo at inflammatory sites through singlet oxygen-mediated oxidations, has recently been identified as an important player in degenerative diseases such as atherosclerosis and Alzheimer's disease. Previous studies from our laboratory and elsewhere have demonstrated the cytotoxic and inflammatory potential of this oxysterol using a wide variety of cell types including primary cells of mammalian origin. In nonimmune cells, such as rat embryonic H9c2 cardiomyocytes, depletion of cellular glutathione and an associated increase in the formation of reactive oxygen species (ROS), particularly H_2O_2, have been suggested to be among the initial events that lead to apoptotic signaling by ChSeco. Herein, we investigated the role of plasma membrane NADPH oxidase system (NOS) as a source of H_2O_2 and related ROS in ChSeco-exposed H9c2 cardiomyocytes. Pretreatment of H9c2 cardiomyocytes with inhibitors of NOS, namely, apocynin (Apo) and diphenyleneiodonium chloride (DPI), lowered the ChSeco-mediated formation of intracellular peroxides or peroxide-like substances (measured based on dichlorofluorescein fluorescence) and H_2O_2, the latter being measured based on intracellular hydrolysis of Amplex red to 3,7-dihydroxyphenoxazine and the subsequent oxidation of 3,7-dihydroxyphenoxazine to resorufin. The superoxide dismutase activity, which was shown to increase in a time-dependent manner in response to ChSeco exposure, increased further when the cardiomyocytes were pretreated with Apo, but not

L. Laynes · A. C. Raghavamenon · R. M. Uppu (✉)
Department of Environmental Toxicology, College of Agriculture,
Southern University and A&M Colleg, Baton Rouge, LA, USA
e-mail: rao_uppu@subr.edu

D. S. Atkins-Ball
Department of Biological Sciences and Chemistry, College of Sciences and Engineering,
Southern University and A&M College, Baton Rouge, LA, USA

N. L. Parinandi, T. J. Hund (eds.), *Cardiovascular Signaling in Health and Disease*, https://doi.org/10.1007/978-3-031-08309-9_7

DPI. The ChSeco-mediated cytotoxicity and lowering of glutathione (GSH) levels were mitigated to some extent when cardiomyocytes were pretreated Apo, while pretreatment with DPI nearly reversed the effects of ChSeco. The NOS inhibitors (Apo and DPI) were effective in lowering lipid peroxidation (formation of thiobarbituric acid-reactive substances) as well as the overexpression of pp38 and SAPK/JNK proteins in the ChSeco-exposed cardiomyocytes. Furthermore, the loss of mitochondrial transmembrane potential (ΔΨm) resulting from exposure to ChSeco was also mitigated by pretreatments with both Apo and DPI, while the loss of ΔΨm induced by rotenone (inhibitor of mitochondrial complex I) was only marginally restored by these two NOS inhibitors. Taken together, we suggest that the plasma membrane NADPH oxidase system contributes to the formation of H_2O_2 and related ROS which may have a secondary effect(s) on mitochondrial function leading to the observed ChSeco cytotoxicity in H9c2 cardiomyocytes.

Keywords Apocynin · Atheronals A and B · Diphenyleneiodonium chloride · Mitochondrial electron transport system · NADPH oxidase system · Oxysterols · Reactive oxygen species

Abbreviations

5-NitroApo	5-Nitroapocynin
Amplex red	10-Acetyl-3,7-dihydroxyphenoxazine
Apo	Apocynin or 4-hydroxy-3-methoxyacetophenone
Apo•	Apocynin phenoxyl radical
Atheronal-B	3β-Hydroxy-5β-hydroxy-B-norcholestan-6β-carboxaldehyde
BCA	Bicinchoninic acid
BSA	Bovine serum albumin
ChSeco	3β-Hydroxy-5-oxo-5,6-secocholestan-6-al (cholesterol secoaldehyde or Atheronal-A)
CMDCF	5-(and-6)-Chloromethyl-2,7-dichloroflurescein
CMDCFDA	5-(and-6)-Chloromethyl-2,7-dichlorodihydrofluorescein
CMH_2DCFDA	5-(and-6)-Chloromethyl-2,7-dichlorodihydro-fluorescein diacetate
DiApo	Diapocynin
DMEM	Dulbecco's modified Eagle's medium
DMSO	Dimethyl sulfoxide
DPI	Diphenyleneiodonium chloride
DTPA	Diethylenetriaminepentaacetic acid
ETS	Electron transport system
FBS	Fetal bovine serum
GST	Glutathione S-transferase
HEPES	4-(2-Hydroxyethy)-1-piperazineethanesulfonic acid
HRP	Horseradish peroxidase

JC-1	5,5′,6,6′-Tetrachloro-1,1′,3,3′-tetraethylbenzimidazolylcarbocyanine iodide
KRH	Krebs-Ringer-HEPES
MCB	Monochlorobimane
MDA	Malondialdehyde
MTS	3-(4,5-Dimethylthiazol-2-yl)-5-(3-carboxymethoxyphenyl)-2-(4-sulfophenyl)-2H-tetrazolium, inner salt
NOS	NADPH oxidase system
PBS	Phosphate-buffered saline
PES	Phenazine ethosulfate
PVDF	Polyvinylidene fluoride
ROS	Reactive oxygen species
SDS	Sodium dodecyl sulfate
SOD	Superoxide dismutase
TBA	Thiobarbituric acid
TBARs	Thiobarbituric acid-reactive substances
TBST	Tris-buffered saline with Tween 20
WST-1	2-(4-Iodophenyl)-3-(4-nitrophenyl)-5-(2,4-disulfophenyl)-2H-tetrazolium, monosodium salt
WST-1	4-[3-(4-Iodophenyl)-2-(4-nitrophenyl)-2H-5-tetrazolio]-1,3-benzenesulfonate disodium salt
$\Delta\Psi m$	Mitochondrial transmembrane potential

Introduction

Oxidized cholesterol species (oxysterols) formed during auto-oxidation or enzymatic reactions have been extensively studied and are shown to produce cytotoxic effects in several cell culture systems [1–3]. In most cases, an increase in intracellular Ca^{2+} levels and excessive formation of reactive oxygen species (ROS) such as H_2O_2 are indicated in oxysterol-mediated cytotoxicity. Once thought to be an 'ozone-specific' oxysterol, 3β-hydroxy-5,6-secocholestan-6-al (cholesterol secoaldehyde, ChSeco, or Atheronal-A; Fig. 1) is now widely recognized to be formed in vivo mainly from Hock cleavage of cholesterol-5α-hydroperoxide [4], the latter being formed in reactions of singlet oxygen with cholesterol and possibly cholesterol 3-acyl esters [5–7]. The natural occurrence of ChSeco and its aldolized product, 3β-hydroxy-5β-hydroxy-B-norcholestane-6β-carboxaldehyde (Atheronal-B; Fig. 1) in atherosclerotic plaques [8], and the demonstration in vitro that ChSeco induces the monocyte maturation to macrophage and causes overexpression of adhesion molecules in endothelial cells [9], have been suggested to indicate a role for this oxysterol in atherosclerosis [10]. In addition to its presence in the atherosclerotic plaque, ChSeco has shown to be present in the brain samples of patients with Alzheimer's disease [11] and Lewy body dementia [12], and promotes the

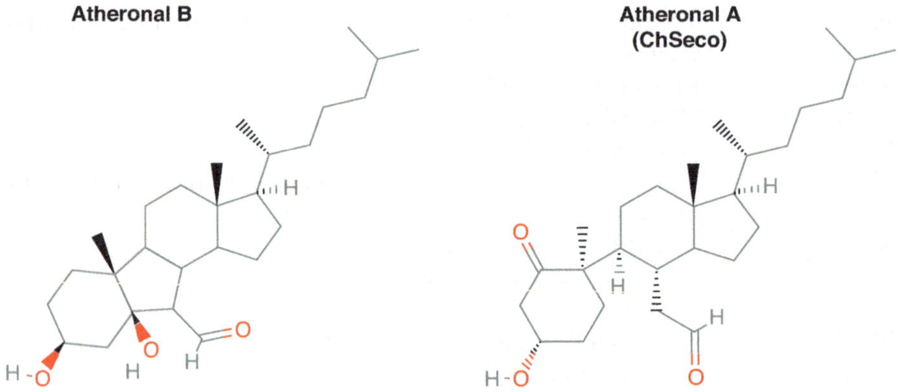

Fig. 1 Structures of Atheronals A and B

formation of amyloid-beta (Aβ) fibrils in rat GT1–7 hypothalamic neurons [13] and in rat primary cortical neurons in vitro [14]. Kelly and his colleagues have shown that aldehydic metabolites of cholesterol (ChSeco) and phospholipids accelerate Alzheimer's amyloidogenesis by a two-step mechanism, Schiff base formation with Aβ peptides followed by downhill polymerization of the Aβ-metabolite aggregates, eliminating the requirement for nucleation [15]. Thus, there is a broad recognition for Atheronals A and B in degenerative diseases [10, 16].

Studies from our laboratory have shown that ChSeco induces apoptosis in rat embryonic H9c2 cardiomyocytes [17, 18] and murine GT1–7 hypothalamic neurons [13, 19] through depletion of cellular glutathione, generation of excessive ROS, and activation of mitochondrial and death receptor pathways. The role of ROS in ChSeco exposures was confirmed by the observations that N-acetyl-L-cysteine (NAC), Trolox (a water-soluble analog of vitamin E), catalase, and deferoxamine (an iron chelator) were all effective in inhibiting most events related to oxidative stress and apoptosis and that H_2O_2 was identified as the major ROS mediating cell death in ChSeco-exposed cardiomyocytes [13, 17–20]. However, the source of H_2O_2 has not yet been identified. Since the mitochondrial electron transport system (ETS) and the plasma membrane NADPH oxidase system (NOS) are two important sources of ROS in the cellular milieu, herein, we investigated the significance of these pathways with known inhibitors of NOS (Apo and DPI) and ETS (rotenone). Additionally, we examined the activities of antioxidant enzymes (catalase and superoxide dismutase), the formation of lipid peroxidation products (thiobarbituric acid-reactive products or TBARs), and activation of MAP Kinases (p38 and SAPK) that could play a role in apoptosis, in general, in ChSeco-exposed H9c2 cardiomyocytes.

Materials and Methods

Chemicals, H9c2 Cardiomyocytes, and Cell Culture Supplies

Apocynin (Apo), bovine serum albumin (BSA), dimethyl sulfoxide (DMSO), diphenyleneiodonium chloride (DPI), Dulbecco's modified Eagle's medium (DMEM), fetal bovine serum (FBS), 4-(2-hydroxyethyl)-1-piperazineethanesulfonic acid (HEPES), menadione, 2-mercaptoethanol, penicillin-streptomycin (10,000 units/mL penicillin and 10 mg/mL streptomycin), phosphate-buffered saline (PBS), protease inhibitor cocktail [4-(2-aminoethyl)-benzenesulfonyl fluoride hydrochloride, pepstatin A, E-64, bestatin, leupeptin, and aprotinin], sodium dodecyl sulfate (SDS), 1,13,3-tetrqaethoxypropane (malondialdehyde precursor), thiobarbituric acid (TBA), Tris base (Trizma), Triton X-100, and trypsin-EDTA (2.5 g EDTA-Na_4 per liter of HBSS) were purchased from Sigma-Aldrich (St. Louis, MO). Other chemicals and reagents were obtained from the following sources: acrylamide/bis-acrylamide (30% w/v in water) and polyvinylidene fluoride (PVDF) membranes from Bio-Rad (Hercules, CA); diethylenetriaminepentaacetic acid (DTPA) was from Acros Organics (Morris Plains, NJ); 5-(and 6)-chloromethyl-2′,7′-dichloro-dihydrofluorescein diacetate (CMH_2DCFDA) from Invitrogen (Carlsbad, CA); bicinchoninic acid (BCA) protein assay kit and SuperSignal™ West Femto chemiluminescent substrate from Pierce (Rockford, IL); CellTiter 96® AQueous one solution from Promega (Madison, WI); kits for the measurement of superoxide dismutase (SOD) activity and reduced glutathione (GSH) and for the fractionation of mitochondria and cytosol from BioVision (Mountain View, CA); MitoCapture kit for the measurement of mitochondrial transmembrane potential ($\Delta\Psi m$) from CalBiochem (La Jolla, CA); the Amplex red assay kit for the determination of H_2O_2 from Invitrogen (Grand Island, NY); and primary antibodies to pp38, pSAPK, and β-actin from Cell Signaling Technology (Danvers, MA). Rat embryonic H9c2 cardiomyocytes were procured from the American Type Culture Collection (ATCC; Rockville, MD). Stock solutions of Apo, ChSeco, DPI, rotenone, and Trolox were prepared in DMSO, ethanol, or PBS, as needed.

Synthesis of Cholesterol Secoaldehyde

ChSeco was synthesized in the laboratory as described previously [17, 21]. Briefly, following ozonation of cholesterol in mixed solvents of dichloromethane and methanol to complete oxidation (monitored by reversed-phase HPLC), the residue of the oxidized cholesterol was dissolved in a minimal volume acetic acid/water (19/1, v/v) and reduced with excess Zn dust (ten-fold molar excess to the amount of cholesterol originally used in the ozonation reaction). The product, ChSeco, was isolated by solvent extraction and the purity tested by reversed-phase HPLC [17]. Small aliquots of the final product were prepared in DMSO at a concentration of 20 mM and stored at −80 °C until use.

Cell Culture and Treatments

H9c2 cardiomyocytes were maintained in DMEM containing 4 mM glutamine, 1.5 g/L sodium bicarbonate, 4.5 g/L glucose, 1.0 mM sodium pyruvate, 10% FBS, 100 units/mL penicillin, and 10 μg/mL streptomycin and incubated at 37 °C in a humidified atmosphere containing 95% air/5% CO_2. The medium was replaced once every 2–3 days, and the cells were split following trypsinization and subcultured to preconfluence (ca. 70%), as needed. For experimental purposes, cardiomyocytes were seeded at a cell density of 4×10^4/well in 24-well plates or 1.5×10^6/well in 6-well plates. At preconfluence, cardiomyocytes were exposed to 10 μM Apo (or DPI or rotenone) for 1 h in DMEM containing 2% FBS (v/v) and antibiotics. Followed by this, cardiomyocytes were exposed to 0–15 μM ChSeco for different time periods and then analyzed for various biochemical parameters including cell proliferation and metabolism, cell viability, antioxidant enzymes, $\Delta\Psi m$, lipid peroxidation, H_2O_2 and other peroxide or peroxide-like substances, and expression of MAP kinases.

Cell Proliferation and Metabolism

The cytotoxicity assay initially performed was based on the metabolic activity of H9c2 cardiomyocytes which relied on phenazine ethosulfate (PES; or 1-methoxy-PES)-mediated reduction of 3-(4,5-dimethylthiazol-2-yl)-5-(3-carboxymethoxyphenyl)-2-(4-sulfophenyl)-2H-tetrazolium (MTS) to MTS formazan. These assays were performed using a CellTiter 96 AQueous one solution as described in our previous publications [16, 18]. Briefly, H9c2 cardiomyocytes were seeded in 24-well plates at a cell density of ca. 4×10^4/well and exposed to 10 μM Apo (or DPI) for 1 h in DMEM containing 2% FBS. Thereafter, the medium was replaced with DMEM containing 0–15 μM ChSeco and 2% FBS and the incubation continued for 24 h. At the end of the incubation period, aliquots (10 μL each) of the AQueous one solution were added directly to cell cultures (the volume of cell culture in each case was 500 μL). The cells were incubated for 1–4 h, and the MTS formazan produced was measured at 490 nm using a BioTek ELx800 microplate reader (Winooski, VT).

Trypan Blue Exclusion Assay

The second assay, which provides a more direct determination of cytotoxicity, was based on Trypan blue exclusion [22]. This assay is based on the principle that live cells with intact cell membranes exclude Trypan blue and, therefore, maintain a clear cytoplasm. However, dead cells have compromised cell membranes that allow entry of Trypan blue into the cell causing their cytoplasm to become blue in color. H9c2 cardiomyocyte suspensions (20 μL each) were mixed with equal volumes of 0.25% Trypan blue in DMEM containing 2% FBS. A 10-μL aliquot of the Trypan blue mixed cardiomyocyte suspension was introduced into a hemocytometer, and the live and dead cells were visually examined and counted using a Nikon inverted microscope.

Measurement of Intracellular Peroxides

The intracellular formation of peroxide or peroxide-like molecules was determined using a membrane-permeable probe, 5-(and 6)-chloromethyl-2,7-dichlorodihydrofluorescein diacetate (CMH_2DCFDA) as described previously [10, 23]. H9c2 cardiomyocytes grown in 24-well plates, at ~70% confluence, were exposed to 10 μM CMH_2DCFDA in DMEM containing 2% FBS for 30 min at 37 °C. After the incubation, the cells were washed twice with KRH buffer and treated first with 10 μM Apo (or DPI) for 1 h and then exposed to 0, 10, or 15 μM ChSeco for up to 9 h. The appearance of 5-(and-6)-chloromethyl-2,7-dichloroflurescein (CMDCF; also referred to as DCF fluorescence) was monitored using a SpectraMax Gemini EM microplate reader (Molecular Devices; Sunnyvale, CA) at excitation and emission wavelengths of 485 and 530 nm, respectively, over the course of ChSeco exposure.

Measurement of Hydrogen Peroxide

10-Acetyl-3-7-dihydroxyphenoxazine (Amplex red), a highly sensitive probe specifically designed for the measurement of H_2O_2, was employed [23]. The assay was based on the intracellular hydrolysis of Amplex red to 3,7-dihydroxyphenoxazine and its subsequent oxidation by H_2O_2 in a peroxidase-catalyzed reaction to resorufin, a highly fluorescent compound (excitation: 571 nm; emission: 585 nm). Briefly, H9c2 cardiomyocytes (ca. 1.5×10^6) were seeded in 6-well plates. At ~70% confluence, cardiomyocytes were initially treated with 10 μM Apo (or DPI) for 1 h. Following this treatment, the medium was removed, the cells were washed with PBS, and then exposed to 0 or 15 μM ChSeco for 0–10 h in DMEM containing 2% FBS. At different incubation periods, cardiomyocytes were harvested by scraping into ice-cold PBS and centrifuging at 3000 rpm for 20 min. The cell pellets were resuspended in 250 μL each of the reaction buffer. Aliquots (50 μL each) of the cardiomyocyte suspensions were placed in individual wells in a 96-well plate and mixed with 50 μL each of the Amplex red/horseradish peroxidase (HRP) working solution (final concentration of Amplex red: 50 μM; final concentration of HRP: 0.1 unit/mL). The reaction was allowed to take place for 30 min at ambient temperature in the absence of light. The formation of resorufin was measured in each case using a SpectraMax Gemini EM microplate reader (excitation: 530–560 nm; emission: 590 nm). All components were provided with the kit. The Amplex red reagent was prepared fresh before use for optimal results.

Thiobarbituric Acid-Reactive Substances (TBARs)

H9c2 cardiomyocytes, at ~70% confluence, were pretreated with Apo or DPI, as described earlier, and then exposed to 15 μM ChSeco for 6 h. At the end of the exposure period, approximately 5×10^6 cells were lysed in 150 μL of cold 0.1 M Tris-HCl buffer, pH 6.80, containing 1% (w/v) SDS and 1% (v/v) protease inhibitor cocktail. The lysate was centrifuged at 15,000 × g for 15 min at 4 °C. Aliquots of the supernatant (125 μL each) were mixed with 50 μL of 10% SDS, 250 μL of 20% (v/v) acetic acid, and 750 μL of thiobarbituric acid reagent (TBA; 0.67% w/v, freshly prepared in deionized water) in different test tubes [24]. The reaction mixtures were incubated for 1 h in boiling water bath. At the end of the incubation period, the mixtures were cooled to room temperature, and 500 μL of deionized water was added to each test tube. Where necessary, the product of MDA (or other aldehydes)-TBA reaction was extracted into pyridine. The organic layer was read against the pyridine blank at 532 nm using a Thermo Spectronic Genesys 10 UV-Vis spectrophotometer (Madison, WI).

Measurement of Reduced Glutathione Levels

This assay was based on glutathione-S-transferase catalyzed conjugation of GSH to monochlorobimane (MCB, nonfluorescent) which results in the formation of fluorescent GS-bimane conjugate with excitation at 380 nm and emission at 460 nm [25–27]. The assay was performed as follows: H9c2 cardiomyocytes were seeded in a 6-well plate at a density of 1×10^6/well. At ~70% confluence, cardiomyocytes were pretreated with 10 μM Apo (or DPI) for 1 h. Thereafter, cardiomyocytes were exposed to 0 or 10 μM ChSeco for 6 h. At the end of the incubation period, cardiomyocytes were harvested and lysed in the lysis buffer (provided as part of the kit). Aliquots (100 μL each) of the cardiomyocyte lysates were mixed with 2 μL of MCB (25 mM) and 2 μL of GST (50 units/mL) reagents, incubated at 37 °C for 30 min. The formation of GS-bimane conjugate was measured using a SpectraMax Gemini EM microplate reader. The protein in the supernatant was measured using the bicinchoninic acid (BCA) kit. The values of fluorescence resulting from the formation of GS-bimane conjugate were calculated per mg protein and compared with those obtained for the corresponding unexposed controls.

Assay of Superoxide Dismutase Activity

H9c2 cardiomyocytes were cultured in 6-well plates (ca. 1×10^6) and treated as mentioned above, first with 10 μM Apo (or DPI) for 1 h and then with 0 or 15 μM ChSeco for 0–12 h. At various periods, cardiomyocytes were harvested by scraping into 1× PBS followed by centrifugation at 600 × g for 5 min at 4 °C. The cell pellets were resuspended in 1.0 mL each of 1× cytosol extraction buffer containing DTT and protease inhibitors (provided as part of Mitochondria/Cytosol fractionation kit). The cells were then homogenized in the cold using a tissue grinder. The homogenates, obtained typically after 30–50 passes, were transferred to a microcentrifuge tube (capacity: 1.5 mL) and centrifuged at 700 × g for 30 min at 4 °C. The supernatant was collected and transferred to a fresh microcentrifuge tube, and the pellet discarded. In a 96-well plate, aliquots (20 μL each) of the supernatant were mixed with 200 μL of a working WST-1 solution (provided by the manufacturer), 20 μL of dilution buffer (for blank only), and 20 μL of the xanthine oxidase enzyme solution. The 96-well microplate was incubated for 20 min at 37 °C and the yield of WST-1 formazan was measured at 450 nm using a BioTek ELx800 microplate reader. The SOD activity was calculated based on inhibition of the formation of WST-1 formazan (see below):

$$\%\text{Inhibition} = \left[\left(\text{blank1} - \text{blank3}\right) - \left(\text{sample} - \text{blank2}\right) / \left(\text{blank1} - \text{blank3}\right)\right] \times 100$$

Measurement of Mitochondrial Transmembrane Potential

The mitochondrial transmembrane potential (ΔΨm) was measured using the MitoCapture kit from CalBiochem as described in our previous publications [18, 26]. The assay is based on the accumulation of a cationic dye, 5,5′,6,6′-tetrachloro-1,1′,3,3′-tetraethylbenzimidazolylcarbo-cyanine iodide (JC-1), in the mitochondria, when the ΔΨm lowers. Briefly, H9c2 cardiomyocytes were initially seeded in 24-well plates at a cell density of ca. 5×10^6/well. At preconfluence, the cells were treated with 10 μM Apo (or DPI or rotenone) for 1 h and later with 0 or 15 μM ChSeco for 8 h. Cardiomyocytes that were incubated with Apo (or DPI or rotenone) for 1 h and not exposed to ChSeco were used as controls. At the end of the exposure period, the JC-1 solution was added at a final concentration of 10 μM and incubated at 37 °C for 30 min. Cardiomyocytes were washed with PBS to remove the unbound JC-1. They were then overlaid with 500 μL of assay buffer (provided as part of the kit), and the fluorescence was measured at excitation and emission wavelengths of 488 and 590 nm, respectively, using a SpectraMax Gemini EM microplate reader.

Western Blot Analysis

H9c2 cardiomyocytes in T-25 cm^2 Corning flasks were pretreated with 10 μM Apo (or DPI) for 1 h and, thereafter, exposed to 0 or 15 μM ChSeco for periods up to 24 h. Following these treatments, cardiomyocytes were rinsed twice with PBS and lysed using 100 μL of 62.5 mM Tris-HCl buffer, pH 6.80, that also contained 1% SDS and 1% protease inhibitor cocktail. A probe sonicator set at 2 Hz for 2 × 10 s each was used to facilitate the lysis of cardiomyocytes. The lysate was centrifugated at 10,000 × g for 10 min at 4 °C, and the protein concentration in the supernatant was determined using the BCA protein assay kit. A known amount of the supernatant protein (50 μg) was resolved by 10% SDS-PAGE and then electrotransferred onto a PVDF membrane. The membrane was then blocked for 1 h using TBST (Tris-HCl buffer, pH 7.50, containing saline and 0.1% (w/v) Tween 20) that also contained 5% (w/v) skim milk powder. Following overnight incubation with primary monoclonal antibodies (against p38, pp38, SAPK/JNK, pSAPK/JNK, and β-actin; 1:1000 dilution) at 4 °C, the PVDF membrane was washed and incubated with HRP-conjugated, species-specific secondary antibody in a blocking solution for 1 h. Thereafter, the membranes were washed with TBST and exposed to SuperSignal West Femto substrate for 5 min, and the chemiluminescence signal was captured using the Alpha imager software system (Alpha Innotech Corporation; San Leandro, CA).

Statistical Analysis

Data analysis was performed using GraphPad Prism software and a comparison was made between treated and their respective controls by means of unpaired Student's t-test and their significance was established by ANOVA. A difference of $p < 0.05$ was considered statistically significant. Values presented represent mean ± SD of three experiments performed in duplicate.

Results

Apocynin Exacerbates While DPI Reduces the ChSeco-Induced Cytotoxic Response in H9c2 Cardiomyocytes

The changes in cell viability manifested in response to the exposure of ChSeco, Apo, and DPI, either singly or in combination, were determined based on the measurement of metabolic activity of live cells (MTS reduction; Fig. 2a) as well as by direct enumeration of live and dead cells (Fig. 2b). When H9c2 cardiomyocytes were exposed to 15 μM ChSeco for 24 h in 2% FBS-containing DMEM, there was a substantial decrease in the formation of MTS formazan (55 ± 7%) compared to the untreated controls. The formation of MTS formazan was reduced further (45 ± 6%) when the cardiomyocytes were pretreated with Apo (10 μM, I h) and then exposed to ChSeco (Fig. 2a). Pretreatment with DPI, on the other hand, resulted in a significantly higher formation of MTS formazan (82 ± 1%) in the ChSeco-exposed cardiomyocytes compared to those treated with ChSeco alone (55 ± 7%; Fig. 2a).

As shown in Fig. 2b, when assessed using the technique of Trypan blue exclusion, the percentage cell viability in ChSeco-exposed (15 μM, 24 h) cardiomyocytes was 60 ± 2%. Pretreatment with Apo (10 μM, 1 h) resulted in a marginal decrease in the viability of ChSeco-exposed cardiomyocytes (56 ± 5%). Cardiomyocytes pretreated with DPI (10 μM, 1 h), and then exposed to ChSeco, exhibited a substantially higher cell viability (75 ± 2%) compared to those exposed to ChSeco alone (60 ± 2%).

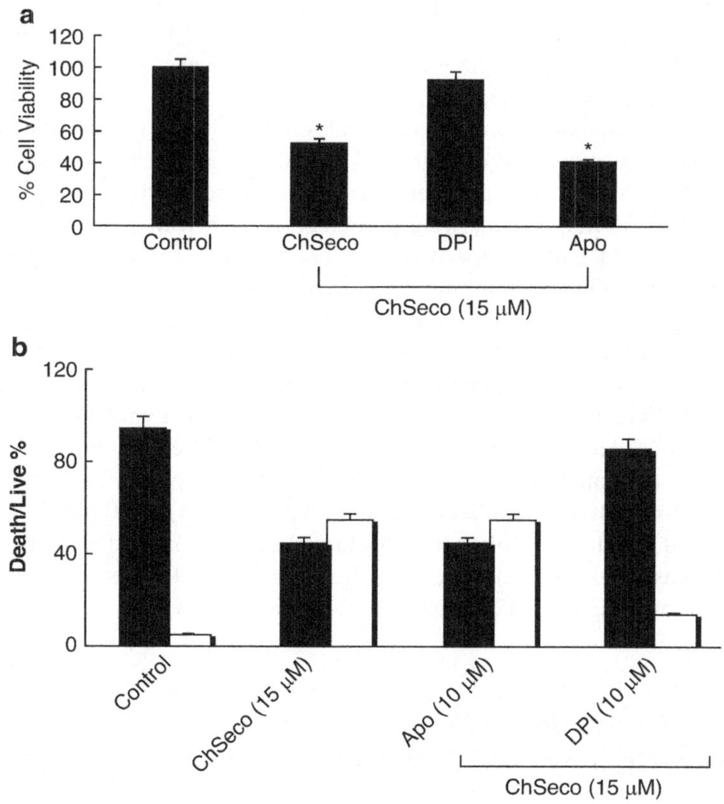

Fig. 2 Cytotoxicity of ChSeco in H9c2 cardiomyocytes pretreated with Apo and DPI. Cardiomyocytes at ca. 70% confluence were pretreated with 10 μM Apo (or DPI) for 1 h. Following pretreatment, the cells were washed twice with PBS and exposed to 15 μM ChSeco for 24 h. The cell viability was then determined based on (**a**) PES-assisted MTS reduction to MTS formazan using the CellTiter 96 AQueous one solution kit, and based on (**b**) actual measurement of live and dead cells (% cell viability) as observed under a Nikon inverted microscope following Trypan blue staining. Data presented are mean ± SD of three different experiments. * Indicates significance at $p < 0.05$ against the untreated (respective) controls

NOS Inhibitors Attenuate the Formation of Peroxide or Peroxide-Like Substances in ChSeco-Exposed H9c2 Cardiomyocytes

There was little or no formation of peroxide or peroxide-like substances in cardiomyocytes that were not exposed to ChSeco. Pretreatments with Apo and DPI, per se, without subsequent exposure to ChSeco also did not result in any increase in the formation of peroxide(s) in cardiomyocytes.

As shown in Fig. 3a, in cardiomyocytes exposed to 10 μM ChSeco for 0–8 h, there was a gradual increase in the formation of peroxide(s) (see, curve marked with closed circles). When pretreated with 10 μM Apo for 1 h and then exposed to 10 μM ChSeco, the formation of peroxide(s) in cardiomyocytes was substantially lower compared to cardiomyocytes that were not pretreated with Apo (see, curved marked with closed triangles). Treatment with 25 μM menadione (positive control) induced a profound increase in the formation of peroxide(s) over a period of 0–8 h (see, curve marked with closed squares).

Figure 3b, curve marked with closed circles, shows the formation of peroxide(s) in H9c2 cardiomyocytes that were treated 15 μM ChSeco (see, curve marked with closed circles). As can be seen, there was a substantial decrease in formation of peroxide when cardiomyocytes were pretreated for 1 h with 10 μM DPI and then exposed to 15 μM ChSeco for 9 h (see, curve marked with closed triangles). Although the magnitude of inhibition of peroxide formation by DPI was found to be smaller in comparison to that observed with Apo, the response to pretreatments with DPI and Apo in ChSeco-exposed cardiomyocytes was found to be similar.

Experiments performed using Amplex red, a sensitive probe specific for measurement of H_2O_2, are shown in Fig. 4. As expected, it was found that H9c2 cardiomyocytes exposed to ChSeco (15 μM) had much higher yields of intracellular H_2O_2 compared to untreated controls (see, curves marked in purple and back colors). Pretreatment with 10 μM Apo (see, curve marked in green) or 10 μM DPI (see, the curve marked in yellow) resulted in a substantial decrease in the formation of H_2O_2. Unlike the case with DCF fluorescence, treatment with DPI was found to be more effective in lowering the ChSeco-induced formation of H_2O_2.

Apocynin Lowers the GSH Levels in ChSeco-Exposed Cardiomyocytes

The levels of GSH in H9c2 cardiomyocytes exposed to 10 μM ChSeco were somewhat lower compared to the unexposed controls (Fig. 5). The depletion of GSH was found to be more pronounced when cardiomyocytes were pretreated with Apo and then exposed to ChSeco. Pretreatment with DPI, on the other hand, mitigated partially the ChSeco-induced decrease in GSH levels.

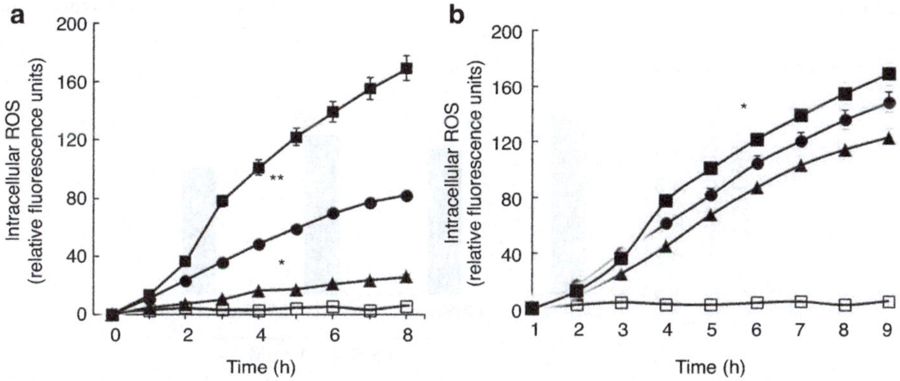

Fig. 3 Effect of NOS inhibitors (Apo and DPI) on ChSeco-induced peroxide generation in H9c2 cardiomyocytes. The formation of intracellular peroxide(s) was measurted in cardiomyocytes pretreated with (**a**) 10 μM Apo or (**b**) 10 μM DPI for 1 h and then exposed to 10 or 15 μM ChSeco, respectively, for 9 h. In these assays, cardiomyocytes (ca. 5×10^4) grown in 24-well plates were incubated with 10 μM CMH_2DCFDA for 30 min at 37 °C. Following removal of the extracellular CMH_2DCFDA, the cells were exposed to ChSeco (10 μM for cells pretreated with Apo and 15 μM for cells pretreated with DPI). The time course for the appearance of CMDCF (DCF fluorescence) was measured as described in the Materials and Methods section. * and ** Indicate significance at $p < 0.05$ and $p < 0.01$ against the untreated (respective) controls: (□--□) none (control); (●--●) cells exposed to 10 or 15 ChSeco; (▲--▲) cells pretreated with Apo (or DPI) and then exposed ChSeco; and (◙--◙) cell exposed to 25 μM menadione (positive control)

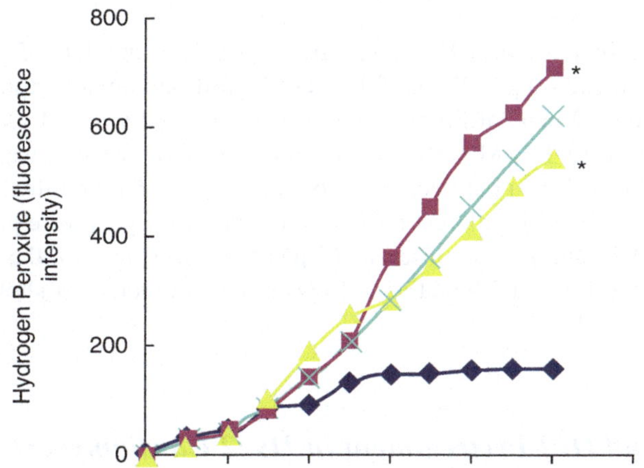

Fig. 4 Effects of pretreatments of Apo and DPI on the formation of H_2O_2 in H9c2 cardiomyocytes exposed to ChSeco. Cardiomyocytes were pretreated with 10 μM of Apo (or DPI) for 1 h, washed with PBS, and then exposed to 15 μM ChSeco for 0–10 h. Each well in the 96-well plate was loaded with 50 μL of cell suspension and 50 μL Amplex red/HRP working solution and incubated for 30 min. The formation of resorufin, the fluorescent product that indicates H_2O_2 formation, was measured using a SpectraMax Gemini EM microplate reader. Other details are as given in the text and Materials and Methods section. Data points indicate mean ± SD of three different experiments performed in duplicates. * Indicates significance at $p < 0.05$ against the untreated (respective) controls

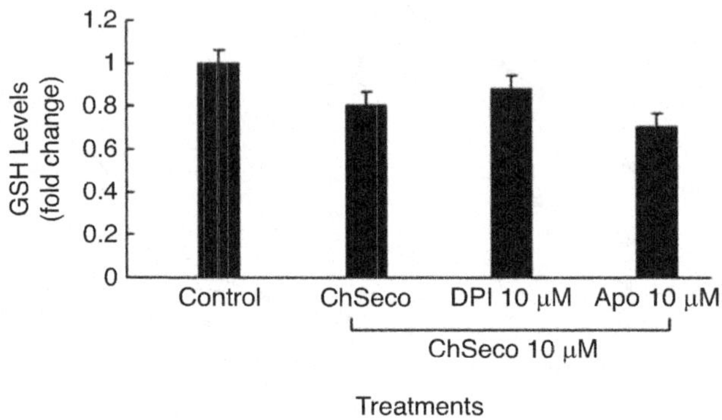

Fig. 5 Pretreatment with DPI, but not Apo, mitigates the ChSeco-induced decrease in the levels of cellular GSH in H9c2 cardiomyocytes. Cardiomyocytes at ca. 70% confluence were pretreated with 10 μM Apo (or DPI) and then exposed to 10 μM ChSeco for 6 h. At the end of the incubation period, cardiomyocytes were lysed in 100 μL of lysis buffer and the lysates mixed with 2 μL each of MCB and GST reagents as described in the Materials and Methods section. Following incubation for 30 min at 37 °C, the product of GSH conjugation to MCB (GS-bimane), which is fluorescent, was measured using a SpectraMax Gemini EM microplate reader

Apocynin But Not DPI Enhances the ChSeco-Induced Activation of SOD Activity in H9c2 Cardiomyocytes

We observed an increase in SOD activity from 18 ± 0.5% at 0 h to 35 ± 2.0% at 2 h, 54 ± 1.5% at 6 h and 60 ± 2.0% at 12 h in H9c2 cardiomyocytes exposed to 15 μM ChSeco (Fig. 6a). When cardiomyocytes were pretreated with 10 μM DPI for 1 h and then exposed to 15 μM ChSeco (Fig. 6b), the SOD activity was found to be higher at 2 h (67 ± 2.0%) but decreased to 53 ± 0.5% at 6 h and finally restored to the original levels at 12 h. Interestingly, when cardiomyocytes were pretreated with Apo (10 μM, 1 h) and then exposed to 15 μM ChSeco (Fig. 6c), the SOD activity increased to 81 ± 1.5% at 2 h and 85 ± 1.0% at 6 h and thereafter showed a decline to 76 ± 1.5% at 12 h.

Apocynin and DPI Pretreatment in H9c2 Cardiomyocytes Reduces TBARs Formed in Response to the Exposure of ChSeco

Following exposure to 15 μM ChSeco for a period of 8 h, there was a 7.0 ± 0.03-fold increase in the levels of TBARs as compared to the unexposed controls (Fig. 7). The excessive production of TBARs in ChSeco exposures was completely mitigated when cardiomyocytes were pretreated with either Apo or DPI. For example, when cardiomyocytes were pretreated with 10 μM Apo for 1 h and then exposed to 15 μM ChSeco, the extent of increase in TBARs was 1.6 ± 0.09 -fold compared to controls that were not exposed to Apo and ChSeco. Similarly, in cardiomyocytes pretreated

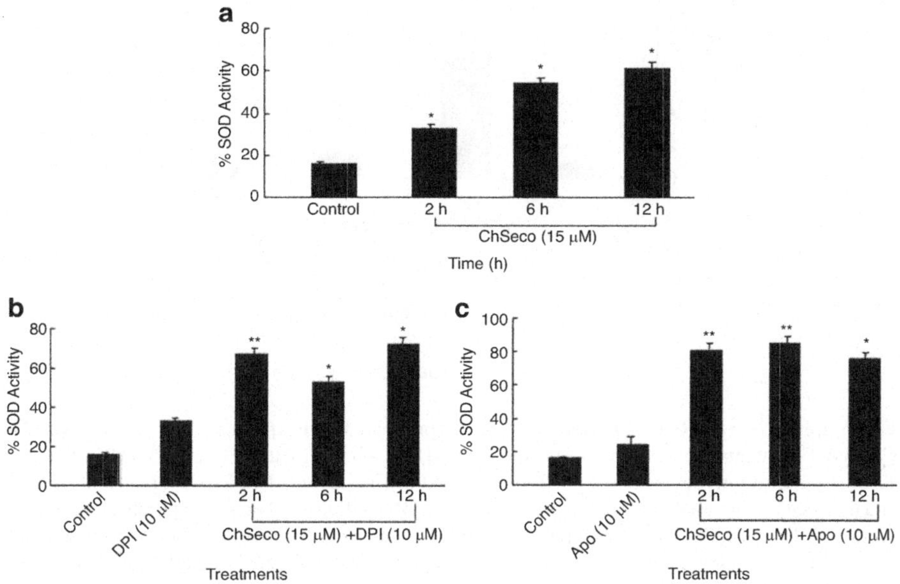

Fig. 6 Effect of pretreatment with Apo and DPI on the activity of SOD in H9c2 cardiomyocytes exposed to ChSeco: (**a**) cardiomyocytes exposed to 15 μM ChSeco for 0–12 h; (**b**) cardiomyocytes pretreated with 10 μM Apo for 1 h and then exposed to 15 μM ChSeco for 0–12 h; and (**c**) cardiomyocytes pretreated with 10 μM DPI for 1 h and then exposed to 15 μM ChSeco for 0–12 h. The SOD activity was measured as described in the Materials and Methods section. Data presented are mean ± SD of three different experiments. * and ** Indicate significance at $p < 0.05$ and $p < 0.01$ (respectively) against untreated controls

with 10 μM DPI and then exposed to 15 μM ChSeco, the extent of increase in TBARs was about 1.0 ± 0.07 -fold compared to the corresponding untreated controls that were exposed to nether DPI nor ChSeco.

ChSeco-Induced Loss in Mitochondrial Transmembrane Potential Is Reversed by Pretreatments with Apocynin and DPI

There was a substantial reduction of ΔΨm (49%) in H9c2 cardiomyocytes exposed to 10 μM ChSeco for 8 h (Fig. 8). Cardiomyocytes pretreated with 10 μM Apo (or DPI or rotenone) and then exposed to 10 μM ChSceco had much higher ΔΨm of 80% or 82% (respectively). The rotenone-induced loss in ΔΨm (79%) was about the same as that observed for Apo (or DPI) treated controls. These values were much smaller when compared to those observed for cardiomyocytes exposed to ChSeco alone. However, the rotenone-induced loss in ΔΨm was about the same when compared to cardiomyocytes exposed to Apo and DPI.

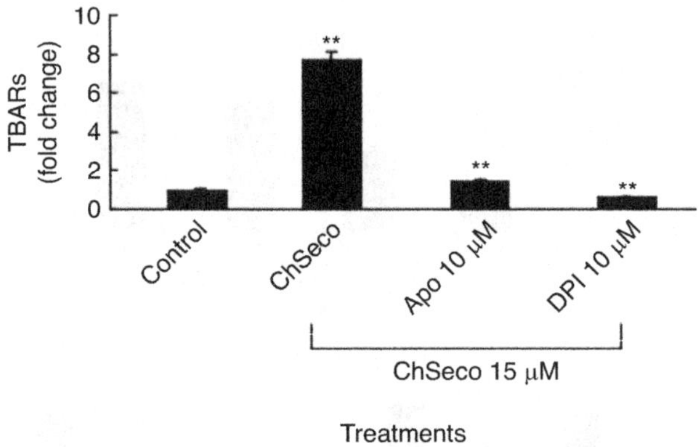

Fig. 7 Levels of TBARs in H9c2 cardiomyocytes pretreated with Apo or DPI and then exposed to ChSeco. Following exposure to 10 μM Apo (or DPI) for 1 h (pretreatment) and 15 μM ChSeco for 12 h, H9c2 cardiomyocytes (ca. 5 × 10^6) were lysed in 150 μL of 0.1 M Tris-HCl buffer, pH 6.80, and the lysates were analyzed for the presence of TBARs as described in the Materials and Methods section. MDA, which is a representative of TBARs, was used for the construction of a standard curve. Data points represent mean ± SD of three experiments performed in duplicate (** Indicates $p < 0.01$ against untreated controls)

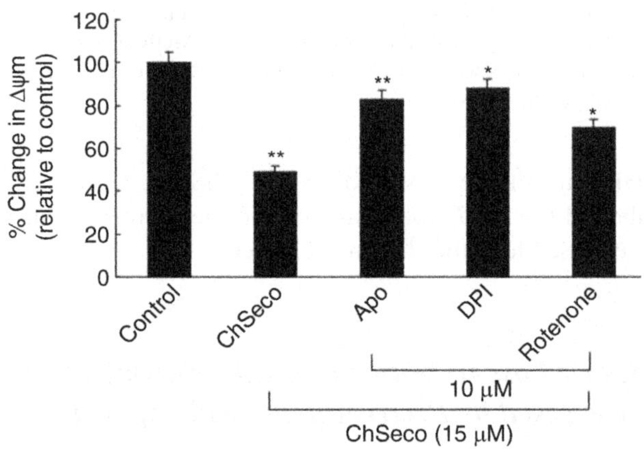

Fig. 8 Changes in the mitochondrial transmembrane potential in H9c2 cardiomyocytes pretreated with Apo, DPI, or rotenone and then exposed to ChSeco. Cardiomyocytes were treated with 10 μM Apo (or DPI or rotenone) for 1 h prior to the exposure to 10 μM ChSeco (6 h). The mitochondrial membrane potential was determined using JC-1 as described in the Materials and Methods section. Data points represent mean ± SD of three experiments performed in duplicates. * and ** Indicate significance at $p < 0.05$ and $p < 0.01$ (respectively) against untreated controls

ChSeco-Induced Overexpression of pp38 and pSAPK in H9c2 Cardiomyocytes Is Mitigated by Pretreatments with Apo and DPI

H9c2 cardiomyocytes exposed to 10 μM ChSeco had higher levels of pSAPK compared to unexposed controls (Fig. 9). Pretreatment with either 10 μM Apo (or DPI) mitigated this rise in pSAPK levels in the ChSeco-exposed cardiomyocytes. A similar phenomenon was observed with regard to the upregulation of pp38 in cardiomyocytes exposed to ChSeco with or without Apo and DPI pretreatments (Fig. 9).

Discussion

This study reports that Apo and DPI reversed the intracellular oxidative stress and the associated decrease in cell viability (assessed based on Trypan blue exclusion) in H9c2 cardiomyocytes exposed to ChSeco and suggests that the primary source of ROS in ChSeco exposures might involve the plasma membrane NADPH oxidase system (NOS). This proposition is supported by the findings that the ChSeco-induced loss in $\Delta\Psi m$ and the lowering of cellular GSH levels in H9c2 cardiomyocytes were mitigated by both Apo and DPI and that the rotenone-induced loss in $\Delta\Psi m$ was not altered. Also, as expected, the ChSeco-induced stress-activated kinase and the levels of phosphorylated p38 in H9c2 cardiomyocytes were found to be reduced by Apo and DPI.

The increase in the ChSeco-induced loss of H9c2 cardiomyocyte viability (assessed based on metabolic activity) by Apo, although seem unexpected, can be explained based on the pro-oxidant potential of Apo phenoxyl radicals (Apo•; 28) and the possible need for Apo preactivation to diapocynin (DiApo) through H_2O_2-dependent, (myelo)peroxidase-mediated oxidations [28–31]. It has been shown that Apo• and similar phenoxyl radicals formed in H_2O_2-dependent,

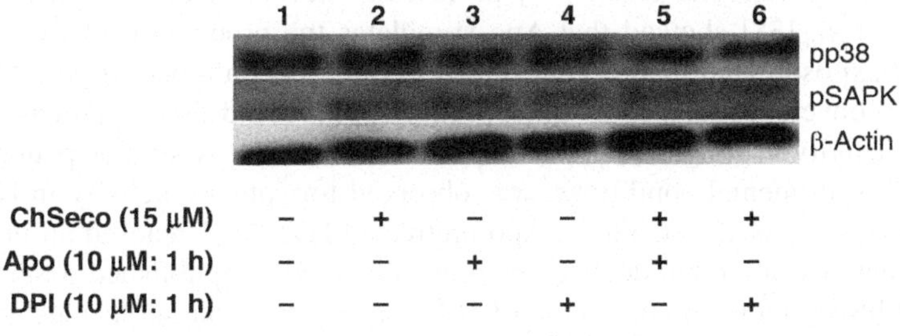

Fig. 9 The involvement of pp38 and pSAPK/JNK in H9c2 cardiomyocyte apoptosis induced by ChSeco. Cardiomyocytes were pretreated with 10 μM Apo (or DPI) and then exposed to 15 μM ChSeco for 2 h. The cells were processed as described in the Materials and Methods section. Along with β-actin, the expression of MAPK was analyzed by immunoblotting

peroxidase-mediated oxidations can facilitate the oxidation of NAD(P)H, GSH, and protein thiols [28–32]. Thus, it appears likely that Apo• could interfere with the PES (or 1-methoxy-PES)-mediated reduction of tetrazolium salts by cellular NAD(P)H and other reducing agents or systems (like the NADPH-ubiquinone oxidoreductase) that form the metabolism-based measurement of cell viability [33, 34]. Another possible explanation for some of the observed discrepancies is that DiApo but not Apo is the most effective inhibitor of NOS [30, 31, 35]. In a study of the flexible docking of Apo, DiApo, and 5-nitroapocynin (5-NitroApo) with 1K4U subsection of human NOS, we found that DiApo had the lowest binding energy (−39.99 eV) as against Apo (−31.70 eV) and 5-NitroApo (−32.36 eV), meaning that DiApo forms a more stable complex with 1K4U subsection [36]. Similarly, in primary cultures of $SOD1^{G93A}$-expressing motor neurons, Beckman and his colleagues have demonstrated that DiApo at a concentration as low as 10 μM provided a significantly greater protection compared to Apo at a concentration of 200 μM against nitric oxide-mediated cell death [30]. Riganti et al. [37] observed that exposure to Apo increases H_2O_2 levels in nonphagocytic cells like the N11 glial cell line and, under conditions of extended exposure, there could be oxidative damage and cytotoxicity. In another nonphagocytic cells like the vascular fibroblast cells, Vejrazka et al. [38] reported exposure to Apo results in increased ROS production.

In our study, similar to the observations made with regard to metabolism-based measurements of cytotoxicity, pretreatment with Apo but not DPI resulted in a further loss of GSH in the ChSeco-exposed H9c2 cardiomyocytes. The effect of Apo is somewhat similar to the observation made by Ximenes et al. [39] who showed that, in alveolar macrophages, GSH levels decreased in response to the treatment with Apo. In another study, Stefanska and Pawliczak [31] demonstrated that Apo exposure causes an increase in intracellular H_2O_2 while the ratio GSH/GSSG shows a decrease.

In this study, it was shown that Apo, but not DPI, increased the ChSeco-induced SOD activity. As discussed above, this is possibly a result of redox cycling of Apo• [28, 32]. For instance, Apo• can oxidize NAD(P)H to NAD(P) pyridinyl radical, NAD(P)• which, in turn, react with O_2 resulting in the formation of superoxide anion radical ($O_2^{\bullet -}$) and $NAD(P)^+$. This explanation is consistent with the observations made by Vejrazka et al. [38] in vascular fibroblast cells exposed to Apo. Vejrazka et al. [38] showed that Apo stimulates the production of ROS in early stages of exposure and this mechanism stops after a certain period possibly due to complete conversion of Apo into other products. This could also explain the decrease in SOD activity at longer hours of exposure to ChSeco. A similar pattern, under identical experimental conditions, was observed for catalase activity in H9c2 cardiomyocytes exposed to ChSeco. Apo pretreated H9c2 cells caused an increase in catalase activity at 2 h but declined significantly at 6 h (unpublished observations).

The ChSeco-induced generation of H_2O_2 or peroxide-like molecules was found to be lowered when the cardiomyocytes were preincubated with either Apo or DPI. This led us to believe that ChSeco may have a direct effect on NOS to generate $O_2^{\bullet -}$. Although not well studied in cardiomyocytes compared to phagocytic cells, the NOS has been suggested to play a major role in cardiomyocytes in conditions of reperfusion injury [40, 41]. Further, we observed that pretreatment with rotenone which inhibits the mitochondrial complex 1 contributed minimally to the lowering of ΔΨm in ChSeco-exposed H9c2 cardiomyocytes (other than plasma membrane, the mitochondrial complex 1 supposedly is the main source of $O_2^{\bullet-}$ in the mitochondrial matrix). It thus appears possible that, in Chseco-exposed cardiomyocytes, H_2O_2 generated from the plasma-membrane NOS may have a direct action on the mitochondrial membrane giving an additive effect. Our observations are in agreement with the findings reported by previous studies [42, 43] which demonstrated that oxysterols, in general, activate the NOS in nonimmune cells but in immune cells (e.g., macrophages).

It was shown by Laynes et al. [20] that there was a loss in cellular antioxidant enzyme defense and subsequent increase in lipid peroxidation in H9c2 cardiomyocytes exposed to ChSeco. In agreement with this, this study not only demonstrated elevated levels of TBARs in ChSeco-exposed cardiomyocytes but also showed that pretreatment with Apo and DPI were effective in bringing down the levels of TBARs.

Phosphorylated p38 and pSAPK play an active role in the induction of apoptosis of H9c2 cells caused by the exposure of ChSeco [20]. Similar results were observed in this study and both Apo and DPI pretreatments suppressed this mitogen kinase phosphorylation. Earlier on, Heumuller et al. [44] demonstrated that the activation of mitogen-activated protein kinases like p38 and extracellular signaling regulated kinases are suppressed by the addition of Apo in response to H_2O_2. It is thus possible that, in ChSeco-exposed cardiomyocytes, oxidative stress may have a direct role in phosphorylation of p38 and SAPK and the subsequent events leading to apoptosis. Apo and DPI which reduce peroxide levels could mitigate the effects of ChSeco and thereby contributes to improved cell survival.

Conclusion

In summary, this study demonstrates that ChSeco-induced oxidative stress in H9c2 cardiomyocytes is mediated primarily through increased generation of H_2O_2 from the NADPH oxidase system. The oxidative stress and further cellular signaling through pp38 and pSAPK in ChSeco-exposed cardiomyocytes are all mitigated to a large extent by NOS inhibitors. Hydrogen peroxide, along with the secondary oxidants (ROS) from it, may have secondary effects on the mitochondria leading to loss of transmembrane potential and subsequent energy crisis. Given the fact that there has been growing evidence that ChSeco is formed naturally through singlet oxygen-mediated oxidations at inflammatory sites [4–8, 12, 16], a detailed understanding of the mechanisms of action of this oxysterol may have important implications to cardiovascular and cerebrovascular diseases.

Acknowledgments The authors acknowledge the support from the National Institutes of Health (NIH) through the National Institute of General Medical Science (NIGMS) Grant 5 P2O GM103424-17 and the US Department of Education (US DoE; Title III, HBGI Part B grant number P031B040030). Its contents are solely the responsibility of the authors and do not represent the official views of NIH, NIGMS, or US DoE.

Financial Interests Declaration The authors declare no competing financial interests.

References

1. Schroepfer GJ Jr. Oxysterols: modulators of cholesterol metabolism and other processes. Physiol Rev. 2000;80:361–554.
2. Lordan S, Mackrill JJ, O'Brien NM. Oxysterols and mechanisms of apoptotic signaling: implications in the pathology of degenerative diseases. J Nutr Biochem. 2009;20:321–36.
3. Vejux A, Lizard G. Cytotoxic effects of oxysterols associated with human diseases: induction of cell death (apoptosis and/or oncosis), oxidative and inflammatory activities, and phospholipidosis. Mol Asp Med. 2009;30:153–70.
4. Brinkhorst J, Nara SJ, Pratt DA. Hock cleavage of cholesterol 5alpha-hydroperoxide: an ozone-free pathway to the cholesterol ozonolysis products identified in arterial plaque and brain tissue. J Am Chem Soc. 2008;130:12224–5.
5. Uemi M, Ronsein GE, Miyamoto S, Medeiros MH, Di Mascio P. Generation of cholesterol carboxyaldehyde by the reaction of singlet molecular oxygen [O2 (1Delta(g))] as well as ozone with cholesterol. Chem Res Toxicol. 2009;22:875–84.
6. Tomono S, Miyoshi N, Sato K, Yoshihiro Ohba Y, Ohshima H. Formation of cholesterol ozonolysis products through an ozone-free mechanism mediated by the myeloperoxidase–H2O2–chloride system. Biochem Biophys Res Commun. 2009;383:222–7.
7. Tomono S, Miyoshi N, Shiokawa H, Iwabuchi T, Aratani Y, Higashi T, Nukaya H, Ohshima H. Formation of cholesterol ozonolysis products in vitro and in vivo through a myeloperoxidase-dependent pathway. J Lipid Res. 2011;52:87–97.
8. Wentworth P Jr, Nieva J, Takeuchi C, Galve R, Wentworth AD, Dilley RB, DeLaria GA, Saven A, Babior BM, Janda KD, Eschenmoser A, Lerner RA. Evidence for ozone formation in human atherosclerotic arteries. Science. 2003;302:1053–6.
9. Takeuchi C, Galve R, Nieva J, Witter DP, Wentworth AD, Troseth RP, Lerner RA, Wentworth P Jr. Proatherogenic effects of the cholesterol ozonolysis products, atheronal-A and atheronal-B. Biochemistry. 2006;45:7162–70.
10. Gao X, Raghavamenon AC, Atkins-Ball DS, Uppu RM. Atherogenic oxoaldehyde of cholesterol induces innate immune response in monocytes and macrophages. Cell Biochem Biophys. 2021;79:649–58.
11. Zhang Q, Powers ET, Nieva J, Huff ME, Dendle MA, Bieschke J, Glabe CG, Eschenmoser A, Wentworth P Jr, Lerner RA, Kelly JW. Metabolite-initiated protein misfolding may trigger Alzheimer's disease. Proc Natl Acad Sci U S A. 2004;101:4752–7.
12. Bosco DA, Fowler DM, Zhang Q, Nieva J, Powers ET, Wentworth P Jr, Lerner RA, Kelly JW. Elevated levels of oxidized cholesterol metabolites in Lewy body disease brains accelerate alpha-synuclein fibrilization. Nat Chem Biol. 2006;2:249–53.
13. Sathishkumar K, Xi X, Martin R, Uppu RM. Cholesterol secoaldehyde, an ozonation product of cholesterol, induces amyloid aggregation and apoptosis in murine GT1-7 hypothalamic neurons. J Alzheimers Dis. 2007;11:261–74.
14. Raghavamenon AC, Gernapudi R, Babu S, D'Auvergne O, Murthy SN, Kadowitz PJ, Uppu RM. Intracellular oxidative stress and cytotoxicity in rat primary cortical neurons exposed to cholesterol secoaldehyde. Biochem Biophys Res Commun. 2009;386:170–4.

15. Bieschke J, Zhang Q, Powers ET, Lerner RA, Kelly JW. Oxidative metabolites accelerate Alzheimer's amyloidogenesis by a two-step mechanism, eliminating the requirement for nucleation. Biochemistry. 2005;44:4977–83.
16. Raghavamenon AC, Gao X, Atkins-Ball DS, Varikuti S, Parinandi NL, Uppu RM. 'Ozone-specific' oxysterols and neuronal cell signaling. In: Berliner L, Parinandi N, editors. Measuring oxidants and oxidative stress in biological systems. 2020. Biol Magn Reson 34, 109, 122.
17. Sathishkumar K, Haque M, Perumal TE, Francis J, Uppu RM. A major ozonation product of cholesterol, 3beta-hydroxy-5-oxo-5,6-secocholestan-6-al, induces apoptosis in H9c2 cardiomyoblasts. FEBS Lett. 2005;579:6444–50.
18. Sathishkumar K, Gao X, Raghavamenon AC, Parinandi N, Pryor WA, Uppu RM. Cholesterol secoaldehyde induces apoptosis in H9c2 cardiomyoblasts through reactive oxygen species involving mitochondrial and death receptor pathways. Free Radic Biol Med. 2009;47:548–58.
19. Sathishkumar K, Murthy SN, Uppu RM. Cytotoxic effects of oxysterols produced during ozonolysis of cholesterol in murine GT1-7 hypothalamic neurons. Free Radic Res. 2007;41:82–8.
20. Laynes L, Raghavamenon AC, D'Auvergne O, Achuthan V, Uppu RM. MAPK signaling in H9c2 cardiomyoblasts exposed to cholesterol secoaldehyde—role of hydrogen peroxide. Biochem Biophys Res Commun. 2011;404:90–5.
21. Sathishkumar K, Raghavamenon AC, Ganeshkumar K, Telaprolu R, Parinandi NL, Uppu RM. Simultaneous analysis of expression of multiple redox-sensitive and apoptotic genes in hypothalamic neurons exposed to cholesterol secoaldehyde. Methods Mol Biol. 2010;610:263–84.
22. Strober W. Trypan blue exclusion test of cell viability. Curr Protoc Immunol. 2001. Appendix 3:Appendix 3B. https://doi.org/10.1002/0471142735.ima03bs21.
23. Karakuzu O, Cruz MR, Liu Y, Garsin DA. Amplex red assay for hydrogen peroxide production from *Caenorhabditis elegans*. Bio Protoc. 2019;9:e3409.
24. Ohkawa H, Ohishi N, Yagi K. Assay for lipid peroxides in animal tissue by thiobarbituric acid reaction. Anal Biochem. 1979;95:351–8.
25. Chatterjee S, Noack H, Possel H, Keilhoff G, Wolf G. Glutathione levels in primary glial cultures: monochlorobimane provides evidence of cell type-specific distribution. Glia. 1999;27:152–61.
26. Sathishkumar K, Gao X, Raghavamenon AC, Murthy SN, Kadowitz PJ, Uppu RM. Determination of glutathione, mitochondrial transmembrane potential, and cytotoxicity in H9c2 cardiomyoblasts exposed to reactive oxygen and nitrogen species. Methods Mol Biol. 2010;610:51–61.
27. Sathishkumar K, Rangan V, Gao X, Uppu RM. Methyl vinyl ketone induces apoptosis in murine GT1-7 hypothalamic neurons through glutathione depletion and the generation of reactive oxygen species. Free Radic Res. 2007;41:469–77.
28. Castor LRG, Locatelli KA, Ximenes VF. Pro-oxidant activity of apocynin radical. Free Radic Biol Med. 2010;48:1636–44.
29. Hart B, Simons J. Metabolic activation of phenols by stimulated neutrophils: a concept for a selective type of anti-inflammatory drug. Biotechnol Ther. 1992;3:119–35.
30. Trumbull KA, McAllister D, Gandelman MM, Fung WY, Lew T, Brennan L, Lopez N, Morré J, Kalyanaraman B, Beckman JS. Diapocynin and apocynin administration fails to significantly extend survival in G93A SOD1 ALS mice. Neurobiol Dis. 2012;45:137–44.
31. Stefanska J, Pawliczak R. Apocynin: molecular aptitudes. Mediat Inflamm. 2008;2008:106507. https://doi.org/10.1155/2008/106507.
32. Babu S, Uppu S, Claville MO, Uppu RM. Prooxidant actions of bisphenol A (BPA) phenoxyl radicals: implications to BPA-related oxidative stress and toxicity. Toxicol Mech Methods. 2013;23:273–80.
33. Mosmann T. Rapid colorimetric assay for cellular growth and survival: application to proliferation and cytotoxicity assays. J Immunol Methods. 1983;65:55–63.
34. https://www.promega.com/-/media/files/resources/protocols/technical-bulletins/0/celltiter-96-aqueous-one-solution-cell-proliferation-assay-system-protocol.pdf
35. Kanegae MP, Condino-Neto A, Pedroza LA, de Almeida AC, Rehder J, da Fonseca LM, Ximenes VF. Diapocynin versus apocynin as pretranscriptional inhibitors of NADPH oxi-

dase and cytokine production by peripheral blood mononuclear cells. Biochem Biophys Res Commun. 2010;393:551–4.
36. Panguruli V, Yang S, Khosravi E, Babu S, Laynes L, Raghavamenon AC, Uppu RM. Docking of apocynin, 5-nitroapocynin, and diapocynin with 1K4U subsection of human neutrophil NADPH oxidase system. Free Radic Biol Med. 2010;49:S203. https://doi.org/10.1016/j.freeradbiomed.2010.10.5.
37. Riganti C, Costamagna A, Bosia A, Ghigo D. The NADPH oxidase inhibitor apocynin (acteovanillone) induces oxidative stress. Toxicol Appl Pharmacol. 2006;212:179–87.
38. Vejražka M, Míček R, Štípek S. Apocynin inhibits NADPH oxidase in phagocytes but stimulates ROS production in non-phagocytic cells. Biochim Biophys Acta. 2005;1722:143–7.
39. Ximenes VF, Kanegae MPP, Rissato SR, Galhiane MS. The oxidation of apocynin catalyzed by myeloperoxidase: proposal for NADPH oxidase inhibition. Arch Biochem Biophys. 2007;457:134–41.
40. Park C, So HS, Shin CH, Baek SH, Moon BS, Shin SH, Lee HS, Lee DW, Park R. Quercetin protects the hydrogen peroxide induced apoptosis via inhibition of mitochondrial dysfunction in H9c2 cardiomyoblast cells. Biochem Pharmacol. 2003;66:1287–95.
41. Turner NA, Xia F, Azhar G, Zhang X, Liu L, Wei JY. Oxidative stress induces DNA fragmentation and caspase activation via the c-Jun NH2-terminal kinase pathway in H9c2 cardiac muscle cells. J Mol Cell Cardiol. 1998;30:1789–801.
42. Biasi F, Mascia C, Astegiano M, Chiarpotto E, Nano M, Vizio B. Pro-oxidant and proapoptotic effects of cholesterol oxidation products on human colonic epithelial cells: a potential mechanism of inflammatory bowel disease progression. Free Radic Biol Med. 2009;47:1731–41.
43. Leonarduzzia G, Gambaa P, Sotteroa B, Kadlb A, Robbesyna F, Calogeroa RA, Biasia F, Chiarpottoa E, Leitingerb N, Sevanian A, Poli G. Oxysterol induced up-regulation of MCP-1 expression and synthesis in macrophage cells. Free Radic Biol Med. 2005;39:1152–61.
44. Heumüller S, Wind S, Barbosa-Sicard E, Harald HH, Schmidt RB, Busse R, Schröder K, Brandes RP. Apocynin is not an inhibitor of vascular NADPH oxidases but an antioxidant. J Hypertens. 2008;51:211–7.

Lipid Mediators in Cardiovascular Physiology and Disease

Diego Hernandez-Saavedra and Kristin I. Stanford

Abstract Bioactive lipids have taken the center stage in the last decade as dynamic mediators of cellular signaling and regulation. Lipids can directly and indirectly modify cellular processes that are independent from their utilization as fuel or structural properties. Among those processes, bioactive lipids are strong mediators of both cardiac and vascular function through diverse mechanisms. One family of lipids includes oxylipins, lipids derived from ω-3 and ω-6 fatty acids such as arachidonic and linoleic acids. Oxylipins play an essential role in whole-body physiology and function including maintenance of cardiac health and vascular homeostasis through direct and indirect mechanisms such as oxylipin receptors or regulation of inflammation. Overactivation or chronic stimulation of oxylipin synthesis has been linked to atherosclerosis, endothelial dysfunction, fibroblast overactivation, myocardial dysfunction, and immune cell activation. Thus, oxylipins are important lipid mediators of cardiovascular physiology and disease.

Keywords Oxylipins · Polyunsaturated fatty acids · Eicosanoids · Prostanoids · Octadecanoids · Cardiovascular disease

---

D. Hernandez-Saavedra
Dorothy M. Davis Heart and Lung Research Institute, The Ohio State University Wexner Medical Center, Columbus, OH, USA

Department of Physiology and Cell Biology, The Ohio State University College of Medicine, Columbus, OH, USA

K. I. Stanford (✉)
Dorothy M. Davis Heart and Lung Research Institute, The Ohio State University Wexner Medical Center, Columbus, OH, USA

Department of Physiology and Cell Biology, The Ohio State University College of Medicine, Columbus, OH, USA

Department of Internal Medicine, Ohio State University College of Medicine, Columbus, OH, USA
e-mail: Kristin.Stanford@osumc.edu

N. L. Parinandi, T. J. Hund (eds.), *Cardiovascular Signaling in Health and Disease*, https://doi.org/10.1007/978-3-031-08309-9_8

Abbreviations

11,12-EET	11,12-Epoxyeicosatrienoic acid
12,13-diHOME	(9Z,12S,13S)-12,13-Dihydroxy-9-octadecenoic acid
13-HODE	13-Hydroxyoctadecadienoic acid
14,15-EET	14,15-Epoxyeicosatrienoic acid
16 (R)-HETE	16(R)-Hydroxyeicosatetraenoic acid
16 (S)-HETE	16(S)-Hydroxyeicosatetraenoic acid
16-HETE	16-Hydroxyeicosatetraenoic acid
17 (R)-HETE	17(R)-Hydroxyeicosatetraenoic acid
17 (S)-HETE	17(S)-Hydroxyeicosatetraenoic acid
17-HETE	17-Hydroxyeicosatetraenoic acid
18 (R)-HETE	18(R)-Hydroxyeicosatetraenoic acid
18-HETE	18-Hydroxyeicosatetraenoic acid
19-HETE	19-Hydroxyeicosatetraenoic acid
20-HETE	19-Hydroxyeicosatetraenoic acid
5,6-EET	5,6-Epoxyeicosatrienoic acid
8,9-EET	8,9-Epoxyeicosatrienoic acid
9-HODE	9-Hydroxyoctadecadienoic acid
9,10-diHOME	(9R,10R,12Z)-9,10-Dihydroxyoctadec-12-enoic acid
AA	Arachidonic acid (20:4)
ACS	Acyl-CoA synthase
ALOX12	Arachidonate 12-lipoxygenase, 12S Type
ALOX12B	Arachidonate 12-lipoxygenase, 12R Type
ALOX15/LOX-1	Arachidonate 15-lipoxygenase
ALOX15B	Arachidonate 15-Lipoxygenase Type B
ALOX5	Arachidonate 5-lipoxygenase
ALOXE3	Arachidonate lipoxygenase 3
ATGL	Adipose tissue triglyceride lipase
BAT	Brown adipose tissue
BKCa	Large-conductance voltage- and Ca^{2+}-activated K^{+} channel
BW755c	Dual lipoxygenase/cyclooxygenase inhibitor
CE	Cholesteryl esters
CEH	Cholesteryl-ester hydrolase
CES1	Carboxylesterase
CHD	Coronary heart disease
COX1	Cyclooxygenase 1
COX2	Cyclooxygenase 2
cPLA2α	Calcium-activated cPLA2
CVD	Cardiovascular disease
CYP	Cytochrome P450
CYP1A	Cytochrome P450 Family 1 Subfamily A
CYP1B1	Cytochrome P450 Family 1 Subfamily B Member 1
CYP2C	Cytochrome P450 Family 2 Subfamily C

CYP2C8	Cytochrome P450 Family 2 Subfamily C Member 8
CYP2C9	Cytochrome P450 Family 2 Subfamily C Member 9
CYP2E	Cytochrome P450 Family 2 Subfamily E
CYP2J	Cytochrome P450 Family 2 Subfamily J
CYP2J2	Cytochrome P450 Family 2 Subfamily J Member 2
CYP4A	Cytochrome P450 Family 4 Subfamily A
CYP4F	Cytochrome P450 Family 4 Subfamily F
CytC	Cytochrome C
DAG	Diacylglycerol
DGLA	Dihomo-γ-linolenic acid
DHA	Docosahexaenoic acid (22:6)
DHETs	Dihydroxyeicosatrienoic acids
diHOME	Dihydroxyoctadecenoic acid
dihomo-PGE2	Dihomo-prostaglandin E2
DP1	Prostanoid receptor D
EDHF	Endothelium-derived hyperpolarizing factors
EET	Epoxyeicosatrienoic acid
EP3	Prostanoid receptor E3
EP4	Prostanoid receptor E4
EPA	Eicosapentaenoic acid (20:5)
EpOME	Epoxyoctadecaenoic acid
ER	Endoplasmic reticulum
ERK	Extracellular-regulated kinases
FA	Fatty acid
FA*sn-1*-lysoPL	*sn-1* Fatty acid lysophospholipid
FOXO	Forkhead box O
FP	Prostanoid receptor F
GPCRs	G-protein-coupled receptors
GLA	γ-Linolenic acid
HETE	Hydroxyeicosatetraenoic acid
HODE	Hydroxyoctadecadienoic acid
HSL	Hormone-sensitive lipase
IL-13	Interleukin-13
IL-1β	Interleukin-1beta
IL-4	Interleukin-4
IP	Prostanoid receptor I
JNK	c-Jun NH2-terminal kinases
LA	Linoleic acid (18:2)
LDAH	Lipid droplet–associated hydrolase
LOX	Lipoxygenase
LTA4	Leukotriene A4
LTB4	Leukotriene B4
LTC4	Leukotriene C4
lysoPL	Lysophospholipids
MAPK	Mitogen-activated protein kinase

MGL	Monoacylglycerol lipase
Na+/K+-ATPase	Sodium/potassium-transporting ATPase
NAFLD	Non-alcoholic fatty liver disease
NCEH1	Neutral cholesterol ester hydrolase 1
NO	Nitric oxide
NOS1	Nitric oxide synthase 1
NSAID	Non-steroidal anti-inflammatory drug
p38	P38 MAP Kinase
PG	Prostaglandins
PGD2	Prostaglandin D2
PGE1	Prostaglandin E1
PGE2	Prostaglandin E2
PGE3	Prostaglandin E3
PGF2a	Prostaglandin F2a
PGG2/PGH2	Prostaglandin G2/H2
PGI2	Prostacyclin
PKC	Protein kinase C
PL	Phospholipids
PLA1	Phospholipase A1
PLA2	Phospholipase A2
PLIN1	Perilipin 1
PMN	Polymorphonuclear cells
PUFAs	Polyunsaturated fatty acids
PUFA*sn-2*-lysoPL	*sn-2* PUFA-lysophospholipid
R-	Rectus
RyR	Ryanodine receptor
S-	Sinister
sEH/EPHX	Soluble epoxide hydrolases/epoxide hydrolase
SERCA2a	Sarcoplasmic reticulum calcium ATPase
sn-1/2	Nucleophilic substitution-1/2
SPMs	Specialized pro-resolvin mediators
TAG	Triacylglycerols
TP	Thromboxane receptor
TXA2	Thromboxane A2
TXB2	Thromboxane B2
ω-3	Omega-3 fatty acid
ω-6	Omega-6 fatty acid

Introduction

Cardiovascular diseases (CVD) are the leading causes of mortality worldwide and in the United States. They comprise a group of diseases that affect the heart and vascular endothelium function including stroke, congenital diseases, atherosclerosis, coronary heart disease (CHD), heart failure (HF), and others [1]. The combined death burden from CVD has been estimated to be around 32% of global deaths, with both heart attacks and stroke accounting for the majority of deaths [2]. Moreover, the economic burden of CVD has skyrocketed in the last decades, and estimates calculate that this will increase to over $1000 billion by the year 2030. Both the deaths by CVD and economic burden are superior to all communicable and non-communicable diseases, thus justifying the need for strategies that can help combat CVD and associated comorbidities.

The etiology of CVD is complex and multifactorial. While CVD has a strong genetic component, lifestyle factors and behaviors appear to have an additive effect that contributes to increased CVD risk [3]. Contributing lifestyle factors include a poor diet, physical inactivity, tobacco and alcohol consumption, among others. These factors can impair cardiovascular health and increase the risk to develop CVD. Thus, given the growing incidence of CVD worldwide, the identification of novel and precise biomarkers that are able to detect and predict the risk of CVD are of the utmost importance. Among such biomarkers are oxylipins, oxygenated by-products of polyunsaturated fatty acids. Oxylipins are modulated by both genetic and lifestyle factors and have been found to play an important role in cardiovascular health and CVD [4–10] and constitute an expanding family of lipids that mediate both intracellular and endocrine signaling. By binding to membrane-bound receptors or enzymes, oxylipins can regulate inflammation and second messenger production to modulate vascular endothelium and cardiac muscle function [10].

Lipid Mediators

Oxylipins are a diverse family of lipids that comprise a wide-ranging group of chemical compounds that possess structural and signaling properties [6, 11]. Recent work has demonstrated the importance of structural lipids such as membrane-residing phospholipids (PL) as fatty acid reservoirs that give rise to lipid mediators such as oxylipins and endocannabinoids. These lipid mediators have emerged as important modulators of multiple physiological systems, affecting both cellular signaling and physiological function.

Oxylipins

Among the wide variety of lipids, oxidized lipids or oxylipins originate from polyunsaturated fatty acids (PUFAs) and have redundant properties, thus activating similar biological processes, or opposing mechanisms, which balance biological responses. Within this growing family of oxylipins, recent advances in detection and quantification have expanded their numbers to >100 oxylipins [10, 12]. Oxylipins have been involved in multiple harmful processes related to thrombosis, hypertension, diabetes, and hemostasis, and on the other hand recent evidence has linked them to improved cardiac function. Altogether, the context of oxylipin production determines their contribution to either disease generation or therapeutic potential.

PUFA Reservoirs

Lipid membranes outline cellular and intracellular compartments and allow the separation of cellular functions and communication among cell organelles. These lipid compartments can be found in the form of lipid bilayers such as the plasma membrane, endoplasmic reticulum (ER), lysosome, mitochondrial membranes, and in lipid monolayers such as lipid droplets. Both lipid mono- and bilayers are conformed by multiple phospholipid species and cholesterol that, given their chemical polarities, help to demarcate the boundaries of intracellular organelles. In the case of lipid droplets, lipid layers allow for the storage of energy in the form of lipids. Thus, phospholipid mono- and bilayers not only facilitate cellular compartmentalization but also provide fatty acid substrates that can be readily utilized as secondary signaling molecules.

Phospholipids as Oxylipin Reservoirs Membrane-bound phospholipids are important regulators of the cytoplasmic membrane biophysical properties and, given the great variation in their acyl moieties, they function as dynamic reservoirs of fatty acids. Following cellular activation, membrane-bound phospholipids are acted on by cytosolic phospholipases that release fatty acids and PUFAs into the cytosol where they can be further metabolized into oxylipins by different enzymes (cytochrome P450 [CYP], lipoxygenases [LOX], cyclooxygenases [COX], Soluble epoxide hydrolases [sEH], etc.) [8, 9]. This cleavage can occur through multiple pathways: cleavage by phospholipase A2 (PLA_2), cholesteryl ester cleavage, phospholipase A1 (PLA_1), and plasmalogen conversion [11]. The first phospholipid cleavage pathway is controlled by calcium-activated $cPLA_2$ ($cPLA_2\alpha$) and involves the lysis of the PUFA from the *sn-2* position within membrane-bound phospholipids to produce PUFAs and *sn-1* fatty acid lysophospholipid ($FA_{sn\text{-}1}$-lysoPL) (Red and yellow pathways, Fig. 1a) [13, 14]. Released PUFAs can then be transformed into oxylipins and subsequently captured by acyl-CoA synthase (ACS) into oxylipin-CoA (Fig. 1a). A second source of oxylipins can be obtained from direct oxygenation of PUFA-cholesteryl esters (CE) into oxylipin-CE, and further cleaved by cholesteryl-ester hydrolase (CEH) (Orange pathway, Fig. 1a). A third pathway of oxylipin synthesis involves the lysis of the fatty acids from the *sn-1* position within membrane-bound phospholipids to produce a fatty acid and *sn-2* PUFA-lysophospholipid ($PUFA_{sn\text{-}2}$-lysoPL) (Green and cyan pathways, Fig. 1a) [15]. Finally, oxylipins can also originate from plasmalogens, *sn-1* vinyl-ether bond-containing phospholipids, that are synthesized within the peroxisomes and are cleaved by cytochrome C (CytC) to generate $PUFA_{sn\text{-}2}$-lysoPL (Purple pathway,

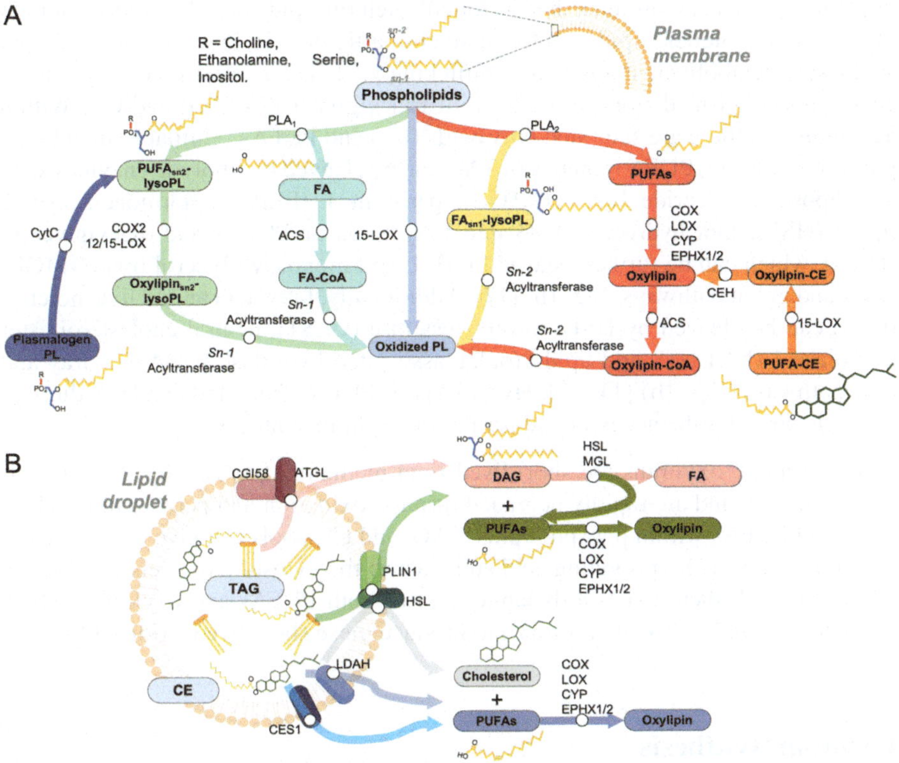

Fig. 1 Oxylipin reservoirs in mammalian cells. (**a**) Membrane-bound phospholipids (PL) are major cellular polyunsaturated fatty acid (PUFA) reservoirs that are cleaved through diverse phospholipase (PLA) enzymes to produce PUFA or fatty acids (FA) and the corresponding lyso PL. PUFAs are subsequently metabolized by various oxygenases (COX, LOX, CYP, sEH) to produce a wide variety of oxylipins. Additionally, oxylipins can be obtained from peroxisome-derived plasmalogens and cholesteryl ester (CE) reservoirs. (**b**) Lipid droplets are important for lipid storage in the form of triglycerides (TAG) and CEs. Diverse lipases (ATGL, HSL, LDAH, CES1, MGL) hydrolyze TAG and CE from lipid droplets to release PUFA that can be utilized to produce oxylipins. PLA, phospholipase A; PUFA, polyunsaturated fatty acid; ACS, acyl-CoA synthase; CE, cholesterylester; CEH, neutral cholesterol ester hydrolase; FA, fatty acid; lysoPL, lysophospholipid; COX, cyclooxygenase; LOX, lipoxygenase; CYP, cytochrome P450; EPHX1/2, soluble epoxyde hydrolase; CytC, cytochrome C; CES1, carboxylesterase 1; LDAH, lipid droplet–associated hydrolase; TAG, triglyceride; DAG, diacylglycerol

Fig. 1a) [16]. All phospholipid-cleaving pathways can be dynamically regulated to provide PUFAs or reciprocally store PUFAs within the cell membrane. While other compartments exist, the phospholipid PUFA reservoir is thought to be the major source of substrate for oxylipin synthesis.

Lipid Droplets as Oxylipin Reservoirs While phospholipids are a major source of oxylipins, several intracellular lipid compartments function as PUFA and oxylipin reservoirs. Lipid droplets are ubiquitous cellular compartments that are composed of phospholipids and proteins that surround a neutral lipid core [17]. Often thought of as inert organelles, lipid droplets actively participate in the regulation of lipid synthesis, metabolism, storage, and trafficking [17]. The neutral lipid cargo stores energy and lipid mediators in the form of triacylglycerols (TAG) and CE. Within acyl moieties, there are saturated fatty acids (FA) and PUFAs. Similar to membrane phospholipids, the PUFA moieties of TAG can be cleaved by lipolytic enzymes such as adipose triglyceride lipase (ATGL), perilipin 1 (PLIN1), hormone-sensitive lipase (HSL), that hydrolyze TAG into fatty acids or PUFAs and diacylglycerol (DAG), which can be further cleaved by HSL and monoacylglycerol lipase (MGL) (pink and green pathways, Fig. 1b) [17]. Additionally, fatty acid and PUFA moieties of CE can be cleaved by HSL, carboxylesterase (CES1), neutral cholesterol ester hydrolase 1 (NCEH1), and lipid droplet–associated hydrolase (LDAH) (Gray and blue pathways, Fig. 1b) [11, 17]. Hydrolysis of PUFAs from TAG and CE through lipolysis provides the necessary substrate for oxylipin synthesis.

Importantly, oxylipins are short-lived lipid mediators that rapidly regulate cellular functions and are rapidly degraded through oxidation and peroxidation. Thus, increased PUFAs within phospholipids, TAG, and CE pools can give rise to greater oxylipin synthesis by providing acyl substrate to the cytosolic oxygenases. Hence, within the cell, there exists a dynamic pool of both PUFAs and oxylipins, which regulate the availability of substrates to produce these short-lived oxidized lipids.

Oxylipin Synthesis

Synthesis of oxylipins occurs in multiple cells of the body and involves the transformation of PUFAs such as arachidonic acid (AA; 20:4), eicosapentaenoic acid (EPA; 20:5), docosahexaenoic acid (DHA; 22:6), and linoleic acid (LA; 18:2). Conversion of AA and EPA by oxylipin synthesis enzymes produces eicosanoids (20-carbon compounds) such as epoxyeicosatrienoic acids (EET), hydroxyeicosatetraenoic acid (HETE), and prostaglandins (PG), whereas conversion of LA by oxylipin synthesis enzymes produces octadecanoids derivatives (18-carbon compounds) that include Epoxyoctadecaenoic acid (EpOME), Dihydroxyoctadecenoic acid (diHOME), and hydroxyoctadecadienoic acid (HODE) [6]. As described above, the oxylipin synthesis process is mediated by oxygenases such as cytochrome P450 (CYP)-epoxygenases, lipoxygenases (LOX), ω-hydroxylases, and cyclooxygenases (COX). Given the nature of oxylipins, their synthesis occurs rapidly to respond to stimuli and their half-life is short, which indicates a tightly monitored process of biotransformation of the available PUFAs.

Cyclooxygenase-Derived Oxylipins

Eicosanoid biosynthesis begins with the conversion of AA (20:4), EPA (20:5), and DHA (docosanoids, 22:6) by CYP, COX, LOX, and other enzymes. COXs are a family of heme-containing enzymes that catalyze oxygenase and peroxidase activities, and these reactions are carried out by either COX-1 or COX-2. COX enzymes reside within multiple membranes of the cell; COX1 and 2 reside within the nuclear membrane as well as the ER and mitochondrial membranes, and COX2 has been found within the Golgi apparatus (Fig. 2) [18]. COX-1 is the constitutively active isoform COX-1 and is ubiquitously expressed across body tissues (vascular endothelium, platelets, immune cells, smooth muscle cells, etc.), while COX-2 is the inducible isoform and responds to inflammation. Recently, a constitutive function has been described for COX-2 in multiple tissues including blood vessels, intestine, and brain, among others [19–21]. COX-1 and COX-2 are localized within the ER and nuclear membranes [22], and COX-2 has also been found within the Golgi membrane [23]. Both COX enzymes—through their dioxygenase activity—catalyze the conversion of AA and EPA to produce diverse prostanoids (i.e., PGE_1,

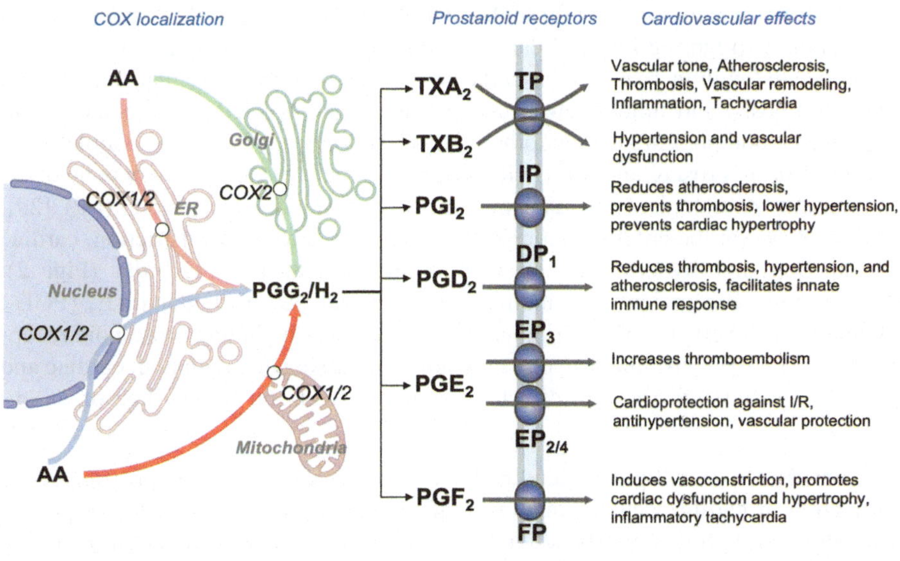

Fig. 2 Effect of prostanoids and thromboxanes on cardiovascular function. Arachidonic acid (AA) can be metabolized by cyclooxygenases (COX1/2) that reside within the nuclear membrane, endoplasmic reticulum, mitochondria, and the Golgi apparatus. AA is converted to the prostaglandin precursor PGG_2/H_2 that gives rise to a wide variety of prostanoids. Each prostanoid acts in an autocrine or paracrine manner through different receptors within the vascular endothelium, myocardium, and immune cells to regulate cardiovascular function. AA, arachidonic acid; COX, cyclooxygenase; DP, receptor for prostanoid D; EP, receptor for prostanoid E; FP, receptor for prostanoid F; IP, receptor for prostanoid I; PGG_2/H_2, prostaglandin G_2/H_2; PGI2, prostacyclin; PGD_2, prostaglandin D_2; PGE_2, prostaglandin E_2; PGF_2, prostaglandin F_2; TP, receptor for prostanoid TxA2; TxA2, thromboxane A2

PGE_3, and dihomo-PGE_2.), thromboxanes (i.e., TXA_2, TXB_2), and leukotrienes (i.e., LTA_4, LTB_4, LTC_4). While the studies on LA-derived oxylipins are fewer than AA oxylipins, LA-derived oxylipins are found in greater amounts than other PUFA-derived oxylipins (octadecanoids) [6, 24–26]. Although the substrate preference for LA is markedly lower than AA, COX enzymes are reported to convert LA into 9-HODE and 13-HODE [27, 28]. Importantly, COX-derived eicosanoids produced from AA, but not LA, have been long associated with cardiovascular effects such as atherosclerosis, clotting, vasoconstriction, and cardiac dysfunction, and drug therapies such as non-steroidal anti-inflammatory drug (NSAID) aspirin target COX-derived oxylipin production. Thus, eicosanoids such as prostanoids and thromboxanes are greatly relevant to both cellular function and disease.

Prostanoids and Cardiovascular Function Prostanoid derivatives are a subclass of eicosanoids that result from the conversion of AA and other PUFAs by COX enzymes—both COX1 and COX2. Following the release of AA and other PUFAs from cellular compartments (Fig. 1), AA is dioxygenated into prostaglandin G_2/H_2 (PGG_2/PGH_2) and is the major common precursor of prostaglandins and prostacyclins (Fig. 2) [4, 11, 18]. Further metabolism of PGG_2/PGH_2 by thromboxane synthase generates diverse thromboxane derivatives such as thromboxane A2 and B2 (TXA2/TXB2), which can act through the thromboxane receptor (TP)-coupled to G protein, thus triggering intracellular signaling [4, 29]. TXA_2 signaling is known to be associated with increases in vascular tone, atherosclerosis, vascular remodeling, pro-thrombosis, and inflammatory tachycardia, whereas TXB_2 is associated with increased blood pressure and endothelial dysfunction (Fig. 2) [8]. In a similar way, metabolism by prostaglandin F (PGF) synthase generates PGF_{2a} which is involved in multiple pathological mechanisms through prostanoid receptor F (FP) [29]. PGF_{2a} effects on the cardiovascular system are related to vasoconstriction, cardiac dysfunction and hypertrophy, as well as inflammatory tachycardia (Fig. 2). Moreover, microsomal prostaglandin E synthase catalyzes prostaglandin E2 (PGE_2) synthesis and exerts its effect through diverse receptors including prostanoid receptor E (EP_3) [29]. Thus, multiple prostaglandins are associated with both cardiac and vascular dysfunction through direct or indirect effects on the cardiac or vascular tissue.

Conversely, several prostaglandins have been associated with improvements or protection of cardiovascular function. PGE_2 signaling through EP_4, in contrast to activation through EP_3, has beneficial effects on cardiovascular function including protection of the heart after ischemia/reperfusion injury, reduced hypertension, and improved endothelial function (Fig. 2). Other examples of beneficial prostaglandins include prostacyclin (PGI_2) and prostaglandin D2 (PGD_2) [8, 29]. PGI_2 is synthesized by PGI synthase and acts through prostanoid receptor I (IP) to reduce atherosclerosis risk, prevent thrombosis, reduce hypertension, and protection against cardiac hypertrophy. In addition, PGD_2 is synthesized by either lipocalin-type PGD synthase or hematopoietic-type PGD synthase and acts through different receptors including prostanoid receptor D (DP_1) [8]. In turn, PGD_2-related mechanisms include reduced thrombosis, lower hypertension, and atherosclerosis, and it is known to aid in the activation of the innate immune response. Altogether, prostaglandins can induce both protective or pathophysiological effects on cardiac and vascular function and are intricately related to the cleavage of AA and other PUFAs.

Lipoxygenase-Derived Oxylipins

Oxylipins can also be produced by the non-heme iron-containing LOX family of enzymes that comprise six members (ALOX12, ALOX12B, ALOX15, ALOX15B, ALOX5, and ALOXE3) that display dioxygenase activities to produce lipid hydroperoxides. Importantly, most LOX enzymes are constitutively active, whereas LOX-1 (ALOX15) is inducible by inflammation (IL-4 and IL-13) [30]. Interestingly, some LOX enzymes are activated by their hydroperoxide products and intracellular calcium, whereas some are irreversibly inactivated by their products [31]. Substrate preference varies among LOX enzymes, while most LOXs have increased preference for AA, other enzymes accept other fatty acids such as EPA, DHA, LA, γ-linolenic acid (GLA), and dihomo-γ-linolenic acid (DGLA). Important oxylipin by-products of LA include HODEs, such as 9- and 13-HODE, and a variety of mid-chain hydroxyeicosatetraenoic acids (HETE; 7, 10, 13, 18, 19-HETE) that form AA, which are involved in stress and inflammation responses. In some instances, they can directly oxidize fatty acid-containing amides [32], lysophospholipids (lysoPL), phospholipids (PL) [14, 33, 34], CE [35], and lipoproteins [36] (Fig. 1). From a wide variety of PUFAs, LOX enzymes can catalyze their conversion to generate a myriad of oxylipins such as leukotrienes and diverse specialized pro-resolvin mediators (SPMs) such as lipoxins, resolvins, protectins, maresins, and others [6]. Given this wide range of lipid mediators, the physiological effect of LOX-derived products is dependent on the context and concentration of the oxylipin. Additionally, LOX enzymes accept a wide variety of PUFAs and lipid products, thus participating in diverse physiological processes.

Mid-chain HETEs and Cardiovascular Function Important by-products of LOX (and CYP1B1) [37] include the hydroxylation of AA to form hydroxyeicosatetraenoic acids (HETEs) in mid-chain positions. These mid-chain HETEs, namely, 7-, 10-, 12-, 13-, 15-, 18-, and 19-HETE, are important in the development of cardiovascular disease. Early research on mid-chain HETEs demonstrated their strong effect on cardiovascular dysfunction due to their effects on smooth muscle cells, endothelial cells, cardiomyocytes, fibroblasts, and immune cells [38–40]. The mechanism by which mid-chain HETEs increase the risk of vascular and cardiac disease involves the activation of multiple receptors that leads to intracellular activation of protein kinase C (PKC) and downstream factors mitogen-activated protein kinase (MAPK), extracellular-regulated kinases (ERK), c-Jun NH2-terminal kinases (JNK), and p38.

Mid-chain HETE, 12-HETE, increases angiotensin II-hypertension by fostering aldosterone secretion; Mechanistically, angiotensin II directly induces 12-HETE production in adrenal cortex (glomerulosa cells) through Ca^{2+}-dependent signaling, and in turn 12-HETE increases vasoconstriction through multiple G-protein-coupled receptors (GPCRs) including GPR31 [41]. This is supported by studies showing that non-specific inhibitors of LOX (BW755c) prevent aldosterone secretion [42–44], and anti-hypertensive therapies that suppress 12-HETE secretion [41, 45]. Similar effects on hypertension have been reported for other mid-chain HETEs 5- and 15-HETE.

15-HETE has been strongly linked to HF, inflammation, fibrosis, and cardiac hypertrophy. Studies have shown that 15-HETE increases isoproterenol sensitivity [41, 46–50], and 15- and 12-HETE are involved in norepinephrine-induced hypertrophy. The mechanism by which mid-chain HETEs such as 15-HETE stimulate chronic cardiac dysfunction involves its incorporation into phospholipids (Fig. 1) and further substitution into DAG species, which can in turn activate PKC. Inhibition of LOX prevents myocardial fibrosis in hypertension models [51] and reduces angiotensin II-mediated collagen deposition. On the other hand, 15-HETE can be converted into lipoxins, which are strong modulators of inflammation resolution [52]. Thus, mid-chain HETEs are associated with both vascular and myocardial dysfunction [41] which involves the stimulation of LOX enzymes or cleavage from their intracellular reservoirs.

HODEs and Cardiovascular Function Oxylipins derived from LA include 9-HODE and 13-HODE which are synthesized by both LOX and CYP epoxygenases. Generally, HODEs are associated with inflammation and oxidative stress; 9-HODE is thought to be pro-oxidative and pro-inflammatory, while 13-HODE is believed to be anti-oxidative and anti-inflammatory.

9-HODE regulates stress and inflammation by activating ER stress and secretion of inflammatory cytokine IL-1β in macrophages [53–55], and stimulates a pro-inflammatory environment in experimental models of wound-healing [56, 57]. Additionally, combined lipidomic and transcriptomic analysis identified 9-HODE as a strong regulator of Forkhead box O (FOXO) transcription factor through JNK-mediated activation, thus increasing inflammation [58]. Altogether, 9-HODE is regarded as a strong activator of inflammation and cellular stress.

Conversely, reports have described multiple beneficial effects of 13-HODE. 13-HODE inhibits platelet adhesion to the vascular endothelium [59, 60], reduces platelet aggregation [61], inhibits tumor cell adhesion to endothelial cells [54], regulates intracellular Ca^{2+} signaling [62], stimulates PGI_2 synthesis [63] (Fig. 2), while reducing neutrophil-mediated LTB_4 release [64] and inhibiting TAG synthesis and release [65]. Similar to other oxylipins, 13-HODE can be incorporated into phospholipids (specifically phosphatidylcholine) from endothelial cells and the heart [66] can be gradually removed through enzymatic cleavage [67] (Fig. 1).

Further metabolism of 9-HODE and 13-HODE produces oxylipins with diverging mechanisms. Degradation products of 9-HODE and 13-HODE are metabolized by hydroxy fatty acid dehydrogenases that generate the keto fatty acid derivatives 9-oxo-ODE and 13-oxo-ODE respectively, with both anti-inflammatory and anti-proliferative properties [54, 68–70]. Thus, despite their similarities in terms of origin and structure, 9-HODE and 13-HODE (and their degradation products) appear to have opposing effects on inflammation and cellular aggregation that alters cardiovascular function.

Cytochrome P450-Derived Oxylipins

Oxylipin synthesis—eicosanoids, docosanoids, and octadecanoids—can also be catalyzed by the superfamily of CYP enzymes. These heme-containing monooxygenase CYPs catalyze an incredible variety of reactions including terminal or mid-chain hydroxylation and epoxidation of PUFAs and degradation of oxylipins. Hydroxylation of PUFAs such as AA produces a great variety of hydroxyeicosatetraenoic acids (HETE) through terminal ω-hydroxylation (16-, 17-, 18-, 19-, 20-HETE) [71–75]. Moreover, CYPs catalyze the formation of epoxyeicosatrienoic acids (EET) from AA, as well as the epoxidation of EPA, DHA, and LA. Importantly, PUFA epoxy-derivatives are oxylipins that have strong effects on vasodilation, vasoconstriction, and inflammation [76], all of which are important for cardiovascular function and disease. Epoxides from PUFAs are further metabolized by epoxide hydrolases (soluble [sEH] and microsomal [mEH]) to produce dihydroxyl products with different physiological effects than EETs. Additionally, CYP epoxygenases metabolize LA into EpOMEs and further metabolism of EpOMEs by cytosolic sEH increases the production of important diHOMEs. Both 9,10-diHOME and 12,13-diHOME are important regulators of metabolic and cardiac function. Thus, CYP and sEH derivatives (EETs and EpOME/diHOMEs) are important oxylipins that modulate oxidative stress and inflammation, as well as vascular and cardiac function.

Terminal HETEs and Cardiovascular Function Terminal hydroxylation of AA by CYP450 (CYP1A, CYP4A, and CYP4F) produces the hydroxyeicosatetraenoic acids (HETE) 16-, 17-, 18-, 19-, and 20-HETE [37]. Hydroxylation within these terminal sites introduces chiral centers or "stereogenic carbons" (carbons with four different substituents) that produce two spatial enantiomers R (*rectus*) and S (*sinister*), that is, 16 (R)-HETE and 16 (S)-HETE. These enantiomers are mirror images of each other and have widely different physiological effects. While 16 (S)-HETE is the less active form, 16 (R)-HETE is produced from polymorphonuclear (PMN) cells following angiotensin II activation and functions as a lipid inhibitor of neutrophil activity [77, 78] and PMN cell aggregation. Additionally, 16 (R)-HETE (and to a lesser extent 16 (S)-HETE) has vasodilation properties, as demonstrated by its effects on renal vasodilation and protection against central hypertension. Evidence exists for 16-HETE incorporation into phospholipids of platelets [79], which could be required for the anti-aggregative properties. 16 (R)-HETE is regarded as a positive biomarker in CVD and other inflammatory diseases such as non-alcoholic fatty liver disease (NAFLD) [80], additional studies will identify the link between 16-HETE and CVD risk.

The effects of 17-HETE and 18-HETE have not been thoroughly investigated [37]. Some studies show that 17-HETE is involved in electrolyte and fluid transport within the kidney, primarily regulating sodium transport [81] and proximal tubule ATPase activity [82]. While the effects have been attributed to 17 (S)-HETE, the 17 (R)-HETE isomer is thought to be inactive. 18 (R)-HETE is associated with vascular insulin resistance, muscle vascularization [83], and stimulating renal vasoconstriction [84]. Both 17- and 18-HETE can be incorporated within phospholipids of the renal cortex and medulla [81], but the long-term effect of the incorporation of 17- and 18-HETE within phospholipids has not been described. Thus, while it is clear that 17- and 18-HETE are involved in cardiac function and CVD, future studies will provide direct evidence for the function of these AA metabolites on vascular and cardiac function.

The role of 19-HETE on cardiovascular physiology has been previously described [37]. While pathologic cardiac hypertrophy is associated with reduced 19-HETE [85], treatment with a racemic mixture of 19-HETE leads to a reduction in 20-HETE and confers protection against angiotensin II-induced cardiac hypertrophy. The effects of 19-HETE on the heart might be mediated through local production through stimulation of the CYP enzymes CYP1A, CYP2E, and CYP4A [86–88]. Moreover, cardiovascular improvements in response to 19 (S)-HETE might be indirectly regulated through stimulation of Na^+/K^+-ATPase within renal proximal tubules [89]. While 19-HETE is regarded as a beneficial AA oxylipin, terminal 20-HETE has been associated with increased CVD risk [10]. Importantly, 19 (R)-HETE is a strong antagonist of pathological 20-HETE [90], and inhibits the effects of 20-HETE on vasoconstriction and endothelial dysfunction [89, 90]. 20-HETE is a potent arterial vasoconstrictor and activator of intracellular Ca^{2+} signaling [91–95]. Importantly, inhibition of 20-HETE production leads to nitric oxide (NO)-mediated vasodilation [96], and addition of 19-HETE prevents the inhibition of NO vasodilation [89], thus providing a direct link between 20-HETE and its vasoactive properties. Given its widespread involvement in arterial vasoconstriction, 20-HETE is intricately linked to ischemic cerebrovascular disease, cardiac ischemia/reperfusion, renal disease, hypertension, diabetes, among others [96].

EETs and Cardiovascular Function EETs are well-described lipid mediators of vascular and cardiac function [7, 10]. Synthesis of EETs from AA is catalyzed by the CYP epoxygenases CYP2C and CYP2J to generate four active regioisomers: 5,6-EET, 8,9-EET, 11,12-EET, and 14,15-EET. Although EETs have structural differences, the biological effects seem to be conserved for all EETs to a similar extent [97]. Classical catabolism of EETs is facilitated by sEHs [97–99] that add a water molecule to the epoxide bond to form dihydroxyeicosatrienoic acids (DHETs), which reduces their bioactivity [7]. Alternative metabolism of EETs involves the shortening by β-oxidation into 16-carbon EETs or elongation into 22-carbon EET derivatives [97]. All EETs can also be incorporated into phospholipids primarily within endothelial cells [100], which can prolong their half-life and bioactivity.

Bioactive properties of EET lipid mediators include vasodilation, anti-inflammation [101, 102], anti-thrombosis, increased fibrinolysis [102, 103], cardioprotection from ischemia/reperfusion [104], amelioration of hypertension, regulation of angiogenesis, among others.

Vasodilation effects of EETs are thought to be mediated through the large-conductance voltage- and Ca^{2+}-activated K^+ channel (BKCa) within smooth muscle cells. This vasodilatory effect can be potentiated by the action of PGI_2 and NO synthesis [103]. EETs also play a role in the regulation of intracellular Ca^{2+} levels by stimulating the expression of the sarcoplasmic reticulum calcium ATPase (SERCA2a) [105]—a major regulator of Ca^{2+} stores. Thus, altogether EETs are well regarded as endothelium-derived hyperpolarizing factors (EDHF) and play a major role in CVD prevention [7, 106, 107]. While therapeutic strategies that increase EETs have shown promising results in patients with CVD, further characterization of the mechanisms of EET-enhancing compounds is needed.

diHOMEs and Cardiovascular Function The LA-derived oxylipins, 9,10-diHOME ((9R,10R,12Z)-9,10-dihydroxyoctadec-12-enoic acid) and 12,13-diHOME ((9Z,12S,13S)-12,13-Dihydroxy-9-octadecenoic acid) have been associated with leukotoxicity [5, 108], but recent studies indicate that 12,13-diHOME has robust therapeutic potential to improve metabolic and cardiac function [109–112]. The synthesis of 12,13-diHOME begins with the metabolism of LA by the CYP450 family of enzymes (CYP2J2, CYP2C8, CYP2C9, CYP1A) [5] that give rise to the linoleic epoxide 12,13-EpOME (and 9,10-EpOME), which in turn is subsequently hydrolyzed by sEH (EPHX1–4) to produce 12,13-diHOME and 9,10-diHOME [5, 113] (Fig. 3). Although high levels of 9,10- and 12,13-diHOME have been associated with pathological conditions such as inflammation and oxidative stress [108, 114–116], endocrine disruption [117, 118], mitogenesis [118, 119], pain [120–122], and cardiotoxicity [123], new studies have dissected novel mechanisms of these diols on metabolism and cardiac function.

Obesity and dyslipidemia are common comorbidities of CVD [124–126]. In obese subjects, the expression levels of CYP and sEH (EPHX) are reduced together with their oxylipin products. In mice, obesity (consumption of a high-fat diet) is associated with disruptions in the oxylipin profile in adipose tissue and decreases important CYP-derived oxylipins such as EpOMEs and EETs, as well as the sEH (EPHX)-products diHOMEs [127]. Additionally, in humans with dyslipidemia, the levels of circulating 12,13-diHOME were reduced, compared to healthy subjects [25]. In human patients with heart disease, circulating levels of 12,13-diHOME are directly associated with increased functional cardiac parameters such as ejection fraction and fractional shortening [109]. Thus, 12,13-diHOME is associated with a reduction of important risk factors for CVD such as obesity and dyslipidemia.

Exercise is an important therapeutic strategy to combat obesity and reduce the risk of CVD [128, 129]. Studies in humans have demonstrated that exercise markedly increases linoleic diol 12,13-diHOME [5, 112, 130, 131] (Fig. 3). We and others have previously shown that 12,13-diHOME functions as a lipokine—a lipid

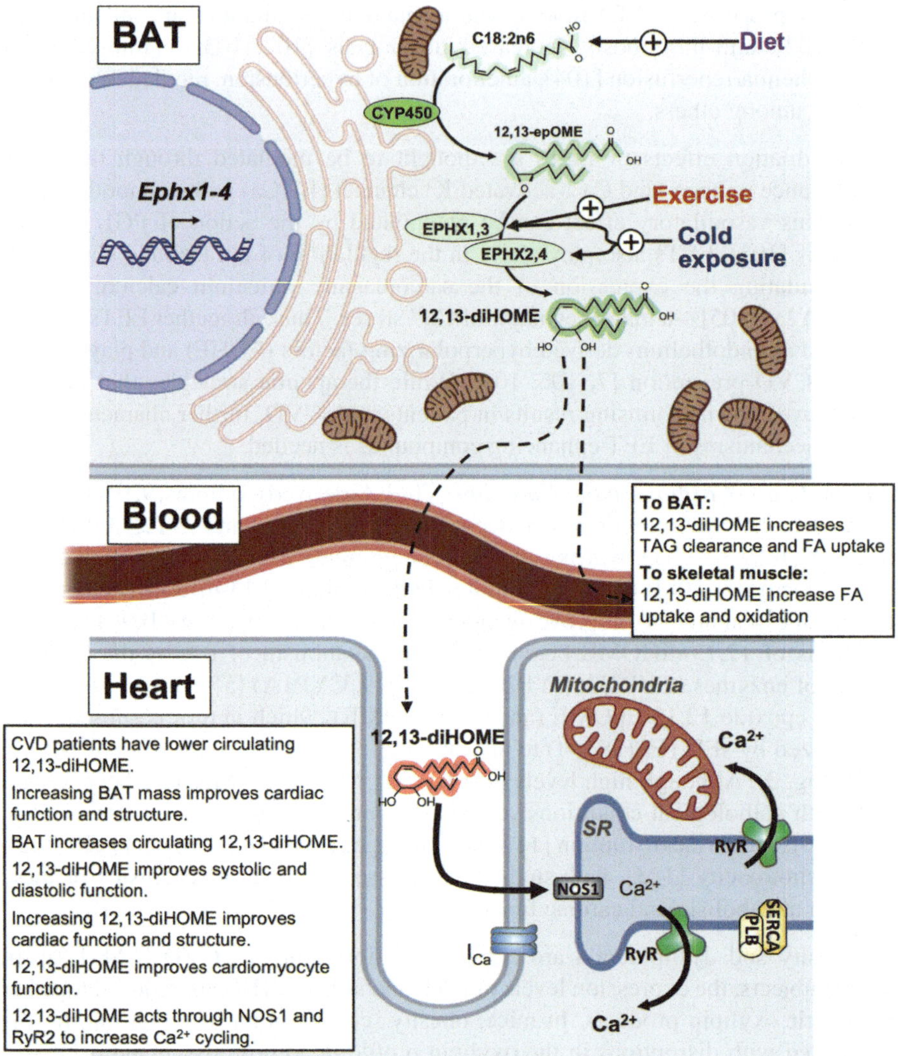

Fig. 3 Effects of 12,13-diHOME on metabolism and cardiovascular function. 12,13-diHOME is an important lipid mediator that has been described as a brown adipose tissue (BAT) lipokine. Dietary linoleic acid (LA) is epoxidized to 12,13-EpOME by CYP450 enzymes, and subsequently converted to 12,13-diHOME by soluble epoxide hydrolases EPHX1–4. Environmental stimuli, such as exercise and cold exposure, increase BAT-released 12,13-diHOME. This lipokine improves metabolism by increasing fatty acid uptake and oxidation in BAT and skeletal muscle. In the heart, 12,13-diHOME improves cardiac function and structure. This effect might be mediated through cell-autonomous improvements in cardiomyocyte contraction kinetics and mitochondrial respiration, which might involve calcium cycling through nitric oxide synthase 1 (NOS1) and the ryanodine receptor (RyR)

mediator released from adipose tissue—that is released in response to exercise and brown adipose tissue (BAT) activation by cold exposure [110, 111, 113, 132, 133] (Fig. 3). Previous studies in mice and humans showed that exercise-stimulated 12,13-diHOME release through the activation of BAT sEH, *Ephx1/2*, and improves metabolic function. The mechanism by which exercise-stimulated 12,13-diHOME improves metabolism involves skeletal muscle fatty acid uptake and oxidation [111, 113, 132]. Moreover, BAT activation through cold exposure and sympathetic activation in rodents and humans increases 12,13-diHOME through the induction of BAT sEH, *Ephx1/2,* and improves metabolic health. The mechanism by which cold-stimulated 12,13-diHOME improves metabolism involves increased uptake of FA into BAT to facilitate mitochondrial heat generation [110, 113, 132]. Thus, 12,13-diHOME is an important lipid mediator that improves metabolic health.

We have recently described the role of the lipokine 12,13-diHOME and BAT in cardiac physiology and CVD risk [109, 134]. Increasing BAT mass by transplantation results in greater circulating levels of the lipokine 12,13-diHOME [109, 134]. We showed that this increase in 12,13-diHOME is associated with improved systolic and diastolic function, as well as cardiac remodeling in mice. We demonstrated that acute injection of 12,13-diHOME recapitulates the improvements in systolic and diastolic function. Utilizing nanotransfection technology, we overexpressed sEH *Ephx1* and *Ephx2*, which resulted in sustained circulating 12,13-diHOME levels and improved systolic and diastolic function in an obese mouse. These improvements in cardiac function were mediated through improvements to cardiomyocyte contraction kinetics and mitochondrial function. Mechanistically, 12,13-diHOME increases cardiomyocyte contraction and mitochondrial respiration via a nitric oxide synthase 1 (NOS1)-dependent mechanism that likely involves the ryanodine receptor (RyR) [109], which dictates calcium cycling. Altogether, 12,13-diHOME is a lipid mediator that is associated with improved metabolic and cardiac health and may hold promise as a therapeutic strategy for the treatment of obesity, type 2 diabetes, and CVD.

Conclusion and Future Perspectives

In summary, lipid mediators play a pivotal role in the regulation of cardiac and vascular physiology. Cardiovascular physiology and CVD are intricately related to the availability of potent bioactive lipids such as oxylipins. Given that oxylipins can be stored and accessed through diverse lipid compartments—such as phospholipids, cholesteryl esters, triglycerides, among others—lipid mediators can be rapidly and dynamically increased in response to physiological and pathological stimuli. Relevant oxylipin species such as prostanoids, eicosanoids, and octadecanoids demonstrate vasoactive and cardioregulatory properties. Additional research is necessary to understand the influence of dietary and environmental factors on oxylipin synthesis and cardiac metabolism. Future studies will investigate the protective and therapeutic effects of oxylipins and their reservoirs on the function of the vascular endothelium and myocardium.

References

1. Benjamin EJ, Muntner P, Alonso A, Bittencourt MS, Callaway CW, Carson AP, et al. Heart disease and stroke statistics-2019 update: a report from the American Heart Association. Circulation. 2019;139(10):e56–e528.
2. Group WCRCW. World Health Organization cardiovascular disease risk charts: revised models to estimate risk in 21 global regions. Lancet Glob Health. 2019;7(10):e1332–e45.
3. Said MA, van de Vegte YJ, Zafar MM, van der Ende MY, Raja GK, Verweij N, et al. Contributions of interactions between lifestyle and genetics on coronary artery disease risk. Curr Cardiol Rep. 2019;21(9):89.
4. Zhu L, Zhang Y, Guo Z, Wang M. Cardiovascular biology of prostanoids and drug discovery. Arterioscler Thromb Vasc Biol. 2020;40(6):1454–63.
5. Hildreth K, Kodani SD, Hammock BD, Zhao L. Cytochrome P450-derived linoleic acid metabolites EpOMEs and DiHOMEs: a review of recent studies. J Nutr Biochem. 2020;86:108484.
6. Gabbs M, Leng S, Devassy JG, Monirujjaman M, Aukema HM. Advances in our understanding of oxylipins derived from dietary PUFAs. Adv Nutr. 2015;6(5):513–40.
7. Tacconelli S, Patrignani P. Inside epoxyeicosatrienoic acids and cardiovascular disease. Front Pharmacol. 2014;5:239.
8. Yuhki K, Kojima F, Kashiwagi H, Kawabe J, Fujino T, Narumiya S, et al. Roles of prostanoids in the pathogenesis of cardiovascular diseases: novel insights from knockout mouse studies. Pharmacol Ther. 2011;129(2):195–205.
9. Caligiuri SPB, Parikh M, Stamenkovic A, Pierce GN, Aukema HM. Dietary modulation of oxylipins in cardiovascular disease and aging. Am J Physiol Heart Circ Physiol. 2017;313(5):H903–H18.
10. Nayeem MA. Role of oxylipins in cardiovascular diseases. Acta Pharmacol Sin. 2018;39(7):1142–54.
11. Hajeyah AA, Griffiths WJ, Wang Y, Finch AJ, O'Donnell VB. The biosynthesis of enzymatically oxidized lipids. Front Endocrinol (Lausanne). 2020;11:591819.
12. Tourdot BE, Ahmed I, Holinstat M. The emerging role of oxylipins in thrombosis and diabetes. Front Pharmacol. 2014;4:176.
13. Liu X, Sims HF, Jenkins CM, Guan S, Dilthey BG, Gross RW. 12-LOX catalyzes the oxidation of 2-arachidonoyl-lysolipids in platelets generating eicosanoid-lysolipids that are attenuated by iPLA2gamma knockout. J Biol Chem. 2020;295(16):5307–20.
14. Liu X, Moon SH, Jenkins CM, Sims HF, Gross RW. Cyclooxygenase-2 mediated oxidation of 2-arachidonoyl-lysophospholipids identifies unknown lipid signaling pathways. Cell Chem Biol. 2016;23(10):1217–27.
15. Liu GY, Moon SH, Jenkins CM, Sims HF, Guan S, Gross RW. Synthesis of oxidized phospholipids by sn-1 acyltransferase using 2-15-HETE lysophospholipids. J Biol Chem. 2019;294(26):10146–59.
16. Jenkins CM, Yang K, Liu G, Moon SH, Dilthey BG, Gross RW. Cytochrome c is an oxidative stress-activated plasmalogenase that cleaves plasmenylcholine and plasmenylethanolamine at the sn-1 vinyl ether linkage. J Biol Chem. 2018;293(22):8693–709.
17. Xu S, Zhang X, Liu P. Lipid droplet proteins and metabolic diseases. Biochim Biophys Acta Mol Basis Dis. 2018;1864(5 Pt B):1968–83.
18. Smith WL, DeWitt DL, Garavito RM. Cyclooxygenases: structural, cellular, and molecular biology. Annu Rev Biochem. 2000;69:145–82.
19. Kirkby NS, Chan MV, Zaiss AK, Garcia-Vaz E, Jiao J, Berglund LM, et al. Systematic study of constitutive cyclooxygenase-2 expression: role of NF-kappaB and NFAT transcriptional pathways. Proc Natl Acad Sci U S A. 2016;113(2):434–9.
20. Kirkby NS, Zaiss AK, Urquhart P, Jiao J, Austin PJ, Al-Yamani M, et al. LC-MS/MS confirms that COX-1 drives vascular prostacyclin whilst gene expression pattern reveals non-vascular sites of COX-2 expression. PLoS One. 2013;8(7):e69524.

21. Therland KL, Stubbe J, Thiesson HC, Ottosen PD, Walter S, Sorensen GL, et al. Cycloxygenase-2 is expressed in vasculature of normal and ischemic adult human kidney and is colocalized with vascular prostaglandin E2 EP4 receptors. J Am Soc Nephrol. 2004;15(5):1189–98.
22. Spencer AG, Woods JW, Arakawa T, Singer II, Smith WL. Subcellular localization of prostaglandin endoperoxide H synthases-1 and -2 by immunoelectron microscopy. J Biol Chem. 1998;273(16):9886–93.
23. Yuan C, Smith WL. A cyclooxygenase-2-dependent prostaglandin E2 biosynthetic system in the Golgi apparatus. J Biol Chem. 2015;290(9):5606–20.
24. Caligiuri SP, Love K, Winter T, Gauthier J, Taylor CG, Blydt-Hansen T, et al. Dietary linoleic acid and alpha-linolenic acid differentially affect renal oxylipins and phospholipid fatty acids in diet-induced obese rats. J Nutr. 2013;143(9):1421–31.
25. Schuchardt JP, Schmidt S, Kressel G, Dong H, Willenberg I, Hammock BD, et al. Comparison of free serum oxylipin concentrations in hyper- vs. normolipidemic men. Prostaglandins Leukot Essent Fatty Acids. 2013;89(1):19–29.
26. Psychogios N, Hau DD, Peng J, Guo AC, Mandal R, Bouatra S, et al. The human serum metabolome. PLoS One. 2011;6(2):e16957.
27. Funk CD, Powell WS. Metabolism of linoleic acid by prostaglandin endoperoxide synthase from adult and fetal blood vessels. Biochim Biophys Acta. 1983;754(1):57–71.
28. Laneuville O, Breuer DK, Xu N, Huang ZH, Gage DA, Watson JT, et al. Fatty acid substrate specificities of human prostaglandin-endoperoxide H synthase-1 and -2. Formation of 12-hydroxy-(9Z, 13E/Z, 15Z)- octadecatrienoic acids from alpha-linolenic acid. J Biol Chem. 1995;270(33):19330–6.
29. Coleman RA, Grix SP, Head SA, Louttit JB, Mallett A, Sheldrick RL. A novel inhibitory prostanoid receptor in piglet saphenous vein. Prostaglandins. 1994;47(2):151–68.
30. Wuest SJ, Crucet M, Gemperle C, Loretz C, Hersberger M. Expression and regulation of 12/15-lipoxygenases in human primary macrophages. Atherosclerosis. 2012;225(1):121–7.
31. Lepley RA, Fitzpatrick FA. Irreversible inactivation of 5-lipoxygenase by leukotriene A4. Characterization of product inactivation with purified enzyme and intact leukocytes. J Biol Chem. 1994;269(4):2627–31.
32. Forsell PK, Brunnstrom A, Johannesson M, Claesson HE. Metabolism of anandamide into eoxamides by 15-lipoxygenase-1 and glutathione transferases. Lipids. 2012;47(8):781–91.
33. Huang LS, Kang JS, Kim MR, Sok DE. Oxygenation of arachidonoyl lysophospholipids by lipoxygenases from soybean, porcine leukocyte, or rabbit reticulocyte. J Agric Food Chem. 2008;56(4):1224–32.
34. Schewe T, Halangk W, Hiebsch C, Rapoport SM. A lipoxygenase in rabbit reticulocytes which attacks phospholipids and intact mitochondria. FEBS Lett. 1975;60(1):149–52.
35. Hutchins PM, Murphy RC. Cholesteryl ester acyl oxidation and remodeling in murine macrophages: formation of oxidized phosphatidylcholine. J Lipid Res. 2012;53(8):1588–97.
36. Belkner J, Wiesner R, Rathman J, Barnett J, Sigal E, Kuhn H. Oxygenation of lipoproteins by mammalian lipoxygenases. Eur J Biochem. 1993;213(1):251–61.
37. Shoieb SM, El-Sherbeni AA, El-Kadi AOS. Subterminal hydroxyeicosatetraenoic acids: crucial lipid mediators in normal physiology and disease states. Chem Biol Interact. 2019;299:140–50.
38. Natarajan R, Bai W, Rangarajan V, Gonzales N, Gu JL, Lanting L, et al. Platelet-derived growth factor BB mediated regulation of 12-lipoxygenase in porcine aortic smooth muscle cells. J Cell Physiol. 1996;169(2):391–400.
39. Natarajan R, Gu JL, Rossi J, Gonzales N, Lanting L, Xu L, et al. Elevated glucose and angiotensin II increase 12-lipoxygenase activity and expression in porcine aortic smooth muscle cells. Proc Natl Acad Sci U S A. 1993;90(11):4947–51.
40. Conrad DJ, Kuhn H, Mulkins M, Highland E, Sigal E. Specific inflammatory cytokines regulate the expression of human monocyte 15-lipoxygenase. Proc Natl Acad Sci U S A. 1992;89(1):217–21.

41. Maayah ZH, El-Kadi AO. The role of mid-chain hydroxyeicosatetraenoic acids in the pathogenesis of hypertension and cardiac hypertrophy. Arch Toxicol. 2016;90(1):119–36.
42. Takai S, Jin D, Kirimura K, Fujimoto Y, Miyazaki M. 12-Hydroxyeicosatetraenoic acid potentiates angiotensin II-induced pressor response in rats. Eur J Pharmacol. 2001;418(1–2):R1–2.
43. Yiu SS, Zhao X, Inscho EW, Imig JD. 12-Hydroxyeicosatetraenoic acid participates in angiotensin II afferent arteriolar vasoconstriction by activating L-type calcium channels. J Lipid Res. 2003;44(12):2391–9.
44. Nadler JL, Natarajan R, Stern N. Specific action of the lipoxygenase pathway in mediating angiotensin II-induced aldosterone synthesis in isolated adrenal glomerulosa cells. J Clin Invest. 1987;80(6):1763–9.
45. Stern N, Golub M, Nozawa K, Berger M, Knoll E, Yanagawa N, et al. Selective inhibition of angiotensin II-mediated vasoconstriction by lipoxygenase blockade. Am J Phys. 1989;257(2 Pt 2):H434–43.
46. Zhang L, Li Y, Chen M, Su X, Yi D, Lu P, et al. 15-LO/15-HETE mediated vascular adventitia fibrosis via p38 MAPK-dependent TGF-beta. J Cell Physiol. 2014;229(2):245–57.
47. Wallukat G, Morwinski R, Kuhn H. Modulation of the beta-adrenergic response of cardiomyocytes by specific lipoxygenase products involves their incorporation into phosphatidylinositol and activation of protein kinase C. J Biol Chem. 1994;269(46):29055–60.
48. Levick SP, Loch DC, Taylor SM, Janicki JS. Arachidonic acid metabolism as a potential mediator of cardiac fibrosis associated with inflammation. J Immunol. 2007;178(2):641–6.
49. Kayama Y, Minamino T, Toko H, Sakamoto M, Shimizu I, Takahashi H, et al. Cardiac 12/15 lipoxygenase-induced inflammation is involved in heart failure. J Exp Med. 2009;206(7):1565–74.
50. Maayah ZH, Althurwi HN, Abdelhamid G, Lesyk G, Jurasz P, El-Kadi AO. CYP1B1 inhibition attenuates doxorubicin-induced cardiotoxicity through a mid-chain HETEs-dependent mechanism. Pharmacol Res. 2016;105:28–43.
51. Kong EK, Yu S, Sanderson JE, Chen KB, Huang Y, Yu CM. A novel anti-fibrotic agent, baicalein, for the treatment of myocardial fibrosis in spontaneously hypertensive rats. Eur J Pharmacol. 2011;658(2–3):175–81.
52. Buckley CD, Gilroy DW, Serhan CN. Proresolving lipid mediators and mechanisms in the resolution of acute inflammation. Immunity. 2014;40(3):315–27.
53. Kramer HJ, Stevens J, Grimminger F, Seeger W. Fish oil fatty acids and human platelets: dose-dependent decrease in dienoic and increase in trienoic thromboxane generation. Biochem Pharmacol. 1996;52(8):1211–7.
54. Honn KV, Nelson KK, Renaud C, Bazaz R, Diglio CA, Timar J. Fatty acid modulation of tumor cell adhesion to microvessel endothelium and experimental metastasis. Prostaglandins. 1992;44(5):413–29.
55. Folcik VA, Cathcart MK. Predominance of esterified hydroperoxy-linoleic acid in human monocyte-oxidized LDL. J Lipid Res. 1994;35(9):1570–82.
56. Hattori T, Obinata H, Ogawa A, Kishi M, Tatei K, Ishikawa O, et al. G2A plays pro-inflammatory roles in human keratinocytes under oxidative stress as a receptor for 9-hydroxyoctadecadienoic acid. J Invest Dermatol. 2008;128(5):1123–33.
57. Obinata H, Izumi T. G2A as a receptor for oxidized free fatty acids. Prostaglandins Other Lipid Mediat. 2009;89(3–4):66–72.
58. Kwon SY, Massey K, Watson MA, Hussain T, Volpe G, Buckley CD, et al. Oxidised metabolites of the omega-6 fatty acid linoleic acid activate dFOXO. Life Sci Alliance. 2020;3(2):e201900356.
59. Theken KN, Schuck RN, Edin ML, Tran B, Ellis K, Bass A, et al. Evaluation of cytochrome P450-derived eicosanoids in humans with stable atherosclerotic cardiovascular disease. Atherosclerosis. 2012;222(2):530–6.
60. Buchanan MR, Haas TA, Lagarde M, Guichardant M. 13-Hydroxyoctadecadienoic acid is the vessel wall chemorepellant factor, LOX. J Biol Chem. 1985;260(30):16056–9.

61. Tloti MA, Moon DG, Weston LK, Kaplan JE. Effect of 13-hydroxyoctadeca-9,11-dienoic acid (13-HODE) on thrombin induced platelet adherence to endothelial cells in vitro. Thromb Res. 1991;62(4):305–17.
62. Stoll LL, Morland MR, Spector AA. 13-HODE increases intracellular calcium in vascular smooth muscle cells. Am J Phys. 1994;266(4 Pt 1):C990–6.
63. De Meyer GR, Bult H, Verbeuren TJ, Herman AG. The role of endothelial cells in the relaxations induced by 13-hydroxy- and 13-hydroperoxylinoleic acid in canine arteries. Br J Pharmacol. 1992;107(2):597–603.
64. Iversen L, Fogh K, Bojesen G, Kragballe K. Linoleic acid and dihomogammalinolenic acid inhibit leukotriene B4 formation and stimulate the formation of their 15-lipoxygenase products by human neutrophils in vitro. Evidence of formation of antiinflammatory compounds. Agents Actions. 1991;33(3–4):286–91.
65. Murthy S, Born E, Mathur S, Field FJ. 13-Hydroxy octadecadienoic acid (13-HODE) inhibits triacylglycerol-rich lipoprotein secretion by CaCo-2 cells. J Lipid Res. 1998;39(6):1254–62.
66. Zhang Z, Emami S, Hennebelle M, Morgan RK, Lerno LA, Slupsky CM, et al. Linoleic acid-derived 13-hydroxyoctadecadienoic acid is absorbed and incorporated into rat tissues. Biochim Biophys Acta Mol Cell Biol Lipids. 1866;2021(3):158870.
67. Fang X, Kaduce TL, Spector AA. 13-(S)-hydroxyoctadecadienoic acid (13-HODE) incorporation and conversion to novel products by endothelial cells. J Lipid Res. 1999;40(4):699–707.
68. Ramsden CE, Ringel A, Feldstein AE, Taha AY, MacIntosh BA, Hibbeln JR, et al. Lowering dietary linoleic acid reduces bioactive oxidized linoleic acid metabolites in humans. Prostaglandins Leukot Essent Fatty Acids. 2012;87(4–5):135–41.
69. Shureiqi I, Wojno KJ, Poore JA, Reddy RG, Moussalli MJ, Spindler SA, et al. Decreased 13-S-hydroxyoctadecadienoic acid levels and 15-lipoxygenase-1 expression in human colon cancers. Carcinogenesis. 1999;20(10):1985–95.
70. Ziboh VA, Miller CC, Cho Y. Significance of lipoxygenase-derived monohydroxy fatty acids in cutaneous biology. Prostaglandins Other Lipid Mediat. 2000;63(1–2):3–13.
71. Brash AR, Boeglin WE, Capdevila JH, Yeola S, Blair IA. 7-HETE, 10-HETE, and 13-HETE are major products of NADPH-dependent arachidonic acid metabolism in rat liver microsomes: analysis of their stereochemistry, and the stereochemistry of their acid-catalyzed rearrangement. Arch Biochem Biophys. 1995;321(2):485–92.
72. Oliw EH. bis-Allylic hydroxylation of linoleic acid and arachidonic acid by human hepatic monooxygenases. Biochim Biophys Acta. 1993;1166(2–3):258–63.
73. Laethem RM, Balazy M, Falck JR, Laethem CL, Koop DR. Formation of 19(S)-, 19(R)-, and 18(R)-hydroxyeicosatetraenoic acids by alcohol-inducible cytochrome P450 2E1. J Biol Chem. 1993;268(17):12912–8.
74. Nebert DW, Wikvall K, Miller WL. Human cytochromes P450 in health and disease. Philos Trans R Soc Lond Ser B Biol Sci. 2013;368(1612):20120431.
75. Wu CC, Gupta T, Garcia V, Ding Y, Schwartzman ML. 20-HETE and blood pressure regulation: clinical implications. Cardiol Rev. 2014;22(1):1–12.
76. Spector AA, Kim HY. Cytochrome P450 epoxygenase pathway of polyunsaturated fatty acid metabolism. Biochim Biophys Acta. 2015;1851(4):356–65.
77. Bednar MM, Gross CE, Balazy MK, Belosludtsev Y, Colella DT, Falck JR, et al. 16(R)-hydroxy-5,8,11,14-eicosatetraenoic acid, a new arachidonate metabolite in human polymorphonuclear leukocytes. Biochem Pharmacol. 2000;60(3):447–55.
78. Reddy YK, Reddy LM, Capdevila JH, Falck JR. Practical, asymmetric synthesis of 16-hydroxyeicosa-5(Z),8(Z), 11(Z),14(Z)-tetraenoic acid (16-HETE), an endogenous inhibitor of neutrophil activity. Bioorg Med Chem Lett. 2003;13(21):3719–20.
79. Zhu Y, Schieber EB, McGiff JC, Balazy M. Identification of arachidonate P-450 metabolites in human platelet phospholipids. Hypertension. 1995;25(4 Pt 2):854–9.
80. Maciejewska D, Ossowski P, Drozd A, Ryterska K, Jamiol-Milc D, Banaszczak M, et al. Metabolites of arachidonic acid and linoleic acid in early stages of non-alcoholic fatty liver disease--a pilot study. Prostaglandins Other Lipid Mediat. 2015;121(Pt B):184–9.

81. Carroll MA, Balazy M, Huang DD, Rybalova S, Falck JR, McGiff JC. Cytochrome P450-derived renal HETEs: storage and release. Kidney Int. 1997;51(6):1696–702.
82. Wang W, Lu M. Effect of arachidonic acid on activity of the apical K+ channel in the thick ascending limb of the rat kidney. J Gen Physiol. 1995;106(4):727–43.
83. Chadderdon SM, Belcik JT, Bader L, Kievit P, Grove KL, Lindner JR. Vasoconstrictor eicosanoids and impaired microvascular function in inactive and insulin-resistant primates. Int J Obes. 2016;40(10):1600–3.
84. Zhang F, Deng H, Kemp R, Singh H, Gopal VR, Falck JR, et al. Decreased levels of cytochrome P450 2E1-derived eicosanoids sensitize renal arteries to constrictor agonists in spontaneously hypertensive rats. Hypertension. 2005;45(1):103–8.
85. El-Sherbeni AA, El-Kadi AO. Alterations in cytochrome P450-derived arachidonic acid metabolism during pressure overload-induced cardiac hypertrophy. Biochem Pharmacol. 2014;87(3):456–66.
86. El-Sherbeni AA, El-Kadi AO. Characterization of arachidonic acid metabolism by rat cytochrome P450 enzymes: the involvement of CYP1As. Drug Metab Dispos. 2014;42(9):1498–507.
87. Poloyac SM, Tortorici MA, Przychodzin DI, Reynolds RB, Xie W, Frye RF, et al. The effect of isoniazid on CYP2E1- and CYP4A-mediated hydroxylation of arachidonic acid in the rat liver and kidney. Drug Metab Dispos. 2004;32(7):727–33.
88. Capdevila JH, Falck JR, Harris RC. Cytochrome P450 and arachidonic acid bioactivation. Molecular and functional properties of the arachidonate monooxygenase. J Lipid Res. 2000;41(2):163–81.
89. Cheng J, Ou JS, Singh H, Falck JR, Narsimhaswamy D, Pritchard KA Jr, et al. 20-Hydroxyeicosatetraenoic acid causes endothelial dysfunction via eNOS uncoupling. Am J Physiol Heart Circ Physiol. 2008;294(2):H1018–26.
90. Alonso-Galicia M, Falck JR, Reddy KM, Roman RJ. 20-HETE agonists and antagonists in the renal circulation. Am J Phys. 1999;277(5):F790–6.
91. Roman RJ. P-450 metabolites of arachidonic acid in the control of cardiovascular function. Physiol Rev. 2002;82(1):131–85.
92. Nayeem MA, Zeldin DC, Boegehold MA, Morisseau C, Marowsky A, Ponnoth DS, et al. Modulation by salt intake of the vascular response mediated through adenosine A(2A) receptor: role of CYP epoxygenase and soluble epoxide hydrolase. Am J Physiol Regul Integr Comp Physiol. 2010;299(1):R325–33.
93. Nayeem MA, Zeldin DC, Boegehold MA, Falck JR. Salt modulates vascular response through adenosine A(2A) receptor in eNOS-null mice: role of CYP450 epoxygenase and soluble epoxide hydrolase. Mol Cell Biochem. 2011;350(1–2):101–11.
94. Nayeem MA, Ponnoth DS, Boegehold MA, Zeldin DC, Falck JR, Mustafa SJ. High-salt diet enhances mouse aortic relaxation through adenosine A2A receptor via CYP epoxygenases. Am J Physiol Regul Integr Comp Physiol. 2009;296(3):R567–74.
95. Nayeem MA, Poloyac SM, Falck JR, Zeldin DC, Ledent C, Ponnoth DS, et al. Role of CYP epoxygenases in A2A AR-mediated relaxation using A2A AR-null and wild-type mice. Am J Physiol Heart Circ Physiol. 2008;295(5):H2068–78.
96. Miyata N, Roman RJ. Role of 20-hydroxyeicosatetraenoic acid (20-HETE) in vascular system. J Smooth Muscle Res. 2005;41(4):175–93.
97. Spector AA. Arachidonic acid cytochrome P450 epoxygenase pathway. J Lipid Res. 2009;50(Suppl):S52–6.
98. Bellien J, Joannides R. Epoxyeicosatrienoic acid pathway in human health and diseases. J Cardiovasc Pharmacol. 2013;61(3):188–96.
99. Wang D, Dubois RN. Epoxyeicosatrienoic acids: a double-edged sword in cardiovascular diseases and cancer. J Clin Invest. 2012;122(1):19–22.
100. Weintraub NL, Fang X, Kaduce TL, VanRollins M, Chatterjee P, Spector AA. Epoxide hydrolases regulate epoxyeicosatrienoic acid incorporation into coronary endothelial phospholipids. Am J Phys. 1999;277(5):H2098–108.

101. Shahabi P, Siest G, Meyer UA, Visvikis-Siest S. Human cytochrome P450 epoxygenases: variability in expression and role in inflammation-related disorders. Pharmacol Ther. 2014;144(2):134–61.
102. Fleming I. Vascular cytochrome p450 enzymes: physiology and pathophysiology. Trends Cardiovasc Med. 2008;18(1):20–5.
103. Deng Y, Theken KN, Lee CR. Cytochrome P450 epoxygenases, soluble epoxide hydrolase, and the regulation of cardiovascular inflammation. J Mol Cell Cardiol. 2010;48(2):331–41.
104. Seubert J, Yang B, Bradbury JA, Graves J, Degraff LM, Gabel S, et al. Enhanced postischemic functional recovery in CYP2J2 transgenic hearts involves mitochondrial ATP-sensitive K+ channels and p42/p44 MAPK pathway. Circ Res. 2004;95(5):506–14.
105. Wang X, Ni L, Yang L, Duan Q, Chen C, Edin ML, et al. CYP2J2-derived epoxyeicosatrienoic acids suppress endoplasmic reticulum stress in heart failure. Mol Pharmacol. 2014;85(1):105–15.
106. Lee CR, North KE, Bray MS, Couper DJ, Heiss G, Zeldin DC. CYP2J2 and CYP2C8 polymorphisms and coronary heart disease risk: the Atherosclerosis Risk in Communities (ARIC) study. Pharmacogenet Genomics. 2007;17(5):349–58.
107. Lee CR, North KE, Bray MS, Fornage M, Seubert JM, Newman JW, et al. Genetic variation in soluble epoxide hydrolase (EPHX2) and risk of coronary heart disease: the Atherosclerosis Risk in Communities (ARIC) study. Hum Mol Genet. 2006;15(10):1640–9.
108. Thompson DA, Hammock BD. Dihydroxyoctadecamonoenoate esters inhibit the neutrophil respiratory burst. J Biosci. 2007;32(2):279–91.
109. Pinckard KM, Shettigar VK, Wright KR, Abay E, Baer LA, Vidal P, et al. A novel endocrine role for the BAT-released lipokine 12,13-diHOME to mediate cardiac function. Circulation. 2021;143(2):145–59.
110. Lynes MD, Leiria LO, Lundh M, Bartelt A, Shamsi F, Huang TL, et al. The cold-induced lipokine 12,13-diHOME promotes fatty acid transport into brown adipose tissue. Nat Med. 2017;23(5):631–7.
111. Stanford KI, Lynes MD, Takahashi H, Baer LA, Arts PJ, May FJ, et al. 12,13-diHOME: an exercise-induced lipokine that increases skeletal muscle fatty acid uptake. Cell Metab. 2018;27(5):1111–20. e3
112. Gollasch B, Dogan I, Rothe M, Gollasch M, Luft FC. Maximal exercise and plasma cytochrome P450 and lipoxygenase mediators: a lipidomics study. Physiol Rep. 2019;7(13):e14165.
113. Hernandez-Saavedra D, Stanford KI. The regulation of lipokines by environmental factors. Nutrients. 2019;11(10):2422.
114. Edwards LM, Lawler NG, Nikolic SB, Peters JM, Horne J, Wilson R, et al. Metabolomics reveals increased isoleukotoxin diol (12,13-DHOME) in human plasma after acute Intralipid infusion. J Lipid Res. 2012;53(9):1979–86.
115. Totani Y, Saito Y, Ishizaki T, Sasaki F, Ameshima S, Miyamori I. Leukotoxin and its diol induce neutrophil chemotaxis through signal transduction different from that of fMLP. Eur Respir J. 2000;15(1):75–9.
116. Viswanathan S, Hammock BD, Newman JW, Meerarani P, Toborek M, Hennig B. Involvement of CYP 2C9 in mediating the proinflammatory effects of linoleic acid in vascular endothelial cells. J Am Coll Nutr. 2003;22(6):502–10.
117. Lecka-Czernik B, Moerman EJ, Grant DF, Lehmann JM, Manolagas SC, Jilka RL. Divergent effects of selective peroxisome proliferator-activated receptor-gamma 2 ligands on adipocyte versus osteoblast differentiation. Endocrinology. 2002;143(6):2376–84.
118. Markaverich BM, Crowley JR, Alejandro MA, Shoulars K, Casajuna N, Mani S, et al. Leukotoxin diols from ground corncob bedding disrupt estrous cyclicity in rats and stimulate MCF-7 breast cancer cell proliferation. Environ Health Perspect. 2005;113(12):1698–704.
119. Fromel T, Jungblut B, Hu J, Trouvain C, Barbosa-Sicard E, Popp R, et al. Soluble epoxide hydrolase regulates hematopoietic progenitor cell function via generation of fatty acid diols. Proc Natl Acad Sci U S A. 2012;109(25):9995–10000.

120. Eskander MA, Ruparel S, Green DP, Chen PB, Por ED, Jeske NA, et al. Persistent nociception triggered by nerve growth factor (NGF) is mediated by TRPV1 and oxidative mechanisms. J Neurosci. 2015;35(22):8593–603.
121. Sisignano M, Angioni C, Park CK, Meyer Dos Santos S, Jordan H, Kuzikov M, et al. Targeting CYP2J to reduce paclitaxel-induced peripheral neuropathic pain. Proc Natl Acad Sci U S A. 2016;113(44):12544–9.
122. Zimmer B, Angioni C, Osthues T, Toewe A, Thomas D, Pierre SC, et al. The oxidized linoleic acid metabolite 12,13-DiHOME mediates thermal hyperalgesia during inflammatory pain. Biochim Biophys Acta Mol Cell Biol Lipids. 2018;1863(7):669–78.
123. Samokhvalov V, Jamieson KL, Darwesh AM, Keshavarz-Bahaghighat H, Lee TYT, Edin M, et al. Deficiency of soluble epoxide hydrolase protects cardiac function impaired by LPS-induced acute inflammation. Front Pharmacol. 2018;9:1572.
124. Bays HE, Toth PP, Kris-Etherton PM, Abate N, Aronne LJ, Brown WV, et al. Obesity, adiposity, and dyslipidemia: a consensus statement from the National Lipid Association. J Clin Lipidol. 2013;7(4):304–83.
125. Cercato C, Fonseca FA. Cardiovascular risk and obesity. Diabetol Metab Syndr. 2019;11:74.
126. Howard BV, Ruotolo G, Robbins DC. Obesity and dyslipidemia. Endocrinol Metab Clin N Am. 2003;32(4):855–67.
127. Wang W, Yang J, Qi W, Yang H, Wang C, Tan B, et al. Lipidomic profiling of high-fat diet-induced obesity in mice: importance of cytochrome P450-derived fatty acid epoxides. Obesity (Silver Spring). 2017;25(1):132–40.
128. Platt C, Houstis N, Rosenzweig A. Using exercise to measure and modify cardiac function. Cell Metab. 2015;21(2):227–36.
129. Roh J, Rhee J, Chaudhari V, Rosenzweig A. The role of exercise in cardiac aging: from physiology to molecular mechanisms. Circ Res. 2016;118(2):279–95.
130. Nieman DC, Shanely RA, Luo B, Meaney MP, Dew DA, Pappan KL. Metabolomics approach to assessing plasma 13- and 9-hydroxy-octadecadienoic acid and linoleic acid metabolite responses to 75-km cycling. Am J Physiol Regul Integr Comp Physiol. 2014;307(1):R68–74.
131. Nieman DC, Shanely RA, Gillitt ND, Pappan KL, Lila MA. Serum metabolic signatures induced by a three-day intensified exercise period persist after 14 h of recovery in runners. J Proteome Res. 2013;12(10):4577–84.
132. Peres Valgas da Silva C, Hernandez-Saavedra D, White JD, Stanford KI. Cold and exercise: therapeutic tools to activate brown adipose tissue and combat obesity. Biology (Basel). 2019;8(1):9.
133. Gavalda-Navarro A, Villarroya J, Cereijo R, Giralt M, Villarroya F. The endocrine role of brown adipose tissue: an update on actors and actions. Rev Endocr Metab Disord. 2022;23(1):31–41.
134. Becher T, Palanisamy S, Kramer DJ, Eljalby M, Marx SJ, Wibmer AG, et al. Brown adipose tissue is associated with cardiometabolic health. Nat Med. 2021;27(1):58–65.

Cardiac Inflammasome and Arrhythmia

Na Li and Dobromir Dobrev

Abstract Cardiac arrhythmias affect the quality of life and are often life threatening. The conventional anti-arrhythmic drugs targeting cardiac ion channels are associated with severe adverse effects including drug-induced cardiac toxicity and are often arrhythmogenic. Elucidating the molecular mechanisms contributing to the fundamental events that promote arrhythmias is a prerequisite for the development of novel and effective therapeutics, as well as better patient care in the different patient populations. Enhanced inflammatory response has been associated with the development of several major forms of cardiovascular diseases including cardiac arrhythmia. Recent studies point to a critical role of the cardiac inflammasome signaling in the pathogenesis of cardiac arrhythmias, especially in the context of atrial fibrillation. In this chapter, we discuss the current understanding of the inflammasome biology, the causative involvement of inflammasome in the development of atrial fibrillation, and the therapeutic potential of targeting the inflammasome pathway in patients at risk for or in atrial fibrillation.

Keywords Arrhythmia · Atrial fibrillation · Inflammasome · NLRP3

N. Li (✉)
Department of Medicine (Section of Cardiovascular Research), Baylor College of Medicine, Houston, TX, USA

Department of Molecular Physiology and Biophysics, Baylor College of Medicine, Houston, TX, USA

Cardiovascular Research Institute, Baylor College of Medicine, Houston, TX, USA
e-mail: nal@bcm.edu

D. Dobrev
Department of Molecular Physiology and Biophysics, Baylor College of Medicine, Houston, TX, USA

Institute of Pharmacology, West German Heart and Vascular Centre, University Duisburg-Essen, Essen, Germany

Montréal Heart Institute and University de Montréal, Medicine and Research Center, QC, Canada

N. L. Parinandi, T. J. Hund (eds.), *Cardiovascular Signaling in Health and Disease*, https://doi.org/10.1007/978-3-031-08309-9_9

Introduction

Inflammation is a double-edged sword. As an essential biological process, inflammation promotes healing upon tissue injury or insult. However, inflammation can adversely affect the physiological function of various organs when it becomes persistent. In 2002, Dr. Tschopp and colleagues described for the first time a multiprotein complex called the inflammasome, which plays a crucial role in modulating innate immunity by facilitating the maturation of caspase-1 and the processing of interleukin-1β (IL-1β) [1–3]. Since that time, inflammasomes have been recognized for their roles not only in the host defense against invading pathogens but also in the development of auto-inflammatory disease, cancer, metabolic, and neurodegenerative diseases [4, 5]. Over the last decade, several animal models have demonstrated a causative role of "NLR family pyrin domain containing 3" (NLRP3) in the pathogenesis of major cardiovascular diseases including atherosclerosis, aortic aneurysm, heart failure, and atrial fibrillation (AF) [2, 6–10]. The levels of IL-1 family cytokines generated upon inflammasome activation are often associated with common cardiovascular diseases. In the Canakinumab Anti-Inflammatory Thrombosis Outcomes Study (CANTOS) clinical trial, the use of the monoclonal IL-1β targeting antibody canakinumab in patients at risk of heart attack showed a clear reduction of overall cardiac events, validating the causative role of NLRP3 and its products in the development of life-threatening cardiovascular events [11]. Collectively, these studies position the targeting of the NLRP3 inflammasome pathway as an effective therapeutic strategy for cardiovascular diseases and open an avenue for further elucidation of the inflammasome biology in cardiovascular diseases with a goal to discover new targets and therapeutics. In this chapter, we will discuss the contribution of inflammasome signaling pathway in the pathophysiology of AF, the most encountered arrhythmia in adulthood.

Cardiac Electrophysiology

Cardiac excitability is one of the basic characteristics of the heart and controls the contractility of the myocardium. Cardiomyocytes generates action potentials (APs) that drive cardiac excitability. AP is developed as a result of the sequential activation and inactivation of a number of voltage-gated ion channels located on the plasma membrane of the cardiomyocyte. The components of these channels leading to an AP vary depending on the location of cells within the cardiac conduction system and the cardiac chamber (atria vs ventricle), especially during the repolarization phase [12]. Due to ionic gradients crossing the plasma membrane, the resting membrane potential (RMP) of cardiomyocytes is around – 80mV. When the cardiomyocyte receives an electrical impulse, AP initiates due to the activation of voltage-gated Na^+ (Nav) channels. The rapid opening of Nav channels produces an inward I_{Na} current which depolarizes the cell and forms the upstroke of the AP (phase 0).

Following this, the activation of transient outward K^+ (I_{to}) channel initiates the repolarization of AP (phase 1). During phase 2, multiple voltage-gated K^+ (Kv) channels are activated. The relative Kv channel expression levels differ in atrial and ventricular cardiomyocytes. While the slow- and rapid-rectifier K^+ (I_{Ks} and I_{Kr}) channels are abundantly expressed in ventricular cardiomyocytes, the K^+ channels forming the ultra-rapid rapid-rectifier K^+ (I_{Kur}) current are selectively expressed in atrial cardiomyocytes, which partially explains the shorter repolarization phase of atrial cardiomyocytes compared to ventricular cardiomyocytes. During phase 2, depolarizing L-type Ca^{2+} currents (LTCCs) are also activated. The counterbalance between outward I_K (I_{Ks}, I_{Kr}, and I_{Kur}) and inward $I_{Ca,L}$ currents with some predominance of $I_{Ca,L}$ causes the characteristic plateau phase of cardiac AP. Once the LTCCs are closed, the repolarization continues as the outward K^+ currents persist until the membrane potential reaches the range of RMP (Phase 3). The inward rectifier K^+ current (I_{K1}) through Kir channels contributes to the AP repolarization (phase 4) and supports the maintenance of the membrane potential at the RMP.

The activation of the $I_{Ca,L}$ current is key for the initiation of excitation-contraction coupling (ECC) in cardiomyocytes [13]. The Ca^{2+} influx mediated by $I_{Ca,L}$ activates the ryanodine receptor type-2 (RyR2) channels, a homotetrameric macromolecular complex located on the sarcoplasmic reticulum (SR). Activated RyR2 channels release Ca^{2+} from SR where the Ca^{2+} is stored in high concentration (high micromolar to low millimolar range). Ca^{2+} release from SR abruptly raises the cytosolic Ca^{2+} level. Cytosolic free Ca^{2+} then binds to troponin C and other Ca^{2+}-sensitive myofibrillar proteins, allowing the actin-myosin cross-bridging, shortening of the sarcomeres and the generation of muscle contraction. When Ca^{2+} release is terminated and Ca^{2+} is removed from the cytoplasm, it results in muscle relaxation. Ca^{2+} removal during relaxation is achieved through several routes: (1) a large part of cytosolic Ca^{2+} are pumped back into SR by the sarco-endoplasmic reticulum Ca^{2+} ATPase type-2a (SERCA2a), (2) Ca^{2+} can also be extruded by the Na^+/Ca^{2+} exchanger (NCX1) located on the plasma membrane and the plasmalemmal Ca^{2+}-ATPase (PMCA), and (3) recent studies suggest that Ca^{2+} can be taken into mitochondria by the mitochondrial Ca^{2+} uniporter [13, 14]. The proper function of these ion channels, transporters and Ca^{2+}-handling proteins are tightly regulated at the transcriptional, posttranscriptional, and posttranslational level. For instance, the activity of RyR2 and SERCA2a are influenced by their binding partners and the steady-state phosphorylation [15, 16]. FK506-binding protein 12.6 (FKBP12.6) and junctophilin 2 (JPH2) are the most known factors that can directly interact with RyR2 and cause inhibitory effect on the RyR2 channel activity. Loss of FKBP12.6 or JPH2 can cause arrhythmogenic spontaneous Ca^{2+} release events (SCaEs) due to the enhanced RyR2 activity. Similarly, the increased phosphorylation of RyR2 due to the enhanced activity of protein kinase A (PKA) and Ca^{2+}/Calmodulin kinase II (CaMKII) or the impaired function of protein phosphates PP1 and PP2A can also promote RyR2-mediated SCaEs and triggered activity [16]. On the other hand, the function of SERCA2a is negatively regulated by its binding partners phospholamban (PLN), sarcolipin (SLN), and myoregulin (MLN) [17]. Loss of PLN or SLN can enhance the SERCA2a activity and increase SR Ca^{2+} content, which may subsequently lead to more Ca^{2+} release from SR. Enhanced phosphorylation of PLN by PKA or CaMKII also can release the inhibitory effect of PLN on SERCA2a, increasing Ca^{2+} uptake into SR [16]. Dysfunctions of any of these ion channels/transporters and Ca^{2+}-handling proteins can lead to electrophysiological disturbance, manifested as different forms of cardiac arrhythmias including AF [18].

Pathophysiology of AF

AF is the most common cardiac arrhythmia with an increasing prevalence worldwide, and it has become a significant public health burden [19, 20]. Although a substantial progress has been made to identify the molecular basis of AF [21–23], the clinical profile of AF is complex [24] and there are substantial challenges in the translation of novel discoveries to clinical application [25, 26].

In a healthy heart, sinoatrial node (SAN) is responsible for the generation of electrical impulses, which then travel through the specialized cardiac conduction system resulting in the sequential depolarization of the atria and the ventricles [27]. When the automaticity in the pulmonary veins is enhanced or focal triggered activity (TA) appears due to the occurrence of early afterdepolarizations (EADs) or delayed afterdepolarizations (DADs), the SAN is no longer the sole source of cardiac impulse, which is termed ectopic firing. When ectopic firing encounters an arrhythmogenic substrate in the atrial, it can promote reentrant conduction or a spiral wave-generating rotor [28], thereby causing AF-maintaining reentry. Although ectopic firing could serve as a trigger of AF-maintaining reentry, it can also sustain AF by producing fibrillatory conduction even in the absence of a proarrhythmic substrate in atria [23, 29]

In 1998, Haissaguerre et al. first discovered that the enhanced automaticity due to the spontaneous firing from a discrete region known as the pulmonary veins can initiate AF [30]. These focal triggers from pulmonary veins are a major source of ectopic firing in AF [31], and have become the main target for the modern AF ablation technology. The rationale for the ablation procedures by applying either high energy (radiofrequency ablation) or cold temperature (cryoablation) is to destroy the tissues surrounding the pulmonary veins, thereby preventing spontaneous pulmonary vein firing-induced fibrillatory conduction to the atria that governs AF [32]. Apart from the pulmonary veins, other regions within atria can also generate ectopic firing due to the Ca^{2+}-handling abnormalities within atrial cardiomyocytes [29, 33–35]. Altered functions of Ca^{2+}-handling proteins in cardiomyocytes can promote both EADs or DADs, which can cause membrane depolarization and initiate the triggered activity during or after a regular AP, respectively. The molecular basis for EADs and DADs is different. The generation of EADs is a consequence of alterations in AP duration (APD). When APD is prolonged, the $I_{Ca,L}$ can be reactivated during phase 2 of the AP and generate EAD [18, 23]. There is little evidence for a role of this type of EADs in AF. However, when APD is shortened, NCX1 can be activated by large amplitude Ca^{2+} transients thereby producing a depolarizing NCX current (I_{NCX}) that can cause late phase-3 EADs [18, 36]. DADs occur after the completion of AP repolarization, primarily because that SR Ca^{2+} overload or RyR2 dysfunction can promote SCaEs that may activate the NCX and produce depolarizing transient inward currents (I_{ti}) [23, 37]. Overall, the enhanced frequency of SCaEs has been largely attributed to the dysfunction of RyR2 channels, as a result of altered posttranslational modifications (e.g., phosphorylation, oxidation), protein-protein interactions, and posttranscriptional regulations [16].

Reentry is recognized as the major proarrhythmic event that maintains AF. The development of reentrant circuits is a result of various anatomical and functional alternations in atrial tissue. An anatomical substrate could develop when the non-excitable necrotic tissue or fibrosis is surrounded by a conductive pathway. A functional substrate could develop when the heterogeneities in excitability or conduction are enhanced, which are often associated with abbreviated refractoriness and slowed conduction [38]. Conceptually, the wavelength of a circuit equals the product of effective refractory period (ERP) and conduction velocity (CV): WL = ERP × CV. The WL of a circuit decreases when ERP is shortened or CV is reduced [39], which are the best-known mechanisms of atrial remodeling. Because the ERP is determined by the cardiomyocyte APD, the shortening of ERP is a direct consequence of APD abbreviation, due to either the increase in repolarizing I_k currents or the reduction in depolarizing $I_{Ca,L}$ [40]. Several repolarizing I_k currents are associated with the APD shortening including I_{Kur}, I_{Ks}, I_{K1}, the constitutively active G-protein coupled inward rectifying K^+ current $I_{K,ACh}$, and the two-pore-domain K^+ currents I_{K2P} [41–46]. On the other hand, CV is determined by a combination of factors including the cellular excitability controlled by the Nav1.5 channel, the cell-to-cell communication via gap junctions, and the heterogeneity of tissue composition. Thus, one or more of the following events – the impaired cardiac Nav channel function [47], the altered expression or distribution of gap junction proteins, and the increased extracellular volume due to interstitial fibrosis and local inflammation [48], can lead to the slowed CV. In general, when atrial tachycardia remodeling due to AF occurs, the APD shortens and thus the reentry circuits become smaller. The smaller the circuit is, the more circuits the atria can accommodate. Meanwhile, if the atria undergo structural enlargement or hypertrophy, it allows the existence of the increased number of wavelets. In summary, abbreviation of ERP, reduction in CV, and enlargement of atria are the best-established arrhythmogenic events supporting the evolution of proarrhythmic substrates for AF [49].

Inflammasome Signaling

The inflammasome is a multimeric protein complex that senses pathogen-associated molecular patterns (PAMPs) or danger-associated molecular patterns (DAMPs) and drives inflammation [3]. The assembly of the inflammasome complex can cause the autocleavage of precursor caspases into mature and active form of pro-inflammatory caspases [50]. More than ten various forms of inflammasomes have

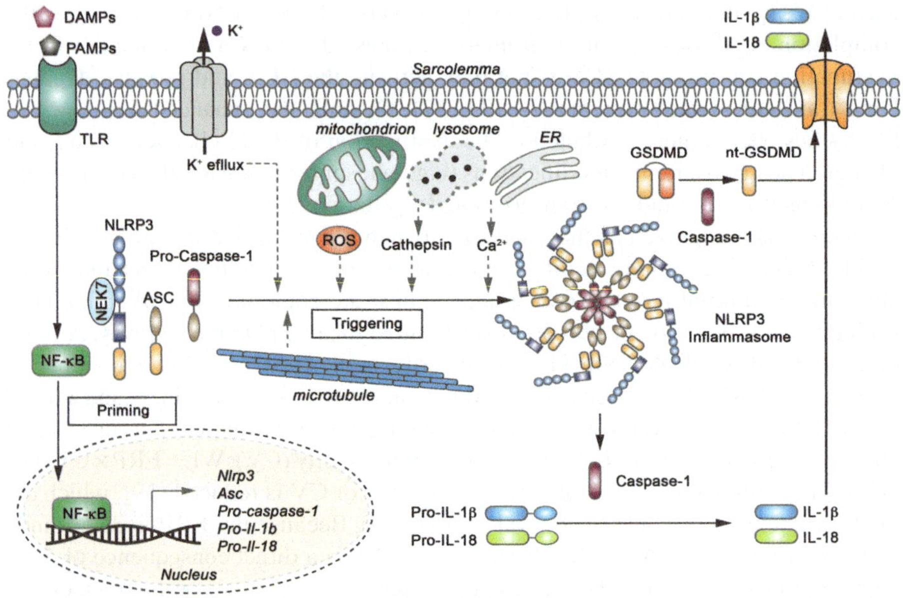

Fig. 1 Mechanisms associated with the activation of NLRP3 inflammasome involving various "priming" and "triggering" signals. The NLRP3 inflammasome is composed of the sensing subunit NLRP3, the adaptor protein known as apoptosis-associated speck-like *protein* containing a CARD (ASC), and the effector protein precursor caspase-1 (pro-caspase-1). NLRP3 protein has three main structures including (1) the N-terminus pyrin domain responsible for the recruitment of ASC, (2) the central nucleotide-binding oligomerization domain that can activate the inflammasome signaling platform, and (3) the C-terminus leucine-rich repeat (LRR) that can sense ligand and autoregulate. Basal level of NLRP3 is relatively low and insufficient for active inflammasome formation, and NLRP3 is kept in an inactive ubiquitinated state until a priming signal evokes de-ubiquitination. (Adapted from the Figure 2 of "The crosstalk between cardiomyocyte calcium and inflammasome signaling pathways in atrial fibrillation" by Wang et al. [49]. Copyright by Springer Nature)

been reported based on the identity of the sensing unit, which mainly include two types of proteins – "nucleotide-binding domain and leucine-rich repeat receptors" (NLRs) or "absent in melanoma 2 (AIM2)-like receptors" (ALRs) [3, 50]. Among them, the NLRP3 inflammasome is the best characterized and most extensively studied inflammasome complex and is unique in a way that it responds to a large diversity of stimuli [2, 3, 50, 51]. Activation of NLRP3 inflammasome is dependent on two types of signals: "priming" and "triggering" (Fig. 1). "Priming" refers to the activation of toll-like receptors (TLRs) by PAMPs/DAMPs or NLRs and cytokine receptors by its respective ligands, which lead to transcription factor "nuclear factor kappa-light-chain-enhancer of activated B cells" (NFκB)-mediated transcription of the genes encoding each inflammasome components *Nlrp3, Asc, pro-caspase-1*, as well as the effector genes *pro-Il-1b* and *pro-Il-18* [52, 53]. Activation of NF-κB also leads to de-ubiquitination of NLRP3 through the deubiquitinating enzyme BRCC3 [54, 55], which allows the system to be ready for quick "triggering"-mediated activation. "Triggering" is the process that promotes the assembly of NLRP3, ASC, and pro-caspase-1 proteins [2, 56–60]. The oligomerization of the multiprotein platform facilitates the autocleavage of pro-aspase-1 into the mature caspase-1. Active caspase-1 is an aspartate-specific cysteine protease that can proteolytically cleave the pro-inflammatory cytokines pro-IL-1β and pro-IL-18 into their mature and active forms IL-1β and IL-18. On the other hand, active caspase-1 also cleaves gasdermin D (GSDMD). The N-terminus fragment of GSDMD (Nt-GSDMD) can form pores on the sarcolemma, which facilitates the release of mature IL-1β and IL-18. In some cell types, pore-forming by Nt-GSDMD may also trigger a lytic, pro-inflammatory cell death, known as pyroptosis [61–64]. There is no consensus on a unified mechanism for the activation of the NLRP3 inflammasome. A number of triggering stimuli have been proposed, including an increase in intracellular Ca^{2+} level, a decrease in cellular AMP level, an increase in K^+-efflux and the resultant decrease in the intracellular K^+ level, uric acid, mitochondrial-associated dysfunction, the production of reactive oxygen species (ROS), and cathepsin B released from the damaged lysosome, among others [50, 65–68]. More recently, studies have shown that the NIMA-related kinase 7 (NEK7)-mediated interaction with NLRP3 is essential for the inflammasome assembly and activation downstream of the K^+-efflux [59, 69], and the spatial arrangement controlled by the microtubule network is also required for the proper activation of NLRP3 [67, 70] (Fig. 1).

Although inflammasomes were initially described in immune cells, emerging evidence point to a wide distribution of the NLRP3 inflammasome in many types of nonimmune cells including endothelial cells, cardiomyocytes, and cardiac fibroblasts [71, 72]. The enhanced function of NLRP3 inflammasome has been associated with the pathogenesis of multiple metabolic and cardiovascular diseases such as diabetes [73], atherosclerosis [10, 74], ischemic cardiomyopathy [71, 75–78], diabetic cardiomyopathy [79], and AF [34, 80, 81]. The innate immune system functions as the primary cardiac defense against pathogens and tissue damage [82]. Myocardial infarction, the most common cause of cardiac injury [83], is associated with acute death of cardiomyocytes. Necrotic cardiac cells can stimulate innate immune response to get rid of dead cell debris in the infarcted area [84, 85]. Endogenous ligands released from the dead cells can serve as DAMPs and activate TLRs, thereby initiating inflammatory responses by activating the NF-κB system and related signals [86]. Inflammatory macrophages and neutrophils can then be recruited to the infarcted area. The immune cells secrete growth factors and cytokines such as transforming growth factor-β (TGF-β) and IL-10 to promote cardiac repair [87, 88]. However, when the inflammatory response becomes overly enhanced, it can cause adverse tissue remodeling. Compared to the well-established canonical function of the NLRP3 inflammasome in the innate immune cells, the putative function of the NLRP3 inflammasome in nonimmune cells including cardiomyocytes and cardiac fibroblasts are less clear despite the recent evidence for a causative role of the NLRP3 inflammasome in the pathophysiology of AF.

Cardiac Inflammasome Signaling and Arrhythmogenesis

Recent evidence points to a potential involvement of the NLRP3 inflammasome in AF pathogenesis and suggests a causal role of upregulated inflammasomes in cardiac arrhythmogenesis. In this section, we will (i) elucidate the putative cell-type specific molecular mechanisms underlying the NLRP3 inflammasome-mediated promotion of AF, (ii) discuss the inflammasome as the potential link between AF risk factors and AF development, and (iii) elaborate the common nodal signaling contributing to the NLRP3 inflammasome activation in the evolution of AF.

Inflammasome-Mediated Pathogenesis of AF

Increases in atrial NLRP3 inflammasome activity have been linked to the development of AF in patients. In patients with a history of paroxysmal AF (pAF; self-terminates within 7 days), enhanced "triggering" signals drive the hyperactivation of NLRP3 inflammasome in atrial tissue. In contrast, in patients with a history of long-lasting persistent AF (perAF), both abnormal "priming" and "triggering" signals contribute to the higher activity of the NLRP3 inflammasome [80]. In patients who are susceptible to the development of postoperative AF (poAF), upregulated "priming" and "triggering" signals contribute to the activation of NLRP3 inflammasome [34] and the formation of a preexisting substrate which promotes the predisposition to poAF. Subsequent studies dissecting the contribution of cardiomyocyte and cardiac fibroblast inflammasome to AF pathogenesis showed that the NLRP3 inflammasome was similarly upregulated in atrial cardiomyocytes of pAF and perAF patients. These data confirm that the cardiomyocyte NLRP3 inflammasome is the primary driver of the hyperactive NLRP3 inflammasome in the atria of AF patients, which may partially be attributed to the fact the cardiomyocytes make up approximately 75% of normal adult myocardial tissue volume [89].

To study the causal role of cardiomyocyte NLRP3 inflammasome in AF development, a cardiomyocyte-specific knockin (CM-KI) mouse model with the cardiomyocyte-specific expression of constitutively active NLRP3 has been established (*Myh6:Nlrp3*$^{A350V/+}$). With this model, we have shown that the restricted activation of the NLRP3 inflammasome in cardiomyocytes is sufficient to promote ectopic activity and atrial ERP shortening, along with secondary fibrosis, thereby enhancing the susceptibility to AF development [34, 80, 81]. The development of ectopic activity in CM-KI mice is associated with an increased incidence of SCaEs through RyR2 channels, as evidenced by the increased frequency of Ca^{2+} sparks which reflect spontaneous Ca^{2+} release via a cluster of RyR2 channels [90]. The shortening of atrial ERP in CM-KI mice is attributed to the increased expression of *Kcna5* (encoding Kv1.5 protein) and the function of Kv1.5 channel (I_{Kur}). Although the CV is unchanged, the secondary development of fibrosis, due to ectopic activity and ERP shortening-induced atrial remodeling, may also enhance the vulnerability

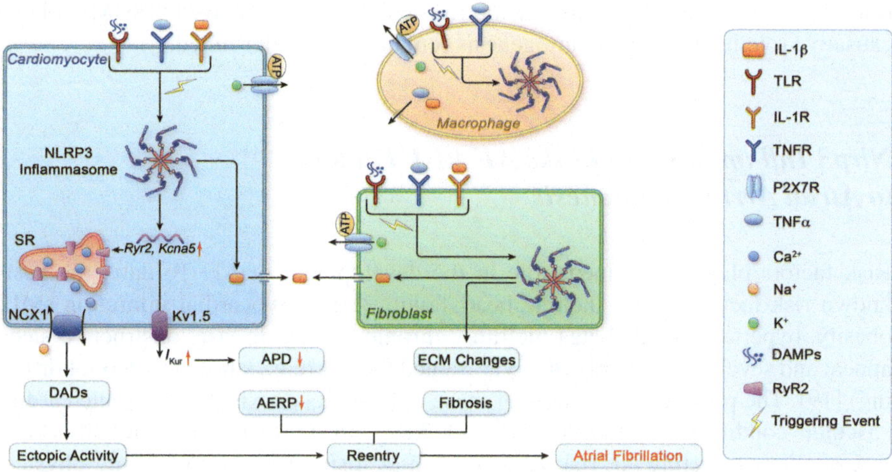

Fig. 2 Mechanisms underlying atrial fibrillation driven by the enhanced activity of NLRP3 inflammasome in different cell types. (Adapted from the Figure 2 of "Inflammasomes and proteostasis novel molecular mechanisms associated with atrial fibrillation" by Li and Brundel [92]. Copyright by Wolters Kluwer)

to AF in CM-KI mice, pointing to a potential contribution of cardiac fibroblast NLRP3 inflammasomes signaling to atrial arrhythmogenesis. Based on evidence in ischemic cardiomyopathy [71, 75–78], the activation of cardiac fibroblast NLRP3 inflammasome could be sufficient to promote atrial fibrosis and may cause atrial myopathy, which are known to enhance atrial arrhythmogenesis. Additionally, it is known that macrophage infiltration is increased in patients with AF, and the enhanced activation of NLRP3 inflammasome and IL-1β signaling could be found in macrophages [91]. Regardless of the source of secretion, the secreted IL-1β, upon binding with its receptor – IL-1R, could subsequently activate the NF-kB-mediated "priming" process in many cell types, ultimately amplifying the NLRP3 activation in cardiomyocytes, cardiac fibroblast, and macrophages [2]. Thus, the autocrine and paracrine crosstalk among different cell types via activated NLRP3 inflammasome signaling may exist in atria, with IL-1β and GSDMD being key effectors (Fig. 2).

For a long time, inflammatory cytokines have been associated with the onset and maintenance of AF, as well as the outcome of AF ablation [92]. There is ample evidence that the proinflammatory cytokine IL-1β is arrhythmogenic. Direct application of IL-1β (40 ng/mL) for 5 min onto human atrial cardiomyocytes can evoke potentially proarrhythmic SCaEs [34]. Similarly, direct application of IL-1β onto rat ventricular cardiomyocytes with low concentration of IL-1β (2 ng/mL) for 3 h is sufficient to increase SR Ca^{2+} and decrease the amplitude of the Ca^{2+} transients along with cell contractility [93]. Incubation of mouse atrial cardiomyocyte cell line HL-1 cells with the IL-1β (40 ng/mL) for 5 min not only induces aberrant SCaEs but also increases the CaMKII-mediated phosphorylation of RyR2, a known culprit mechanism of dysfunctional RyR2 channels [16, 34]. The potential arrhythmogenic role of IL-1β is emerging; whether the full-length and the cleaved GSDMD play a causative role in atrial arrhythmogenesis remains to be determined.

Nlrp3 Inflammasome Links AF Risk Factors to Atrial Arrhythmogenesis

Risk factors play a prominent role in the development of AF. To date, the best-known risk factors of AF include heart failure (HF), myocardial infarction (MI), obesity, hypertension, diabetes mellitus, chronic kidney disease, obstructive sleep apnea, and several modifiable lifestyle-related factors (e.g., smoking, alcohol drinking) [19]. The prevalence of many of these risk factors coexist with chronic cardiovascular conditions including HF and MI, and is increasing globally [19]. Interventions targeting AF risk factors, such as losing weight, increasing physical activity, and modifying lifestyle-related factors, have shown benefits for sinus rhythm maintenance [94, 95]. In addition to rate and rhythm control, and anticoagulation treatments, risk factor management has been suggested as a new pillar in AF care [94]. Although a great deal of knowledge has been obtained for some risk factors, the precise mechanisms of arrhythmogenesis remain poorly understood. Interestingly, several of these risk factors provide established signals for NLRP3 inflammasome activation, such as enhanced endoplasmic reticulum (ER) stress, elevated intracellular Ca^{2+} level, and increased ROS production. The enhancement in these inflammasome activation signals positions the NLRP3 inflammasome as a critical link between the risk factors and AF pathogenesis. The connection between heart failure and AF development has been thoroughly reviewed previously [96–98]. In light of very recent findings, in this section, we will discuss the role of specific signals associated with NLRP3 inflammasome activation in the context of obesity, gut dysbiosis, and chronic kidney disease.

Obesity Overweight and obesity are established risk factors of AF and are associated with lower efficacy of rhythm control strategies [94, 99–102]. Animal models of obesity exhibit reduced cardiac conduction and atrial enlargement, two established proarrhythmic substrates. The overfeeding-induced obesity in sheep promotes heterogeneous activation, reduced conduction velocity, spontaneous AF development, and increased AF burden [103]. In the pig model of high-fat diet-induced obesity, reduced ERP in the pulmonary veins contributes to the development of sustained AF [104]. The weight gain in obese sheep correlates with inflammatory infiltration, fibrosis, and enhanced activation of the TGF-β1 and the platelet-derived growth factor pathways [105]. Additionally, in the large animal models of obesity, the increased pericardial fat volume along with the fat infiltration into the atrial myocardium can form a unique substrate for AF, which is generally absent in the small animal models of obesity. Although obesity and inflammation are two independent risk factors of AF, obesity is frequently associated with enhanced inflammatory signaling [48, 106–111]. The activity of NLRP3 inflammasome is enhanced in atrial tissue of overweight and obese patients, obese sheep, and obese mice subjected to high-fat diet [81]. The increased AF susceptibility in obese mice is primarily driven by the NLRP3 inflammasome because whole-body genetic inhibition of NLRP3 prevents the development of abnormal SCaEs-mediated triggered activity and the reentrant substrate including ERP shortening and fibrosis, thereby preventing obesity-induced atrial arrhythmogenesis in obese mice [81]. The specific contribution of NLRP3 inhibition in different cell types in preventing the obesity-induced AF remains to be determined.

The high-calorie western diet is a major cause of obesity in modern societies and often rich in saturated fatty acids content. The saturated fatty acids can trigger the NLRP3-inflammasome activation via the ER stress pathway in macrophages [112]. Transcriptional upregulation of the NLRP3 inflammasome effector cytokines IL-1β and IL-18 is mediated through inositol-requiring enzyme 1 (IRE1α) upon dissociation of heat shock protein 5A (HSP5A, also known as the glucose-regulated protein of 78 kDa, GRP78) [113]. HSP5A normally resides near the ER, acting as a master controller of the unfolded protein response (UPR) [114]. In the atria of obese mice due to high-fat diet and obese patients, the expression of HSP5A is elevated, which may enhance the transcription of UPR target genes within atrial cardiomyocytes and drive the NLRP3-inflammasome activity. Conversely, inhibition of NLRP3 also blunts the diet-induced increase in HSPA5 level, suggesting a putative feedforward loop between NLRP3-inflammasome upregulation and ER-stress activation, which should be further dissected and validated in future studies [81]. Overall, the molecular mechanisms of obesity-mediated atrial arrhythmogenesis remain incompletely understood. Future research should also address the role of paracrine communication via adipocytokines between peripheral adipose tissue and atria and determine the direct local effects of epicardial adipose tissue on AF-promoting atrial remodeling.

Gut dysbiosis Emerging evidence suggest that the disruption in the gut microbiota composition, also known as "gut dysbiosis," is a potential new contributor of AF pathogenesis [115, 116]. Several gut microbiota-derived metabolites and endotoxins have been associated with the increased AF risk including trimethylamine N-oxide (TMAO, the product of trimethylamine, derived from the dietary choline and carnitine) [117, 118], indoxyl sulfate (the most common uremic toxin, derived from the dietary tryptophan or protein-based food) [119–121], and lipopolysaccharide (LPS, an endotoxin found in the out-layer of Gram-negative bacteria) [122, 123]. Gut dysbiosis is also associated with many AF risk factors, such as aging, obesity, hypertension, diabetes, obstructive sleep apnea, heart failure, and atherosclerosis [116, 124–131]. Gut dysbiosis could enhance the AF arrhythmogenesis by promoting the abovementioned risk factors of AF, or by increasing gut metabolites and endotoxins that can act on the atria thereby promoting the evolution of an arrhythmic substrate. In the doxorubicin-induced cardiac fibrosis model, TMAO can exacerbate fibrosis by activating the NLRP3 inflammasome in cardiac fibroblasts [132]. One study by Zhang et al. showed that aging-associated gut dysbiosis alters the gut barrier and increases the levels of circulating LPS and glucose, which then jointly activate the NLRP3 inflammasome in atrial myofibroblasts to cause atrial fibrosis, thereby enhancing AF susceptibility [133]. For detailed review on the causal relationship between gut dysbiosis and AF, we refer the readers to a recent comprehensive review by Gawalko et al. [115]. The direct effect of other gut metabolites such as bile acids and indoxyl sulfate on NLRP3 activation and arrhythmogenesis remain poorly understood and requires extensive investigation in future work.

Chronic kidney disease Chronic kidney disease (CKD) is a major public health problem with an estimated 26 million Americans affected. The high risk of cardiovascular events in CKD represents a major cause of morbidity and mortality in this population [134]. In the large population-based "Atherosclerosis Risk in Communities" (ARIC) study, impaired kidney function was strongly associated with the incidence of AF independently of other risk factors [134]. Classification of different stages of CKD is based on the level of the estimated glomerular filtration rate (eGFR) [135]. The prevalence of AF markedly increases with the severity/stage of CKD and is about 10- to 20-fold elevated in stage 5 CKD patients on hemodialysis compared with patients with normal kidney function [136, 137]. Despite its high prevalence in CKD patients, the pathogenesis of AF in this patient population is not well understood. The causal association between CKD and AF is complex and may result from fluctuations in electrolyte levels, abnormal sympathetic nervous system activity, and changes in the renin-angiotensin system that appear to promote AF in CKD [138, 139]. In the nephrectomy-induced CKD rat model, the increased AF susceptibility is associated with left atrial enlargement, atrial fibrosis, and increased activity of the profibrotic signaling pathway (e.g., TGF-β1/Smad2/3 pathway), and the enhanced inflammatory response due to activation of NLRP3 inflammasome. However, the mechanisms underlying the activation of NLRP3 in the CKD model are currently unknown. It has been postulated that hyperuricemia associated with CKD can cause activation of the NLRP3 inflammasome [140]. Serum uric acid and the salt or ester forms of uric acid – urate can activate the NLRP3 inflammasome in macrophage via the purinergic receptor P2X7R [141]. Fibroblasts can internalize the uric acid crystal (UAC) via the uric acid transporter (UAT), which stimulates ROS production and increases the Ca^{2+} levels in ER. The latter signals can trigger the NLRP3 inflammasome activation in fibroblasts. In addition, the elevated ROS in fibroblasts may also activate the Ca^{2+}-permeable transient receptor potential melastatin-related type-7 channels (TRPM7). Enhanced function of TRPM7 channels that contributes to the TGFβ1-induced fibroblast differentiation have been implicated in the AF-related development of fibrosis in patients [142]. Uric acid may also directly affect the electrophysiology of cardiomyocytes. In the mouse atrial myocyte like HL-1 cells, uric acid intake via the UATs upregulates *Kcna* expression by activating the extracellular signal-regulated kinase (ERK) signaling pathway [143]. The increased protein level of Kv1.5 may increase the amplitude of I_{Kur}, which can abbreviate the atrial APD and ERP. The increased K^+-efflux due to the enhanced I_{Kur} may, in turn, activate the NLRP3 inflammasome in cardiomyocytes, forming a feedforward loop.

Nodal Signaling Points of Nlrp3 Activation

In addition to the molecular signatures unique to various AF risk factors, several nodal signaling points commonly existing in many cardiovascular diseases appear to be key for the regulation of NLRP3 inflammasome activation. For instance, CaMKII and ROS are key contributors to AF pathophysiology and established enhancers of the NLRP3 inflammasome activity [144–146], which position them as potential targets for therapeutic interventions in AF.

CaMKII CaMKII, a multifunctional serine-threonine protein kinase, is abundant in various tissues. Among four homologous isoforms CaMKII α, β, δ, and γ, CaMKIIδ and CaMKIIγ are the predominant isoforms in the heart [147, 148]. CaMKII exists as a holoenzyme that consists of two hexamers [149]. Within each monomer, it contains an N-terminal catalytic domain, a central regulatory domain, and a C-terminal association domain. In the inactive state, the pseudo-substrate region of the regulatory domain inhibits the catalytic domain of each CaMKII subunit by sterically blocking the substrate to interact with the ATP-binding pocket. When intracellular Ca^{2+} level increases, Ca^{2+} can bind to calmodulin (CaM). Subsequently, the Ca^{2+}/CaM complex can activate CaMKII by interacting with the C-terminal region of the regulatory domain, causing a conformational change with allosteric displacement of the pseudo-substrate region, and allowing the substrate and ATP to access the catalytic domain [150]. In the presence of ATP, sustained increases in the Ca^{2+}/CaM level can result in autophosphorylation of Threonine 287 in the autoinhibitory region of the regulatory domain, promoting a conformational change that prevents the association between the catalytic and regulatory domains. This mode of Ca^{2+}-independent activation of CaMKII is known as autophosphorylation [149]. In 2008, Mark E. Anderson's group discovered that oxidation of Methionine 281 & Methionine 282 can also cause a Ca^{2+}-independent activation of CaMKII holoenzyme [151]. To date, although other types of posttranslational modifications such as O-linked glycosylation of Serine 280, NO-dependent nitrosylation of Cysteine 116/273/290 have been reported to cause autonomous CaMKII activity [152, 153], the autophosphorylation and oxidation are still the most known mechanisms responsible for the enhanced CaMKII activity in many physiological and pathological conditions of the heart. In cardiomyocytes, CaMKII can modulate many diverse targets by phosphorylation of certain serine or threonine residues. The CaMKII target proteins include ion channels Cav1.2 (alpha subunit of the L-type Ca^{2+} current), Nav1.5 (alpha subunit of voltage-dependent Na^{2+} current), and Kv4.3 (alpha subunit of transient outward K^+ current), as well as Ca^{2+} handling proteins like RyR2 and PLN [154–159]. Because these targets are critically involved in shaping the AP morphology and mediating ECC, CaMKII acts as an important nodal signaling in many aspects of cardiac diseases.

There is ample evidence that the enhanced function of CaMKII promotes atrial arrhythmogenesis by targeting a variety of proteins involved in cellular electrophysiology, Ca^{2+} homeostasis, inflammation, cell survival, etc. [150]. The most established role for CaMKII in AF development is the promotion of triggered activity by phosphorylation of RyR2 at Ser2814 and subsequent enhancement in frequency of SCaEs [35, 160, 161]. Additionally, abnormal CaMKII can facilitate electrical remodeling in perAF by contributing to ion channel dysfunction [40, 159]. Increased activity of CaMKII may secondarily contribute to AF-related structural remodeling by increasing the death of cardiomyocytes, causing fibrosis and extracellular matrix remodeling, and enhancing the transcription of inflammatory genes such as tumor necrosis factor alpha (TNFα) [148, 162–164].

Recent studies highlight the presence of a CaMKII/NLRP3 nexus that promotes the formation of a preexisting substrate vulnerable for the development of poAF [34]. In atrial cardiomyocytes of poAF patients, enhanced activity of CaMKII and NLRP3 signaling are concomitant and associated with enhancements of IL-1β-mediated SR Ca^{2+} leak, SCaEs, and DADs [34]. Acute application of IL-1β elicits abnormal SCaEs in human atrial cardiomyocytes and in HL-1 cells, which is accompanied by an increased phosphorylation of RyR2 and PLN by CaMKII [34]. On the other hand, in murine models of cardiac remodeling induced by angiotensin II-infusion or transverse aortic constriction, cardiomyocyte-specific deletion of CaMKII diminishes the activation of NLRP3 inflammasome [144, 165]. It has been shown that the cardiomyocyte CaMKIIδc can phosphorylate the inhibitor of NFκB (IκB kinase) and subsequently cause the activation of NFκB [162, 164], which controls the transcription of inflammasome genes. Although not universally accepted, CaMKII could also increase Ca^{2+} entry in the mitochondria by phosphorylating the mitochondrial Ca^{2+} uniporter, which can subsequently increase ROS production [166]. Thus, the CaMKII-mediated activation of the NLRP3 inflammasome in cardiomyocytes may be a consequence of both increases in ROS production and SR Ca^{2+} leak.

ROS It is well known that the elevated level of atrial ROS/reactive nitrogen species (RNS) is associated with AF [167–171]. Both the increased production of mitochondrial ROS, primarily due to the functionally deficient electron transport chain, and the enhanced function of NADPH oxidase (NOX) located on the cell membrane can contribute to the overall increase in the ROS/RNS production in atria [167–169, 172]. In a dog model of AF induced by atrial tachypacing, NOX2 inhibition by the short hairpin RNA-mediated knockdown prevents the onset and maintenance of AF [173]. ROS acting as an NLRP3-activating trigger is supported by the evidence that inhibition of NOX-derived ROS prevents the caspase-1 and IL-1β maturation in alveolar macrophages [145, 146]. Similarly, the knockdown of the $p22^{Phox}$ subunit of NOX can decrease IL-1β release from human monocyte THP1 cells [174]. The crystal structure of NLRP3 reveals that the pyrin domain and the nucleotide-binding site domain are connected via a highly conserved disulfide bond. This suggests that NLRP3 could be very sensitive to alterations in redox states [175]. Additionally, it is well-established that ROS/RNS signaling can directly modulate the activity of several Ca^{2+} handling proteins and CaMKII in cardiomyocytes independently [176, 177]. Oxidized or S-nitrosylated RyR2 channels also exhibit increased Ca^{2+} release function during diastole [178]. All of these actions can elevate the intracellular Ca^{2+}

level, which is an established trigger of the NLRP3 inflammasome [179]. In macrophages, enhanced Ca^{2+} signaling clearly contributes to NLRP3 inflammasome activation [180], likely by facilitating the interaction between NLRP3 and ASC [181], thereby enhancing the oligomerization of the inflammasome complex. The altered Ca^{2+} homeostasis due to either ROS/RNS or CaMKII activity may also promote ER stress, another known "triggering" signal activating the NLRP3 inflammasome [182]. Although there is a correlation between intracellular Ca^{2+} and NLRP3 inflammasome activation, the precise mechanisms linking Ca^{2+} fluxes with NLRP3-inflammasome activation remain elusive. On the other hand, the increased cytosolic Ca^{2+} level may enhance the uptake of Ca^{2+} into the mitochondria and increase the mitochondrial ROS production, subsequently activating NLRP3 inflammasome [180, 183, 184]. Thus, it is plausible that ROS, Ca^{2+}, and CaMKII signaling can interact with each other, synergistically triggering the activation of NLRP3 inflammasome in AF.

Therapeutic Potential of Targeting the Cardiac NLRP3 Inflammasome

Currently, the standard treatments of AF include rhythm and rate control, as well as anticoagulation. Despite clear evidence that innate inflammatory signaling pathways may play a causal role in AF development, use of conventional anti-inflammatory medications such as steroids and nonsteroidal anti-inflammatory drugs (NSAIDs) resulted in mixed results in AF. Clinical studies have shown that dexamethasone and oral administration of prednisone have no or limited benefits in preventing poAF and recurrent AF post-ablation, despite the reduction of circulating inflammatory cytokines [185, 186]. Whether a selective NLRP3 inhibition constitutes a more effective anti-AF option requires direct investigation. Currently, the available strategies to reduce the overall inflammatory burden target either the effectors of the NLRP3 inflammasome (caspase-1, IL-1β, and IL-1R) or its upstream stimuli (NFκB, TLR, and purinergic receptor P2X7R) [92]. Several "biologicals" including neutralizing antibodies and recombined decoy proteins targeting IL-1β or IL-1R are available [92], and many of these FDA-approved orphan drugs are indicated for autoinflammatory and autoimmune diseases like rheumatoid arthritis. Their potential therapeutic effect on cardiac diseases, neurogenerative disease, the coronavirus disease 2019 (COVID-19)-related respiratory disease, and other diseases with enhanced inflammatory response are being investigated (Table 1). There is also a growing interest in the development of selective inhibitors targeting NLRP3 [187]. If the NLRP3-specific inhibitors (i.e., DFV890 and OLT1177) [188, 189] are safe in patients, their efficacy in preventing AF would require additional clinical investigations.

Table 1 Ongoing non-cancer-related clinical trials involving the drugs targeting NLRP3 inflammasome pathway

Drug	Mechanism of action	Clinical trial phase	Indication	Identifier (clinicaltrial.gov)
DFV890	NLRP3 inhibitor	Phase II	Knee osteoarthritis	NCT04886258
		Phase II	Familial cold auto-inflammatory syndrome	NCT04868968
Dapansutrile (OLT1177)	NLRP3 inhibitor	Phase II	COVID-19, cytokine release syndrome	NCT04540120
		Phase II	Schnitzler syndrome	NCT03595371
Belnacasan (VX-765)	Casapse-1 inhibitor	Phase II	COVID-19	NCT05164120
Canakinumab (Ilaris®)	Neutralizing anti-IL-1β antibody	Phase III	Cryopyrin-associated periodic syndrome Familial cold autoinflammatory syndrome	NCT01576367
		Phase II	Alcoholic hepatitis	NCT03775109
		Phase II	Knee osteoarthritis	NCT04814368
		Phase I & II	Duchenne muscular dystrophy	NCT03936894
		Phase II	Mild cognitive impairment Alzheimer disease	NCT04795466
Rilonacept (Arcalyst®)	IL-1 decoy receptor	Phase III	Recurrent pericarditis	NCT03737110 NCT04687358
Anakinra (Kineret®)	IL-1R antagonist	Phase II	Epidermolysis bullosa	NCT03468322
		Early phase I	Endometriosis	NCT03991520
		Phase I & II	Kawasaki disease	NCT02179853
		Phase II	Respiratory failure by COVID-19	NCT04357366 NCT04680949 NCT04643678
		Phase II	Multiple myeloma	NCT04099901
		Phase II & III	Gout and chronic kidney disease	NCT02578394
		Phase I	Macrophage activation syndrome	NCT02780583
		Phase II & III	Acute myocarditis	NCT03018834
		Phase II	Type 2 diabetes	NCT04227769
		Phase I & II	Multiple sclerosis	NCT04025554
		Phase II	Systolic heart failure	NCT03797001
		Phase II	Cardiac sarcoidosis	NCT04017936
sc-rAAV2.5IL-1Ra	Gene transfer vector expressing IL-1Ra	Phase I	Knee osteoarthritis	NCT02790723
IL-1Ra	IL-1R antagonist	Phase III	Subarachnoid hemorrhage	NCT03249207

Summary and Future Perspectives

Over the past two decades, we witness an explosion of information on inflammasome biology. As the function of inflammasomes has been critically established in many systems, there is an increasing interest in understanding the function of inflammasomes and their disease-causal role in cardiovascular diseases including coronary artery disease, cardiomyopathy, and cardiac arrhythmias. Despite the great progress, many knowledge gaps and issues remain. Here are some areas of research priorities for preclinical and clinical investigations: (i) to define the precise function of individual inflammasome effectors in cardiac electrophysiology and AF pathophysiology, (ii) to determine the upstream molecular signals including the noncanonical activation pathways activating NLRP3 inflammasome in AF, (iii) to elucidate the NLRP3-driven interactions between cardiomyocytes, cardiac fibroblasts, adipocytes, and immune cells via diverse modes (autocrine, paracrine, and direct cell contacts), (iv) to evaluate the clinical efficacy of NLRP3-specific inhibitors in cardiac arrhythmias including AF, and (v) to explore the roles of other inflammasome complexes (e.g., AIM2) and test their therapeutic potential in AF patients.

Acknowledgment D.D. is supported by the National Institute of Health (R01HL136389, R01HL131517, and R01HL089598), the German Research Foundation (DFG, Do 769/4-1), and the European Union (large-scale integrative project MEASTRIA, No. 965286). N.L. is supported by the grants from National Institute of Health (R01HL136389, R01HL147108, R01HL163277).

References

1. Broz P, Dixit VM. Inflammasomes: mechanism of assembly, regulation and signalling. Nat Rev Immunol. 2016;16:407–20.
2. Scott L Jr, Li N, Dobrev D. Role of inflammatory signaling in atrial fibrillation. Int J Cardiol. 2019;287:195–200.
3. Schroder K, Tschopp J. The inflammasomes. Cell. 2010;140:821–32.
4. Man SM, Kanneganti T-D. Regulation of inflammasome activation. Immunol Rev. 2015;265:6–21.
5. De Nardo D, Latz E. NLRP3 inflammasomes link inflammation and metabolic disease. Trends Immunol. 2011;32:373–9.
6. Sharma BR, Kanneganti TD. NLRP3 inflammasome in cancer and metabolic diseases. Nat Immunol. 2021;22:550–9.
7. Ren P, Wu D, Appel R, Zhang L, Zhang C, Luo W, Robertson AAB, Cooper MA, Coselli JS, Milewicz DM, Shen YH, LeMaire SA. Targeting the NLRP3 inflammasome with inhibitor MCC950 prevents aortic aneurysms and dissections in mice. J Am Heart Assoc. 2020;9:e014044.
8. Wu D, Ren P, Zheng Y, Zhang L, Xu G, Xie W, Lloyd EE, Zhang S, Zhang Q, Curci JA, Coselli JS, Milewicz DM, Shen YH, LeMaire SA. NLRP3 (Nucleotide oligomerization domain-like receptor family, pyrin domain containing 3)-caspase-1 inflammasome degrades contractile proteins: implications for aortic biomechanical dysfunction and aneurysm and dissection formation. Arterioscler Thromb Vasc Biol. 2017;37:694–706.
9. Abbate A, Toldo S, Marchetti C, Kron J, Van Tassell BW, Dinarello CA. Interleukin-1 and the inflammasome as therapeutic targets in cardiovascular disease. Circ Res. 2020;126:1260–80.

10. Baldrighi M, Mallat Z, Li X. NLRP3 inflammasome pathways in atherosclerosis. Atherosclerosis. 2017;267:127–38.
11. Ridker PM, Everett BM, Thuren T, MacFadyen JG, Chang WH, Ballantyne C, Fonseca F, Nicolau J, Koenig W, Anker SD, Kastelein JJP, Cornel JH, Pais P, Pella D, Genest J, Cifkova R, Lorenzatti A, Forster T, Kobalava Z, Vida-Simiti L, Flather M, Shimokawa H, Ogawa H, Dellborg M, Rossi PRF, Troquay RPT, Libby P, Glynn RJ, CANTOS Trial Group. Antiinflammatory therapy with canakinumab for atherosclerotic disease. N Engl J Med. 2017;377:1119–31.
12. Nerbonne JM, Kass RS. Molecular physiology of cardiac repolarization. Physiol Rev. 2005;85:1205–53.
13. Bers DM. Cardiac excitation-contraction coupling. Nature. 2002;415:198–205.
14. Garbincius JF, Elrod JW. Mitochondrial calcium exchange in physiology and disease. Physiol Rev. 2022;102:893–992.
15. Heijman J, Dewenter M, El-Armouche A, Dobrev D. Function and regulation of serine/threonine phosphatases in the healthy and diseased heart. J Mol Cell Cardiol. 2013;64:90–8.
16. Lahiri SK, Aguilar-Sanchez Y, Wehrens XHT. Mechanisms underlying pathological Ca(2+) handling in diseases of the heart. Pflugers Arch. 2021;473:331–47.
17. MacLennan DH, Asahi M, Tupling AR. The regulation of SERCA-type pumps by phospholamban and sarcolipin. Ann N Y Acad Sci. 2003;986:472–80.
18. Landstrom AP, Dobrev D, Wehrens XHT. Calcium signaling and cardiac arrhythmias. Circ Res. 2017;120:1969–93.
19. Kornej J, Borschel CS, Benjamin EJ, Schnabel RB. Epidemiology of atrial fibrillation in the 21st century: novel methods and new insights. Circ Res. 2020;127:4–20.
20. Rahman F, Kwan GF, Benjamin EJ. Global epidemiology of atrial fibrillation. Nat Rev Cardiol. 2014;11:639–54.
21. Nattel S, Dobrev D. Electrophysiological and molecular mechanisms of paroxysmal atrial fibrillation. Nat Rev Cardiol. 2016;13:575–90.
22. Dobrev D, Aguilar M, Heijman J, Guichard JB, Nattel S. Postoperative atrial fibrillation: mechanisms, manifestations and management. Nat Rev Cardiol. 2019;16:417–36.
23. Nattel S, Heijman J, Zhou L, Dobrev D. Molecular basis of atrial fibrillation pathophysiology and therapy: a translational perspective. Circ Res. 2020;127:51–72.
24. Andrade J, Khairy P, Dobrev D, Nattel S. The clinical profile and pathophysiology of atrial fibrillation: relationships among clinical features, epidemiology, and mechanisms. Circ Res. 2014;114:1453–68.
25. Heijman J, Guichard JB, Dobrev D, Nattel S. Translational challenges in atrial fibrillation. Circ Res. 2018;122:752–73.
26. Nattel S, Sager PT, Huser J, Heijman J, Dobrev D. Why translation from basic discoveries to clinical applications is so difficult for atrial fibrillation and possible approaches to improving it. Cardiovasc Res. 2021;117:1616–31.
27. Christoffels VM, Smits GJ, Kispert A, Moorman AF. Development of the pacemaker tissues of the heart. Circ Res. 2010;106:240–54.
28. Pandit SV, Jalife J. Rotors and the dynamics of cardiac fibrillation. Circ Res. 2013;112:849–62.
29. Heijman J, Voigt N, Nattel S, Dobrev D. Cellular and molecular electrophysiology of atrial fibrillation initiation, maintenance, and progression. Circ Res. 2014;114:1483–99.
30. Haissaguerre M, Jais P, Shah DC, Takahashi A, Hocini M, Quiniou G, Garrigue S, Le Mouroux A, Le Metayer P, Clementy J. Spontaneous initiation of atrial fibrillation by ectopic beats originating in the pulmonary veins. N Engl J Med. 1998;339:659–66.
31. European Heart Rhythm Association, European Cardiac Arrhythmia Scoiety, American College of Cardiology, American Heart Association, Society of Thoracic Surgeons, Calkins H, Brugada J, Packer DL, Cappato R, Chen SA, Crijns HJ, Damiano RJ Jr, Davies DW, Haines DE, Haissaguerre M, Iesaka Y, Jackman W, Jais P, Kottkamp H, Kuck KH, Lindsay BD, Marchlinski FE, PM MC, Mont JL, Morady F, Nademanee K, Natale A, Pappone C, Prystowsky E, Raviele A, Ruskin JN, Shemin RJ. HRS/EHRA/ECAS expert Consensus

Statement on catheter and surgical ablation of atrial fibrillation: recommendations for personnel, policy, procedures and follow-up. A report of the Heart Rhythm Society (HRS) Task Force on catheter and surgical ablation of atrial fibrillation. Heart Rhythm. 2007;4:816–61.
32. Andrade JG, Champagne J, Dubuc M, Deyell MW, Verma A, Macle L, Leong-Sit P, Novak P, Badra-Verdu M, Sapp J, Mangat I, Khoo C, Steinberg C, Bennett MT, Tang ASL, Khairy P, Investigators C-DS. Cryoballoon or radiofrequency ablation for atrial fibrillation assessed by continuous monitoring: a randomized clinical trial. Circulation. 2019;140:1779–88.
33. Schweikert RA, Perez Lugones A, Kanagaratnam L, Tomassoni G, Beheiry S, Bash D, Pisano E, Saliba W, Tchou PJ, Natale A. A simple method of mapping atrial premature depolarizations triggering atrial fibrillation. Pacing Clin Electrophysiol. 2001;24:22–7.
34. Heijman J, Muna AP, Veleva T, Molina CE, Sutanto H, Tekook M, Wang Q, Abu-Taha IH, Gorka M, Kunzel S, El-Armouche A, Reichenspurner H, Kamler M, Nikolaev V, Ravens U, Li N, Nattel S, Wehrens XHT, Dobrev D. Atrial myocyte NLRP3/CaMKII nexus forms a substrate for postoperative atrial fibrillation. Circ Res. 2020;127:1036–55.
35. Voigt N, Li N, Wang Q, Wang W, Trafford AW, Abu-Taha I, Sun Q, Wieland T, Ravens U, Nattel S, Wehrens XH, Dobrev D. Enhanced sarcoplasmic reticulum Ca2+ leak and increased Na+-Ca2+ exchanger function underlie delayed afterdepolarizations in patients with chronic atrial fibrillation. Circulation. 2012;125:2059–70.
36. Burashnikov A, Antzelevitch C. Late-phase 3 EAD. A unique mechanism contributing to initiation of atrial fibrillation. Pacing Clin Electrophysiol. 2006;29:290–5.
37. Wakili R, Voigt N, Kaab S, Dobrev D, Nattel S. Recent advances in the molecular pathophysiology of atrial fibrillation. J Clin Invest. 2011;121:2955–68.
38. Waks JW, Josephson ME. Mechanisms of atrial fibrillation – reentry, rotors and reality. Arrhythm Electrophysiol Rev. 2014;3:90–100.
39. Comtois P, Kneller J, Nattel S. Of circles and spirals: bridging the gap between the leading circle and spiral wave concepts of cardiac reentry. Europace. 2005;7(Suppl 2):10–20.
40. Christ T, Boknik P, Wohrl S, Wettwer E, Graf EM, Bosch RF, Knaut M, Schmitz W, Ravens U, Dobrev D. L-type Ca2+ current downregulation in chronic human atrial fibrillation is associated with increased activity of protein phosphatases. Circulation. 2004;110:2651–7.
41. Caballero R, de la Fuente MG, Gomez R, Barana A, Amoros I, Dolz-Gaiton P, Osuna L, Almendral J, Atienza F, Fernandez-Aviles F, Pita A, Rodriguez-Roda J, Pinto A, Tamargo J, Delpon E. In humans, chronic atrial fibrillation decreases the transient outward current and ultrarapid component of the delayed rectifier current differentially on each atria and increases the slow component of the delayed rectifier current in both. J Am Coll Cardiol. 2010;55:2346–54.
42. Dobrev D, Friedrich A, Voigt N, Jost N, Wettwer E, Christ T, Knaut M, Ravens U. The G protein-gated potassium current I(K,ACh) is constitutively active in patients with chronic atrial fibrillation. Circulation. 2005;112:3697–706.
43. Voigt N, Trausch A, Knaut M, Matschke K, Varro A, Van Wagoner DR, Nattel S, Ravens U, Dobrev D. Left-to-right atrial inward rectifier potassium current gradients in patients with paroxysmal versus chronic atrial fibrillation. Circ Arrhythm Electrophysiol. 2010;3:472–80.
44. Schmidt C, Wiedmann F, Voigt N, Zhou XB, Heijman J, Lang S, Albert V, Kallenberger S, Ruhparwar A, Szabo G, Kallenbach K, Karck M, Borggrefe M, Biliczki P, Ehrlich JR, Baczko I, Lugenbiel P, Schweizer PA, Donner BC, Katus HA, Dobrev D, Thomas D. Upregulation of K(2P)3.1 K+ current causes action potential shortening in patients with chronic atrial fibrillation. Circulation. 2015;132:82–92.
45. Makary S, Voigt N, Maguy A, Wakili R, Nishida K, Harada M, Dobrev D, Nattel S. Differential protein kinase C isoform regulation and increased constitutive activity of acetylcholine-regulated potassium channels in atrial remodeling. Circ Res. 2011;109:1031–43.
46. Voigt N, Maguy A, Yeh YH, Qi X, Ravens U, Dobrev D, Nattel S. Changes in I K, ACh single-channel activity with atrial tachycardia remodelling in canine atrial cardiomyocytes. Cardiovasc Res. 2008;77:35–43.

47. Sossalla S, Kallmeyer B, Wagner S, Mazur M, Maurer U, Toischer K, Schmitto JD, Seipelt R, Schondube FA, Hasenfuss G, Belardinelli L, Maier LS. Altered Na(+) currents in atrial fibrillation effects of ranolazine on arrhythmias and contractility in human atrial myocardium. J Am Coll Cardiol. 2010;55:2330–42.
48. Hu YF, Chen YJ, Lin YJ, Chen SA. Inflammation and the pathogenesis of atrial fibrillation. Nat Rev Cardiol. 2015;12:230–43.
49. Wang X, Chen X, Dobrev D, Li N. The crosstalk between cardiomyocyte calcium and inflammasome signaling pathways in atrial fibrillation. Pflugers Arch. 2021;473:389–405.
50. Martinon F, Burns K, Tschopp J. The inflammasome: a molecular platform triggering activation of inflammatory caspases and processing of proIL-beta. Mol Cell. 2002;10:417–26.
51. Gross O, Thomas CJ, Guarda G, Tschopp J. The inflammasome: an integrated view. Immunol Rev. 2011;243:136–51.
52. Liu T, Zhang L, Joo D, Sun SC. NF-kappaB signaling in inflammation. Signal Transduct Target Ther. 2017;2:17023.
53. Ylikoski J, Pirvola U, Happola O, Panula P, Virtanen I. Immunohistochemical demonstration of neuroactive substances in the inner ear of rat and guinea pig. Acta Otolaryngol. 1989;107:417–23.
54. Py Bénédicte F, Kim M-S, Vakifahmetoglu-Norberg H, Yuan J. Deubiquitination of NLRP3 by BRCC3 critically regulates inflammasome activity. Mol Cell. 2013;49:331–8.
55. Qiao Y, Wang P, Qi J, Zhang L, Gao C. TLR-induced NF-κB activation regulates NLRP3 expression in murine macrophages. FEBS Lett. 2012;586:1022–6.
56. Kanneganti TD. The inflammasome: firing up innate immunity. Immunol Rev. 2015;265:1–5.
57. Chen G, Chelu MG, Dobrev D, Li N. Cardiomyocyte inflammasome signaling in cardiomyopathies and atrial fibrillation: mechanisms and potential therapeutic implications. Front Physiol. 2018;9:1115.
58. Sharif H, Wang L, Wang WL, Magupalli VG, Andreeva L, Qiao Q, Hauenstein AV, Wu Z, Nunez G, Mao Y, Wu H. Structural mechanism for NEK7-licensed activation of NLRP3 inflammasome. Nature. 2019;570:338–43.
59. He Y, Zeng MY, Yang D, Motro B, Nunez G. NEK7 is an essential mediator of NLRP3 activation downstream of potassium efflux. Nature. 2016;530:354–7.
60. Swanson KV, Deng M, Ting JPY. The NLRP3 inflammasome: molecular activation and regulation to therapeutics. Nat Rev Immunol. 2019;19:477–89.
61. Fink SL, Cookson BT. Caspase-1-dependent pore formation during pyroptosis leads to osmotic lysis of infected host macrophages. Cell Microbiol. 2006;8:1812–25.
62. Shi J, Zhao Y, Wang K, Shi X, Wang Y, Huang H, Zhuang Y, Cai T, Wang F, Shao F. Cleavage of GSDMD by inflammatory caspases determines pyroptotic cell death. Nature. 2015;526:660–5.
63. Kayagaki N, Stowe IB, Lee BL, O'Rourke K, Anderson K, Warming S, Cuellar T, Haley B, Roose-Girma M, Phung QT, Liu PS, Lill JR, Li H, Wu J, Kummerfeld S, Zhang J, Lee WP, Snipas SJ, Salvesen GS, Morris LX, Fitzgerald L, Zhang Y, Bertram EM, Goodnow CC, Dixit VM. Caspase-11 cleaves gasdermin D for non-canonical inflammasome signalling. Nature. 2015;526:666–71.
64. Liu X, Zhang Z, Ruan J, Pan Y, Magupalli VG, Wu H, Lieberman J. Inflammasome-activated gasdermin D causes pyroptosis by forming membrane pores. Nature. 2016;535:153–8.
65. Gong T, Yang Y, Jin T, Jiang W, Zhou R. Orchestration of NLRP3 inflammasome activation by ion fluxes. Trends Immunol. 2018;39:393–406.
66. Zhou R, Yazdi AS, Menu P, Tschopp J. A role for mitochondria in NLRP3 inflammasome activation. Nature. 2011;469:221–5.
67. Misawa T, Takahama M, Kozaki T, Lee H, Zou J, Saitoh T, Akira S. Microtubule-driven spatial arrangement of mitochondria promotes activation of the NLRP3 inflammasome. Nat Immunol. 2013;14:454–60.
68. Hornung V, Latz E. Critical functions of priming and lysosomal damage for NLRP3 activation. Eur J Immunol. 2010;40:620–3.

69. Niu T, De Rosny C, Chautard S, Rey A, Patoli D, Groslambert M, Cosson C, Lagrange B, Zhang Z, Visvikis O, Hacot S, Hologne M, Walker O, Wong J, Wang P, Ricci R, Henry T, Boyer L, Petrilli V, Py BF. NLRP3 phosphorylation in its LRR domain critically regulates inflammasome assembly. Nat Commun. 2021;12:5862.
70. Magupalli VG, Negro R, Tian Y, Hauenstein AV, Di Caprio G, Skillern W, Deng Q, Orning P, Alam HB, Maliga Z, Sharif H, Hu JJ, Evavold CL, Kagan JC, Schmidt FI, Fitzgerald KA, Kirchhausen T, Li Y, Wu H. HDAC6 mediates an aggresome-like mechanism for NLRP3 and pyrin inflammasome activation. Science. 2020;369:eaas8995.
71. Kawaguchi M, Takahashi M, Hata T, Kashima Y, Usui F, Morimoto H, Izawa A, Takahashi Y, Masumoto J, Koyama J, Hongo M, Noda T, Nakayama J, Sagara J, Taniguchi S, Ikeda U. Inflammasome activation of cardiac fibroblasts is essential for myocardial ischemia/reperfusion injury. Circulation. 2011;123:594–604.
72. Takahashi M. NLRP3 in myocardial ischaemia-reperfusion injury: inflammasome-dependent or -independent role in different cell types. Cardiovasc Res. 2013;99:4–5.
73. Vandanmagsar B, Youm YH, Ravussin A, Galgani JE, Stadler K, Mynatt RL, Ravussin E, Stephens JM, Dixit VD. The NLRP3 inflammasome instigates obesity-induced inflammation and insulin resistance. Nat Med. 2011;17:179–88.
74. Duewell P, Kono H, Rayner KJ, Sirois CM, Vladimer G, Bauernfeind FG, Abela GS, Franchi L, Nunez G, Schnurr M, Espevik T, Lien E, Fitzgerald KA, Rock KL, Moore KJ, Wright SD, Hornung V, Latz E. NLRP3 inflammasomes are required for atherogenesis and activated by cholesterol crystals. Nature. 2010;464:1357–61.
75. Mezzaroma E, Toldo S, Farkas D, Seropian IM, Van Tassell BW, Salloum FN, Kannan HR, Menna AC, Voelkel NF, Abbate A. The inflammasome promotes adverse cardiac remodeling following acute myocardial infarction in the mouse. Proc Natl Acad Sci U S A. 2011;108:19725–30.
76. Sandanger O, Ranheim T, Vinge LE, Bliksoen M, Alfsnes K, Finsen AV, Dahl CP, Askevold ET, Florholmen G, Christensen G, Fitzgerald KA, Lien E, Valen G, Espevik T, Aukrust P, Yndestad A. The NLRP3 inflammasome is up-regulated in cardiac fibroblasts and mediates myocardial ischaemia-reperfusion injury. Cardiovasc Res. 2013;99:164–74.
77. Toldo S, Marchetti C, Mauro AG, Chojnacki J, Mezzaroma E, Carbone S, Zhang S, Van Tassell B, Salloum FN, Abbate A. Inhibition of the NLRP3 inflammasome limits the inflammatory injury following myocardial ischemia-reperfusion in the mouse. Int J Cardiol. 2016;209:215–20.
78. Liu Y, Lian K, Zhang L, Wang R, Yi F, Gao C, Xin C, Zhu D, Li Y, Yan W, Xiong L, Gao E, Wang H, Tao L. TXNIP mediates NLRP3 inflammasome activation in cardiac microvascular endothelial cells as a novel mechanism in myocardial ischemia/reperfusion injury. Basic Res Cardiol. 2014;109:415.
79. Fender AC, Kleeschulte S, Stolte S, Leineweber K, Kamler M, Bode J, Li N, Dobrev D. Thrombin receptor PAR4 drives canonical NLRP3 inflammasome signaling in the heart. Basic Res Cardiol. 2020;115:10.
80. Yao C, Veleva T, Scott L Jr, Cao S, Li L, Chen G, Jeyabal P, Pan X, Alsina KM, Abu-Taha ID, Ghezelbash S, Reynolds CL, Shen YH, LeMaire SA, Schmitz W, Muller FU, El-Armouche A, Tony Eissa N, Beeton C, Nattel S, Wehrens XHT, Dobrev D, Li N. Enhanced cardiomyocyte NLRP3 inflammasome signaling promotes atrial fibrillation. Circulation. 2018;138:2227–42.
81. Scott L Jr, Fender AC, Saljic A, Li L, Chen X, Wang X, Linz D, Lang J, Hohl M, Twomey D, Pham TT, Diaz-Lankenau R, Chelu MG, Kamler M, Entman ML, Taffet GE, Sanders P, Dobrev D, Li N. NLRP3 inflammasome is a key driver of obesity-induced atrial arrhythmias. Cardiovasc Res. 2021;117:1746–59.
82. Askevold ET, Gullestad L, Dahl CP, Yndestad A, Ueland T, Aukrust P. Interleukin-6 signaling, soluble glycoprotein 130, and inflammation in heart failure. Curr Heart Fail Rep. 2014;11:146–55.
83. Jennings RB, Murry CE, Steenbergen C Jr, Reimer KA. Development of cell injury in sustained acute ischemia. Circulation. 1990;82:II2–12.

84. Opie LH, Commerford PJ, Gersh BJ, Pfeffer MA. Controversies in ventricular remodelling. Lancet. 2006;367:356–67.
85. Pfeffer MA, Braunwald E. Ventricular remodeling after myocardial infarction. Experimental observations and clinical implications. Circulation. 1990;81:1161–72.
86. Lawrence T. The nuclear factor NF-kappaB pathway in inflammation. Cold Spring Harb Perspect Biol. 2009;1:a001651.
87. Frangogiannis NG. Targeting the transforming growth factor (TGF)-β cascade in the remodeling heart: benefits and perils. J Mol Cell Cardiol. 2014;76:169–71.
88. Kaur K, Dhingra S, Slezak J, Sharma AK, Bajaj A, Singal PK. Biology of TNFα and IL-10, and their imbalance in heart failure. Heart Fail Rev. 2009;14:113–23.
89. Camelliti PL, Borg TK, Kohl P. Structural and functional characterisation of cardiac fibroblasts. Cardiovasc Res. 2005;65(1):40–51.
90. Cheng H, Lederer WJ. Calcium sparks. Physiol Rev. 2008;88:1491–545.
91. He G, Tan W, Wang B, Chen J, Li G, Zhu S, Xie J, Xu B. Increased M1 macrophages infiltration is associated with thrombogenesis in rheumatic mitral stenosis patients with atrial fibrillation. PLoS One. 2016;11:e0149910.
92. Li N, Brundel B. Inflammasomes and proteostasis novel molecular mechanisms associated with atrial fibrillation. Circ Res. 2020;127:73–90.
93. Duncan DJ, Yang Z, Hopkins PM, Steele DS, Harrison SM. TNF-alpha and IL-1beta increase Ca2+ leak from the sarcoplasmic reticulum and susceptibility to arrhythmia in rat ventricular myocytes. Cell Calcium. 2010;47:378–86.
94. Lau DH, Nattel S, Kalman JM, Sanders P. Modifiable risk factors and atrial fibrillation. Circulation. 2017;136:583–96.
95. Pathak RK, Elliott A, Middeldorp ME, Meredith M, Mehta AB, Mahajan R, Hendriks JM, Twomey D, Kalman JM, Abhayaratna WP, Lau DH, Sanders P. Impact of CARDIOrespiratory FITness on arrhythmia recurrence in obese individuals with atrial fibrillation: the CARDIO-FIT study. J Am Coll Cardiol. 2015;66:985–96.
96. Carlisle MA, Fudim M, DeVore AD, Piccini JP. Heart failure and atrial fibrillation, like fire and fury. JACC Heart Fail. 2019;7:447–56.
97. Husti Z, Varro A, Baczko I. Arrhythmogenic remodeling in the failing heart. Cell. 2021;10:3203.
98. Heijman J, Voigt N, Abu-Taha IH, Dobrev D. Rhythm control of atrial fibrillation in heart failure. Heart Fail Clin. 2013;9:407–15, vii–viii.
99. Goudis CA, Korantzopoulos P, Ntalas IV, Kallergis EM, Ketikoglou DG. Obesity and atrial fibrillation: a comprehensive review of the pathophysiological mechanisms and links. J Cardiol. 2015;66:361–9.
100. Guglin M, Maradia K, Chen R, Curtis AB. Relation of obesity to recurrence rate and burden of atrial fibrillation. Am J Cardiol. 2011;107:579–82.
101. Tsang TS, Barnes ME, Miyasaka Y, Cha SS, Bailey KR, Verzosa GC, Seward JB, Gersh BJ. Obesity as a risk factor for the progression of paroxysmal to permanent atrial fibrillation: a longitudinal cohort study of 21 years. Eur Heart J. 2008;29:2227–33.
102. Chung MK, Eckhardt LL, Chen LY, Ahmed HM, Gopinathannair R, Joglar JA, Noseworthy PA, Pack QR, Sanders P, Trulock KM, American Heart Association Electrocardiography and Arrhythmias Committee and Exercise, Cardiac Rehabilitation, and Secondary Prevention Committee of the Council on Clinical Cardiology; Council on Arteriosclerosis, Thrombosis and Vascular Biology; Council on Cardiovascular and Stroke Nursing; and Council on Lifestyle and Cardiometabolic Health. Lifestyle and risk factor modification for reduction of atrial fibrillation: a scientific statement from the American Heart Association. Circulation. 2020;141:e750–e72.
103. Mahajan R, Lau DH, Brooks AG, Shipp NJ, Manavis J, Wood JP, Finnie JW, Samuel CS, Royce SG, Twomey DJ, Thanigaimani S, Kalman JM, Sanders P. Electrophysiological, electroanatomical, and structural remodeling of the atria as consequences of sustained obesity. J Am Coll Cardiol. 2015;66:1–11.

104. Okumura Y, Watanabe I, Nagashima K, Sonoda K, Sasaki N, Kogawa R, Takahashi K, Iso K, Ohkubo K, Nakai T, Takahashi R, Taniguchi Y, Mitsumata M, Nikaido M, Hirayama A. Effects of a high-fat diet on the electrical properties of porcine atria. J Arrhythm. 2015;31:352–8.
105. Abed HS, Samuel CS, Lau DH, Kelly DJ, Royce SG, Alasady M, Mahajan R, Kuklik P, Zhang Y, Brooks AG, Nelson AJ, Worthley SG, Abhayaratna WP, Kalman JM, Wittert GA, Sanders P. Obesity results in progressive atrial structural and electrical remodeling: implications for atrial fibrillation. Heart Rhythm. 2013;10:90–100.
106. Cheng T, Wang XF, Hou YT, Zhang L. Correlation between atrial fibrillation, serum amyloid protein A and other inflammatory cytokines. Mol Med Rep. 2012;6:581–4.
107. Wang H, Yan HM, Tang MX, Wang ZH, Zhong M, Zhang Y, Deng JT, Zhang W. Increased serum levels of microvesicles in nonvalvular atrial fibrillation determinated by ELISA using a specific monoclonal antibody AD-1. Clin Chim Acta. 2010;411:1700–4.
108. Luan Y, Guo Y, Li S, Yu B, Zhu S, Li S, Li N, Tian Z, Peng C, Cheng J, Li Q, Cui J, Tian Y. Interleukin-18 among atrial fibrillation patients in the absence of structural heart disease. Europace. 2010;12:1713–8.
109. Galea R, Cardillo MT, Caroli A, Marini MG, Sonnino C, Narducci ML, Biasucci LM. Inflammation and C-reactive protein in atrial fibrillation: cause or effect? Tex Heart Inst J. 2014;41:461–8.
110. Zhang Y, Wang YT, Shan ZL, Guo HY, Guan Y, Yuan HT. Role of inflammation in the initiation and maintenance of atrial fibrillation and the protective effect of atorvastatin in a goat model of aseptic pericarditis. Mol Med Rep. 2015;11:2615–23.
111. Issac TT, Dokainish H, Lakkis NM. Role of inflammation in initiation and perpetuation of atrial fibrillation: a systematic review of the published data. J Am Coll Cardiol. 2007;50:2021–8.
112. Robblee MM, Kim CC, Porter Abate J, Valdearcos M, Sandlund KL, Shenoy MK, Volmer R, Iwawaki T, Koliwad SK. Saturated fatty acids engage an IRE1alpha-dependent pathway to activate the NLRP3 inflammasome in myeloid cells. Cell Rep. 2016;14:2611–23.
113. Tufanli O, Telkoparan Akillilar P, Acosta-Alvear D, Kocaturk B, Onat UI, Hamid SM, Cimen I, Walter P, Weber C, Erbay E. Targeting IRE1 with small molecules counteracts progression of atherosclerosis. Proc Natl Acad Sci U S A. 2017;114:E1395–E404.
114. Kopp MC, Larburu N, Durairaj V, Adams CJ, Ali MMU. UPR proteins IRE1 and PERK switch BiP from chaperone to ER stress sensor. Nat Struct Mol Biol. 2019;26:1053–62.
115. Gawalko M, Agbaedeng TA, Saljic A, Muller DN, Wilck N, Schnabel R, Penders J, Rienstra M, van Gelder I, Jespersen T, Schotten U, Crijns H, Kalman JM, Sanders P, Nattel S, Dobrev D, Linz D. Gut microbiota, dysbiosis and atrial fibrillation. Arrhythmogenic mechanisms and potential clinical implications. Cardiovasc Res. 2021:cvab292. https://doi.org/10.1093/cvr/cvab292.
116. Linz D, Gawalko M, Sanders P, Penders J, Li N, Nattel S, Dobrev D. Does gut microbiota affect atrial rhythm? Causalities and speculations. Eur Heart J. 2021;42:3521–5.
117. Nguyen BO, Meems LMG, van Faassen M, Crijns H, van Gelder IC, Kuipers F, Rienstra M. Gut-microbe derived TMAO and its association with more progressed forms of AF: results from the AF-RISK study. Int J Cardiol Heart Vasc. 2021;34:100798.
118. Yu L, Meng G, Huang B, Zhou X, Stavrakis S, Wang M, Li X, Zhou L, Wang Y, Wang M, Wang Z, Deng J, Po SS, Jiang H. A potential relationship between gut microbes and atrial fibrillation: trimethylamine N-oxide, a gut microbe-derived metabolite, facilitates the progression of atrial fibrillation. Int J Cardiol. 2018;255:92–8.
119. Aoki K, Teshima Y, Kondo H, Saito S, Fukui A, Fukunaga N, Nawata T, Shimada T, Takahashi N, Shibata H. Role of indoxyl sulfate as a predisposing factor for atrial fibrillation in renal dysfunction. J Am Heart Assoc. 2015;4:e002023.
120. Chen WT, Chen YC, Hsieh MH, Huang SY, Kao YH, Chen YA, Lin YK, Chen SA, Chen YJ. The uremic toxin indoxyl sulfate increases pulmonary vein and atrial arrhythmogenesis. J Cardiovasc Electrophysiol. 2015;26:203–10.

121. Yamagami F, Tajiri K, Doki K, Hattori M, Honda J, Aita S, Harunari T, Yamasaki H, Murakoshi N, Sekiguchi Y, Homma M, Takahashi N, Aonuma K, Nogami A, Ieda M. Indoxyl sulphate is associated with atrial fibrillation recurrence after catheter ablation. Sci Rep. 2018;8:17276.
122. Chen YY, Sun ZW, Jiang JP, Kang XD, Wang LL, Shen YL, Xie XD, Zheng LR. α-Adrenoceptor-mediated enhanced inducibility of atrial fibrillation in a canine system inflammation model. Mol Med Rep. 2017;15:3767–74.
123. Okazaki R, Iwasaki YK, Miyauchi Y, Hirayama Y, Kobayashi Y, Katoh T, Mizuno K, Sekiguchi A, Yamashita T. Lipopolysaccharide induces atrial arrhythmogenesis via down-regulation of L-type Ca2+ channel genes in rats. Int Heart J. 2009;50:353–63.
124. Ganesh BP, Nelson JW, Eskew JR, Ganesan A, Ajami NJ, Petrosino JF, Bryan RM Jr, Durgan DJ. Prebiotics, probiotics, and acetate supplementation prevent hypertension in a model of obstructive sleep apnea. Hypertension. 2018;72:1141–50.
125. Shi H, Zhang B, Abo-Hamzy T, Nelson JW, Ambati CSR, Petrosino JF, Bryan RM Jr, Durgan DJ. Restructuring the gut microbiota by intermittent fasting lowers blood pressure. Circ Res. 2021;128:1240–54.
126. Bowerman KL, Rehman SF, Vaughan A, Lachner N, Budden KF, Kim RY, Wood DLA, Gellatly SL, Shukla SD, Wood LG, Yang IA, Wark PA, Hugenholtz P, Hansbro PM. Disease-associated gut microbiome and metabolome changes in patients with chronic obstructive pulmonary disease. Nat Commun. 2020;11:5886.
127. Jonsson AL, Backhed F. Role of gut microbiota in atherosclerosis. Nat Rev Cardiol. 2017;14:79–87.
128. Marques FZ, Mackay CR, Kaye DM. Beyond gut feelings: how the gut microbiota regulates blood pressure. Nat Rev Cardiol. 2018;15:20–32.
129. Maruvada P, Leone V, Kaplan LM, Chang EB. The human microbiome and obesity: moving beyond associations. Cell Host Microbe. 2017;22:589–99.
130. Tang WHW, Li DY, Hazen SL. Dietary metabolism, the gut microbiome, and heart failure. Nat Rev Cardiol. 2019;16:137–54.
131. Yang G, Wei J, Liu P, Zhang Q, Tian Y, Hou G, Meng L, Xin Y, Jiang X. Role of the gut microbiota in type 2 diabetes and related diseases. Metabolism. 2021;117:154712.
132. Li X, Geng J, Zhao J, Ni Q, Zhao C, Zheng Y, Chen X, Wang L. Trimethylamine N-oxide exacerbates cardiac fibrosis via activating the NLRP3 inflammasome. Front Physiol. 2019;10:866.
133. Zhang Y, Zhang S, Li B, Luo Y, Gong Y, Jin X, Zhang J, Zhou Y, Zhuo X, Wang Z, Zhao X, Han X, Gao Y, Yu H, Liang D, Zhao S, Sun D, Wang D, Xu W, Qu G, Bo W, Li D, Wu Y, Li Y. Gut microbiota dysbiosis promotes age-related atrial fibrillation by lipopolysaccharide and glucose-induced activation of NLRP3-inflammasome. Cardiovasc Res. 2022;118:785–97.
134. Alonso A, Lopez FL, Matsushita K, Loehr LR, Agarwal SK, Chen LY, Soliman EZ, Astor BC, Coresh J. Chronic kidney disease is associated with the incidence of atrial fibrillation: the Atherosclerosis Risk in Communities (ARIC) study. Circulation. 2011;123:2946–53.
135. Levey AS, Coresh J, Balk E, Kausz AT, Levin A, Steffes MW, Hogg RJ, Perrone RD, Lau J, Eknoyan G, National Kidney Foundation. National Kidney Foundation practice guidelines for chronic kidney disease: evaluation, classification, and stratification. Ann Intern Med. 2003;139:137–47.
136. Stamellou E, Floege J. Novel oral anticoagulants in patients with chronic kidney disease and atrial fibrillation. Nephrol Dial Transplant. 2018;33:1683–9.
137. Naser N, Dilic M, Durak A, Kulic M, Pepic E, Smajic E, Kusljugic Z. The impact of risk factors and comorbidities on the incidence of atrial fibrillation. Mater Sociomed. 2017;29:231–6.
138. Mert KU, Mert GO, Basaran O, Beton O, Dogan V, Tekinalp M, Aykan AC, Kalaycioglu E, Bolat I, Tasar O, Safak O, Kalcik M, Yaman M, Kirma C, Biteker M, Investigators R. Real-world stroke prevention strategies in nonvalvular atrial fibrillation in patients with renal impairment. Eur J Clin Investig. 2017;47:428–38.
139. Reinecke H, Brand E, Mesters R, Schabitz WR, Fisher M, Pavenstadt H, Breithardt G. Dilemmas in the management of atrial fibrillation in chronic kidney disease. J Am Soc Nephrol. 2009;20:705–11.

140. Li N, Dobrev D. Hyperuricemia: a causal player or a bystander linking inflammatory signaling and atrial fibrillation? Int J Cardiol. 2017;231:177–8.
141. Gicquel T, Robert S, Loyer P, Victoni T, Bodin A, Ribault C, Gleonnec F, Couillin I, Boichot E, Lagente V. IL-1beta production is dependent on the activation of purinergic receptors and NLRP3 pathway in human macrophages. FASEB J. 2015;29:4162–73.
142. Du J, Xie J, Zhang Z, Tsujikawa H, Fusco D, Silverman D, Liang B, Yue L. TRPM7-mediated Ca2+ signals confer fibrogenesis in human atrial fibrillation. Circ Res. 2010;106:992–1003.
143. Maharani N, Ting YK, Cheng J, Hasegawa A, Kurata Y, Li P, Nakayama Y, Ninomiya H, Ikeda N, Morikawa K, Yamamoto K, Makita N, Yamashita T, Shirayoshi Y, Hisatome I. Molecular mechanisms underlying urate-induced enhancement of Kv1.5 channel expression in HL-1 atrial myocytes. Circ J. 2015;79:2659–68.
144. Suetomi T, Willeford A, Brand CS, Cho Y, Ross RS, Miyamoto S, Brown JH. Inflammation and NLRP3 inflammasome activation initiated in response to pressure overload by Ca(2+)/calmodulin-dependent protein kinase II delta signaling in cardiomyocytes are essential for adverse cardiac remodeling. Circulation. 2018;138:2530–44.
145. Cruz CM, Rinna A, Forman HJ, Ventura AL, Persechini PM, Ojcius DM. ATP activates a reactive oxygen species-dependent oxidative stress response and secretion of proinflammatory cytokines in macrophages. J Biol Chem. 2007;282:2871–9.
146. Panday A, Sahoo MK, Osorio D, Batra S. NADPH oxidases: an overview from structure to innate immunity-associated pathologies. Cell Mol Immunol. 2015;12:5–23.
147. Tobimatsu T, Fujisawa H. Tissue-specific expression of four types of rat calmodulin-dependent protein kinase II mRNAs. J Biol Chem. 1989;264:17907–12.
148. Kreusser MM, Lehmann LH, Keranov S, Hoting MO, Oehl U, Kohlhaas M, Reil JC, Neumann K, Schneider MD, Hill JA, Dobrev D, Maack C, Maier LS, Grone HJ, Katus HA, Olson EN, Backs J. Cardiac CaM kinase II genes delta and gamma contribute to adverse remodeling but redundantly inhibit calcineurin-induced myocardial hypertrophy. Circulation. 2014;130:1262–73.
149. Hudmon A, Schulman H. Structure-function of the multifunctional Ca2+/calmodulin-dependent protein kinase II. Biochem J. 2002;364:593–611.
150. Mesubi OO, Anderson ME. Atrial remodelling in atrial fibrillation: CaMKII as a nodal proarrhythmic signal. Cardiovasc Res. 2016;109:542–57.
151. Erickson JR, Joiner ML, Guan X, Kutschke W, Yang J, Oddis CV, Bartlett RK, Lowe JS, O'Donnell SE, Aykin-Burns N, Zimmerman MC, Zimmerman K, Ham AJ, Weiss RM, Spitz DR, Shea MA, Colbran RJ, Mohler PJ, Anderson ME. A dynamic pathway for calcium-independent activation of CaMKII by methionine oxidation. Cell. 2008;133:462–74.
152. Erickson JR, Pereira L, Wang L, Han G, Ferguson A, Dao K, Copeland RJ, Despa F, Hart GW, Ripplinger CM, Bers DM. Diabetic hyperglycaemia activates CaMKII and arrhythmias by O-linked glycosylation. Nature. 2013;502:372–6.
153. Gutierrez DA, Fernandez-Tenorio M, Ogrodnik J, Niggli E. NO-dependent CaMKII activation during beta-adrenergic stimulation of cardiac muscle. Cardiovasc Res. 2013;100:392–401.
154. Anderson ME, Braun AP, Schulman H, Premack BA. Multifunctional Ca2+/calmodulin-dependent protein kinase mediates Ca(2+)-induced enhancement of the L-type Ca2+ current in rabbit ventricular myocytes. Circ Res. 1994;75:854–61.
155. El-Haou S, Balse E, Neyroud N, Dilanian G, Gavillet B, Abriel H, Coulombe A, Jeromin A, Hatem SN. Kv4 potassium channels form a tripartite complex with the anchoring protein SAP97 and CaMKII in cardiac myocytes. Circ Res. 2009;104:758–69.
156. Kranias EG. Regulation of Ca2+ transport by cyclic 3′,5′-AMP-dependent and calcium-calmodulin-dependent phosphorylation of cardiac sarcoplasmic reticulum. Biochim Biophys Acta. 1985;844:193–9.
157. Wagner S, Dybkova N, Rasenack EC, Jacobshagen C, Fabritz L, Kirchhof P, Maier SK, Zhang T, Hasenfuss G, Brown JH, Bers DM, Maier LS. Ca2+/calmodulin-dependent protein kinase II regulates cardiac Na+ channels. J Clin Invest. 2006;116:3127–38.

158. Wehrens XH, Lehnart SE, Reiken SR, Marks AR. Ca2+/calmodulin-dependent protein kinase II phosphorylation regulates the cardiac ryanodine receptor. Circ Res. 2004;94:e61–70.
159. Tessier S, Karczewski P, Krause EG, Pansard Y, Acar C, Lang-Lazdunski M, Mercadier JJ, Hatem SN. Regulation of the transient outward K(+) current by Ca(2+)/calmodulin-dependent protein kinases II in human atrial myocytes. Circ Res. 1999;85:810–9.
160. Chelu MG, Sarma S, Sood S, Wang S, van Oort RJ, Skapura DG, Li N, Santonastasi M, Muller FU, Schmitz W, Schotten U, Anderson ME, Valderrabano M, Dobrev D, Wehrens XH. Calmodulin kinase II-mediated sarcoplasmic reticulum Ca2+ leak promotes atrial fibrillation in mice. J Clin Invest. 2009;119:1940–51.
161. Li N, Wang T, Wang W, Cutler MJ, Wang Q, Voigt N, Rosenbaum DS, Dobrev D, Wehrens XH. Inhibition of CaMKII phosphorylation of RyR2 prevents induction of atrial fibrillation in FKBP12.6 knockout mice. Circ Res. 2012;110:465–70.
162. Beckendorf J, van den Hoogenhof MMG, Backs J. Physiological and unappreciated roles of CaMKII in the heart. Basic Res Cardiol. 2018;113:29.
163. Gray CB, Suetomi T, Xiang S, Mishra S, Blackwood EA, Glembotski CC, Miyamoto S, Westenbrink BD, Brown JH. CaMKIIdelta subtypes differentially regulate infarct formation following ex vivo myocardial ischemia/reperfusion through NF-kappaB and TNF-alpha. J Mol Cell Cardiol. 2017;103:48–55.
164. Ling H, Gray CB, Zambon AC, Grimm M, Gu Y, Dalton N, Purcell NH, Peterson K, Brown JH. Ca2+/calmodulin-dependent protein kinase II delta mediates myocardial ischemia/reperfusion injury through nuclear factor-kappaB. Circ Res. 2013;112:935–44.
165. Willeford A, Suetomi T, Nickle A, Hoffman HM, Miyamoto S, Heller Brown J. CaMKIIdelta-mediated inflammatory gene expression and inflammasome activation in cardiomyocytes initiate inflammation and induce fibrosis. JCI Insight. 2018;3:e97054.
166. Joiner ML, Koval OM, Li J, He BJ, Allamargot C, Gao Z, Luczak ED, Hall DD, Fink BD, Chen B, Yang J, Moore SA, Scholz TD, Strack S, Mohler PJ, Sivitz WI, Song LS, Anderson ME. CaMKII determines mitochondrial stress responses in heart. Nature. 2012;491:269–73.
167. Antoniades C, Demosthenous M, Reilly S, Margaritis M, Zhang MH, Antonopoulos A, Marinou K, Nahar K, Jayaram R, Tousoulis D, Bakogiannis C, Sayeed R, Triantafyllou C, Koumallos N, Psarros C, Miliou A, Stefanadis C, Channon KM, Casadei B. Myocardial redox state predicts in-hospital clinical outcome after cardiac surgery effects of short-term pre-operative statin treatment. J Am Coll Cardiol. 2012;59:60–70.
168. Dudley SC Jr, Hoch NE, McCann LA, Honeycutt C, Diamandopoulos L, Fukai T, Harrison DG, Dikalov SI, Langberg J. Atrial fibrillation increases production of superoxide by the left atrium and left atrial appendage: role of the NADPH and xanthine oxidases. Circulation. 2005;112:1266–73.
169. Kim YM, Guzik TJ, Zhang YH, Zhang MH, Kattach H, Ratnatunga C, Pillai R, Channon KM, Casadei B. A myocardial Nox2 containing NAD(P)H oxidase contributes to oxidative stress in human atrial fibrillation. Circ Res. 2005;97:629–36.
170. Simon JN, Ziberna K, Casadei B. Compromised redox homeostasis, altered nitroso-redox balance, and therapeutic possibilities in atrial fibrillation. Cardiovasc Res. 2016;109:510–8.
171. Dobrev D, Dudley SC. Oxidative stress: a bystander or a causal contributor to atrial remodelling and fibrillation? Cardiovasc Res. 2021;117:2291–3.
172. Mighiu AS, Recalde A, Ziberna K, Carnicer R, Tomek J, Bub G, Brewer AC, Verheule S, Shah AM, Simon JN, Casadei B. Inducibility, but not stability, of atrial fibrillation is increased by NOX2 overexpression in mice. Cardiovasc Res. 2021;117:2354–64.
173. Yoo S, Pfenniger A, Hoffman J, Zhang W, Ng J, Burrell A, Johnson DA, Gussak G, Waugh T, Bull S, Benefield B, Knight BP, Passman R, Wasserstrom JA, Aistrup GL, Arora R. Attenuation of oxidative injury with targeted expression of NADPH oxidase 2 short hairpin RNA prevents onset and maintenance of electrical remodeling in the canine atrium: a novel gene therapy approach to atrial fibrillation. Circulation. 2020;142:1261–78.
174. Dostert C, Petrilli V, Van Bruggen R, Steele C, Mossman BT, Tschopp J. Innate immune activation through Nalp3 inflammasome sensing of asbestos and silica. Science. 2008;320:674–7.

175. Bae JY, Park HH. Crystal structure of NALP3 protein pyrin domain (PYD) and its implications in inflammasome assembly. J Biol Chem. 2011;286:39528–36.
176. Yoo S, Aistrup G, Shiferaw Y, Ng J, Mohler PJ, Hund TJ, Waugh T, Browne S, Gussak G, Gilani M, Knight BP, Passman R, Goldberger JJ, Wasserstrom JA, Arora R. Oxidative stress creates a unique, CaMKII-mediated substrate for atrial fibrillation in heart failure. JCI Insight. 2018;3:e120728.
177. Purohit A, Rokita AG, Guan X, Chen B, Koval OM, Voigt N, Neef S, Sowa T, Gao Z, Luczak ED, Stefansdottir H, Behunin AC, Li N, El-Accaoui RN, Yang B, Swaminathan PD, Weiss RM, Wehrens XH, Song LS, Dobrev D, Maier LS, Anderson ME. Oxidized Ca(2+)/calmodulin-dependent protein kinase II triggers atrial fibrillation. Circulation. 2013;128:1748–57.
178. Xie W, Santulli G, Reiken SR, Yuan Q, Osborne BW, Chen BX, Marks AR. Mitochondrial oxidative stress promotes atrial fibrillation. Sci Rep. 2015;5:11427.
179. Yang X, An N, Zhong C, Guan M, Jiang Y, Li X, Zhang H, Wang L, Ruan Y, Gao Y, Liu N, Shang H, Xing Y. Enhanced cardiomyocyte reactive oxygen species signaling promotes ibrutinib-induced atrial fibrillation. Redox Biol. 2020;30:101432.
180. Murakami T, Ockinger J, Yu J, Byles V, McColl A, Hofer AM, Horng T. Critical role for calcium mobilization in activation of the NLRP3 inflammasome. Proc Natl Acad Sci. 2012;109:11282.
181. Lee G-S, Subramanian N, Kim AI, Aksentijevich I, Goldbach-Mansky R, Sacks DB, Germain RN, Kastner DL, Chae JJ. The calcium-sensing receptor regulates the NLRP3 inflammasome through Ca2+ and cAMP. Nature. 2012;492:123–7.
182. Menu P, Mayor A, Zhou R, Tardivel A, Ichijo H, Mori K, Tschopp J. ER stress activates the NLRP3 inflammasome via an UPR-independent pathway. Cell Death Dis. 2012;3:e261.
183. Horng T. Calcium signaling and mitochondrial destabilization in the triggering of the NLRP3 inflammasome. Trends Immunol. 2014;35:253–61.
184. Liu Q, Zhang D, Hu D, Zhou X, Zhou Y. The role of mitochondria in NLRP3 inflammasome activation. Mol Immunol. 2018;103:115–24.
185. Yared JP, Bakri MH, Erzurum SC, Moravec CS, Laskowski DM, Van Wagoner DR, Mascha E, Thornton J. Effect of dexamethasone on atrial fibrillation after cardiac surgery: prospective, randomized, double-blind, placebo-controlled trial. J Cardiothorac Vasc Anesth. 2007;21:68–75.
186. Iskandar S, Reddy M, Afzal MR, Rajasingh J, Atoui M, Lavu M, Atkins D, Bommana S, Umbarger L, Jaeger M, Pimentel R, Dendi R, Emert M, Turagam M, Di Biase L, Natale A, Lakkireddy D. Use of oral steroid and its effects on atrial fibrillation recurrence and inflammatory cytokines post ablation – the steroid AF study. J Atr Fibrillation. 2017;9:1604.
187. Coll RC, Robertson AA, Chae JJ, Higgins SC, Munoz-Planillo R, Inserra MC, Vetter I, Dungan LS, Monks BG, Stutz A, Croker DE, Butler MS, Haneklaus M, Sutton CE, Nunez G, Latz E, Kastner DL, Mills KH, Masters SL, Schroder K, Cooper MA, O'Neill LA. A small-molecule inhibitor of the NLRP3 inflammasome for the treatment of inflammatory diseases. Nat Med. 2015;21:248–55.
188. Aliaga J, Bonaventura A, Mezzaroma E, Dhakal Y, Mauro AG, Abbate A, Toldo S. Preservation of contractile reserve and diastolic function by inhibiting the NLRP3 inflammasome with OLT1177((R)) (dapansutrile) in a mouse model of severe ischemic cardiomyopathy due to non-reperfused anterior wall myocardial infarction. Molecules. 2021;26:3534.
189. Elsayed MS, Abu-Elsaad NM, Nader MA. The NLRP3 inhibitor dapansutrile attenuates folic acid induced nephrotoxicity via inhibiting inflammasome/caspase-1/IL axis and regulating autophagy/proliferation. Life Sci. 2021;285:119974.

Myocardial Fibrosis: Cell Signaling and In Vitro Modeling

Caleb Graham and Palaniappan Sethu

Abstract Heart disease is the leading cause of death in the US and worldwide. In addition to alterations in cardiomyocyte structure and cellular composition, the adverse ventricular remodeling seen post-myocardial infarction, in various types of cardiomyopathies, and in response to pressure and volume overload usually includes myocardial fibrosis. This condition is characterized by the deposition of excess extracellular matrix (ECM) proteins and often begins as an adaptive response to injury or hemodynamic stress, but then persists and transitions into a pathological process. Presence and amount of fibrosis have proven to be reliable negative prognostic indicators in the setting of heart failure, and there is mounting evidence to suggest that fibrosis directly worsens disease outcomes. While 2D cell monoculture experiments have proven invaluable in understanding some of the triggers and signaling dynamics involved in the development of myocardial fibrosis and its downstream effects, these disease models do not accurately recapitulate pathophysiological remodeling seen in the in vivo setting. Despite advances in the development of 3D cell culture and tissue engineering techniques over the past few decades, as well as the discovery and utilization of induced pluripotent stem cells, this complex pathological process has proven difficult to faithfully model in vitro. This complex and dynamic process of fibrotic remodeling relies on a multitude of cellular and extracellular signals originating from within the myocardium and from systemic interactions with the immune and endocrine systems. The goal of this chapter is to provide an understanding of fibrosis in the context of the myocardium, detail efforts to model myocardial fibrosis in vitro, discuss limitations of available in vitro models and, finally, highlight how in vitro models have made critical contributions to our understanding of this condition by elucidating cell-cell and cell-ECM interactions and signaling pathways involved in its development.

C. Graham (✉) · P. Sethu
Division of Cardiovascular Disease, University of Alabama at Birmingham, Birmingham, AL, USA

Department of Biomedical Engineering, University of Alabama at Birmingham, Birmingham, AL, USA
e-mail: cgraha5@uab.edu; psethu@uabmc.edu

N. L. Parinandi, T. J. Hund (eds.), *Cardiovascular Signaling in Health and Disease*, https://doi.org/10.1007/978-3-031-08309-9_10

Keywords Myocardial Fibrosis · Cardiac Fibrosis · Heart · Tissue Engineering · Myofibroblast · Cell Culture · In Vitro Models · Disease Modeling · Tissue Chip

Abbreviations

AngII	Angiotensin II
AT1R	Angiotensin II Type 1 Receptor
AV	Atrioventricular
CF	Cardiac Fibroblast
CM	Cardiomyocyte
CTGF	Connective Tissue Growth Factor
DDR2	Discoidin Domain Receptor-2
EC	Endothelial Cell
ECM	Extracellular Matrix
ERK	Extracellular Signal-Regulated Kinase
ESC	Embryonic Stem Cell
ET-1	Endothelin-1
FAK	Focal Adhesion Kinase
FB	Fibroblast
GelMA	Gelatin Methacrylate
HF	Heart Failure
HSKMs	Human Skeletal Muscle Myotubes
IBZ	Infarct Border Zone
IGF1	Insulin-Like Growth Factor 1
IL	Interleukin
iPSC	Induced Pluripotent Stem Cell
JNK	c-JUN N-Terminal Kinase
LAD	Left Anterior Descending
LV	Left Ventricle
LVH	Left Ventricular Hypertrophy
MAPK	Mitogen-activated Protein Kinase
MI	Myocardial Infarction
MMP	Matrix Metalloproteinase
MSC	Mesenchymal Stem Cell
MVEC	Microvascular Endothelial Cell
NO	Nitric Oxide
PC	Pericyte
PDMS	Polydimethylsiloxane
PEGDA	Poly(ethylene glycol) Diacrylate
ROCK	Rho-kinase
SA	Sinoatrial
SMA	Alpha-Smooth Muscle Actin
SMC	Smooth Muscle Cell

TAC	Transaortic Constriction
TGF-β	Transforming Growth Factor-β
TIMP	Tissue Inhibitor of Metalloproteinase
TNF	Tumor Necrosis Factor
VEGF	Vascular Endothelial Growth Factor

Introduction

Heart disease accounted for over a fifth of deaths in the United States in 2018 [1] and cardiac issues specifically related to ischemia accounted for 16% of mortality worldwide in 2019 [2]. In total, heart conditions lead to $219.6 billion in direct and indirect costs from 2016 to 2017 in the US, almost 8% of the country's total healthcare expenditures in that time period [1]. While mortality rates from acute myocardial infarction (MI) are lower than ever [3] due to advances in percutaneous intervention and medicinal regimens, past MI is the leading cause of heart failure (HF) [4]. Following MI and during the course of many types of cardiovascular disease, the heart (particularly the left ventricle) undergoes adaptive remodeling to maintain myocardial function and structural integrity in response to increased blood pressure, volume, neurohormonal activation, inflammation, and/or cardiomyocyte (CM) death [5]. This remodeling, which is compensatory initially, involves changes in cellular and interstitial composition and gene expression, which lead to alterations in the size, shape, and function of the organ. Two hallmarks of cardiac tissue remodeling, scar formation (in the case of infarction) and CM hypertrophy, are consequences of the heart muscle cells' extremely limited proliferative capacity [6]. Though CMs can adapt to increased hemodynamic loads via hypertrophy and enhancement of their contractile capabilities through the addition of sarcomeres, these changes are frequently accompanied by further modifications to the myocardium that augment its structural integrity. Aside from reparative fibrosis, which is responsible for scar formation at sites of focal cardiomyocyte necrosis, other patterns of fibrosis, broadly defined as the deposition of excess extracellular matrix (ECM) [7], also frequently develop during cardiac remodeling [8]. While supplemental ECM material, which predominantly consists of cross-linked collagen types I and III [9], is necessary following infarction to prevent ventricular wall rupture and could serve a mechanistic compensatory role in cases of hemodynamic overload [10, 11], the fibrotic remodeling caused by prolonged dysregulation of ECM deposition and degradation throws the highly organized architecture of the myocardium into disarray, replacing contractile myocardium with noncontractile scar, and its presence has been associated with decreased heart function [12, 13], adverse outcomes in a variety of cardiac conditions [14–17], and a predisposition toward arrhythmia [18].

Documented observations of myocardial fibrosis date back at least as far as the turn of the twentieth century [19, 20], and its association with several different cardiac and systemic pathologies, such as myocardial ischemia [21, 22] or infarction

[23–25], pressure overload due to outflow issues [26–28] or hypertension [29], volume overload due to aortic regurgitation [28], HF [30, 31], cardiomyopathies [32–36], diabetes [37–39], different forms of myocarditis [40–42], connective tissue disorders [43], aging [44], and drug [45, 46] and radiation toxicity [47, 48], had been observed through animal experiments and human observational and postmortem studies by the 1980s. As suggested by its association with a wide range of conditions, myocardial fibrosis is a dynamic and multifactorial process involving multiple cell types (both resident and circulating populations), ECM remodeling, and numerous cell-signaling pathways mediated through direct cell-cell and cell-ECM interactions, as well as via autocrine, paracrine, and endocrine mechanisms. This chapter will highlight the main signaling pathways, cells, and ECM factors involved in the development of myocardial fibrosis, examine the difficulties involved in modeling this condition, and discuss in vitro, tissue engineering-based studies that have furthered our understanding of this process and have helped identify and assess the safety and effectiveness of anti-fibrotic therapeutics.

The Normal Myocardium

Though the heart is composed of three tissue layers, the intermediate muscular myocardial layer is responsible for generating the cyclic contraction and relaxation that propels blood through the circulatory system. During the diastolic (filling) phase of the cardiac cycle, the heart chambers relax and fill with blood, with the increased pressure in the atria leading to opening of the atrioventricular (AV) valves and blood flow into the ventricles. An action potential from the sinoatrial (SA) node then triggers the atria to contract simultaneously, further filling the ventricles with blood. During systole (contraction), the electrical signal originating in the SA node makes its way through the atrioventricular (AV) node, the bundle of His, and the Purkinje fibers, which triggers coordinated myocardial contraction. The ventricles (particularly the left) perform significantly more contractile work than the atria and have much thicker myocardial tissue. Unsurprisingly, the cellular composition of the two chamber types differ significantly: in the adult human, atria are comprised of 30.1% cardiomyocytes (CMs), 24.3% cardiac fibroblasts (CFs), 17.1% mural cells, such as pericytes (PCs) and smooth muscle cells (SMCs), 12.2% endothelial cells (ECs), and 10.4% immune cells, whereas the ventricles consist of 49.2% CMs, 21.2% mural cells, 15.5% CFs, 7.8% ECs, and 5.3% immune cells [49]. These cells do not perform their functions in isolation, but rather communicate with themselves and surrounding cells via autocrine and paracrine mechanisms and direct cell-cell contacts to maintain homeostasis [50]. Cells can also signal to one another by remodeling or exerting tension upon the surrounding ECM and secreting dormant signaling mediators into the matrix. The dynamic myocardial ECM is composed primarily of fibrillar collagens I and III [9], as well as non-fibrillary components like proteoglycans, glycoproteins (e.g. laminin, fibronectin, tenascins, thrombospondins), and glycosaminoglycans [51, 52]. This matrix surrounds and has robust connections with the cells of the myocardium, serving numerous purposes, the most important of which include providing strength and elasticity to the heart and acting as a plastic scaffold that cells can attach to, migrate along, and remodel [53].

Cardiomyocytes are striated contractile cells that form a functional syncytium, propagating action potentials through gap junctions composed of connexins [54] and exerting force on neighboring CMs and ECM through intercalated disks, adherens junctions, desmosomes, and costameres, which are associated with integrins and focal adhesions [55, 56]. CMs have also been shown to exert paracrine signaling through secretion of molecules such as vascular endothelial growth factors (VEGFs), fibroblast growth factors (FGFs), and natriuretic peptides [57, 58], and can also generate a variety of pro-fibrotic and inflammatory mediators [59].

Cardiac fibroblasts are mesenchymal cells that reside both within the interstitial ECM as well as in close proximity to other resident myocardial cells, and during homeostasis they are responsible for ECM maintenance, replenishing degraded collagen [59, 60]. Through their activation and differentiation into myofibroblasts (myoFBs), these become the principal effector cells of myocardial fibrosis.

Endothelial cells line the lumens of the vasculature within the myocardium, protecting the tissue from clot-induced ischemia or infarction by shielding the underlying collagen from the bloodstream and secreting anti-thrombotic factors [61]. ECs can also influence CMs and other neighboring cells through the secretion of numerous soluble mediators, such as nitric oxide (NO), endothelin-1 (ET-1), growth factors, and angiotensin II (AngII) [50]. They also express cell surface receptors integral for white blood cell recruitment during inflammation [62].

Regarding the mural cells, SMCs surround the ECs of the coronary arteries to the downstream arterioles, allowing for local control over blood supply through vasoconstriction and dilation [63]. PCs, on the other hand, are found around capillary ECs, where they contribute to the barrier function of the vessel [64] and are involved in angiogenesis [65].

Resident cardiac macrophages are thought to play roles in immune homeostasis and clearance of apoptotic cells during homeostasis [66], but are mostly replaced by circulating monocytes/macrophages in inflammatory scenarios [6]. Little is known regarding the function of resident lymphocytes or mast cells during normal state homeostasis.

Myocardial Fibrosis

Though the known triggers of myocardial fibrosis exhibit several distinct differences in pathophysiology, the resulting fibrotic remodeling appears to be mediated through the same cell populations and similar signaling pathways [6]. MyoFBs are a contractile, secretory cell type that are highly active in ECM remodeling through deposition of structural and nonstructural proteins and secretion of matrix metalloproteinases (MMPs), tissue inhibitor of metalloproteinases (TIMPs), and pro-inflammatory and pro-fibrotic soluble mediators [9] (Fig. 1). Their presence is

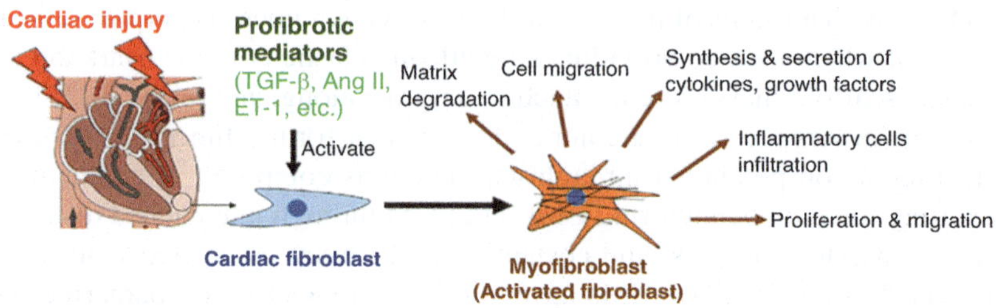

Fig. 1 The development of myocardial fibrosis. Following an insult to the heart, CFs are activated and differentiated into myoFBs by a variety of mediators. MyoFBs are highly active in ECM remodeling, synthesis of pro-fibrotic mediators, and recruitment of immune cells. (Reproduced with permission from [67]. This work was published under a CC BY license (Creative Commons Attribution 4.0 International License; https://creativecommons.org/licenses/by/4.0/))

required for the development of fibrosis [68], and though some have suggested the contribution of various cell types to the myoFB pool (reviewed in [69]), recent studies have reported they are mostly derived from activated CFs [68, 70]. Quiescent CFs do not form actin-associated cell-cell or cell-ECM adhesion complexes [71], but upon activation these cells begin to exhibit actin stress fibers and form associated adhesion complexes to allow for migration [72]. Activated CFs also undergo rapid proliferation, and once they reach their destination they begin to secrete substantial amounts of ECM proteins such as collagens, laminin, and fibronectin, as well as express specific cadherins, integrins, and the contractile protein alpha-smooth muscle actin (SMA), through which they exert contractile forces on surrounding cells and ECM [73–75].

Due to their central importance in myocardial fibrosis, understanding the signaling leading to CF activation and differentiation into myoFBs as well as that involved in their pro-fibrotic activities is important (Fig. 2). As this chapter will only focus on a handful of mediators and pathways, please see [59, 76, 77] for recent reviews. Transforming growth factor-β1 (TGF- β1) has been shown to be a potent activator of CFs in animals [78], human hearts [79], and in vitro [80]. In the canonical pathway, TGF- β1 binds to Type II serine/threonine kinase receptors, which transphosphorylate the kinase domain of the Type I receptor [77]. This leads to activation of SMAD2/3, which form a complex with SMAD4 and translocate to the nucleus, where the complex upregulates transcription of collagens (type I and III, predominantly), TIMPs, TGF-β1, integrins, and the pro-fibrotic matricellular protein connective tissue growth factor (CTGF) [9, 59]. After TGF-β1 expression is induced in activated CFs, it serves as an important pro-fibrotic paracrine and autocrine mediator. There are numerous sources of TGF-β1 in myocardial fibrosis, including many other resident myocardial cells, inflammatory cells recruited from the circulation, and latent stores in the ECM [78] which can be activated by cell binding through αvβ5 integrin [81].

Angiotensin II also activates CFs, both directly and indirectly. CFs and myoFBs express AngII type 1 receptor (AT1R), through which AngII mediates increased expression of fibronectin, collagens, and TGF-β1 [82, 83]. AngII has been shown in

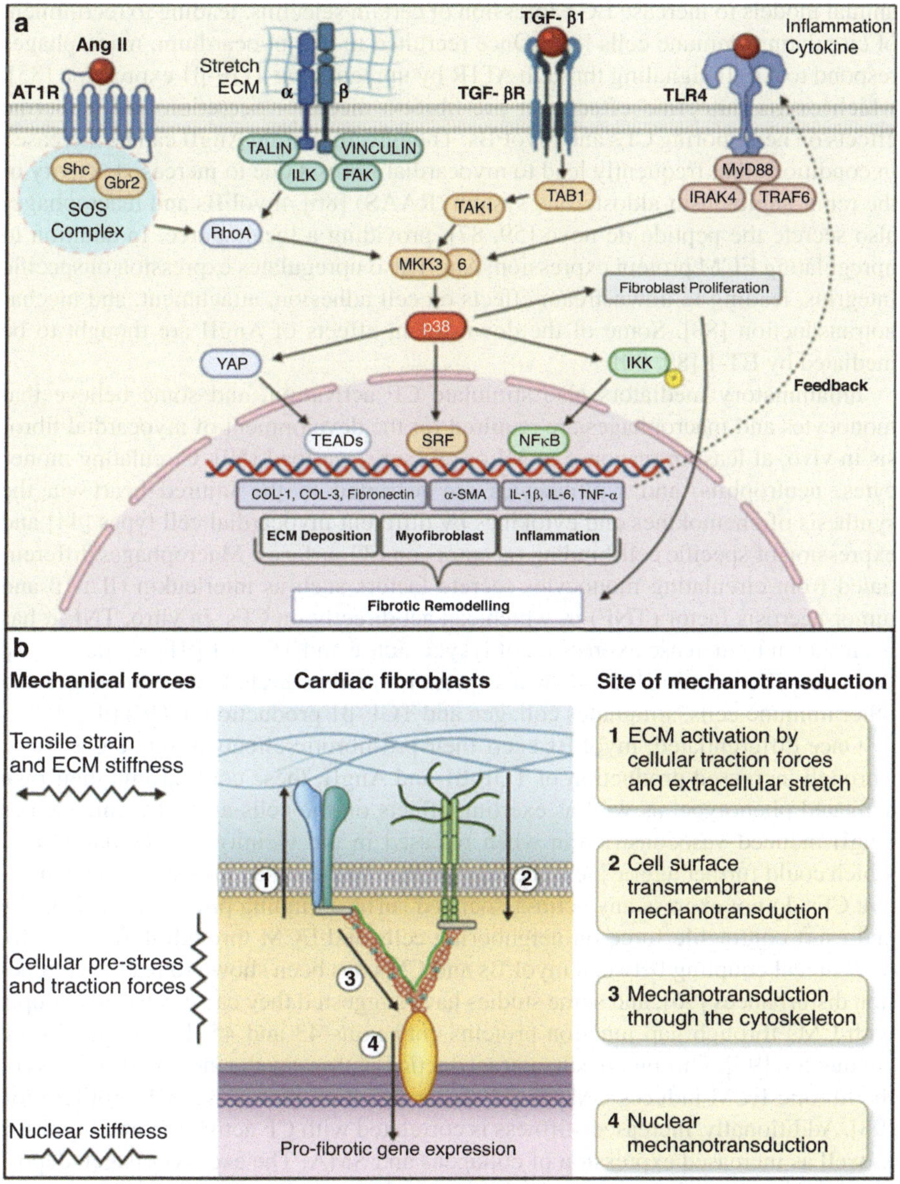

Fig. 2 (**a**) Summary of pro-fibrotic signal transduction mediated by AngII, mechanotransduction, TGF-β1, and inflammatory cytokines leading to activation of CFs and differentiation into myoFBs, expression of ECM components, contractile proteins, and pro-inflammatory signals. (Reproduced with permission from [77]. This work was published under a CC BY license (Creative Commons Attribution 4.0 International License; https://creativecommons.org/licenses/by/4.0/)). (**b**) Mechanotransductive signaling in CF. Steps 2–4 highlight ECM-to-cell signaling through cell surface transmembrane receptors, leading to downstream effects on the cytoskeleton and, eventually gene expression. Step 1 demonstrates cell-to-ECM signaling. (Reproduced with permission from [95]. This work was published under a CC BY license (Creative Commons Attribution 4.0 International License; https://creativecommons.org/licenses/by/4.0/))

animal models to increase EC expression of certain selectins, leading to recruitment of circulating immune cells [84]. Once recruited to the myocardium, macrophages respond to AngII signaling through AT1R by upregulating TGF-β1 expression [85], which exerts autocrine effects on pro-fibrotic mediator secretion and paracrine effects on neighboring CFs and myoFBs. Though circulating AngII can be increased in conditions that frequently lead to myocardial fibrosis due to increased activity of the renin-angiotensin-aldosterone system (RAAS) [86], myoFBs and macrophages also secrete the peptide de novo [59, 87], providing a local source. In addition to upregulating ECM protein expression, AngII also upregulates expression of specific integrins, leading to downstream effects on cell adhesion, attachment, and mechanotransduction [88]. Some of the downstream effects of AngII are thought to be mediated by ET-1 [89, 90].

Inflammatory mediators also stimulate CF activation, and some believe that monocytes and macrophages are required for the development of myocardial fibrosis in vivo, at least in response to MI or pressure overload [59]. Circulating monocytes, neutrophils, and lymphocytes are recruited to the injured heart via the synthesis of chemokines and cytokines by different myocardial cell types [44] and expression of specific cell-binding receptors on EC surfaces. Macrophages differentiated from circulating monocytes secrete factors such as interleukin (IL)-1β and tumor necrosis factor (TNF)-α, which can act directly on CFs. In vitro, TNF-α has been shown to increase expression of lysyl oxidase and TGF-β1 [91], while (IL)-1β increases CF migration [92]. IL-6, also predominantly secreted by macrophages and other immune cells, stimulates collagen and TGF-β1 production in CFs [93, 94].

Once differentiated, myoFBs exert their pro-fibrotic effects in numerous ways. Through increased production of TGF-β1 and AngII, these cells sustain their own activated phenotype, as well as exerting effects on the cells around them, such as AngII-induced vasoconstriction when released in the vicinity of vascular SMCs, which could further aggravate ischemia and lead to increased fibrosis. Though inactive CFs do not express any actin-associated surface binding proteins, myoFBs do, and exert contractile force on neighboring cells and ECM through these. In vitro, mechanical coupling between myoFBs and CMs has been shown to lead to conduction disturbances [96], and some studies have suggested they can electrically couple with CMs through gap junction proteins connexins 43 and 45, leading to slowed conduction [97]. Through mechanotransduction pathways, the increased stiffness of the fibrotic ECM induces CM-CF paracrine signaling that causes CF proliferation [98]. Additionally, increased stiffness is correlated with CF activation into myoFBs, as well as increased expression of collagens and SMA. The excess collagen deposited by myoFBs can also surround and isolate CMs, leading to conduction and contractile disturbances [9]. MyoFBs also secrete matricellular proteins such as thrombospondins, osteopontin, periostin, and tenascin-C, which exert mostly pro-fibrotic effects on the surrounding environment, such as further activation of CFs [59], increasing myoFB survival [99], and activation of latent growth factors in the ECM [100]. Though myoFBs have been shown to either apoptose or enter a quiescent phase following dermal wound healing [101, 102], all of the stimulating factors above incite a positive feedback loop in the myocardium that leads to further CF-to-myoFB differentiation and sustained activation [103].

With a basic understanding of the major signaling pathways involved in CF activation and differentiation into myoFBs and the downstream pro-fibrotic effects of these effector cells, the subsequent sections highlight in vitro tissue-engineered models of myocardial fibrosis to replicate and gain further insights into the drivers of pathophysiological fibrotic tissue remodeling.

Modeling Myocardial Fibrosis

Due to the complexity of both the myocardium and the development and progression of fibrosis, creation of physiologically relevant, reproducible, predictive, and high throughput in vitro models of myocardial fibrosis has proceeded at a slower pace than other areas of tissue-engineered disease modeling, hampering efforts to find pharmaceutical solutions [104]. Though animal models have proven invaluable for myocardial fibrosis disease modeling and drug discovery, they are expensive, lack tunability, are considered by some researchers and members of the general public to be unethical [105], and are frequently not predictive of safety or efficacy in humans due to interspecies differences in physiology [106–108]. This section will review many of the existing tissue engineering models of myocardial fibrosis, which have served useful in examining the roles of several different mechanical, soluble, ECM-, and cell-mediated signaling pathways involved in the pathophysiology of fibrotic tissue remodeling. Some of these models have been shown to react appropriately to cardiotoxins [109], fibrogenic compounds [110], and antifibrotic drugs [111], offering a glimpse of the role tissue-engineered myocardial fibrosis models could play in drug screening and discovery.

An ideal model of myocardial fibrosis would (1) include all of the resident cell types of the normal heart in their appropriate proportions as well as the circulating cell types implicated in the development or perpetuation of the pathological process; (2) recapitulate in vivo-like organization of cells surrounded by ECM that mimics normal or fibrotic tissue in terms of mechanical properties and the presence of cell-binding ligands and growth factors; (3) have the capacity for pacing and monitoring of calcium handling and electrophysiological behavior; and (4) incorporate the ability to subject the tissues to physiologically-/pathophysiologically-relevant amounts of strain and pressure and the capability to assay stiffness and contractile forces. While not exhaustive, this list should give readers an idea of the difficulties involved in modeling this disease process.

Optimizing the cellular composition of myocardial fibrosis models is challenging for several reasons. First, with recent studies coming to significantly different conclusions [49, 112], there is no consensus on the proportion of resident heart cell populations. Additionally, many cells of the heart are either difficult to procure or show phenotypic changes when harvested or cultured on traditional rigid tissue culture plastics. Immortalized cell lines offer advantages over primary cells, such as ease of culture and substantial proliferative capacity, but they behave differently than their primary counterparts [113, 114] and often lack critical functional components [113, 115], thus making primary cells more desirable. Primary cells, unfortunately, tend to have low proliferative capacity and exhibit significant phenotypic changes due to passaging or substrate stiffness. While primary adult ventricular myocytes are the ideal CM cell source for modeling fibrosis, they show changes in shape and ultrastructure during early passages in 2D culture [116]. CFs, which serve important homeostatic roles in the normal heart and differentiate into myoFBs during myocardial fibrosis as explained above, show passage-dependent differentiation towards myoFBs in 2D culture [117–120]. Due to the scarcity of primary human cells and the difficulty associated with their culture, primary cells of animal origin have been preferred. However, these often differ physiologically from human cells [106, 107], and may not be able to faithfully model human disease. The maturation status and tissue origin of cells are also important factors that need to be considered with model development. Induced pluripotent stem cell (iPSC) derived CMs have an immature phenotype and undergo many modifications during maturation, including changes in morphology, sarcomere organization, T-tubular and mitochondrial distribution, and calcium handling capabilities [113, 121, 122], which lead to changes in conduction and contractility, as well as differences in susceptibility to various compounds. Though there have been advancements in maturation protocols for human iPSC-CMs (as reviewed in [123]), nothing approaching adult-stage development has been achieved thus far. In regard to the importance of tissue source, fibroblasts (FBs) from different locations in the body have been shown to have divergent gene expression profiles [124, 125], and CFs express unique cardiogenic genes demonstrated to be required for development and healing [126], thus making them the most appropriate source of cells for myocardial fibrosis modeling.

When designing a tissue-engineered model, it is important to tailor the cell density and ECM composition to produce physiologically relevant cell-cell and cell-ECM interactions. The cell density of the heart is around 1.0×10^8 cells/cm^3 [127], or 100 million cells/mL, and different cell densities can lead to divergent tissue properties and functional capabilities in myocardial fibrosis models [128]. Aside from the importance of cell density, the incorporation of multiple resident cell types, such as CFs and ECs, in addition to CMs, increases the in vivo-like quality of the model and leads to changes in performance [110, 129]. Additionally, differing proportions of cocultured cells can affect model behavior [110, 129–131]. Incorporating more cell types, while increasing the physiological relevancy of the model, can also complicate downstream assays. Coculture of cells from the same species may require sorting prior to gene expression analysis in order to decipher the expression patterns attributable to each cell type. Other issues include the difficulty of reliably seeding reproducible numbers of cells in coculture experiments as well as finding a common media suitable for all incorporated cell types. In regard to ECM, options most suitable for myocardial fibrosis modeling should possess tunable mechanical properties that can achieve the stiffness of normal and fibrotic heart (which have been reported as 6–8 kPa [132, 133] and 17–137 kPa [132, 134], respectively, though different estimates can be found in [135]). ECM should also contain binding peptides for cells, be able to undergo remodeling, and allow for electrical conductance. Though in vivo ECM also contains stores of latent growth factors that would be useful to incorporate into models, matrices containing these factors have issues with batch-to-batch variability, so most experimenters use purified matrices or growth-factor-reduced substances, some of which can be impregnated with dormant molecules. A notable exception to this is the use of decellularized tissue, which retains the natural structure and proteins bound to ECM for use in tissue engineering. None of the models discussed here utilize this source, but its processing and application in various tissue models is reviewed in [136]. Most experimenters in this review employed natural ECM derivatives rather than synthetic alternatives. In 2D experiments, many researchers have taken advantage of cells' preference for adhering to these natural substrates by utilizing micropatterning, the process of printing small patterns of ECM or other attachment factors on top of another surface, to either organize specific cell types in different spaces [96, 137] or to control cell shape [138]. Though 2D models can offer increased control over cellular organization and are less complicated in regard to fabrication and assays, cells cultured on flat surfaces have been shown to behave differently than those in 3D culture [109], making the latter method preferable. In the realm of 3D myocardial fibrosis modeling, most experimenters prefer the use of hydrogels of natural ECM derivatives, like fibrin, Matrigel, and gelatin methacrylate (GelMA), where the degree of cross-linking or gelation can be modulated. Cells can be seeded into these gels prior to polymerization, and can then proliferate, migrate, and organize due to the presence of cell-binding ligands in the matrices and their inherent ability to be remodeled. Hydrogels are also preferred due to the ease of casting suspensions into molds of varying shapes, as well as their hydrophilicity, which leads to enhanced diffusion of oxygen, nutrients, and waste products [139]. Though these natural hydrogels are not as tunable as synthetic options with regard to mechanical properties and conductivity (reviewed in [140]), they are widely available and require no further processing prior to use.

Because myocardial fibrosis is believed to alter the electrical properties of the myocardium [9] and predispose the heart to arrhythmias [141], the ability to pace cells and to monitor parameters associated with calcium handling, action potential propagation, and other conductive properties would ideally be included in models of this process. Pacing can be performed through field stimulation via electrodes immersed in the culture media and various functional outputs, such as impulse propagation via optical mapping with fluorescent voltage-sensitive dyes [96, 137, 142], calcium handling via the use of calcium sensitive dyes [142, 143], contraction rate analysis via video processing, and field potential analysis via multi-electrode array [110], can be monitored. In addition to being an important component of these models, pacing has also been shown to elicit improvements in human iPSC-CM maturity [144, 145], and different pacing regimens have been shown to alter outcomes in regard to excitation contraction coupling [128].

Lastly, ideal models of myocardial fibrosis should allow for the experimenter to subject cells or tissue to the same amounts of strain and pressure they would experience in the body and to assay passive and active mechanical properties like stiffness and force of contraction. In vivo, myocardial fibers in normal human hearts are estimated to undergo strains, calculated as $\frac{\text{End diastolic length} - \text{End systolic length}}{\text{End diastolic length}}$, of 15 to 22% [146–148]. In compromised, fibrotic hearts, the strain values differ based on disease process [149, 150]. Strain can be exerted on model systems in a variety of ways. In 2D studies, experimenters can culture cells on flexible substrates, such as polydimethylsiloxane (PDMS), and exert static or cyclic strain by applying vacuum to the bottom aspect of the substrate, a method employed by Flexcell® (Flexcell® International Corporation, Burlington, NC, USA) in their tension systems. Additionally, static perpendicular force can be applied to monolayer cultures using magnetic beads [117, 118], inducing strain when the culture surface is rigid. Strain can be imposed on 3D constructs as well, usually by situating posts or clamps at the longitudinal ends of the tissues, which are deflected due to the movement of a flexible membrane that they are either anchored to [151] or are in periodic contact with [152], or are actuated directly [153]. Additionally, when using an incompressible hydrogel, strain can be induced by compressing a construct against a non-deformable substrate [154]. Though none of the models discussed here incorporate this parameter, intraventricular pressure could play roles in fibrosis beyond its downstream effects on other forces acting upon the myocardium. Pressure in cardiac models can be applied and tuned by employing hydrostatic fluid columns, one-way valves, and pneumatic pumps [151, 155, 156]. In addition to allowing for the mimicking of mechanical cues seen in vivo, an ideal system should also allow one to assay the mechanical properties of the cell or tissue construct as well as the force it exerts. The stiffness of tissues can be determined via nondestructive methods like atomic force microscopy or monitoring the passive stiffness via deflection of posts or wires of known elasticity [111, 131] or by assays like destructive tensile testing [157]. The contractile forces generated in 2D cultures can be quantified by tracking the movement of fluorescent beads within a deformable culture substrate [138], a method called traction force microscopy. These forces can be monitored in 3D systems through analysis of the movement of posts or wires of known elasticity [111, 131] or by utilization of force actuators and transducers attached to either the rods or clamps holding the tissue in place [153] or by direct application to the tissue [142].

With these parameters of an optimal platform in mind, this section will review many of the tissue-engineered myocardial fibrosis models of the past few decades and describe how researchers have incorporated various culture and assay methods to recapitulate important aspects of this disease process. This survey will cover models ranging in complexity and physiological relevance, from 2D monocultures of animal-derived cells on micropatterned substrates to fully humanized 3D triculture models capable of pacing and monitoring of electrical and contractile function. We will explore the available tissue-engineered models based on their suitability for modeling and examining the signaling pathways involved in and/or downstream effects of the following phenomena involved in myocardial fibrosis: mechanotransduction, cell-ECM and/or cell-cell interactions (aside from those integral to mechanotransduction), and pro-fibrotic soluble mediators.

As detailed in this section, due to their central role in myocardial fibrosis, CFs and/or myoFBs are incorporated into virtually all models of this condition. The presence of myoFBs is often assayed by the upregulated gene expression or positive staining of SMA, which endows these cells with their contractile phenotype [158]. As myoFBs secrete numerous ECM proteins (e.g., collagens I and III, fibronectin, osteopontin, etc.), factors that affect ECM degradation such as MMPS and TIMPs, and pro-fibrotic autocrine and paracrine factors (e.g., TGF-β1, AngII), the presence of these molecules is also usually assayed through gene expression studies, immunofluorescent imaging, and/or direct quantification of protein levels and activity.

In addition to monitoring molecular changes in cells and ECM, many of the studies below also examine changes in electrophysiology, contractile function, and other mechanical properties of fibrotic cardiac tissue constructs. Models often incorporate functional assays such as these along with molecular assays, allowing investigation of relationships between synthetic and behavioral phenomena to enhance predictive power.

Mechanotransduction

In the body, cells sense mechanical input from neighboring cells and the ECM through cell surface receptors and respond to them via a variety of signaling pathways in a process known as mechanotransduction [159]. Understanding the role of mechanotransduction in myocardial fibrosis is of paramount importance, as the condition is often triggered or worsened by increased hemodynamic loads [26–31], exerting pathophysiological amounts of force onto the myocardium. Also, myoFBs are known to exert tension on neighboring cells and ECM through increased expression of SMA and to directly affect ECM composition though increased expression of MMPs, TIMPs, and various ECM constituents like collagen I and III, further highlighting the need to understand how various mechanical properties and forces affect the cells and ECM of the heart.

To recapitulate the downstream effects of volume overload, the condition where end-diastolic volume exceeds physiological levels due to valvular regurgitation or arteriovenous fistula, models can be designed to allow the experimenter to subject cells to stretch. Though these phenomena can be more holistically mimicked using perfusion systems, flexible culture chambers, and pumps [151], thus far researchers focused on myocardial fibrosis have mostly focused on direct application of strain. Strain can be applied in 2D models either through the stretching of a flexible culture substrate [160–163] or by exposing cells cultured on rigid substrates to downward force through the use of magnetic beads coating the upper aspect of the monolayer [117, 118]. Groups have employed both methods to study the effects of stretch on CF proliferation and differentiation into myoFBs. In flexible substrate models, cyclic strain has been associated with increased myoFB differentiation based on SMA staining and gene expression [160, 163], along with increased ECM deposition [160, 162, 163] and degradation [160, 162]. Interestingly, Butt and Bishop did not see any effect of strain on CF proliferation in their model [160], while Dalla Costa et al. saw a doubling in Ki67 positive proliferative CFs undergoing stretch [162]. This discrepancy could be due to differences in strain magnitude, frequency, substrate coating, and the developmental stage of the cells. When subjected to static strain exerted by collagen I coated magnetic beads, both neonatal and adult myoFBs, which were differentiated through growth on rigid polystyrene, decreased their expression of SMA, indicating that the phenotype is reversible [117, 118]. On the other hand, when early passage, predominantly SMA negative CFs were subjected to the same force, SMA gene expression was upregulated, suggesting that myoFBs and less activated CFs respond differently to mechanical cues [118]. 2D studies have also allowed for examination of the combinatorial roles played by various ECM components, integrins, and intracellular pathways in strain mechanotransduction. c-JUN N-terminal kinase (JNK), extracellular signal-regulated kinase (ERK), and p38 mitogen-activated protein kinases (MAPKs) have been implicated in the development of myocardial fibrosis [164–166], and are involved in both mechanotransductive [167] and TGF-β [168] signaling. Activation of these pathways through mechanotransduction has been shown to rely upon integrin-triggered activation of focal adhesion kinase (FAK) [164]. MacKenna et al. explored the contributions of static stretch and different ECM substrates to the activation of MAPK pathways, finding that stretch led to activation of JNK and ERK when CFs were allowed to deposit their own ECM, but that stretch-induced pathway activation was variable when cells were plated on other ECM proteins [161]. When subjected to strain via collagen I coated magnetic beads, CFs show differing MAPK pathway activation patterns based on whether they have been previously differentiated into myoFBs. Nonactivated neonatal rat CFs upregulate ERK activation in response to force, whereas neonatal rat myoFBs display p38 activation [118]. In a separate study, when adult rat myoFBs were subjected to force via magnetic beads coated with bovine serum albumin (BSA) or poly-L-lysine, activation of p38 was not seen, nor was the force-induced decrease in SMA, providing evidence that MAPK signaling and specific ECM-integrin interactions play a role in myoFB mechanotransduction [117]. Further supporting the role of integrins in this process, stretch-induced myoFB differentiation was associated with FAK activation, and culture with RGD peptide (acting as an integrin blocker) or knockdown of FAK expression via siRNA decreased SMA expression [162].

As 3D models increase the amount of contact between cells and ECM, they more accurately mimic the cell-cell and cell-ECM communications involved in stretch-related mechanotransduction. However, thus far these studies have come to different conclusions regarding the effect of stretch on important aspects of fibrosis, such as myoFB activation. When feline CFs were seeded into a nylon scaffold and subjected to static stretch, they expressed less SMA and collagen I [169]. However, when researchers cultured neonatal rat CFs in GelMA with physiological stiffness and subjected these constructs to varying levels of strain via the use of a pneumatic pump and culture chambers composed of a rigid glass top and flexible PDMS bottom, increased strain was correlated with greater expression of SMA, collagen I, TGF-β, and fibronectin [120].

In addition to strain, CFs also experience force due to interstitial flow, which has been shown to trigger integrin-mediated mechanotransducive signaling in other organs [170]. One group has designed a bioreactor that has a media inlet tube at the bottom of a deformable chamber, which is stretched and compressed along a perpendicular axis, allowing application of strain and interstitial flow at the same time [171]. In this model, when neonatal rat CFs were exposed to flow, they showed increased expression of collagen III, TGF-β, and SMA [172]. Activation of Smad2, a transcription factor and downstream mediator of TGF-β known to play a role in fibrosis [173], was also seen, and inhibition of TGF-β inhibited pro-fibrotic gene upregulation. Interestingly, when combined with flow, strain significantly decreased SMA expression.

The impact of ECM stiffness on CMs and CFs has also been explored in several 2D and 3D models of myocardial fibrosis. Zhao et al. used photolithography to pattern poly(ethylene glycol) diacrylate (PEGDA) substrates with normal or fibrotic stiffness into alternating bands [119] and saw that CF proliferation and differentiation into myoFBs correlated with culture on stiff substrate. In addition to pro-fibrotic gene expression changes, patchy collagen deposition was detected on stiff substrates with Sirius red stain. Another group looked at the synergistic roles played by matrix stiffness and TGF-β signaling in fibrosis development by culturing adult human CFs on PDMS of various elasticity and exposing some samples to TGF-β [174], finding that TGF-β signaling much more significantly impacted pro-fibrotic gene changes. Exploring the role of ECM stiffness on CFs in 3D, Galie, Westfall, and Stegemann reported that culture in confined collagen I gels (higher stiffness) led to more SMA expression than culture in free-floating gels [175]. Models have also been designed to explore the effect of ECM stiffness on CM behavior. Both the effects of substrate elasticity and cell shape were explored by McCain et al., who used photolithography to pattern polyacrylamide gel of normal and high stiffness onto a culture substrate and then micropatterned fibronectin rectangles of various aspect ratios on top [138]. Through traction force microscopy they saw that CMs exerted peak systolic work at 7:1 aspect ratio when cultured on normal stiffness gel, whereas those on fibrotic gel worked most at 2:1, suggesting that changes in ECM stiffness led to contractile adaptations in CMs. In a separate study, mouse iPSC-CMs showed disorganized sarcomeric structure and irregular beating when cultured on polystyrene or polyacrylamide gel with increased stiffness (123 kPa), while cultures on soft (12 kPa) and medium (30 kPa) gels showed good organization and synchronous contraction [176]. Worke et al. cultured embryonic mouse CMs on collagen I matrices of different stiffnesses and showed that lower stiffness was correlated with discrete and synchronized calcium peaks, whereas calcium transients on stiffer collagen beds were flat and asynchronous [143].

Lastly, one study has examined the differing mechanotransductive effects of CFs and myoFBs on CMs [96]. In this 2D model, when added to micropatterned cardiomyocytes, myoFBs led to conduction disturbances. Researchers sought to understand whether this was due to electrical or mechanical coupling between CMs and myoFBs, and found that while inhibiting connexin 43 did not change the deleterious effect myoFBs had on conduction, inhibiting contraction with blebbistatin did, suggesting that the tension exerted by these cells on neighboring CMs can have negative effects [96].

Cell-Cell and Cell-ECM Interactions

In addition to reacting to extracellular mechanical stimuli and exerting tension on their surroundings, myoFBs and other cells of the myocardium signal to and act upon one another in a variety of ways, including direct cell-cell interactions, autocrine signaling, and paracrine signaling [177]. Heart cells also experience nonmechanical signaling from the ECM that contributes to pro-fibrotic changes which in turn results in progressive ECM remodeling [178]. In vitro models have proven useful for dissecting the effects that specific cell-cell and cell-ECM interactions have on the development of myocardial fibrosis, as well as understanding the anti-fibrotic therapeutic mechanisms of cell therapy.

Beginning with 2D models, Chang et al. used fibronectin micropatterning and cell coculture to recreate structural features of an infarct border zone (IBZ) and the surrounding healthy heart tissue [137]. Human skeletal muscle myotubes (HSKMs) were used in coculture with neonatal rat CMs to simulate the healed IBZ because HSKMs lack gap junctions, exhibit linear morphology similar to fibrous collagen, and disturb the orientation of neighboring cells. Coculture of 20–30% HSKMs with 70–80% CMs resulted in lower conduction velocities in comparison to the neighboring CM only "healthy heart tissue". Upon rapid pacing, reentrant wave fronts emerged in these models, which were terminated by nitrendipine, an L-type calcium channel blocker. Microscopy showed HSKMs disturbing CM organization in the coculture region, suggesting that decreased cell-cell contact between CMs and non-uniform anisotropic architecture led to the conduction abnormalities. Focusing on cell-ECM interactions, Watson et al. investigated the differential effects of plating human primary CFs on collagen I, IV, VI, and laminin, and showed that collagens I and VI actually upregulated collagen I gene expression in the absence of any other experimental variables [163]. When exposed to stretch and/or TGF-β, CFs showed variable expression of pro-fibrotic genes correlating with the substrate they were plated on, further suggesting that the ECM composition affects CF synthetic activity.

Aside from modulating the expression of collagen, changes in ECM composition have also been shown to affect the contractile phenotype of CFs [75]. In a study investigating how AngII mediates its pro-fibrotic effects, researchers examined the roles of osteopontin, an ECM structural protein, and β3 integrin, a membrane receptor subunit found in many cell-ECM attachments, in the contraction of 3D collagen gels seeded with adult rat CFs. These specific molecules were chosen for study because they were upregulated by AngII in 2D culture. Increased gel contraction was seen when osteopontin was added to the construct, whereas RGD peptide and β3 integrin monoclonal antibodies blocked the effect, suggesting a causal link between AngII, myoFB osteopontin and integrin upregulation, and increased contractile phenotype in fibrotic tissue.

Several groups have used 3D models with varying proportions of myocardial cells to examine the relative contributions of each cell type to development of fibrosis and the effects the cell types have on one another, as well as to model the interface between fibrotic and normal tissue. Lee et al. created spheroids containing different ratios of human embryonic stem cell (hESC)-derived CMs and mesenchymal stem cells (MSCs), a progenitor of CFs, showing that increasing amounts of MSCs (up to 20%) correlated with increased spheroid compaction and more regular and rapid spontaneous contractile activity measured by microelectrode array [110]. Interestingly, after coculture MSCs expressed discoidin domain receptor-2 (DDR2), a CF marker, which they did not express prior to the experiment, suggesting that cell-cell and/or cell-ECM signaling encouraged the differentiation of CFs. Additionally, when the constructs were treated with the fibrogenic compounds bisphenol A, aldosterone, and metoprolol, cocultured constructs demonstrated greater levels of CM apoptosis than CM-only spheroids. Overall, the coculture constructs demonstrated appropriate responses to these fibrogenic compounds as determined by expression of pro-fibrotic and TGF-β-responsive genes, demonstrating their potential in drug screening applications. Mainardi et al. explored whether the presence of CMs would inhibit CF activation in 3D hydrogel constructs seeded between posts and subjected to uniaxial stretch [152] (see Fig. 3a). They created CF-only constructs as well as those containing neonatal rat CM:CF ratios of 80:20, 50:50, 20:80 and examined various parameters after five days of cyclical stretch. The constructs containing greater numbers of CFs unsurprisingly demonstrated more SMA, but interestingly, CMs expressed more SMA the greater the number of CFs they were cultured with, suggesting that activated CFs can increase the resting contractile tone of CMs. Constructs with higher numbers of CFs also showed higher expression of collagen I and fibronectin, but only those created with 50% or fewer CFs could be paced, making constructs containing larger amounts of CFs less useful for studying electrical phenomena. Wang et al. similarly explored the impact of cell ratios in a 3D hydrogel model, seeding constructs of 75:25 and 25:75 human iPSC-CM to human adult CF ratios in fibrin and casting them into custom dumbbell-shaped PDMS wells with a flexible fluorescent polymer wire at each end, allowing for convenient tracking of deflection for force calculations [131]. Beyond finding the expected trends in ECM deposition and SMA-positive cells between constructs, they also made a construct with one half composed of "normal" (75:25 CM:CF) and the other composed of "fibrotic" (25:75 CM:CF) tissue that maintained the morphological hallmarks of the interface between healthy and fibrotic myocardium for weeks based on collagen and SMA staining. When subjected to pacing, the normal half of the construct demonstrated diminished impulse propagation velocity in comparison to "normal" constructs as well as arrhythmia development, presumably due to conduction block created by the fibrotic half. When the normal half was excised and compared to "normal" constructs, it exhibited worsened contractile parameters, suggesting that fibrotic tissue exerts paracrine effects on neighboring healthy tissue.

In addition to comparing the contributions of different cell types to tissue remodeling and function, some researchers have also explored the effects that FBs from different sources have on these parameters. Li, Asfour, and Bursac created 3D

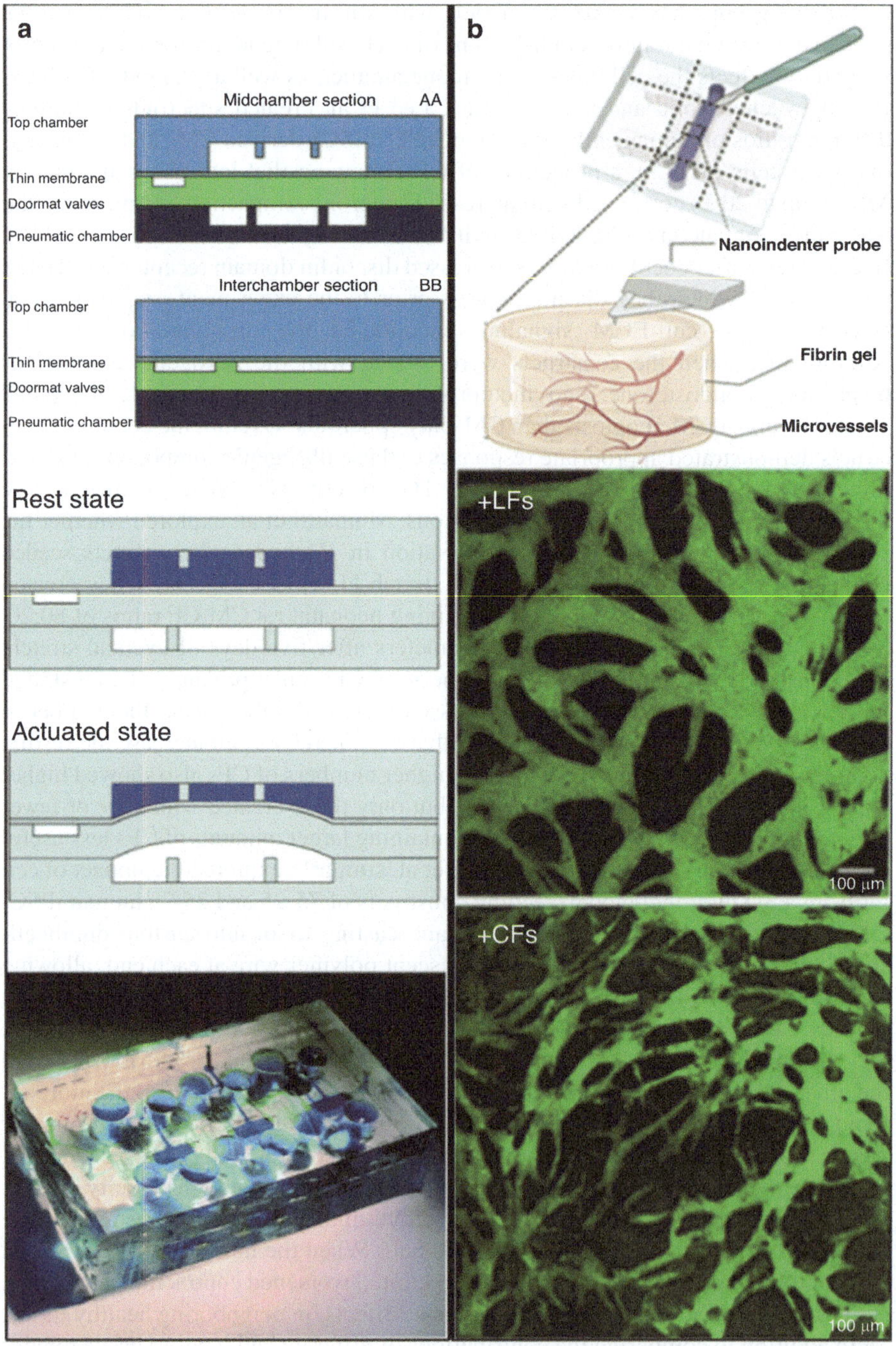

Fig. 3 (**a**) Design schematics for a 4-layer, PDMS myocardial fibrosis platform composed of a top chamber to house the tissue constructs suspended between posts hanging from a rigid upper aspect

hydrogel constructs containing a 0.3:1 ratio of either fetal or adult mouse CFs and neonatal rat CMs and suspended them in PDMS troughs surrounded by a rigid frame, generating 3D cylindrical bundles [142]. They found that bundles containing adult CFs had slower conduction velocities and prolonged action potential durations, as well as lower contractile forces and higher passive stiffness, in comparison to those containing neonatal CFs. The adult CF bundles also demonstrated increased collagen, fibronectin, CTGF, and SMA expression, along with decreased expression of genes associated with mechanical function, calcium handling, and connexin 43, suggesting that aging has detrimental effects on cardiac function regardless of the health of the cells. Some of the same investigators used this model to evaluate the impact of CFs harvested from sham or transaortic constriction (TAC) model mice on CF:CM 3D coculture constructs. Due to phenotypic differences observed in Thymocyte Differentiation Antigen 1 (Thy1) positive and negative FBs in other organs [179, 180], these researchers set out to learn if this marker correlated with different CF phenotypes, and whether or not these associations were due to preexisting disease. Coculture constructs containing $Thy1^-$ CFs from TAC mice generated lower contractile force and demonstrated greater contractile rise time and calcium transient amplitudes than the other constructs, as well as increased collagen staining, suggesting that this specific subset of CFs is particularly pro-fibrotic in diseased hearts but phenotypically similar to $Thy1^+$ CFs in the healthy heart. A separate group created spheroids containing a 4:1 ratio of hiPSC-CMs to CFs derived from either healthy patients or those with HF and saw that the HF spheroids demonstrated a faster contractile rate and shortened relaxation phase in comparison to the healthy spheroids [181]. Lastly, one group investigated the effects of human primary CFs and lung FBs on human iPSC-derived endothelial cells (ECs) [182] (see Fig. 3b). Using a 5:1 EC:FB ratio, the cells were seeded in hydrogel into a PDMS device bonded to glass. The device contained a central chamber for the tissue construct and two neighboring channels for media. In comparison to lung FBs, CFs led to worsened perfusion as assayed via FITC-labeled dextran and also decreased barrier function, leading to increased permeability.

Myocardial fibrosis models have also been used to evaluate the therapeutic properties of cell therapy. Repurposing their bioreactor that incorporates both interstitial flow and cyclic uniaxial strain, Galie and Stegemann created a myocardial fibrosis

Fig. 3 (continued) made of thick PDMS (blue), a thin membrane (gray), doormat valves (green), and a pneumatic chamber (navy). At rest, the posts are not deflected, but when the membrane is actuated, it deflects upward and contacts the posts, pushing them apart. The bottom aspect shows the realized device. (Reproduced with permission from [152]. This work was published under a CC BY license (Creative Commons Attribution 4.0 International License; https://creativecommons.org/licenses/by/4.0/)). (**b**) Model concept composed of a tri-channel PDMS culture chamber bonded to a glass coverslip. ECs +/− CFs or lung FBs were seeded with fibrin in the central channel, and the side channels were filled with culture media. Tissue stiffness was assayed via nanoindentation. The lower panels show the differential effects of LFs (lung FBs) and CFs on microvascular network formation. (Reproduced with permission from [182]. This work was published under a CC BY license (Creative Commons Attribution 4.0 International License; https://creativecommons.org/licenses/by/4.0/))

tissue model consisting of neonatal rat CFs in collagen gel and injected rat MSCs into them before exposing them to flow, strain, or hypoxia [183]. Though increased amounts of vascular endothelial growth factor (VEGF) and insulin-like growth factor 1 (IGF1) were seen in the supernatant of the coculture models, both of which have been associated with the paracrine therapeutic effects of MSCs, the responses of CFs 250 μm away from the MSC injection site based on the expression of SMA and other pro-fibrotic genes was variable amongst the different conditions, and the researchers suspected that the CFs were not sufficiently activated into myoFBs based on SMA stain. Another group cultured human fetal CFs in 3D GelMA constructs and stimulated them with TGF-β to differentiate them into myoFBs leading to an increase in the expression of SMA and collagen I [184]. When these fibrotic constructs were placed in a Transwell, which prevents cell migration but allows for media transport, into a dish with cardiac progenitor cells, they showed decreased expression of pro-fibrotic genes, an effect not seen in coculture with human microvascular ECs, suggesting that cardiac progenitor cells exert paracrine antifibrotic effects.

Pro-Fibrotic Soluble Mediators

A number of soluble mediators are known to play stimulatory roles in the development of myocardial fibrosis [8, 59, 168, 185, 186], however, the majority of in vitro myocardial fibrosis models have focused on the downstream effects of TGF-β, with many only using the compound due to its reliability in triggering CF to myoFB differentiation without trying to parse any new information regarding its effects on different cell types or the mechanisms by which it mediates its pro-fibrotic effects [96, 109, 111, 152, 154, 160, 184]. Here, the focus will be on models that have been used to further understand the role of TGF-β signaling in the development of myocardial fibrosis, as well as highlight models investigating other soluble mediators.

First, revisiting a model discussed previously, Wang, Seth, and McCulloch saw that when adult rat myoFBs were exposed to static strain via collagen I coated magnetic beads, SMA expression decreased [117]. However, when AngII was added to these experiments, the decrease in SMA was not observed, indicating that AngII can perpetuate the myoFB phenotype in the face of mechanotransductive signals encouraging a quiescent phenotype. In another experiment evaluating the relative contributions of mechanotransduction and soluble mediators on fibrosis development, Cho, Razipour, and McCain plated primary human CFs on PDMS of different stiffnesses with or without TGF-β [174]. Though SMA expression was upregulated by stiffer matrices, none demonstrated greater than 50% SMA positivity as observed with staining, whereas all of the conditions including TGF-β demonstrated greater than 50% SMA positivity, suggesting that TGF-β contributes more to myocardial fibrosis than ECM stiffness. Nunohiro et al. saw that AngII upregulated the expression of ECM components and specific integrins in 2D culture of adult rat CFs, and then used a collagen contraction assay in combination with AngII, osteopontin, and integrin blockers to dissect how AngII exerts its pro-fibrotic effects, concluding that its upregulation of osteopontin results in an increased rate of matrix remodeling that is mediated through specific upregulated integrins [75]. In a study exposing neonatal rat CFs to interstitial flow in a 3D collagen gel, flow was correlated with increased collagen III, TGF-β, and SMA expression [172]. Inhibiting either TGF-β or AngII signaling diminished the pro-fibrotic gene upregulation, indicating that both are involved in flow-induced mechanotransduction signaling. When neonatal rat CFs were seeded in GelMA of physiological stiffness and subjected to increasing magnitudes of cyclic strain via a pneumatically-actuated flexible PDMS and glass device, increased strain correlated with increased expression of TGF-β, which the experimenters hypothesized was related to traction-induced release and activation of dormant TGF-β in the ECM [120]. When these constructs were exposed to strain in the presence of TGF-β inhibitor, pro-fibrotic gene expression was decreased, again suggesting that TGF-β plays an important role in mediating pro-fibrotic strain signaling. Santos et al. seeded neonatal rat or human primary CFs and collagen I into ring-shaped molds and monitored changes in compaction (via cross-sectional area) and stiffness (via destructive tensile testing) when the tissues were treated with a Rho-kinase (ROCK) inhibitor and/or TGF-β. Rho-kinase is a downstream mediator of the small GTP-binding protein RhoA, and it plays roles in cell motility, proliferation, and contraction through interactions with the actin cytoskeleton [187]. In this model, ROCK inhibition led to decreases in tissue compaction and stiffness, which were countered by simultaneous treatment with TGF-β.

In addition to exploring the role of TGF-β in exerting or counteracting mechanotransductive signaling in myocardial fibrosis, others have sought to understand its importance in sustaining the fibrotic phenotype and its effect on cells other than CFs. Blyszczuk et al. created spheroids containing either human fetal CFs alone, hiPSC-CMs alone, or the two in a 4:1 CM:CF ratio [181]. When treated with TGF-β the CF-containing constructs grew and increased expression of collagen I, whereas CM-only constructs demonstrated no changes. Interestingly, the CM:CF constructs showed decreased expression of cardiac troponin T and connexin 43 in response to TGF-β, which are integral in contraction and action potential propagation, respectively. To determine whether the pro-fibrotic changes were due to continued presence of TGF-β or its effect on CF to myoFB differentiation alone, myoFBs were differentiated beforehand with TGF-β and then cultured in constructs without continued TGF-β administration. All of changes associated with TGF-β treatment were replicated, highlighting the central role of myoFBs in myocardial fibrosis. This group also examined the effects of TGF-β on CM:CF constructs containing either healthy adult CFs or those sourced from HF patients. TGF-β caused increased contraction rates in 4 of 6 healthy constructs, but this was observed in only 1 of 6 constructs made with unhealthy CFs, indicating that CFs from diseased hearts lose some capacity to respond to TGF-β signaling.

Lastly, exosomes, which are 30–200 nm membrane-bound extracellular vesicles containing proteins and nucleic acids [188, 189], are secreted into the ECM by many different cell types and have been implicated in the development of various forms of heart disease [190–192] and have been suspected to play a role in fibrotic

ECM remodeling. These vesicles harbor microRNAs, which are single-stranded RNA fragments that inhibit gene expression by targeting complementary RNA strands for destruction or otherwise impeding their translation [193]. After using either doxorubicin or left anterior descending artery (LAD) ligation to elicit myocardial fibrosis in rats, Yang et al. found that miR-208a was increased in both models [188] and sought to understand if and how it elicited pro-fibrotic signaling in vitro. When rat CMs or CFs were exposed to hypoxia or AngII in monocultures, CMs upregulated expression of the microRNA, whereas CFs didn't. However, when conditioned media from the CM experiments were added to CF cultures, CFs upregulated miR-208a expression, as well as that of collagens I, III, SMA, thrombospondin, and CTGF, indicating pro-fibrotic paracrine signaling was involved in mediating the miR effect, as well as positive feedback. The researchers utilized fluorescent nanoparticle tracking to visually capture entry of the exosomal content into the CFs, confirming that they could be responsible for the observed changes in gene expression. When exosomes were removed from the media or miR-208a was blocked through addition of an antagomir, expression of miR-208a and pro-fibrotic genes decreased in the CF cultures. After these in vitro experiments, the group induced MI in rats through LAD ligation and treated them with miR-208a antagomir. Rats that received this treatment showed increased cardiac performance and decreased fibrosis weeks after surgery, providing further evidence for the role of CM-derived exosomal miR-208a in the development of myocardial fibrosis. Please see Table 1 for a chronological listing of the models referenced in this section.

Conclusion

As detailed above, myocardial fibrosis is a dynamic, multicellular process that relies upon signaling originating in the heart, circulating immune cells, or distant organs. Due in large part to this complexity, there are relatively fewer tissue-engineered in vitro models of myocardial fibrosis than many other disease processes. However, as described in this chapter, important signaling aspects of myocardial fibrosis can be recapitulated in vitro, and tissue-engineered disease models have helped elucidate the contributions of specific cell-cell, cell-ECM, and soluble mediator signaling pathways to this maladaptive process, as well as the effect of ECM stiffness on myocardial cell phenotype.

Earlier, the parameters of an ideal tissue-engineered model of myocardial fibrosis were detailed (see Box 1). These design recommendations are geared toward recapitulating the in vivo setting as accurately as possible, thus leading to greater physiological relevance and reliability of information generated from these models. Due to limitations associated with current tissue engineering and assay technologies, achieving these goals is still challenging, and tradeoffs are often made to focus on a handful of signaling phenomena of importance and to maintain moderate throughput of the platform. While no model thus far has managed to achieve all of these aspirations, the recommendation that is furthest from being realized is that

Table 1 In vitro tissue-engineered myocardial fibrosis models

Publication Year	Cell Origin/Maturation Status	Cellular/Tissue Organization	Mechano-transduction	Cell-Cell/ Cell-ECM communication	Soluble Mediators	Citation
1997	Fetal rat CFs	2D		-		[160]
1998	Adult rat CFs	2D		-	-	[161]
1999	Adult rat CFs	3D	-			[75]
2000	Adult rat CFs	2D		-		[117]
2003	Neonatal rat CFs	2D		-	-	[118]
2006	Feline CFs	3D		-	-	[169]
2009	Neonatal rat CFs + HSKMs	2D	-		-	[137]
2009	Neonatal rat CFs	2D		-	-	[162]
2011	Neonatal rat CFs	3D		-	-	[175]
2011	Neonatal rat CFs + CMs	2D		-	-	[96]
2012	Neonatal rat CFs	3D		-		[172]
2014	Neonatal rat CFs + rat MSCs	3D	-		-	[183]
2014	Neonatal rat CMs	2D		-	-	[138]
2014	Adult human CFs	2D				[163]
2014	Adult rat CFs + NIH 3T3 rat FBs	2D and 3D		-	-	[119]
2017	Neonatal rat CMs + neonatal or adult mouse CFs	3D	-		-	[142]
2017	Neonatal rat CFs + CMs	3D	-	-		[154]
2017	Embryonic mouse CMs	2D		-	-	[143]
2018	Adult human CFs	2D		-		[174]
2018	Adult rat CFs + CMs	2D	-			[188]
2019	Fetal human CFs + cardiac progenitor cells or ECs	3D	-		-	[184]
2019	Adult mouse CFs + mouse iPSC-CMs	2D		-	-	[176]
2019	Neonatal rat CFs	3D		-		[120]
2019	Human embryonic stem cell-derived CMs + MSCs	3D	-		-	[110]
2019	Neonatal rat CFs +/- CMs or adult human CFs alone	3D	-	-		[157]
2019	Human iPSC-CMs + adult human CFs	3D	-		-	[131]
2020	Human iPSC-CMs + fetal human CFs, healthy adult human CFs, or HF adult human CFs	3D	-			[181]
2020	Neonatal rat CMs + adult mouse CFs of different phenotypes	3D	-		-	[199]
2020	Human iPSC-CMs + adult human CFs	3D	-	-	-	[111]
2021	Human iPSC-ECs +/- adult human CFs or lung FBs	3D	-			[182]
2021	Neonatal rat CFs + CMs	3D			-	[152]

Blue highlight indicates the platform was used to study associated phenomena

Box 1. Parameters of an Ideal Tissue-Engineered Myocardial Fibrosis Model:

1. Include all of the resident cell types of the normal heart in their appropriate proportions as well as the circulating cell types implicated in the development or perpetuation of the pathological process.
2. Recapitulate in vivo-like organization of cells surrounded by ECM that mimics normal or fibrotic tissue in terms of mechanical properties and the presence of cell-binding ligands and growth factors.
3. Have the capacity for pacing and monitoring of calcium handling and electrophysiological behavior.
4. Incorporate the ability to subject the tissues to physiologically-/pathophysiologically-relevant amounts of strain and pressure and the capability to assay stiffness and contractile forces.

regarding the cellular composition of the tissue within these platforms. Though many of the models above included CMs and CFs or myoFBs, none of them included inflammatory cells of any kind. While the importance of resident macrophages and lymphocytes to the development of myocardial fibrosis appears to be limited beyond initial recruitment of circulating immune cells [44, 194], these circulating monocytes are thought to be required for development of the condition in response to MI and pressure overload [59], and lymphocytes recruited to the heart also exert pro-fibrotic effects [195]. Additionally, only two studies examined the role of ECs in myocardial fibrosis, with one examining their paracrine effects on myoFBs [184] and another investigating the effects that direct coculture with either CFs or lung FBs would have on microvascular network formation, perfusability, and permeability [182]. In addition to their involvement in pro-fibrotic signaling through the secretion of soluble mediators and recruitment of circulating immune cells [62], myocardial fibrosis has been associated with microvascular rarefaction [62], or diminished capillary density, in some HF patients [196–198], and further understanding of the interplay between dysfunctional ECs and fibrotic ventricular remodeling would be of use. Due to their proximity to microvascular ECs and their known effects on barrier function, PCs could also be important to include in models exploring this relationship. With regard to other mural cells, though no studies incorporated vascular SMCs, these would only be of particular importance in models examining changes in arteriolar tone or the potential contribution of these cells to the myoFB pool. Also with regard to the cellular composition of the reviewed models, many of the studies not only utilized animal cells, but those from early stages in development, both of which limit their physiological relevance due to interspecies and maturation-dependent differences in physiology [106–108, 113, 121, 122]. These cell source issues may have contributed to some of the contradictory phenomena observed in a few of the models reviewed. Though progress remains to be made on this front, multiple models reviewed here advanced knowledge in regard to the effect of CM to stromal cell ratios on the behavior of cardiac tissue models [110,

131, 152], and also worked to establish the effects of cell seeding density on various cardiac functional parameters and ECM composition [128].

The reviewed models were more consistent in satisfying the remaining three criteria (see Box 1), though none achieved them all, and here some of the most physiologically relevant and high throughput platforms discussed will be highlighted, drawing lessons from them for future application. To learn more about the effects of cyclic strain magnitude and TGF-β signaling on CF activation and ECM remodeling, Kong et al. created a bioreactor composed of a bottom pneumatic chamber, a culture chamber with a flexible membrane, and a rigid glass top [120]. Five different levels of strain were achieved within this model by fabricating interconnected pneumatic chambers of different diameters, with larger diameters leading to greater strain with the same applied pressure. Each culture chamber contained a cylindrical post, upon which the 3D tissue constructs were positioned, and with each pneumatic pulse, these constructs were compressed against the glass top to induce strain. Combining neonatal rat CFs and 7% GelMA, which has physiologically relevant stiffness, the researchers seeded all five culture chambers of the device simultaneously through one microfluidic channel, and then used photolithography and UV exposure to only polymerize the areas of the cell gel suspension that were atop the cylindrical posts, washing away the remaining non-polymerized gel. Though the use of neonatal rat cells and the lack of other myocardial cell types limits the physiological relevance of this model, the ability to test five levels of strain within a device that is seeded all at once increases the throughput of the model, and by combining treatment with TGF-β or an inhibitor, the experimenters were able to learn about the combinatorial effects of both mechanotransduction and soluble mediator signaling on CF activation and tissue remodeling. Mainardi et al. similarly employed a multi-chamber, interconnected device capable of pneumatic activation to investigate the effects of differing neonatal rat CM:CF ratios, cyclic stretch, and TGF-β signaling on CF activation, profibrotic gene expression, and electrophysiology using 3D fibrin-based tissue constructs [152]. The bioreactor employed contained six culture chambers, three static and three dynamic, that were interconnected via doormat valves and could be seeded simultaneously (see Fig. 3a). Two deflectable posts hung from the top of each culture chamber, and when the flexible membrane at the bottom of the chamber was pneumatically activated, it bulged upward, pushing the posts apart and exerting 10% strain on the tissue. Following each experiment, the tissue constructs were also externally paced, and their behavior was examined via motion tracking analysis. Lastly, Wang et al. utilized the Biowire™ II device (TARA Biosystems, Inc., New York City, NY, USA), which is composed of a dumbbell-shaped trough for cell suspension seeding and two flexible polymer wires at each end, upon which the tissue constructs exert force following gelation [131]. By doping the polymer with fluorescent molecules, the displacement of the wires was easily monitored, allowing for calculations of passive stiffness and contractile force of the tissue constructs. The group seeded either "normal" fibrin-based constructs, composed of 75:25 ratio of human iPSC-CMs and primary human CFs, "fibrotic" constructs with the opposite ratio, or half "normal", half "fibrotic" constructs, within which gels containing both CM:CF ratios were seeded on opposing sides of the culture chamber, mimicking the border between healthy and fibrotic tissue. External pacing was applied to these constructs for 6 weeks, and the group assayed contractile force, passive stiffness, profibrotic gene expression changes, and electrophysiological behavior, observing significant differences in function and composition between the constructs.

Though tissue-engineered in vitro models of myocardial fibrosis have helped elucidate the signaling pathways involved in this condition and some have responded appropriately to fibrogenic and anti-fibrotic compounds, further improvements to increase translational relevance could make them more useful in preclinical studies. With regard to cell composition, experimenters should seek to use human cells, and in regard to CMs, they should be matured toward the adult phenotype as much as possible to allow for recapitulation of the electrophysiological behavior and calcium handling seen in vivo. Pertaining to CFs, one should pay particular attention to their differentiation status in culture, ensuring that cells are quiescent prior to experimentation, unless one is seeking to model existing fibrosis. As discussed above, other resident cell types, especially ECs, should be more frequently included in models when possible. Immune cells in the bloodstream could also be incorporated into future platforms, either by introducing cells into the media surrounding constructs in static culture, or by creating a perfusion system that allows for circulation of the cells. Experimenters could also better replicate the cardiac cycle by incorporating both cyclic stretch and electrical pacing simultaneously. Pitoulis et al. have designed a bioreactor that allows for application of both, which involves suspending tissue in a culture chamber between clamps, one of which is attached to a force sensor, and the other to an actuator [153]. In this perfusion model, pacing is applied, and the force sensor records the first part of the resultant contractile force. This information is then relayed to a computer model that predicts what the end systolic fiber length would be in vivo, and then this output is communicated to the actuator clamp, which moves to this position. Though the experimenters were employing animal heart tissue slices, this technology could be utilized in tissue-engineered models of myocardial fibrosis. Lastly, due to the frequent association of myocardial fibrosis with pressure overload conditions, experimenters could mimic pathophysiological intraventricular pressures through the use of pneumatic pumps. Though the ability to meet all four parameters of an ideal tissue-engineered myocardial fibrosis model may be extremely challenging with current cell culture technology, acting upon the above considerations should increase the physiological relevance of future models.

Funding Sources NIH Grants T32EB023872 & R01HL148462.

References

1. Virani SS, Alonso A, Aparicio HJ, Benjamin EJ, Bittencourt MS, Callaway CW, et al. Heart disease and stroke statistics--2021 update. Circulation. 2021;143(8):e254–743.

2. The top 10 causes of death: World Health Organization; 2020 [updated December 9, 2020; cited 2021 October 21]. Available from: https://www.who.int/news-room/fact-sheets/detail/the-top-10-causes-of-death.
3. Krumholz HM, Normand S-LT, Wang Y. Twenty-year trends in outcomes for older adults with acute myocardial infarction in the United States. JAMA Netw Open. 2019;2(3):e191938–e.
4. Roger VL. Epidemiology of heart failure. Circ Res. 2013;113(6):646–59.
5. Cohn JN, Ferrari R, Sharpe N. Cardiac remodeling—concepts and clinical implications: a consensus paper from an international forum on cardiac remodeling. J Am Coll Cardiol. 2000;35(3):569–82.
6. Kong P, Christia P, Frangogiannis NG. The pathogenesis of cardiac fibrosis. Cell Mol Life Sci. 2014;71(4):549–74.
7. Rockey DC, Bell PD, Hill JA. Fibrosis — a common pathway to organ injury and failure. N Engl J Med. 2015;372(12):1138–49.
8. Hinderer S, Schenke-Layland K. Cardiac fibrosis – a short review of causes and therapeutic strategies. Adv Drug Deliv Rev. 2019;146:77–82.
9. Weber KT, Sun Y, Bhattacharya SK, Ahokas RA, Gerling IC. Myofibroblast-mediated mechanisms of pathological remodelling of the heart. Nat Rev Cardiol. 2013;10(1):15–26.
10. Berk BC, Fujiwara K, Lehoux S. ECM remodeling in hypertensive heart disease. J Clin Invest. 2007;117(3):568–75.
11. Travers JG, Kamal FA, Robbins J, Yutzey KE, Blaxall BC. Cardiac Fibrosis. Circ Res. 2016;118(6):1021–40.
12. Baicu CF, Stroud JD, Livesay VA, Hapke E, Holder J, Spinale FG, et al. Changes in extracellular collagen matrix alter myocardial systolic performance. Am J Physiol Heart Circ Physiol. 2003;284(1):H122–32.
13. Burlew BS, Weber KT. Cardiac fibrosis as a cause of diastolic dysfunction. Herz. 2002;27(2):92–8.
14. Kato S, Saito N, Kirigaya H, Gyotoku D, Iinuma N, Kusakawa Y, et al. Prognostic significance of quantitative assessment of focal myocardial fibrosis in patients with heart failure with preserved ejection fraction. Int J Cardiol. 2015;191:314–9.
15. Wong TC, Piehler KM, Kang IA, Kadakkal A, Kellman P, Schwartzman DS, et al. Myocardial extracellular volume fraction quantified by cardiovascular magnetic resonance is increased in diabetes and associated with mortality and incident heart failure admission. Eur Heart J. 2014;35(10):657–64.
16. Aoki T, Fukumoto Y, Sugimura K, Oikawa M, Satoh K, Nakano M, et al. Prognostic impact of myocardial interstitial fibrosis in non-ischemic heart failure. -comparison between preserved and reduced ejection fraction heart failure. Circ J. 2011;75(11):2605–13.
17. Schelbert EB, Fridman Y, Wong TC, Daya HA, Piehler KM, Kadakkal A, et al. Temporal relation between myocardial fibrosis and heart failure with preserved ejection fraction: association with baseline disease severity and subsequent outcome. JAMA Cardiol. 2017;2(9):995–1006.
18. Nguyen TP, Qu Z, Weiss JN. Cardiac fibrosis and arrhythmogenesis: the road to repair is paved with perils. J Mol Cell Cardiol. 2014;70:83–91.
19. Maccallum JB. A contribution to the knowledge of the pathology of fragmentation and segmentation, and fibrosis of the myocardium. J Exp Med. 1899;4(3–4):409–24.
20. Evans RD, Nuzum FR. Fibrosis of the myocardium. Cal West Med. 1929;30(1):11–6.
21. White FC, Sanders M, Peterson T, Bloor CM. Ischemic myocardial injury after exercise stress in the pressure-overloaded heart. Am J Pathol. 1979;97(3):473–88.
22. Lee RJ, Zaidi IH, Baky SH. Pathophysiology of the atherosclerotic rabbit. Environ Health Perspect. 1978;26:225–31.
23. Judd JT, Wexler BC, Williamson G, Bickers M, Springs M. Myocardial connective tissue metabolism in response to injury: histological and chemical studies of mucopolysaccharide and collagen in rat hearts after isoproterenol-induced infarction. Circ Res. 1969;25(2):201–14.
24. Opie LH. Acute metabolic response in myocardial infarction. Br Heart J. 1971;33(Suppl):Suppl:129–37.

25. Fishbein MC, Maclean D, Maroko PR. Experimental myocardial infarction in the rat: qualitative and quantitative changes during pathologic evolution. Am J Pathol. 1978;90(1):57–70.
26. Meerson FZ, Alekhina GM, Aleksandrov PN, Bazardjan AG. Dynamics of nucleic acid and protein synthesis of the myocardium in compensatory hyperfunction and hypertrophy of the heart. Am J Cardiol. 1968;22(3):337–48.
27. Grove D, Zak R, Nair KG, Aschenbrenner V. Biochemical correlates of cardiac hypertrophy: IV. observations on the cellular organization of growth during myocardial hypertrophy in the rat. Circ Res. 1969;25(4):473–85.
28. Oldershaw PJ, Brooksby IA, Davies MJ, Coltart DJ, Jenkins BS, Webb-Peploe MM. Correlations of fibrosis in endomyocardial biopsies from patients with aortic valve disease. Br Heart J. 1980;44(6):609–11.
29. Averill DB, Ferrario CM, Tarazi RC, Sen S, Bajbus R. Cardiac performance in rats with renal hypertension. Circ Res. 1976;38(4):280–8.
30. Krayenbuehl HP, Hess OM, Schneider J, Turina M. Physiologic or pathologic hypertrophy. Eur Heart J. 1983;4 Suppl A:29–34.
31. Dunnigan A, Staley NA, Smith SA, Ella Pierpont M, Judd D, Benditt DG, et al. Cardiac and skeletal muscle abnormalities in cardiomyopathy: comparison of patients with ventricular tachycardia or congestive heart failure. J Am Coll Cardiol. 1987;10(3):608–18.
32. OakleyC.Ventricularhypertrophyincardiomyopathy.BrHeartJ.1971;33(Suppl):Suppl:179–86.
33. Goodwin JF. Alcohol and the heart: alcoholic cardiomyopathy. J R Coll Physicians Lond. 1977;12(1):5–11.
34. St John Sutton MG, Lie JT, Anderson KR, O'Brien PC, Frye RL. Histopathological specificity of hypertrophic obstructive cardiomyopathy. Myocardial fibre disarray and myocardial fibrosis. Br Heart J. 1980;44(4):433–43.
35. Nakayama Y, Shimizu G, Hirota Y, Saito T, Kino M, Kitaura Y, et al. Functional and histopathologic correlation in patients with dilated cardiomyopathy: an integrated evaluation by multivariate analysis. J Am Coll Cardiol. 1987;10(1):186–92.
36. Bortone AS, Hess OM, Chiddo A, Gaglione A, Locuratolo N, Caruso G, et al. Functional and structural abnormalities in patients with dilated cardiomyopathy. J Am Coll Cardiol. 1989;14(3):613–23.
37. Hamby RI, Zoneraich S, Sherman L. Diabetic cardiomyopathy. JAMA. 1974;229(13):1749–54.
38. Regan TJ, Lyons MM, Ahmed SS, Levinson GE, Oldewurtel HA, Ahmad MR, et al. Evidence for cardiomyopathy in familial diabetes mellitus. J Clin Invest. 1977;60(4):884–99.
39. Seneviratne BI. Diabetic cardiomyopathy: the preclinical phase. Br Med J. 1977;1(6074):1444–6.
40. Dragatakis LN, Klassen J, Hüttner I, Fraser DG, Poirier NL, Klassen GA. Autoimmune myocarditis: a clinical entity. Can Med Assoc J. 1979;120(3):317–21.
41. Billingham ME, Tazelaar HD. The morphological progression of viral myocarditis. Postgrad Med J. 1986;62(728):581–4.
42. Matsumori A, Kawai C. Viral myocarditis and cardiomyopathy. Jpn J Med. 1989;28(3):416–8.
43. Bulkley BH, Ridolfi RL, Salyer WR, Hutchins GM. Myocardial lesions of progressive systemic sclerosis. A cause of cardiac dysfunction. Circulation (New York, NY). 1976;53(3):483–90.
44. Swirski FK, Nahrendorf M. Cardioimmunology: the immune system in cardiac homeostasis and disease. Nat Rev Immunol. 2018;18(12):733–44.
45. Olson HM, Young DM, Prieur DJ, LeRoy AF, Reagan RL. Electrolyte and morphologic alterations of myocardium in adriamycin-treated rabbits. Am J Pathol. 1974;77(3):439–54.
46. Wilcox RG, James PD, Toghill PJ. Endomyocardial fibrosis associated with daunorubicin therapy. Br Heart J. 1976;38(8):860–3.
47. Khan MY. Radiation-induced cardiomyopathy. I. An electron microscopic study of cardiac muscle cells. Am J Pathol. 1973;73(1):131–46.
48. O'Donnell L, O'Neill T, Toner M, O'Briain S, Graham I. Myocardial hypertrophy, fibrosis and infarction following exposure of the heart to radiation for Hodgkin's disease. Postgrad Med J. 1986;62(733):1055–8.

49. Litviňuková M, Talavera-López C, Maatz H, Reichart D, Worth CL, Lindberg EL, et al. Cells of the adult human heart. Nature (London). 2020;588(7838):466–72.
50. Tirziu D, Giordano FJ, Simons M. Cell Communications in the Heart. Circulation. 2010;122(9):928–37.
51. Silva AC, Pereira C, Fonseca ACRG, Pinto-do-Ó P, Nascimento DS. Bearing my heart: the role of extracellular matrix on cardiac development, homeostasis, and injury response. Front Cell Dev Biol. 2021;8(1705):621644.
52. Rienks M, Papageorgiou A-P, Frangogiannis NG, Heymans S. Myocardial Extracellular Matrix. Circ Res. 2014;114(5):872–88.
53. Howard CM, Baudino TA. Dynamic cell-cell and cell-ECM interactions in the heart. J Mol Cell Cardiol. 2014;70:19–26.
54. Martins-Marques T, Hausenloy DJ, Sluijter JPG, Leybaert L, Girao H. Intercellular communication in the heart: therapeutic opportunities for cardiac ischemia. Trends Mol Med. 2021;27(3):248–62.
55. Samarel AM. Costameres, focal adhesions, and cardiomyocyte mechanotransduction. Am J Physiol Heart Circ Physiol. 2005;289(6):H2291–H301.
56. Shoykhet M, Trenz S, Kempf E, Williams T, Gerull B, Schinner C, et al. Cardiomyocyte adhesion and hyperadhesion differentially require ERK1/2 and plakoglobin. JCI. Insight. 2020;5(18):e140066.
57. Talman V, Kivelä R. Cardiomyocyte—endothelial cell interactions in cardiac remodeling and regeneration. Front Cardiovasc Med. 2018;5:101.
58. Forte M, Madonna M, Schiavon S, Valenti V, Versaci F, Zoccai GB, et al. Cardiovascular pleiotropic effects of natriuretic peptides. Int J Mol Sci. 2019;20(16):3874.
59. Frangogiannis NG. Cardiac fibrosis: cell biological mechanisms, molecular pathways and therapeutic opportunities. Mol Aspects Med. 2019;65:70–99.
60. Weber K, Janicki J, Shroff S, Pick R, Chen R, Bashey R. Collagen remodeling of the pressure-overloaded, hypertrophied nonhuman primate myocardium. Circ Res. 1988;62(4):757–65.
61. Yau JW, Teoh H, Verma S. Endothelial cell control of thrombosis. BMC Cardiovasc Disord. 2015;15(1):130.
62. Sun X, Nkennor B, Mastikhina O, Soon K, Nunes SS. Endothelium-mediated contributions to fibrosis. Semin Cell Dev Biol. 2020;101:78–86.
63. Touyz RM, Alves-Lopes R, Rios FJ, Camargo LL, Anagnostopoulou A, Arner A, et al. Vascular smooth muscle contraction in hypertension. Cardiovasc Res. 2018;114(4):529–39.
64. Murray IR, Baily JE, Chen WCW, Dar A, Gonzalez ZN, Jensen AR, et al. Skeletal and cardiac muscle pericytes: functions and therapeutic potential. Pharmacol Ther. 2017;171:65–74.
65. Díaz-Flores L, Gutiérrez R, Madrid JF, Varela H, Valladares F, Acosta E, et al. Pericytes. Morphofunction, interactions and pathology in a quiescent and activated mesenchymal cell niche. Histol Histopathol. 2009;24(7):909–69.
66. Sansonetti M, Waleczek FJG, Jung M, Thum T, Perbellini F. Resident cardiac macrophages: crucial modulators of cardiac (patho)physiology. Basic Res Cardiol. 2020;115(6):77.
67. Parichatikanond W, Luangmonkong T, Mangmool S, Kurose H. Therapeutic targets for the treatment of cardiac fibrosis and cancer: focusing on TGF-β signaling. Front Cardiovasc Med. 2020;7:34.
68. Kanisicak O, Khalil H, Ivey MJ, Karch J, Maliken BD, Correll RN, et al. Genetic lineage tracing defines myofibroblast origin and function in the injured heart. Nat Commun. 2016;7:12260.
69. Aujla PK, Kassiri Z. Diverse origins and activation of fibroblasts in cardiac fibrosis. Cell Signal. 2021;78:109869.
70. Fu X, Khalil H, Kanisicak O, Boyer JG, Vagnozzi RJ, Maliken BD, et al. Specialized fibroblast differentiated states underlie scar formation in the infarcted mouse heart. J Clin Invest. 2018;128(5):2127–43.
71. Tomasek JJ, Gabbiani G, Hinz B, Chaponnier C, Brown RA. Myofibroblasts and mechano-regulation of connective tissue remodelling. Nat Rev Mol Cell Biol. 2002;3(5):349–63.

72. Mitchell MD, Laird RE, Brown RD, Long CS. IL-1β stimulates rat cardiac fibroblast migration via MAP kinase pathways. Am J Physiol Heart Circ Physiol. 2007;292(2):H1139–H47.
73. Bowler MA, Bersi MR, Ryzhova LM, Jerrell RJ, Parekh A, Merryman WD. Cadherin-11 as a regulator of valve myofibroblast mechanobiology. Am J Physiol Heart Circ Physiol. 2018;315(6):H1614–H26.
74. Gibb AA, Lazaropoulos MP, Elrod JW. Myofibroblasts and fibrosis. Circ Res. 2020;127(3):427–47.
75. Nunohiro T, Ashizawa N, Graf K, Hsueh W, Yano K. Angiotensin II promotes integrin-mediated collagen gel contraction by adult rat cardiac fibroblasts. Jpn Heart J. 1999;40(4):461–9.
76. Kurose H. Cardiac fibrosis and fibroblasts. Cells. 2021;10(7):1716.
77. Umbarkar P, Ejantkar S, Tousif S, Lal H. Mechanisms of fibroblast activation and myocardial fibrosis: lessons learned from FB-specific conditional mouse models. Cell. 2021;10(9):2412.
78. Dobaczewski M, Chen W, Frangogiannis NG. Transforming growth factor (TGF)-β signaling in cardiac remodeling. J Mol Cell Cardiol. 2011;51(4):600–6.
79. Li R-K, Li G, Mickle DA, Weisel RD, Merante F, Luss H, et al. Overexpression of transforming growth factor-β1 and insulin-like growth factor-I in patients with idiopathic hypertrophic cardiomyopathy. Circulation. 1997;96(3):874–81.
80. Petrov VV, Fagard RH, Lijnen PJ. Stimulation of collagen production by transforming growth factor-beta1 during differentiation of cardiac fibroblasts to myofibroblasts. Hypertension. 2002;39(2):258–63.
81. Perrucci GL, Barbagallo VA, Corlianò M, Tosi D, Santoro R, Nigro P, et al. Integrin ανβ5 in vitro inhibition limits pro-fibrotic response in cardiac fibroblasts of spontaneously hypertensive rats. J Transl Med. 2018;16(1):352.
82. Villarreal FJ, Kim NN, Ungab GD, Printz MP, Dillmann WH. Identification of functional angiotensin II receptors on rat cardiac fibroblasts. Circulation. 1993;88(6):2849–61.
83. Lee AA, Dillmann WH, McCulloch AD, Villarreal FJ. Angiotensin II stimulates the autocrine production of transforming growth factor-beta 1 in adult rat cardiac fibroblasts. J Mol Cell Cardiol. 1995;27(10):2347–57.
84. Jia L, Li Y, Xiao C, Du J. Angiotensin II induces inflammation leading to cardiac remodeling. Front Biosci (Landmark Ed). 2012;17:221–31.
85. Sun Y, Zhang JQ, Zhang J, Ramires FJ. Angiotensin II, transforming growth factor-beta1 and repair in the infarcted heart. J Mol Cell Cardiol. 1998;30(8):1559–69.
86. Weber KT. Fibrosis in hypertensive heart disease: focus on cardiac fibroblasts. J Hypertens. 2004;22(1):47–50.
87. Katwa LC, Campbell SE, Tyagi SC, Lee SJ, Cicila GT, Weber KT. Cultured Myofibroblasts generate angiotensin peptidesde novo. J Mol Cell Cardiol. 1997;29(5):1375–86.
88. Chen C, Li R, Ross RS, Manso AM. Integrins and integrin-related proteins in cardiac fibrosis. J Mol Cell Cardiol. 2016;93:162–74.
89. Hokimoto S, Yasue H, Fujimoto K, Yamamoto H, Nakao K, Kaikita K, et al. Expression of angiotensin-converting enzyme in remaining viable myocytes of human ventricles after myocardial infarction. Circulation. 1996;94(7):1513–8.
90. Leask A. Potential therapeutic targets for cardiac fibrosis: TGFβ, angiotensin, endothelin, CCN2, and PDGF, partners in fibroblast activation. Circ Res. 2010;106(11):1675–80.
91. Voloshenyuk TG, Hart AD, Khoutorova E, Gardner JD. TNF-α increases cardiac fibroblast lysyl oxidase expression through TGF-β and PI3Kinase signaling pathways. Biochem Biophys Res Commun. 2011;413(2):370–5.
92. Palmer JN, Hartogensis WE, Patten M, Fortuin FD, Long CS. Interleukin-1 beta induces cardiac myocyte growth but inhibits cardiac fibroblast proliferation in culture. J Clin Invest. 1995;95(6):2555–64.
93. Zhang Y, Wang J-H, Zhang Y-Y, Wang Y-Z, Wang J, Zhao Y, et al. Deletion of interleukin-6 alleviated interstitial fibrosis in streptozotocin-induced diabetic cardiomyopathy of mice through affecting TGFβ1 and miR-29 pathways. Sci Rep. 2016;6(1):1–11.

94. Datta R, Bansal T, Rana S, Datta K, Datta Chaudhuri R, Chawla-Sarkar M, et al. Myocyte-derived Hsp90 modulates collagen upregulation via biphasic activation of STAT-3 in fibroblasts during cardiac hypertrophy. Mol Cell Biol. 2017;37(6):e00611–6.
95. Herum KM, Lunde IG, McCulloch AD, Christensen G. The soft- and hard-heartedness of cardiac fibroblasts: mechanotransduction signaling pathways in fibrosis of the heart. J Clin Med. 2017;6(5):53.
96. Thompson SA, Copeland CR, Reich DH, Tung L. Mechanical coupling between myofibroblasts and cardiomyocytes slows electric conduction in fibrotic cell monolayers. Circulation. 2011;123(19):2083–93.
97. Rohr S. Myofibroblasts in diseased hearts: new players in cardiac arrhythmias? Heart Rhythm. 2009;6(6):848–56.
98. Herum KM, Choppe J, Kumar A, Engler AJ, McCulloch AD. Mechanical regulation of cardiac fibroblast profibrotic phenotypes. Mol Biol Cell. 2017;28(14):1871–82.
99. Zohar R, Zhu B, Liu P, Sodek J, McCulloch C. Increased cell death in osteopontin-deficient cardiac fibroblasts occurs by a caspase-3-independent pathway. Am J Physiol Heart Circ Physiol. 2004;287(4):H1730–H9.
100. Adams JC, Lawler J. The thrombospondins. Int J Biochem Cell Biol. 2004;36(6):961–8.
101. Desmoulière A, Badid C, Bochaton-Piallat ML, Gabbiani G. Apoptosis during wound healing, fibrocontractive diseases and vascular wall injury. Int J Biochem Cell Biol. 1997;29(1):19–30.
102. Hinz B, Mastrangelo D, Iselin CE, Chaponnier C, Gabbiani G. Mechanical tension controls granulation tissue contractile activity and myofibroblast differentiation. Am J Pathol. 2001;159(3):1009–20.
103. D'Urso M, Kurniawan NA. Mechanical and physical regulation of fibroblast-Myofibroblast transition: from cellular Mechanoresponse to tissue pathology. Front Bioeng Biotechnol. 2020;8:609653.
104. Palano G, Foinquinos A, Müllers E. In vitro assays and imaging methods for drug discovery for cardiac fibrosis. Front Physiol. 2021;12:697270.
105. Portillo-Lara R, Spencer AR, Walker BW, Shirzaei Sani E, Annabi N. Biomimetic cardiovascular platforms for in vitro disease modeling and therapeutic validation. Biomaterials. 2019;198:78–94.
106. Kaese S, Verheule S. Cardiac electrophysiology in mice: a matter of size. Front Physiol. 2012;3:345.
107. Sala L, Bellin M, Mummery CL. Integrating cardiomyocytes from human pluripotent stem cells in safety pharmacology: has the time come? Br J Pharmacol. 2017;174(21):3749–65.
108. Downey JM, Cohen MV. Why do we still not have cardioprotective drugs? Circ J. 2009;73(7):1171–7.
109. Figtree GA, Bubb KJ, Tang O, Kizana E, Gentile C. Vascularized cardiac spheroids as novel 3D in vitro models to study cardiac fibrosis. Cells Tissues Organs. 2017;204(3–4):191–8.
110. Lee M-O, Jung KB, Jo S-J, Hyun S-A, Moon K-S, Seo J-W, et al. Modelling cardiac fibrosis using three-dimensional cardiac microtissues derived from human embryonic stem cells. J Biol Eng. 2019;13(1):15.
111. Mastikhina O, Moon B-U, Williams K, Hatkar R, Gustafson D, Mourad O, et al. Human cardiac fibrosis-on-a-chip model recapitulates disease hallmarks and can serve as a platform for drug testing. Biomaterials. 2020;233:119741.
112. Pinto AR, Ilinykh A, Ivey MJ, Kuwabara JT, D'Antoni ML, Debuque R, et al. Revisiting cardiac cellular composition. Circ Res. 2016;118(3):400–9.
113. Peter AK, Bjerke MA, Leinwand LA. Biology of the cardiac myocyte in heart disease. Mol Biol Cell. 2016;27(14):2149–60.
114. Kaur G, Dufour JM. Cell lines: valuable tools or useless artifacts. Spermatogenesis. 2012;2(1):1–5.
115. Pan C, Kumar C, Bohl S, Klingmueller U, Mann M. Comparative proteomic phenotyping of cell lines and primary cells to assess preservation of cell type-specific functions. Mol Cell Proteomics. 2009;8(3):443–50.

116. Louch WE, Sheehan KA, Wolska BM. Methods in cardiomyocyte isolation, culture, and gene transfer. J Mol Cell Cardiol. 2011;51(3):288–98.
117. Wang J, Seth A, McCulloch CAG. Force regulates smooth muscle actin in cardiac fibroblasts. Am J Physiol Heart Circ Physiol. 2000;279(6):2776–85.
118. Wang J, Chen H, Seth A, McCulloch CA. Mechanical force regulation of myofibroblast differentiation in cardiac fibroblasts. Am J Physiol Heart Circ Physiol. 2003;285(5):1871–81.
119. Zhao H, Li X, Zhao S, Zeng Y, Zhao L, Ding H, et al. Microengineered in vitro model of cardiac fibrosis through modulating myofibroblast mechanotransduction. Biofabrication. 2014;6(4):045009.
120. Kong M, Lee J, Yazdi IK, Miri AK, Lin Y-D, Seo J, et al. Cardiac fibrotic remodeling on a chip with dynamic mechanical stimulation. Adv Healthc Mater. 2019;8(3):e1801146-e.
121. Yang X, Pabon L, Murry CE. Engineering adolescence: maturation of human pluripotent stem cell-derived cardiomyocytes. Circ Res. 2014;114(3):511–23.
122. Leonard A, Bertero A, Powers JD, Beussman KM, Bhandari S, Regnier M, et al. Afterload promotes maturation of human induced pluripotent stem cell derived cardiomyocytes in engineered heart tissues. J Mol Cell Cardiol. 2018;118:147–58.
123. Karbassi E, Fenix A, Marchiano S, Muraoka N, Nakamura K, Yang X, et al. Cardiomyocyte maturation: advances in knowledge and implications for regenerative medicine. Nat Rev Cardiol. 2020;17(6):341–59.
124. Chang HY, Chi JT, Dudoit S, Bondre C, van de Rijn M, Botstein D, et al. Diversity, topographic differentiation, and positional memory in human fibroblasts. Proc Natl Acad Sci U S A. 2002;99(20):12877–82.
125. Zhang H, Tian L, Shen M, Tu C, Wu H, Gu M, et al. Generation of quiescent cardiac fibroblasts from human induced pluripotent stem cells for in vitro modeling of cardiac fibrosis. Circ Res. 2019;125(5):552–66.
126. Furtado MB, Costa MW, Pranoto EA, Salimova E, Pinto AR, Lam NT, et al. Cardiogenic genes expressed in cardiac fibroblasts contribute to heart development and repair. Circ Res. 2014;114(9):1422–34.
127. Guo Y, Zhang XZ, Wei Y, Guo C, Li RX, Zeng QC, et al. Culturing of ventricle cells at high density and construction of engineered cardiac cell sheets without scaffold. Int Heart J. 2009;50(5):653–62.
128. Zhao Y, Rafatian N, Wang EY, Feric NT, Lai BFL, Knee-Walden EJ, et al. Engineering microenvironment for human cardiac tissue assembly in heart-on-a-chip platform. Matrix Biol. 2020;85-86:189–204.
129. van Spreeuwel ACC, Bax NAM, van Nierop BJ, Aartsma-Rus A, Goumans MTH, Bouten CVC. Mimicking cardiac fibrosis in a dish: fibroblast density rather than collagen density weakens Cardiomyocyte function. J Cardiovasc Transl Res. 2017;10(2):116–27.
130. Iseoka H, Miyagawa S, Fukushima S, Saito A, Masuda S, Yajima S, et al. Pivotal role of non-cardiomyocytes in electromechanical and therapeutic potential of induced pluripotent stem cell-derived engineered cardiac tissue. Tissue Eng Part A. 2018;24(3–4):287–300.
131. Wang EY, Rafatian N, Zhao Y, Lee A, Lai BFL, Lu RX, et al. Biowire model of interstitial and focal cardiac fibrosis. ACS Cent Sci. 2019;5(7):1146–58.
132. Zile MR, Baicu CF, Ikonomidis JS, Stroud RE, Nietert PJ, Bradshaw AD, et al. Myocardial stiffness in patients with heart failure and a preserved ejection fraction: contributions of collagen and titin. Circulation. 2015;131(14):1247–59.
133. Arani A, Arunachalam SP, Chang ICY, Baffour F, Rossman PJ, Glaser KJ, et al. Cardiac MR elastography for quantitative assessment of elevated myocardial stiffness in cardiac amyloidosis. J Magn Reson Imaging. 2017;46(5):1361–7.
134. Chaturvedi RR, Herron T, Simmons R, Shore D, Kumar P, Sethia B, et al. Passive stiffness of myocardium from congenital heart disease and implications for diastole. Circulation. 2010;121(8):979–88.
135. Chen Q-Z, Harding SE, Ali NN, Lyon AR, Boccaccini AR. Biomaterials in cardiac tissue engineering: ten years of research survey. Mater Sci Eng R: Rep. 2008;59(1):1–37.

136. McCrary MW, Bousalis D, Mobini S, Song YH, Schmidt CE. Decellularized tissues as platforms for in vitro modeling of healthy and diseased tissues. Acta Biomater. 2020;111:1–19.
137. Chang MG, Zhang Y, Chang CY, Xu L, Emokpae R, Tung L, et al. Spiral waves and reentry dynamics in an in vitro model of the healed infarct border zone. Circ Res. 2009;105(11):1062–71.
138. McCain ML, Yuan H, Pasqualini FS, Campbell PH, Parker KK. Matrix elasticity regulates the optimal cardiac myocyte shape for contractility. Am J Physiol Heart Circ Physiol. 2014;306(11):H1525–39.
139. Bracco Gartner TCL, Stein JM, Muylaert DEP, Bouten CVC, Doevendans PA, Khademhosseini A, et al. Advanced in vitro modeling to study the paradox of mechanically induced cardiac fibrosis. Tissue Eng Part C Methods. 2021;27(2):100–14.
140. Unal AZ, West JL. Synthetic ECM: bioactive synthetic hydrogels for 3D tissue engineering. Bioconjug Chem. 2020;31(10):2253–71.
141. Pellman J, Zhang J, Sheikh F. Myocyte-fibroblast communication in cardiac fibrosis and arrhythmias: mechanisms and model systems. J Mol Cell Cardiol. 2016;94:22–31.
142. Li Y, Asfour H, Bursac N. Age-dependent functional crosstalk between cardiac fibroblasts and cardiomyocytes in a 3D engineered cardiac tissue. Acta Biomater. 2017;55:120–30.
143. Worke LJ, Barthold JE, Seelbinder B, Novak T, Main RP, Harbin SL, et al. Densification of type I collagen matrices as a model for cardiac fibrosis. Adv Healthc Mater. 2017;6(22)
144. Ronaldson-Bouchard K, Ma SP, Yeager K, Chen T, Song L, Sirabella D, et al. Advanced maturation of human cardiac tissue grown from pluripotent stem cells. Nature. 2018;556(7700):239–43.
145. Guo Y, Pu WT. Cardiomyocyte maturation. Circ Res. 2020;126(8):1086–106.
146. Potter E, Marwick TH. Assessment of left ventricular function by echocardiography: the case for routinely adding global longitudinal strain to ejection fraction. JACC Cardiovasc Imaging. 2018;11(2 Pt 1):260–74.
147. Wang VY, Casta C, Zhu YM, Cowan BR, Croisille P, Young AA, et al. Image-based investigation of human in vivo Myofibre strain. IEEE Trans Med Imaging. 2016;35(11):2486–96.
148. MacGowan GA, Shapiro EP, Azhari H, Siu CO, Hees PS, Hutchins GM, et al. Noninvasive measurement of shortening in the fiber and cross-fiber directions in the normal human left ventricle and in idiopathic dilated cardiomyopathy. Circulation. 1997;96(2):535–41.
149. Steen H, Giusca S, Montenbruck M, Patel AR, Pieske B, Florian A, et al. Left and right ventricular strain using fast strain-encoded cardiovascular magnetic resonance for the diagnostic classification of patients with chronic non-ischemic heart failure due to dilated, hypertrophic cardiomyopathy or cardiac amyloidosis. J Cardiovasc Magn Reson. 2021;23(1):45.
150. Marwick TH, Shah SJ, Thomas JD. Myocardial strain in the assessment of patients with heart failure: a review. JAMA Cardiol. 2019;4(3):287–94.
151. Rogers AJ, Miller JM, Kannappan R, Sethu P. Cardiac tissue chips (CTCs) for Modeling cardiovascular disease. IEEE Trans Biomed Eng. 2019;66(12):3436–43.
152. Mainardi A, Carminati F, Ugolini GS, Occhetta P, Isu G, Robles Diaz D, et al. A dynamic microscale mid-throughput fibrosis model to investigate the effects of different ratios of cardiomyocytes and fibroblasts. Lab Chip. 2021;21(21):4177–95.
153. Pitoulis FG, Nunez-Toldra R, Xiao K, Kit-Anan W, Mitzka S, Jabbour RJ, et al. Remodelling of adult cardiac tissue subjected to physiological and pathological mechanical load in vitro. Cardiovasc Res. 2021;118:814.
154. Sadeghi AH, Shin SR, Deddens JC, Fratta G, Mandla S, Yazdi IK, et al. Engineered 3D cardiac fibrotic tissue to study fibrotic remodeling. Adv Healthc Mater. 2017;6(11) https://doi.org/10.1002/adhm.201601434.
155. Rogers AJ, Kannappan R, Abukhalifeh H, Ghazal M, Miller JM, El-Baz A, et al. Hemodynamic stimulation using the biomimetic cardiac tissue model (BCTM) enhances maturation of human induced pluripotent stem cell-derived cardiomyocytes. Cells Tissues Organs. 2018;206(1–2):82–94.

156. Giridharan GA, Nguyen M-D, Estrada R, Parichehreh V, Hamid T, Ismahil MA, et al. Microfluidic cardiac cell culture model (μCCCM). Anal Chem. 2010;82(18):7581–7.
157. Santos GL, Hartmann S, Zimmermann WH, Ridley A, Lutz S. Inhibition of Rho-associated kinases suppresses cardiac myofibroblast function in engineered connective and heart muscle tissues. J Mol Cell Cardiol. 2019;134:13–28.
158. Pakshir P, Noskovicova N, Lodyga M, Son DO, Schuster R, Goodwin A, et al. The myofibroblast at a glance. J Cell Sci. 2020;133(13):jcs227900.
159. Martino F, Perestrelo AR, Vinarský V, Pagliari S, Forte G. Cellular Mechanotransduction: from tension to function. Front Physiol. 2018;9:824.
160. Butt RP, Bishop JE. Mechanical load enhances the stimulatory effect of serum growth factors on cardiac fibroblast procollagen synthesis. J Mol Cell Cardiol. 1997;29(4):1141–51.
161. MacKenna DA, Dolfi F, Vuori K, Ruoslahti E. Extracellular signal-regulated kinase and c-Jun NH2-terminal kinase activation by mechanical stretch is integrin-dependent and matrix-specific in rat cardiac fibroblasts. J Clin Invest. 1998;101(2):301–10.
162. Dalla Costa AP, Clemente CFMZ, Carvalho HF, Carvalheira JB, Nadruz W Jr, Franchini KG. FAK mediates the activation of cardiac fibroblasts induced by mechanical stress through regulation of the mTOR complex. Cardiovasc Res. 2009;86(3):421–31.
163. Watson CJ, Phelan D, Collier P, Horgan S, Glezeva N, Cooke G, et al. Extracellular matrix sub-types and mechanical stretch impact human cardiac fibroblast responses to transforming growth factor beta. Connect Tissue Res. 2014;55(3):248–56.
164. Zhang S, Weinheimer C, Courtois M, Kovacs A, Zhang CE, Cheng AM, et al. The role of the Grb2-p38 MAPK signaling pathway in cardiac hypertrophy and fibrosis. J Clin Invest. 2003;111(6):833–41.
165. Li X, Wang G, QiLi M, Liang H, Li T, E X, et al. Aspirin reduces cardiac interstitial fibrosis by inhibiting Erk1/2-Serpine2 and P-Akt signalling pathways. Cell Physiol Biochem. 2018;45(5):1955–65.
166. Ichihara S, Li P, Mise N, Suzuki Y, Izuoka K, Nakajima T, et al. Ablation of aryl hydrocarbon receptor promotes angiotensin II-induced cardiac fibrosis through enhanced c-Jun/HIF-1α signaling. Arch Toxicol. 2019;93(6):1543–53.
167. Lammerding J, Kamm RD, Lee RT. Mechanotransduction in cardiac myocytes. Ann N Y Acad Sci. 2004;1015:53–70.
168. Frangogiannis NG. Transforming growth factor–β in tissue fibrosis. J Exp Med. 2020;217(3):e20190103.
169. Poobalarahi F, Baicu CF, Bradshaw AD. Cardiac myofibroblasts differentiated in 3D culture exhibit distinct changes in collagen I production, processing, and matrix deposition. Am J Physiol Heart Circ Physiol. 2006;291(6):2924–32.
170. Ng CP, Swartz MA. Fibroblast alignment under interstitial fluid flow using a novel 3-D tissue culture model. Am J Physiol Heart Circ Physiol. 2003;284(5):H1771–H7.
171. Galie PA, Stegemann JP. Simultaneous application of interstitial flow and cyclic mechanical strain to a three-dimensional cell-seeded hydrogel. Tissue Eng Part C Methods. 2011;17(5):527–36.
172. Galie PA, Russell MW, Westfall MV, Stegemann JP. Interstitial fluid flow and cyclic strain differentially regulate cardiac fibroblast activation via AT1R and TGF-β1. Exp Cell Res. 2012;318(1):75–84.
173. Khalil H, Kanisicak O, Prasad V, Correll RN, Fu X, Schips T, et al. Fibroblast-specific TGF-β–Smad2/3 signaling underlies cardiac fibrosis. J Clin Invest. 2017;127(10):3770–83.
174. Cho N, Razipour SE, McCain ML. Featured article: TGF-β1 dominates extracellular matrix rigidity for inducing differentiation of human cardiac fibroblasts to myofibroblasts. Exp Biol Med (Maywood). 2018;243(7):601–12.
175. Galie PA, Westfall MV, Stegemann JP. Reduced serum content and increased matrix stiffness promote the cardiac myofibroblast transition in 3D collagen matrices. Cardiovasc Pathol. 2011;20(6):325–33.
176. Heras-Bautista CO, Mikhael N, Lam J, Shinde V, Katsen-Globa A, Dieluweit S, et al. Cardiomyocytes facing fibrotic conditions re-express extracellular matrix transcripts. Acta Biomater. 2019;89:180–92.

177. Fountoulaki K, Dagres N, Iliodromitis EK. Cellular Communications in the Heart. Card Fail Rev. 2015;1(2):64–8.
178. Rienks M, Papageorgiou A-P, Frangogiannis NG, Heymans S. Myocardial extracellular matrix an ever-changing and diverse entity. Circ Res. 2014;114(5):872–88.
179. Croft AP, Campos J, Jansen K, Turner JD, Marshall J, Attar M, et al. Distinct fibroblast subsets drive inflammation and damage in arthritis. Nature. 2019;570(7760):246–51.
180. Phipps RP, Penney DP, Keng P, Quill H, Paxhia A, Derdak S, et al. Characterization of two major populations of lung fibroblasts: distinguishing morphology and discordant display of Thy 1 and class II MHC. Am J Respir Cell Mol Biol. 1989;1(1):65–74.
181. Błyszczuk P, Zuppinger C, Costa A, Nurzynska D, Di Meglio FD, Stellato M, et al. Activated cardiac fibroblasts control contraction of human fibrotic cardiac microtissues by a β-Adrenoreceptor-dependent mechanism. Cell. 2020;9(5):1270.
182. Akinbote A, Beltran-Sastre V, Cherubini M, Visone R, Hajal C, Cobanoglu D, et al. Classical and non-classical fibrosis phenotypes are revealed by lung and cardiac like microvascular tissues on-chip. Front Physiol. 2021;12:735915.
183. Galie PA, Stegemann JP. Injection of mesenchymal stromal cells into a mechanically stimulated in vitro model of cardiac fibrosis has paracrine effects on resident fibroblasts. Cytotherapy. 2014;16(7):906–14.
184. Bracco Gartner TCL, Deddens JC, Mol EA, Magin Ferrer M, van Laake LW, Bouten CVC, et al. Anti-fibrotic effects of cardiac progenitor cells in a 3D-model of human cardiac fibrosis. Front Cardiovasc Med. 2019;6:52.
185. Park S, Nguyen NB, Pezhouman A, Ardehali R. Cardiac fibrosis: potential therapeutic targets. Transl Res. 2019;209:121–37.
186. Li R, Frangogiannis NG. Chemokines in cardiac fibrosis. Curr Opin Physio. 2021;19:80–91.
187. Shimizu T, Liao JK. Rho kinases and cardiac remodeling. Circ J. 2016;80(7):1491–8.
188. Yang J, Yu X, Xue F, Li Y, Liu W, Zhang S. Exosomes derived from cardiomyocytes promote cardiac fibrosis via myocyte-fibroblast cross-talk. Am J Transl Res. 2018;10(12):4350–66.
189. Gurunathan S, Kang MH, Jeyaraj M, Qasim M, Kim JH. Review of the isolation, characterization, biological function, and multifarious therapeutic approaches of exosomes. Cell. 2019;8(4):307.
190. Ranjan P, Kumari R, Verma SK. Cardiac fibroblasts and cardiac fibrosis: precise role of exosomes. Front Cell Dev Biol. 2019;7:318.
191. Bang C, Batkai S, Dangwal S, Gupta SK, Foinquinos A, Holzmann A, et al. Cardiac fibroblast-derived microRNA passenger strand-enriched exosomes mediate cardiomyocyte hypertrophy. J Clin Invest. 2014;124(5):2136–46.
192. Tian C, Gao L, Zimmerman MC, Zucker IH. Myocardial infarction-induced microRNA-enriched exosomes contribute to cardiac Nrf2 dysregulation in chronic heart failure. Am J Physiol Heart Circ Physiol. 2018;314(5):H928–h39.
193. Mohr AM, Mott JL. Overview of microRNA biology. Semin Liver Dis. 2015;35(1):3–11.
194. Heidt T, Courties G, Dutta P, Sager HB, Sebas M, Iwamoto Y, et al. Differential contribution of monocytes to heart macrophages in steady-state and after myocardial infarction. Circ Res. 2014;115(2):284–95.
195. Blanton RM, Carrillo-Salinas FJ, Alcaide P. T-cell recruitment to the heart: friendly guests or unwelcome visitors? Am J Physiol Heart Circ Physiol. 2019;317(1):H124–H40.
196. Hirakawa K, Yamamuro M, Uemura T, Takashio S, Kaikita K, Utsunomiya D, et al. Correlation between microvascular dysfunction and B-type natriuretic peptide levels in non-ischemic heart failure patients with cardiac fibrosis. Int J Cardiol. 2017;228:881–5.
197. Camici PG, Tschöpe C, Di Carli MF, Rimoldi O, Van Linthout S. Coronary microvascular dysfunction in hypertrophy and heart failure. Cardiovasc Res. 2020;116(4):806–16.
198. Trenson S, Hermans H, Craps S, Pokreisz P, de Zeeuw P, Van Wauwe J, et al. Cardiac microvascular endothelial cells in pressure overload-induced heart disease. Circ Heart Fail. 2021;14(1):e006979.
199. Li Y, Song D, Mao L, Abraham DM, Bursac N. Lack of Thy1 defines a pathogenic fraction of cardiac fibroblasts in heart failure. Biomaterials. 2020;236:119824.

Neural Regulation of Cardiac Rhythm

Crystal M. Ripplinger

Abstract The autonomic nervous system (ANS) regulates cardiac function, including chronotropy, inotropy, lusitropy, and dromotropy. The cardiac nervous system includes afferent, efferent, and interconnecting neurons, which span the intrinsic cardiac nervous system (ICNS: within the pericardium), extracardiac ganglia, and central nervous system (CNS: spinal cord and brain). This chapter focuses primarily on efferent sympathetic and parasympathetic neurotransmission and resulting end-organ (i.e., heart) electrophysiological responses. The signaling mechanisms of primary neurotransmitters are discussed, as well as the contribution of co-transmitters. Finally, several in vitro experimental models for the study of neural control of cardiac rhythm are highlighted, including key physiological findings revealed by these experimental models.

Keywords Autonomic · sympathetic · parasympathetic · cardiac electrophysiology · heart rate · arrhythmia

Introduction

The autonomic nervous system (ANS) regulates and fine-tunes nearly every aspect of cardiac physiology, including chronotropy (heart rate), inotropy (contractility), lusitropy (rate of relaxation), and dromotropy (conduction). The cardiac nervous system includes afferent, efferent, and interconnecting neurons, with processing and nested feedback loops occurring at multiple levels, which span the intrinsic cardiac nervous system (ICNS: within the pericardium), extracardiac ganglia, and central nervous system (CNS: spinal cord and brain) [1]. Sensory afferent neurons transmit signals for processing both within the ICNS and to higher levels, which then feedback to alter efferent sympathetic and parasympathetic outflow. Although the hierarchical organization and processing within these nested feedback loops is of utmost interest, this chapter will focus primarily on efferent sympathetic and

C. M. Ripplinger (✉)
Department of Pharmacology, UC Davis School of Medicine, Davis, CA, USA
e-mail: cripplinger@ucdavis.edu

N. L. Parinandi, T. J. Hund (eds.), *Cardiovascular Signaling in Health and Disease*, https://doi.org/10.1007/978-3-031-08309-9_11

parasympathetic neurotransmission and resulting end-organ (i.e., heart) electrophysiological responses.

Neuronal Anatomy and Organization

Sympathetic preganglionic nerves in the spinal cord project to sympathetic postganglionic neurons primarily located in the stellate and cranial thoracic sympathetic chains, as well as the middle and superior cervical ganglia. However, there is also evidence of a population of postganglionic sympathetic neurons located within the ICNS [4]. Parasympathetic preganglionic neurons originate for the most part in the nucleus ambiguous and project to the heart via the vagus nerve. Parasympathetic postganglionic neurons are located in the ICNS (Fig. 1). Within the ICNS are several ganglionated plexi (GP: clusters of neuronal cell bodies and fibers), which serve as integration centers for the extrinsic and intrinsic nervous systems. Most GPs are located in supraventricular regions on the epicardium or in epicardial fat pads [5–8]. The precise locations of GPs are somewhat species-dependent, and only larger mammals (i.e., non-rodents) have ganglia located in the ventricles [9–12].

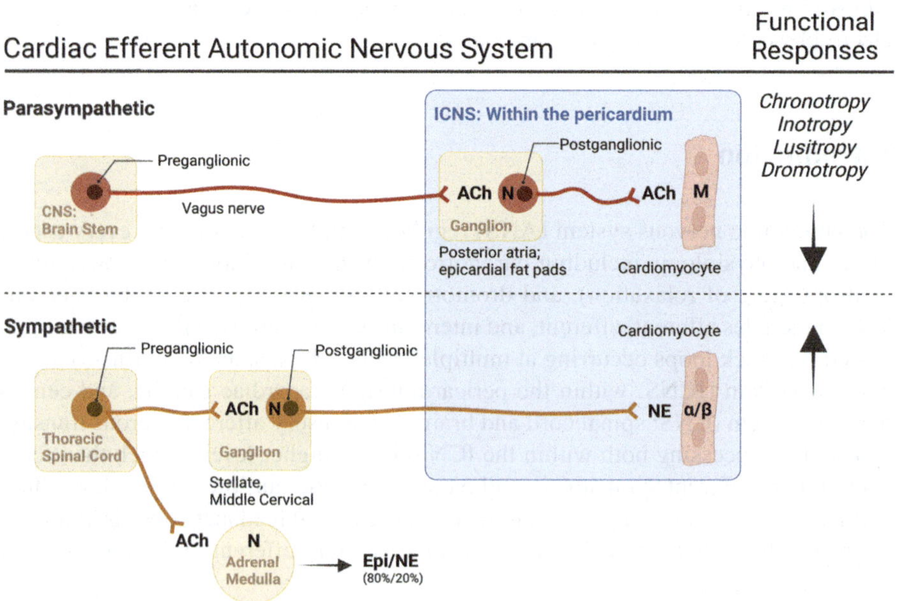

Fig. 1 Simplified schematic of the cardiac efferent ANS, major neurotransmitters and receptors involved, and functional responses. CNS central nervous system; ICNS intrinsic cardiac nervous system; ACh acetylcholine; Epi epinephrine; NE norepinephrine; α/β α- or β-adrenergic receptor; M muscarinic receptor; N nicotinic receptor. (Figure created with Biorender.com)

The distribution of both sympathetic and parasympathetic nerve fibers is non-uniform throughout the heart. Sympathetic nerve density is generally greater in the atria compared to the ventricles, greater at the base of the heart than the apex, and greater at the epi- versus endocardium [13–16]. Despite the long-held belief that the ventricles have very little parasympathetic innervation, there is now abundant evidence refuting this notion and significant data demonstrating dense innervation of the ventricles by parasympathetic fibers (for an excellent discussion of this misconception and a survey of the evidence, see [17]). There are, however, mixed reports on parasympathetic fiber density in the atria compared to ventricles, with greater atrial density found in human hearts [13], but greater ventricular density observed in porcine hearts [18]. There is no difference in parasympathetic fiber density from base to apex (although nerve thickness is greater at the base), and density is generally higher in the endocardium compared to epicardium [13, 18].

The sympathetic postganglionic neurons affect the heart primarily through the release of norepinephrine (NE or noradrenalin). Epinephrine (Epi or adrenalin) is synthesized, stored, and released by the adrenal medulla (chromaffin cells) and impacts the heart via the circulation. NE and Epi bind to adrenergic receptors (α-AR and β-AR) on cardiac myocytes to produce functional responses (Fig. 1). Parasympathetic neurons primarily release acetylcholine (ACh), which can bind to two types of cholinergic receptors: nicotinic receptors (N) and muscarinic receptors (predominantly M2 on cardiomyocytes). Of note, nicotinic receptors are located on the postganglionic parasympathetic and postganglionic sympathetic neurons. Nicotinic receptors are ligand-gated ion channels whose activation causes depolarization and excitation, but due to the lack of specificity of nicotinic receptors between the sympathetic and parasympathetic nervous system, the postganglionic neurons of the heart are difficult to target pharmacologically.

Cardiac Responses to Neurochemicals

Adrenergic Signaling

Cardiac myocytes have both α- and β-ARs, with β-ARs accounting for about 90% of all ARs [19]. Both β1- and β2-AR subtypes are found in the heart, but under normal conditions, the ratio of β1: β2 is about 80:20% [20], although β2 expression is known to increase under pathological conditions, including heart failure (HF) [20, 21]. All ARs are G-protein-coupled receptors (GPCRs).

Both NE and Epi bind β1-ARs to produce prototypical sympathetic responses, including positive chronotropy, inotropy, lusitropy, and dromotropy. The β1-AR signaling cascade involves activation of Gs, which stimulates adenylyl cyclase (AC) to produce cyclic AMP (cAMP) (Fig. 2). Increases in cAMP increase the funny current (I_f, via direct interaction with the channel) to increase the rate of diastolic depolarization in pacemaker cells of the sino-atrial node (SAN), which contributes to an

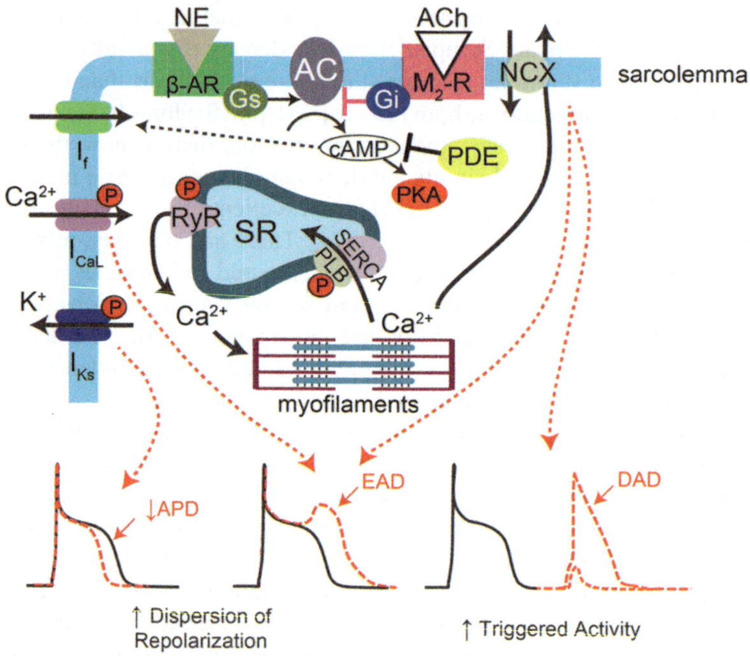

Fig. 2 Cardiomyocyte signaling cascades and electrophysiological responses. AC adenylyl cyclase; ACh acetylcholine; APD action potential duration; β-AR β-adrenergic receptor; cAMP cyclic adenosine monophosphate; DAD delayed afterdepolarization; EAD early afterdepolarization; M2 muscarinic receptor; NCX Na^+/Ca^{2+} exchanger; NE norepinephrine; P phosphorylated; PDE phosphodiesterase; PKA protein kinase A; PLB phospholamban; RyR ryanodine receptor; SERCA sarco/endoplasmic reticulum Ca^{2+} ATPase; SR sarcoplasmic reticulum. (Figure adapted with permission from [3])

increase in heart rate (positive chronotropy) [22]. cAMP also activates protein kinase A (PKA), which phosphorylates several downstream targets, including L-type Ca^{2+} channels, ryanodine receptors (RyRs), and phospholamban (PLB), the latter of which relieves inhibition of the sarco-endoplasmic reticulum Ca^{2+} ATPase (SERCA) pump. Collectively, this results in an increase in L-type Ca^{2+} current (I_{CaL}) and an increase in sarcoplasmic reticulum (SR) Ca^{2+} content, leading to more Ca^{2+} available for contraction and positive inotropy [23–27].

Increased SR Ca^{2+} content and phosphorylation of RyR may also contribute to diastolic Ca^{2+} leak from the SR [26, 28, 29]. The electrogenic Na-Ca^{2+} exchanger (NCX) extrudes this Ca^{2+} in exchange for 3 Na^+ ions, resulting in a net depolarizing inward current. In SAN cells, this process (known as the 'Ca^{2+} clock') contributes to diastolic depolarization and an increase in heart rate [30–32]. In non-nodal cells, however, this diastolic inward current can trigger delayed and/or early afterdepolarizations (DADs, EADs), which can underlie focal arrhythmias and premature contractions [33, 34] (Fig. 2).

Phosphorylation of PLB relieves inhibition on the SERCA pump, which accelerates Ca^{2+} reuptake into the SR and contributes to positive lusitropy (faster relaxation) [35]. The affinity of contractile proteins for Ca^{2+} is also reduced by PKA phosphorylation of troponin I, which results in a decrease in the affinity of troponin C for Ca^{2+} [36]. The decreased affinity for Ca^{2+} further contributes to the positive lusitropic effect by increasing the rate at which Ca^{2+} dissociates from the contractile proteins. By itself, this decrease in affinity would be expected to decrease the force of contraction, but it is outweighed by the increase in Ca^{2+} available for contraction.

Positive dromotropy (faster conduction) within the SAN and atrio-ventricular node (AVN) is primarily due to increased I_{CaL}, which is the main contributor to the action potential upstroke in nodal cells [37]. In ventricular myocardium, increased conduction velocity may be due to PKA-mediated phosphorylation of Na^+ channels (which potentiates I_{Na}) and/or modulation of the gap junction protein, connexin 43 (Cx43) [38, 39]. β1-AR stimulation can impact Cx43 via both short-term changes in phosphorylation and assembly, or via longer-term alterations in expression [40–42].

To accommodate faster heart rates during β1-AR stimulation, the action potential duration (APD) typically shortens. In larger mammals (i.e., non-rodents), this is achieved primarily via PKA-mediated phosphorylation of I_{Ks}, which augments the current and counterbalances the increase in depolarizing I_{CaL} [37, 43–45]. Rodent hearts lack I_{Ks}, however, and recent studies from our group indicate that APD in normal mouse hearts is typically prolonged (at least transiently) in response to sympathetic nerve stimulation (likely via the predominant effect on I_{CaL}) [2].

β2-ARs can couple to either Gs or Gi, depending on the condition or pathological state [46]. Gs-mediated effects are similar to those described above, but β2-ARs are predominately found on the T-tubule membrane co-localized with L-type Ca^{2+} channels, which allows for more compartmentalized signaling compared to β1 activation [21, 47]. Acute β2-AR activation typically produces positive tropic responses via Gs, whereas longer-term Gi activation may activate cell survival pathways to improve cardiac function [46]. Of note, the major nerve-released neurotransmitter, NE, has very low affinity for β2-AR, whereas circulating Epi stimulates both β1- and β2-ARs.

Cardiomyocytes also express α1-ARs, although at much lower levels than β-ARs. α1-ARs classically couple to Gq and the phospholipase C pathway, and α1 activation has been shown to prevent pathological remodeling in heart failure [19]. Interestingly, in heterologous expression systems as well as rat cardiomyocytes, α1-AR activation acutely decreases the transient outward K^+ current (I_{to}) via the Gs-cAMP-PKA pathway [48, 49]. This is thought to occur via co-localization of a subset of I_{to} channel-forming proteins (Kv4.2 and Kv4.3) with α1-ARs in caveolae (reviewed in [50]). Both NE and Epi stimulate α1-ARs; however, NE has a higher affinity than Epi. Because β1-ARs vastly outnumber both β2- and α-ARs, the acute effects of either NE or Epi on cardiac rhythm is largely mediated by β1-AR effects.

Muscarinic Signaling

Muscarinic receptors are also GPCRs, and the predominant cardiac receptor, M2, is coupled to Gi. Parasympathetic nerve-released ACh binds to M2 receptors on cardiomyocytes and therefore opposes the β-AR Gs-mediated signaling mechanisms discussed above by inhibiting AC (Fig. 2). This generally results in negative chronotropy, inotropy, lusitropy, and dromotropy. In addition, M2 stimulation also increases the ACh-activated K^+ current (I_{KACh}) via Gβγ [51]. Increased I_{KACh} hyperpolarizes the membrane of pacemaking cells. Therefore, increased I_{KACh}, along with decreased I_f and slowing of the Ca^{2+} clock, likely together mediate the parasympathetic-induced decrease in heart rate [52]. Atrial myocytes also have I_{KACh} channels, and parasympathetic activity shortens atrial APD via this mechanism [37].

Although cardiomyocytes mainly express M2 receptors, there is also evidence for M3 receptor expression [53, 54], and recent studies have implicated both M2 and M3 in mediating beneficial alterations to ventricular Ca^{2+} handling in response to ACh [55]. Specifically, M2-Gi signaling can lead to PKG activation, which was found to phosphorylate RyR at Ser-2808; whereas M3 stimulation dephosphorylated RyR at Ser-2818 via decreased reactive oxygen species (ROS)-dependent activation of CaMKII. This reciprocal phosphorylation of RyR resulted in improved Ca^{2+} cycling efficiency via an increase in systolic Ca^{2+} release at low SR Ca^{2+} load without any associated SR Ca^{2+} leak [55, 56]. These exciting data suggest that the beneficial effects of parasympathetic activity may go beyond canonical signaling mechanisms, and these pathways may therefore represent important therapeutic opportunities.

Co-transmission

Although NE and ACh are the primary neurotransmitters released during cardiac sympathetic and parasympathetic nerve activation, respectively, co-release of neuropeptides and other signaling molecules may also occur. Some of these co-transmitters are known to directly impact cardiomyocytes, whereas others may act on the prejunctional nerve terminals to regulate their activity. In addition to NE, sympathetic neurons can release neuropeptide Y (NPY), galanin, and adenosine triphosphate (ATP). Parasympathetic neurons, on the other hand, can also release vasoactive intestinal peptide (VIP), nitric oxide (NO), and ATP [57].

NPY NPY release during sympathetic activity is perhaps best known for inhibiting ACh release from parasympathetic neurons via binding to Y_2 receptors on cholinergic nerve terminals [58, 59]. Indeed, NPY is known to underlie the inhibition of vagal-induced bradycardia that occurs following sympathetic stimulation [59]. Recent evidence also points to a direct pro-arrhythmic role of NPY via binding to Y_1 receptors on cardiomyocytes, which has been shown to impact intracellular Ca^{2+} handling, including increased Ca^{2+} transient amplitude, shortened Ca^{2+} transient duration, and an increased incidence of ventricular arrhythmias in isolated hearts [60]. Notably, NPY expression is increased in patients with chronic HF and following acute myocardial infarction (MI) [60, 61], and high levels of NPY following MI are associated with increased incidence of ventricular tachycardia (VT) or fibrillation (VF) [60]. These exciting findings point not only to potentially novel anti-arrhythmic targets, but perhaps a novel biomarker for risk assessment.

Galanin The cardiac sympathetic nerves can also co-release galanin, although at significantly lower levels than NPY [62]. The effects of galanin are similar to those of NPY, whereby it inhibits parasympathetic ACh release via galanin receptors ($GalR_1$) on cardiac parasympathetic neurons. Specific inhibition of GalR can partially reverse reductions in vagal bradycardia following sympathetic stimulation [62].

VIP VIP is co-released with ACh from parasympathetic neurons. In contrast to ACh, exogenous application of VIP results in tachycardia via activation of the Gs signaling cascade by binding to VIP receptors on cardiomyocytes (VPAC1, VPAC2) [63–65]. However, the actions of VIP may not be limited to direct cardiomyocyte effects. Some reports indicate that VIP-mediated tachycardia does not occur in the absence of muscarinic antagonism [66]. Further, application of a VIP receptor antagonist prior to vagal stimulation amplifies vagal-induced bradycardia, but the effects are abolished with the ganglionic blocker hexamethonium [67]. These latter findings suggest that VIP may also act at the parasympathetic preganglionic–postganglionic synapse to alter cholinergic signaling.

NO Nitric oxide (NO) is a ubiquitous intra- and intercellular signaling molecule released by cardiac parasympathetic neurons (in addition to other sources of NO, including cardiomyocytes and nearby endothelial cells). NO can directly impact cardiomyocyte electrophysiology by binding to soluble guanylyl cyclase to cause cyclic GMP (cGMP) production. cGMP signaling involves many targets, including protein kinase G (PKG) and several phosphodiesterases (PDEs) [68]. However, downstream effects are difficult to predict, as NO may also activate Gs or Gi (which may be concentration-dependent). For example, NO has been shown to both increase and decrease I_{CaL} and activate I_f, resulting in either an increase or decrease in heart rate [68, 69]. Moreover, data from a decentralized innervated rabbit heart preparation suggest that NO release during vagal nerve stimulation is anti-arrhythmic and contributes to flattening of the APD restitution curve [70]. In addition to these direct effects of NO on cardiomyocytes, NO also inhibits NE release from sympathetic neurons and potentiates ACh release from parasympathetic neurons, and in vivo data suggest that NO effects on the nervous system may dominate [71, 72].

ATP Both sympathetic and parasympathetic cardiac neurons co-release ATP, but due to its short biological half-life, the direct impact of nerve-released ATP on cardiomyocytes is not fully understood [73]. Cardiomyocytes possess the two major classes of purinergic receptors, P1 (adenosine receptors A_1, $A_{2A/B}$, and A_3) and P2

(P2X and P2Y), and the effects of adenosine (resulting from ATP breakdown) and ATP on the heart have been reviewed extensively [74, 75]. The direct effect of adenosine or ATP application is to slow pacemaking activity and AVN conduction, primarily via Gi-coupled A_1 receptors [76, 77]. Moreover, in vivo administration of ATP, but not adenosine, triggers a strong vagal reflex via activation of P2X receptors on vagal sensory nerves within the wall of the left ventricle [78]. In summary, although the actions of ATP and adenosine have been examined in vitro and in vivo, the role of cardiac nerve-released ATP in normal and pathological states remains an important area for future study.

In vitro Models to Study Neural Regulation of Cardiac Rhythm

The most translational experiments to assess neural control of cardiac electrophysiology are of course in vivo animal or human studies, in which branches of the ANS are stimulated and resulting cardiac responses are recorded with ECG-based or mapping-based approaches [79–81]. These experiments have the added benefit of intact reflex responses, higher-order processing of afferent and efferent signals, and physiological levels of circulating neuro-endocrine factors. However, detailed cellular signaling mechanisms and individual contributing factors to complex multiorgan responses can be difficult to dissect in vivo. Therefore, at the other end of the spectrum, isolated cardiomyocytes are commonly used to assess the electrophysiological and signaling responses to exogenous neurochemicals. This approach has the advantage of eliminating a multitude of confounding factors compared to the in vivo milieu, but the concentration, kinetics, and spatial distribution of nerve-released chemicals can be difficult to replicate. Therefore, various in vitro neuro-cardiac models have been developed, and these approaches have been particularly useful in revealing detailed aspects of the neuro-cardiac junction, neuron-cardiomyocyte signaling, and tissue- and organ-level electrophysiological responses.

Co-cultures

A variety of neuron-cardiomyocyte co-culture systems have been developed to assess the neuro-cardiac junction and cellular-level responses to nerve activity. Some of the earlier co-culture experiments used neonatal cardiomyoyctes co-cultured with sympathetic neurons isolated from cervical or stellate ganglia [82, 83], or co-cultured with parasympathetic neurons isolated from sacral cord explants or ciliary muscle [84, 85]. Functional neurocardiac junctions were confirmed with expected changes in cardiomyocyte beat rate (either increase or decrease) upon stimulation of the neurons. Neuronal stimulation was historically performed with nicotine application to activate neuronal nicotinic receptors, but optogenetic approaches have now been employed, which allow for precise optical control of neuronal activation [86]. More recent co-culture experiments have turned to iPSC-derived neurons (either sympathetic or parasympathetic) together with iPSC-derived cardiomyocytes [87, 88]. This exciting development has opened the door to studying human neuron-cardiomyocyte co-cultures and patient-specific genetic mutations, and may also allow for optogenetic control of neuron activation [88].

Sympathetic Neuron-Cardiomyocyte Co-cultures Studies using neonatal cardiomyocyte co-cultures revealed that the presence of adrenergic neurons promotes the development of distinct signaling domains in adjacent cardiomyocytes. These domains are enriched with β1-ARs, the scaffold proteins SAP97 and AKAP150, and deficient in caveolin-3 [83]. Upon neuronal stimulation, there is also a loss of β2-AR in these regions. These intriguing results suggest that the sympathetic nerves play an active role in organizing signaling domains in adjacent cardiomyocytes. Co-cultures also revealed that NE release at the neuro-cardiac junction is extremely diffusion restricted, with high NE concentration in the cleft (estimated at ~100 nM), and activation of only those β-ARs directly adjacent to varicosities on sympathetic axons (sites of NE release) [86].

The individual roles of neurons versus cardiomyocytes in contributing to observed pathological phenotypes have also been investigated in co-culture systems. For example, Larsen et al. showed that altered cardiomyocyte cAMP signaling provoked by sympathetic neuron activation in co-cultures from hypertensive rats could be rescued by culturing the hypertensive cardiomyocytes with sympathetic neurons from normal healthy rats [89]. Likewise, pathological cAMP signaling could be induced in normal cardiomyocytes by co-culturing with sympathetic neurons from hypertensive rats. More recently, iPSC-derived sympathetic neuron-myocyte co-cultures from two patients with long QT type 1 (LQT1) syndrome were developed by Winbo et al. Interestingly, they reported significant neuronal hyperactivity and increased firing rate of the differentiated sympathetic neurons from both patients harboring loss-of-function KCNQ1 mutations (which encodes Kv7.1, I_{Ks}). This neuronal hyperactivity may play an important (and previously unrecognized) role in LQT-related arrhythmogenesis [87].

Parasympathetic Neuron-Cardiomyocyte Co-cultures There are somewhat fewer reports of parasympathetic neuron-cardiomyocyte co-cultures (compared to sympathetic), but early studies of embryonic chick heart myocytes showed that chronotropic responses to pharmacological muscarinic stimulation occurred when embryonic myocytes were co-cultured with ciliary ganglia, but not when cultured alone [84]. Interestingly, muscarinic responsiveness also developed when myocytes were cultured in media conditioned by myocyte-ciliary ganglia co-cultures, suggesting that soluble factors may be involved. Increased muscarinic responsiveness of cardiomyocytes was associated with increased expression of Gi subunits, indicating that parasympathetic innervation may play an important role in coupling cardiomyocyte muscarinic receptors to downstream physiological activity [84].

Interestingly, parasympathetic neurons may also respond to phenotypic changes in the cells they are innervating. Flannery and Brusés reported that expression of N-cadherin in CHO cells induced differentiation of cholinergic presynaptic terminals in co-cultured brainstem neurons [90]. Whether similar mechanisms are involved in the cardiac parasympathetic neurocardiac junction has yet to be determined.

Isolated Heart and Tissue Preparations

Because adult primary cardiomyocytes are difficult to maintain in culture, the co-culture experiments described above are typically performed with embryonic, neonatal, or iPSC-derived cardiomyocytes, all of which may have different electrophysiological properties compared to adult cardiomyocytes. Cardiac tissue and whole-heart preparations have therefore been used to assess electrophysiological responses to nerve activity, and the signaling mechanisms involved, in a variety of animal models. It is important to note, however, that traditional isolated, perfused hearts and cardiac tissues (e.g., Langendorff-perfused or working hearts) are not necessarily 'denervated' per se. Rather, these preparations are 'decentralized', as isolated hearts and tissues still contain many components of the ICNS, including GPs, afferent and interconnecting neurons, and postganglionic efferent parasympathetic and sympathetic fibers. Intrinsic efferent neurons can be stimulated via electrical, chemical, or optical approaches to determine functional electrophysiological effects.

Electrical stimulation of the GPs or chemical stimulation of hearts with nicotine will stimulate both sympathetic and parasympathetic responses. This is because the ICNS and GPs contain both adrenergic and cholinergic neurons [91], and both types are activated via electrical stimulation or via nicotinic receptors. Indeed, electrical stimulation of GPs in Langendorff-perfused mouse and rabbit hearts has been shown to produce bradycardia, tachycardia, or more complex biphasic brady-tachycardia [91, 92]. Similar results were observed with local nicotine application, and for both modes of stimulation, tachycardia was prevented with a beta-blocker and bradycardia prevented with a muscarinic antagonist. The magnitude and direction of heart rate responses depend on the location of the ganglion stimulated, but parasympathetic responses tend to be more dominant. This is consistent with histological studies indicating that the majority of intrinsic cardiac neurons are cholinergic [93, 94]. Indeed, in Langendorff-perfused mouse hearts, Jungen et al. found that targeted atrial application of nicotine resulted in a decrease in cAMP levels in both atrial and ventricular regions [95]. Moreover, they found that resection of atrial GPs reduced ventricular refractory periods and increased susceptibility to ventricular arrhythmias, underscoring the fact that atrial GPs contain fibers projecting to the ventricles and that the ICNS plays an active role in ventricular electrophysiology, even in decentralized Langendorff-perfused hearts [95]. Chemical sympathetic activation can be achieved via application of tyramine, which causes the release of NE from adrenergic nerve terminals. Our group has demonstrated that electrophysiological responses to tyramine in Langendorff-perfused hearts closely resemble those provoked by electrical sympathetic nerve stimulation [2].

New developments in optogenetics now allow for more precise optical stimulation of the ICNS, and these approaches have also been applied in isolated hearts. One of the first such studies employed a mouse model expressing channelrhodopsin-2 (ChR2) in catecholaminergic neurons [96]. Photoactivation of the neurons in Langendorff-perfused hearts demonstrated prototypical increases in heart rate, contractility, and increased arrhythmia susceptibility. ChR2 has also been expressed in cholinergic neurons, where reductions in heart rate, along with changes to the p-wave duration were observed in isolated hearts [97]. Although the focus here is on in vitro experimental systems, it should be noted that optical stimulation of adrenergic and cholinergic cardiac neurons has also been performed in vivo [86], and such experiments are an exciting step forward in experimental modulation of neural control of the heart.

In addition to stimulation of ICNS components, several in vitro heart and tissue preparations have been developed in which one or more components of the extrinsic cardiac nervous system (ECNS) remain intact for stimulation and assessment of cardiac responses. Some of the earliest such experiments used 'innervated atrial' preparations, in which atria were isolated with intact vagal nerves or intact sympathetic (stellate) ganglia [98–100]. Electrical stimulation of the left or right vagus or stellate was performed to assess SAN function and/or atrial contractility, electrophysiology, and signaling mechanisms involved. Some of these studies were the first to assess SAN responses to nerve stimulation compared to exogenously applied neurotransmitters [98, 100], and posited distinct roles and signaling mechanisms for junctional (i.e., stimulated via nerve-released neurochemicals) versus extrajunctional (i.e., stimulated via circulating neurochemicals) adrenergic and muscarinic receptors. Notably, some of these concepts remain unresolved and are still topics of lively investigation and debate [52, 101].

In 2001, Ng et al. developed a novel Langendorff-perfused rabbit heart with intact sympathetic and parasympathetic innervation [102]. In this model, hearts are isolated with the spinal cord intact, which allows for stimulation of the sympathetic thoracic ganglia, as well as intact right and left vagus nerves. Early studies with this model recorded cardiac responses via left ventricular pressure, ECG, and monophasic action potentials [102]. Later experiments employed optical mapping approaches for more detailed assessment of ventricular electrophysiological responses [103, 104]. Indeed, Ng and colleagues were the first to report distinct changes in repolarization dynamics with sympathetic nerve activation, including heterogeneity of restitution and reversal of the repolarization sequence, which may play important roles in sympathetic-mediated arrhythmias [103, 104]. Our group later extended the innervated heart model to the mouse, allowing for detailed cross-species comparison of sympathetic responses (Fig. 3) and for assessment of genetically modified mouse models [2, 16, 105, 106].

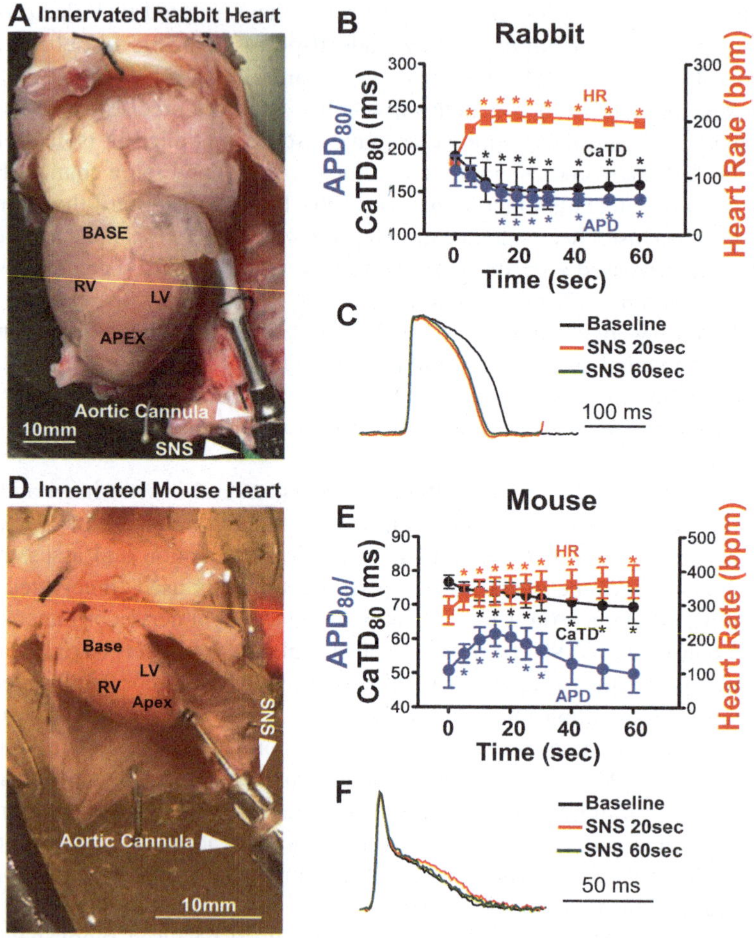

Fig. 3 In vitro innervated heart preparations allow for detailed assessment of electrophysiological responses to electrical sympathetic nerve stimulation (SNS) in rabbit (**a**–**c**) and mouse (**d**–**f**) hearts. Action potential duration (APD) monotonically decreases in the rabbit, whereas transient prolongation of APD is observed in the mouse. In both species, heart rate (HR) monotonically increases and Ca^{2+} transient duration (CaTD) monotonically decreases. RV right ventricle; LV left ventricle. (Modified with permission from [2])

Although the innervated heart model described above allows for assessment of efferent sympathetic and/or parasympathetic responses, it is devoid of central reflexes, which may play a key role in normal and abnormal cardiac rhythm. To assess reflex responses, a working heart–brainstem preparation was developed in the mouse heart in 1996 by Paton [107], and in 2013, it was extended by Ashton and colleagues to include cardiac electrophysiological assessments [108]. Using a rat model of the working heart–brainstem, Ashton et al. assessed, for the first time, baroreflex- and chemoreflex-mediated shifts of the leading pacemaker [109]. These exciting studies open the door to further exploration of hierarchical processing, feedback, and autonomic control of the heart.

Conclusions

Despite the incredible progress over the past several decades in furthering our understanding of the signaling mechanisms involved in autonomic control of heart rhythm, there is still much to be discovered. This chapter focused on cardiomyocyte responses to efferent ANS activity and neurochemicals, primarily in the normal, undiseased heart. Cardiovascular disease dramatically alters the structure and function of the ANS, underlying cardiomyocyte electrophysiological function, as well as how cardiomyocytes respond to autonomic inputs [3, 110]. In order to develop and test novel neuromodulatory approaches to delay disease progression and prevent arrhythmias, multidisciplinary, multiscale, and multiorgan approaches will be needed. Novel experimental approaches will be essential, including some of the in vitro models described here, and collaboration between neuroscientists, cardiovascular researchers, and clinicians will be key to advancing novel anti-arrhythmic therapies.

References

1. Fukuda K, Kanazawa H, Aizawa Y, Ardell JL, Shivkumar K. Cardiac innervation and sudden cardiac death. Circ Res. 2015;116(12):2005–19.
2. Wang L, Morotti S, Tapa S, Francis Stuart SD, Jiang Y, Wang Z, et al. Different paths, same destination: divergent action potential responses produce conserved cardiac fight-or-flight response in mouse and rabbit hearts. J Physiol. 2019;597(15):3867–83.
3. Ripplinger CM, Noujaim SF, Linz D. The nervous heart. Prog Biophys Mol Biol. 2016;120(1-3):199–209.
4. Butler CK, Smith FM, Cardinal R, Murphy DA, Hopkins DA, Armour JA. Cardiac responses to electrical stimulation of discrete loci in canine atrial and ventricular ganglionated plexi. Am J Phys. 1990;259(5 Pt 2):H1365–73.
5. Armour JA, Murphy DA, Yuan BX, Macdonald S, Hopkins DA. Gross and microscopic anatomy of the human intrinsic cardiac nervous system. Anat Rec. 1997;247(2):289–98.
6. Singh S, Johnson PI, Lee RE, Orfei E, Lonchyna VA, Sullivan HJ, et al. Topography of cardiac ganglia in the adult human heart. J Thorac Cardiovasc Surg. 1996;112(4):943–53.
7. Rysevaite K, Saburkina I, Pauziene N, Noujaim SF, Jalife J, Pauza DH. Morphologic pattern of the intrinsic ganglionated nerve plexus in mouse heart. Heart Rhythm. 2011;8(3):448–54.
8. Saburkina I, Gukauskiene L, Rysevaite K, Brack KE, Pauza AG, Pauziene N, et al. Morphological pattern of intrinsic nerve plexus distributed on the rabbit heart and interatrial septum. J Anat. 2014;224(5):583–93.
9. Batulevicius D, Pauziene N, Pauza DH. Topographic morphology and age-related analysis of the neuronal number of the rat intracardiac nerve plexus. Ann Anat. 2003;185(5):449–59.
10. Gagliardi M, Randall WC, Bieger D, Wurster RD, Hopkins DA, Armour JA. Activity of in vivo canine cardiac plexus neurons. Am J Phys. 1988;255(4 Pt 2):H789–800.

11. Pauza DH, Skripka V, Pauziene N, Stropus R. Morphology, distribution, and variability of the epicardiac neural ganglionated subplexuses in the human heart. Anat Rec. 2000;259(4):353–82.
12. Saburkina I, Rysevaite K, Pauziene N, Mischke K, Schauerte P, Jalife J, et al. Epicardial neural ganglionated plexus of ovine heart: anatomic basis for experimental cardiac electrophysiology and nerve protective cardiac surgery. Heart Rhythm. 2010;7(7):942–50.
13. Kawano H, Okada R, Yano K. Histological study on the distribution of autonomic nerves in the human heart. Heart Vessel. 2003;18(1):32–9.
14. Ito M, Zipes DP. Efferent sympathetic and vagal innervation of the canine right ventricle. Circulation. 1994;90(3):1459–68.
15. Crick SJ, Anderson RH, Ho SY, Sheppard MN. Localisation and quantitation of autonomic innervation in the porcine heart II: endocardium, myocardium and epicardium. J Anat. 1999;195(Pt 3):359–73.
16. Francis Stuart SD, Wang L, Woodard WR, Ng GA, Habecker BA, Ripplinger CM. Age-related changes in cardiac electrophysiology and calcium handling in response to sympathetic nerve stimulation. J Physiol. 2018;596(17):3977–91.
17. Coote JH. Myths and realities of the cardiac vagus. J Physiol. 2013;591(17):4073–85.
18. Ulphani JS, Cain JH, Inderyas F, Gordon D, Gikas PV, Shade G, et al. Quantitative analysis of parasympathetic innervation of the porcine heart. Heart Rhythm. 2010;7(8):1113–9.
19. O'Connell TD, Jensen BC, Baker AJ, Simpson PC. Cardiac alpha1-adrenergic receptors: novel aspects of expression, signaling mechanisms, physiologic function, and clinical importance. Pharmacol Rev. 2014;66(1):308–33.
20. Bristow MR, Ginsburg R, Umans V, Fowler M, Minobe W, Rasmussen R, et al. Beta 1- and beta 2-adrenergic-receptor subpopulations in nonfailing and failing human ventricular myocardium: coupling of both receptor subtypes to muscle contraction and selective beta 1-receptor down-regulation in heart failure. Circ Res. 1986;59(3):297–309.
21. Nikolaev VO, Moshkov A, Lyon AR, Miragoli M, Novak P, Paur H, et al. Beta2-adrenergic receptor redistribution in heart failure changes cAMP compartmentation. Science. 2010;327(5973):1653–7.
22. Bucchi A, Baruscotti M, Robinson RB, DiFrancesco D. Modulation of rate by autonomic agonists in SAN cells involves changes in diastolic depolarization and the pacemaker current. J Mol Cell Cardiol. 2007;43(1):39–48.
23. Bers DM. Cardiac excitation-contraction coupling. Nature. 2002;415(6868):198–205.
24. Reuter H. Localization of beta adrenergic receptors, and effects of noradrenaline and cyclic nucleotides on action potentials, ionic currents and tension in mammalian cardiac muscle. J Physiol. 1974;242(2):429–51.
25. Kirchberber MA, Tada M, Katz AM. Phospholamban: a regulatory protein of the cardiac sarcoplasmic reticulum. Recent Adv Stud Cardiac Struct Metab. 1975;5:103–15.
26. Shannon TR, Ginsburg KS, Bers DM. Potentiation of fractional sarcoplasmic reticulum calcium release by total and free intra-sarcoplasmic reticulum calcium concentration. Biophys J. 2000;78(1):334–43.
27. Viatchenko-Karpinski S, Györke S. Modulation of the Ca(2+)-induced Ca(2+) release cascade by beta-adrenergic stimulation in rat ventricular myocytes. J Physiol. 2001;533(Pt 3):837–48.
28. Marx SO, Reiken S, Hisamatsu Y, Jayaraman T, Burkhoff D, Rosemblit N, et al. PKA phosphorylation dissociates FKBP12.6 from the calcium release channel (ryanodine receptor): defective regulation in failing hearts. Cell. 2000;101(4):365–76.
29. Valdivia HH, Kaplan JH, Ellis-Davies GC, Lederer WJ. Rapid adaptation of cardiac ryanodine receptors: modulation by Mg2+ and phosphorylation. Science. 1995;267(5206):1997–2000.
30. Vinogradova TM, Lyashkov AE, Zhu W, Ruknudin AM, Sirenko S, Yang D, et al. High basal protein kinase A-dependent phosphorylation drives rhythmic internal Ca2+ store oscillations and spontaneous beating of cardiac pacemaker cells. Circ Res. 2006;98(4):505–14.
31. Bogdanov KY, Maltsev VA, Vinogradova TM, Lyashkov AE, Spurgeon HA, Stern MD, et al. Membrane potential fluctuations resulting from submembrane Ca2+ releases in rabbit sino-

atrial nodal cells impart an exponential phase to the late diastolic depolarization that controls their chronotropic state. Circ Res. 2006;99(9):979–87.
32. Lakatta EG, Maltsev VA, Vinogradova TM. A coupled SYSTEM of intracellular Ca2+ clocks and surface membrane voltage clocks controls the timekeeping mechanism of the heart's pacemaker. Circ Res. 2010;106(4):659–73.
33. Pogwizd SM, Schlotthauer K, Li L, Yuan W, Bers DM. Arrhythmogenesis and contractile dysfunction in heart failure: roles of sodium-calcium exchange, inward rectifier potassium current, and residual beta-adrenergic responsiveness. Circ Res. 2001;88(11):1159–67.
34. Pogwizd SM, Bers DM. Cellular basis of triggered arrhythmias in heart failure. Trends Cardiovasc Med. 2004;14(2):61–6.
35. Li L, Desantiago J, Chu G, Kranias EG, Bers DM. Phosphorylation of phospholamban and troponin I in beta-adrenergic-induced acceleration of cardiac relaxation. Am J Physiol Heart Circ Physiol. 2000;278(3):H769–79.
36. Kentish JC, McCloskey DT, Layland J, Palmer S, Leiden JM, Martin AF, et al. Phosphorylation of troponin I by protein kinase A accelerates relaxation and crossbridge cycle kinetics in mouse ventricular muscle. Circ Res. 2001;88(10):1059–65.
37. Bartos DC, Grandi E, Ripplinger CM. Ion channels in the heart. Compr Physiol. 2015;5(3):1423–64.
38. Campbell AS, Johnstone SR, Baillie GS, Smith G. beta-Adrenergic modulation of myocardial conduction velocity: Connexins vs. sodium current. J Mol Cell Cardiol. 2014;77:147–54.
39. Herren AW, Bers DM, Grandi E. Post-translational modifications of the cardiac Na channel: contribution of CaMKII-dependent phosphorylation to acquired arrhythmias. Am J Physiol Heart Circ Physiol. 2013;305(4):H431–45.
40. Somekawa S, Fukuhara S, Nakaoka Y, Fujita H, Saito Y, Mochizuki N. Enhanced functional gap junction neoformation by protein kinase A-dependent and Epac-dependent signals downstream of cAMP in cardiac myocytes. Circ Res. 2005;97(7):655–62.
41. TenBroek EM, Lampe PD, Solan JL, Reynhout JK, Johnson RG. Ser364 of connexin43 and the upregulation of gap junction assembly by cAMP. J Cell Biol. 2001;155(7):1307–18.
42. Salameh A, Frenzel C, Boldt A, Rassler B, Glawe I, Schulte J, et al. Subchronic alpha- and beta-adrenergic regulation of cardiac gap junction protein expression. FASEB J. 2006;20(2):365–7.
43. Kagan A, Melman YF, Krumerman A, McDonald TV. 14-3-3 amplifies and prolongs adrenergic stimulation of HERG K+ channel activity. EMBO J. 2002;21(8):1889–98.
44. Marx SO, Kurokawa J, Reiken S, Motoike H, D'Armiento J, Marks AR, et al. Requirement of a macromolecular signaling complex for beta adrenergic receptor modulation of the KCNQ1-KCNE1 potassium channel. Science. 2002;295(5554):496–9.
45. Chiamvimonvat N, Chen-Izu Y, Clancy CE, Deschenes I, Dobrev D, Heijman J, et al. Potassium currents in the heart: functional roles in repolarization, arrhythmia and therapeutics. J Physiol. 2017;595(7):2229–52.
46. Xiao RP. Beta-adrenergic signaling in the heart: dual coupling of the beta2-adrenergic receptor to G(s) and G(i) proteins. Sci STKE. 2001;2001(104):re15.
47. Davare MA, Avdonin V, Hall DD, Peden EM, Burette A, Weinberg RJ, et al. A beta2 adrenergic receptor signaling complex assembled with the Ca2+ channel Cav1.2. Science. 2001;293(5527):98–101.
48. Gallego M, Casis O. Regulation of cardiac transient outward potassium current by norepinephrine in normal and diabetic rats. Diabetes Metab Res Rev. 2001;17(4):304–9.
49. Gallego M, Setien R, Puebla L, Boyano-Adanez Mdel C, Arilla E, Casis O. alpha1-Adrenoceptors stimulate a Galphas protein and reduce the transient outward K+ current via a cAMP/PKA-mediated pathway in the rat heart. Am J Physiol Cell Physiol. 2005;288(3):C577–85.
50. Gallego M, Alday A, Alonso H, Casis O. Adrenergic regulation of cardiac ionic channels: role of membrane microdomains in the regulation of kv4 channels. Biochim Biophys Acta. 2014;1838(2):692–9.

51. Krapivinsky G, Krapivinsky L, Wickman K, Clapham DE. G beta gamma binds directly to the G protein-gated K+ channel, IKACh. J Biol Chem. 1995;270(49):29059–62.
52. Han SY, Bolter CP. Effects of tertiapin-Q and ZD7288 on changes in sinoatrial pacemaker rhythm during vagal stimulation. Auton Neurosci. 2015;193:117–26.
53. Dhein S, van Koppen CJ, Brodde OE. Muscarinic receptors in the mammalian heart. Pharmacol Res. 2001;44(3):161–82.
54. Wang Z, Shi H, Wang H. Functional M3 muscarinic acetylcholine receptors in mammalian hearts. Br J Pharmacol. 2004;142(3):395–408.
55. Ho HT, Belevych AE, Liu B, Bonilla IM, Radwanski PB, Kubasov IV, et al. Muscarinic stimulation facilitates sarcoplasmic reticulum ca release by modulating ryanodine receptor 2 phosphorylation through protein Kinase G and Ca/Calmodulin-dependent protein Kinase II. Hypertension. 2016;68(5):1171–8.
56. Baine S, Thomas J, Bonilla I, Ivanova M, Belevych A, Li J, et al. Muscarinic-dependent phosphorylation of the cardiac ryanodine receptor by protein kinase G is mediated by PI3K-AKT-nNOS signaling. J Biol Chem. 2020;295(33):11720–8.
57. Burnstock G. Autonomic neurotransmission: 60 years since sir Henry Dale. Annu Rev Pharmacol Toxicol. 2009;49:1–30.
58. Smith-White MA, Herzog H, Potter EK. Role of neuropeptide Y Y(2) receptors in modulation of cardiac parasympathetic neurotransmission. Regul Pept. 2002;103(2-3):105–11.
59. Herring N, Lokale MN, Danson EJ, Heaton DA, Paterson DJ. Neuropeptide Y reduces acetylcholine release and vagal bradycardia via a Y2 receptor-mediated, protein kinase C-dependent pathway. J Mol Cell Cardiol. 2008;44(3):477–85.
60. Kalla M, Hao G, Tapoulal N, Tomek J, Liu K, Woodward L, et al. The cardiac sympathetic co-transmitter neuropeptide Y is pro-arrhythmic following ST-elevation myocardial infarction despite beta-blockade. Eur Heart J. 2019;41:2168.
61. Ajijola OA, Chatterjee NA, Gonzales MJ, Gornbein J, Liu K, Li D, et al. Coronary sinus neuropeptide Y levels and adverse outcomes in patients with stable chronic heart failure. JAMA Cardiol. 2019;5:318.
62. Herring N, Cranley J, Lokale MN, Li D, Shanks J, Alston EN, et al. The cardiac sympathetic co-transmitter galanin reduces acetylcholine release and vagal bradycardia: implications for neural control of cardiac excitability. J Mol Cell Cardiol. 2012;52(3):667–76.
63. Christophe J, Waelbroeck M, Chatelain P, Robberecht P. Heart receptors for VIP, PHI and secretin are able to activate adenylate cyclase and to mediate inotropic and chronotropic effects. Species variations and physiopathology. Peptides. 1984;5(2):341–53.
64. Rigel DF. Effects of neuropeptides on heart rate in dogs: comparison of VIP, PHI, NPY, CGRP, and NT. Am J Phys. 1988;255(2 Pt 2):H311–7.
65. De Neef P, Robberecht P, Chatelain P, Waelbroeck M, Christophe J. The in vitro chronotropic and inotropic effects of vasoactive intestinal peptide (VIP) on the atria and ventricular papillary muscle from Cynomolgus monkey heart. Regul Pept. 1984;8(3):237–44.
66. Markos F, Snow HM. An investigation into the physiological relevance of the vagal tachycardia in the anaesthetized dog. Acta Physiol (Oxf). 2006;186(3):179–84.
67. Hogan K, Markos F. Vasoactive intestinal polypeptide receptor antagonism enhances the vagally induced increase in cardiac interval of the rat atrium in vitro. Exp Physiol. 2006;91(3):641–6.
68. Fischmeister R, Castro L, Abi-Gerges A, Rochais F, Vandecasteele G. Species- and tissue-dependent effects of NO and cyclic GMP on cardiac ion channels. Comp Biochem Physiol A Mol Integr Physiol. 2005;142(2):136–43.
69. Massion PB, Feron O, Dessy C, Balligand JL. Nitric oxide and cardiac function: ten years after, and continuing. Circ Res. 2003;93(5):388–98.
70. Brack KE, Patel VH, Coote JH, Ng GA. Nitric oxide mediates the vagal protective effect on ventricular fibrillation via effects on action potential duration restitution in the rabbit heart. J Physiol. 2007;583(Pt 2):695–704.
71. Chowdhary S, Vaile JC, Fletcher J, Ross HF, Coote JH, Townend JN. Nitric oxide and cardiac autonomic control in humans. Hypertension. 2000;36(2):264–9.

72. Paterson D. Nitric oxide and the autonomic regulation of cardiac excitability. The G.L. Brown Prize Lecture. Exp Physiol. 2001;86(1):1–12.
73. Herring N. Autonomic control of the heart: going beyond the classical neurotransmitters. Exp Physiol. 2015;100(4):354–8.
74. Burnstock G. Purinergic signaling in the cardiovascular system. Circ Res. 2017;120(1):207–28.
75. Headrick JP, Ashton KJ, Rose'meyer RB, Peart JN. Cardiovascular adenosine receptors: expression, actions and interactions. Pharmacol Ther. 2013;140(1):92–111.
76. Belardinelli L, Giles WR, West A. Ionic mechanisms of adenosine actions in pacemaker cells from rabbit heart. J Physiol. 1988;405:615–33.
77. West GA, Belardinelli L. Correlation of sinus slowing and hyperpolarization caused by adenosine in sinus node. Pflugers Arch. 1985;403(1):75–81.
78. Katchanov G, Xu J, Hurt CM, Pelleg A. Electrophysiological-anatomic correlates of ATP-triggered vagal reflex in the dog. III. Role of cardiac afferents. Am J Phys. 1996;270(5 Pt 2):H1785–90.
79. Ajijola OA, Vaseghi M, Zhou W, Yamakawa K, Benharash P, Hadaya J, et al. Functional differences between junctional and extrajunctional adrenergic receptor activation in mammalian ventricle. Am J Physiol Heart Circ Physiol. 2013;304(4):H579–88.
80. Ajijola OA, Yagishita D, Patel KJ, Vaseghi M, Zhou W, Yamakawa K, et al. Focal myocardial infarction induces global remodeling of cardiac sympathetic innervation: neural remodeling in a spatial context. Am J Physiol Heart Circ Physiol. 2013;305(7):H1031–40.
81. Vaseghi M, Lux RL, Mahajan A, Shivkumar K. Sympathetic stimulation increases dispersion of repolarization in humans with myocardial infarction. Am J Physiol Heart Circ Physiol. 2012;302(9):H1838–46.
82. Lockhart ST, Turrigiano GG, Birren SJ. Nerve growth factor modulates synaptic transmission between sympathetic neurons and cardiac myocytes. J Neurosci. 1997;17(24):9573–82.
83. Shcherbakova OG, Hurt CM, Xiang Y, Dell'Acqua ML, Zhang Q, Tsien RW, et al. Organization of beta-adrenoceptor signaling compartments by sympathetic innervation of cardiac myocytes. J Cell Biol. 2007;176(4):521–33.
84. Barnett JV, Taniuchi M, Yang MB, Galper JB. Co-culture of embryonic chick heart cells and ciliary ganglia induces parasympathetic responsiveness in embryonic chick heart cells. Biochem J. 1993;292(Pt 2):395–9.
85. Marvin WJ Jr, Atkins DL, Chittick VL, Lund DD, Hermsmeyer K. In vitro adrenergic and cholinergic innervation of the developing rat myocyte. Circ Res. 1984;55(1):49–58.
86. Prando V, Da Broi F, Franzoso M, Plazzo AP, Pianca N, Francolini M, et al. Dynamics of neuroeffector coupling at cardiac sympathetic synapses. J Physiol. 2018;596(11):2055–75.
87. Winbo A, Ramanan S, Eugster E, Jovinge S, Skinner JR, Montgomery JM. Functional coculture of sympathetic neurons and cardiomyocytes derived from human-induced pluripotent stem cells. Am J Physiol Heart Circ Physiol. 2020;319(5):H927–H37.
88. Takayama Y, Kushige H, Akagi Y, Suzuki Y, Kumagai Y, Kida YS. Selective induction of human autonomic neurons enables precise control of Cardiomyocyte beating. Sci Rep. 2020;10(1):9464.
89. Larsen HE, Lefkimmiatis K, Paterson DJ. Sympathetic neurons are a powerful driver of myocyte function in cardiovascular disease. Sci Rep. 2016;6:38898.
90. Flannery RJ, Bruses JL. N-cadherin induces partial differentiation of cholinergic presynaptic terminals in heterologous cultures of brainstem neurons and CHO cells. Front Synaptic Neurosci. 2012;4:6.
91. Zarzoso M, Rysevaite K, Milstein ML, Calvo CJ, Kean AC, Atienza F, et al. Nerves projecting from the intrinsic cardiac ganglia of the pulmonary veins modulate sinoatrial node pacemaker function. Cardiovasc Res. 2013;99(3):566–75.
92. Allen E, Coote JH, Grubb BD, Batten TFC, Pauza DH, Ng GA, et al. Electrophysiological effects of nicotinic and electrical stimulation of intrinsic cardiac ganglia in the absence of extrinsic autonomic nerves in the rabbit heart. Heart Rhythm. 2018;15(11):1698–707.
93. Rajendran PS, Challis RC, Fowlkes CC, Hanna P, Tompkins JD, Jordan MC, et al. Identification of peripheral neural circuits that regulate heart rate using optogenetic and viral vector strategies. Nat Commun. 2019;10(1):1944.

94. Hoover DB, Ganote CE, Ferguson SM, Blakely RD, Parsons RL. Localization of cholinergic innervation in guinea pig heart by immunohistochemistry for high-affinity choline transporters. Cardiovasc Res. 2004;62(1):112–21.
95. Jungen C, Scherschel K, Eickholt C, Kuklik P, Klatt N, Bork N, et al. Disruption of cardiac cholinergic neurons enhances susceptibility to ventricular arrhythmias. Nat Commun. 2017;8:14155.
96. Wengrowski AM, Wang X, Tapa S, Posnack NG, Mendelowitz D, Kay MW. Optogenetic release of norepinephrine from cardiac sympathetic neurons alters mechanical and electrical function. Cardiovasc Res. 2015;105(2):143–50.
97. Moreno A, Endicott K, Skancke M, Dwyer MK, Brennan J, Efimov IR, et al. Sudden heart rate reduction upon optogenetic release of acetylcholine from cardiac parasympathetic neurons in perfused hearts. Front Physiol. 2019;10:16.
98. Choate JK, Edwards FR, Hirst GD, O'Shea JE. Effects of sympathetic nerve stimulation on the sino-atrial node of the guinea-pig. J Physiol. 1993;471:707–27.
99. Bramich NJ, Brock JA, Edwards FR, Hirst GD. Responses to sympathetic nerve stimulation of the sinus venosus of the toad. J Physiol. 1993;461:403–30.
100. Campbell GD, Edwards FR, Hirst GD, O'Shea JE. Effects of vagal stimulation and applied acetylcholine on pacemaker potentials in the guinea-pig heart. J Physiol. 1989;415:57–68.
101. Hutter OF. The vagus and the heart: revisiting an early contribution to a still on-going dispute. J Physiol. 2012;590(10):2535–6. author reply 7
102. Ng GA, Brack KE, Coote JH. Effects of direct sympathetic and vagus nerve stimulation on the physiology of the whole heart--a novel model of isolated Langendorff perfused rabbit heart with intact dual autonomic innervation. Exp Physiol. 2001;86(3):319–29.
103. Ng GA, Mantravadi R, Walker WH, Ortin WG, Choi BR, de Groat W, et al. Sympathetic nerve stimulation produces spatial heterogeneities of action potential restitution. Heart Rhythm. 2009;6(5):696–706.
104. Mantravadi R, Gabris B, Liu T, Choi BR, de Groat WC, Ng GA, et al. Autonomic nerve stimulation reverses ventricular repolarization sequence in rabbit hearts. Circ Res. 2007;100(7):e72–80.
105. Tapa S, Wang L, Francis Stuart SD, Wang Z, Jiang Y, Habecker BA, et al. Adrenergic supersensitivity and impaired neural control of cardiac electrophysiology following regional cardiac sympathetic nerve loss. Sci Rep. 2020;10(1):18801.
106. Wang L, Olivas A, Francis Stuart SD, Tapa S, Blake MR, Woodward WR, et al. Cardiac sympathetic nerve transdifferentiation reduces action potential heterogeneity after myocardial infarction. Am J Physiol Heart Circ Physiol. 2020;318(3):H558–H65.
107. Paton JF. A working heart-brainstem preparation of the mouse. J Neurosci Methods. 1996;65(1):63–8.
108. Ashton JL, Paton JF, Trew ML, LeGrice IJ, Smaill BH. A working heart-brainstem preparation of the rat for the study of reflex mediated autonomic influences on atrial arrhythmia development. Annu Int Conf IEEE Eng Med Biol Soc. 2013;2013:3785–8.
109. Ashton JL, Trew ML, LeGrice IJ, Paterson DJ, Paton JF, Gillis AM, et al. Shift of leading pacemaker site during reflex vagal stimulation and altered electrical source-to-sink balance. J Physiol. 2019;597(13):3297–313.
110. Habecker BA, Anderson ME, Birren SJ, Fukuda K, Herring N, Hoover DB, et al. Molecular and cellular neurocardiology: development, and cellular and molecular adaptations to heart disease. J Physiol. 2016;594(14):3853–75.

Part II
Vascular Signaling

Mechanisms of Lipoproteins and Reverse Cholesterol Transport in Atherosclerotic Cardiovascular Disease

Holly C. Sucharski and Sara N. Koenig

Abstract Coronary heart disease (CHD) makes up approximately 42.1% of all cardiovascular disease deaths in the United States. Cholesterol deposition in the arteries from LDL-C, or the "bad cholesterol," increases the risk of CHD, atherosclerosis, myocardial infarction, and stroke. However, increased HDL-C, "good cholesterol," has been associated with lower risk of CHD and plays an important role in the reverse cholesterol transport (RCT) pathway. The RCT pathway is the process of cholesterol efflux from peripheral cells and tissues by HDL, and transported to the liver for uptake, excretion, and recycling. Pathogenic variants in key players within the RCT pathway, like *SCARB1, ApoA-I, ABCA1/ABCG1,* are associated with atherosclerosis and coronary artery disease. Thus, understanding RCT mechanisms is of significant scientific interest. In this chapter, we discuss lipoproteins, with particular emphasis on the known mechanisms of RCT, disease-associated variants, and current therapies.

Keywords High density lipoprotein · HDL-C · SR-BI · RCT · coronary artery disease · atherosclerosis

H. C. Sucharski · S. N. Koenig (✉)
Department of Physiology and Cell Biology, College of Medicine, The Ohio State University, Columbus, OH, USA

Dorothy M. Davis Heart and Lung Research Institute, College of Medicine, The Ohio State University Wexner Medical Center, Columbus, OH, USA

Bob and Corinne Frick Center for Heart Failure and Arrhythmia, The Ohio State University Wexner Medical Center, Columbus, OH, USA

Division of Cardiovascular Medicine, Department of Internal Medicine, College of Medicine, The Ohio State University, Columbus, OH, USA
e-mail: Holly.Sucharski@osumc.edu; Sara.Koenig@osumc.edu

N. L. Parinandi, T. J. Hund (eds.), *Cardiovascular Signaling in Health and Disease*, https://doi.org/10.1007/978-3-031-08309-9_12

Introduction

Cardiovascular disease (CVD) is the leading cause of death in the United States, claiming more lives than cancer and chronic lung disease combined [1]. Coronary Heart Disease (CHD) accounts for 42.1% of all CVD deaths, with an estimated economic cost of over $200 billion per year in the United States [1]. Arterial cholesterol deposition increases risk of CHD, atherosclerosis, heart attack, and stroke. Elevated low-density lipoprotein-associated cholesterol (LDL-C), termed "bad cholesterol," is a CVD risk factor that increases the risk of arterial cholesterol deposition. The "good cholesterol," high-density lipoprotein-associated cholesterol (HDL-C), is associated with improved outcomes in CHD patients [2].

The Framingham Heart study first identified a correlation between elevated plasma HDL-C and decreased risk of atherosclerotic CVDs, including coronary artery disease (CAD), leading to the investigation of the mechanism of athero-protective HDL-C [3] and the discovery of reverse cholesterol transport (RCT) [4]. RCT is the mechanism where cholesterol is transported by HDL from cells in peripheral tissues to the liver to be recycled or excreted in bile or feces. Several clinical trials have been successful in raising HDL-C and lowering LDL-C with CETP inhibition; however, they were not successful in altering outcomes of CAD [5–8]. Elevated HDL-C is widely accepted as a biomarker for decreased risk of CVD, but the mechanisms of RCT and HDL functions are still not fully understood. Despite setbacks from CETP inhibition clinical trials, cholesterol metabolism, RCT, and cell-specific uptake offer novel therapeutic targets to reduce plasma cholesterol levels.

Environmental and nutritional factors that influence the risk of atherosclerosis development, such as tobacco smoke, obesity, sedentary lifestyle, diabetes mellitus, poor diet, and hypertension [9], all of which can be controlled to an extent to reduce atherosclerosis risk. An in-depth analysis from 2015 discussed the athero-protective contribution of different nutrients, particularly found within the "Mediterranean diet" [9–15] . While the Mediterranean diet is athero-protective, there are also foods that contribute to atherosclerosis and CVD risk, including trans-fats [16–18] and processed sugars [19–23]. In addition to diet, exercise is an important lifestyle factor that is associated with risk of CVD [24]. In 2004, 52 countries were represented in a case-control study that determined that physical inactivity has a population attributable risk of 12.2% for myocardial infarction [24, 25]. Exercise is a non-medical treatment for increasing HDL-C levels and improving RCT function (Fig. 1) [26].

While environmental and nutritional factors play a substantial role in CVD risk, genetic risk factors are becoming more apparent. Genome-wide association studies (GWAS) have identified genetic markers located in chromosome 9p21.3 that are associated with CHD and myocardial infarction (MI) in European-derived populations from the Atherosclerotic Disease, Vascular Function, and Genetic Epidemiology (ADVANCE) study [1, 27], with 50% of the European-derived population estimated to harbor one risk allele and 23% harbors two risk alleles [1, 28]. In addition, a multi-ethnic extension of the ADVANCE study determined that SNPs in this region

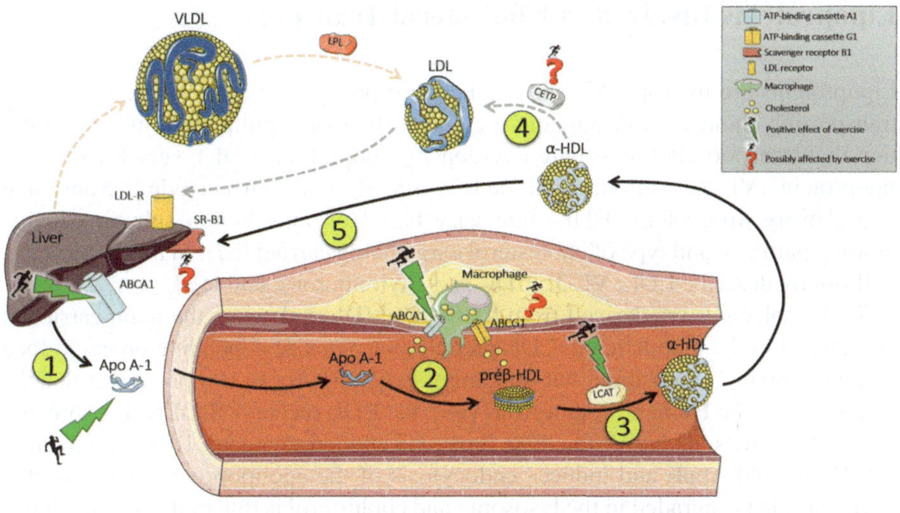

Fig. 1 Schematic depicting the RCT pathway in relation to arterial lipid deposition and the positive impact of exercise. Abbreviations: VLDL very low density lipoprotein, LPL lipoprotein lipase, LDL low density lipoprotein, CETP cholesterol ester transport protein, HDL high density lipoprotein, ApoA-1 apolipoprotein A-1, SR-BI scavenger receptor BI. (Designed and originally presented by [26])

were not associated with CAD in African American subjects, but there was an increased odds ratio for US Hispanics and East Asian subjects [29]. Over 2000 genetic variants within 57 loci have been significantly associated with CAD involving pathways in blood vessel morphogenesis, lipid metabolism, nitric oxide signaling, and inflammation [30, 31]. Patient variants in lipid regulators that are associated with CVD have highlighted the importance of cholesterol transport proteins (ex: *LDLR*, *APOE*, *SCARB1*).

In patients with high cholesterol and atherosclerotic disease, statins are commonly used to reduce LDL-C and greatly reduce risk of death by CHD. Poor adherence and chronic use of statins significantly increases the risk of adverse cardiovascular events [32]. PCSK9 inhibitors are an effective therapy at reducing CHD death. However, the exorbitant price of PCKS9 inhibitors contributes to medical, financial, and economic burdens, ultimately leading to a high prescription abandonment rate and resulting in medical disparity with increased risk of CVD, disproportionately affecting women, racial minorities, and low-income groups [33–35]. Despite the success of statins and PCSK9 inhibition, alternative therapies are warranted to improve patient outcomes. Exploring and enhancing RCT mechanisms could help identify novel therapeutics for CVD. Pathogenic variants in SR-BI, a key player in RCT, were recently identified as causing inherited CAD, demonstrating the potential therapeutic power of this pathway [36].

In this book chapter, we will highlight the known mechanisms in cholesterol signaling that contribute to CVD, go in depth on HDL metabolism with a focus on the RCT pathway, and discuss the genetic predisposition that underlies CVD as well as current therapies.

Lipoproteins Involved in Cholesterol Transport

Lipoproteins are hydrophobic carriers made of proteins and lipids that are vital to transporting cholesterol throughout the body. The main apolipoproteins in circulation that transport cholesterol are low-density lipoprotein (LDL), very low-density lipoprotein (VLDL), intermediate-density lipoprotein (IDL), chylomicrons, and high-density lipoprotein (HDL) that vary based on apo-protein subunit density, binding partners, and type of cholesterol cargo being carried [37]. In this section, we will briefly describe LDL, VLDL, IDL and chylomicrons, and HDL.

LDL molecules are derived from VLDL and IDL and act as the main carrier of cholesterol in the bloodstream. LDL particles vary in size depending on cargo load and apoprotein composition, but contain at least one ApoB-100 protein molecule. According to the Human Protein Atlas [38], the LDL receptor (LDLR) is expressed in various tissues, including the liver, lung, and endocrine organs. LDLR recognizes ApoB-100 and ApoE and induces endocytosis of the apolipoprotein, whereby the LDL particle is degraded in the lysosome and cholesterol is released. Elevated intracellular cholesterol decreases the activity of hydroxy-3-methylglutaryl-CoA (HMG-CoA) reductase that converts HMG-CoA to mevalonate, a cholesterol precursor and the rate-limiting step in cholesterol biosynthesis. HMG-CoA reductase reduction decreases the expression of LDLR to decrease cholesterol uptake [39]. Statins competitively block HMG-CoA reductase and can reduce proprotein convertase subtilisin/kexin type (PCSK9) that helps reduce cholesterol production while increasing LDLR expression, in turn reducing serum LDL-C. The catalytic unit of PCSK9 binds to LDLR and facilitates transport to the lysosome for degradation [40–42], therefore inhibition drives LDLR membrane retention. Smaller LDL particles are considered pro-atherosclerotic due to lower binding affinity for LDLR, leading to increased circulation retention and increased risk of arterial wall deposition [37].

VLDL particles are produced by the liver and mainly carry triglycerides in circulation. Each VLDL contains one ApoB-100 as the core structural protein, but can also contain other apolipoproteins such as Apo B-100, C-I, C-II, C-III, and E. The size of VLDL can vary based on triglyceride load, triglyceride production, and apolipoprotein composition. When triglyceride production is increased, triglyceride production VLDL particles grow based on demand, but remain smaller than chylomicrons. IDL particles are formed when adipose or muscle tissue removes the triglycerides from VLDL, leaving the particle enriched with cholesterol and Apo B-100 expression. IDL contains Apo B-100 expression, but may also contain ApoE and ApoC, depending on the originating VLDL [37].

Chylomicrons are low-density particles produced by the liver and enriched for triglycerides, forming the largest lipoprotein. Chylomicrons may contain Apo B-48, C, E, A-1, A-II, A-IV, but have Apo B-48 as its core structural protein. Based on diet and triglyceride production, chylomicron size will be altered: a high fat meal will produce large chylomicrons, where fasting leads to small chylomicrons with less triglycerides [37].

HDL particles are enriched for cholesterol esters (CE) and phospholipids and are considered cardio-protective, since HDL-C is inversely correlated with CVD outcomes [3, 37, 43]. The biogenesis of HDL occurs mostly in the liver [44], but can also occur in the intestine [45], where different subspecies are formed that vary based on size, density, and apolipoprotein composition (Fig. 2). Originally determined by centrifugation, the HDL subspecies include: pre-β or "nascent" HDL that is discoidal in shape and lipid-free/lipid-poor, alpha-1 (α-1 or HDL2) which is considered the "mature" form of HDL that is spherical and is carrying the most cholesterol, and α-II and α-III (HDL3 particles) which are smaller/denser, spherical, ApoA-I containing HDL that accept cholesterol [46, 47]. HDL particles vary in size and may contain Apo A-I, A-II, C, or E apolipoproteins. ApoA-1 is the major HDL structural protein, accounting for up to 70% of HDL composition that also aids in recognition of HDL-C for transporters (ABCA1: liver, ABCG1: peripheral cells), and receptors (scavenger receptor BI (SR-BI)) interactions. To develop the HDL core of cholesterol esters, ApoA1 functions as a cofactor for lecithin: cholesterol acyltransferase (LCAT) that esterifies free cholesterol to cholesterol esters. When lipid levels are low, the liver secretes ApoA-I that acquires additional cholesterol

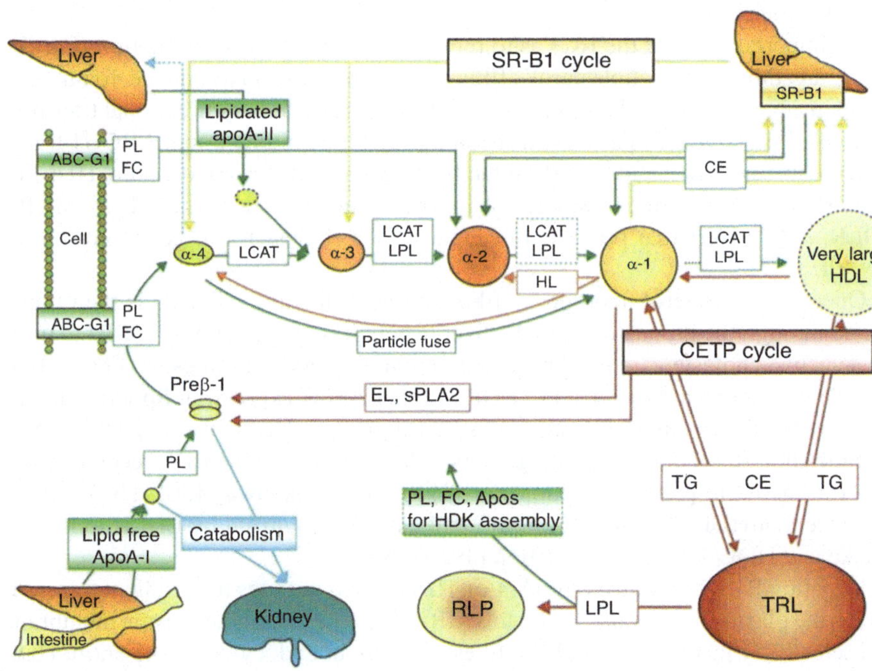

Fig. 2 HDL metabolism/remodeling in RCT. The composition of HDL-C dictates its size and next steps in the RCT process. Abbreviations: PL phospholipids, FC free cholesterol, TG triglycerides, CE cholesterol ester, ABCG1 ATP binding cassette transporter G1, EL endothelial lipase, LCAT Lecithin-Cholesterol Acyltransferase, sPLA2 secretory phospholipase A2, TRL TG-rich lipoprotein, SR-BI scavenger receptor B1. (Originally adapted from [51] by [46], presented here with permission)

and phospholipids from the ATP binding cassette transporter 1, ABCA1 or ABCG1, transporter to produce nascent HDL particles [48]. ApoA-1 has also been identified as an anti-atherosclerotic marker, because it acquires free cholesterol and phospholipids that are effluxed by hepatocytes and enterocytes [37, 49, 50] at the beginning of the RCT pathway that will be described in the next section.

The Reverse Cholesterol Transport Pathway

The RCT pathway is the process of cholesterol efflux from peripheral cells and tissues by HDL, and transported through the circulation, to the liver where the cholesterol is taken up and recycled, excreted in bile or metabolized as bile salts prior to excretion. A major component of the RCT pathway is the lipid carrier, HDL [47]. Nascent HDL particles are a discoidal shape [52], which allows them to acquire more lipid particles from peripheral cell types, such as macrophages or vascular smooth muscle cells [53], through ABCA1-mediated efflux. The HDL-C particle then becomes a more spherical shape with the addition of cholesterol esters and is ready for transport [47].

An important role of the RCT is to remove cholesterol from circulation (Fig. 3). There are three ways cholesterol efflux can occur from macrophage foam cells found in atherosclerotic plaques [54]: (1) ABCA1-mediated unidirectional transport to ApoA-1 on HDL; (2) passive diffusion through the unidirectional ABCG1 transporter to mature HDL; and (3) SR-BI facilitated passive diffusion to mature HDL. To do this, there are various cholesterol receptors and transporters such as SR-BI, LDLR, ABCA1/G1 that provide a mechanism for cholesterol to go into or out of the cell.

Once the cholesterol undergoes efflux and HDL loading, HDL-C is then transported through the bloodstream to the liver, where it is recognized by SR-BI to initiate selective cholesterol ester uptake. Any remaining cholesterol esters in the HDL particle may then be transferred to LDL or other ApoB-expressing lipoprotein particles by CETP for direct hepatic uptake through the LDL receptor (LDLR) [55]. Hepatic cholesterol uptake can occur in two direct ways: (1) the LDL receptor pathway (discussed in previous section); or (2) the SR-BI pathway [56]. There is also evidence that HDL-C expressing ApoE that can be recognized by LDLR to initiate uptake and HDL/ApoA-I degradation [49, 57, 58].

Hepatic SR-BI recognizes ApoA-1 on HDL-C and initiates the gradient transport across the plasma membrane. To do this, SR-BI binds HDL at the plasma membrane and selectively takes in the cholesterol esters without endocytosing the entire HDL particle [47]. The cholesterol esters transported this way will then be hydrolyzed by hormone-sensitive lipase (HSL-1) to free cholesterol. The free cholesterol can then be used for the cell's structure, transported to bile, or used as a steroidal precursor in endocrine tissues [56]. In 2013, a crystal structure of the lysosomal integral membrane protein (LIMP2), which has sequence homology to SR-BI and is in the class B scavenger receptor family with SR-BI, identified a "hydrophobic channel" in the

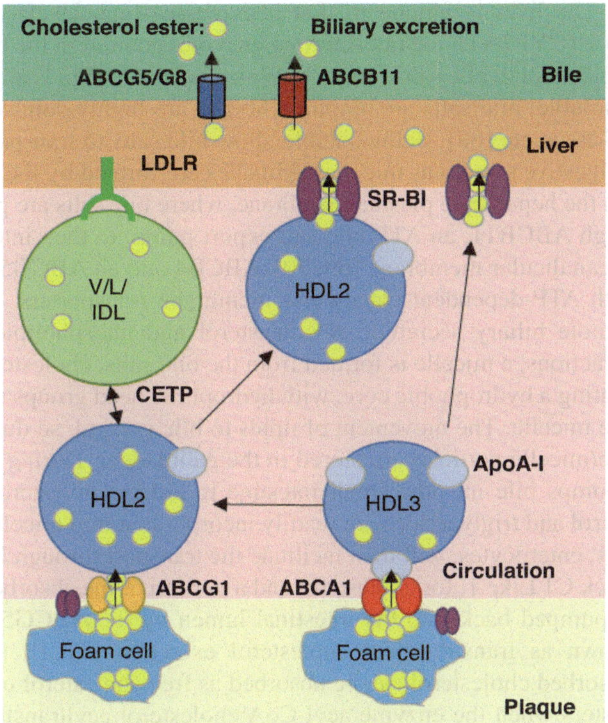

Fig. 3 Key steps in RCT. Briefly, RCT starts with the removal of cholesterol from peripheral cells, such as arterial foam cells from either vascular smooth muscle cell (V-mac) or macrophage origin (left panel). Since this process requires the efflux of cholesterol to cholesterol acceptors, such as nascent or mature HDL, and macrophage migration out of the plaque, efflux is considered the rate-limiting step of RCT. Once cholesterol is transported to the liver, there are direct and indirect ways of cholesterol uptake and HDL-C can undergo remodeling to adapt to the most efficient uptake method (middle panel). The last step of RCT is the cholesterol excretion through biliary excretion or transintestinal cholesterol efflux (TICE) for feces excretion (right panel). Abbreviations: LCAT lecithin:cholesterol acyltransferase, LXR liver X receptors, OSBP oxysterol-binding protein. (Adapted with permission from [53])

extracellular domain that would help explain the mechanics of selective cholesterol ester uptake from HDL-C [59]. Nonpolar cholesterol esters move freely down the gradient, through the suggested SR-B1 "channel," whereas more polar phospholipids will take longer [54]. SR-BI is bidirectional, meaning it allows for cholesterol uptake and promotes efflux. The importance of SR-BI is highlighted in SR-BI knockout mice that exhibit increased amount and size of HDL-C with accelerated atherosclerosis and decreased cholesterol secreted in bile [60–62]. Additionally, SR-BI overexpression mice have decreased HDL-C and reduced atherosclerosis [60–65].

After cholesterol is taken up by hepatocytes, a portion of it will be enzymatically converted to bile salts, whereas another fraction can be used for lipid membrane maintenance. The liver is unique in that it expresses high levels of cholesterol 7α-hydroxylase (CYP7A1), the rate-limiting enzyme required in the multistep conversion of cholesterol to bile salts [66, 67]. While cholesterol/lipids are hydrophobic and water insoluble, bile salts are amphiphilic and are highly soluble due to their ability to self-associate [68]. Being soluble allow bile salts to transport cholesterol through the digestive system as micelles. Micelles are formed by the interaction of bile salts with the hepatocyte plasma membrane, where bile salts are pumped out of the cell through ABCB11, an ATP bile salt export pump, to then interact with the extracellular canalicular membrane [69, 70]. ABCB4 and an ABCG5/ABCG8 heterodimer, both ATP-dependent canalicular membrane transporters, are then activated to promote biliary secretion of cholesterol and phospholipids [67]. With complex interactions, a micelle is formed from the bile salts, cholesterol, and phospholipids creating a hydrophobic core, with hydrophilic head groups making up the exterior of the micelle. The movement of lipids to bile is rapid, so during rest from digestion, bile/micelle particles are stored in the gallbladder. During digestion, the gallbladder pumps bile into the small intestine to aid in cholesterol metabolism since cholesterol and triglycerides are readily incorporated into micelles. The small intestinal cells, enterocytes, will then facilitate the transport through NPC1L1 [71], or Nieman-Pick C1 Like 1, for cholesterol and a portion of the absorbed cholesterol will then be pumped back into the intestinal lumen by the ABCG5/ABCG8 heterodimer known as transintestinal cholesterol excretion (TICE) [67, 72]. The remaining absorbed cholesterol can be absorbed as free cholesterol or converted to cholesterol esters when the enzyme acyl-CoA:cholesterol acyltransferase (ACAT) covalently attaches a fatty acid to the free hydroxyl group of the cholesterol particle [73]. The free cholesterol, triglycerides, and cholesterol esters are then combined within the enterocyte and added to apolipoprotein B48 to form chylomicrons to be secreted into the lymph to be transported back to the plasma for liver processing, known as enterohepatic circulation [74]. About 1–2% of the bile salts will escape this cycle and be excreted in feces, as the liver cholesterol production and rate of absorption are tightly regulated [67] (Fig. 4).

Mechanisms of Diseases Associated with Atherosclerotic CVD

Under genetic predisposition or exposure to environmental risk factors, arterial cholesterol deposition can occur and start the atherosclerotic process. Lipid deposition leads to the recruitment of monocytes that attempt to clear the lipid deposition by increasing cholesterol uptake. The intracellular cholesterol esters are then packaged into lipid droplets. The monocytes then become macrophage foam cells that are cholesterol-laden, efflux defective, and pro-inflammatory cells that leak cholesterol that contribute to the arterial plaque. There are also vascular smooth muscle cells that can become foam cells as well [75–79]. To be released from the cell, cholesterol esters must undergo hydrolysis to be converted to free, unesterified cholesterol. The cholesterol esters in the lipid droplets consistently go through cycles of hydrolysis

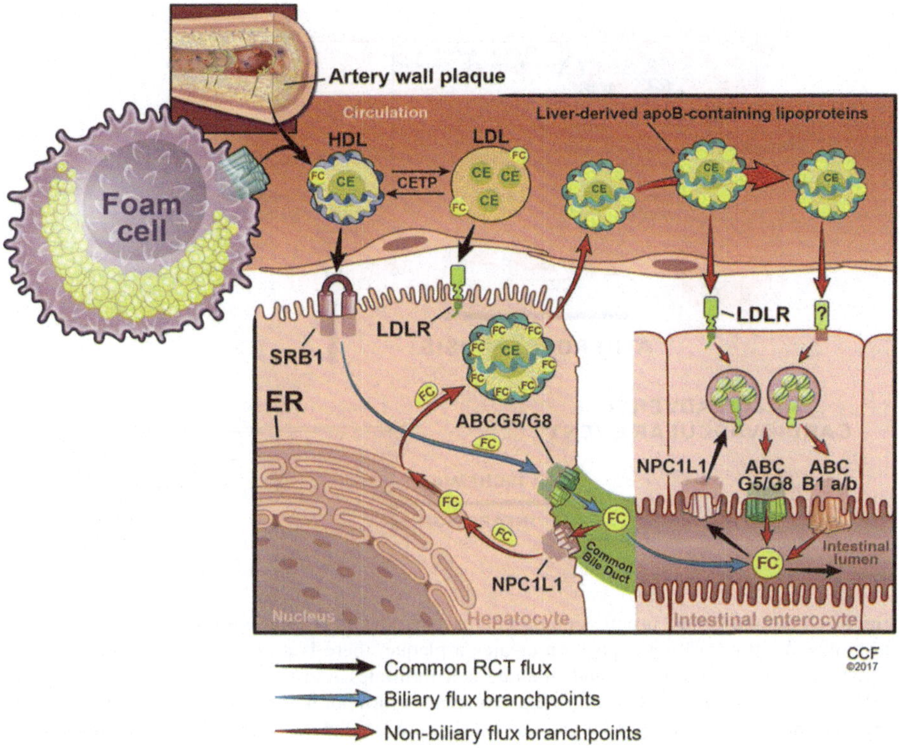

Fig. 4 Important steps for cholesterol excretion after RCT processes. Discussed in detail in the text, but briefly there is biliary flux that converts free cholesterol to bile salts to be stored in the gallbladder, non-biliary flux that transports cholesterol to the small intestine for transintestinal cholesterol excretion in feces or reabsorption for enterohepatic circulation in the lymph. (Presented with permission from [72])

and re-esterification, create a cholesterol-laden foam cell, if the cholesterol is re-esterified by ER-resident protein acyl-CoA:cholesterol acyltransferase (ACAT) [80, 81], instead of undergoing efflux. If cholesterol efflux is disrupted, hydrolyzed free cholesterol cannot be efficiently released, preventing efflux and RCT mechanisms from clearing the lipid deposition. Buildup of cholesterol and foam cells lead to damage and inflammation in the endothelium, leading to infiltration of additional cell types such as macrophages, fibroblasts, and smooth muscle cells that contribute to plaque formation. The atherosclerotic plaques result in vessel wall thinning and restricted blood flow, gradually leading to the development of atherosclerosis, CAD, or heart failure [53]. Plaques are also at risk of bursting, resulting in activated clotting factors and can lead to stroke, myocardial infarction, or activation of other clotting disorders that restrict blood flow (Fig. 5) [82, 83].

Historically, HDL-C has been considered atheroprotective since HDL-C was found to inversely correlate with CVD risk [3, 43, 53]. Recent evidence suggests that HDL-C levels are not indicative of RCT efficiency since HDL-C

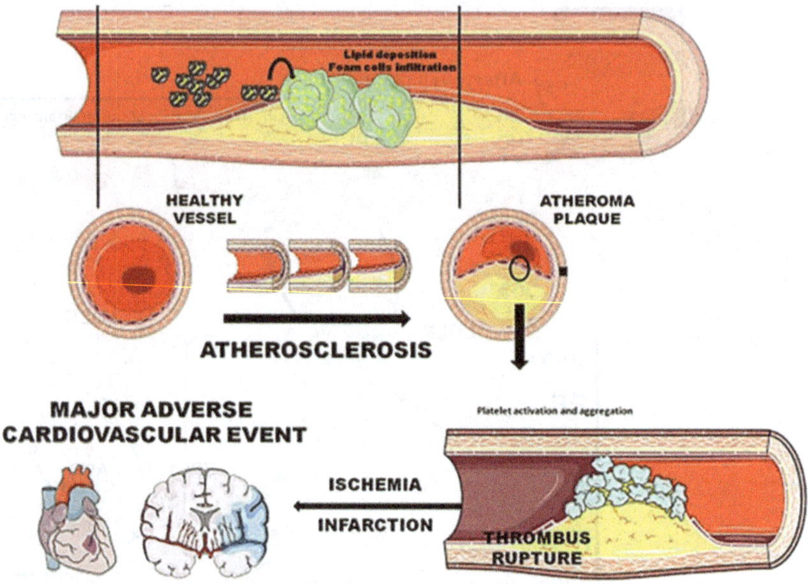

Fig. 5 Development and progression of atherosclerosis with arterial lipid deposition and cellular infiltrates. As arterial lipid deposition creates a plaque, there is an increase in proinflammatory signals that drive macrophage and immune cell infiltrates into the intima. These cells become efflux-deficient foam cells that contribute to thrombus formation. If the thrombus blocks the entire vessel, ischemia or coronary artery disease leads to myocardial infarction. (Presented with permission from [88])

replacement and CETP inhibition increased HDL-C, lowered LDL-C, but overall, had no effect on CVD outcomes [5–8, 84]. HDL has many contributing factors to athero-protection, including its main function in RCT/excretion to prevent plaque formation and its anti-inflammatory properties, such as inhibiting adhesion molecules and MCP-1 on endothelial cells from binding circulating monocytes that are associated with inflammation and the initiation of atherosclerotic disease [85–87].

The oxidation hypothesis of atherosclerosis suggests that HDL-C aids in mediating the oxidation of phospholipids in LDL-C [89, 90]. Pro-inflammatory, oxidized phospholipids result from oxidation of LDL phospholipids that contain arachidonic acid and subsequently be recognized by the innate immune system [55, 91]. During an acute phase response, HDL becomes pro-inflammatory to aid the immune response during injury or infection, while decreasing RCT functions. After healing and homeostasis are restored, HDL returns to anti-inflammatory functions and resumes higher RCT activity [91]. In the case of poor diet and exercise or genetic predisposition, the acute phase response to oxidative phospholipids has potential to become chronic, rendering HDL pro-inflammatory and less likely to perform RCT and anti-inflammatory functions [92], highlighting the potential link between RCT and the oxidation hypothesis in atherosclerosis [93, 94].

Genetic Predisposition to Cholesterol-Driven Cardiovascular Disease

While environmental and nutritional factors play a large role in CVD, genetic predisposition may shed light on causes of dyslipidemia-associated diseases. Fortunately, GWAS has brought the world one step closer to personalized medicine and is helping to uncover other genetic predisposition variants in key lipid regulating genes.

For example, GWAS has identified 56 new loci associated with CAD, most are pleiotropic, meaning they are also associated with other diseases. In a discovery analysis of common SNPs, six new loci were associated with CAD, such as SNPs from the *SCARB1, CETP, LRP1, C2, KCNJ13-GIGYF2, MRVI1-CTR20-9* genes [1, 95]. CAD has been shown to be associated with variants in the main lipid modulators/transporters such as *LDLR, APOB, PCSK9, SCARB1* and *CETP* [96–98] that will be highlighted in this section. This section will also go into more detail on dyslipidemia-related variants and recent research findings.

Historically, familial variants help lead to the identification of different cholesterol transport proteins such as LDLR [96, 99]. Familial hypercholesterolemia (FH) is defined by genetic predisposition to elevated serum LDL-C, where the most common variants are *LDLR, APOB, PCSK9,* as previously reviewed [97]. Variants in *LDLR* were first found in homozygous patients with severe FH, where even heterozygous patients had reduced LDLR protein expression and increased LDL-C [99]. LDLR knockout mice have been extensively studied and developed as an in vivo atherosclerosis model due to the increase in weight gain and cholesterol levels when fed a high cholesterol diet, especially in an ApoE knockout background [100–103]. Mutations in *APOB*, that encodes the LDLR ligand, ApoB, have also been associated with FH and CAD [97, 104–107]. Interestingly, protein-truncating variants in *APOB* have been associated with familial hypobetalipoproteinemia and have been shown to decrease LDL-C, triglycerides, and lower risk of CHD [108, 109]. Since 2004, over 30 gain-of-function and loss-of-function *PCSK9* mutations have been associated with FH [110, 111]. PCSK9 binds LDLR and leads to lysosomal degradation of the complex, so a gain-of-function mutation in *PCSK9* would increase LDLR degradation, increasing serum LDL-C levels [40, 97, 111–113].

The majority of current hyperlipidemia therapies target mostly the LDL-C pathway, but HDL-C pathway variants have provided a unique challenge for therapeutic development and enhancement of RCT. Variants have been identified in proteins responsible for RCT that are associated with CAD and dyslipidemia, such as *SCARB1.* While the severity of disease associated with *SCARB1* variants has been shown to be dependent on its location and the impact on function of SR-BI [95, 114], variants in *SCARB1* have been directly linked to increased HDL-C and increased risk of CAD [36, 115–117], although risk of CAD was not observed in a large Icelandic population study with *SCARB1* variants [118]. The first reported mutation of SR-BI in humans (p.P297S mutation) was associated with reduced capacity for cholesterol efflux from macrophages and increased HDL-C levels, but overall did not have increased severity of atherosclerosis based on the small sample size of carriers [119] and was later linked to issues with uptake for steroidogenesis [120].

SR-BI is also highly expressed in steroidogenic tissues such as the adrenal glands, placenta, ovary, and testis that require cholesterol for steroidogenesis [56]. Cholesterol esters needed for steroid synthesis is obtained by (a) de novo cholesterol synthesis within the cell; (b) pulling from lipid droplet reservoirs; and (c) cholesterol esters from either LDLR-mediated endocytic uptake or the selective uptake via SR-BI [56]. In *SCARB1*$^{-/-}$ and *apoA-1* $^{-/-}$ null mice, there was almost no cholesterol ester accumulation in adrenal cells suggesting that SR-BI-mediated selective cholesterol ester uptake provides the majority of cholesterol needed for steroidogenesis in mouse adrenal cells [121]. In humans, variants in *SCARB1* have also been associated with diminished adrenal steroid production [119, 120].

More recently, compound heterozygous variants (p.G319V and c.754_755delinsC variant) were identified in a family with severe, early-onset CAD and dyslipidemia following Mendelian inheritance. Although the p.G319V mutation was previously associated with elevated HDL-C but not with CAD [118], the p.G319V mutation was pathogenic in the presence of the null c.754_755delinsC variant. In further support of pathogenicity of this variant, homozygous p.G319V mouse model that had >95% lethality, and heterozygous knock-in mice have elevated total cholesterol in the heterozygous p.G319V mice similar to that of SR-BI knockout mice [36].

To model atherosclerotic CVDs in the lab, in vivo mouse models are utilized with deficient expression of lipoprotein components and binding partners such as LDLR, ApoE, and SR-BI in combination with diets of altered cholesterol and/or fat [122]. LDLR knock-out mice develop hypercholesterolemia with elevated serum cholesterol levels under normal conditions (200–400 mg/dL) and significantly elevated cholesterol levels (>2000 mg/dL) on high fat diet [123]. The ApoE knock-out mouse model develops arterial lesions starting at three months of age and have elevated serum cholesterol under normal conditions with the reduced ability to remove excess cholesterol esters from the blood [124–127]. Both of these models have been used to recapitulate atherosclerosis in mice, although the mechanisms and lipoprotein composition vary greatly from human atherosclerosis [128]. On a "Western" diet, the LDLR/ApoE double knock-out mouse model develops obstructive CAD and MI associated with hyperlipidemia [129]. Interestingly, SR-BI knockout mice have increased HDL-C and total cholesterol compared to wild-type mice and experience homozygous lethality [61]. These mice also have abnormally large HDL particles and increased risk of atherosclerosis [61, 130, 131]. A double knockout of SR-BI/ApoE mouse model develops atherosclerotic lesions in the coronary arteries starting at 4–5 weeks of age and die of severe CAD by 8 weeks [132].

Therapeutic Approaches Targeting Lipoproteins

Ever since the correlation between increased LDL-C with increased risk of CVD [133], and increased HDL-C with decreased risk of CVD was established [3, 43], research has been done to identify therapeutic targets involved in the cholesterol transport pathways that has led to current drug therapies for atherosclerotic CVDs.

LDL-C-Targeted Therapies

Since the low-density and intermediate-density lipoproteins contribute to arterial lipid deposition and atherosclerotic CVDs, therapies targeting these pathways have been successful in decreasing LDL-C and reducing CVD deaths [33, 34]. Statin therapy was first approved in 1987 [134] and acts by inhibiting the enzyme, HMG-CoA reductase, which converts HMG-CoA to mevalonate (rate-limiting cholesterol precursor) in hepatocytes. Statins actively compete with normal substrates as well as reversibly altering the conformation of the enzyme after binding to the active site [135, 136]. Inhibiting HMG-CoA reductase with statins reduces cholesterol biosynthesis and increases LDLR expression for increased cholesterol uptake: effectively reducing serum LDL-C levels. Although statin therapy is the primary form of therapy for CVD, clinical trials have shown statin toxicity, resistance, and intolerance in up to 15% of patients leading to different disorders such as developing type II diabetes, muscle disorder, and myopathies [32, 137], and inter-patient efficacy variability has been observed [138, 139].

In addition to lowering cholesterol production, statins also decrease PCSK9, in turn reducing LDL-C levels. The catalytic unit of PCSK9 binds to LDLR and facilitates transport to the lysosome for degradation. More recently, PCSK9 inhibitors have been developed that inhibit PCSK9 and prevent LDLR degradation, increasing the bioavailability of functionally active LDLR at the membrane to ultimately decrease serum LDL-C [33, 140, 141]. In the Further Cardiovascular Outcomes Research with PCSK9i in Subjects with Elevated Risk (FOURIER) trial, Evolocumab, a monoclonal antibody targeting PCSK9, was used in a statin-treated background and demonstrated a significant reduction in LDL-C and associated CVD outcomes [141]. Other PCSK9i have also been developed such as alirocumab, bococizumab, and RG7652 that were shown to be effective at reducing LDL-C by over 30% and reduced CVD events compared to placebo groups [41]. Although PCSK9i was shown to be effective at decreasing LDL-C levels, the fiscal impact of only treating patients with severe atherosclerotic CVD was predicted to be $1.5 billion over three years [33], causing the budget impact of treating all patients with hyperlipidemia to be substantial, so PCSK9i are currently limited to those with familial hyperlipidemia.

There are many ongoing clinical trials for new therapies to address statin resistance, such as Orlistat or Vupanorsen that target intestinal lipases to prevent metabolism and uptake of cholesterol, that can be viewed on clinicaltrials.gov and was reviewed in 2016 by Shapiro and Fazio [142].

HDL-C-Targeted Therapies

While the LDL-C lowering therapies have been successful, these therapies may not work for all patients due to [143–146], so studies to raise HDL-C while lowering LDL-C have been ongoing. Since HDL and RCT are anti-atherosclerotic, clinical trials investigating therapeutics to increase HDL-C have proven to be complex. Since ApoA-I comprises the largest portion of the HDL protein composition, some studies showed elevation in HDL-C after supplementing ApoA-I resulting in reduced atherosclerotic plaque size [55, 143]. The first ApoA-I variant identified resulted in an arginine to cysteine mutation at residue 173, named ApoA-IMilano, that was associated with low HDL-C. Despite lower HDL-C, heterozygous patients for ApoA-IMilano displayed increased longevity and had less carotid intimal wall thickness compared to patients with similarly low levels of HDL-C [147], leading to clinical trials with the *Escherichia coli* recombinant ApoA-1Milano protein, ETC-216 [148]. In the phase-II clinical trial of ETC-216, a 10.9% reduction was observed in atheroma volume [149], which unfortunately was less than what was identified in the pre-clinical animal studies and not further pursued [150].

In 2007, an Australian-based company, Commonwealth Serum Laboratories, purified ApoA-I from human plasma and reconstituted with soybean phosphatidylcholine to create CSL-111 and initiated the ERASE trial. Several pre-clinical studies with CSL-111 displayed a similar effect on cholesterol efflux and endothelial function as ApoA-IMilano reconstituted HDL [151, 152], but clinical trials did not show a significant change in plaque size. Furthermore, the higher dose trial group was halted early due to an increase in liver function test abnormalities [153]. The Medicines Company also conducted a clinical trial in 2018 to further explore the findings from ApoA-IMilano, but these trials did not result in plaque regression among patients with acute coronary syndrome already treated with statins [154]. Isolating ApoA-I and HDL-C for supplementation is complex, so researchers developed ApoA-I mimetics that showed increased efflux from ABCA1 in vitro [155] and protected ApoE knockout mice from atherosclerosis [156]. Unfortunately, clinical trials did not recapitulate the in vivo and ex vivo data in humans and increased inflammatory markers, CRP and IL-6 [157, 158].

Renal filtration is suggested to be a big contributor to reducing supplemental HDL/ApoA-I effectiveness, and Roche developed a trimeric ApoA-I to increase its size in order to prevent glomerular filtration. ApoA-I is usually associated with HDL but can dissociate and be filtered and excreted out of the body. The trimeric ApoA-I has been shown to enhance cholesterol efflux from ABCA1 [159], but did not have a significant effect on lipid levels or lesion area in cholesterol-fed LDLR knockout mice treated with phospholipid-reconstituted trimeric ApoA-I [158, 159].

There have not been clinical trials using trimeric ApoA-I due to the limited data available and minimal benefit in animal models.

In 2004, a pre-clinical trial was conducted with African Green monkeys where the plasma underwent selective HDL delipidation and was reinfused. This study demonstrated the capacity to change the lipid concentration of HDL-C and promote RCT in vivo [160, 161].

Increasing HDL-C by supplementation or replacement studies has not provided major therapeutic benefit to be marketed for widespread use. This does not negate the importance of HDL-C and RCT in preventing atherosclerosis, but does warrant new therapies to be developed that will enhance these processes in humans. There have been great strides in learning about the anti-atherosclerotic properties of HDL-C and defining novel targets in the RCT pathway will help to improve clinical outcomes.

CETP inhibition has also been a consideration in drug development to treat dyslipidemia, but has come with trials and tribulations reviewed here [84]. The primary role of CETP is transferring cholesterol esters from HDL to LDL or VLDL that can then be recognized by LDLR. Inhibition of CETP leads to more cholesterol ester retention with HDL and therefore increasing HDL-C while lowering LDL-C [162, 163]. Unfortunately, three compounds failed in phase III clinical trials due to drug toxicity citing high blood pressure, hyperaldosteronism [5], and increased response to endothelin in the vasculature [164], leading to higher mortality and halting trials. The CETP inhibitor, Anacetrapib, was found to be moderately effective at decreasing adverse coronary events when added to statin therapy in the REVEAL trial but is still associated with mildly increased blood pressure and with minimal, tolerable side effects that need to be tightly monitored in high-risk patients [84]. CETP inhibitors used as a monotherapy still need to be considered and determine the off-target effects to better determine alternatives for drug development. While CETP inhibition has been shown to be a direct mechanism to increase HDL-C and decrease LDL-C, this mechanism likely does not contribute to enhanced cholesterol uptake and efflux, so it underscores the impact of viewing high HDL-C as athero-protective when efflux and uptake may offer additional benefits [5–8].

References

1. Virani SS, Alonso A, Aparicio HJ, Benjamin EJ, Bittencourt MS, Callaway CW, et al. Heart disease and stroke statistics - 2021 update. Circulation. 2021;143(8):e254–743.
2. Pencina MJ, D'Agostino RB, Larson MG, Massaro JM, Vasan RS. Predicting the 30-year risk of cardiovascular disease. Circulation. 2009;119(24):3078–84.
3. Wilson PW, Garrison RJ, Castelli WP, Feinleib M, McNamara PM, Kannel WB. Prevalence of coronary heart disease in the Framingham offspring study: role of lipoprotein cholesterols. Am J Cardiol. 1980;46(4):649–54.
4. Glomset JA, Janssen ET, Kennedy R, Dobbins J. Role of plasma lecithin:cholesterol acyltransferase in the metabolism of high density lipoproteins. J Lipid Res. 1966;7(5):638–48.

5. Barter PJ, Caulfield M, Eriksson M, Grundy SM, Kastelein JJ, Komajda M, et al. Effects of torcetrapib in patients at high risk for coronary events. N Engl J Med. 2007;357(21):2109–22.
6. Cannon CP, Shah S, Dansky HM, Davidson M, Brinton EA, Gotto AM, et al. Safety of Anacetrapib in patients with or at high risk for coronary heart disease. N Engl J Med. 2010;363(25):2406–15.
7. Nicholls SJ, Brewer HB, Kastelein JJ, Krueger KA, Wang MD, Shao M, et al. Effects of the CETP inhibitor evacetrapib administered as monotherapy or in combination with statins on HDL and LDL cholesterol: a randomized controlled trial. JAMA. 2011;306(19):2099–109.
8. Schwartz GG, Olsson AG, Abt M, Ballantyne CM, Barter PJ, Brumm J, et al. Effects of Dalcetrapib in patients with a recent acute coronary syndrome. N Engl J Med. 2012;367(22):2089–99.
9. Torres N, Guevara-Cruz M, Velázquez-Villegas LA, Tovar AR. Nutrition and atherosclerosis. Arch Med Res. 2015;46(5):408–26.
10. Massaro M, Scoditti E, Carluccio MA, De Caterina R. Nutraceuticals and prevention of atherosclerosis: focus on ω-3 polyunsaturated fatty acids and Mediterranean diet polyphenols. Cardiovasc Ther. 2010;28(4):e13–e9.
11. Estruch R, Ros E, Salas-Salvadó J, Covas M-I, Corella D, Arós F, et al. Primary prevention of cardiovascular disease with a Mediterranean diet. N Engl J Med. 2013;368(14):1279–90.
12. de Lorgeril M, Renaud S, Salen P, Monjaud I, Mamelle N, Martin JL, et al. Mediterranean alpha-linolenic acid-rich diet in secondary prevention of coronary heart disease. Lancet. 1994;343(8911):1454–9.
13. Korre M, Tsoukas MA, Frantzeskou E, Yang J, Kales SN. Mediterranean diet and workplace health promotion. Curr Cardiovasc Risk Rep. 2014;8(12):416.
14. Vilahur G, Badimon L. Antiplatelet properties of natural products. Vasc Pharmacol. 2013;59(3–4):67–75.
15. Saita E, Kondo K, Momiyama Y. Anti-inflammatory diet for atherosclerosis and coronary artery disease: antioxidant foods. Clin Med Insights Cardiol. 2015;8(Suppl 3):61–5.
16. Mozaffarian D, Benjamin EJ, Go AS, Arnett DK, Blaha MJ, Cushman M, et al. Heart disease and stroke statistics--2015 update: a report from the American Heart Association. Circulation. 2015;131(4):e29–322.
17. Mozaffarian D, Katan MB, Ascherio A, Stampfer MJ, Willett WC. Trans fatty acids and cardiovascular disease. N Engl J Med. 2006;354(15):1601–13.
18. Correction to: 2019 ACC/AHA Guideline on the Primary Prevention of Cardiovascular Disease: A Report of the American College of Cardiology/American Heart Association Task Force on Clinical Practice Guidelines. Circulation. 2019;140(11):e649–e50.
19. Temple NJ. Fat, sugar, whole grains and heart disease: 50 years of confusion. Nutrients. 2018;10(1):39.
20. Anand SS, Hawkes C, de Souza RJ, Mente A, Dehghan M, Nugent R, et al. Food consumption and its impact on cardiovascular disease: importance of solutions focused on the globalized food system: a report from the workshop convened by the World Heart Federation. J Am Coll Cardiol. 2015;66(14):1590–614.
21. Narain A, Kwok CS, Mamas MA. Soft drinks and sweetened beverages and the risk of cardiovascular disease and mortality: a systematic review and meta-analysis. Int J Clin Pract. 2016;70(10):791–805.
22. Yang Q, Zhang Z, Gregg EW, Flanders WD, Merritt R, Hu FB. Added sugar intake and cardiovascular diseases mortality among US adults. JAMA Intern Med. 2014;174(4):516–24.
23. DiNicolantonio JJ, JH OK. Added sugars drive coronary heart disease via insulin resistance and hyperinsulinaemia: a new paradigm. Open. Heart. 2017;4(2):e000729.
24. Brown JC, Gerhardt TE, Kwon E. Risk factors for coronary artery disease. StatPearls. Treasure Island (FL): StatPearls Publishing. Copyright © 2021, StatPearls Publishing LLC; 2021.
25. Yusuf S, Hawken S, Ounpuu S, Dans T, Avezum A, Lanas F, et al. Effect of potentially modifiable risk factors associated with myocardial infarction in 52 countries (the INTERHEART study): case-control study. Lancet. 2004;364(9438):937–52.

26. Marques LR, Diniz TA, Antunes BM, Rossi FE, Caperuto EC, Lira FS, et al. Reverse cholesterol transport: molecular mechanisms and the non-medical approach to enhance HDL cholesterol. Front Physiol. 2018;9:526.
27. Helgadottir A, Thorleifsson G, Manolescu A, Gretarsdottir S, Blondal T, Jonasdottir A, et al. A common variant on chromosome 9p21 affects the risk of myocardial infarction. Science. 2007;316(5830):1491–3.
28. Palomaki GE, Melillo S, Bradley LA. Association between 9p21 genomic markers and heart disease: a meta-analysis. JAMA. 2010;303(7):648–56.
29. Assimes TL, Knowles JW, Basu A, Iribarren C, Southwick A, Tang H, et al. Susceptibility locus for clinical and subclinical coronary artery disease at chromosome 9p21 in the multiethnic ADVANCE study. Hum Mol Genet. 2008;17(15):2320–8.
30. Nikpay M, Goel A, Won H-H, Hall LM, Willenborg C, Kanoni S, et al. A comprehensive 1000 genomes–based genome-wide association meta-analysis of coronary artery disease. Nat Genet. 2015;47(10):1121–30.
31. Nelson CP, Goel A, Butterworth AS, Kanoni S, Webb TR, Marouli E, et al. Association analyses based on false discovery rate implicate new loci for coronary artery disease. Nat Genet. 2017;49(9):1385–91.
32. Ward NC, Watts GF, Eckel RH. Statin toxicity. Circ Res. 2019;124(2):328–50.
33. Ko DT, Khan AM, Kotrri G, Austin PC, Wijeysundera HC, Koh M, et al. Eligibility, clinical outcomes, and budget impact of PCSK9 inhibitor adoption: the CANHEART PCSK9 study. J Am Heart Assoc. 2018;7(21):e010007.
34. Schmidt AF, Carter JP, Pearce LS, Wilkins JT, Overington JP, Hingorani AD, et al. PCSK9 monoclonal antibodies for the primary and secondary prevention of cardiovascular disease. Cochrane Database Syst Rev. 2020;10(10):CD011748.
35. Myers KD, Farboodi N, Mwamburi M, Howard W, Staszak D, Gidding S, et al. Effect of access to prescribed PCSK9 inhibitors on cardiovascular outcomes. Circ Cardiovasc Qual Outcomes. 2019;12(8):e005404.
36. Koenig SN, Sucharski HC, Jose EM, Dudley EK, Madiai F, Cavus O, et al. Inherited variants in *SCARB1* cause severe early-onset coronary artery disease. Circ Res. 2021;129(2):296–307.
37. Feingold KR. Introduction to lipids and lipoproteins. In: Feingold KR, Anawalt B, Boyce A, Chrousos G, de Herder WW, Dhatariya K, et al., editors. Endotext. South Dartmouth (MA): MDText.com, Inc. Copyright © 2000–2021, MDText.com, Inc.; 2000.
38. Uhlén M, Fagerberg L, Hallström BM, Lindskog C, Oksvold P, Mardinoglu A, et al. Tissue-based map of the human proteome. Science. 2015;347(6220):1260419. https://www.proteinatlas.org/ENSG00000130164-LDLR
39. Friesen JA, Rodwell VW. The 3-hydroxy-3-methylglutaryl coenzyme-A (HMG-CoA) reductases. Genome Biol. 2004;5(11):248.
40. Shapiro MD, Tavori H, Fazio S. PCSK9. Circ Res. 2018;122(10):1420–38.
41. Poirier S, Mayer G, Benjannet S, Bergeron E, Marcinkiewicz J, Nassoury N, et al. The Proprotein convertase PCSK9 induces the degradation of low density lipoprotein receptor (LDLR) and its closest family members VLDLR and ApoER2*. J Biol Chem. 2008;283(4):2363–72.
42. Brown MS, Anderson RG, Goldstein JL. Recycling receptors: the round-trip itinerary of migrant membrane proteins. Cell. 1983;32(3):663–7.
43. Barr DP, Russ EM, Eder HA. Protein-lipid relationships in human plasma. II. In atherosclerosis and related conditions. Am J Med. 1951;11(4):480–93.
44. Basso F, Freeman L, Knapper CL, Remaley A, Stonik J, Neufeld EB, et al. Role of the hepatic ABCA1 transporter in modulating intrahepatic cholesterol and plasma HDL cholesterol concentrations. J Lipid Res. 2003;44(2):296–302.
45. Brunham LR, Kruit JK, Iqbal J, Fievet C, Timmins JM, Pape TD, et al. Intestinal ABCA1 directly contributes to HDL biogenesis in vivo. J Clin Investig. 2006;116(4):1052–62.
46. Asztalos BF, Tani M, Schaefer EJ. Metabolic and functional relevance of HDL subspecies. Curr Opin Lipidol. 2011;22(3):176–85.

47. Kulkarni KR, Marcovina SM, Krauss RM, Garber DW, Glasscock AM, Segrest JP. Quantification of HDL2 and HDL3 cholesterol by the Vertical Auto Profile-II (VAP-II) methodology. J Lipid Res. 1997;38(11):2353–64.
48. Nofer JR, Remaley AT. Tangier disease: still more questions than answers. Cell Mol Life Sci. 2005;62(19–20):2150–60.
49. Bailey A, Mohiuddin SS. Biochemistry, high density lipoprotein. StatPearls. Treasure Island (FL): StatPearls Publishing Copyright © 2021, StatPearls Publishing LLC; 2021.
50. Voight BF, Peloso GM, Orho-Melander M, Frikke-Schmidt R, Barbalic M, Jensen MK, et al. Plasma HDL cholesterol and risk of myocardial infarction: a mendelian randomisation study. Lancet. 2012;380(9841):572–80.
51. Asztalos BF, Schaefer EJ, Horvath KV, Yamashita S, Miller M, Franceschini G, et al. Role of LCAT in HDL remodeling: investigation of LCAT deficiency states. J Lipid Res. 2007;48(3):592–9.
52. Oram JF, Vaughan AM. ATP-binding cassette cholesterol transporters and cardiovascular disease. Circ Res. 2006;99(10):1031–43.
53. Ouimet M, Barrett TJ, Fisher EA. HDL and reverse cholesterol transport. Circ Res. 2019;124(10):1505–18.
54. Ono K. Current concept of reverse cholesterol transport and novel strategy for atheroprotection. J Cardiol. 2012;60(5):339–43.
55. Remaley AT, Amar M, Sviridov D. HDL-replacement therapy: mechanism of action, types of agents and potential clinical indications. Expert Rev Cardiovasc Ther. 2008;6(9):1203–15.
56. Shen WJ, Azhar S, Kraemer FB. SR-B1: a unique multifunctional receptor for cholesterol influx and efflux. Annu Rev Physiol. 2018;80:95–116.
57. Charles MA, Kane JP. New molecular insights into CETP structure and function: a review. J Lipid Res. 2012;53(8):1451–8.
58. Rosenson RS, Brewer HB Jr, Davidson WS, Fayad ZA, Fuster V, Goldstein J, et al. Cholesterol efflux and atheroprotection: advancing the concept of reverse cholesterol transport. Circulation. 2012;125(15):1905–19.
59. Neculai D, Schwake M, Ravichandran M, Zunke F, Collins RF, Peters J, et al. Structure of LIMP-2 provides functional insights with implications for SR-BI and CD36. Nature. 2013;504(7478):172–6.
60. Varban ML, Rinninger F, Wang N, Fairchild-Huntress V, Dunmore JH, Fang Q, et al. Targeted mutation reveals a central role for SR-BI in hepatic selective uptake of high density lipoprotein cholesterol. Proc Natl Acad Sci U S A. 1998;95(8):4619–24.
61. Rigotti A, Trigatti BL, Penman M, Rayburn H, Herz J, Krieger M. A targeted mutation in the murine gene encoding the high density lipoprotein (HDL) receptor scavenger receptor class B type I reveals its key role in HDL metabolism. Proc Natl Acad Sci U S A. 1997;94(23):12610–5.
62. Van Eck M, Twisk J, Hoekstra M, Van Rij BT, Van der Lans CA, Bos IS, et al. Differential effects of scavenger receptor BI deficiency on lipid metabolism in cells of the arterial wall and in the liver. Int J Biol Chem. 2003;278(26):23699–705.
63. Kozarsky KF, Donahee MH, Rigotti A, Iqbal SN, Edelman ER, Krieger M. Overexpression of the HDL receptor SR-BI alters plasma HDL and bile cholesterol levels. Nature. 1997;387(6631):414–7.
64. Ueda Y, Royer L, Gong E, Zhang J, Cooper PN, Francone O, et al. Lower plasma levels and accelerated clearance of high density lipoprotein (HDL) and non-HDL cholesterol in scavenger receptor class B type I transgenic mice. Int J Biol Chem. 1999;274(11):7165–71.
65. Wang N, Arai T, Ji Y, Rinninger F, Tall AR. Liver-specific overexpression of scavenger receptor BI decreases levels of very low density lipoprotein ApoB, low density lipoprotein ApoB, and high density lipoprotein in transgenic mice. J Biol Chem. 1998;273(49):32920–6.
66. Chiang JY. Regulation of bile acid synthesis: pathways, nuclear receptors, and mechanisms. J Hepatol. 2004;40(3):539–51.

67. Cohen DE. Balancing cholesterol synthesis and absorption in the gastrointestinal tract. J Clin Lipidol. 2008;2(2):S1–3.
68. Carey MC, Lamont JT. Cholesterol gallstone formation. 1. Physical-chemistry of bile and biliary lipid secretion. Prog Liver Dis. 1992;10:139–63.
69. Zakim D, Boyer TD. Hepatology: a textbook of liver disease. Saunders; 1990.
70. Oude Elferink RP, Paulusma CC, Groen AK. Hepatocanalicular transport defects: pathophysiologic mechanisms of rare diseases. Gastroenterol. 2006;130(3):908–25.
71. Altmann SW, Davis HR Jr, Zhu LJ, Yao X, Hoos LM, Tetzloff G, et al. Niemann-pick C1 like 1 protein is critical for intestinal cholesterol absorption. Science. 2004;303(5661):1201–4.
72. Brown JM, Temel RE, Graf GA. Para-bile-osis establishes a role for nonbiliary macrophage to Feces reverse cholesterol transport. Arterioscler Thromb Vasc Biol. 2017;37(5):738–9.
73. Wang DQ. Regulation of intestinal cholesterol absorption. Annu Rev Physiol. 2007;69:221–48.
74. Hussain MM. A proposed model for the assembly of chylomicrons. Atherosclerosis. 2000;148(1):1–15.
75. Rong JX, Shapiro M, Trogan E, Fisher EA. Transdifferentiation of mouse aortic smooth muscle cells to a macrophage-like state after cholesterol loading. Proc Natl Acad Sci U S A. 2003;100(23):13531–6.
76. Vengrenyuk Y, Nishi H, Long X, Ouimet M, Savji N, Martinez FO, et al. Cholesterol loading reprograms the microRNA-143/145-myocardin axis to convert aortic smooth muscle cells to a dysfunctional macrophage-like phenotype. Arterioscler Thromb Vasc Biol. 2015;35(3):535–46.
77. Shankman LS, Gomez D, Cherepanova OA, Salmon M, Alencar GF, Haskins RM, et al. KLF4-dependent phenotypic modulation of smooth muscle cells has a key role in atherosclerotic plaque pathogenesis. Nat Med. 2015;21(6):628–37.
78. Feil S, Fehrenbacher B, Lukowski R, Essmann F, Schulze-Osthoff K, Schaller M, et al. Transdifferentiation of vascular smooth muscle cells to macrophage-like cells during atherogenesis. Circ Res. 2014;115(7):662–7.
79. Allahverdian S, Chehroudi AC, McManus BM, Abraham T, Francis GA. Contribution of intimal smooth muscle cells to cholesterol accumulation and macrophage-like cells in human atherosclerosis. Circulation. 2014;129(15):1551–9.
80. Brown MS, Goldstein JL, Krieger M, Ho YK, Anderson RG. Reversible accumulation of cholesteryl esters in macrophages incubated with acetylated lipoproteins. J Cell Biol. 1979;82(3):597–613.
81. McGookey DJ, Anderson RG. Morphological characterization of the cholesteryl ester cycle in cultured mouse macrophage foam cells. J Cell Biol. 1983;97(4):1156–68.
82. Gallino A, Aboyans V, Diehm C, Cosentino F, Stricker H, Falk E, et al. Non-coronary atherosclerosis. Eur Heart J. 2014;35(17):1112–9.
83. Ross R. Atherosclerosis--an inflammatory disease. N Engl J Med. 1999;340(2):115–26.
84. Tall AR, Rader DJ. Trials and tribulations of CETP inhibitors. Circ Res. 2018;122(1):106–12.
85. Barter PJ, Rye KA. High density lipoproteins and coronary heart disease. Atherosclerosis. 1996;121(1):1–12.
86. Barter PJ, Baker PW, Rye K-A. Effect of high-density lipoproteins on the expression of adhesion molecules in endothelial cells. Curr Opin Lipidol. 2002;13(3):285–8.
87. Barter PJ, Rye KA. Relationship between the concentration and antiatherogenic activity of high-density lipoproteins. Curr Opin Lipidol. 2006;17(4):399–403.
88. Badimon L, Chiva-Blanch G. Chapter 24 - lipid metabolism in Dyslipidemia and familial hypercholesterolemia. In: Patel VB, editor. The molecular nutrition of fats. Academic Press; 2019. p. 307–22.
89. Navab M, Ananthramaiah GM, Reddy ST, Van Lenten BJ, Ansell BJ, Fonarow GC, et al. The oxidation hypothesis of atherogenesis: the role of oxidized phospholipids and HDL. J Lipid Res. 2004;45(6):993–1007.

90. Terasaka N, Wang N, Yvan-Charvet L, Tall AR. High-density lipoprotein protects macrophages from oxidized low-density lipoprotein-induced apoptosis by promoting efflux of 7-ketocholesterol via ABCG1. Proc Natl Acad Sci U S A. 2007;104(38):15093–8.
91. Barter PJ, Nicholls S, Rye KA, Anantharamaiah GM, Navab M, Fogelman AM. Antiinflammatory properties of HDL. Circ Res. 2004;95(8):764–72.
92. Navab M, Anantharamaiah GM, Fogelman AM. The role of high-density lipoprotein in inflammation. Trends Cardiovasc Med. 2005;15(4):158–61.
93. Ansell BJ, Watson KE, Fogelman AM, Navab M, Fonarow GC. High-density lipoprotein function recent advances. J Am Coll Cardiol. 2005;46(10):1792–8.
94. Navab M, Ananthramaiah GM, Reddy ST, Van Lenten BJ, Ansell BJ, Hama S, et al. The double jeopardy of HDL. Ann Med. 2005;37(3):173–8.
95. Webb TR, Erdmann J, Stirrups KE, Stitziel NO, Masca NGD, Jansen H, et al. Systematic evaluation of pleiotropy identifies 6 further loci associated with coronary artery disease. J Am Coll Cardiol. 2017;69(7):823–36.
96. Hobbs HH, Russell DW, Brown MS, Goldstein JL. The LDL receptor locus in familial hypercholesterolemia: mutational analysis of a membrane protein. Annu Rev Genet. 1990;24:133–70.
97. Meshkov A, Ershova A, Kiseleva A, Zotova E, Sotnikova E, Petukhova A, et al. The LDLR, APOB, and PCSK9 variants of index patients with familial hypercholesterolemia in Russia. Genes (Basel). 2021;12(1):66.
98. Inazu A, Brown ML, Hesler CB, Agellon LB, Koizumi J, Takata K, et al. Increased high-density lipoprotein levels caused by a common cholesteryl-ester transfer protein gene mutation. N Engl J Med. 1990;323(18):1234–8.
99. Goldstein JL, Brown MS. The LDL receptor. Arterioscler Thromb Vasc Biol. 2009;29(4):431–8.
100. Lewis MJ, Malik TH, Ehrenstein MR, Boyle JJ, Botto M, Haskard DO. Immunoglobulin M is required for protection against atherosclerosis in low-density lipoprotein receptor-deficient mice. Circulation. 2009;120(5):417–26.
101. Wouters K, Shiri-Sverdlov R, van Gorp PJ, van Bilsen M, Hofker MH. Understanding hyperlipidemia and atherosclerosis: lessons from genetically modified apoe and ldlr mice. Clin Chem Lab Med. 2005;43(5):470–9.
102. Kennedy DJ, Kuchibhotla SD, Guy E, Park YM, Nimako G, Vanegas D, et al. Dietary cholesterol plays a role in CD36-mediated atherogenesis in LDLR-knockout mice. Arterioscler Thromb Vasc Biol. 2009;29(10):1481–7.
103. Langheinrich AC, Michniewicz A, Sedding DG, Walker G, Beighley PE, Rau WS, et al. Correlation of Vasa Vasorum neovascularization and plaque progression in aortas of Apolipoprotein $E^{-/-}$ low-density lipoprotein$^{-/-}$ double knockout mice. Arterioscler Thromb Vasc Biol. 2006;26(2):347–52.
104. Semenova AE, Sergienko IV, García-Giustiniani D, Monserrat L, Popova AB, Nozadze DN, et al. Verification of underlying genetic cause in a Cohort of Russian patients with familial hypercholesterolemia using targeted next generation sequencing. J Cardiovasc Dev Dis. 2020;7(2):16.
105. Pogoda T, Metelskaya V, Perova N, Limborska S. Detection of the apoB-3500 mutation in a Russian family with coronary heart disease. Hum Hered. 1998;48(5):291–2.
106. Shakhtshneider E, Ivanoshchuk D, Orlov P, Timoshchenko O, Voevoda M. Analysis of the Ldlr, Apob, Pcsk9 and Ldlrap1 genes variability in patients with familial hypercholesterolemia in West Siberia using targeted high throughput resequencing. Atherosclerosis. 2019;287:e285.
107. Sadananda SN, Foo JN, Toh MT, Cermakova L, Trigueros-Motos L, Chan T, et al. Targeted next-generation sequencing to diagnose disorders of HDL cholesterol. J Lipid Res. 2015;56(10):1993–2001.

108. Peloso GM, Nomura A, Khera AV, Chaffin M, Won H-H, Ardissino D, et al. Rare protein-truncating variants in *APOB*, lower low-density lipoprotein cholesterol, and protection against coronary heart disease. Circ Genom Precis Med. 2019;12(5):e002376.
109. Rabacchi C, Simone ML, Pisciotta L, Di Leo E, Bocchi D, Pietrangelo A, et al. In vitro functional characterization of splicing variants of the APOB gene found in familial hypobetalipoproteinemia. J Clin Lipidol. 2019;13(6):960–9.
110. Leren TP. Mutations in the PCSK9 gene in Norwegian subjects with autosomal dominant hypercholesterolemia. Clin Genet. 2004;65(5):419–22.
111. Guo Q, Feng X, Zhou Y. PCSK9 variants in familial hypercholesterolemia: a comprehensive synopsis. Front Genet. 2020;11:1020.
112. Di Taranto MD, Giacobbe C, Fortunato G. Familial hypercholesterolemia: a complex genetic disease with variable phenotypes. Eur J Med Genet. 2020;63(4):103831.
113. Ragusa R, Basta G, Neglia D, De Caterina R, Del Turco S, Caselli C. PCSK9 and atherosclerosis: looking beyond LDL regulation. Eur J Clin. 2021;51(4):e13459.
114. Kent AP, Stylianou IM. Scavenger receptor class B member 1 protein: hepatic regulation and its effects on lipids, reverse cholesterol transport, and atherosclerosis. Hepat Med. 2011;3:29–44.
115. Xie L, Lv X, Sun Y, Tong Y, Zhang S, Deng Y. Association of rs5888 SNP in SCARB1 gene with coronary artery disease. Herz. 2019;44(7):644–50.
116. Wu DF, Yin RX, Cao XL, Chen WX, Aung LH, Wang W, et al. Scavenger receptor class B type 1 gene rs5888 single nucleotide polymorphism and the risk of coronary artery disease and ischemic stroke: a case-control study. Int J Med Sci. 2013;10(12):1771–7.
117. Ayhan H, Gormus U, Isbir S, Yilmaz SG, Isbir T. SCARB1 gene polymorphisms and HDL subfractions in coronary artery disease. In vivo (Athens, Greece). 2017;31(5):873–6.
118. Helgadottir A, Sulem P, Thorgeirsson G, Gretarsdottir S, Thorleifsson G, Jensson B, et al. Rare SCARB1 mutations associate with high-density lipoprotein cholesterol but not with coronary artery disease. Eur Heart J. 2018;39(23):2172–8.
119. Vergeer M, Korporaal SJ, Franssen R, Meurs I, Out R, Hovingh GK, et al. Genetic variant of the scavenger receptor BI in humans. N Engl J Med. 2011;364(2):136–45.
120. Ljunggren SA, Levels JH, Hovingh K, Holleboom AG, Vergeer M, Argyri L, et al. Lipoprotein profiles in human heterozygote carriers of a functional mutation P297S in scavenger receptor class B1. Biochim Biophys Acta. 2015;1851(12):1587–95.
121. Williams DL, Temel RE, Connelly MA. Roles of scavenger receptor BI and APO A-I in selective uptake of HDL cholesterol by adrenal cells. Endocr Res. 2000;26(4):639–51.
122. Schreyer SA, Wilson DL, LeBoeuf RC. C57BL/6 mice fed high fat diets as models for diabetes-accelerated atherosclerosis. Atherosclerosis. 1998;136(1):17–24.
123. Ishibashi S, Brown MS, Goldstein JL, Gerard RD, Hammer RE, Herz J. Hypercholesterolemia in low density lipoprotein receptor knockout mice and its reversal by adenovirus-mediated gene delivery. J Clin Investig. 1993;92(2):883–93.
124. Lo Sasso G, Schlage WK, Boué S, Veljkovic E, Peitsch MC, Hoeng J. The Apoe(−/−) mouse model: a suitable model to study cardiovascular and respiratory diseases in the context of cigarette smoke exposure and harm reduction. J Transl Med. 2016;14(1):146.
125. Meir KS, Leitersdorf E. Atherosclerosis in the Apolipoprotein Edeficient mouse. Arterioscler Thromb Vasc Biol. 2004;24(6):1006–14.
126. Zhang SH, Reddick RL, Piedrahita JA, Maeda N. Spontaneous hypercholesterolemia and arterial lesions in mice lacking apolipoprotein E. Science. 1992;258(5081):468–71.
127. Plump AS, Smith JD, Hayek T, Aalto-Setälä K, Walsh A, Verstuyft JG, et al. Severe hypercholesterolemia and atherosclerosis in apolipoprotein E-deficient mice created by homologous recombination in ES cells. Cell. 1992;71(2):343–53.
128. Getz GS, Reardon CA. Do the *Apoe*$^{-/-}$ and *Ldlr*$^{-/-}$ mice yield the same insight on Atherogenesis? Arterioscler Thromb Vasc Biol. 2016;36(9):1734–41.

129. Caligiuri G, Levy B, Pernow J, Thorén P, Hansson GK. Myocardial infarction mediated by endothelin receptor signaling in hypercholesterolemic mice. Proc Natl Acad Sci U S A. 1999;96(12):6920–4.
130. Contreras-Duarte S, Santander N, Birner-Gruenberger R, Wadsack C, Rigotti A, Busso D. High density lipoprotein cholesterol and proteome in SR-B1 KO mice: lost in precipitation. J Transl Med. 2018;16(1):309.
131. Trigatti B, Rayburn H, Viñals M, Braun A, Miettinen H, Penman M, et al. Influence of the high density lipoprotein receptor SR-BI on reproductive and cardiovascular pathophysiology. Proc Natl Acad Sci U S A. 1999;96(16):9322–7.
132. Tsukamoto K, Mani DR, Shi J, Zhang S, Haagensen DE, Otsuka F, et al. Identification of apolipoprotein D as a cardioprotective gene using a mouse model of lethal atherosclerotic coronary artery disease. Proc Natl Acad Sci U S A. 2013;110(42):17023–8.
133. Report of the National Cholesterol Education Program Expert Panel on Detection, Evaluation, and Treatment of High Blood Cholesterol in Adults. The Expert Panel. Arch Intern Med. 1988;148(1):36–69.
134. Harrington RA. Statins—Almost 30 years of use in the United States and still not quite there. JAMA Cardiol. 2017;2(1):66.
135. Stancu C, Sima A. Statins: mechanism of action and effects. J Cell Mol Med. 2001;5(4):378–87.
136. Corsini A, Bellosta S, Baetta R, Fumagalli R, Paoletti R, Bernini F. New insights into the pharmacodynamic and pharmacokinetic properties of statins. Pharmacol Ther. 1999;84(3):413–28.
137. Jayatilaka S, Desai K, Rijal S, Zimmerman D. Statin-induced autoimmune necrotizing myopathy. J Prim Care Community Health. 2021;12:21501327211028714.
138. Postmus I, Trompet S, Deshmukh HA, Barnes MR, Li X, Warren HR, et al. Pharmacogenetic meta-analysis of genome-wide association studies of LDL cholesterol response to statins. Nat Commun. 2014;5(1):5068.
139. Thompson PD, Panza G, Zaleski A, Taylor B. Statin-associated side effects. J Am Coll Cardiol. 2016;67(20):2395–410.
140. Giugliano RP, Sabatine MS. Are PCSK9 inhibitors the next breakthrough in the cardiovascular field? J Am Coll Cardiol. 2015;65(24):2638–51.
141. Sabatine MS, Giugliano RP, Keech AC, Honarpour N, Wiviott SD, Murphy SA, et al. Evolocumab and clinical outcomes in patients with cardiovascular disease. N Engl J Med. 2017;376(18):1713–22.
142. Shapiro MD, Fazio S. From lipids to inflammation: new approaches to reducing atherosclerotic risk. Circ Res. 2016;118(4):732–49.
143. Chhabria MT, Suhagia BN, Brahmkshatriya PS. HDL elevation and lipid lowering therapy: current scenario and future perspectives. Recent Adv Cardiovasc Drug Discov. 2007;2(3):214–27.
144. Wong NC. Novel therapies to increase apolipoprotein AI and HDL for the treatment of atherosclerosis. Curr Opin Investig Drugs (London, England : 2000). 2007;8(9):718–28.
145. Sacks FM. The relative role of low-density lipoprotein cholesterol and high-density lipoprotein cholesterol in coronary artery disease: evidence from large-scale statin and fibrate trials. Am J Cardiol. 2001;88(12a):14n–8n.
146. Baigent C, Keech A, Kearney PM, Blackwell L, Buck G, Pollicino C, et al. Efficacy and safety of cholesterol-lowering treatment: prospective meta-analysis of data from 90,056 participants in 14 randomised trials of statins. Lancet (London, England). 2005;366(9493):1267–78.
147. Franceschini G, Sirtori CR, Bosisio E, Gualandri V, Orsini GB, Mogavero AM, et al. Relationship of the phenotypic expression of the A-IMilano apoprotein with plasma lipid and lipoprotein patterns. Atherosclerosis. 1985;58(1–3):159–74.
148. Franceschini G, Sirtori CR, Capurso A 2nd, Weisgraber KH, Mahley RW. A-IMilano apoprotein. Decreased high density lipoprotein cholesterol levels with significant lipoprotein modifications and without clinical atherosclerosis in an Italian family. J Clin Investig. 1980;66(5):892–900.

149. Nicholls SJ, Tuzcu EM, Sipahi I, Schoenhagen P, Crowe T, Kapadia S, et al. Relationship between atheroma regression and change in lumen size after infusion of apolipoprotein A-I Milano. J Am Coll Cardiol. 2006;47(5):992–7.
150. Ibanez B, Vilahur G, Cimmino G, Speidl WS, Pinero A, Choi BG, et al. Rapid change in plaque size, composition, and molecular footprint after recombinant apolipoprotein A-I Milano (ETC-216) administration: magnetic resonance imaging study in an experimental model of atherosclerosis. J Am Coll Cardiol. 2008;51(11):1104–9.
151. Spieker LE, Sudano I, Hürlimann D, Lerch PG, Lang MG, Binggeli C, et al. High-density lipoprotein restores endothelial function in hypercholesterolemic men. Circulation. 2002;105(12):1399–402.
152. Nanjee MN, Cooke CJ, Garvin R, Semeria F, Lewis G, Olszewski WL, et al. Intravenous apoA-I/lecithin discs increase pre-beta-HDL concentration in tissue fluid and stimulate reverse cholesterol transport in humans. J Lipid Res. 2001;42(10):1586–93.
153. Tardif JC, Grégoire J, L'Allier PL, Ibrahim R, Lespérance J, Heinonen TM, et al. Effects of reconstituted high-density lipoprotein infusions on coronary atherosclerosis: a randomized controlled trial. JAMA. 2007;297(15):1675–82.
154. Nicholls SJ, Puri R, Ballantyne CM, Jukema JW, Kastelein JJP, Koenig W, et al. Effect of infusion of high-density lipoprotein mimetic containing recombinant Apolipoprotein A-I Milano on coronary disease in patients with an acute coronary syndrome in the MILANO-PILOT trial: a randomized clinical trial. JAMA Cardiol. 2018;3(9):806–14.
155. Remaley AT, Thomas F, Stonik JA, Demosky SJ, Bark SE, Neufeld EB, et al. Synthetic amphipathic helical peptides promote lipid efflux from cells by an ABCA1-dependent and an ABCA1-independent pathway. J Lipid Res. 2003;44(4):828–36.
156. Anantharamaiah G, Navab M, Reddy ST, Garber DW, Datta G, Gupta H, et al. Synthetic peptides: managing lipid disorders. Curr Opin Lipidol. 2006;17(3):233–7.
157. Watson CE, Weissbach N, Kjems L, Ayalasomayajula S, Zhang Y, Chang I, et al. Treatment of patients with cardiovascular disease with L-4F, an apo-A1 mimetic, did not improve select biomarkers of HDL function. J Lipid Res. 2011;52(2):361–73.
158. Sethi AA, Amar M, Shamburek RD, Remaley AT. Apolipoprotein AI mimetic peptides: possible new agents for the treatment of atherosclerosis. Curr Opin Investig Drugs (London, England : 2000). 2007;8(3):201–12.
159. Graversen JH, Laurberg JM, Andersen MH, Falk E, Nieland J, Christensen J, et al. Trimerization of apolipoprotein A-I retards plasma clearance and preserves antiatherosclerotic properties. J Cardiovasc Pharmacol. 2008;51(2):170–7.
160. Brewer HBAP, Kostner G, et al. Selective plasma HDL delipidation and reinfusion: a new approach for acute HDL therapy in the treatment of cardiovascular disease. Circulation. 2004;110:51–2.
161. Sacks FM, Rudel LL, Conner A, Akeefe H, Kostner G, Baki T, et al. Selective delipidation of plasma HDL enhances reverse cholesterol transport in vivo. J Lipid Res. 2009;50(5):894–907.
162. Whitlock ME, Swenson TL, Ramakrishnan R, Leonard MT, Marcel YL, Milne RW, et al. Monoclonal antibody inhibition of cholesteryl ester transfer protein activity in the rabbit. Effects on lipoprotein composition and high density lipoprotein cholesteryl ester metabolism. J Clin Investig. 1989;84(1):129–37.
163. Ikewaki K, Rader DJ, Sakamoto T, Nishiwaki M, Wakimoto N, Schaefer JR, et al. Delayed catabolism of high density lipoprotein apolipoproteins A-I and A-II in human cholesteryl ester transfer protein deficiency. J Clin Investig. 1993;92(4):1650–8.
164. Simic B, Hermann M, Shaw SG, Bigler L, Stalder U, Dörries C, et al. Torcetrapib impairs endothelial function in hypertension. Eur Heart J. 2012;33(13):1615–24.

Atherosclerotic Plaque Regression: Future Perspective

Indu M. Suseela, Jose Padikkala, Thekkekara D. Babu, Rao M. Uppu, and Achuthan C. Raghavamenon

Abstract Atherosclerotic plaque is a complex environment in which cholesterol, phospholipids, protein, and their oxidized molecules co-exist. Many of these oxidized species have been shown to undergo auto-oxidation generating downstream stable lipid carbonyls and their cyclized products which, independently or complexed with proteins, may have plaque destabilizing effects. Reverse cholesterol transport involving the efflux and transport of cholesterol and phospholipids from peripheral tissues to the liver for metabolism helps to maintain lipid homeostasis. The same process is also expected to reduce plaque burden and thereby cardiac incidents. High-density lipoprotein cholesterol (HDL) is crucial in this process. Through HDL mimetics, drugs that enhance functional HDL, dietary modifications, and exercise, we can achieve only 10–30% plaque burden. However, none of these molecules are reported to scavenge or quench oxidized forms of the trapped lipid moieties or their decomposition products. Molecules that scavenge/quench lipid carbonyls can prevent carbonyl adduct formation and may provide additional benefits. Improved plaque regression therefore could be possible with molecules that enhance functional HDL as well as scavenge lipid carbonyls.

Keywords Plaque regression · atherosclerosis · reverse cholesterol transport

Introduction

Atherosclerotic plaque, the core lesion at the subendothelial space of arteries is composed of macrophage foam cells rich in proteins; protein-bound carbonyls, phospholipids, and oxidized cholesterol [1, 2]. In advanced lesions, the components

I. M. Suseela · J. Padikkala · T. D. Babu · A. C. Raghavamenon (✉)
Department of Biochemistry, Amala Cancer Research Centre, Thrissur, Kerala, India
e-mail: raghav@amalaims.org

R. M. Uppu
Division of Environmental Toxicology, Southern University A&M College, Baton Rouge, LA, USA

N. L. Parinandi, T. J. Hund (eds.), *Cardiovascular Signaling in Health and Disease*, https://doi.org/10.1007/978-3-031-08309-9_13

of the plaque are contained within a fibrous cap rich in collagen, which helps in plaque stabilization and prevents access of its thrombogenic core to the bloodstream. This cap itself is continually remodeled, with simultaneous removal and replacement of collagen [3]. The plaque clogs up the artery, thereby disrupting the flow of blood around the body. Any reduction in the strength of the fibrous cap causes plaque rupture resulting in life-threatening conditions such as myocardial infarction and stroke.

American Heart Association (AHA) has established criteria by which atherosclerotic plaques are classified (Table 1) according to content and structure [4]. The initial lesion (type I) contains enough atherogenic lipoproteins to elicit an increase in macrophages and the formation of scattered macrophage foam cells. Type II lesions consist primarily of layers of macrophage foam cells and lipid-laden smooth muscle cells and include lesions grossly designated as fatty streaks. Type III lesions contain scattered collections of extracellular lipid droplets and particles that disrupt the coherence of some intimal smooth muscle cells. These lipid droplets are immediate precursors of the larger, confluent, and more disruptive core that characterizes type IV lesions. During the fourth decade of life, lesions having a lipid core can also contain thick layers of fibrous connective tissue (type V lesion) and/or fissure, hematoma, and thrombus (type VI lesion). Some type V lesions may be largely calcified (type Vb), and some consist mainly of fibrous connective tissue and little or no accumulated lipid or calcium (type Vc).

Development of Atherosclerotic Plaque and Formation of Primary Oxidation Products

Atherosclerotic plaque formation is considered a unidirectional process that starts in early childhood and progresses with age. This concept has been modified and it is now viewed that plaque development is a dynamic process that can be slowed, stopped, or even reversed. The initial event is a fatty streak formation at the

Table 1 Classification of atherosclerotic plaques by American Heart Association (AHA)

Stages	Events
I	Initial with foam cell
II	Fatty streak with multiple foam cells
III	Preatheroma with extracellular lipid pools
IV	Atheroma with confluent extracellular lipids
V	Fibroatheroma
VI	Complex plaque with surface defect, hemorrhage
VII	Calcified Plaque
VIII	Fibrotic plaque without lipid core

sub-endothelial space in which oxidized LDL contributes significantly. LDL oxidation is a slow process, occurring in the sub-intimal space, by reactive oxygen species generated by vascular cells. The oxidation process generates a huge variety of lipid peroxidation products like MDA, 4-HNE, acrolein or glyoxal [5, 6]. Interaction of these aldehydes with lysine residues in the apolipoprotein B-100 moiety of LDL renders the lipoprotein molecule more negative charge. This modification results in decreased affinity for LDL to its receptor and increased affinity for scavenger receptor-bearing cells like macrophages, which are progressively transformed into foam cells [7]. The accumulation of foam cells thus formed leads to the fatty streaks that are characteristic of the early atherosclerotic lesions.

When the macrophage engulfs ox LDL, it causes the loss of pH homeostasis within the lysosome, affecting the activity of lysosomal lipase, an enzyme responsible for the hydrolysis of the cholesterol esters. By loss of its activity, the lipids cannot be processed and therefore accumulate within the cells. Myeloperoxidase–hydrogen peroxide reaction within the macrophage foam cells can generate hypochlorous acid that also can modify the oxidized LDL [8]. Guanosine triphosphate–cyclohydrolase activity on GTP produces Dihydroneopterin within the macrophages. Dihydroneopterin is converted to neopterin by the same hypochlorous acid action. Neopterin is a stable molecule within the plaque whose values are raised in cardiovascular diseases [9]. The MPO-H_2O_2 system also results in the production of dityrosine from two tyrosine radicals. Dityrosine is a stable protein identified within the atherosclerotic plaque acting as a marker for free radical damage [10]. In addition, the cholesterols undergo enzymatic or non-enzymatic oxidation to form oxysterols. Oxysterols are known to contribute to plaque growth as this material cannot be detoxified and removed [11].

Role of Protein Carbonyls in Atherosclerotic Plaque Progression

Protein carbonyls form a major constituent in atherosclerotic plaque. The oxidative theory of atherosclerosis postulates that 4-hydroxynonenal (4-HNE) and malondialdehyde (MDA)-modified LDLs are directly involved in the mechanisms of fatty streak formation, an early step in atherogenesis [12, 13]. A hallmark in atherosclerosis is the detection of 4-HNE and MDA adducts and oxidized LDLs (ox LDLs) within the plaque [14, 15]. Most of these oxidized species are shown to be pro-inflammatory in nature.

Suvarna et al. (1995) [16] have reported that the human atherosclerotic plaque contained free cholesterol, cholesterol oleate (Ch 18:1), and cholesteryl linoleate (Ch18:2) of which 30% are in oxidized form. In addition, fatty hydroxides (17%), ketones (12%), and hydroperoxides (1%) have been documented. Above all, the authors detected endogenous vitamin E (43%), vitamin C (87%), and alpha-tocopherol and coenzyme Q in their oxidized forms, tocopheryl quinone and ubiquinone 10. Therefore, it has been assumed that atherosclerotic plaque observes very little oxidation process. However, the primary oxidized lipids in the plaque can undergo non-enzymatic conversion at physiological environment over time, generating secondary aldehydes and acids. Raghavamenon et al. [17] have reported that vitamin E and alpha tocopherols are incapable of preventing the decomposition of already oxidized fatty acids at physiological conditions. The lower level of lipid hydroperoxide detected in human atherosclerotic plaque by Susan et al. (1995) [16] indicates this autoxidation reaction. Parthasarathy et al. (2008) [18] have opined that differences from early fatty lesion where primary oxidation products are detected, advanced lesion might have experienced the steps involved in the decomposition of peroxidized lipids into aldehydes and their further oxidation into carboxylic acids and that these steps may not be responsive to attenuation by antioxidants. Antioxidants might actually counter the stabilization of plaque by preventing the formation of carboxylic acids which are anti-inflammatory in nature. The formation of such dicarboxylic acids may also be conducive to plaque stabilization by trapping calcium. It is thus clear that several molecules with biological activity are present within the plaque environment that contributes to the development of vulnerable plaque. Removals of these molecules from the plaques are necessary for better reduction in plaque burden and plaque stabilization.

Factors Affecting Plaque Regression

Plaque regression is defined as the return of the arterial wall to its initial state. The mechanisms of plaque regression are distinct and composed of three important processes: (1) the reduction or clearance of necrotic and extracellular material from the tunica intima; (2) repair of endothelium and their regeneration and return to homeostasis; and (3) the cessation of smooth muscle cell proliferation [19]. The most important step in plaque regression is the conversion of the inert pool of extracellular lipid in plaque to a metabolically active intracellular pool and subsequent clearance by the high-density lipoprotein-mediated reverse cholesterol transport system [20]. Reductions of LDL by using lipid-lowering drugs like statins and prevention of oxidation of LDL by using antioxidants like niacin are the conventional treatment options for atherosclerosis. Newer approaches for the prevention of plaque progression and promotion of plaque regression are achieved by increasing functional HDL and the use of HDL mimetics [21].

Increasing HDL Cholesterol

High-density lipoprotein cholesterol (HDL) is crucial in removing extrahepatic cholesterol and phospholipids to the liver for metabolism and excretion through a process designated as reverse cholesterol transport (RCT). Several studies have indicated that transport of arterial lipid deposition through the RCT pathway reduces plaque burden and thereby cardiac incidents [21, 22]. Additionally, paraoxonase enzyme seen associated with HDL molecule prevent oxidation of LDL and HDL itself and reduce atherogenesis [23]. Thus the atheroprotective effect of HDL at the molecular level seems to be through the processes of reducing adhesion of inflammatory cells to the endothelium, preventing their migration into the arterial intima and reducing inflammation within the artery wall [24, 25]. The human paraoxonase enzyme associated with HDL interferes at various levels of atherosclerosis progression by reducing the oxidative stress, preventing LDL oxidation and its uptake by macrophages [26]. It also reduces macrophage cholesterols synthesis and increases cholesterol efflux. Therefore, achieving enhanced transport of excess cholesterol from arterial lipid-rich lesions has emerged as an important approach in anti-atherosclerotic drug development. Molecules that can improve functional HDL, as well as molecules involved in the RCT pathway, have a high appreciation as a candidate drug.

Reverse Cholesterol Transport

The concept of RCT was first proposed by Glomset in 1968 and it represents the most widely accepted mechanism underlying the HDL hypothesis which proposes that pharmacological intervention to raise HDL will reduce cardiovascular risk. The first step in reverse cholesterol transport is the efflux of cholesterol from macrophages. The cholesterol efflux is mediated by ATP binding cassette transporter A1 (ABCA1), ABCG1, Scavenger receptor B1 (SR-B1) present in macrophages or by passive diffusion. ABCA1 transfers free cholesterol to nascent HDL containing Apo A1, whereas ABCG1 and SR-B1 transfer free cholesterol to mature HDL. Within the mature HDL, the free cholesterol transferred is esterified by an enzyme, lecithin cholesterol acyl transferase (LCAT), to form cholesteryl ester (CE). Cholesteryl esters are transported to the liver by direct or indirect pathways. In direct pathway, the HDL with CE gets attached to the SR-B1 receptor. In an indirect pathway, the HDL transfers CE in exchange with phospholipid to Apo-B containing lipoproteins like VLDL and LDL with subsequent uptake in the liver via the low-density lipoprotein receptor (LDLR). In the circulation, the exchange of CE and phospholipid is mediated by Cholesteryl ester transfer protein (CETP) and Phospholipid transfer protein (PLTP). The CE taken up by the liver will be acted upon by the enzyme hepatic lipase and the cholesterol will be metabolized and excreted through bile or feces (Fig. 1).

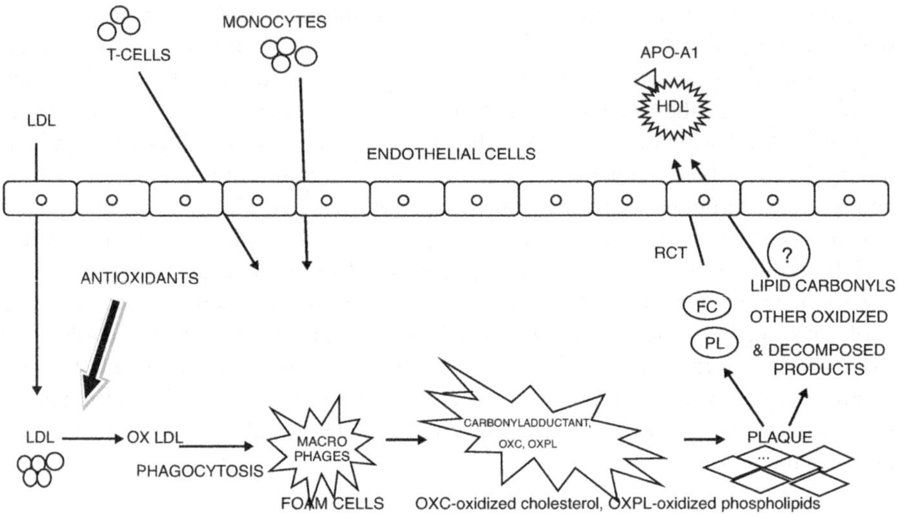

Fig. 1 Schematic representation of Reverse Cholesterol Transport

Plaque Regression: Current Approaches

Increasing the Efflux of Cholesterol From Macrophages

Presently several studies are underway for increasing Apo A1, ABCA-1, ABCG-1 and SR-B1, thereby increasing the cholesterol efflux from the macrophages as a potential mechanism to increase RCT. HDL cholesterol increases ABCA1 and ABCG1 expression through the micro RNA [27]. Studies have shown that the administration of ProAlgaZyme (PAZ) and its subfractions alter mRNA levels of ABCA1, ApoA1, SRB1, and CETP [28]. Transcription of ApoA1, ABCA1, and SRB1 genes was upregulated, whereas transcription of the gene encoding CETP was downregulated after 4 weeks of dietary intervention with PAZ.

Reservelogix-208 (RVX-208) is a small molecule that increases the endogenous synthesis of ApoAI. Oral administration of RVX-208 resulted in increased levels of plasma Apo A-I and HDL [29]. Serum from these animals has been shown to mediate enhanced cholesterol efflux from J774 macrophages through the ABCA1, ABCG1, and SR-BI-dependent pathways.

Synthetic LXR agonists including LXR α/β are known to induce the transcription of ABCA1 and ABCG1. As potent activators of the cellular cholesterol efflux, these compounds have been found to raise HDL levels and to reduce atherosclerosis in transgenic mouse models [30]. Thus, LXR agonist's activation may be a promising pharmacologic target for the treatment of dyslipidemia and atherosclerosis. The LXR agonist LXR-623 is associated with increased expression of ABCA1 and ABCG1 in cells [31].

Increasing Transport of Cholesterol From Macrophages Through the Plasma to the Liver

A second essential determinant of efficient cholesterol elimination from macrophage foam cells is the number of acceptors, principally apoA-I and HDL, present in the circulation. Overexpression of human apoA-I in mice resulted in more cholesterol being removed from macrophages and deposited in the liver via the RCT pathway which eventually degrades to bile acids.

Recombinant apoA-I, when delivered by intravenous infusion, has been shown to promote regression of atherosclerosis lesion to a good extent [32]. Ibanez et al. [33] in their studies has shown that recombinant HDL apoA-I_{Milano} exerts good anti-inflammatory and plaque stabilizing properties. Intravenous infusion of autologous delipidated HDL is a novel approach to raise HDL. The preclinical evaluation showed a significant 6.9% reduction in aortic atheroma volume [34].

LCAT Activity Modulators Early studies for the treatment of atherosclerosis and CVD, by raising HDL through plasma LCAT enzyme activity, were initiated by Zhou et al. [35] in a rabbit model. Data concluded that recombinant LCAT administration may represent a novel approach for the treatment of atherosclerosis and dyslipidemia associated with low HDL. A preclinical mouse study of human recombinant LCAT (rLCAT) was reported [36]. An rLCAT (rLCAT, AlphaCore Pharma, Ann Arbor, MI, USA) when injected into LCAT-null mice was found to reverse the abnormal lipoprotein profile by increasing HDL to near-normal levels for several days. Intravenous infusion of human rLCAT in rabbits was found to raise HDL, to increase fecal secretion of cholesterol, and to reduce atherosclerosis. LCAT promotes the maturation of HDL particles in plasma and facilitates reverse cholesterol transport by maintaining a concentration gradient for the diffusion of cellular unesterified cholesterol to HDL. Studies have shown that the methanol fraction of *Aconitum heterophyllum* could activate LCAT and thereby increase the HDL levels [37].

CETP inhibitors like dalcetrapib [38] used in clinical trials have been shown to increase HDL by 30–40%, without changing LDL and blood pressure. In hamsters, which naturally express CETP, treatment with the potent CETP inhibitor torceptrapib or anacetrapib to some extent improved the movement of cholesterol from macrophages in the peritoneal cavity to the feces [39]. Evacetrapib is a benzazepine compound (LY248595) and a potent and selective inhibitor of CETP both in vitro and in vivo. Clinical trials with evacetrapib showed substantially increased HDL (54–129%) and decreased LDL (14–36%) across a dose range of evacetrapib in 398 dyslipidemic patients [40].

Increasing Uptake of Cholesterol by the Liver for Metabolism and Excretion

Following transport through the plasma, the final step in RCT is the delivery of cholesterol to the liver. Hepatic SR-BI is the key receptor responsible for the selective uptake of CEs from HDL into the liver, and hepatic SR-BI has been recognized as a positive regulator of RCT. Consistent with the effects on experimental atherosclerosis, hepatic SR-BI overexpression resulted in more macrophage-derived cholesterol being excreted into the feces [41].

Carbonyls Scavengers

Antioxidants are capable of inhibiting reactive oxygen species and lipid peroxidation products, but are less efficient in quenching reactive carbonyl substances (RCS) before adducts are formed on proteins. Carbonyl scavengers are agents capable of removing the protein carbonyls and preventing the formation of protein carbonyl adducts. Studies are underway in developing molecules that efficiently interfere with different phases of the reaction cascades, such as by directly trapping RCS, by acting as antioxidants or by chelating metal ions (Fig. 2). Among the various targets, direct trapping of reactive aldehydes seems to be the most promising therapeutic approach. The, β unsaturated aldehydes like 4-hydroxynonenal (HNE), acrolein, and dicarbonyls methylglyoxal (MG) are the most abundant and toxic lipid-derived RCS. Studies have shown that agents like epigallocatechin-3-gallate (EGCG) prevent protein glycation by competing with lysine or arginine and can rapidly trap methylglyoxal (MG) at neutralizing or alkaline conditions [42]. Besides EGCG, catechin, epicatechin, theaflavin, proanthocyanidins, phloretin, phloridzin,

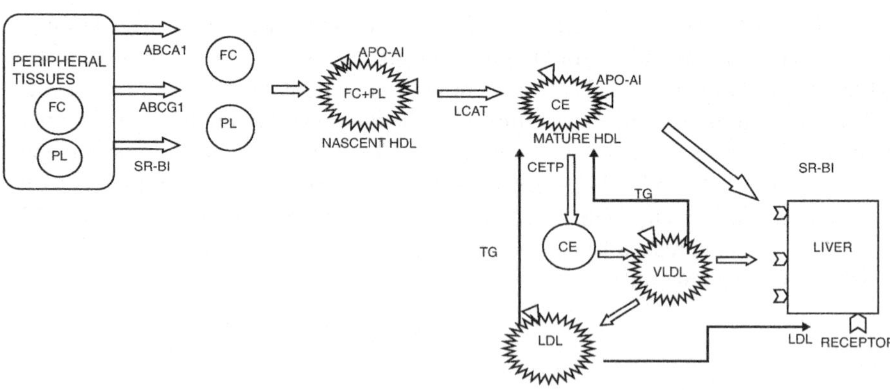

Fig. 2 Do lipid poor HDL, transport oxidized and decomposed products, carbonyls and carbonyl adducts efficiently?

curcumin, a pharmaceutical agent like aminoguanidine, etc., can effectively trap MG [43]. Sulforaphane (SF), the main active isothiocyanate component in cruciferous vegetables, has been proven as a carbonyl scavenger [44] and a persuasive protector against oxidative damage by Nrf-2 mediated induction of phase 2 detoxification enzymes. Studies recommend that SF has the potential to reduce the risk of various types of cancers, diabetes, atherosclerosis, respiratory diseases, neurodegenerative disorders, ocular disorders, and cardiovascular diseases.

Therefore, these compounds represent a new group of 1, 2 dicarbonyl scavenging agents. Studies have also shown that hydrazine derivatives [45] like hydralazine, isoniazid are capable of preventing the protein glycation by reacting with 4-hydroxynonenal (HNE), thereby acting as an excellent carbonyl scavenger.

.Conclusion

Atherosclerotic plaque is rich in cholesterol, phospholipids, and their oxidized forms. The currently achieved reduction in plaque burden by increasing functional HDL and by improving molecules of reverse cholesterol transport pathway is only 30–40%. There is no literature available to ensure that other forms of oxidized and decomposed lipids from plaques are being removed by RCT. The removal of lipid carbonyls from the plaque likely provides a better reduction in plaque volume in addition to enhancing RCT. Several plant extracts are good carbonyl scavengers that can reduce oxidized lipid species accumulation in the plaque and to a certain extent can reduce lipid carbonyls of the plaque microenvironments. It is thus suggested that investigation for newer candidate drugs to improve plaque regression needs to stress on molecules that enhance functional HDL as well as carbonyl scavengers. More research on these lines is indeed essential in the development of the new class of anti-atherosclerotic drugs.

Acknowledgments The authors would like to acknowledge the financial support of the Science Engineering Research Board Department of Science and Technology, India (DST/SERB-sanction No.SR/SO/HS-0240/2012).

Conflict of Interest The authors declare that there are no conflicts of interest.

References

1. Vaya J. The association between biomarkers in the blood and carotid plaque composition-focusing on oxidized lipids, oxysterols and plaque status. Biochem Pharmacol. 2013;86(1):15–8.
2. Shen L, Yamamoto T, Tan XW, Ogata K, Ando E, Ozeki E, et al. Identification and visualization of oxidized lipids in atherosclerotic plaques by microscopic imaging mass spectrometry-based metabolomics. Atherosclerosis. 2020;311:1–12.

3. Tschoepe D, Stratmann B. Plaque stability and plaque regression: new insights. Eur Heart J Suppl. 2006;8(suppl_F):F34–F9.
4. Stary HC, Chandler AB, Dinsmore RE, Fuster V, Glagov S, Insull W Jr, et al. A definition of advanced types of atherosclerotic lesions and a histological classification of atherosclerosis. A report from the Committee on Vascular Lesions of the Council on Arteriosclerosis, American Heart Association. Circulation. 1995;92(5):1355–74.
5. Esterbauer H, Schaur RJ, Zollner H. Chemistry and biochemistry of 4-hydroxynonenal, malonaldehyde and related aldehydes. Free Radic Biol Med. 1991;11(1):81–128.
6. Negre-Salvayre A, Coatrieux C, Ingueneau C, Salvayre R. Advanced lipid peroxidation end products in oxidative damage to proteins. Potential role in diseases and therapeutic prospects for the inhibitors. Br J Pharmacol. 2008;153(1):6–20.
7. Steinbrecher UP. Receptors for oxidized low density lipoprotein. Biochim Biophys Acta. 1999;1436(3):279–98.
8. Carr AC, McCall MR, Frei B. Oxidation of LDL by myeloperoxidase and reactive nitrogen species: reaction pathways and antioxidant protection. Arterioscler Thromb Vasc Biol. 2000;20(7):1716–23.
9. Gieseg SP, Crone EM, Flavall EA, Amit Z. Potential to inhibit growth of atherosclerotic plaque development through modulation of macrophage neopterin/7,8-dihydroneopterin synthesis. Br J Pharmacol. 2008;153(4):627–35.
10. Tien M. Myeloperoxidase-catalyzed oxidation of tyrosine. Arch Biochem Biophys. 1999;367(1):61–6.
11. Larsson H, Böttiger Y, Iuliano L, Diczfalusy U. In vivo interconversion of 7beta-hydroxycholesterol and 7-ketocholesterol, potential surrogate markers for oxidative stress. Free Radic Biol Med. 2007;43(5):695–701.
12. Steinberg D. Low density lipoprotein oxidation and its pathobiological significance. J Biol Chem. 1997;272(34):20963–6.
13. Stocker R, Keaney JF Jr. Role of oxidative modifications in atherosclerosis. Physiol Rev. 2004;84(4):1381–478.
14. Palinski W, Rosenfeld ME, Ylä-Herttuala S, Gurtner GC, Socher SS, Butler SW, et al. Low density lipoprotein undergoes oxidative modification in vivo. Proc Natl Acad Sci U S A. 1989;86(4):1372–6.
15. Petersen DR, Doorn JA. Reactions of 4-hydroxynonenal with proteins and cellular targets. Free Radic Biol Med. 2004;37(7):937–45.
16. Suarna C, Dean RT, May J, Stocker R. Human atherosclerotic plaque contains both oxidized lipids and relatively large amounts of alpha-tocopherol and ascorbate. Arterioscler Thromb Vasc Biol. 1995;15(10):1616–24.
17. Parthasarathy S, Raghavamenon A, Garelnabi MO, Santanam N. Oxidized low-density lipoprotein. Methods Mol Biol (Clifton, NJ). 2010;610:403–17.
18. Parthasarathy S, Litvinov D, Selvarajan K, Garelnabi M. Lipid peroxidation and decomposition--conflicting roles in plaque vulnerability and stability. Biochim Biophys Acta. 2008;1781(5):221–31.
19. Francis AA, Pierce GN. An integrated approach for the mechanisms responsible for atherosclerotic plaque regression. Exp Clin Cardiol. 2011;16(3):77–86.
20. Schwartz CJ, Valente AJ, Sprague EA, Kelley JL, Nerem RM. The pathogenesis of atherosclerosis: an overview. Clin Cardiol. 1991;14(2 Suppl 1):I1–16.
21. Hafiane A, Genest J. HDL, atherosclerosis, and emerging therapies. Cholesterol. 2013;2013:891403.
22. Krause BR, Auerbach BJ. Reverse cholesterol transport and future pharmacological approaches to the treatment of atherosclerosis. Curr Opin Investig Drugs (London, England : 2000). 2001;2(3):375–81.
23. Khera AV, Cuchel M, de la Llera-Moya M, Rodrigues A, Burke MF, Jafri K, et al. Cholesterol efflux capacity, high-density lipoprotein function, and atherosclerosis. N Engl J Med. 2011;364(2):127–35.

24. Stein O, Stein Y. Atheroprotective mechanisms of HDL. Atherosclerosis. 1999;144(2):285–301.
25. Camont L, Chapman MJ, Kontush A. Biological activities of HDL subpopulations and their relevance to cardiovascular disease. Trends Mol Med. 2011;17(10):594–603.
26. Chistiakov DA, Melnichenko AA, Orekhov AN, Bobryshev YV. Paraoxonase and atherosclerosis-related cardiovascular diseases. Biochimie. 2017;132:19–27.
27. Moore KJ, Rayner KJ, Suárez Y, Fernández-Hernando C. The role of microRNAs in cholesterol efflux and hepatic lipid metabolism. Annu Rev Nutr. 2011;31:49–63.
28. Geamanu A, Saadat N, Goja A, Wadehra M, Ji X, Gupta SVJN, et al. ProAlgaZyme and its subfractions increase plasma HDL cholesterol via upregulation of ApoA1, ABCA1, and SRB1, and inhibition of CETP in hypercholesterolemic hamsters. Nutr Metlab. 2012;4:17–24.
29. Bailey D, Jahagirdar R, Gordon A, Hafiane A, Campbell S, Chatur S, et al. RVX-208: a small molecule that increases apolipoprotein A-I and high-density lipoprotein cholesterol in vitro and in vivo. J Am Coll Cardiol. 2010;55(23):2580–9.
30. Yasuda T, Grillot D, Billheimer JT, Briand F, Delerive P, Huet S, et al. Tissue-specific liver X receptor activation promotes macrophage reverse cholesterol transport in vivo. Arterioscler Thromb Vasc Biol. 2010;30(4):781–6.
31. Quinet EM, Basso MD, Halpern AR, Yates DW, Steffan RJ, Clerin V, et al. LXR ligand lowers LDL cholesterol in primates, is lipid neutral in hamster, and reduces atherosclerosis in mouse. J Lipid Res. 2009;50(12):2358–70.
32. Nissen SE, Tsunoda T, Tuzcu EM, Schoenhagen P, Cooper CJ, Yasin M, et al. Effect of recombinant ApoA-I Milano on coronary atherosclerosis in patients with acute coronary syndromes: a randomized controlled trial. JAMA. 2003;290(17):2292–300.
33. Ibanez B, Giannarelli C, Cimmino G, Santos-Gallego CG, Alique M, Pinero A, et al. Recombinant HDL(Milano) exerts greater anti-inflammatory and plaque stabilizing properties than HDL(wild-type). Atherosclerosis. 2012;220(1):72–7.
34. Sacks FM, Rudel LL, Conner A, Akeefe H, Kostner G, Baki T, et al. Selective delipidation of plasma HDL enhances reverse cholesterol transport in vivo. J Lipid Res. 2009;50(5):894–907.
35. Zhou M, Sawyer J, Kelley K, Fordstrom P, Chan J, Tonn G, et al. Lecithin cholesterol acyltransferase promotes reverse cholesterol transport and attenuates atherosclerosis progression in New Zealand white rabbits. Am Heart Assoc. 2009;120:S1175.
36. Rousset X, Vaisman B, Auerbach B, Krause BR, Homan R, Stonik J, et al. Effect of recombinant human lecithin cholesterol acyltransferase infusion on lipoprotein metabolism in mice. J Pharmacol Exp Ther. 2010;335(1):140–8.
37. Subash AK, Augustine A. Hypolipidemic effect of methanol fraction of Aconitum heterophyllum wall ex Royle and the mechanism of action in diet-induced obese rats. J Adv Pharm Technol Res. 2012;3(4):224–8.
38. Schwartz GG, Olsson AG, Abt M, Ballantyne CM, Barter PJ, Brumm J, et al. Effects of dalcetrapib in patients with a recent acute coronary syndrome. N Engl J Med. 2012;367(22):2089–99.
39. Briand F, Thieblemont Q, André A, Ouguerram K, Sulpice T. CETP inhibitor torcetrapib promotes reverse cholesterol transport in obese insulin-resistant CETP-ApoB100 transgenic mice. Clin Transl Sci. 2011;4(6):414–20.
40. Nicholls SJ, Brewer HB, Kastelein JJ, Krueger KA, Wang M-D, Shao M, et al. Effects of the CETP inhibitor evacetrapib administered as monotherapy or in combination with statins on HDL and LDL cholesterol: a randomized controlled trial. JAMA. 2011;306(19):2099–109.
41. Zhang Y, Da Silva JR, Reilly M, Billheimer JT, Rothblat GH. Rader DJJT Joci. Hepatic expression of scavenger receptor class B type I (SR-BI) is a positive regulator of macrophage reverse cholesterol transport in vivo. J Clin Invest. 2005;115(10):2870–4.
42. Shin ER, Jung W, Kim MK, Chong YJABC. Identification of (−)-epigallocatechin (EGC) as a methylglyoxal (MGO)-trapping agent and thereby as an inhibitor of advanced glycation end product (AGE) formation. Appl Biol Chem. 2018;61(5):587–91.
43. Jiang H, Li D. Polyphenols as reactive carbonyl species scavengers-the solution to the current puzzle of polyphenols' health effects. Med Hypotheses. 2020;142:110144.

44. Galvani S, Coatrieux C, Elbaz M, Grazide MH, Thiers JC, Parini A, et al. Carbonyl scavenger and antiatherogenic effects of hydrazine derivatives. Free Radic Biol Med. 2008;45(10):1457–67.
45. Angeloni C, Leoncini E, Malaguti M, Angelini S, Hrelia P, Hrelia S. Modulation of phase II enzymes by sulforaphane: implications for its cardioprotective potential. J Agric Food Chem. 2009;57(12):5615–22.

Role of Bioactive Lipid, Phosphatidic Acid, in Hypercholesterolemia Drug-Induced Myotoxicity: Statin-Induced Phospholipase D (PLD) Lipid Signaling in Skeletal Muscle Cells

Eric M. Tretter, Patrick J. Oliver, Sainath R. Kotha, Travis O. Gurney, Drew M. Nassal, Jodi C. McDaniel, Thomas J. Hund, and Narasimham L. Parinandi

Abstract Cardiovascular diseases (CVDs) are among the leading causes of mortality in the United States and worldwide. Cholesterol at high levels in circulation has been established as a major risk factor for CVDs in humans. Statins have been widely used for lowering and controlling the endogenous levels of cholesterol to prevent or treat CVDs. Statins lower cholesterol by inhibiting 3-hydroxy-3-methylglutaryl-CoA reductase (HMG-CoA reductase), the rate-limiting enzyme in the biosynthetic pathway of endogenous cholesterol. Despite their efficacy for cardiovascular indications, statins may induce undesired side effects, including cause skeletal muscle damage (myotoxicity and myalgia). However, the mechanisms and treatment of statin-induced myotoxicity and myalgia are not well known. Phospholipase D (PLD) is a ubiquitous membrane phospholipid-hydrolyzing enzyme involved in mediating lipid signaling in mammalian cells, including skeletal muscle cells (myocytes). Therefore, we hypothesized that statins would mediate skeletal muscle myocyte damage through activation of the PLD-mediated lipid signaling, and inhibition of PLD activation would protect against the statin-induced

This work was submitted as honors dissertation by the author, Eric M. Tretter for partial fulfillment of the requirements for graduation "With Honors Research Distinction" in the College of Arts and Sciences of The Ohio State University in 2015 under the supervision of Dr. Narasimham L. Parinandi.

E. M. Tretter · P. J. Oliver · S. R. Kotha · T. O. Gurney · D. M. Nassal · J. C. McDaniel · T. J. Hund · N. L. Parinandi (✉)
Department of Internal Medicine, Division of Pulmonary, Critical Care, and Sleep Medicine, Department of Biomedical Engineering, College of Engineering, College of Nursing Center on Healthy Aging, Self-Management and Complex Care, Dorothy M. Davis Heart & Lung Research Institute, The Ohio State University Wexner Medical Center, Columbus, OH, USA
e-mail: narasimham.parinandi@osumc.edu

N. L. Parinandi, T. J. Hund (eds.), *Cardiovascular Signaling in Health and Disease*, https://doi.org/10.1007/978-3-031-08309-9_14

myotoxicity. In order to test our hypothesis, we used the well-established C2C12 mouse skeletal muscle myoblast cell line and studied the PLD activation and cytotoxicity in vitro induced by two different widely prescribed statins (mevastatin and simvastatin). Our studies revealed that the statins caused dose- (10–100 μM) and time-dependent (4–24 h) activation of PLD (as determined by the [^{32}]P-labeling of cells and thin-layer chromatography of phosphatidylbutanol formation) and cytotoxicity and mitochondrial dysfunction (as determined by the release of intracellular lactate dehydrogenase, suppression of MTT reduction, and alterations in cell morphology) in the C2C12 myoblast cells. Our results also showed that cholesterol replenishment protected against the statin-induced toxicity to the C2C12 cells. Furthermore, our results showed that the novel PLD-specific inhibitor, 5-fluoro-2-indolyl des-chlorohalopemide (FIPI) inhibited the statin-induced PLD activation and cytotoxicity in the C2C12 myoblast cells. For the first time, our study demonstrated the role of endogenous cellular cholesterol depletion and PLD-mediated lipid signaling in statin-induced skeletal muscle myocyte damage and emphasized the importance of PLD inhibition in attenuating the statin-induced myotoxicity and myalgia in CVD patients consuming statins to lower the elevated levels of endogenous cholesterol.

Keywords Statins · Cholesterol-lowering drugs · Skeletal muscle cells · PLD · Phosphatidic acid signaling

Abbreviations

ATP	Adenosine 5-phosphate
BCA	Bicinchoninic acid
BSA	Bovine serum albumin
CHD	Chronic heart disease
CoQ10	Coenzyme Q10
CVD	Cardiovascular disease
DAG	Diacylglycerol
DMEM	Dulbecco's modified Eagle medium
DMSO	Dimethylsulfoxide
EDTA	Ethylenediaminetetraacetic acid
FBS	Fetal bovine serum
FIPI	5-fluoro-2-indolyl des-chlorohalopemide hydrochloride hydrate
HDL	High-density lipoprotein
HMG-CoA reductase	3-hydroxy-3-methyl-glutaryl-CoA reductase
LDH	Lactate dehydrogenase
LDL-C	Low-density lipoprotein-cholesterol
LPA	Lysophosphatidic acid
MAPK	Mitogen-activated protein kinase

MβCD	Methyl-β-cyclodextrin
MEM	Minimal essential medium
MTT	3-[4,5-dimethylthiazol-2-yl]-2,5-diphenyl tetrazolium bromide
PA	Phosphatidic acid
PBS	Phosphate-buffered saline
PBS-T	Phosphate-buffered saline-Tween-20
PBt	Phosphatidylbutanol
PKC	Protein kinase C
PLA_2	Phospholipase A_2
PLD	Phospholipase D
ROS	Reactive oxygen species
TLC	Thin-layer chromatography

Introduction

Cardiovascular Disease

In developed countries across the world, and particularly in the United States, cardiovascular disease (CVD), including the cerebrovascular diseases, remain the number one cause of death [1, 2]. CVD is the result of restricted blood vessels that lead to chest pain, stroke, heart attack, or other painful heart conditions [3–7]. A host of risk factors have been identified for CVD, including a diet high in fat and cholesterol, lack of exercise, use of tobacco, drugs, and alcohol, and a stressful environment [8–15]. High concentrations of circulating blood cholesterol can cause the deposition of this cholesterol in blood vessels and eventually lead to plaque formation in the blood vessel and ultimate obstruction of the circulation to vital organs, including the heart, brain, and kidney [16–19]. Prescription of drugs that lower endogenously synthesized cholesterol in order to lower the risk for cardiovascular disease have become increasingly common to modify CVD risk [20].

Cholesterol and Cardiovascular Diseases

Cholesterol is the major component of the cell membrane lipid backbone that regulates membrane fluidity, structure, and function [21, 22]. Cholesterol occurs in circulation in two different types: high-density lipoproteins (HDLs) and low-density lipoproteins (LDLs) [23, 24]. HDLs have been found to reduce the risk for heart disease when in high concentrations, and thus are often designated as "good cholesterol" [25]. In contrast, high concentrations of LDLs are associated with coronary artery disease, earning them the distinction of "bad cholesterol" [26]. Having high concentrations of LDL-cholesterol (LDL-C) is referred to as hypercholesterolemia, and can have genetic or dietary roots [26].

Familial hypercholesterolemia is an inherited condition caused by a defect in LDL-C receptor expression/function [27]. Dysfunction in LDL-C receptors results in insufficient uptake of LDL-C into cells, leading to increased circulating LDL-C in the bloodstream that can accumulate on the vessel walls. In addition, individuals with familial hypercholesterolemia typically demonstrate a loss of the normal feedback inhibition that stops the synthesis of cholesterol by inhibiting 3-hydroxy-3-methyl-glutaryl-CoA reductase (HMG-CoA reductase), a rate-limiting enzyme in cholesterol synthesis (Schema 1), when high concentrations are detected in the blood [28–30] (Schema 1). Dietary hypercholesterolemia results from a diet high in saturated fats and cholesterol, more than recommended daily allowance. The excessive amount of cholesterol consumed finds its way into the blood stream and can accumulate on vessel walls leading to plaque formation and atherosclerotic lesions

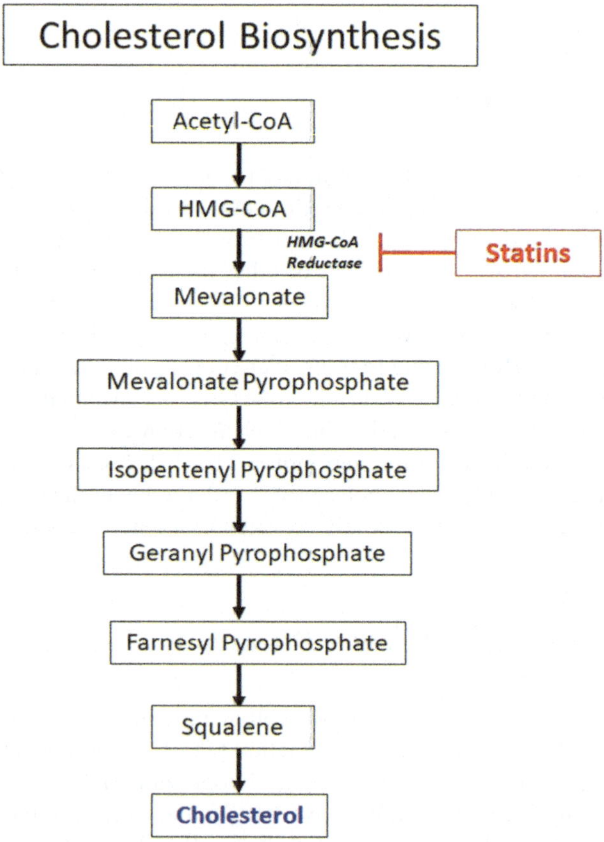

Schema 1 Pathway of cellular cholesterol synthesis in mammalian cells. HMG-CoA reductase is the rate-limiting enzyme that converts HMG-CoA to mevalonate. HMG-CoA reductase is inhibited by the statins to cause decrease/reduction in the de novo synthesized cellular cholesterol

in the blood vessel that would obstruct blood flow leading to the CVDs and cerebrovascular disorders.

Statins as the Endogenous Cholesterol-Lowering Drugs

Current therapeutic intervention for hypercholesterolemia involves the use of HMG-CoA reductase-specific inhibitors, known as statins (Schema 2), to attenuate or suppress the synthesis of endogenous cholesterol [28–30]. Suppressing or inhibiting

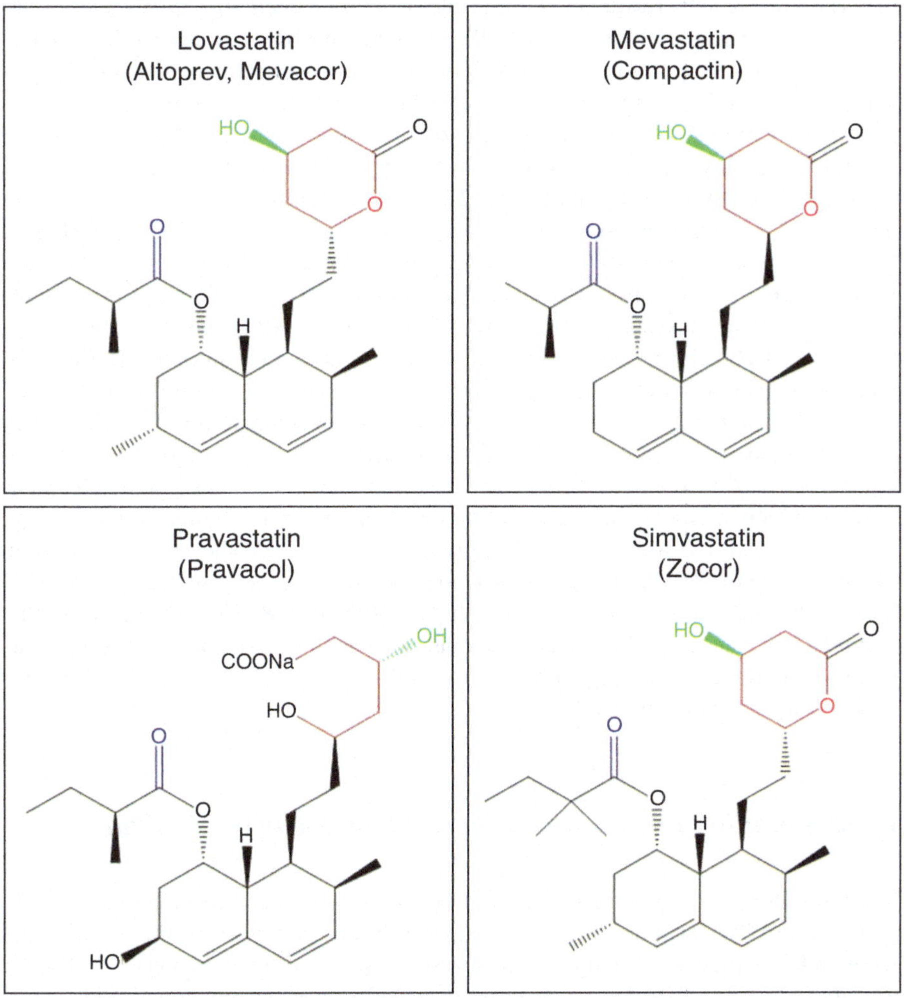

Schema 2 Chemical structures of different statins used to lower endogenous cholesterol by inhibiting HMG-CoA reductase in mammalian cells

endogenous cholesterol synthesis lowers the levels of LDL-C in the blood circulation. Among the most common statins in clinical use are atorvastatin (Lipitor), pravastatin (Pravachol), simvastatin (Zocor), mevastatin (Compactin), and lovastatin (Altoprev) [31–33].

Although statins have dramatically altered the landscape for CVD risk, adverse effects have been identified, yet are not well understood. Myopathy is a disorder encompassing an array of ailments that impact the skeletal muscles, and is one of the most common and significant side effects of statin myotoxicity. Conditions developed can vary from the milder myalgia or myositis to the possibly life-threatening rhabdomyolysis [34–40]. Symptoms of statin-induced myopathy include muscle pain, weakness, and fatigue [35]. Studies have shown that statins use in certain individuals results in statin intolerance associated with myotoxicity [34–42]. The exact rate of incidence is still disputed. In observational studies, statin-associated muscle symptoms (SAMS) have been seen in between 10% and 30% of statin users [35]. In other randomized control studies, 9.4% of patients taking statins experienced myalgia compared to the 4.6% of placebo patients that experienced myalgia [35]. However, a recent study from Jordan found that overall incidence of myopathy in patients taking statins was 27.8%; specifically, incidence was 31.4% in males, 22.6% in females, and 34% among patients $\geq$60 years old [36]. According to a separate study of patients experiencing statin myopathy, 13% were hospitalized for treatment for rhabdomyolysis [37]. More importantly, though, all of the patients in the study who ceased using statins were able to successfully recover from statin-associated myopathy within an average of 2.3 months after stopping statin treatment, with over 50% reporting resolution of muscle symptoms within 1 month [37]. Statins have pleiotropic effects in the body, but the mechanism through which they cause myalgia or other adverse effects remains unclear. If understood, the safety and effectiveness of statins could increase. It seems there exist multiple mechanisms through which statins cause myotoxicity, one being the mitochondrial mechanism, which holds valid to some extent [43]. Previous studies have shown that statins can cause damage to skeletal myocyte mitochondria, which reduces the energy metabolism of the cell and can ultimately result in cell death [38, 39, 44]. Based on this previous work, we hypothesized that statins cause membrane lipid signaling perturbation and mitochondrial dysfunction through cholesterol depletion, which ultimately leads to statin-induced myalgia.

Lipid Signaling and Statin-Induced Myotoxicity or Myalgia

The phospholipid bilayer forms a barrier between the cell and the external environment. Membrane phospholipids, protein channels, and cholesterol compose much of the membrane bilayer of living cells across species, including in humans [45–47]. An essential function of the bilayer is to interact with the environment and relay information to the cell wherein the membrane lipid signaling accomplishes these goals [48]. Phospholipases are the house-keeping enzymes that hydrolyze membrane phospholipids to support turnover/maintenance of the membrane, leading to the formation of a host of bioactive lipid molecules, each with an associated membrane lipid signal as interpreted by the cell. Four major phospholipases exist, including the phospholipase A1, A2, C, and D, and each catalyzes a specific hydrolysis of the membrane phospholipid within the cell [49–54]. Specifically, phospholipase D (PLD) hydrolyzes the membrane phospholipid (phosphatidylcholine, PC) releasing phosphatidic acid (PA), a potent bioactive cell signaling mediator [53, 55]. The cell can further convert the PLD-generated PA to potent bioactive lipid signal mediators such as lysophosphatidic acid (LPA) or diacylglycerol (DAG) upon the actions of phospholipase A1 (PLA1) or PLA2, which can cause myotoxicity (Schema 3) [56–61]. It has been demonstrated that lowering of cell membrane cholesterol in the vascular endothelial cells leads to the activation of PLD and generation of PA [62]. Therefore, here it is hypothesized that statins cause cholesterol depletion in the membranes of the skeletal muscle cells through the inhibition of HMG-CoA reductase,

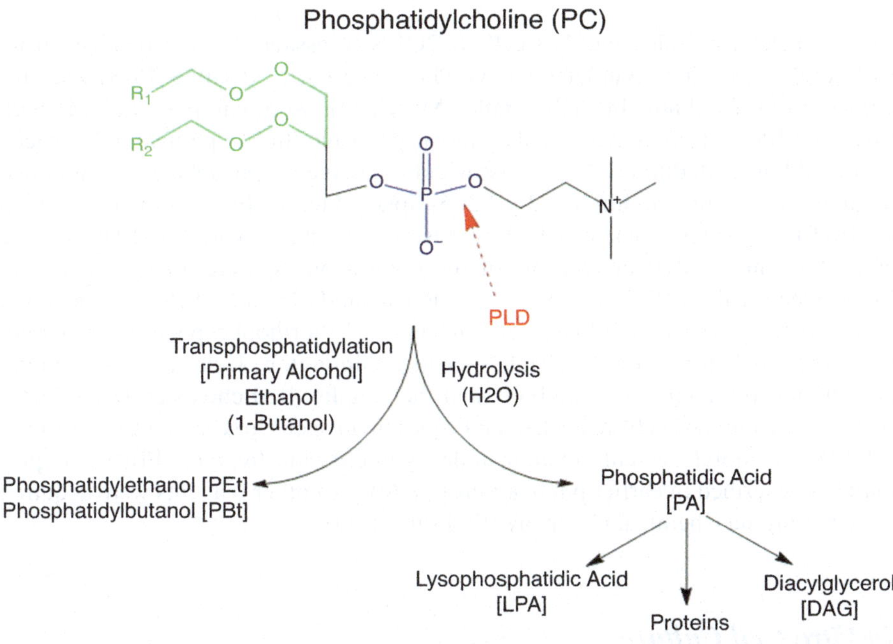

Schema 3 Mechanism of the phospholipase D (PLD)-mediated hydrolysis of the membrane phospholipids (e.g., phosphatidylcholine, PC) in mammalian cells. PLD hydrolyzes PC at the head group and forms the bioactive lipid signal mediator, phosphatidic acid (PA). PA can further be converted to potent bioactive lipid signal mediators such as the lysophosphatidic acid (LPA) by the action of phospholipases A1/A2 (PLA1 or PLA2) and diacylglycerol (DAG) by the action of lipid phosphate phosphatase. Both LPA and DAG are potent lipid mediators of cell signaling. PA can also directly modulate cellular proteins and cause functional alterations in the cells. One salient feature of PLD is that enzyme is capable of using stereospecifically a primary alcohol instead of water such as ethanol, 1-butanol, and 1-propanol during the hydrolysis of the membrane phospholipid to convert the PLD-generated PA into the corresponding phosphatidylalcohols (e.g., phosphatidylethanol [PEt], phosphatidylbutanol [PBt], and phosphatidylpropanol [PProp]). Hence, the phosphatidylalcohols serve as the indices of PLD activity in cells in situ

leading to the activation of PLD that generates the bioactive lipid signal mediator (PA), resulting in mitochondrial damage and myotoxicity as a mechanistic basis of the statin-induced myalgia or myotoxicity. In order to test our hypothesis, in the current study, we chose the well-established skeletal muscle cell model, C2C12 myoblast cells. Our studies revealed that the two widely used statins (HMG-CoA reductase inhibitors), mevastatin and simvastatin, caused PLD activation and generation of the bioactive lipid signal mediator (PA), leading to the mitochondrial dysfunction and cytotoxicity in the C2C12 myoblast cells through cholesterol depletion [63].

Materials and Methods

Materials

Mouse skeletal muscle myoblast cells (C2C12s) (passage 2) were obtained from Cell Applications Inc. (San Diego, CA). Phosphate-buffered saline (PBS) was purchased from Biofluids Inc (Rockville, Maryland). Minimal Essential Medium (MEM), FBS, trypsin, nonessential amino acids, penicillin/streptomycin, Dulbecco Modified Eagle Medium (DMEM) tissue culture reagents, phosphate-free modified medium, 3-[4,5-dimethylthiazol-2-yl]-2, 5-diphenyl tetrazolium bromide reduction kit (MTT assay kit), lactate dehydrogenase cytotoxicity assay kit (LDH release assay kit), and analytical reagents of the highest purity were all obtained from Sigma Chemical Co (St Louis, Missouri). Phosphatidylbutanol (PBt), was acquired from Avanti Polar Lipids (Alabaster, Alabama). [^{32}P]orthophosphate (carrier-free) was purchased from New England Nuclear (Wilmington, Delaware). Anti-rabbit AlexaFluor 488-conjugated antibody and the Amplex Red cholesterol determination kit were purchased from Molecular Probes Invitrogen Co (Carlsbad, California). 5-Fluoro-2-indolyl des-chlorohalopemide hydrochloride hydrate (FIPI) was prepared as described in earlier publications [59, 60]. All other reagents were acquired from the Sigma Chemical Company (St. Louis, MO).

In Vitro Cell Culture

The C2C12 myoblast cells were cultured in Dulbecco's Modified Eagle Medium (DMEM) supplemented with 10% FBS and antibiotics up to 90–100% confluence in sterile 35-mm or 60-mm dishes and in 96-well plates under a sterile and humidified atmosphere of 95% air–5% CO_2 at 37 °C. C2C12s were used up to passage 20 for experiments.

Assay of Phospholipase D (PLD) Activation

PLD activity in the C2C12 myoblast cells was determined according to our previously published procedure [59, 60]. C2C12 myoblast cells cultured in 35-mm dishes were labeled with [^{32}P]orthophosphate (5 mCi/ml) in DMEM phosphate-free medium containing 2% (vol/vol) fetal bovine serum for 14 h. Following the experimental treatments for the chosen periods of time, [^{32}P]-labeled phosphatidylbutanol ([^{32}P]PBt), formed from the PLD activation and transphosphatidylation reaction in cellular lipid extracts as an index of PLD activity in intact cells, was separated by thin-layer chromatography (TLC). Radioactivity associated with the [^{32}P]PBt was quantified by liquid scintillation counting and data were expressed as DPM normalized to the total [^{32}P] in the lipid extract of the cells in the dish.

Lactate Dehydrogenase (LDH) Release Assay of Cytotoxicity

The C2C12 myoblast cells were grown up to 90–100% confluence in sterile 15.5-mm dishes (24-well culture plate) and treated with DMEM alone or DMEM containing the chosen concentrations of statins and/or FIPI at designated time points. At the end of the incubation period, the supernatant was removed, and the level of lactate dehydrogenase (LDH) activity was measured spectrophotometrically according to the manufacturer's protocol (Sigma Chemical Co., St. Louis, MO).

MTT Cell Proliferation Assay

The C2C12 myoblast cells were grown up to 90–100% confluence in sterile 15.5-mm dishes (24-well culture plate) and treated with DMEM alone or DMEM containing the chosen concentrations of statins and/or FIPI at designated time points. At the end of the incubation period, the supernatant was removed, and the extent of MTT reduction was measured spectrophotometrically according to the manufacturer's protocols (Cayman Chemical Co., Ann Arbor, MI).

Cellular Morphology

Morphological changes in the C2C12 myoblast cells grown in the sterile 35-mm dishes up to 90–100% confluence, following their exposure to the chosen concentrations of statins and/or FIPI for different time periods, were examined under light microscope as an index of cytotoxicity. Images of cell morphology were digitally captured using the Zeiss Axioskop 200 with Zen 2011 software at 20× magnification.

Cholesterol Determination

The C2C12 myoblast cells were grown up to 90–100% confluence in sterile 60-mm dishes, and treated with DMEM alone or DMEM containing the chosen concentrations of statins and/or FIPI for the chosen periods of time. Cells were then liberated from dish, reconstituted in PBS, and protein was determined following the protein determination by the BCA assay. Cholesterol contents in the samples was normalized to the total cellular protein (1 mg) according to our previously published method [64].

Phospholipase D1 (PLD1) Phosphorylation Visualization by Confocal Immunofluorescence Microscopy

The C2C12 myoblast cells were grown on sterile glass cover slips (~90% confluence) and treated with DMEM alone or DMEM containing chosen concentrations of statins for 12 h. Cover slips were then rinsed three times with PBS, and fixed with 3.7% formaldehyde in PBS for 10 min. at room temperature. The cells were permeabilized in 0.25% Triton X-100 prepared in PBS containing 0.01% Tween-20 (PBS-T) for 5 min. The cells were again washed three times with PBS-T, and treated with PBS-T containing 1% BSA blocking buffer for 30 min. at room temperature. Cover slips were then incubated overnight at room temperature with the primary antibody [phospho-PLD1 (1:150 dilution)] in 1% BSA solution. After rinsing three times with PBS-T, the cells were labeled with secondary AlexaFluor 488 (1:100 dilution) in 1% BSA in PBS-T for 1 h. Finally, the cells were washed three times with PBS-T, mounted, and examined under Zeiss LSM 710 Confocal/Multiphoton Microscope powered by Argon-2 laser with 500–550 BP filter. The images were captured digitally, and the average fluorescent intensity of triplicate samples was determined using ImageJ.

Preparation of Solutions Containing Pharmacological Agents for Treatment of Cells

All water-soluble pharmacological agent solutions were freshly formulated in DMEM for treatment of cells. The stock solutions of lipophilic pharmacological agents, including all statins and FIPI, were freshly assembled in DMSO and then diluted in DMEM for treatment of cells. The final DMSO concentration in the cell treatment medium did not exceed 0.1% (vol/vol) and did not appear to have any influence on experimental outcomes.

Statistical Analysis

All experiments were completed in triplicate. Results were reported as mean ± standard deviation (SD). Statistical analysis of data was accomplished by one-way analysis of variance (ANOVA) by use of the SigmaStat (Jandel Scientific, San Rafael, California). The statistical significance level was taken as $p \leq 0.05$.

Results

Statins induce PLD activation in C2C12 myoblast cells Statins are HMG-CoA reductase inhibitors and lower endogenous levels of cholesterol by inhibiting cellular cholesterol synthesis [65, 66]. We have previously shown that the cholesterol-depleting agent such as the methyl-β-cyclodextrin (MβCD) lowers cholesterol levels and causes activation of PLD in the vascular endothelial cells [62, 64]. Taking these as the premise, here we hypothesized that statin-induced decrease of cellular cholesterol would lead to activation of PLD in C2C12 myoblast cells. To test our hypothesis, we treated C2C12 cells with two selected statins, mevastatin and simvastatin, for different periods of time (4–24 h) and assayed the activation of PLD. We determined the activity of PLD by analyzing the intracellular formation of phosphatidylbutanol as the transphosphatidylation reaction product of PLD, which is an established and widely used assay of cellular PLD activity [67]. Both mevastatin and simvastatin (10 μM) caused significant time-dependent activation of PLD in the C2C12 myoblast cells (Fig. 1a, b). Although the PLD activity peaked at 12 h for mevastatin-treated cells, simvastatin caused a significant linear increase of the PLD activity from 4 h up to 24 h compared to the untreated control cells (Fig. 1a, b). At 12 h, mevastatin caused a 4.5-fold increase of PLD activation in the C2C12 cells compared to control untreated and mevastatin-treated cells at 4 h. PLD activity remained significantly elevated at 24 h compared to control, although not to the level observed at 12 h. On the other hand, simvastatin caused a significant and approximately sevenfold increase in PLD activity at 24 h as compared to the same in the control untreated and simvastatin-treated cells at 4 h (Fig. 1b). One noteworthy response of the control untreated cells was that the basal PLD activity (formation of PA without statin treatment) also linearly increased with time (Fig. 1a, b). These results revealed that both mevastatin and simvastatin caused a significant activation of PLD in the C2C12 myoblast cells and simvastatin apparently was more potent than mevastatin in inducing the activation of PLD and generating the bioactive lipid signal mediator (PA) in cells.

FIPI, the PLD-specific inhibitor attenuates statin-induced PLD activation in C2C12 cells In previous studies, we have demonstrated efficacy of FIPI, the only available PLD-specific pharmacological inhibitor, in PLD inhibition in cell culture models [59, 60]. Here, we used FIPI (*i*) to investigate its inhibitory action on the

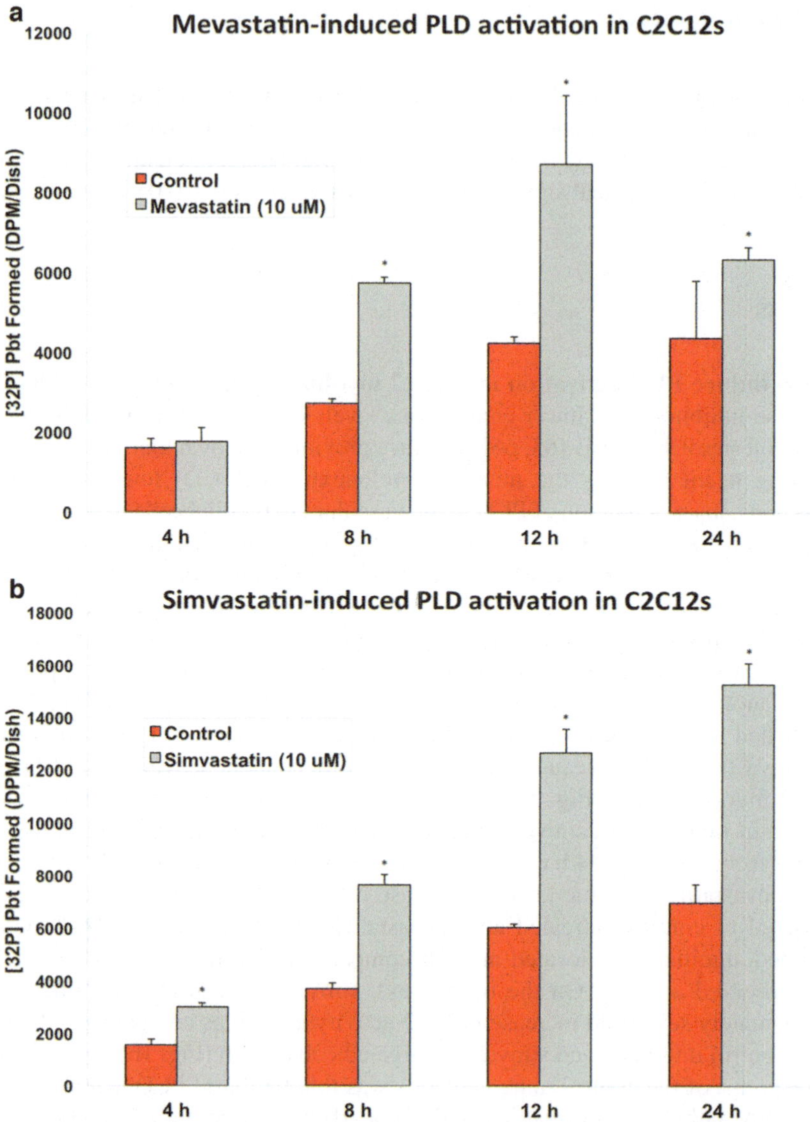

Fig. 1 (**a**) Mevastatin induces PLD activation in C2C12 cells. C2C12 myoblast cells (2×10^5 cells) were pre-labeled for 12 h with carrier free [^{32}P] orthophosphate in complete medium, following which cells were treated with mevastatin (10 μM) for different time periods (4–24 h) in 95% air–5% CO_2 under humidified sterile environment at 37 °C. Under identical conditions, appropriate controls were established without the statin treatment. At the end of the experiment for the designated time, lipids were extracted under acidic conditions with 2:1 chloroform-methanol (vol/vol) and separated by thin-layer chromatography (TLC) as described in the section "Materials and Methods". Phosphatidylbutanol (PBt) as the product of PLD activity generated by the transphosphatidylation reaction was identified by iodination on the TLC plate with authentic PBt standard as described in the section "Materials and Methods". The PBt spots on the TLC plates were

statin-induced PLD activation and (*ii*) to confirm that statins indeed activate PLD in the C2C12 myoblast cells. By virtue of its PLD-specific inhibitory action, FIPI (0.1–1 μM, 12 h of pretreatment) significantly and drastically inhibited the basal PLD activity in the control untreated C2C12 myoblast cells at 24 h (Fig. 2a, b). At 1 μM concentration, FIPI caused ~85–92% decrease of basal PLD activation in the control untreated cells (Fig. 2a, b). Furthermore, FIPI, in a dose-dependent manner (0.1–1 μM), significantly and robustly attenuated the statin-induced PLD activation in cells treated with both mevastatin (10 μM) and simvastatin (10 μM) for 24 h as compared with the cells treated with statins alone (Fig. 2a, b). FIPI (1 μM) significantly attenuated the statin-induced PLD activation by ~90–95% in cells treated with both mevastatin (10 μM) and simvastatin (10 μM) for 24 h as compared to the same in the C2C12 myoblast cells treated with statins alone (Fig. 2a, b). These results clearly revealed that (*i*) FIPI was a potent inhibitor of the statin-induced PLD activation at doses ranging between 0.1–1 μM and the 1 μM dose was the most efficacious in causing effective inhibition of the statin-induced PLD activation and (*ii*) indeed, statins, induced the activation of PLD in the C2C12 myoblast cells.

Statins induce cytotoxicity in C2C12 cells It is becoming increasingly evident that the lipid-lowering and cholesterol-depleting drugs such as the lipophilic statins cause cytotoxicity to mammalian cells including the normal and malignant cells [68–71]. Also, our earlier studies revealed that cyclodextrin-induced cellular cholesterol depletion causes cytotoxicity in the vascular endothelial cells in culture [62, 64]. Based on these findings, we investigated whether statins (simvastatin and mevastatin) induce cytotoxicity in C2C12 cells as determined by lactate dehydrogenase (LDH) release. Our results revealed that both simvastatin and mevastatin (10–100 μM) significantly induced LDH release from cells in a dose-dependent fashion at 24 h of exposure as compared to control untreated cells (Fig. 3a, b). Both

Fig. 1 (continued) scrapped and [^{32}P] radioactivity was determined on a liquid scintillation counter and normalized to the total lipid phosphorus [^{32}P] of the cells and expressed as DPM/cells in the dish. Each histogram is an average of results obtained from three independent experiments under identical conditions with ± S.D. *Significantly different from the untreated control cells at $p \leq 0.05$. (**b**) Simvastatin induces PLD activation in C2C12 cells. C2C12 myoblast cells (2×10^5 cells) were pre-labeled for 12 h with carrier free [^{32}P] orthophosphate in complete medium, following which cells were treated with simvastatin (10 μM) for different time periods (4–24 h) under a humidified sterile atmosphere of 95% air–5% CO_2 at 37 °C. Under identical conditions, appropriate controls were established without the statin treatment. At the end of the experiment for the designated time, lipids were extracted under acidic conditions with 2:1 chloroform-methanol (vol/vol) and separated by thin-layer chromatography (TLC) as described in the section "Materials and Methods". Phosphatidylbutanol (PBt) as the product of PLD activity generated by the transphosphatidylation reaction was identified by iodination on the TLC plate with authentic PBt standard as described in the section "Materials and Methods". The PBt spots on the TLC plates were scrapped and [^{32}P] radioactivity was determined on a liquid scintillation counter and normalized to the total lipid phosphorus [^{32}P] of the cells and expressed as DPM/cells in the dish. Each histogram is an average of results obtained from three independent experiments under identical conditions with ± S.D. *Significantly different from the untreated control cells at $p \leq 0.05$

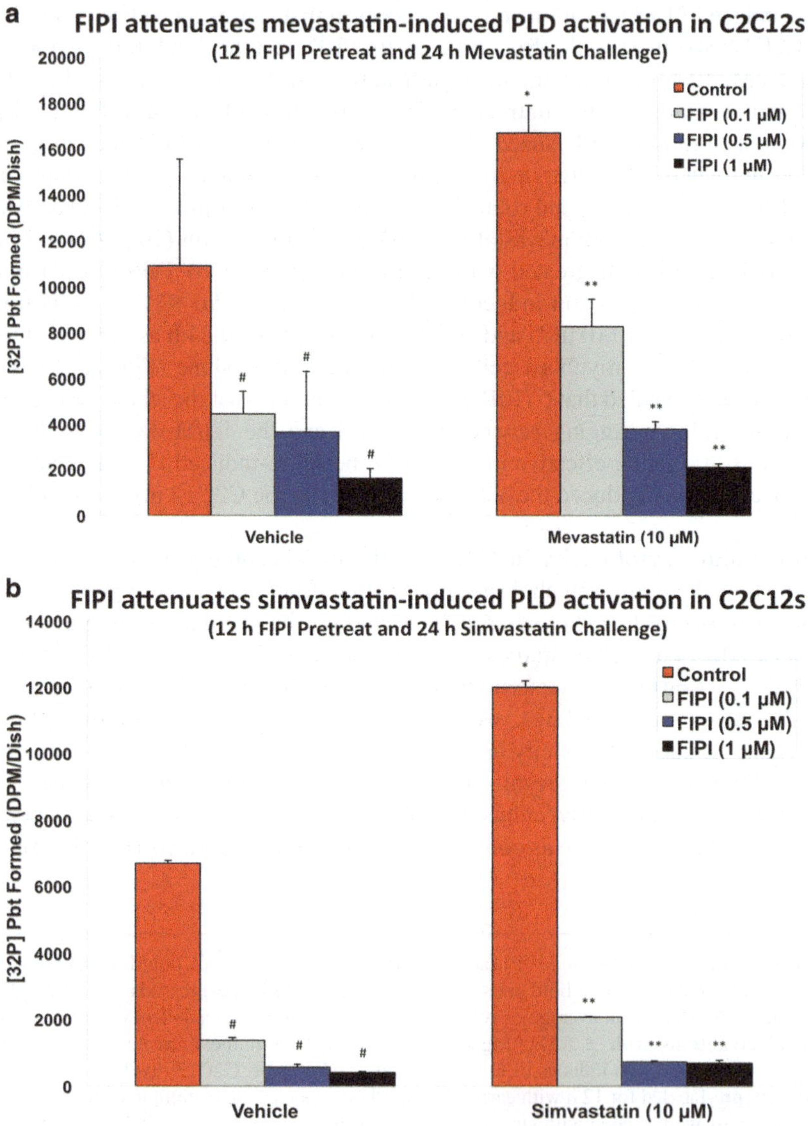

Fig. 2 (**a**) FIPI, the PLD-specific inhibitor, attenuates mevastatin-induced PLD activation in C2C12 cells. C2C12 myoblast cells (2×10^5 cells) were pre-labeled for 12 h with carrier free [^{32}P] orthophosphate in complete medium, following which cells were first treated with the PLD-specific pharmacological inhibitor, FIPI (0.1–1 μM) for 12 h and then treated with mevastatin (10 μM) for 24 h in absence and presence of FIPI under a humidified sterile atmosphere of 95% air–5% CO_2 at 37 °C. Under identical conditions, appropriate controls were established without FIPI and the statin treatment and with FIPI treatments alone. At the end of the experiment for the designated time, lipids were extracted under acidic conditions with 2:1 chloroform-methanol (vol/vol) and separated by thin-layer chromatography (TLC) as described in the section "Materials and Methods". Phosphatidylbutanol (PBt) as the product of PLD activity generated by the transphosphatidylation reaction was identified by iodination on the TLC plate with authentic PBt standard

statins at 10 μM dose caused robust and significant increase in the LDH release from cells (twofold increase by simvastatin; 3.7-fold increase by mevastatin) and further increase in release of the intracellular LDH was not markedly enhanced by increasing the dose of statins at 24 h of treatment of the C2C12 cells (Fig. 3a, b). Therefore, our current results revealed that statins (simvastatin and mevastatin) induced cytotoxicity in the C2C12 cells even at 10 μM dose as demonstrated by the release of intracellular LDH, the standard mammalian cytotoxicity assay. Furthermore, mevastatin was more potent than simvastatin at 10 μM dose in causing cytotoxicity in the C2C12 myoblast cells (Fig. 3a, b).

Statins decrease cholesterol in C2C12 cells As statins are established to lower or decrease the cellular levels of cholesterol [70], here we investigated whether statins would decrease/lower the cholesterol levels in the C2C12 myoblast cells. All the three tested statins (simvastatin, mevastatin, and lovastatin) at 10 μM dose caused significant decrease of intracellular cholesterol (25%, 21%, and 46%, respectively) as compared to that in the control untreated cells at 24 h of treatment (Fig. 4a). Among all the chosen statins, lovastatin was the most effective statin in decreasing the intracellular cholesterol levels in the C2C12 myoblast cells. Therefore, these results revealed that statins caused significant decrease of the intracellular levels of cholesterol in the C2C12 myoblast cells.

Cholesterol replenishment protects against statin-induced cytotoxicity in C2C12 cells Our earlier studies demonstrated that cholesterol replenishment offered protection against the cyclodextrin-induced cytotoxicity mediated through

Fig. 2 (continued) as described in the section "Materials and Methods". The PBt spots on the TLC plates were scrapped and [^{32}P] radioactivity was determined on a liquid scintillation counter and normalized to the total lipid phosphorus [^{32}P] of the cells and expressed as DPM/cells in the dish. Each histogram is an average of results obtained from three independent experiments under identical conditions with ± S.D. *Significantly different from the untreated control cells at $p \leq 0.05$. **Significantly different from the statin-treated cells at $p \leq 0.05$. (**b**) FIPI, the PLD-specific inhibitor, attenuates simvastatin-induced PLD activation in C2C12 cells. C2C12 myoblast cells (2×10^5 cells) were pre-labeled for 12 h with carrier free [^{32}P] orthophosphate in complete medium, following which cells were first treated with the PLD-specific pharmacological inhibitor, FIPI (0.1–1 μM) for 12 h and then treated with simvastatin (10 μM) for 24 h in absence and presence of FIPI under a humidified sterile atmosphere of 95% air–5% CO_2 at 37 °C. Under identical conditions, appropriate controls were established without FIPI and the statin treatment and with FIPI treatments alone. At the end of the experiment for the designated time, lipids were extracted under acidic conditions with 2:1 chloroform-methanol (vol/vol) and separated by thin-layer chromatography (TLC) as described in the section "Materials and Methods". Phosphatidylbutanol (PBt) as the product of PLD activity generated by the transphosphatidylation reaction was identified by iodination on the TLC plate with authentic PBt standard as described in the section "Materials and Methods". The PBt spots on the TLC plates were scrapped and [^{32}P] radioactivity was determined on a liquid scintillation counter and normalized to the total lipid phosphorus [^{32}P] of the cells and expressed as DPM/cells in the dish. Each histogram is an average of results obtained from three independent experiments under identical conditions with ± S.D. *Significantly different from the untreated control cells at $p \leq 0.05$. **Significantly different from the statin-treated cells at $p \leq 0.05$

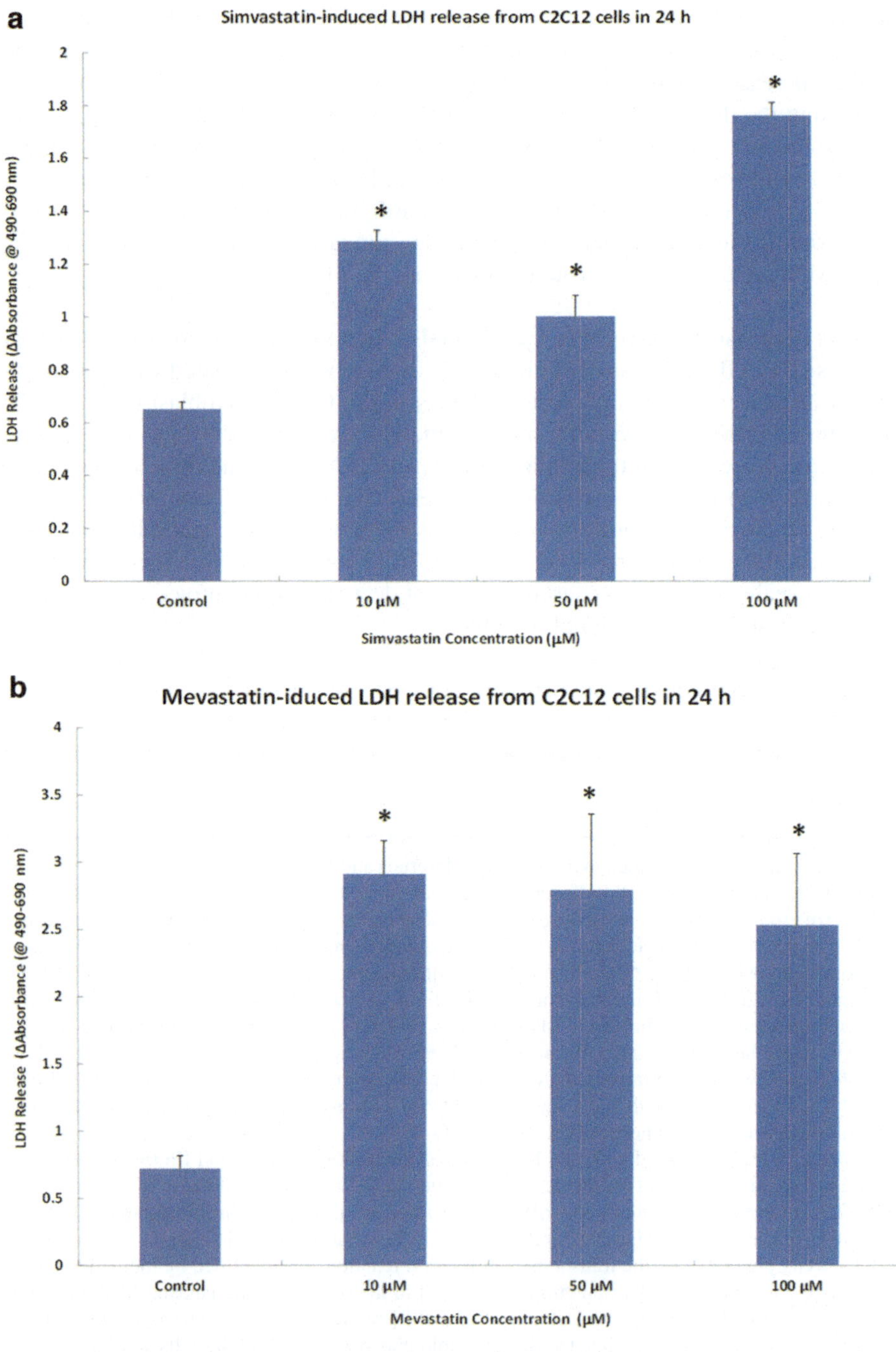

Fig. 3 (**a**) Simvastatin induces lactate dehydrogenase (LDH) release from C2C12 cells. C2C12 myoblast cells were cultured up to ~90–100% confluence in 15.5-mm sterile dishes (24-well cell culture plate) in complete medium and then treated with complete medium or medium containing

the intracellular cholesterol depletion in the vascular endothelial cells [64]. In the current study, we showed that statins cause both cytotoxicity (LDH release) and cholesterol depletion. Therefore, we investigated whether cholesterol replenishment would offer protection against the statin-induced cytotoxicity (LDH release) in the C2C12 myoblast cells. Our results revealed that both mevastatin (10 μM) and simvastatin (10 μM) caused significant cytotoxicity as revealed by the intracellular LDH release assay at 24 h of treatment with statins as compared to the control untreated cells (Fig. 4b, c). Furthermore, treatment of cells with the water-soluble (methylcyclodextrin-conjugated) cholesterol (1%) offered significant protection of the statin-induced cytotoxicity (attenuation of release of intracellular LDH) (Fig. 4b, c). However, cholesterol at 2% dose was not effective in lowering the mevastatin-induced release of intracellular LDH and failed to protect against the mevastatin-induced cytotoxicity in the C2C12 cells as opposed to the protective action offered by cholesterol at 1% dose (Fig. 4b). On the other hand, cholesterol at 2% dose, although effective in significantly protecting against the simvastatin-induced cytotoxicity in the C2C12 cells, its protective action was less effective than that was offered by cholesterol at 1% dose (Fig. 4c). Overall, these results revealed that (*i*) cholesterol replenishment offered protection against the statin-induced cytotoxicity in the C2C12 cells as revealed by the intracellular LDH assay and (*ii*) lower dose of cholesterol (1%) was more effective in significantly protecting against the statin-induced cytotoxicity.

FIPI, the PLD-specific inhibitor attenuates statin-induced cytotoxicity in C2C12 cells We have earlier reported that the PLD-specific pharmacological inhibitor, FIPI protected against the oxidant- and drug-induced cytotoxicity mediated by the PLD-depended bioactive lipid signaling in the vascular endothelial cells [59, 60]. Therefore, here we investigated to show whether FIPI would offer protec-

Fig. 3 (continued) different concentrations of simvastatin (10–100 μM) for 24 h under a humidified sterile atmosphere of 95% air–5% CO_2 at 37 °C. Appropriate controls without the statin treatments were established simultaneously under identical conditions. At the end of the treatment, the supernatant was removed, and the extent of release of LDH was determined on a plate reader (visible) by the commercially available LDH spectrophotometric assay kit (Sigma Chemical Co., St. Louis, MO) according to the manufacturer's recommendations as described in the section "Materials and Methods". Each histogram is an average of results obtained from three independent experiments under identical conditions with ± S.D. *Significantly different from the untreated control cells at $p \leq 0.05$. (**b**) Mevastatin induces lactate dehydrogenase (LDH) release from C2C12 cells. C2C12 myoblast cells were cultured up to ~90–100% confluence in 15.5-mm sterile dishes (24-well cell culture plate) in complete medium and then treated with complete medium or medium containing different concentrations of mevastatin (10–100 μM) for 24 h under a humidified sterile atmosphere of 95% air–5% CO_2 at 37 °C. Appropriate controls without the statin treatments were established simultaneously under identical conditions. At the end of the treatment, the supernatant was removed, and the extent of release of LDH was determined on a plate reader (visible) by the commercially available LDH spectrophotometric assay kit (Sigma Chemical Co., St. Louis, MO) according to the manufacturer's recommendations as described in the section "Materials and Methods". Each histogram is an average of results obtained from three independent experiments under identical conditions with ± S.D. *Significantly different from the untreated control cells at $p \leq 0.05$

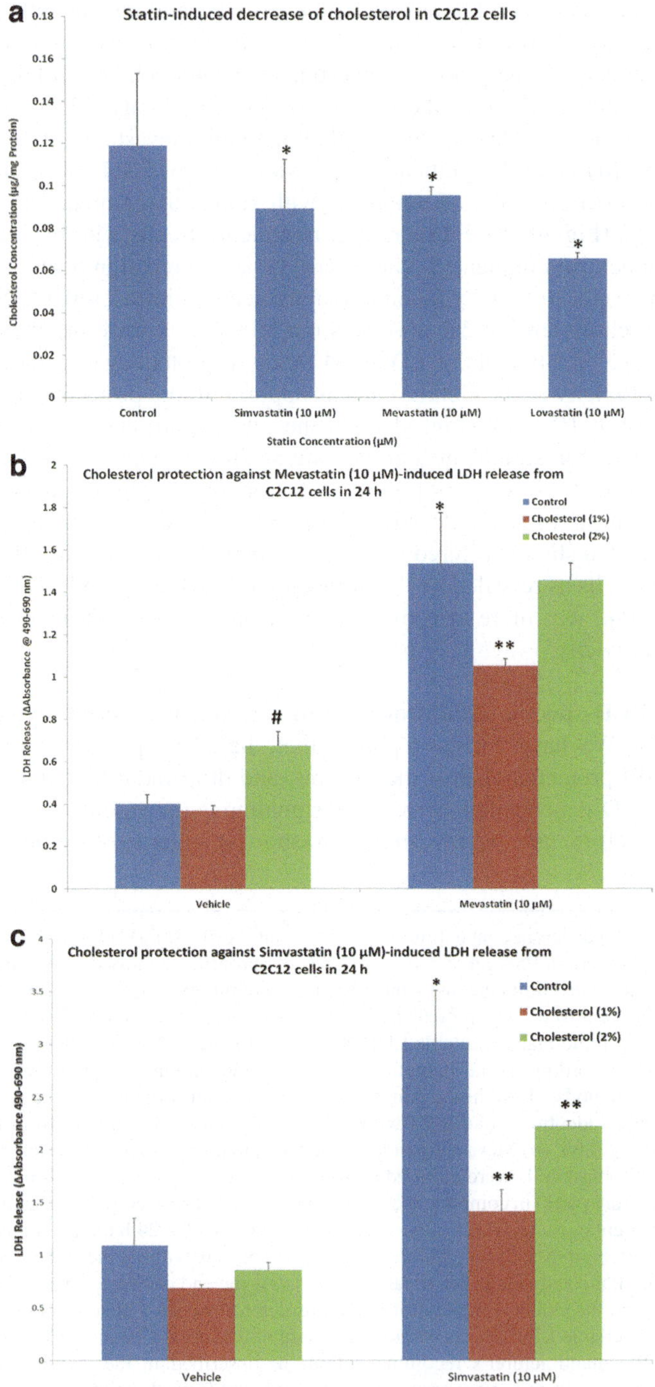
a
Statin-induced decrease of cholesterol in C2C12 cells
Cholesterol Concentration (µg/mg Protein)
0.18
0.16
0.14
0.12
0.1
0.08
0.06
0.04
0.02
0
*
*
*
Control
Simvastatin (10 µM)
Mevastatin (10 µM)
Lovastatin (10 µM)
Statin Concentration (µM)
b
Cholesterol protection against Mevastatin (10 µM)-induced LDH release from C2C12 cells in 24 h
LDH Release (ΔAbsorbance @ 490-690 nm)
2
1.8
1.6
1.4
1.2
1
0.8
0.6
0.4
0.2
0
Control
Cholesterol (1%)
Cholesterol (2%)
*
**
#
Vehicle
Mevastatin (10 µM)
c
Cholesterol protection against Simvastatin (10 µM)-induced LDH release from C2C12 cells in 24 h
LDH Release (ΔAbsorbance 490-690 nm)
4
3.5
3
2.5
2
1.5
1
0.5
0
Control
Cholesterol (1%)
Cholesterol (2%)
*
**
**
Vehicle
Simvastatin (10 µM)

tion against the statin-induced cytotoxicity (intracellular LDH release) in the C2C12 myoblast cells to demonstrate the role of PLD therein, since in the current study, it was revealed that statins caused significant activation of PLD which was significantly attenuated by FIPI in the C2C12 myoblast cells (Figs. 1 and 2). FIPI (1 μM, pretreatment for 12 h) offered significant protection against simvastatin (10 μM)- and mevastatin (10 μM)-induced cytotoxicity at 24 h as demonstrated by the release of intracellular LDH (Fig. 5). Overall, the current study demonstrated that (*i*) the PLD-specific pharmacological inhibitor, FIPI offered significant protection of the statin-induced cytotoxicity and (*ii*) the PLD-mediated bioactive lipid signaling also was involved in the statin-induced cytotoxicity in the C2C12 myoblast cells.

Fig. 4 (**a**) Statins induce decrease of cholesterol in C2C12 cells. C2C12 myoblast cells were cultured in complete medium up to ~90–100% confluence in 60-mm sterile culture dishes and treated with different statins at 10 μM concentration for 24 h under a humidified sterile atmosphere of 95% air–5% CO_2 at 37 °C. Appropriate controls were established simultaneously under identical condition without statin treatments. At the end of the treatment, the cholesterol content in the cells was then determined spectrofluorometrically on a plate reader (fluorescence) with the commercially available cholesterol determination kit according to the manufacturer's recommendation (Molecular Probes – Invitrogen Detection Technologies, Grand Island, NY) as described in the section "Materials and Methods". The cellular levels of cholesterol were normalized to protein and expressed as μg/mg protein. Each histogram is an average of results obtained from three independent experiments under identical conditions with ± S.D. *Significantly different from the untreated control cells at $p \leq 0.05$. (**b**) Cholesterol replenishment protects against mevastatin-induced lactate dehydrogenase (LDH) release from C2C12 Cells. C2C12 myoblast cells were cultured up to ~90–100% confluence in 15.5-mm sterile dishes (24-well cell culture plate) in complete medium and then treated with complete medium or medium containing different concentrations of water-soluble cholesterol alone (1–2%) or mevastatin (10 μM) alone or water-soluble cholesterol (1–2%) + mevastatin (10 μM) for 24 h under a humidified sterile atmosphere of 95% air–5% CO_2 at 37 °C. Appropriate controls without the water-soluble cholesterol and statin treatments were established simultaneously under identical conditions. At the end of the treatment, the supernatant was removed, and the extent of release of LDH was determined on a plate reader (visible) by the commercially available LDH spectrophotometric assay kit (Sigma Chemical Co., St. Louis, MO) according to the manufacturer's recommendations as described in the section "Materials and Methods". Each histogram is an average of results obtained from three independent experiments under identical conditions with ± S.D. *Significantly different from the untreated control cells at $p \leq 0.05$. **Significantly different from the statin-treated cells at $p \leq 0.05$. #Significantly different from the untreated control cells at $p \leq 0.05$. (**c**) Cholesterol replenishment protects against simvastatin-induced lactate dehydrogenase (LDH) release from C2C12 Cells. C2C12 myoblast cells were cultured up to ~90–100% confluence in 15.5-mm sterile dishes (24-well cell culture plate) in complete medium and then treated with complete medium or medium containing different concentrations of water-soluble cholesterol alone (1–2%) or simvastatin (10 μM) alone or water-soluble cholesterol (1–2%) + simvastatin (10 μM) for 24 h under a humidified sterile atmosphere of 95% air–5% CO_2 at 37 °C. Appropriate controls without the water-soluble cholesterol and statin treatments were established simultaneously under identical conditions. At the end of the treatment, the supernatant was removed, and the extent of release of LDH was determined on a plate reader (visible) by the commercially available LDH spectrophotometric assay kit (Sigma Chemical Co., St. Louis, MO) according to the manufacturer's recommendations as described in the section "Materials and Methods". Each histogram is an average of results obtained from three independent experiments under identical conditions with ± S.D. *Significantly different from the untreated control cells at $p \leq 0.05$. **Significantly different from the statin-treated cells at $p \leq 0.05$

Statins induce mitochondrial dysfunction in C2C12 cells It is rapidly emerging that statins cause mitochondrial dysfunction and damage in different organs/cells, including the skeletal muscle, especially during the statin-induced myopathy [34, 72]. Although MTT reduction by the mitochondria is used as a valid indicator of cell proliferation and cytotoxicity, the assay is also utilized to ascertain the mitochondrial function [73]. Therefore, we utilized MTT reduction by the cellular mitochondria in the C2C12 cells to demonstrate the statin-induced mitochondrial dysfunction, since MTT is reduced by a mitochondria-specific dehydrogenase [73]. Both statins, simvastatin and mevastatin, caused significant dose-dependent decrease of MTT reduction in the C2C12 cells at 24 h of treatment as compared to the control untreated cells (Fig. 6a, b). Simvastatin (10 μM) and mevastatin (10 μM) significantly decreased MTT reduction, by 51% and 60%, respectively, by the C2C12 myoblast cells treated for 24 h as compared to the same in the control untreated cells (Fig. 6a, b). Therefore, these studies demonstrated that both mevastatin and simvastatin

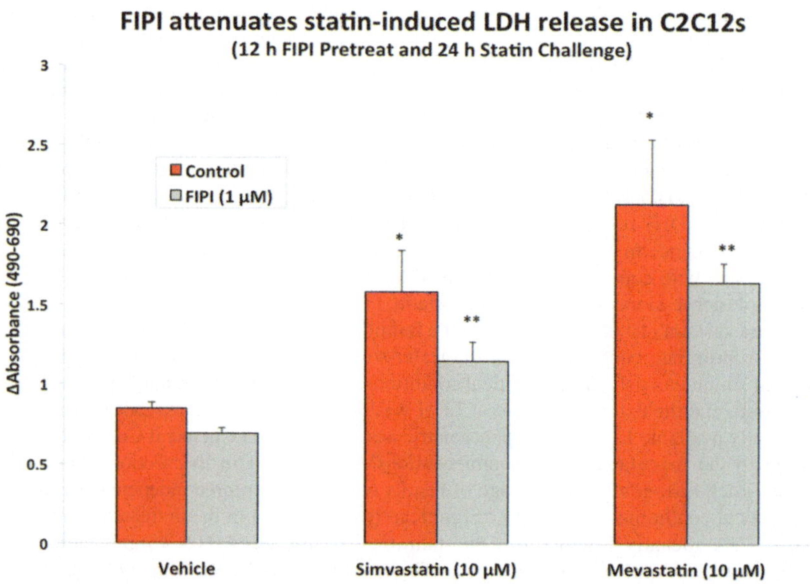

Fig. 5 FIPI, the PLD-specific inhibitor, attenuates statin-induced lactate dehydrogenase (LDH) release from C2C12 cells. C2C12 myoblast cells were cultured up to ~90–100% confluence in 15.5-mm sterile dishes (24-well cell culture plate) in complete medium and then pretreated for 12 h with FIPI, and then treated with simvastatin (10 μM) and mevastatin (10 μM) for 24 h under a humidified sterile atmosphere of 95% air–5% CO_2 at 37 °C. Appropriate controls without and with FIPI and statins alone were established simultaneously under identical conditions. At the end of the treatment, the supernatant was removed, and the extent of release of LDH was determined on a plate reader (visible) by the commercially available LDH spectrophotometric assay kit (Sigma Chemical Co., St. Louis, MO) according to the manufacturer's recommendations as described in the section "Materials and Methods". Each histogram is an average of results obtained from three independent experiments under identical conditions with ± S.D. *Significantly different from the untreated control cells at $p \leq 0.05$. **Significantly different from the statin-treated cells at $p \leq 0.05$

induced the mitochondrial dysfunction as revealed by the decrease in the MTT reduction by the C2C12 cells.

FIPI, the PLD-specific inhibitor attenuates statin-induced mitochondrial dysfunction by C2C12 cells In the earlier experiments of the current study, we had shown that (*i*) statins induced PLD activation; (*ii*) FIPI, the PLD-specific pharmacological inhibitor, attenuated the statin-induced PLD activation; and (*iii*) FIPI offered protection against the statin-induced cytotoxicity (intracellular LDH release) in the C2C12 myoblasts. Hence, here we investigated whether FIPI would attenuate the statin-induced mitochondrial dysfunction as assayed by the decrease of MTT reduction by the C2C12 cells to establish a link between the statin-induced PLD activation and bioactive lipid signal mediator (PA) generation and mitochondrial dysfunction. FIPI (1 μM, 12 h of pretreatment) significantly and markedly attenuated the simvastatin (10 μM)- and mevastatin (10 μM)-induced decrease of MTT reduction by the C2C12 cells (Fig. 7a). Furthermore, the protective action of FIPI against the simvastatin-induced decrease of MTT reduction by the C2C12 cells was more pronounced than that induced by mevastatin (Fig. 7a). Overall, these studies demonstrated that FIPI significantly attenuated the statin-induced mitochondrial dysfunction in the C2C12 myoblast cells as demonstrated by the mitochondria-specific MTT reduction assay, suggesting the role of PLD-generated bioactive lipid signal mediators therein.

FIPI, the PLD-specific inhibitor attenuates statin-induced morphological alterations in C2C12 Cells Our results so far revealed that statins (mevastatin and simvastatin) induced cytotoxicity and mitochondrial dysfunction that was mediated by the PLD-generated bioactive lipid signaling and cholesterol depletion in the C2C12 myoblast cells. Cell morphology alterations serve as an index of cytotoxicity induced by toxic stresses [73, 74]. Here, we sought to test whether statins would induce morphological alterations in the C2C12 myoblast cells, which might be protected by the PLD-specific pharmacological inhibitor, FIPI. Simvastatin, mevastatin, and lovastatin at 10 μM dose caused severe cell morphological alteration at 24 h of treatment characterized by the light microscopic examinations, including elongated myoblast cells turning into round and circular cells (Fig. 7b). This indicated the drastic changes in the morphological nature of the cells upon the statin treatment. However, FIPI pre-treatment (1 μM for 12 h) offered almost complete protection of the statin-induced morphological alterations in the C2C12 myoblast cells (Fig. 7b). These results further confirmed that the statin-induced cell morphological alterations (cytotoxicity) was (*i*) protected by the PLD-specific pharmacological inhibitor, FIPI and (*ii*) PLD-generated bioactive lipid signaling through PA formation played a role in the statin-induced cell morphology alterations in the C2C12 cells.

Statins induce threonine phosphorylation of PLD1 in C2C12 cells Our earlier reports revealed that oxidant stress and heavy metal toxicity cause the protein kinase-mediated serine-threonine phosphorylation of PLD isoenzymes upstream of

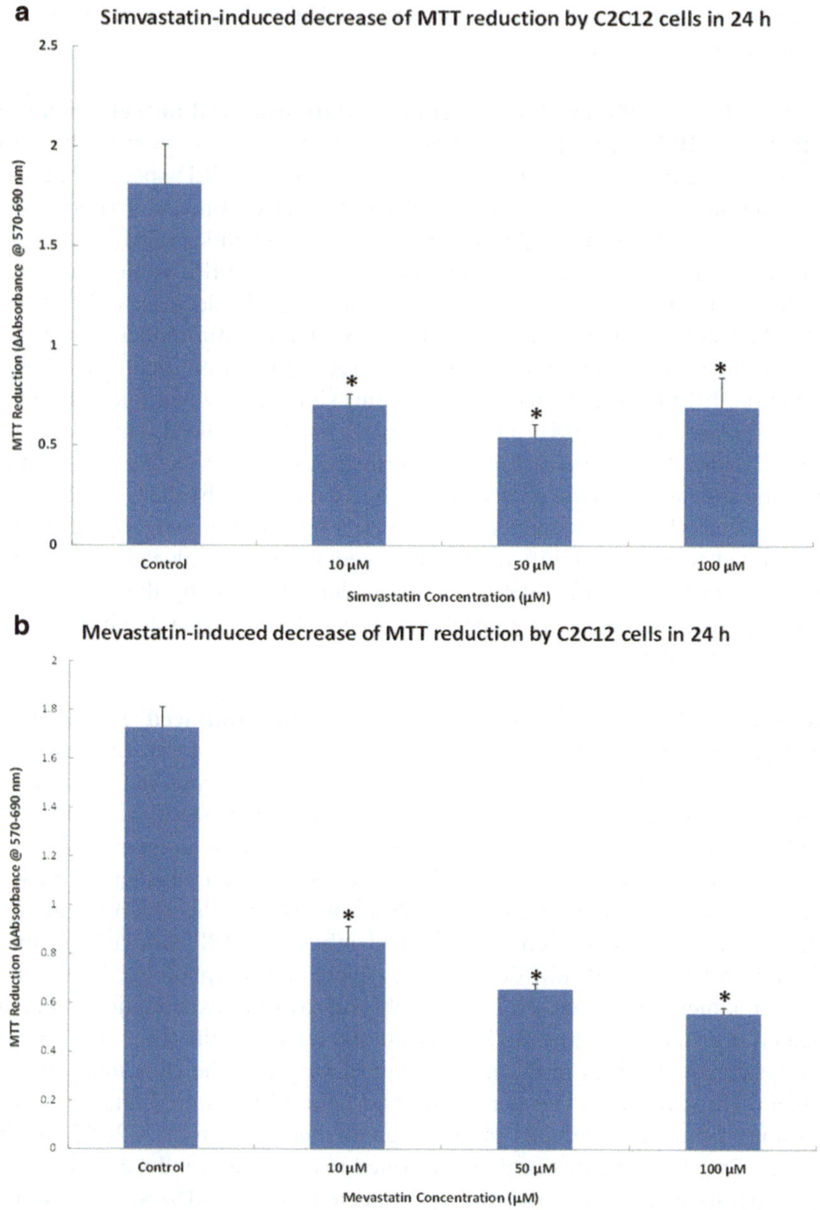

Fig. 6 (**a**) Simvastatin induces decrease of MTT reduction by C2C12 cells. C2C12 myoblast cells were cultured up to ~90–100% confluence in 15.5-mm sterile dishes (24-well cell culture plate) in complete medium and then treated with complete medium or medium containing different concentrations of simvastatin (10–100 μM) for 24 h under a humidified sterile atmosphere of 95% air–5% CO_2 at 37 °C. Appropriate controls without the statin treatments were established simultaneously under identical conditions. At the end of the treatment, the supernatant was removed, and the extent of MTT reduction by the cells was determined on a plate reader (visible) by the commercially

PLD activation [60, 67]. Since the protein kinase-mediated serine-threonine phosphorylation regulates the activity of PLD isoenzymes, in this study, we investigated whether statins would induce threonine phosphorylation of PLD1 isoenzyme in the C2C12 myoblast cells as examined by the confocal immunofluorescence microscopy with the aid of PLD-phosphothreonine-specific antibody immunostaining. Both mevastatin and simvastatin at 10 μM dose induced intense phosphorylation of PLD1 isoenzyme in intact C2C12 myoblast cells at 12 h of treatment of cells with the statins (upstream of maximal PLD activation at 24 h) as compared to the same in the control untreated cells (Fig. 8a, b). This study demonstrated that statins induce serine phosphorylation of PLD1 isoenzyme upstream of the activation of the enzyme.

Discussion

High cholesterol levels (hypercholesterolemia) in circulation have been considered as a risk factor for human vascular disorders, cardiovascular diseases (CVDs), cerebrovascular diseases, and chronic heart diseases (CHD). Therefore, aggressive therapy with cholesterol-lowering drugs to lower the endogenously synthesized cholesterol is common clinical practice [28–33]. Among the most successful and widely prescribed cholesterol-lowering drugs are the statins that target the HMG-CoA reductase as a rate-limiting enzyme in the cholesterol biosynthetic pathway [28–30] (Schema 1). However, it is becoming increasingly evident that statins are associated with adverse effects such as statin myalgia or statin myotoxicity [41, 42]. Mitochondrial dysfunction, apoptosis, and coenzyme-Q10 (CoQ10) depletion have been identified as potential cellular mediators of statin-induced dysfunction, but the precise mechanism remains unknown [34, 43, 44, 70]. In the current study, we hypothesized that statins would cause cholesterol depletion in the membranes of the skeletal muscle cells through the inhibition of HMG-CoA reductase (Schema 1), leading to the activation of PLD that could generate the bioactive lipid signal

Fig. 6 (continued) available MTT reduction spectrophotometric assay kit (Cayman Chemical Co., Ann Arbor, MI) according to the manufacturer's recommendations as described in the section "Materials and Methods". Each histogram is an average of results obtained from three independent experiments under identical conditions with ± S.D. *Significantly different from the untreated control cells at $p \leq 0.05$. (**b**) Mevastatin induces decrease of MTT reduction by C2C12 cells. C2C12 myoblast cells were cultured up to ~90–100% confluence in 15.5-mm sterile dishes (24-well cell culture plate) in complete medium and then treated with complete medium or medium containing different concentrations of mevastatin (10–100 μM) for 24 h under a humidified sterile atmosphere of 95% air–5% CO_2 at 37 °C. Appropriate controls without the statin treatments were established simultaneously under identical conditions. At the end of the treatment, the supernatant was removed, and the extent of MTT reduction by the cells was determined on a plate reader (visible) by the commercially available MTT reduction spectrophotometric assay kit (Cayman Chemical Co., Ann Arbor, MI) according to the manufacturer's recommendations as described in the section "Materials and Methods". Each histogram is an average of results obtained from three independent experiments under identical conditions with ± S.D. *Significantly different from the untreated control cells at $p \leq 0.05$

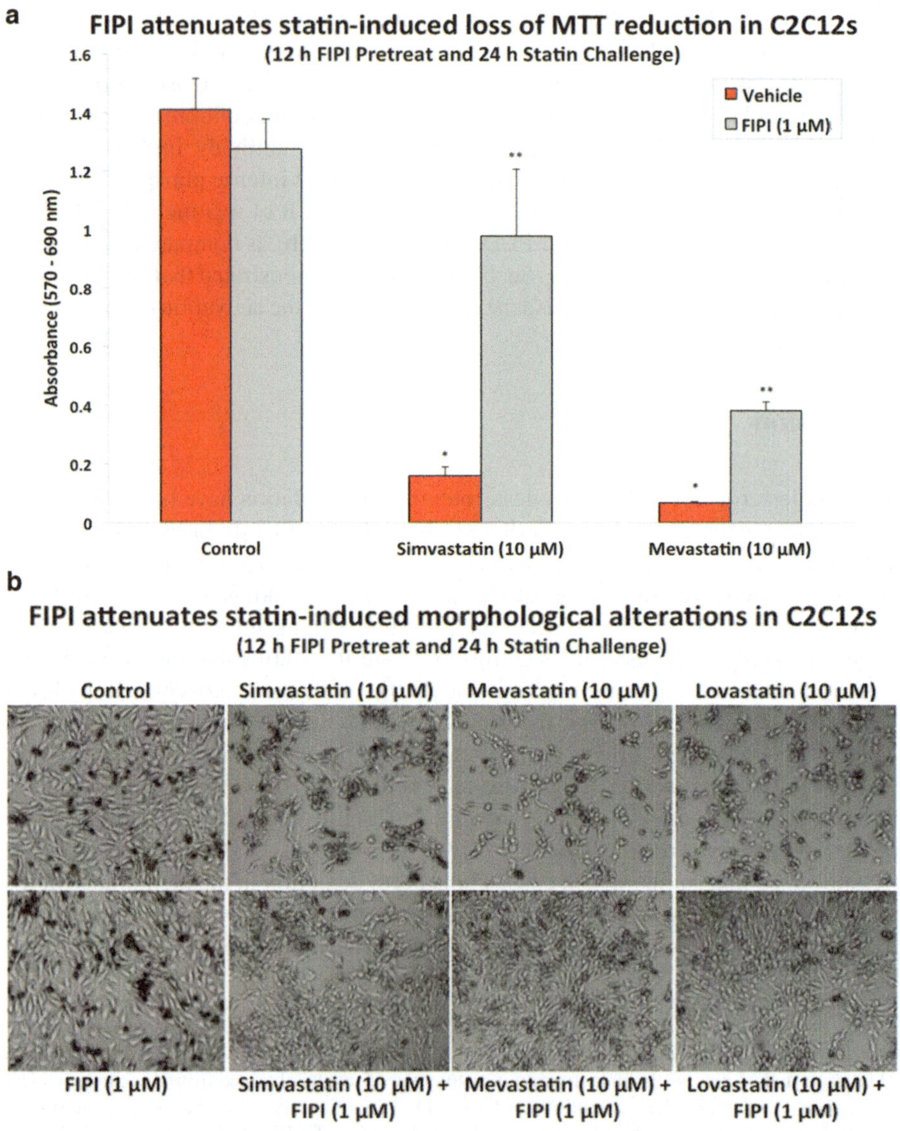

Fig. 7 (**a**) FIPI, the PLD-specific inhibitor, attenuates statin-induced decrease of MTT reduction by C2C12 cells. C2C12 myoblast cells were cultured up to ~90–100% confluence in 15.5-mm sterile dishes (24-well cell culture plate) in complete medium, pre-treated with the PLD-specific pharmacological inhibitor (FIPI, 1 μM) for 12 h, and then treated with complete medium alone or medium containing FIPI alone or medium containing simvastatin (10 μM) and mevastatin (10 μM) for 24 h under a humidified sterile atmosphere of 95% air–5% CO_2 at 37 °C. Appropriate controls without FIPI treatment and the statin treatments were established simultaneously under identical conditions. At the end of the treatment, the supernatant was removed, and the extent of MTT reduction by the cells was determined on a plate reader (visible) by the commercially available MTT reduction spectrophotometric assay kit (Cayman Chemical Co., Ann Arbor, MI) according to the

mediator (PA) (Schema 3), which in turn induces mitochondrial damage and myotoxicity as a mechanistic basis of the statin-induced myalgia or myotoxicity. To test our hypothesis, we utilized the well-established skeletal muscle cell model, the C2C12 myoblast cells. The current study demonstrated that the two widely used statins, mevastatin and simvastatin, induced PLD activation and formation of the bioactive lipid signal mediator (PA) causing mitochondrial dysfunction and cytotoxicity in a skeletal muscle cell model (C2C12 myoblast cells) through the decrease of endogenous cholesterol (Schema 4).

By virtue of their ability to selectively inhibit the rate-limiting enzyme in the cholesterol biosynthetic pathway in the mammalian cells, statins drastically decrease or deplete the endogenous levels of cholesterol. On the other hand, cholesterol is an important lipidic component of the cell membranes that regulates the membrane structure and function [21, 22]. Our earlier studies demonstrated that the widely used experimental cholesterol-depleting agent, methyl-β-cyclodextrin (MβCD) causes cytotoxicity to vascular endothelial cells through cellular cholesterol depletion, which can be reversed with cholesterol replenishment, indicating that cellular cholesterol plays a crucial role in maintaining the viability of the mammalian cells [62, 64]. We propose that statin-induced cytotoxicity in C2C12 myoblast cells observed in the current study was caused by a similar depletion of cellular cholesterol in the skeletal muscle cell model. This was also confirmed by replenishing the cells with the water-soluble cholesterol which rescued the cells from statin-induced cytotoxicity. Together, these data suggest that statin-mediated cellular cholesterol depletion promotes statin-induced myotoxicity and myalgia.

Phospholipase D (PLD) is ubiquitously present in mammalian cells and belongs to the family of phospholipases that drive membrane phospholipid hydrolysis (Schema 3) [56–61]. Although PLD is a house-keeping enzyme involved in the metabolism of membrane phospholipids, the enzyme is known to be activated by a variety of agonists, including hormones, reactive oxygen species, heavy metals, toxins, and metabolic stresses [56–61]. PLD exists in cells as two isoforms, PLD1 and PLD2 [56–61]. Protein kinases such as mitogen-activated protein kinases (MAPKs), protein tyrosine kinases, protein kinase C, G-protein coupled receptors, and

Fig. 7 (continued) manufacturer's recommendations as described in the section "Materials and Methods". Each histogram is an average of results obtained from three independent experiments under identical conditions with ± S.D. *Significantly different from the untreated control cells at $p \leq 0.05$. **Significantly different from the statin-treated cells at $p \leq 0.05$. (**b**) FIPI, the PLD-specific inhibitor, attenuates statin-induced morphological alterations in C2C12 cells. C2C12 myoblast cells were cultured in sterile 35-mm dishes up to ~90–100% confluence in complete medium under a humidified atmosphere of 95% air–5% CO_2 at 37 °C, and then pre-treated with the PLD-specific pharmacological inhibitor, FIPI (1 μM) alone for 12 h, following which the cells were treated with medium alone or medium containing simvastatin (10 μM), mevastatin (10 μM), and lovastatin (10 μM), or FIPI (1 μM) + statin (10 μM) for 24 h. Appropriate controls were established with cells treated with medium alone or FIPI alone under identical conditions. At the end of the experiment, the images of cellular morphology were digitally captured using the Zeiss Axioskop 200 with Zen 2011 software at 20× magnification. Each photomicrograph is a typical representative of at least three independent observations from three different experiments conducted under identical conditions

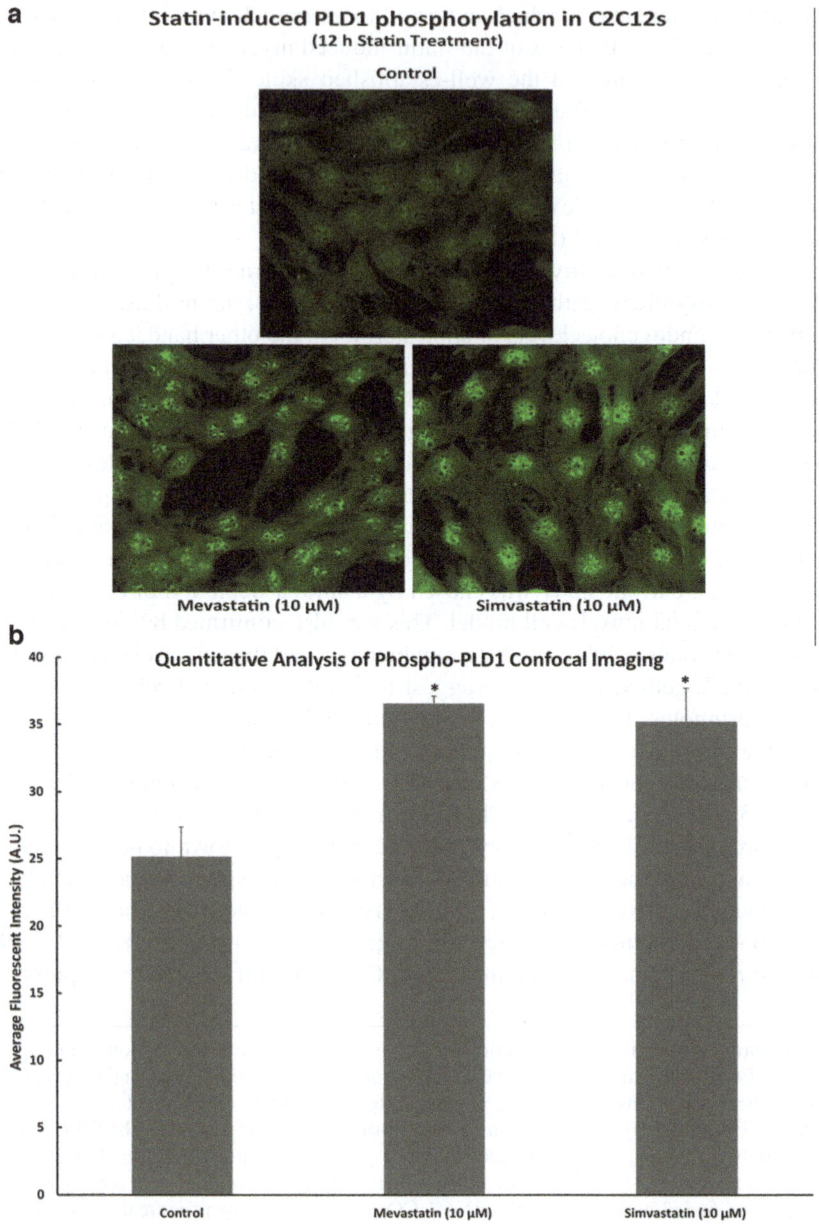

Fig. 8 (**a**) Statins induce threonine phosphorylation of PLD1 in C2C12 cells. C2C12s were cultured on sterile glass cover slips (90% confluence) in complete medium under a humidified atmosphere of 95% air–5% CO_2 at 37 °C and then treated with complete medium alone or medium containing chosen concentrations of statins for 12 h. As described in the section "Materials and Methods", the cells on the cover slips were treated overnight at room temperature with the primary antibody [phosphothreonine-PLD1 (1:150 dilution)] in 1% BSA solution. Following that, the cells were labeled with the secondary AlexaFluor 488 (1:100 dilution) for 1 h and then the cells were

receptor-mediated kinases phosphorylate the PLD isoforms at the corresponding amino acid residues (serine or threonine or tyrosine), leading to their translocation and activation [56–61]. Upon activation, PLD hydrolyzes cell membrane phospholipids (e.g., phosphatidylcholine) to form PA, a highly potent bioactive signal lipid linked to cellular cytoskeletal alterations and cytotoxicity [59, 60]. Our earlier studies demonstrated that cholesterol depletion mediated by MβCD in the vascular endothelial cells activates PLD, causing cytotoxicity through the generation of the lipid signal mediator, PA [62]. Along these lines, in the current study, it was demonstrated for the first time that statins induced the activation of PLD in the C2C12 cells which was attenuated by the PLD-specific inhibitor, FIPI. Furthermore, the current study also showed that the PLD-specific inhibitor, FIPI effectively protected against the statin-induced cytotoxicity. The results of the current study also revealed that statins induced the upstream threonine phosphorylation of PLD1 isoenzyme in the C2C12 cells that could have been mediated by the MAPKs or PKC, leading to the translocation and activation of the enzyme as previously observed in other cellular models. Thus, a connection between statin-induced PLD activation and statin cytotoxicity in the C2C12 muscle cell model offers a PLD-dependent mechanism of statin myotoxicity and myalgia, suggesting a role of the PLD-generated bioactive lipids such as PA, LPA, and DAG therein.

In the current study, it was shown that statins caused the mitochondrial dysfunction as determined by the MTT reduction ability in the C2C12 cells [73]. The statin-induced mitochondrial dysfunction (decrease in MTT reduction catalyzed by the mitochondrial dehydrogenase) and cytotoxic cell morphology alterations were attenuated by the PLD-specific inhibitor. This suggested a reasonable connection or association with the PLD activation and subsequent generation of the bioactive signal lipid mediators (PA, LPA, and DAG) and the mitochondrial dysfunction and cytotoxicity in the C2C12 myoblast cells. However, other mitochondria-driven mechanisms of the statin-induced myotoxicity such as apoptosis, CoQ10 loss, decline of ATP production, and reactive oxygen species generation are not ruled out in mediating the statin-induced cytotoxicity in the C2C12 cells.

Fig. 8 (continued) examined under the Zeiss LSM 710 Confocal/Multiphoton Microscope powered by Argon-2 laser with 500–550 BP filter. The images were captured digitally. Each confocal fluorescence micrograph is a typical representative of three independent observations from three different experiments conducted under identical conditions. The bright green fluorescence in situ depicts the threonine phosphorylation of PLD1 isoform. (**b**) Quantitative Analysis of Phospho-PLD1 by Confocal Fluorescence Microscopy. C2C12s were cultured on sterile glass cover slips (90% confluence) in complete medium under a humidified atmosphere of 95% air–5% CO_2 at 37 °C and then treated with complete medium alone or medium containing chosen concentrations of statins for 12 h. As described in the section "Materials and Methods", the cells on the cover slips were treated overnight at room temperature with the primary antibody [phosphothreonine-PLD1 (1:150 dilution)] in 1% BSA solution. Following that, the cells were labeled with the secondary AlexaFluor 488 (1:100 dilution) for 1 h and then the cells were examined under the Zeiss LSM 710 Confocal/Multiphoton Microscope powered by Argon-2 laser with 500–550 BP filter. The images were captured digitally. Each histogram is an average intensity of fluorescence of three independent confocal fluorescence micrographs of untreated control cells and treatments in triplicates with ± standard deviations. *Significantly different from the untreated control cells at $p \leq 0.05$. **Significantly different from the statin-treated cells at $p \leq 0.05$

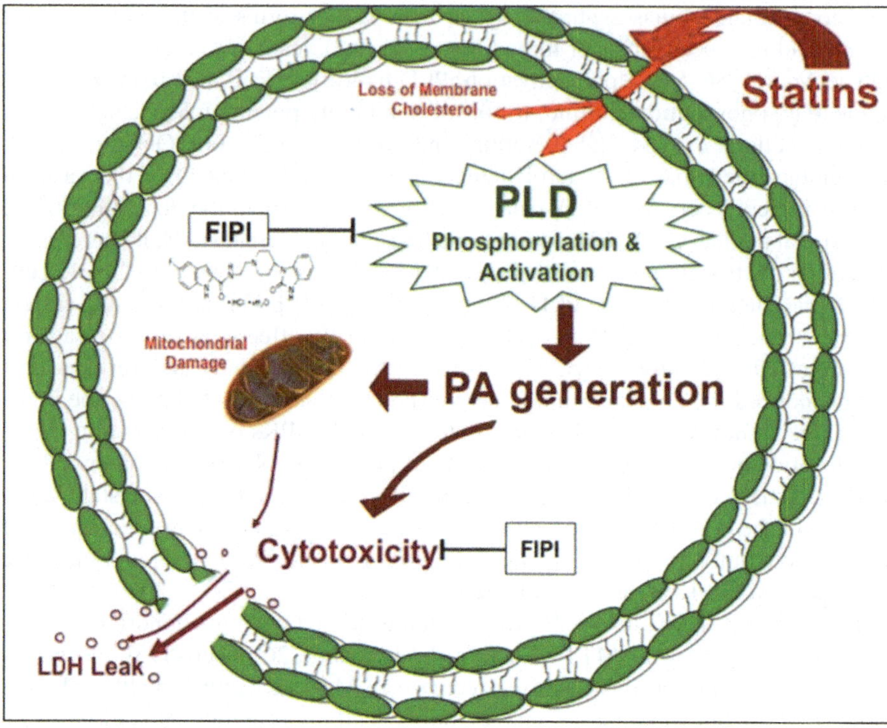

Schema 4 Proposed mechanism of statin-induced PLD activation leading to cytotoxicity through depletion of endogenous cholesterol in C2C12 myoblast cells. In addition to loss of cellular cholesterol, the PLD-generated lipid signal mediator, phosphatidic acid (PA) apparently is responsible for causing the statin-induced myotoxicity in the cells. Also, mitochondrial dysfunction, as observed by the loss of MTT reduction appears to be a critical player in statin-induced cytotoxicity in the C2C12 myoblast cells. Statin-induced PLD activation appears to be mediated by the upstream threonine phosphorylation of PLDs that may be regulated by the statin-mediated loss of membrane cholesterol and subsequent activation of the serine-threonine protein kinases. Both cholesterol replenishment and the pharmacological inhibition of PLD by the PLD-specific inhibitor, FIPI, offer protection against the statin-induced cytotoxicity in C2C12 myoblast cells. Overall, the results of the present study suggest that the statin-induced myotoxicity is mediated by the upstream cholesterol loss and the associated PLD activation leading to generation of the bioactive lipid signal mediator (PA) and also the mitochondrial dysfunction

Overall, the current study demonstrated that the cholesterol-lowering HMG-CoA inhibitor drugs, statins, caused the cellular cholesterol depletion leading to the activation of PLD, which in turn caused the mitochondrial dysfunction and cytotoxicity in the skeletal muscle cell model, C2C12 cells (Schema 4). Thus, it is highly reasonable to ascertain that PLD activation and formation of the PLD-generated bioactive lipids could act as potential players in the statin-induced myalgia and myotoxicity. Therefore, PLD could be a pharmacological target for combating the statin myotoxicity.

References

1. Stock EO, Redberg R. Cardiovascular disease in women. Curr Probl Cardiol. 2012;37(11):450–526. https://doi.org/10.1016/j.cpcardiol.2012.07.001.
2. Sowers JR, Epstein M, Frohlich ED. Diabetes, hypertension, and cardiovascular disease: an update. Hypertension. 2001;37(4):1053–9. https://doi.org/10.1161/01.hyp.37.4.1053.
3. Essop MR, Peters F. Contemporary issues in rheumatic fever and chronic rheumatic heart disease. Circulation. 2014;130(24):2181–8. https://doi.org/10.1161/CIRCULATIONAHA.114.009857.
4. Granger DN, Vowinkel T, Petnehazy T. Modulation of the inflammatory response in cardiovascular disease. Hypertension. 2004;43(5):924–31. https://doi.org/10.1161/01.HYP.0000123070.31763.55.
5. Chen K, Keaney JF Jr. Evolving concepts of oxidative stress and reactive oxygen species in cardiovascular disease. Curr Atheroscler Rep. 2012;14(5):476–83. https://doi.org/10.1007/s11883-012-0266-8.
6. Varghese MJ. Familial hypercholesterolemia: a review. Ann Pediatr Cardiol. 2014;7(2):107–17. https://doi.org/10.4103/0974-2069.132478.
7. Xiong J, Miller VM, Li Y, Jayachandran M. Microvesicles at the crossroads between infection and cardiovascular diseases. J Cardiovasc Pharmacol. 2012;59(2):124–32. https://doi.org/10.1097/FJC.0b013e31820c6254.
8. Nasr VG, Kussman BD. Advances in the care of adults with congenital heart disease. Semin Cardiothorac Vasc Anesth. 2015;19(3):175–86. https://doi.org/10.1177/1089253214563989.
9. Singh S, Schwarz K, Horowitz J, Frenneaux M. Cardiac energetic impairment in heart disease and the potential role of metabolic modulators: a review for clinicians. Circ Cardiovasc Genet. 2014;7(5):720–8. https://doi.org/10.1161/CIRCGENETICS.114.000221.
10. Lamarche B, Couture P. It is time to revisit current dietary recommendations for saturated fat. Appl Physiol Nutr Metab. 2014;39(12):1409–11. https://doi.org/10.1139/apnm-2014-0141.
11. Gielen S, Laughlin MH, O'Conner C, Duncker DJ. Exercise training in patients with heart disease: review of beneficial effects and clinical recommendations. Prog Cardiovasc Dis. 2015;57(4):347–55. https://doi.org/10.1016/j.pcad.2014.10.001.
12. Buttar HS, Li T, Ravi N. Prevention of cardiovascular diseases: role of exercise, dietary interventions, obesity and smoking cessation. Exp Clin Cardiol. 2005;10(4):229–49.
13. Salabei JK, Conklin DJ. Cardiovascular autophagy: crossroads of pathology, pharmacology and toxicology. Cardiovasc Toxicol. 2013;13(3):220–9. https://doi.org/10.1007/s12012-013-9200-8.
14. Katsiki N, Tziomalos K, Mikhailidis DP. Alcohol and the cardiovascular system: a double-edged sword. Curr Pharm Des. 2014;20(40):6276–88. https://doi.org/10.2174/1381612820666140620125741.
15. Hartley L, Mavrodaris A, Flowers N, Ernst E, Rees K. Transcendental meditation for the primary prevention of cardiovascular disease. Cochrane Database Syst Rev. 2014;12:CD010359. https://doi.org/10.1002/14651858.CD010359.pub2.
16. Badimon L, Vilahur G. LDL-cholesterol versus HDL-cholesterol in the atherosclerotic plaque: inflammatory resolution versus thrombotic chaos. Ann N Y Acad Sci. 2012;1254:18–32. https://doi.org/10.1111/j.1749-6632.2012.06480.x.
17. Cozma A, Orăşan O, Sâmpelean D, Fodor A, Vlad C, Negrean V, Rednic N, Zdrenghea D. Endothelial dysfunction in metabolic syndrome. Rom J Intern Med. 2009;47(2):133–40.
18. Sander GE, Giles TD. Hypertension and lipids: lipid factors in the hypertension syndrome. Curr Hypertens Rep. 2002;4(6):458–63. https://doi.org/10.1007/s11906-002-0026-1.
19. Madden JA. Role of the vascular endothelium and plaque in acute ischemic stroke. Neurology. 2012;79(13 Suppl 1):S58–62. https://doi.org/10.1212/WNL.0b013e3182695836.
20. Ewang-Emukowhate M, Wierzbicki AS. Lipid-lowering agents. J Cardiovasc Pharmacol Ther. 2013;18(5):401–11. https://doi.org/10.1177/1074248413492906.

21. Cortes VA, Busso D, Maiz A, Arteaga A, Nervi F, Rigotti A. Physiological and pathological implications of cholesterol. Front Biosci (Landmark Ed). 2014;19(3):416–28. https://doi.org/10.2741/4216.
22. Owen DM, Gaus K. Imaging lipid domains in cell membranes: the advent of super-resolution fluorescence microscopy. Front Plant Sci. 2013;4:503. https://doi.org/10.3389/fpls.2013.00503.
23. Najam O, Ray KK. Familial hypercholesterolemia: a review of the natural history, diagnosis, and management. Cardiol Ther. 2015;4(1):25–38. https://doi.org/10.1007/s40119-015-0037-z.
24. Scanu AM, Wisdom C. Serum lipoproteins structure and function. Annu Rev Biochem. 1972;41:703–30. https://doi.org/10.1146/annurev.bi.41.070172.003415.
25. Rached FH, Chapman MJ, Kontush A. HDL particle subpopulations: focus on biological function. Biofactors. 2015;41(2):67–77. https://doi.org/10.1002/biof.1202.
26. Elshourbagy NA, Meyers HV, Abdel-Meguid SS. Cholesterol: the good, the bad, and the ugly – therapeutic targets for the treatment of dyslipidemia. Med Princ Pract. 2014;23(2):99–111. https://doi.org/10.1159/000356856.
27. Santos RD, Maranhao RC. What is new in familial hypercholesterolemia? Curr Opin Lipidol. 2014;25(3):183–8. https://doi.org/10.1097/MOL.0000000000000073.
28. Trapani L, Segatto M, Incerpi S, Pallottini V. 3-Hydroxy-3-methylglutaryl coenzyme A reductase regulation by antioxidant compounds: new therapeutic tools for hypercholesterolemia? Curr Mol Med. 2011;11(9):790–7. https://doi.org/10.2174/156652411798062403.
29. McPherson PA. From fungus to pharmaceuticals--the chemistry of statins. Mini Rev Med Chem. 2012;12(12):1250–60. https://doi.org/10.2174/138955712802762103.
30. Salakhutdinov NF, Rogoza LN, Tolstikov GA. Hypercholesterolemia: chemical aspect of approach. Curr Med Chem. 2011;18(26):4076–105. https://doi.org/10.2174/092986711796957248.
31. Malinowski JM. Atorvastatin: a hydroxymethylglutaryl-coenzyme A reductase inhibitor. Am J Health Syst Pharm. 1998;55(21):2253–303. https://doi.org/10.1093/ajhp/55.21.2253.
32. Moghadasian MH, Mancini GB, Frohlich JJ. Pharmacotherapy of hypercholesterolaemia: statins in clinical practice. Expert Opin Pharmacother. 2000;1(4):683–95. https://doi.org/10.1517/14656566.1.4.683.
33. Opie LH. Present status of statin therapy. Trends Cardiovasc Med. 2015;25(3):216–25. https://doi.org/10.1016/j.tcm.2014.10.002.
34. Stroes ES, Thompson PD, Corsini A, Vladutiu GD, Raal FJ, Ray KK, Roden M, Stein E, Tokgözoğlu L, Nordestgaard BG, Bruckert E, De Backer G, Krauss RM, Laufs U, Santos RD, Hegele RA, Hovingh GK, Leiter LA, Mach F, März W, European Atherosclerosis Society Consensus Panel, et al. Statin-associated muscle symptoms: impact on statin therapy-European Atherosclerosis Society Consensus Panel Statement on Assessment, Aetiology and Management. Eur Heart J. 2015;36(17):1012–22. https://doi.org/10.1093/eurheartj/ehv043.
35. Vinci P, Panizon E, Tosoni LM, Cerrato C, Pellicori F, Mearelli F, Biasinutto C, Fiotti N, Di Girolamo FG, Biolo G. Statin-associated myopathy: emphasis on mechanisms and targeted therapy. Int J Mol Sci. 2021;22(21):11687. https://doi.org/10.3390/ijms222111687.
36. Abed W, Abujbara M, Batieha A, Ajlouni K. Statin induced myopathy among patients attending the National Center for Diabetes, Endocrinology, & Genetics. Ann Med Surg. 2022;74(2012):103304. https://doi.org/10.1016/j.amsu.2022.103304.
37. Hansen KE, Hildebrand JP, Ferguson EE, Stein JH. Outcomes in 45 patients with statin-associated myopathy. Arch Intern Med. 2005;165(22):2671–6. https://doi.org/10.1001/archinte.165.22.2671.
38. Ward NC, Watts GF, Eckel RH. Statin toxicity. Circ Res. 2019;124(2):328–50. https://doi.org/10.1161/CIRCRESAHA.118.312782.
39. Turner RM, Pirmohamed M. Statin-related myotoxicity: a comprehensive review of pharmacokinetic, pharmacogenomic and muscle components. J Clin Med. 2019;9(1):22. https://doi.org/10.3390/jcm9010022.
40. Sathasivam S, Lecky B. Statin induced myopathy. BMJ. 2008;337:a2286. https://doi.org/10.1136/bmj.a2286.

41. Ganga HV, Slim HB, Thompson PD. A systematic review of statin-induced muscle problems in clinical trials. Am Heart J. 2014;168(1):6–15. https://doi.org/10.1016/j.ahj.2014.03.019.
42. Ahmad Z. Statin intolerance. Am J Cardiol. 2014;113(10):1765–71. https://doi.org/10.1016/j.amjcard.2014.02.033.
43. Sirvent P, Mercier J, Lacampagne A. New insights into mechanisms of statin-associated myotoxicity. Curr Opin Pharmacol. 2008;8(3):333–8. https://doi.org/10.1016/j.coph.2007.12.010.
44. Dirks AJ, Jones KM. Statin-induced apoptosis and skeletal myopathy. Am J Physiol Cell Physiol. 2006;291(6):C1208–12. https://doi.org/10.1152/ajpcell.00226.2006.
45. Korn ED. Current concepts of membrane structure and function. Fed Proc. 1969;28(1):6–11.
46. Siekevitz P. Biological membranes: the dynamics of their organization. Annu Rev Physiol. 1972;34:117–40. https://doi.org/10.1146/annurev.ph.34.030172.001001.
47. Suetsugu S, Kurisu S, Takenawa T. Dynamic shaping of cellular membranes by phospholipids and membrane-deforming proteins. Physiol Rev. 2014;94(4):1219–48. https://doi.org/10.1152/physrev.00040.2013.
48. Slone EA, Fleming SD. Membrane lipid interactions in intestinal ischemia/reperfusion-induced injury. Clin Immunol. 2014;153(1):228–40. https://doi.org/10.1016/j.clim.2014.04.018.
49. Richmond GS, Smith TK. Phospholipases A_1. Int J Mol Sci. 2011;12(1):588–612. https://doi.org/10.3390/ijms12010588.
50. Tani K, Kogure T, Inoue H. The intracellular phospholipase A1 protein family. Biomol Concepts. 2012;3(5):471–8. https://doi.org/10.1515/bmc-2012-0014.
51. Ghosh M, Tucker DE, Burchett SA, Leslie CC. Properties of the Group IV phospholipase A2 family. Prog Lipid Res. 2006;45(6):487–510. https://doi.org/10.1016/j.plipres.2006.05.003.
52. Krizaj I. Roles of secreted phospholipases A_2 in the mammalian immune system. Protein Pept Lett. 2014;21(12):1201–8.
53. Zhang Y, Frohman MA. Cellular and physiological roles for phospholipase D1 in cancer. J Biol Chem. 2014;289(33):22567–74. https://doi.org/10.1074/jbc.R114.576876.
54. Follo MY, Manzoli L, Poli A, McCubrey JA, Cocco L. PLC and PI3K/Akt/mTOR signalling in disease and cancer. Adv Biol Regul. 2015;57:10–6. https://doi.org/10.1016/j.jbior.2014.10.004.
55. Gomez-Cambronero J. Phospholipase D in cell signaling: from a myriad of cell functions to cancer growth and metastasis. J Biol Chem. 2014;289(33):22557–66. https://doi.org/10.1074/jbc.R114.574152.
56. Parinandi NL, Scribner WM, Vepa S, Shi S, Natarajan V. Phospholipase D activation in endothelial cells is redox sensitive. Antioxid Redox Signal. 1999;1(2):193–210. https://doi.org/10.1089/ars.1999.1.2-193.
57. Parinandi NL, Roy S, Shi S, Cummings RJ, Morris AJ, Garcia JG, Natarajan V. Role of Src kinase in diperoxovanadate-mediated activation of phospholipase D in endothelial cells. Arch Biochem Biophys. 2001;396(2):231–43. https://doi.org/10.1006/abbi.2001.2609.
58. Cummings R, Parinandi N, Wang L, Usatyuk P, Natarajan V. Phospholipase D/phosphatidic acid signal transduction: role and physiological significance in lung. Mol Cell Biochem. 2002;234–235(1–2):99–109.
59. Patel RB, Kotha SR, Sherwani SI, Sliman SM, Gurney TO, Loar B, Butler SO, Morris AJ, Marsh CB, Parinandi NL. Pulmonary fibrosis inducer, bleomycin, causes redox-sensitive activation of phospholipase D and cytotoxicity through formation of bioactive lipid signal mediator, phosphatidic acid, in lung microvascular endothelial cells. Int J Toxicol. 2011;30(1):69–90. https://doi.org/10.1177/1091581810388850.
60. Secor JD, Kotha SR, Gurney TO, Patel RB, Kefauver NR, Gupta N, Morris AJ, Haley BE, Parinandi NL. Novel lipid-soluble thiol-redox antioxidant and heavy metal chelator, N,N′-bis(2-mercaptoethyl)isophthalamide (NBMI) and phospholipase D-specific inhibitor, 5-fluoro-2-indolyl des-chlorohalopemide (FIPI) attenuate mercury-induced lipid signaling leading to protection against cytotoxicity in aortic endothelial cells. Int J Toxicol. 2011;30(6):619–38. https://doi.org/10.1177/1091581811422413.
61. Varadharaj S, Steinhour E, Hunter MG, Watkins T, Baran CP, Magalang U, Kuppusamy P, Zweier JL, Marsh CB, Natarajan V, Parinandi NL. Vitamin C-induced activation of phospho-

lipase D in lung microvascular endothelial cells: regulation by MAP kinases. Cell Signal. 2006;18(9):1396–407. https://doi.org/10.1016/j.cellsig.2005.10.019.
62. Kline MA, O'Connor Butler ES, Hinzey A, Sliman S, Kotha SR, Marsh CB, Uppu RM, Parinandi NL. A simple method for effective and safe removal of membrane cholesterol from lipid rafts in vascular endothelial cells: implications in oxidant-mediated lipid signaling. Methods Mol Biol. 2010;610:201–11. https://doi.org/10.1007/978-1-60327-029-8_12.
63. Tretter EM. Statins, the cholesterol-lowering drugs cause myotoxicity through phospholipase D-mediated lipid signaling in skeletal C2C12 myoblast cells. Honors Research Dissertation for Graduation "With Honors Research Distinction," The Ohio State University. The Ohio State University Knowledge Bank. 2015. http://hdl.handle.net/1811/86177
64. Hinzey AH, Kline MA, Kotha SR, Sliman SM, Butler ES, Shelton AB, Gurney TR, Parinandi NL. Choice of cyclodextrin for cellular cholesterol depletion for vascular endothelial cell lipid raft studies: cell membrane alterations, cytoskeletal reorganization and cytotoxicity. Indian J Biochem Biophys. 2012;49(5):329–41.
65. Okuyama H, Langsjoen PH, Hamazaki T, Ogushi Y, Hama R, Kobayashi T, Uchino H. Statins stimulate atherosclerosis and heart failure: pharmacological mechanisms. Expert Rev Clin Pharmacol. 2015;8(2):189–99. https://doi.org/10.1586/17512433.2015.1011125.
66. Bockorny B, Dasanu CA. HMG-CoA reductase inhibitors as adjuvant treatment for hematologic malignancies: what is the current evidence? Ann Hematol. 2015;94(1):1–12. https://doi.org/10.1007/s00277-014-2236-4.
67. Natarajan V, Scribner WM, Morris AJ, Roy S, Vepa S, Yang J, Wadgaonkar R, Reddy SP, Garcia JG, Parinandi NL. Role of p38 MAP kinase in diperoxovanadate-induced phospholipase D activation in endothelial cells. Am J Physiol Lung Cell Mol Physiol. 2001;281(2):L435–49. https://doi.org/10.1152/ajplung.2001.281.2.L435.
68. Costa RA, Fernandes MP, de Souza-Pinto NC, Vercesi AE. Protective effects of l-carnitine and piracetam against mitochondrial permeability transition and PC3 cell necrosis induced by simvastatin. Eur J Pharmacol. 2013;701(1–3):82–6. https://doi.org/10.1016/j.ejphar.2013.01.001.
69. Abdoli N, Heidari R, Azarmi Y, Eghbal MA. Mechanisms of the statins cytotoxicity in freshly isolated rat hepatocytes. J Biochem Mol Toxicol. 2013;27(6):287–94. https://doi.org/10.1002/jbt.21485.
70. Kotamraju S, Williams CL, Kalyanaraman B. Statin-induced breast cancer cell death: role of inducible nitric oxide and arginase-dependent pathways. Cancer Res. 2007;67(15):7386–94. https://doi.org/10.1158/0008-5472.CAN-07-0993.
71. Kanugula AK, Gollavilli PN, Vasamsetti SB, Karnewar S, Gopoju R, Ummanni R, Kotamraju S. Statin-induced inhibition of breast cancer proliferation and invasion involves attenuation of iron transport: intermediacy of nitric oxide and antioxidant defence mechanisms. FEBS J. 2014;281(16):3719–38. https://doi.org/10.1111/febs.12893.
72. Abdoli N, Azarmi Y, Eghbal MA. Protective effects of N-acetylcysteine against the statins cytotoxicity in freshly isolated rat hepatocytes. Adv Pharm Bull. 2014;4(3):249–54. https://doi.org/10.5681/apb.2014.036.
73. Sliman SM, Eubank TD, Kotha SR, Kuppusamy ML, Sherwani SI, Butler ES, Kuppusamy P, Roy S, Marsh CB, Stern DM, Parinandi NL. Hyperglycemic oxoaldehyde, glyoxal, causes barrier dysfunction, cytoskeletal alterations, and inhibition of angiogenesis in vascular endothelial cells: aminoguanidine protection. Mol Cell Biochem. 2010;333(1–2):9–26. https://doi.org/10.1007/s11010-009-0199-x.
74. Patel RB, Kotha SR, Sauers LA, Malireddy S, Gurney TO, Gupta NN, Elton TS, Magalang UJ, Marsh CB, Haley BE, Parinandi NL. Thiol-redox antioxidants protect against lung vascular endothelial cytoskeletal alterations caused by pulmonary fibrosis inducer, bleomycin: comparison between classical thiol-protectant, N-acetyl-L-cysteine, and novel thiol antioxidant, N,N′-bis-2-mercaptoethyl isophthalamide. Toxicol Mech Methods. 2012;22(5):383–96. https://doi.org/10.3109/15376516.2012.673089.

Cell-Cell Communication in the Vascular Endothelium

D. Ryan King, Louisa Mezache, Meghan Sedovy, Przemysław B. Radwański, Scott R. Johnstone, and Rengasayee Veeraraghavan

Abstract The vasculature is a body-wide organ responsible for mass and energy transport throughout the body. In broad terms, all vessels consist of a layer of endothelial cells, which act as the interface with blood, pericytes, which modulate their function, and in the case of resistance arteries and larger vessels, additional layers of smooth muscle cells, which modulate vascular tone. Dynamic regulation of vascular function in response to biological and physiological cues is a vital process in both health and disease. These responses as well as their synchronization along vessels are critically dependent upon communication between vascular cells, which utilizes a plethora of direct cell-to-cell, paracrine, and autocrine mechanisms. Here, we review these mechanisms, focusing especially on endothelial cells, the structural and molecular underpinnings thereof, and the mechanisms by which they are regulated.

D. Ryan King
Dorothy M. Davis Heart and Lung Research Institute, College of Medicine, The Ohio State University Wexner Medical Center, Columbus, OH, USA

L. Mezache · R. Veeraraghavan (✉)
Dorothy M. Davis Heart and Lung Research Institute, College of Medicine, The Ohio State University Wexner Medical Center, Columbus, OH, USA

Department of Biomedical Engineering, The Ohio State University, Columbus, OH, USA
e-mail: veeraraghavan.12@osu.edu

M. Sedovy · S. R. Johnstone (✉)
The Fralin Biomedical Research Institute Center for Heart and Reparative Medicine Research, Virginia Tech Carilion, Roanoke, VA, USA
e-mail: scottrj@vtc.vt.edu

P. B. Radwański (✉)
Dorothy M. Davis Heart and Lung Research Institute, College of Medicine, The Ohio State University Wexner Medical Center, Columbus, OH, USA

Division of Outcomes and Translational Sciences, College of Pharmacy, The Ohio State University, Columbus, OH, USA
e-mail: radwanski.2@osu.edu

N. L. Parinandi, T. J. Hund (eds.), *Cardiovascular Signaling in Health and Disease*, https://doi.org/10.1007/978-3-031-08309-9_15

Keywords Cell-cell communication · Gap junctions · Pannexins · Paracrine signaling · Vascular endothelium

On the grand scale, the mammalian circulatory system can be subdivided into three circuits: the systemic circuit, the pulmonary circuit, and the coronary circuit. Within each circuit exists both arterial and venous blood vessels that carry blood away from and toward the heart, respectively. The arterial vasculature can be largely divided, by vessel diameter and structure, into large conduit arteries, intermediate contractile arteries, arterioles (also referred to as resistance arteries), and capillaries [1, 2]. The tunica intima, or innermost layer of conduit arteries, contractile arteries, and arterioles, is composed of a continuous endothelium and connective tissue bound by an internal elastic membrane. The tunica media of these arteries and arterioles comprise smooth muscle cells and elastic laminae that allow for contraction regulation in these systems. Together, the artery and arteriole network carry oxygenated blood to tissues, including the myocardium, and function to regulate coronary vascular resistance. Outside of these vessels, capillaries are composed of a single layer of endothelial cells, arranged as continuous or fenestrated layers. The capillaries are devoid of smooth muscle cell coverage but have direct pericyte contact. The capillary beds act as functional sites of nutrient and gas exchange between the vascular network and the surrounding interstitum [3].

The venous vasculature can be largely divided into venules and veins. Veins have thin, elastic, walls relative to arteries [4]. This allows veins to hold a large amount of blood at relatively low pressures and at any one point, nearly three-fourths of the entire systemic blood supply are housed within the venous vasculature. A notable feature of veins is that they also have one-way flow valves that permit blood flow only in the forward (toward the heart) direction [4].

Structural/Signaling Components

The vascular endothelium is derived from the mesoderm and forms the innermost layer of the entire vasculature system across all circulatory and lymphatic divisions [5]. Endothelial cells within the blood vessel are directionally polarized, with unique protein expression occurring on the luminal versus basal side. Endothelial cells are mechanically coupled to one another via tight junctions. Further, several lines of evidence suggest that endothelial cells are also electrically coupled to one another via gap junctions [6, 7].

Tight Junctions

As noted above, endothelial cells are mechanically coupled by tight junctions, which are named for the ability to provide a robust paracellular barrier capable of preventing even water movement across the endothelium. Tight junctions are junctional complexes composed of at least 40 different proteins consisting of transmembrane proteins that connect apposing cell membranes, as well as associated cytoplasmic proteins [8]. The extracellular domains of this multiprotein complex are joined directly with one another to form a network of sealing strands, each operating independently of another, that function to control epithelial and endothelial paracellular permeability and regulate apical-basolateral diffusion of membrane proteins, maintaining cell surface polarity [8]. The degree to which a junction prevents solute and protein transfer can depend on the number of sealing strands or the properties of the proteins in the strands [9]. For instance, leaky epithelial cells of the gallbladder have on average two strands, while the tight endothelia of the blood-brain barrier possess at minimum five sealing strands [10]. There are three major transmembrane proteins that determine tight junction selectivity: claudins, occludin, and junction adhesion molecule (JAM) proteins.

Claudins comprise a large family of low molecular weight (21–28 kDa) membrane proteins consisting of four transmembrane domains, two extracellular loops, one cytoplasmic loop, and both N- and C- termini located intracellularly. Claudins are considered the backbone of tight junctions, as they are essential in controlling the extent to which the paracellular space is sealed. Occludin protein, with a molecular weight of 60 kDa, has a similar structure to claudins, and serves to regulate the opening and closing of tight junctions [9]. JAM proteins belong to the immunoglobulin superfamily. They possess a single transmembrane domain and are generally known to play a role in tight junction regulation and signaling. These three major proteins – claudins, occludin, and JAM proteins – are anchored to the cell's actin filaments via association with cytoplasmic proteins like zonula occludens-1 (ZO-1). While the general structure of tight junctions is very much alike in both endothelia across all organ and tissue types, the focus of this chapter will be tight junctions in vascular endothelial cells.

Endothelial tight junctions are Ca^{2+}-dependent adhesion structures and thus, highly sensitive to changes in extracellular Ca^{2+} changes. However, under most physiological and pathophysiological conditions, they are regulated through changes in intracellular Ca^{2+} and signaling mechanisms within endothelial cells, in turn modulated by external stimuli and conditions. These processes are governed through the action of a myriad of receptors and channels, including connexin hemichannels, gap junction channels, pannexin channels (Fig. 1), which are discussed in the following sections.

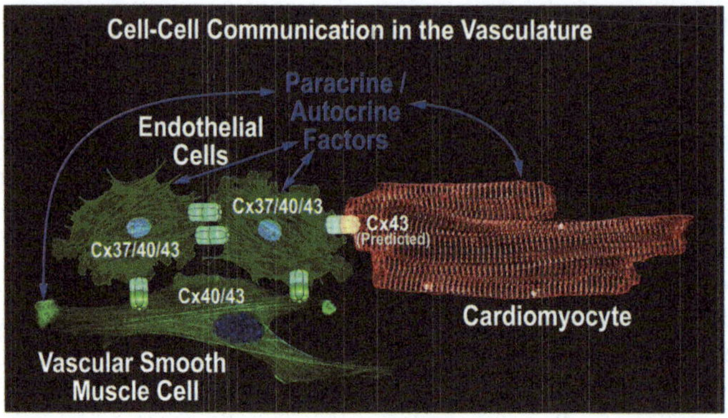

Fig. 1 Direct and indirect modes of communication between endothelial cells and other cell types

Gap Junctions

Endothelial cells are electrically coupled by gap junctions, which provide a conduit for intercellular communication between neighboring cells. A single gap junction is composed of two protein hemichannels (or connexons), each comprising six transmembrane proteins, or connexins (Cxs). Cx proteins contain four transmembrane domains, two extracellular loops, an intracellular loop, and intracellular N- and C-terminus. Within the vasculature, there are five primarily expressed connexin isoforms: Cx32 [11–13], Cx37 [14–16], Cx40 [14, 16, 17], Cx43 [18, 19], Cx45 [20], though a functional role for Cx32 and Cx45 in the vasculature has not been well defined. Connexin hemichannels are assembled prior to being trafficked to the cell membrane where they align and dock with a neighboring cell's hemichannel to form gap junctions before being accrued into closely packed arrays known as gap junction plaques. In addition to providing cell-to-cell electrical and chemical coupling as gap junctions, connexins can also function as undocked hemichannels, which provide a large conductance, non-specific pathway for electrochemical exchange between the endothelial cell and its extracellular environment. Connexin hemichannels participate in purinergic signaling and other key regulatory processes that govern barrier function and angiogenesis, have emerged as important determinants of endothelial health and function, and are attractive targets for therapy [21].

Pannexin Channels

The Pannexin (Panx) family of single-membrane channels contains three members, Panx1–3. The channels allow for nonselective passage of ions and small molecules between the cytosol and extracellular space. Similar to Cx hemichannels, Panx channels possess four transmembrane domains. Multiple studies have identified Panx1 channels to be assembled as heptamers prior to trafficking to the cell membrane [22–25]. Based on sequence similarities, it is suggested that Panx2 and Panx3 may also assemble in a similar manner, although heptamer assembly and plasma membrane localization has not been robustly demonstrated for either. Expression of Panx2 and Panx3 appears to be limited, but identification has been limited by a lack of available specific reagents, including antibodies. Panx1 shows abundant expression in endothelial cells and is significantly regulated under physiological and pathophysiological conditions. The Panx1 channels play an important role in purinergic signaling, regulating cellular ATP release (similar roles to connexin hemichannels), and in mediating inflammation through control of intracellular Ca^{2+}.

Purinergic Signaling in Endothelial Cells

In 1929, Sir Alan Nigel Drury and Sir Albert Szent-Györgyi provided the first evidence that exogenous adenine compounds affect the heart [26]. Though, it would not be until 43 years later, in 1972, that Geoffrey Burnstock would propose purinergic signaling as an endogenous biological mechanism [27]. The early study of endogenous purinergic signaling focused on the role of adenosine triphosphate (ATP) as a neurotransmitter, though the subsequent decades have seen the field expand to nearly all mammalian biological systems. As ATP and the enzymes responsible for ATP synthesis are ubiquitously expressed in cells, it is of no surprise that purinergic signaling has been demonstrated in multi-systemic fashion. Of particular interest to this chapter, several lines of evidence have demonstrated that purinergic signaling is a principal regulator of vasomotor function in the cardiovascular system.

Between the late 1970s and early 1990s, the purinergic receptors for adenosine (P1) and adenosine di- and tri-phosphate (P2) were discovered, cloned, and characterized [28–30]. The P1 receptors are G-coupled protein receptors (GPCRs) and, to date, four subtypes (A_1, A_{2A}, A_{2B}, and A_3) have been identified [31–33]. The A_1 and A_3 receptors are G_i coupled, whereas the A_{2A} and A_{2B} receptors are G_s coupled. The P2 receptors require further subdivision into the ion channel receptor subtype, P2X, and the GPCR subtype, P2Y. To date, seven P2X receptors ($P2X_{1-7}$) have been identified. The $P2X_{1-7}$ receptors all respond to ATP as their activating ligand and assemble into homo- or heterotrimers that form non-selective cation channels; though, it is worth noting that $P2X_5$ has shown some permeability to anions [34]. To date, eight P2Y receptors ($P2Y_1$, $P2Y_2$, $P2Y_4$, $P2Y_6$, $P2Y_{11}$, $P2Y_{12}$, $P2Y_{13}$, $P2Y_{14}$) have been identified [35]. The P2Y receptors form hetero-oligomers and bind extracellular ATP, ADP, UTP, and UDP as activating ligands [36]. The $P2Y_{1,2,4,6,11}$ receptors are G_q coupled, the $P2Y_{11}$ receptor is G_s coupled, and the $P2Y_{12,13,14}$ receptors are G_i coupled [35]. The activation of the G_q coupled receptors leads to the activation of PLCβ, increases in IP_3, and subsequent release of calcium (Ca^{2+}) from intracellular stores [37]. The activation of the G_s coupled receptor leads to the activation of adenylyl cyclase and subsequent cAMP production [37, 38]. Lastly, G_i coupled receptors inhibit adenylyl cyclase and antagonize cAMP production [37, 38]. Notably, $P2Y_{11}$ is both a G_q and G_s coupled receptor, meaning $P2Y_{11}$ activation results in increases in intracellular Ca^{2+} and cAMP.

There is evidence for all four P1 receptor subtypes in both the myocardium and the coronary endothelium, though it must be noted that specific P1 receptor expression ratios has been reported to vary by species [39, 40]. P2 receptors are also found in cardiomyocytes and coronary endothelium. A_1 and A_3 receptors are G_i coupled and their activation results in inhibition of adenylyl cyclase. Conversely, A_{2A} and A_{2B} receptors are G_s coupled and their activation results in activation of adenylyl cyclase, giving the two groups of P1 receptors seemingly antagonistic actions. Specifically, cardiomyocytes have been shown to express all seven P2X ($P2X_{1-7}$) receptors and five P2Y ($P2Y_1$, $P2Y_2$, $P2Y_4$, $P2Y_6$, $P2Y_{11}$) receptors [41–51]. Vascular endothelial cells primarily express the $P2X_4$ receptor and variably express P2Y receptors. For instance, all P2Y receptors have been identified within lung microvascular endothelial cells [52]; however, a 2011 study by Lyubchenko et al. demonstrate that $P2Y_6$, $P2Y_{11}$, and $P2Y_{12}$ are not expressed in pulmonary artery vasa vasorum endothelial cells [53]. To our knowledge, no one has yet to detail the expression pattern of P2X receptors or P2Y receptors specifically within the coronary microvasculature. Further, the P1 and P2X/P2Y receptors are functionally linked through the action of the cell surface enzymes ectonucleoside triphosphate diphosphohydrolase-1 (CD39) and ecto-5′-nucleotidase (CD73), which function to hydrolyze ATP and ADP to AMP, and AMP to adenosine, respectively [54–57]. In short, ATP can bind, and activate, P2X/P2Y receptors and then be subsequently hydrolyzed to adenosine and bind P1 receptors to affect further change.

In cardiomyocytes, adenosine activation of A_1 receptors has been demonstrated to have a negative chronotropic and dromotropic effect [58]. Further, A_1 activation is antagonistic to α- and β- adrenergic stimulation [59] and results in PKC [60] and PKB [61] activation. Knockout of the A_{2A} receptor gene resulted in a blunted adenosine-mediated NO release in the mouse heart [59], and a further study showed the $A_{2A/B}$ agonist CGS21680 resulted in a marked increase in nitric oxide production in cardiomyocytes [62].

Given the variety of effects conferred by P1 receptor activation and the long-standing discovery that adenine compounds affect the heart, it is of no surprise that several pre-clinical and clinical lines of inquiry have investigated P1 activation in the treatment of acute heart failure [63, 64], chronic heart failure [64–66], myocardial infarction [67], ischemia-reperfusion injury [68–70], angina [71], and coronary artery disease [72]. These studies, however, have not come without controversy and contradiction to the effect of P1 receptor activation. For instance, despite the seemingly antagonistic actions of $A_{1/3}$ and $A_{2A/2B}$, activation of all four has been reported to improve outcomes of myocardial ischemia [73–77]. Further, studies have demonstrated that selective A_3 receptor activation is cardioprotective from ischemic damage [78–81], but that an A_3 receptor gene deletion also results in an ischemia-tolerant phenotype [80, 82]. Though the precise mechanism remains contested, there is a general improvement observed following adenosine treatment in the context of heart failure [83, 84] and ischemia-reperfusion injury [85–87].

Several lines of evidence have demonstrated that shear stress-induced vasodilation occurs via P2X or P2Y activation in the endothelium [88]. In response to shear stress, endothelial cells generate and release ATP for paracrine and autocrine signaling, leading in turn to P2X activation and subsequent Ca^{2+} entry through P2X. As with P1 receptors, there are conflicting reports to which specific subtype is responsible for the effect. On the one hand, $P2X_4$ knockdown in human umbilical vein endothelial cells (HUVECs) results in decreased shear stress-induced Ca^{2+} influx, and $P2X_4$ deficient mice show global hypertension and reduced nitric oxide production [89–92]. Though, other studies suggest that $P2Y_2$, not $P2X_4$, is responsible for shear stress-induced Ca^{2+} influx in HUVECs [93, 94]. In general, research on the specific function of P2 receptors is limited by a lack of specific inhibitors with high subtype-specific fidelity and remains an interesting area of inquiry.

In the myocardium, alterations in expression and activation of both P2X and P2Y receptors have been associated with the development and progression of disease. In patients with heart failure, studies have reported increases in mRNA expression of the $P2X_1$, $P2X_6$, and $P2Y_2$ receptors [41, 95]. During acute ischemic episodes, activation or overexpression of $P2X_4$, $P2Y_2$, or $P2Y_4$ was associated with cardioprotection [96–99] from hypoxic damage, whereas activation of P2X7 led to apoptotic death [100]. As both the coronary endothelium and myocardium are responsible for local environment ATP and UTP release, it is understandable that a tight regulation of the interplay between the two may drive health and disease [97].

Ca^{2+} Signaling

As in many cell types, Ca^{2+} homeostasis is an integral part of cell regulation in endothelial cells. Intracellular Ca^{2+} levels control and modulate a wide array of regulatory pathways that govern vascular luminal pressure, endothelial cell–cell adhesion (and thereby, vascular barrier function), and secretion of signaling molecules, including the aforementioned purines. Ca^{2+} permeable channels in endothelial cells include both transmembrane Ca^{2+} channels, which enable Ca^{2+} movement across the cell membrane, and intracellular Ca^{2+} channels, which carry Ca^{2+} between the cytosol and intracellular Ca^{2+} stores.

Transmembrane Ca^{2+} Channels

Voltage-Gated Ca^{2+} Channels (VGCCs) These consist of long opening L-type Ca^{2+} channels (LCCs) and transient opening T-type Ca^{2+} channels (TCCs), each with characteristic biophysical properties [101]. Functional LCCs and TCCs both consist of heteromeric protein complexes, including a pore-forming α_1 subunit and optional auxiliary β, α_2/δ, and γ subunits, which modulate their behavior [102].

Both types have been identified in large arterial endothelial cells [103–105] (but not yet in smaller vessels), though their functional roles in vivo remain unclear.

Na^+-Ca^{2+} exchanger (NCX) First identified in cardiac muscle, NCX is a transmembrane protein that exchanges 1 Ca^{2+} ion for 3 Na^+ ions in forward (Ca^{2+} efflux) or reverse (Ca^{2+} influx) mode, depending on electrochemical gradients [106]. NCX is expressed by various cell types throughout the body, including vascular endothelial cells and vascular smooth muscle cells [107, 108] with key roles in EC-mediated vasodilation in vascular endothelial cells [108–110]. Lillo et al. recently demonstrated abolition of ACh-induced vasodilation in rat mesenteric arteries following pharmacological NCX inhibition. Additionally, siRNA knockdown of NCX expression blunted Ach-induced rise in intracellular Ca^{2+}, indicating an obligate role for NCX in process [109].

Transient receptor potential (TRP) channels These are a superfamily of cation channels widely expressed within the vasculature with 28 members subdivided into six families based on sequence homology: TRPC, TRPV, TRPM, TRPA, TRPML, and TRPP channels [111]. Vascular endothelial cells express several of these, including TRPC1, TRPC3, TRPC4, TRPC5, TRPC6, TRPC7, TRPV1, TRPV2, TRPV4, TRPP1, TRPP2, TRPA1, TRPM1, TRPM2, TRPM3, TRPM4, TRPM6, TRPM7, and TRPM8 channels; though, their expression is specific to species and vascular bed [111–114]. TRP channels play roles in angiogenesis [111], vascular barrier function [54], EC Ca^{2+} entry [113].

Intracellular Ca^{2+} Channels

Sarcoplasmic Reticulum Ca^{2+} ATPase (SERCA) This is a P-type ATPase that replenishes intracellular Ca^{2+} stores by moving Ca^{2+} from the cytosol into the sarcoplasmic reticulum against its electrochemical gradient. Of the 3 isoforms (SERCA1–3) identified in mammals, only SERCA2 (including the splice variants, SERCA2a and SERCA2b) and SERCA3 are expressed in vascular endothelial cells [115]. Vascular endothelial cells also express phospholamban (PLB), a small protein (52 amino acids) that inhibits SERCA function in its native state [116, 117]. Phosphorylation of PLB at serine 10, serine 16, and threonine 17 by PKC, PKG, and CaMKII, respectively [118], relieves PLB's inhibitory function, thereby disinhibiting SERCA function and accelerating sarcoplasmic reticulum Ca^{2+} uptake.

Ryanodine Receptor Ca^{2+} Release Channels (RyR) Three isoforms of RyRs (RyR1–3) are expressed in mammals [119] and form tetrameric channels capable of rapidly releasing Ca^{2+} from the sarcoplasmic reticulum into the cytosol with their activity regulated by both cytosolic and sarcoplasmic reticulum Ca^{2+} levels. Lesh et al. first identified RyRs in endothelial cells followed by Ziegelstein and colleagues, who demonstrated their ability to regulate intracellular Ca^{2+} storage in cultured human and bovine endothelial cells [120, 121]. Although various roles have since been suggested for RyR in regulating vascular EC Ca^{2+} [122–124], their expression/functional roles in resistance artery endothelial cells remain unclear. Whereas Kohler et al. demonstrated the presence of caffeine-inducible Ca^{2+} currents in human third order mesenteric arteries and identified expression of the RyR3 gene [125, 126], Ledoux et al. failed to find either RyR mRNA or a functional effect with caffeine in mesenteric arteries isolated from C57BL6 mice [127].

Inositol 1,4,5-trisphosphate receptor (IP_3R) These six transmembrane domain proteins form tetrameric IP_3R Ca^{2+} release channels [128], which can also rapidly release Ca^{2+} from intracellular stores into the cytosol in IP_3 and to $[Ca^{2+}]$–dependent fashion, although the exact mechanisms of regulation remain to be clarified [129–131]. IP_3 is widely viewed as a global cellular messenger, given its ability to diffuse much faster than Ca^{2+} (280 vs. 38 $\mu m^2/s$) and suitability for long-range signaling [132, 133]. The G_q protein-mediated phospholipase C (PLC), activated by norepinephrine stimulation of α_1ARs, metabolizes membrane phospholipids (phosphatidylinositol 4,5-bisphosphate; PIP_2) to produce diacylglycerol (DAG) and inositol 1,4,5-trisphosphate (IP_3). DAG is linked to TRPC6 channel activation and can elicit Ca^{2+} entry, independent of protein kinase C (PKC) [134, 135], while IP_3 can modulate Ca^{2+} release from intracellular stores via IP_3Rs.

Store-operated Ca^{2+} entry (SOCE) When intracellular Ca^{2+} stores are depleted, SOCE is triggered to replenish them. This process, which has been suggested to modulate vascular smooth muscle cells proliferation, differentiation, and contraction [136] and vascular EC signaling [137, 138], is primarily mediated by the function and interactions of two proteins: STIM1 and Orai1. Orai1 channels permit extracellular Ca^{2+} intro the cytosol, which is then conducted into the sarcoplasmic reticulum by closely aligned STIM1 channels. This can involve either complexation of STIM1 and Orai1, triggered by sarcoplasmic reticulum Ca^{2+} depletion, and/or immediate activation of STIM1L, a long splice variant which is complexed with Orai1 at baseline. Additionally, recent work indicates potential roles for Orai2 and Orai3 [139] as well as TRPC channels [140–143] in SOCE, although unanswered questions remain about their functional roles [144].

Local Ca^{2+} Movements in Endothelial Cells Recent studies suggest that physiological increases in intracellular Ca^{2+} may occur in a localized, rather than cell-wide, manner [127, 145] involving multiple distinct Ca^{2+} signaling events within vascular cells. Spontaneous EC Ca^{2+} events, termed Ca^{2+} pulsars, resemble Ca^{2+} sparks in vascular smooth muscle cells. However, Ca^{2+} pulsars primarily occur around the nucleus and at myoendothelial junctional sites with broader spatiotemporal profiles. IP_3 channels open secondary to GPCR activation to release Ca^{2+} from the sarcoplasmic reticulum [127, 146], in turn activating nearby IK and SK channels and inducing hyperpolarization of EC membrane and dilation of the arteries [147, 148].

A Look Ahead

A wide array of pharmaceutical therapies used to treat a diverse range of diseases rely on the vasculature as the "highway system" for delivery, underscoring the translational importance understanding vascular biology and physiology. More importantly, we are only beginning to realize how attractive and potent a therapeutic target the vasculature itself is in many disease. For instance, ACE inhibitors have proven particularly effective in symptom mitigation and disease recovery for cardiovascular disease. While it is recognized that blood vessels are ubiquitously embedded within all organ systems, little work exists exploring the heterocellular interplay between the vascular endothelium and resident cells of each organ system. In the heart, for example, endothelial cells are thought to outnumber cardiomyocytes 3 to 1 and that every cardiomyocyte is neighbored by at least one endothelial cell. Given that cardiomyocytes and endothelial cells share myriad systems for direct cell-to-cell communication and paracrine signaling (pannexins, hemichannels, purinergic signaling), it stands to reason that they can modulate each other's behavior. Indeed, not only is vascular dysfunction associated with arrhythmias such as atrial fibrillation but serum markers of such dysfunction have been shown to return to baseline levels upon restoration of sinus rhythm [149]. There is even evidence suggesting direct coupling via gap junctions exists between cardiomyocytes and endothelial cells, though it is unclear which isoforms may be involved or whether or not the channels are conducting under normal physiological conditions. While such heterocellular gap junctional communication is known to occur in vitro, direct evidence has yet to be provided for such phenomena in the mammalian heart. Encouragingly for this line of inquiry, myoendothelial junctions are known to exist between endothelial cells and smooth muscle cells [150], and there is a growing body of evidence demonstrating heterocellular gap junctional communication between cardiomyocytes and non-myocyte cells (fibroblasts, macrophages) [151–154]. As we learn more about how endothelial cells communicate with other resident cell types in various organ systems, new opportunities will arise for the advancement of treatments for a wide range of pathologies.

Conclusion

In summary, vascular endothelial cells communicate with each other using a variety of mechanisms over different timescales to regulate their biology and physiology in response to evolving needs and cues in health and disease. As we continue to uncover these mechanisms, more and more therapeutic strategies are developed to target cell–cell communication in the vasculature, and these approaches have proven immensely successful in a variety of pathologies.

References

1. Reese DE, Mikawa T, Bader DM. Development of the coronary vessel system. Circ Res. 2002;91:761–8.
2. Taqueti VR, Di Carli MF. Coronary microvascular disease pathogenic mechanisms and therapeutic options: JACC state-of-the-art review. J Am Coll Cardiol. 2018;72:2625–41.
3. Rhodin JA. The ultrastructure of mammalian arterioles and precapillary sphincters. J Ultrastruct Res. 1967;18:181–223.
4. Pugsley MK, Tabrizchi R. The vascular system. An overview of structure and function. J Pharmacol Toxicol Methods. 2000;44:333–40.
5. Dyer LA, Patterson C. Development of the endothelium: an emphasis on heterogeneity. Semin Thromb Hemost. 2010;36:227–35.
6. Okamoto T, Suzuki K. The role of gap junction-mediated endothelial cell-cell interaction in the crosstalk between inflammation and blood coagulation. Int J Mol Sci. 2017;18:2254.
7. Segal SS. Integration and modulation of intercellular signaling underlying blood flow control. J Vasc Res. 2015;52:136–57.
8. Furuse M. Molecular basis of the core structure of tight junctions. Cold Spring Harb Perspect Biol. 2010;2:a002907.
9. Szaszi K, Amoozadeh Y. New insights into functions, regulation, and pathological roles of tight junctions in kidney tubular epithelium. Int Rev Cell Mol Biol. 2014;308:205–71.
10. Weber CR. Dynamic properties of the tight junction barrier. Ann N Y Acad Sci. 2012;1257:77–84.
11. Okamoto T, Akita N, Hayashi T, Shimaoka M, Suzuki K. Endothelial connexin 32 regulates tissue factor expression induced by inflammatory stimulation and direct cell-cell interaction with activated cells. Atherosclerosis. 2014;236:430–7.
12. Okamoto T, Akiyama M, Takeda M, Gabazza EC, Hayashi T, Suzuki K. Connexin32 is expressed in vascular endothelial cells and participates in gap-junction intercellular communication. Biochem Biophys Res Commun. 2009;382:264–8.
13. Okamoto T, Kawamoto E, Takagi Y, Akita N, Hayashi T, Park EJ, Suzuki K, Shimaoka M. Gap junction-mediated regulation of endothelial cellular stiffness. Sci Rep. 2017;7:6134.
14. Allagnat F, Dubuis C, Lambelet M, Le Gal L, Alonso F, Corpataux JM, Deglise S, Haefliger JA. Connexin37 reduces smooth muscle cell proliferation and intimal hyperplasia in a mouse model of carotid artery ligation. Cardiovasc Res. 2017;113:805–16.
15. Figueroa XF, Duling BR. Dissection of two Cx37-independent conducted vasodilator mechanisms by deletion of Cx40: electrotonic versus regenerative conduction. Am J Physiol Heart Circ Physiol. 2008;295:H2001–7.
16. Meens MJ, Alonso F, Le Gal L, Kwak BR, Haefliger JA. Endothelial Connexin37 and Connexin40 participate in basal but not agonist-induced NO release. Cell Commun Signal. 2015;13:34.
17. Little TL, Beyer EC, Duling BR. Connexin 43 and connexin 40 gap junctional proteins are present in arteriolar smooth muscle and endothelium in vivo. Am J Phys. 1995;268:H729–39.
18. Hakim CH, Jackson WF, Segal SS. Connexin isoform expression in smooth muscle cells and endothelial cells of hamster cheek pouch arterioles and retractor feed arteries. Microcirculation. 2008;15:503–14.
19. Yuan D, Sun G, Zhang R, Luo C, Ge M, Luo G, Hei Z. Connexin 43 expressed in endothelial cells modulates monocyteendothelial adhesion by regulating cell adhesion proteins. Mol Med Rep. 2015;12:7146–52.
20. Schmidt VJ, Jobs A, von Maltzahn J, Worsdorfer P, Willecke K, de Wit C. Connexin45 is expressed in vascular smooth muscle but its function remains elusive. PLoS One. 2012;7:e42287.
21. King DR, Sedovy MW, Leng X, Xue J, Lamouille S, Koval M, Isakson BE, Johnstone SR. Mechanisms of connexin regulating peptides. Int J Mol Sci. 2021;22:10186.

22. Ruan Z, Orozco IJ, Du J, Lu W. Structures of human pannexin 1 reveal ion pathways and mechanism of gating. Nature. 2020;584:646–51.
23. Deng Z, He Z, Maksaev G, Bitter RM, Rau M, Fitzpatrick JAJ, Yuan P. Cryo-EM structures of the ATP release channel pannexin 1. Nat Struct Mol Biol. 2020;27:373–81.
24. Michalski K, Syrjanen JL, Henze E, Kumpf J, Furukawa H, Kawate T. The Cryo-EM structure of pannexin 1 reveals unique motifs for ion selection and inhibition. Elife. 2020;9:e54670.
25. Mou L, Ke M, Song M, Shan Y, Xiao Q, Liu Q, Li J, Sun K, Pu L, Guo L, Geng J, Wu J, Deng D. Structural basis for gating mechanism of Pannexin 1 channel. Cell Res. 2020;30:452–4.
26. Drury AN, Szent-Gyorgyi A. The physiological activity of adenine compounds with especial reference to their action upon the mammalian heart. J Physiol. 1929;68:213–37.
27. Burnstock G. Purinergic nerves. Pharmacol Rev. 1972;24:509–81.
28. Burnstock G. Purinergic nerves and receptors. Prog Biochem Pharmacol. 1980;16:141–54.
29. Burnstock G. Purine and pyrimidine receptors. Cell Mol Life Sci. 2007;64:1471–83.
30. Ralevic V, Burnstock G. Receptors for purines and pyrimidines. Pharmacol Rev. 1998;50:413–92.
31. Burnstock G. Purinergic signaling in the cardiovascular system. Circ Res. 2017;120:207–28.
32. Headrick JP, Ashton KJ, Rose'meyer RB, Peart JN. Cardiovascular adenosine receptors: expression, actions and interactions. Pharmacol Ther. 2013;140:92–111.
33. Zhang YY, Bu YF, Liu JZ. Production of L-ornithine from sucrose and molasses by recombinant Corynebacterium glutamicum. Folia Microbiol (Praha). 2015;60:393–8.
34. Coddou C, Yan Z, Obsil T, Huidobro-Toro JP, Stojilkovic SS. Activation and regulation of purinergic P2X receptor channels. Pharmacol Rev. 2011;63:641–83.
35. Harden TK, Sesma JI, Fricks IP, Lazarowski ER. Signalling and pharmacological properties of the P2Y receptor. Acta Physiol (Oxf). 2010;199:149–60.
36. von Kugelgen I, Hoffmann K. Pharmacology and structure of P2Y receptors. Neuropharmacology. 2016;104:50–61.
37. Erb L, Weisman GA. Coupling of P2Y receptors to G proteins and other signaling pathways. Wiley Interdiscip Rev Membr Transp Signal. 2012;1:789–803.
38. Navarro G, Cordomi A, Casado-Anguera V, Moreno E, Cai NS, Cortes A, Canela EI, Dessauer CW, Casado V, Pardo L, Lluis C, Ferre S. Evidence for functional pre-coupled complexes of receptor heteromers and adenylyl cyclase. Nat Commun. 2018;9:1242.
39. Kapicka CL, Montamat SC, Olson RD, Musser B, Mudumbi RV, Vestal RE. Species comparison of adenosine A1 receptors in isolated mammalian atrial and ventricular myocardium. Life Sci. 2003;72:2825–38.
40. Zhang Y, Wernly B, Cao X, Mustafa SJ, Tang Y, Zhou Z. Adenosine and adenosine receptor-mediated action in coronary microcirculation. Basic Res Cardiol. 2021;116:22.
41. Banfi C, Ferrario S, De Vincenti O, Ceruti S, Fumagalli M, Mazzola A, D'Ambrosi N, Volonte C, Fratto P, Vitali E, Burnstock G, Beltrami E, Parolari A, Polvani G, Biglioli P, Tremoli E, Abbracchio MP. P2 receptors in human heart: upregulation of P2X6 in patients undergoing heart transplantation, interaction with TNFalpha and potential role in myocardial cell death. J Mol Cell Cardiol. 2005;39:929–39.
42. Bogdanov Y, Rubino A, Burnstock G. Characterisation of subtypes of the P2X and P2Y families of ATP receptors in the foetal human heart. Life Sci. 1998;62:697–703.
43. Burnstock G, Knight GE. Cellular distribution and functions of P2 receptor subtypes in different systems. Int Rev Cytol. 2004;240:31–304.
44. Garcia-Guzman M, Soto F, Laube B, Stuhmer W. Molecular cloning and functional expression of a novel rat heart P2X purinoceptor. FEBS Lett. 1996;388:123–7.
45. Garcia-Guzman M, Stuhmer W, Soto F. Molecular characterization and pharmacological properties of the human P2X3 purinoceptor. Brain Res Mol Brain Res. 1997;47:59–66.
46. Hansen MA, Bennett MR, Barden JA. Distribution of purinergic P2X receptors in the rat heart. J Auton Nerv Syst. 1999;78:1–9.
47. Musa H, Tellez JO, Chandler NJ, Greener ID, Maczewski M, Mackiewicz U, Beresewicz A, Molenaar P, Boyett MR, Dobrzynski H. P2 purinergic receptor mRNA in rat and

human sinoatrial node and other heart regions. Naunyn Schmiedeberg's Arch Pharmacol. 2009;379:541–9.
48. Rassendren F, Buell GN, Virginio C, Collo G, North RA, Surprenant A. The permeabilizing ATP receptor, P2X7. Cloning and expression of a human cDNA. J Biol Chem. 1997;272:5482–6.
49. Soto F, Garcia-Guzman M, Gomez-Hernandez JM, Hollmann M, Karschin C, Stuhmer W. P2X4: an ATP-activated ionotropic receptor cloned from rat brain. Proc Natl Acad Sci U S A. 1996;93:3684–8.
50. Surprenant A, Rassendren F, Kawashima E, North RA, Buell G. The cytolytic P2Z receptor for extracellular ATP identified as a P2X receptor (P2X7). Science. 1996;272:735–8.
51. Vassort G. Adenosine 5′-triphosphate: a P2-purinergic agonist in the myocardium. Physiol Rev. 2001;81:767–806.
52. Batori R, Kumar S, Bordan Z, Cherian-Shaw M, Kovacs-Kasa A, MacDonald JA, Fulton DJR, Erdodi F, Verin AD. Differential mechanisms of adenosine- and ATPgammaS-induced microvascular endothelial barrier strengthening. J Cell Physiol. 2019;234:5863–79.
53. Lyubchenko T, Woodward H, Veo KD, Burns N, Nijmeh H, Liubchenko GA, Stenmark KR, Gerasimovskaya EV. P2Y1 and P2Y13 purinergic receptors mediate Ca2+ signaling and proliferative responses in pulmonary artery vasa vasorum endothelial cells. Am J Physiol Cell Physiol. 2011;300:C266–75.
54. Maier-Begandt D, Comstra HS, Molina SA, Kruger N, Ruddiman CA, Chen YL, Chen X, Biwer LA, Johnstone SR, Lohman AW, Good ME, DeLalio LJ, Hong K, Bacon HM, Yan Z, Sonkusare SK, Koval M, Isakson BE. A venous-specific purinergic signaling cascade initiated by Pannexin 1 regulates TNFalpha-induced increases in endothelial permeability. Sci Signal. 2021;14:eaba2940.
55. Marcus AJ, Broekman MJ, Drosopoulos JH, Olson KE, Islam N, Pinsky DJ, Levi R. Role of CD39 (NTPDase-1) in thromboregulation, cerebroprotection, and cardioprotection. Semin Thromb Hemost. 2005;31:234–46.
56. Ohta M, Toyama K, Gutterman DD, Campbell WB, Lemaitre V, Teraoka R, Miura H. Ecto-5′-nucleotidase, CD73, is an endothelium-derived hyperpolarizing factor synthase. Arterioscler Thromb Vasc Biol. 2013;33:629–36.
57. Zimmermann H. 5′-Nucleotidase: molecular structure and functional aspects. Biochem J. 1992;285(Pt 2):345–65.
58. Ikeda U, Kurosaki K, Ohya K, Shimada K. Adenosine stimulates nitric oxide synthesis in vascular smooth muscle cells. Cardiovasc Res. 1997;35:168–74.
59. Teng B, Ledent C, Mustafa SJ. Up-regulation of A 2B adenosine receptor in A 2A adenosine receptor knockout mouse coronary artery. J Mol Cell Cardiol. 2008;44:905–14.
60. Henry P, Demolombe S, Puceat M, Escande D. Adenosine A1 stimulation activates delta-protein kinase C in rat ventricular myocytes. Circ Res. 1996;78:161–5.
61. Germack R, Griffin M, Dickenson JM. Activation of protein kinase B by adenosine A1 and A3 receptors in newborn rat cardiomyocytes. J Mol Cell Cardiol. 2004;37:989–99.
62. Xu Z, Park SS, Mueller RA, Bagnell RC, Patterson C, Boysen PG. Adenosine produces nitric oxide and prevents mitochondrial oxidant damage in rat cardiomyocytes. Cardiovasc Res. 2005;65:803–12.
63. Vlasov AV. Comparative evaluation of late results of the Hoffmeister-Finsterer gastrectomy and the method of valve anastomoses in peptic ulcer. Klin Med (Mosk). 1989;67:84–7.
64. Lee JE, Bokoch G, Liang BT. A novel cardioprotective role of RhoA: new signaling mechanism for adenosine. FASEB J. 2001;15:1886–94.
65. Dinh W, Albrecht-Kupper B, Gheorghiade M, Voors AA, van der Laan M, Sabbah HN. Partial adenosine A1 agonist in heart failure. Handb Exp Pharmacol. 2017;243:177–203.
66. Sabbah HN, Gupta RC, Kohli S, Wang M, Rastogi S, Zhang K, Zimmermann K, Diedrichs N, Albrecht-Kupper BE. Chronic therapy with a partial adenosine A1-receptor agonist improves left ventricular function and remodeling in dogs with advanced heart failure. Circ Heart Fail. 2013;6:563–71.

67. da Silva JS, Gabriel-Costa D, Sudo RT, Wang H, Groban L, Ferraz EB, Nascimento JH, Fraga CA, Barreiro EJ, Zapata-Sudo G. Adenosine A2A receptor agonist prevents cardiac remodeling and dysfunction in spontaneously hypertensive male rats after myocardial infarction. Drug Des Devel Ther. 2017;11:553–62.
68. Tian Y, Piras BA, Kron IL, French BA, Yang Z. Adenosine 2B receptor activation reduces myocardial reperfusion injury by promoting anti-inflammatory macrophages differentiation via PI3K/Akt pathway. Oxidative Med Cell Longev. 2015;2015:585297.
69. Albrecht-Kupper BE, Leineweber K, Nell PG. Partial adenosine A1 receptor agonists for cardiovascular therapies. Purinergic Signal. 2012;8:91–9.
70. Auchampach JA, Ge ZD, Wan TC, Moore J, Gross GJ. A3 adenosine receptor agonist IB-MECA reduces myocardial ischemia-reperfusion injury in dogs. Am J Physiol Heart Circ Physiol. 2003;285:H607–13.
71. Tendera M, Gaszewska-Zurek E, Parma Z, Ponikowski P, Jankowska E, Kawecka-Jaszcz K, Czarnecka D, Krzeminska-Pakula M, Bednarkiewicz Z, Sosnowski M, Ochan Kilama M, Agrawal R. The new oral adenosine A1 receptor agonist capadenoson in male patients with stable angina. Clin Res Cardiol. 2012;101:585–91.
72. Paganelli F, Gaudry M, Ruf J, Guieu R. Recent advances in the role of the adenosinergic system in coronary artery disease. Cardiovasc Res. 2021;117:1284–94.
73. Liu GS, Richards SC, Olsson RA, Mullane K, Walsh RS, Downey JM. Evidence that the adenosine A3 receptor may mediate the protection afforded by preconditioning in the isolated rabbit heart. Cardiovasc Res. 1994;28:1057–61.
74. Liu GS, Thornton J, Van Winkle DM, Stanley AW, Olsson RA, Downey JM. Protection against infarction afforded by preconditioning is mediated by A1 adenosine receptors in rabbit heart. Circulation. 1991;84:350–6.
75. Schlack W, Schafer M, Uebing A, Schafer S, Borchard U, Thamer V. Adenosine A2-receptor activation at reperfusion reduces infarct size and improves myocardial wall function in dog heart. J Cardiovasc Pharmacol. 1993;22:89–96.
76. Woolfson RG, Patel VC, Neild GH, Yellon DM. Inhibition of nitric oxide synthesis reduces infarct size by an adenosine-dependent mechanism. Circulation. 1995;91:1545–51.
77. Yang Z, Cerniway RJ, Byford AM, Berr SS, French BA, Matherne GP. Cardiac overexpression of A1-adenosine receptor protects intact mice against myocardial infarction. Am J Physiol Heart Circ Physiol. 2002;282:H949–55.
78. Fabritz L, Kirchhof P, Fortmuller L, Auchampach JA, Baba HA, Breithardt G, Neumann J, Boknik P, Schmitz W. Gene dose-dependent atrial arrhythmias, heart block, and bradycardiomyopathy in mice overexpressing A(3) adenosine receptors. Cardiovasc Res. 2004;62:500–8.
79. Ge ZD, Peart JN, Kreckler LM, Wan TC, Jacobson MA, Gross GJ, Auchampach JA. Cl-IB-MECA [2-chloro-N6-(3-iodobenzyl)adenosine-5′-N-methylcarboxamide] reduces ischemia/reperfusion injury in mice by activating the A3 adenosine receptor. J Pharmacol Exp Ther. 2006;319:1200–10.
80. Harrison GJ, Cerniway RJ, Peart J, Berr SS, Ashton K, Regan S, Paul Matherne G, Headrick JP. Effects of A(3) adenosine receptor activation and gene knock-out in ischemic-reperfused mouse heart. Cardiovasc Res. 2002;53:147–55.
81. Wan TC, Tosh DK, Du L, Gizewski ET, Jacobson KA, Auchampach JA. Polyamidoamine (PAMAM) dendrimer conjugate specifically activates the A3 adenosine receptor to improve post-ischemic/reperfusion function in isolated mouse hearts. BMC Pharmacol. 2011;11:11.
82. Ashton KJ, Nilsson U, Willems L, Holmgren K, Headrick JP. Effects of aging and ischemia on adenosine receptor transcription in mouse myocardium. Biochem Biophys Res Commun. 2003;312:367–72.
83. Bulluck H, Sirker A, Loke YK, Garcia-Dorado D, Hausenloy DJ. Clinical benefit of adenosine as an adjunct to reperfusion in ST-elevation myocardial infarction patients: an updated meta-analysis of randomized controlled trials. Int J Cardiol. 2016;202:228–37.

84. Rolland-Turner M, Goretti E, Bousquenaud M, Leonard F, Nicolas C, Zhang L, Maskali F, Marie PY, Devaux Y, Wagner D. Adenosine stimulates the migration of human endothelial progenitor cells. Role of CXCR4 and microRNA-150. PLoS One. 2013;8:e54135.
85. Ely SW, Berne RM. Protective effects of adenosine in myocardial ischemia. Circulation. 1992;85:893–904.
86. Shin EY, Wang L, Zemskova M, Deppen J, Xu K, Strobel F, Garcia AJ, Tirouvanziam R, Levit RD. Adenosine production by biomaterial-supported mesenchymal stromal cells reduces the innate inflammatory response in myocardial ischemia/reperfusion injury. J Am Heart Assoc. 2018;7:e006949.
87. Singh L, Kulshrestha R, Singh N, Jaggi AS. Mechanisms involved in adenosine pharmacological preconditioning-induced cardioprotection. Korean J Physiol Pharmacol. 2018;22:225–34.
88. Ralevic V. P2X receptors in the cardiovascular system and their potential as therapeutic targets in disease. Curr Med Chem. 2015;22:851–65.
89. Sathanoori R, Rosi F, Gu BJ, Wiley JS, Muller CE, Olde B, Erlinge D. Shear stress modulates endothelial KLF2 through activation of P2X4. Purinergic Signal. 2015;11:139–53.
90. Yamamoto K, Korenaga R, Kamiya A, Ando J. Fluid shear stress activates Ca(2+) influx into human endothelial cells via P2X4 purinoceptors. Circ Res. 2000;87:385–91.
91. Yamamoto K, Korenaga R, Kamiya A, Qi Z, Sokabe M, Ando J. P2X(4) receptors mediate ATP-induced calcium influx in human vascular endothelial cells. Am J Physiol Heart Circ Physiol. 2000;279:H285–92.
92. Yamamoto K, Sokabe T, Matsumoto T, Yoshimura K, Shibata M, Ohura N, Fukuda T, Sato T, Sekine K, Kato S, Isshiki M, Fujita T, Kobayashi M, Kawamura K, Masuda H, Kamiya A, Ando J. Impaired flow-dependent control of vascular tone and remodeling in P2X4-deficient mice. Nat Med. 2006;12:133–7.
93. Raqeeb A, Sheng J, Ao N, Braun AP. Purinergic P2Y2 receptors mediate rapid Ca(2+) mobilization, membrane hyperpolarization and nitric oxide production in human vascular endothelial cells. Cell Calcium. 2011;49:240–8.
94. Sathanoori R, Bryl-Gorecka P, Muller CE, Erb L, Weisman GA, Olde B, Erlinge D. P2Y2 receptor modulates shear stress-induced cell alignment and actin stress fibers in human umbilical vein endothelial cells. Cell Mol Life Sci. 2017;74:731–46.
95. Hou M, Malmsjo M, Moller S, Pantev E, Bergdahl A, Zhao XH, Sun XY, Hedner T, Edvinsson L, Erlinge D. Increase in cardiac P2X1-and P2Y2-receptor mRNA levels in congestive heart failure. Life Sci. 1999;65:1195–206.
96. Erlinge D, Harnek J, van Heusden C, Olivecrona G, Jern S, Lazarowski E. Uridine triphosphate (UTP) is released during cardiac ischemia. Int J Cardiol. 2005;100:427–33.
97. Mazzola A, Amoruso E, Beltrami E, Lecca D, Ferrario S, Cosentino S, Tremoli E, Ceruti S, Abbracchio MP. Opposite effects of uracil and adenine nucleotides on the survival of murine cardiomyocytes. J Cell Mol Med. 2008;12:522–36.
98. Sonin D, Zhou SY, Cronin C, Sonina T, Wu J, Jacobson KA, Pappano A, Liang BT. Role of P2X purinergic receptors in the rescue of ischemic heart failure. Am J Physiol Heart Circ Physiol. 2008;295:H1191–7.
99. Wee S, Peart JN, Headrick JP. P2 purinoceptor-mediated cardioprotection in ischemic-reperfused mouse heart. J Pharmacol Exp Ther. 2007;323:861–7.
100. Cosentino S, Banfi C, Burbiel JC, Luo H, Tremoli E, Abbracchio MP. Cardiomyocyte death induced by ischaemic/hypoxic stress is differentially affected by distinct purinergic P2 receptors. J Cell Mol Med. 2012;16:1074–84.
101. Catterall WA. Voltage-gated calcium channels. Cold Spring Harb Perspect Biol. 2011;3:a003947.
102. Bodi I, Mikala G, Koch SE, Akhter SA, Schwartz A. The L-type calcium channel in the heart: the beat goes on. J Clin Invest. 2005;115:3306–17.
103. Bossu JL, Elhamdani A, Feltz A. Voltage-dependent calcium entry in confluent bovine capillary endothelial cells. FEBS Lett. 1992;299:239–42.

104. Bossu JL, Feltz A, Rodeau JL, Tanzi F. Voltage-dependent transient calcium currents in freshly dissociated capillary endothelial cells. FEBS Lett. 1989;255:377–80.
105. Gilbert G, Courtois A, Dubois M, Cussac LA, Ducret T, Lory P, Marthan R, Savineau JP, Quignard JF. T-type voltage gated calcium channels are involved in endothelium-dependent relaxation of mice pulmonary artery. Biochem Pharmacol. 2017;138:61–72.
106. King DR, Padget RL, Perry J, Hoeker G, Smyth JW, Brown DA, Poelzing S. Elevated perfusate [Na(+)] increases contractile dysfunction during ischemia and reperfusion. Sci Rep. 2020;10:17289.
107. Graier WF, Paltauf-Doburzynska J, Hill BJ, Fleischhacker E, Hoebel BG, Kostner GM, Sturek M. Submaximal stimulation of porcine endothelial cells causes focal Ca2+ elevation beneath the cell membrane. J Physiol. 1998;506(Pt 1):109–25.
108. Teubl M, Groschner K, Kohlwein SD, Mayer B, Schmidt K. Na(+)/Ca(2+) exchange facilitates Ca(2+)-dependent activation of endothelial nitric-oxide synthase. J Biol Chem. 1999;274:29529–35.
109. Lillo MA, Gaete PS, Puebla M, Ardiles NM, Poblete I, Becerra A, Simon F, Figueroa XF. Critical contribution of Na(+)-Ca(2+) exchanger to the Ca(2+)-mediated vasodilation activated in endothelial cells of resistance arteries. FASEB J. 2018;32:2137–47.
110. Schneider JC, El Kebir D, Chereau C, Mercier JC, Dall'Ava-Santucci J, Dinh-Xuan AT. Involvement of Na(+)/Ca(2+) exchanger in endothelial NO production and endothelium-dependent relaxation. Am J Physiol Heart Circ Physiol. 2002;283:H837–44.
111. Smani T, Gomez LJ, Regodon S, Woodard GE, Siegfried G, Khatib AM, Rosado JA. TRP channels in angiogenesis and other endothelial functions. Front Physiol. 2018;9:1731.
112. Kwan HY, Huang Y, Yao X. TRP channels in endothelial function and dysfunction. Biochim Biophys Acta. 2007;1772:907–14.
113. Ottolini M, Hong K, Sonkusare SK. Calcium signals that determine vascular resistance. Wiley Interdiscip Rev Syst Biol Med. 2019;11:e1448.
114. Ottolini M, Sonkusare SK. The calcium signaling mechanisms in arterial smooth muscle and endothelial cells. Compr Physiol. 2021;11:1831–69.
115. Somlyo AP, Somlyo AV. The sarcoplasmic reticulum: then and now. Novartis Found Symp. 2002;246:258–68; discussion 268–71, 272–6.
116. Koss KL, Kranias EG. Phospholamban: a prominent regulator of myocardial contractility. Circ Res. 1996;79:1059–63.
117. Sutliff RL, Hoying JB, Kadambi VJ, Kranias EG, Paul RJ. Phospholamban is present in endothelial cells and modulates endothelium-dependent relaxation. Evidence from phospholamban gene-ablated mice. Circ Res. 1999;84:360–4.
118. Simmerman HK, Collins JH, Theibert JL, Wegener AD, Jones LR. Sequence analysis of phospholamban. Identification of phosphorylation sites and two major structural domains. J Biol Chem. 1986;261:13333–41.
119. Meissner G. The structural basis of ryanodine receptor ion channel function. J Gen Physiol. 2017;149:1065–89.
120. Lesh RE, Marks AR, Somlyo AV, Fleischer S, Somlyo AP. Anti-ryanodine receptor antibody binding sites in vascular and endocardial endothelium. Circ Res. 1993;72:481–8.
121. Ziegelstein RC, Spurgeon HA, Pili R, Passaniti A, Cheng L, Corda S, Lakatta EG, Capogrossi MC. A functional ryanodine-sensitive intracellular Ca2+ store is present in vascular endothelial cells. Circ Res. 1994;74:151–6.
122. Corda S, Spurgeon HA, Lakatta EG, Capogrossi MC, Ziegelstein RC. Endoplasmic reticulum Ca2+ depletion unmasks a caffeine-induced Ca2+ influx in human aortic endothelial cells. Circ Res. 1995;77:927–35.
123. Echeverri D, Montes FR, Cabrera M, Galan A, Prieto A. Caffeine's vascular mechanisms of action. Int J Vasc Med. 2010;2010:834060.
124. Evangelista AM, Thompson MD, Weisbrod RM, Pimental DR, Tong X, Bolotina VM, Cohen RA. Redox regulation of SERCA2 is required for vascular endothelial growth factor-induced signaling and endothelial cell migration. Antioxid Redox Signal. 2012;17:1099–108.

125. Kohler R, Brakemeier S, Kuhn M, Degenhardt C, Buhr H, Pries A, Hoyer J. Expression of ryanodine receptor type 3 and TRP channels in endothelial cells: comparison of in situ and cultured human endothelial cells. Cardiovasc Res. 2001;51:160–8.
126. Kohler R, Degenhardt C, Kuhn M, Runkel N, Paul M, Hoyer J. Expression and function of endothelial Ca(2+)-activated K(+) channels in human mesenteric artery: a single-cell reverse transcriptase-polymerase chain reaction and electrophysiological study in situ. Circ Res. 2000;87:496–503.
127. Ledoux J, Taylor MS, Bonev AD, Hannah RM, Solodushko V, Shui B, Tallini Y, Kotlikoff MI, Nelson MT. Functional architecture of inositol 1,4,5-trisphosphate signaling in restricted spaces of myoendothelial projections. Proc Natl Acad Sci U S A. 2008;105:9627–32.
128. Foskett JK, White C, Cheung KH, Mak DO. Inositol trisphosphate receptor Ca2+ release channels. Physiol Rev. 2007;87:593–658.
129. Adkins CE, Taylor CW. Lateral inhibition of inositol 1,4,5-trisphosphate receptors by cytosolic Ca(2+). Curr Biol. 1999;9:1115–8.
130. Mak DO, McBride S, Foskett JK. Inositol 1,4,5-trisphosphate [correction of tris-phosphate] activation of inositol trisphosphate [correction of tris-phosphate] receptor Ca2+ channel by ligand tuning of Ca2+ inhibition. Proc Natl Acad Sci U S A. 1998;95:15821–5.
131. Taylor CW, Laude AJ. IP3 receptors and their regulation by calmodulin and cytosolic Ca2+. Cell Calcium. 2002;32:321–34.
132. Allbritton NL, Meyer T, Stryer L. Range of messenger action of calcium ion and inositol 1,4,5-trisphosphate. Science. 1992;258:1812–5.
133. Leybaert L. IP3, still on the move but now in the slow lane. Sci Signal. 2016;9:fs17.
134. Large WA, Saleh SN, Albert AP. Role of phosphoinositol 4,5-bisphosphate and diacylglycerol in regulating native TRPC channel proteins in vascular smooth muscle. Cell Calcium. 2009;45:574–82.
135. Saleh SN, Albert AP, Peppiatt CM, Large WA. Angiotensin II activates two cation conductances with distinct TRPC1 and TRPC6 channel properties in rabbit mesenteric artery myocytes. J Physiol. 2006;577:479–95.
136. Leung FP, Yung LM, Yao X, Laher I, Huang Y. Store-operated calcium entry in vascular smooth muscle. Br J Pharmacol. 2008;153:846–57.
137. Kwan HY, Huang Y, Yao X. Store-operated calcium entry in vascular endothelial cells is inhibited by cGMP via a protein kinase G-dependent mechanism. J Biol Chem. 2000;275:6758–63.
138. Norwood N, Moore TM, Dean DA, Bhattacharjee R, Li M, Stevens T. Store-operated calcium entry and increased endothelial cell permeability. Am J Physiol Lung Cell Mol Physiol. 2000;279:L815–24.
139. Yang Y, Sohma Y, Nourian Z, Ella SR, Li M, Stupica A, Korthuis RJ, Davis MJ, Braun AP, Hill MA. Mechanisms underlying regional differences in the Ca2+ sensitivity of BK(Ca) current in arteriolar smooth muscle. J Physiol. 2013;591:1277–93.
140. Ambudkar IS, de Souza LB, Ong HL. TRPC1, Orai1, and STIM1 in SOCE: friends in tight spaces. Cell Calcium. 2017;63:33–9.
141. Lindsey SH, Tribe RM, Songu-Mize E. Cyclic stretch decreases TRPC4 protein and capacitative calcium entry in rat vascular smooth muscle cells. Life Sci. 2008;83:29–34.
142. Shi J, Ju M, Abramowitz J, Large WA, Birnbaumer L, Albert AP. TRPC1 proteins confer PKC and phosphoinositol activation on native heteromeric TRPC1/C5 channels in vascular smooth muscle: comparative study of wild-type and TRPC1-/- mice. FASEB J. 2012;26:409–19.
143. Thebault S, Zholos A, Enfissi A, Slomianny C, Dewailly E, Roudbaraki M, Parys J, Prevarskaya N. Receptor-operated Ca2+ entry mediated by TRPC3/TRPC6 proteins in rat prostate smooth muscle (PS1) cell line. J Cell Physiol. 2005;204:320–8.
144. Avila-Medina J, Mayoral-Gonzalez I, Dominguez-Rodriguez A, Gallardo-Castillo I, Ribas J, Ordonez A, Rosado JA, Smani T. The complex role of store operated calcium entry pathways and related proteins in the function of cardiac, skeletal and vascular smooth muscle cells. Front Physiol. 2018;9:257.

145. Sonkusare SK, Bonev AD, Ledoux J, Liedtke W, Kotlikoff MI, Heppner TJ, Hill-Eubanks DC, Nelson MT. Elementary Ca2+ signals through endothelial TRPV4 channels regulate vascular function. Science. 2012;336:597–601.
146. Nausch LW, Bonev AD, Heppner TJ, Tallini Y, Kotlikoff MI, Nelson MT. Sympathetic nerve stimulation induces local endothelial Ca2+ signals to oppose vasoconstriction of mouse mesenteric arteries. Am J Physiol Heart Circ Physiol. 2012;302:H594–602.
147. Brahler S, Kaistha A, Schmidt VJ, Wolfle SE, Busch C, Kaistha BP, Kacik M, Hasenau AL, Grgic I, Si H, Bond CT, Adelman JP, Wulff H, de Wit C, Hoyer J, Kohler R. Genetic deficit of SK3 and IK1 channels disrupts the endothelium-derived hyperpolarizing factor vasodilator pathway and causes hypertension. Circulation. 2009;119:2323–32.
148. Si H, Heyken WT, Wolfle SE, Tysiac M, Schubert R, Grgic I, Vilianovich L, Giebing G, Maier T, Gross V, Bader M, de Wit C, Hoyer J, Kohler R. Impaired endothelium-derived hyperpolarizing factor-mediated dilations and increased blood pressure in mice deficient of the intermediate-conductance Ca2+-activated K+ channel. Circ Res. 2006;99:537–44.
149. Seko Y, Nishimura H, Takahashi N, Ashida T, Nagai R. Serum levels of vascular endothelial growth factor and transforming growth factor-beta1 in patients with atrial fibrillation undergoing defibrillation therapy. Jpn Heart J. 2000;41:27–32.
150. Sandow SL, Hill CE. Incidence of myoendothelial gap junctions in the proximal and distal mesenteric arteries of the rat is suggestive of a role in endothelium-derived hyperpolarizing factor-mediated responses. Circ Res. 2000;86:341–6.
151. Givvimani S, Qipshidze N, Tyagi N, Mishra PK, Sen U, Tyagi SC. Synergism between arrhythmia and hyperhomo-cysteinemia in structural heart disease. Int J Physiol Pathophysiol Pharmacol. 2011;3:107–19.
152. Leuschner F, Nahrendorf M. Novel functions of macrophages in the heart: insights into electrical conduction, stress, and diastolic dysfunction. Eur Heart J. 2020;41:989–94.
153. Narmoneva DA, Vukmirovic R, Davis ME, Kamm RD, Lee RT. Endothelial cells promote cardiac myocyte survival and spatial reorganization: implications for cardiac regeneration. Circulation. 2004;110:962–8.
154. Rook MB, van Ginneken AC, de Jonge B, el Aoumari A, Gros D, Jongsma HJ. Differences in gap junction channels between cardiac myocytes, fibroblasts, and heterologous pairs. Am J Phys. 1992;263:C959–77.

Lysophosphatidic Acid Regulates Endothelial Barrier Integrity

Jing Zhao, Sarah J. Taleb, Heather Wang, and Yutong Zhao

Abstract The main function of vasculature is to serve as a vessel network for blood circulation between lungs and other organs. The endothelium is a major component of blood vessels, lining the inside of vessels and playing a central role in maintenance of vascular integrity. The endothelial barrier prevents blood component leakage into perivascular tissues. Increases in vascular permeability result in tissue edema, which is a hallmark of acute inflammatory diseases. Lysophosphatidic acid (LPA) is a simple phospholipid that exerts many physiopathological functions in various cell types including endothelial cells (ECs). LPA levels are detectable in plasma. Abnormal changes in LPA levels are correlated to diseases. LPA has been shown to regulate endothelial barrier integrity differently in different types of ECs. This chapter will summarize the current knowledge of the effect of LPA on endothelial barrier function and discuss how different ECs respond to LPA and molecular mechanisms underlying LPA-regulated EC barrier functions.

Keywords Lysophosphatidic acid · Endothelial cells · Vascular permeability · Edema · Signal pathway

J. Zhao · S. J. Taleb · H. Wang
Department of Physiology and Cell Biology, Dorothy M. Davis Heart and Lung Research Institute, The Ohio State University, Columbus, OH, USA

Y. Zhao (✉)
Department of Physiology and Cell Biology, Dorothy M. Davis Heart and Lung Research Institute, The Ohio State University, Columbus, OH, USA

Department of Physiology and Cell Biology and Internal Medicine, The Ohio State University, Columbus, OH, USA
e-mail: yutong.zhao@osumc.edu

N. L. Parinandi, T. J. Hund (eds.), *Cardiovascular Signaling in Health and Disease*, https://doi.org/10.1007/978-3-031-08309-9_16

Introduction

The blood vessel network circulates blood in the pulmonary and systemic circulatory systems. Oxygenated blood is delivered from the lungs to the left heart, and is then pumped to systemic tissues. Deoxygenated blood is circulated back to the lungs through the right side of heart. Blood vessels exhibit distinct properties and play different physiological roles dependent on their localization in different tissues, if they carry oxygenated or deoxygenated blood, and diameter (reviewed in [1–4]).

Endothelial cells (ECs) form a monolayer and line the interior surface of blood vessels. The major function of ECs is to maintain vessel architecture and prevent blood component leakage into perivessels. The vascular endothelium exhibits a semi-permeable function. In normal physiological conditions, ECs allow solute and certain molecules smaller than 40 kDa to extravasate to surrounding tissues. The barrier formed by the EC monolayer keeps larger molecules and blood cells in circulation in the vessels. During pathological conditions, to a certain extent, EC barrier integrity is disrupted, leading to leakage of blood components including plasma into tissues, causing tissue edema. EC barrier integrity varies between sources of ECs (reviewed in [5–8]). Regulation of EC barrier integrity in blood-air barrier and the blood-brain barrier (BBB) has been well studied (reviewed in [9–15]). ECs express both adherens and tight junctions which control barrier integrity and mediate intracellular signals.

Lysophospholipids belong to a group of bio-active phospholipids that regulates intracellular signals and exerts biological functions through G protein-coupled receptors. Lysophosphatidic acid (LPA) is the simplest glycerolphospholipid, which is considered to be a growth factor in plasma (reviewed in [16–22]). Increase in LPA levels in biological fluids including bronchoalveolar lavage (BAL) has been detected in a variety of diseases including acute lung injury [23–25] and lung fibrosis [25, 26]. Several reports demonstrate the distinct effects of LPA on EC barrier integrity. In this chapter, we will summarize these findings and discuss potential mechanisms by which LPA regulates EC junctions and permeability.

Vascular Endothelial Hyperpermeability in Diseases

Vascular vessels are closed and continuous tubes that carry blood components including blood cells and provide nutrition to tissues. The microvascular EC barrier is a dynamic and complex interface between the blood and the surrounding tissues. Due to its semi-permeability of EC barrier, certain small molecules and solute may pass through the microvascular EC barrier. EC junctions are a major component of the anatomical barrier. In addition to cell–cell junctions, transcellular permeability and specific transporters also control small molecules through EC barrier (reviewed in [6, 7, 9, 10, 13, 14]). The tightness of the microvascular EC barrier is largely dependent on tissues. The central nervous system (CNS) needs a stabled and controlled microenvironment. The BBB strictly controls influx and efflux of essential substances and promotes the normal physiological functions of the CNS. BBB dysfunction is observed in many CNS diseases, including multiple sclerosis, epilepsy, stroke, and Alzheimer's disease (reviewed [9, 14, 15]). Systemic inflammation caused by sepsis also leads to BBB disruption; in turn, BBB breakdown causes brain tissue edema and neuron inflammation and damage. Another well-studied EC barrier is the pulmonary microvascular EC barrier in the blood-air barrier. The lungs' major function is to facilitate gas exchange between the environment and the bloodstream (reviewed in [10, 11]). The lung epithelial barrier prevents inhaled microbes, allergens, and particulate matters from entering into the bloodstream (reviewed in [27–29]), while the pulmonary microvascular EC barrier limits blood component leakage into alveolar or interstitial tissue to prevent edema. Edema is a condition caused by excess fluid in the lungs due to EC barrier dysfunction (reviewed in [10–12]). The fluid in the alveolar space interferes with air exchange and leads to shortness of breath and death. Local infection by bacterial or virus (such as SARS-Cov2) or systemic inflammation (such as sepsis) causes pulmonary microvascular EC barrier dysfunction, leading to inflammatory cells and protein-rich fluid influx into alveolar spaces (reviewed in [11, 30–32]). Maintaining pulmonary microvascular EC barrier integrity is a novel therapeutic strategy for acute respiratory distress syndrome (ARDS).

EC monolayer barrier integrity was measured by several techniques, including transwell leakage assay, impedance-based cell monitoring, and immunostaining of cell–cell junctions. Measurement of protein levels in BAL and Evans Blue dye leakage in the tissues are common methods to determine vascular barrier integrity in in vivo studies.

EC Cell–Cell Junctions Regulate EC Barrier Integrity

Adherens and tight junctions are two major intercellular junctions that connect neighbor cells together including ECs. Adherens junction has been considered to play a critical role in initiation and stabilization of cell junctions (reviewed in [33–35]). VE-cadherin, also called CDH5 and CD144, is an endothelial-specific adhesion protein located at vascular adherens junction. VE-cadherin belongs to a cadherin family which consists of E-, N-, and P-cadherin. Extracellular domains of cadherins from adjacent cells interact each other in a calcium dependent manner. The intracellular domain of cadherins cross-links with the cytoskeleton. In addition to maintenance of cell–cell junctions, cadherins also mediate intracellular signaling (reviewed in [36–38]). In this chapter, we will focus on discussing the molecular regulation of VE-cadherin and the role of VE-cadherin in maintenance of vascular barrier integrity. VE-cadherin is phosphorylated on several tyrosine (tyr) residues including tyr658 and tyr731 in response to lipopolysaccharide (LPS) and TNFα. These phosphorylations have been reported to modulate endothelial permeability through regulation of VE-cadherin shedding, internalization, degradation, and dis-association of VE-cadherin with its associated proteins, including p120, α-catenin, and β-catenin [39–41] (reviewed in [36, 37, 42]). Phosphatase SHP-2 and protein-tyrosine phosphatase nonreceptor 14 (PTPN14) negatively regulate VE-cadherin phosphorylation and promote restoration of endothelial integrity [41, 43].

Tight junctions between vascular endothelial cells mostly occur on apical and basolateral junctional complexes. Claudins and occludin are major tight junction transmembrane components. Similar to adherens junctions, claudins and occludin from adjacent cells interact each other and form a strict intercellular seal. Claudins and occluding are four transmembrane proteins, while VE-cadherin is a single transmembrane protein (reviewed in [44–47]). Claudins, including claudin-3, -5, and -12, have been reported in the endothelium (reviewed in [48, 49]). Among them, claudin-5 is well studied (reviewed in [50, 51]). Claudin-5 deficient mice demonstrate an increase in BBB permeability [49]. Knockdown of claudin-5 attenuated simvastatin-induced rescue of lung endothelial barrier integrity [52]. Occludin levels are downregulated in response to endothelial barrier disruption stimuli such as LPS and hypoxia [53–57]. Phosphorylation of occludin by protein kinase Cβ (PKCβ) in response to vascular endothelial growth factor (VEGF) leads to occludin ubiquitination and increase in endothelial permeability [58, 59]. Zonula occludens-1 (ZO-1) is a claudin and occludin adaptor protein. ZO-1 links claudins and occludin with the actin cytoskeleton [60, 61] (reviewed in [48, 62]). ZO-1 depletion reduces tight junctions and leads to stress fiber formation [60]. Angiotensin II is reported to downregulate ZO-1 expression and disrupt endothelial tight junctions [60].

Rho Family of GTPases Regulate EC Barrier Integrity

The Rho family of GTPases belongs to small G protein superfamily. Rho family members are activated after binding to GTP, while the GDP-bound form is in an inactive state. RhoA, Cdc42, and Rac1 are major members of the Rho family (reviewed in [63–65]). The distinct roles of Rho family members in the regulation of EC barrier integrity are dependent on the effects of their activation on reorganization of the actin cytoskeleton (reviewed in [65–67]). RhoA activation leads to myosin light chain (MLC) phosphorylation and promotes stress fiber formation, resulting in cell contraction and EC barrier disruption. RhoA-induced MLC phosphorylation is mediated by Rho-associated kinase (ROCK)/MLC phosphatase [68–70]. In addition to regulation of the rearrangement of the cytoskeleton, RhoA/ROCK pathway promotes downregulation of VE-cadherin, claudins, and occludin; thus RhoA plays a central role in the regulation of EC barrier function through disrupting both adherens and tight junctions and promoting cell contraction (reviewed in [71]). The effect of RhoA on corneal EC barrier restores and repairs after hyperosmotic stress also has been reported [72]. We will discuss the effect of LPA on RhoA activation in ECs in the chapter.

In contrast to RhoA, activation of Rac1 and Cdc42 preserve EC barrier integrity. Rac1 possesses a coordinating antagonism with RhoA [73, 74]. Rac1 is reported to be activated by extracellular adenosine [75], activated protein C [76], and others [77]. Inhibition of Rac1 reduced EC permeability and intercellular gap formation [77–79]. Notably, the role of Rac1 in disruption of lung epithelial cell barrier integrity has been reported. DiPaolo, B.C. et al. demonstrated that Rac1 inhibitor attenuated stretch-induced increases in alveolar epithelial cell permeability [80]. Cdc42 promotes VE-cadherin-mediated adherens junction assembly [81]. Expression of a dominant active mutant of Cdc42 in endothelial cells reduced LPS-induced EC barrier disruption [82].

LPA Production

LPA, naturally presented in plasma and cells, possesses multiple biological functions, including cell growth and proliferation. LPA is a phospholipid derivative that consists of a glycerol backbone, a fatty acid chain, and a phosphate. According to the different fatty acids, LPA exists in different species, such as 16:0, 18:1, and 22:6 LPA (reviewed in [16, 83, 84]). LPA is generated both intracellularly and extracellularly. Intracellular LPA is synthesized from monoacylglycerol by a monoglycerol kinase (MGK) or converted from phosphatidic acid (PA) by phospholipase A2s (PLA2s) (reviewed in [20, 21]). The role of intracellular LPA in the regulation of EC barrier integrity has not been reported. Most studies regarding the effect of LPA on EC barrier integrity are focusing on extracellular LPA that stimulates cells through LPA receptors (LPARs). Extracellular LPA is generated from lysophophatidylcholine (LPC) by autotaxin (ATX, also called lysoPLD, ENPP2) [85, 86]. LPC is detectable in plasma and bronchoalveolar lavage [87–89]. ATX heterozygous knockout mice show a 50% reduction of plasma LPA levels [90]. Platelets have been shown to release LPA, suggesting that at least part of plasma LPA is from platelets [91, 92]; however, the mechanisms by which activated platelets release LPA have not been reported. LPA also is reported to be generated from phosphoatidylserine-(PS)exposed blood cells by a secretory PLA2 in the pathological conditions [93].

Increases in LPA levels in BAL fluid have been reported in murine models of acute lung injury. Except 18:0LPA, LPA species including 16:0, 16:1, 18:1, 18:2, 20:4, 20:3LPA are increased in murine BAL fluids after intratracheal LPS challenge for 24 h [23]. Mouratis, M-A. et al. examined the time course of LPA generation and found that LPS challenge increased LPA levels in BAL after 12 h and LPA levels remained at similar levels up to 48 h [24]. Increases in ATX activity and protein levels in BAL are correlated with LPA levels. However, bronchial epithelium- or myeloid-specific ATX deletion or inhibition of ATX had minor effects on lung injury [24]; thus, the role of BAL LPA in the pathogenesis of lung injury is unclear. Intratracheal instillation of LPA displays a protective role in LPS-induced lung injury. The protective effect of LPA possibly occurs through enhancing lung epithelial barrier integrity [94]. Increase in systemic ATX worsened LPS-induced lung injury, suggesting that systemic LPA, not local LPA, contributes to the pathogenesis of lung injury [24]. Increases in ATX and LPA species (18:1, 16:0, 18:0, 20:4, 22:6LPA) in plasma were observed following ischemia and reperfusion (I/R) [95]. Vascular endothelial cells are targets of systemic LPA.

LPARs' Expression in Endothelial Cells

The effects of extracellular LPA on the cellular responses occur through its ligation and activation of a group of G protein-coupled receptors (GPCRs) on the cell surface. LPARs are divided into two groups based on sequence similarity. LPAR1–3 belong to endothelial cell differentiation gene (EDG) family of GPCRs. Other GPCRs, including GPR23/P2Y9/LPAR4, GPR92/LPAR5, P2Y5/LPAR6, and P2Y10, were identified as putative LPARs. LPARs coupled with distinct heterotrimeric G proteins [18, 20, 21, 83]. LPARs are expressed at distinct levels in different endothelial cells. Data from different groups reveal distinct expression patterns of LPARs. For example, Gupte R. et al. reported that LPAR5 is the predominant LPAR in human umbilical vein cells (HUVECs) [96], while Yokiura H. et al. showed that LPAR6 is highly expressed in HUVECs [97]. Other studies revealed the expression of LPAR1 and LPAR3 in HUVECs [98, 99]. The expression of LPA receptors in human pulmonary ECs has been reported. Ren Y. et al. demonstrated that LPAR2 and LPAR6 subtypes are highly expressed in both human pulmonary arterial (HPAEC) and microvascular (HLMVEC) ECs [100]. Cai J. et al. detected LPAR1 protein expression in HLMVECs [101]. Overall, LPARs levels in HLMVECs are much lower compared to their expression in human bronchial epithelial cells (unpublished data). In brain endothelial cells, LPAR6 is determined as the predominant LPAR [102]. On NH. et al. reported expression of LPAR1–3 in human brain capillary ECs and expression of LPAR1–5 in the capillary fraction from mouse brain homogenate [103]. LPAR1, not LPAR3, was detected in cerebral microvessels in rat brain by immunofluorescence staining [104].

LPA in Endothelial Barrier Function in Lungs

Pulmonary microvascular EC barrier integrity is responsible for maintaining the blood-air barrier. Disruption of the blood-air barrier is a hallmark of lung injury caused by inhaled pathogens (such as bacterial and SARS-Cov2) or systemic inflammatory diseases (such as sepsis). EC barrier dysfunction leads to protein-rich fluid influx into alveolar spaces, resulting in reduction of air exchange between the blood stream and atmosphere [10, 12, 30–32]. Brp-LPA, a pan LPA receptor inhibitor, reduced LPS injection-induced endothelial barrier disruption in mouse lungs [105], suggesting that LPA plays a critical role in lung EC barrier dysfunction. The effects of LPA on HLMVECs and HPAECs have been reported. Ren Y et al. showed that LPA (0.1–30.0 μM) slightly reduced transendothelial electrical resistance (TEER) in HLMVECs using an electric cell-substrate impedance sensing (ECIS) system. The reduction was mild and transwell leakage assay with dextran-FITC did not confirm the phenomenon [100]. In another study, Cai J. et al. showed that LPA (5 μM) rapidly and significantly reduced TEER. The reduction was reversed back to normal levels after 2 h [101], suggesting a role of LPA in HLMVEC barrier disruption occurring in a short time frame. To compare the effect of LPA with lipopolysaccharide (LPS) on HLMVEC barrier disruption, we treated HLMVECs with LPA (1 μM) and LPS (200 ng/ml). Consistent with the study from Cai et al., LPA induced a rapid and significant reduction of TEER, while LPS induced a delayed reduction of TEER. The peak of reduction from LPS occurred after 8 h, and the TEER returned to basal level after 20 h (Y. Zhao, "unpublished data"). LPA is short-lived; 70% of extracellular LPA is degraded by lipid phosphate phosphatase (LPPs) in 2 h [106]. Interestingly, a metabolically stabilized analog of LPA (OMPT) induced a rapid and more severe reduction of TEER. The reduced TEER returned to basal levels after 20 h (Y. Zhao, "unpublished data"). OMPT is also a specific agonist of LPAR3 [107]. The data suggest that activation of LPAR3 results in HLMVEC barrier disruption. Future studies will be focused on examining the effect of LPAR3 in EC barrier dysfunction in murine models of lung injury.

HPAECs are useful cell models for investigating pulmonary EC barrier integrity. The data from different studies are not consistent. It has been reported that LPA treatment increased TEER in bovine pulmonary arterial ECs [108, 109], while Ren Y et al. showed that LPA reduced TEER in a dose-dependent manner in HPAECs [100]. Munoz NM. et al. and our unpublished data showed that LPA (1 μM) had no effect on TEER of HPACEs [110].

As we discussed, the HLMVECs and HPAECs respond to LPA in terms of EC barrier function differently in different studies. Due to the contrasting findings in the data, it is difficult to conclude the effect of LPA on EC barrier function in pulmonary lung ECs. HLMVECs and HPAECs are primary cells, LPARs expression pattern may be distinct from different donors and different passages. Generation of EC-specific LPARs-deficient mice will be helpful to determine the role of LPA/LPARs in EC barrier function in lung disease models.

LPA in Endothelial Barrier Function in BBB

Homeostasis of the brain microenvironment is important for neuronal activity. The BBB functions as a strict and selective barrier between blood stream and brain tissues to remains homeostasis of the brain microenvironment (reviewed in [13, 14, 95]). Using gadolinium diethylenetriaminepentaacetate (Gd-DTPA) contrast-enhanced MRI, the BBB disruption effect of LPA was observed in mice [103]. The conclusion was confirmed in Wistar rats and Sprague-Dawley rats by using fluorescent dye, sulforhodamine B [111] or Evans blue dye for BBB integrity [112]. LPA treatment of porcine brain capillary ECs in both apical and basolateral side of transwell reduced TEER [113], while incubation of LPA in apical side, not in basolateral side of transwell filter cultured with rat cerebral microvascular ECs, increased transendothelial flux [114]. The localization of LPARs on the basal and basolateral plasma membrane in brain ECs cultured in a transwell chamber in these two cell types was not determined. It is possible that LPARs have different localization patterns in cells from different species. The effects of LPA on increases in permeability have been reported in porcine [115], bovine [103], rat [102], and human brain microvascular EC [100] by independent studies. Recently, Nah, S-Y.'s group demonstrated that gintonin from ginseng reduces BBB through activation of LPAR1/3 [116–118]. All these studies support that LPA increases permeability of brain microvascular ECs. In contrast to this conclusion, LPA increased TEER in corneal ECs isolated from New Zealand White rabbits [119]. Together, these studies indicate that, unlike the controversial conclusions of the effect of LPA on blood-air barrier integrity, BBB is sensitive to LPA. Though the effect of LPA on brain microvascular EC is to increase permeability, the clinical applications of LPA in the brain diseases have not been well studied. Since LPA-increased brain microvascular EC permeability is rapid and transient, it may provide a supplemental therapeutic strategy to increase drugs delivery to brain. Choi S-H. et al. showed that coadministration of gintonin, an LPAR ligand, with donepezil, a potential medicine for Alzheimer disease, increased donepezil concentration in cerebral spinal fluid [118]. The benefit of administration of LPA in delivery of drug to brains needs further evaluation.

LPA in Endothelial Barrier Function in Other Systems

An earlier study demonstrated that LPA decreased permeability in bovine aortic ECs [92]. However, the expression profile of LPA receptors and LPA-mediated signal pathways in bovine aortic ECs have not been determined. HUVECs are used as an EC model for the study of EC functions such as proliferation, cell death, inflammation, and barrier function. The effect of LPA on HUVEC barrier integrity has been studied. LPA treatment of HUVECs increased permeability as evidenced by an increase in leakage of horseradish peroxidase (HRP) [120] and FITC-labeled dextran in the transwell permeability assay [97]. The EC barrier disruption effect of LPA occurred through LPA ligation to LPAR6 [97]. Neidlinger NA et al. showed that LPA treatment of HUVEC monolayers induced cell retraction, increased gaps, and cell detachment, indicating that LPA induces HUVEC barrier disruption [93]. As discussed above, extracellular LPA can be generated by activation of ATX. Incubation of HUVECs with ATX and its substrate LPC increased LPA levels, as well as EC gap formation and permeability. Thus, the data regarding the effect of LPA on HUVEC barrier disruption is consistent in the different independent studies. Interestingly, Hisano Y et al. found that LPA rapidly and transiently increased TEER in LPAR1 overexpressing HUVECs [121], indicating that LPAR isotypes play distinct roles in LPA-altered HUVEC barrier integrity.

Molecular Mechanisms of LPA-Modulated Barrier Function

To investigate the molecular mechanisms by which LPA induces EC barrier disruption, most studies have been focused on the role of LPA in activation of Rho GTPases, cytoskeleton rearrangement, and regulation of cell–cell junctions. Ridley AJ and Hall A found that LPA rapidly induced the formation of focal adhesions and actin stress fibers through activation of Rho in fibroblast cells [122]. This is the initial study that reports LPA activation of Rho and regulation of cytoskeleton rearrangement. Cross MJ et al. were the first to investigate the role of Rho activation in LPA-induced stress fiber formation in ECs. They showed that inhibition of Rho by C3 exotoxin attenuated LPA-induced stress fiber formation in porcine aortic ECs [123]. Further, activation of phospholipase D was identified to activate Rho upon LPA treatment [123]. Though this study did not directly determine the role of LPA/Rho in EC barrier disruption, the data provided indirect evidence that LPA induces EC barrier disruption through PLD/Rho activation. Masago K et al. revealed that Rho regulates LPA-induced BBB dysfunction [102]. The role of Rho in LPA-induced EC barrier disruption has been further confirmed by other groups [112, 116, 120, 124]. MLC phosphorylation and cytoskeleton rearrangement by Rho and Rho kinase upon LPA treatment was shown to play a critical role in the EC barrier disruption [120]. LPA has been reported to regulate Rac1 and Cdc42 in lung epithelial cells (reviewed in [125]); however, the role of Rac1 and Cdc42 in LPA-altered EC barrier function has not been reported.

In addition to regulation of cytoskeleton rearrangement, LPA disrupted tight junctions by altering structural integrity of claudin-5, occludin, and ZO-1 in brain microvascular ECs [102]. LPA also induced phosphorylation of VE-cadherin in HLMVECs [101], which has been shown to be involved in VE-cadherin disruption and internalization, resulting in EC barrier dysfunction. ATP release, and an increase in intracellular calcium have been shown to regulate LPA-induced cytoskeleton rearrangement in HUVECs [126], suggesting that LPA-induced EC barrier disruption may be regulated by ATP release and increase in intracellular calcium.

Summary

LPA is a bioactive lysophospholipid, which induces cellular responses through ligation to a group of LPA receptors on the cell surface. Synthesis and metabolism of LPA occur in both intracellular and extracellular fractions. It has been shown that LPA regulates multiple physiological functions and pathological processes. The effect of LPA on EC barrier integrity has been demonstrated in different types of ECs (Fig. 1.). The EC barrier disruptive effect of LPA has been confirmed in HUVEC model and brain microvascular ECs. In lungs, there are controversial data regarding the role of LPA in lung arterial and microvascular EC barrier integrity. Generation of EC-specific LPA receptor deficient mice will help to understand the

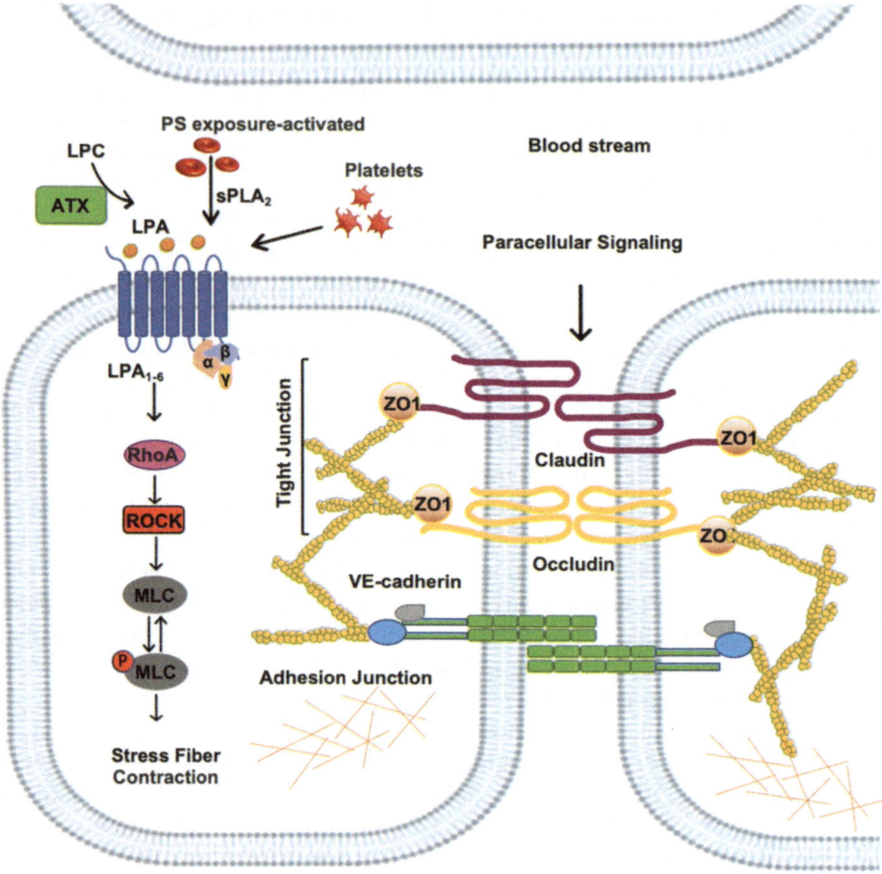

Fig. 1 LPA regulates EC barrier integrity. Extracellular LPA activates intracellular RhoA/ROCK/p-MLC pathway and regulates EC barrier integrity through ligation to LPARs on the cell surface of ECs

role of LPA/LPARs in the regulation of the blood-air barrier. As most studies are focused on extracellular LPA, ATX, and LPA receptors, the role of intracellular LPA in EC barrier integrity has not been investigated. The LPA-derived biolipids such as oxidized LPA need be further focused to investigate their role in EC barrier function. The development of in vivo lipidomic techniques may be useful to determine the local changes of LPA in circulation during disease progress.

Acknowledgement This work was supported by grants from National Institutes of Health (R01 HL131663, R01 HL136294, and R01 HL157164 to Y.Z.; R01 GM115389, R01HL151513 to J.Z.).

Conflict of Interest The authors declare no competing financial interests.

References

1. Tennant M, McGeachie JK. Blood vessel structure and function: a brief update on recent advances. Aust N Z J Surg. 1990;60(10):747–53.
2. Waltemath CL. Oxygen, uptake, transport, and tissue utilization. Anesth Analg. 1970;49(1):184–203.
3. Gutterman DD, et al. The human microcirculation: regulation of flow and beyond. Circ Res. 2016;118(1):157–72.
4. Berne RM. Regulation of coronary blood flow. Physiol Rev. 1964;44:1–29.
5. Deanfield JE, Halcox JP, Rabelink TJ. Endothelial function and dysfunction: testing and clinical relevance. Circulation. 2007;115(10):1285–95.
6. Malik AB, Lynch JJ, Cooper JA. Endothelial barrier function. J Invest Dermatol. 1989;93(2 Suppl):62S–7S.
7. Michiels C. Endothelial cell functions. J Cell Physiol. 2003;196(3):430–43.
8. Rosen H, Gordon S. Current status review: adhesion molecules and myelomonocytic cell-endothelial interactions. Br J Exp Pathol. 1989;70(3):385–94.
9. Daneman R, Prat A. The blood-brain barrier. Cold Spring Harb Perspect Biol. 2015;7(1):a020412.
10. Malik AB. Pulmonary microembolism and lung vascular injury. Eur Respir J Suppl. 1990;11:499s–506s.
11. Jacobson JR, Garcia JG. Novel therapies for microvascular permeability in sepsis. Curr Drug Targets. 2007;8(4):509–14.
12. Borok Z, Verkman AS. Lung edema clearance: 20 years of progress: invited review: role of aquaporin water channels in fluid transport in lung and airways. J Appl Physiol (1985). 2002;93(6):2199–206.
13. Ronaldson PT, Davis TP. Regulation of blood-brain barrier integrity by microglia in health and disease: a therapeutic opportunity. J Cereb Blood Flow Metab. 2020;40(1_suppl):S6–S24.
14. Deli MA, et al. Permeability studies on in vitro blood-brain barrier models: physiology, pathology, and pharmacology. Cell Mol Neurobiol. 2005;25(1):59–127.
15. Engelhardt S, Patkar S, Ogunshola OO. Cell-specific blood-brain barrier regulation in health and disease: a focus on hypoxia. Br J Pharmacol. 2014;171(5):1210–30.
16. Moolenaar WH. Lysophosphatidic acid, a multifunctional phospholipid messenger. J Biol Chem. 1995;270(22):12949–52.
17. Chun J, Contos JJ, Munroe D. A growing family of receptor genes for lysophosphatidic acid (LPA) and other lysophospholipids (LPs). Cell Biochem Biophys. 1999;30(2):213–42.
18. Hla T, et al. Lysophospholipids--receptor revelations. Science. 2001;294(5548):1875–8.
19. Pyne NJ, Dubois G, Pyne S. Role of sphingosine 1-phosphate and lysophosphatidic acid in fibrosis. Biochim Biophys Acta. 2013;1831(1):228–38.
20. Zhao Y, Natarajan V. Lysophosphatidic acid (LPA) and its receptors: role in airway inflammation and remodeling. Biochim Biophys Acta. 2013;1831(1):86–92.
21. Zhao Y, Natarajan V. Lysophosphatidic acid signaling in airway epithelium: role in airway inflammation and remodeling. Cell Signal. 2009;21(3):367–77.
22. Hao Y, et al. Lysophospholipids and their G-coupled protein signaling in Alzheimer's disease: from physiological performance to pathological impairment. Front Mol Neurosci. 2020;13:58.
23. Zhao J, et al. Lysophosphatidic acid receptor 1 modulates lipopolysaccharide-induced inflammation in alveolar epithelial cells and murine lungs. Am J Physiol Lung Cell Mol Physiol. 2011;301(4):L547–56.
24. Mouratis MA, et al. Autotaxin and endotoxin-induced acute lung injury. PLoS One. 2015;10(7):e0133619.
25. Black KE, et al. Autotaxin activity increases locally following lung injury, but is not required for pulmonary lysophosphatidic acid production or fibrosis. FASEB J. 2016;30(6):2435–50.

26. Tager AM, et al. The lysophosphatidic acid receptor LPA1 links pulmonary fibrosis to lung injury by mediating fibroblast recruitment and vascular leak. Nat Med. 2008;14(1):45–54.
27. Guillot L, et al. Alveolar epithelial cells: master regulators of lung homeostasis. Int J Biochem Cell Biol. 2013;45(11):2568–73.
28. Glover LE, Colgan SP. Epithelial barrier regulation by hypoxia-inducible factor. Ann Am Thorac Soc. 2017;14(Supplement_3):S233–6.
29. Carlier FM, de Fays C, Pilette C. Epithelial barrier dysfunction in chronic respiratory diseases. Front Physiol. 2021;12:691227.
30. Habashi NM, et al. Functional pathophysiology of SARS-CoV-2-induced acute lung injury and clinical implications. J Appl Physiol (1985). 2021;130(3):877–91.
31. Gustafson D, et al. Overcoming barriers: the endothelium as a linchpin of coronavirus disease 2019 pathogenesis? Arterioscler Thromb Vasc Biol. 2020;40(8):1818–29.
32. Martin MA, Silverman HJ. Gram-negative sepsis and the adult respiratory distress syndrome. Clin Infect Dis. 1992;14(6):1213–28.
33. Dejana E, Corada M, Lampugnani MG. Endothelial cell-to-cell junctions. FASEB J. 1995;9(10):910–8.
34. Lampugnani MG, et al. The molecular organization of endothelial cell to cell junctions: differential association of plakoglobin, beta-catenin, and alpha-catenin with vascular endothelial cadherin (VE-cadherin). J Cell Biol. 1995;129(1):203–17.
35. Reglero-Real N, et al. Endothelial cell junctional adhesion molecules: role and regulation of expression in inflammation. Arterioscler Thromb Vasc Biol. 2016;36(10):2048–57.
36. Giannotta M, Trani M, Dejana E. VE-cadherin and endothelial adherens junctions: active guardians of vascular integrity. Dev Cell. 2013;26(5):441–54.
37. Vestweber D. VE-cadherin: the major endothelial adhesion molecule controlling cellular junctions and blood vessel formation. Arterioscler Thromb Vasc Biol. 2008;28(2):223–32.
38. Yuan SY. Signal transduction pathways in enhanced microvascular permeability. Microcirculation. 2000;7(6 Pt 1):395–403.
39. Nwariaku FE, et al. Tyrosine phosphorylation of vascular endothelial cadherin and the regulation of microvascular permeability. Surgery. 2002;132(2):180–5.
40. Angelini DJ, et al. TNF-alpha increases tyrosine phosphorylation of vascular endothelial cadherin and opens the paracellular pathway through fyn activation in human lung endothelia. Am J Physiol Lung Cell Mol Physiol. 2006;291(6):L1232–45.
41. Fu P, et al. Phospholipase D2 restores endothelial barrier function by promoting PTPN14-mediated VE-cadherin dephosphorylation. J Biol Chem. 2020;295(22):7669–85.
42. Chan YH, et al. Differential regulation of LPS-mediated VE-cadherin disruption in human endothelial cells and the underlying signaling pathways: a mini review. Front Cell Dev Biol. 2019;7:280.
43. Wessel F, et al. Leukocyte extravasation and vascular permeability are each controlled in vivo by different tyrosine residues of VE-cadherin. Nat Immunol. 2014;15(3):223–30.
44. Cong X, Kong W. Endothelial tight junctions and their regulatory signaling pathways in vascular homeostasis and disease. Cell Signal. 2020;66:109485.
45. Boitano S, et al. Cell-cell interactions in regulating lung function. Am J Physiol Lung Cell Mol Physiol. 2004;287(3):L455–9.
46. Zhang Y, Yang WX. Tight junction between endothelial cells: the interaction between nanoparticles and blood vessels. Beilstein J Nanotechnol. 2016;7:675–84.
47. Cummins PM. Occludin: one protein, many forms. Mol Cell Biol. 2012;32(2):242–50.
48. Gawdi R, Emmady, PD. Physiology, blood brain barrier. In: StatPearls. Treasure Island (FL); 2021.
49. Nitta T, et al. Size-selective loosening of the blood-brain barrier in claudin-5-deficient mice. J Cell Biol. 2003;161(3):653–60.
50. Greene C, Hanley N, Campbell M. Claudin-5: gatekeeper of neurological function. Fluids Barriers CNS. 2019;16(1):3.

51. Jia W, et al. The role of claudin-5 in blood-brain barrier (BBB) and brain metastases (review). Mol Med Rep. 2014;9(3):779–85.
52. Chen W, et al. Role of claudin-5 in the attenuation of murine acute lung injury by simvastatin. Am J Respir Cell Mol Biol. 2014;50(2):328–36.
53. Qin LH, et al. LPS induces occludin dysregulation in cerebral microvascular endothelial cells via MAPK signaling and augmenting MMP-2 levels. Oxidative Med Cell Longev. 2015;2015:120641.
54. Wu J, et al. Betaine attenuates LPS-induced downregulation of Occludin and Claudin-1 and restores intestinal barrier function. BMC Vet Res. 2020;16(1):75.
55. Peng LY, et al. Madecassoside protects against LPS-induced acute lung injury via inhibiting TLR4/NF-kappaB activation and blood-air barrier permeability. Front Pharmacol. 2020;11:807.
56. Luo PL, et al. Hypoxia-induced hyperpermeability of rat glomerular endothelial cells involves HIF-2alpha mediated changes in the expression of occludin and ZO-1. Braz J Med Biol Res. 2018;51(7):e6201.
57. Ma X, et al. Hypoxia/Aglycemia-induced endothelial barrier dysfunction and tight junction protein downregulation can be ameliorated by citicoline. PLoS One. 2013;8(12):e82604.
58. Murakami T, et al. Protein kinase cbeta phosphorylates occludin regulating tight junction trafficking in vascular endothelial growth factor-induced permeability in vivo. Diabetes. 2012;61(6):1573–83.
59. Harhaj NS, et al. VEGF activation of protein kinase C stimulates occludin phosphorylation and contributes to endothelial permeability. Invest Ophthalmol Vis Sci. 2006;47(11):5106–15.
60. Tornavaca O, et al. ZO-1 controls endothelial adherens junctions, cell-cell tension, angiogenesis, and barrier formation. J Cell Biol. 2015;208(6):821–38.
61. Van Itallie CM, et al. ZO-1 stabilizes the tight junction solute barrier through coupling to the perijunctional cytoskeleton. Mol Biol Cell. 2009;20(17):3930–40.
62. Musch MW, Walsh-Reitz MM, Chang EB. Roles of ZO-1, occludin, and actin in oxidant-induced barrier disruption. Am J Physiol Gastrointest Liver Physiol. 2006;290(2):G222–31.
63. Hall A. Rho family GTPases. Biochem Soc Trans. 2012;40(6):1378–82.
64. Narumiya S, Thumkeo D. Rho signaling research: history, current status and future directions. FEBS Lett. 2018;592(11):1763–76.
65. Hall A. Rho GTPases and the actin cytoskeleton. Science. 1998;279(5350):509–14.
66. Aspenstrom P. The Rho GTPases have multiple effects on the actin cytoskeleton. Exp Cell Res. 1999;246(1):20–5.
67. Ridley AJ. Rho family proteins and regulation of the actin cytoskeleton. Prog Mol Subcell Biol. 1999;22:1–22.
68. Yao L, et al. The role of RhoA/Rho kinase pathway in endothelial dysfunction. J Cardiovasc Dis Res. 2010;1(4):165–70.
69. van Nieuw Amerongen GP, et al. Activation of RhoA by thrombin in endothelial hyperpermeability: role of Rho kinase and protein tyrosine kinases. Circ Res. 2000;87(4):335–40.
70. Koizumi N, et al. Rho-associated kinase inhibitor eye drop treatment as a possible medical treatment for Fuchs corneal dystrophy. Cornea. 2013;32(8):1167–70.
71. Spindler V, Schlegel N, Waschke J. Role of GTPases in control of microvascular permeability. Cardiovasc Res. 2010;87(2):243–53.
72. Ortega MC, et al. Activation of Rac1 and RhoA preserve corneal endothelial barrier function. Invest Ophthalmol Vis Sci. 2016;57(14):6210–22.
73. Comunale F, et al. Rac1 and RhoA GTPases have antagonistic functions during N-cadherin-dependent cell-cell contact formation in C2C12 myoblasts. Biol Cell. 2007;99(9):503–17.
74. Kim SY, et al. Coordinated balance of Rac1 and RhoA plays key roles in determining phagocytic appetite. PLoS One. 2017;12(4):e0174603.
75. Kovacs-Kasa A, et al. Extracellular adenosine-induced Rac1 activation in pulmonary endothelium: molecular mechanisms and barrier-protective role. J Cell Physiol. 2018;233(8):5736–46.

76. Finigan JH, et al. Activated protein C mediates novel lung endothelial barrier enhancement: role of sphingosine 1-phosphate receptor transactivation. J Biol Chem. 2005;280(17):17286–93.
77. Natarajan V, et al. Sphingosine-1-phosphate, FTY720, and sphingosine-1-phosphate receptors in the pathobiology of acute lung injury. Am J Respir Cell Mol Biol. 2013;49(1):6–17.
78. Waschke J, et al. Requirement of Rac activity for maintenance of capillary endothelial barrier properties. Am J Physiol Heart Circ Physiol. 2004;286(1):H394–401.
79. Garcia JG, et al. Sphingosine 1-phosphate promotes endothelial cell barrier integrity by Edg-dependent cytoskeletal rearrangement. J Clin Invest. 2001;108(5):689–701.
80. Dipaolo BC, et al. Rac1 pathway mediates stretch response in pulmonary alveolar epithelial cells. Am J Physiol Lung Cell Mol Physiol. 2013;305(2):L141–53.
81. Broman MT, et al. Cdc42 regulates adherens junction stability and endothelial permeability by inducing alpha-catenin interaction with the vascular endothelial cadherin complex. Circ Res. 2006;98(1):73–80.
82. Ramchandran R, et al. Critical role of Cdc42 in mediating endothelial barrier protection in vivo. Am J Physiol Lung Cell Mol Physiol. 2008;295(2):L363–9.
83. Gaits F, et al. Lysophosphatidic acid as a phospholipid mediator: pathways of synthesis. FEBS Lett. 1997;410(1):54–8.
84. Geraldo LHM, et al. Role of lysophosphatidic acid and its receptors in health and disease: novel therapeutic strategies. Signal Transduct Target Ther. 2021;6(1):45.
85. Umezu-Goto M, et al. Autotaxin has lysophospholipase D activity leading to tumor cell growth and motility by lysophosphatidic acid production. J Cell Biol. 2002;158(2):227–33.
86. Aoki J, et al. Serum lysophosphatidic acid is produced through diverse phospholipase pathways. J Biol Chem. 2002;277(50):48737–44.
87. Niewoehner DE, et al. Injurious effects of lysophosphatidylcholine on barrier properties of alveolar epithelium. J Appl Physiol (1985). 1987;63(5):1979–86.
88. Sekas G, et al. Origin of plasma lysophosphatidylcholine: evidence for direct hepatic secretion in the rat. J Lab Clin Med. 1985;105(2):190–4.
89. Zhao Z, et al. Plasma lysophosphatidylcholine levels: potential biomarkers for colorectal cancer. J Clin Oncol. 2007;25(19):2696–701.
90. van Meeteren LA, et al. Autotaxin, a secreted lysophospholipase D, is essential for blood vessel formation during development. Mol Cell Biol. 2006;26(13):5015–22.
91. Eichholtz T, et al. The bioactive phospholipid lysophosphatidic acid is released from activated platelets. Biochem J. 1993;291(Pt 3):677–80.
92. Alexander JS, et al. Platelet-derived lysophosphatidic acid decreases endothelial permeability in vitro. Am J Phys. 1998;274(1):H115–22.
93. Neidlinger NA, et al. Hydrolysis of phosphatidylserine-exposing red blood cells by secretory phospholipase A2 generates lysophosphatidic acid and results in vascular dysfunction. J Biol Chem. 2006;281(2):775–81.
94. He D, et al. Lysophosphatidic acid enhances pulmonary epithelial barrier integrity and protects endotoxin-induced epithelial barrier disruption and lung injury. J Biol Chem. 2009;284(36):24123–32.
95. Bhattarai S, et al. Disrupted blood-brain barrier and mitochondrial impairment by autotaxin-lysophosphatidic acid axis in postischemic stroke. J Am Heart Assoc. 2021;10(18):e021511.
96. Gupte R, et al. Benzyl and naphthalene methylphosphonic acid inhibitors of autotaxin with anti-invasive and anti-metastatic activity. ChemMedChem. 2011;6(5):922–35.
97. Yukiura H, et al. LPP3 localizes LPA6 signalling to non-contact sites in endothelial cells. J Cell Sci. 2015;128(21):3871–7.
98. Lin CI, et al. Lysophosphatidic acid regulates inflammation-related genes in human endothelial cells through LPA1 and LPA3. Biochem Biophys Res Commun. 2007;363(4):1001–8.
99. Hwang SH, et al. Effects of gintonin on the proliferation, migration, and tube formation of human umbilical-vein endothelial cells: involvement of lysophosphatidic-acid receptors and vascular-endothelial-growth-factor signaling. J Ginseng Res. 2016;40(4):325–33.

100. Ren Y, et al. Comparing the differential effects of LPA on the barrier function of human pulmonary endothelial cells. Microvasc Res. 2013;85:59–67.
101. Cai J, et al. AM966, an antagonist of lysophosphatidic acid receptor 1, increases lung microvascular endothelial permeability through activation of rho signaling pathway and phosphorylation of VE-cadherin. Mediat Inflamm. 2017;2017:6893560.
102. Masago K, et al. Lysophosphatidic acid receptor, LPA6, regulates endothelial blood-brain barrier function: implication for hepatic encephalopathy. Biochem Biophys Res Commun. 2018;501(4):1048–54.
103. On NH, et al. Rapid and reversible enhancement of blood-brain barrier permeability using lysophosphatidic acid. J Cereb Blood Flow Metab. 2013;33(12):1944–54.
104. Banks DB, et al. Lysophosphatidic acid and amitriptyline signal through LPA1R to reduce P-glycoprotein transport at the blood-brain barrier. J Cereb Blood Flow Metab. 2018;38(5):857–68.
105. Panchatcharam M, et al. Mice with targeted inactivation of ppap2b in endothelial and hematopoietic cells display enhanced vascular inflammation and permeability. Arterioscler Thromb Vasc Biol. 2014;34(4):837–45.
106. Zhao Y, et al. Lipid phosphate phosphatase-1 regulates lysophosphatidic acid-induced calcium release, NF-kappaB activation and interleukin-8 secretion in human bronchial epithelial cells. Biochem J. 2005;385(Pt 2):493–502.
107. Qian L, et al. Phosphorothioate analogues of alkyl lysophosphatidic acid as LPA3 receptor-selective agonists. ChemMedChem. 2006;1(3):376–83.
108. Minnear FL, et al. Platelet lipid(s) bound to albumin increases endothelial electrical resistance: mimicked by LPA. Am J Physiol Lung Cell Mol Physiol. 2001;281(6):L1337–44.
109. Gainor JP, et al. Platelet-conditioned medium increases endothelial electrical resistance independently of cAMP/PKA and cGMP/PKG. Am J Physiol Heart Circ Physiol. 2001;281(5):H1992–2001.
110. Munoz NM, et al. Group V phospholipase A(2) increases pulmonary endothelial permeability through direct hydrolysis of the cell membrane. Pulm Circ. 2012;2(2):182–92.
111. Sarker MH, Hu DE, Fraser PA. Regulation of cerebromicrovascular permeability by lysophosphatidic acid. Microcirculation. 2010;17(1):39–46.
112. Yu Y, et al. Role of Rho kinase in lysophosphatidic acid-induced altering of blood-brain barrier permeability. Int J Mol Med. 2014;33(3):661–9.
113. Nitz T, et al. Serum-derived factors weaken the barrier properties of cultured porcine brain capillary endothelial cells in vitro. Brain Res. 2003;981(1–2):30–40.
114. Hudson N, et al. Differential apicobasal VEGF signaling at vascular blood-neural barriers. Dev Cell. 2014;30(5):541–52.
115. Schulze C, et al. Lysophosphatidic acid increases tight junction permeability in cultured brain endothelial cells. J Neurochem. 1997;68(3):991–1000.
116. Kim DG, et al. Gintonin, a ginseng-derived exogenous lysophosphatidic acid receptor ligand, enhances blood-brain barrier permeability and brain delivery. Int J Biol Macromol. 2018;114:1325–37.
117. Jang M, et al. Ginseng gintonin attenuates the disruptions of brain microvascular permeability and microvascular endothelium junctional proteins in an APPswe/PSEN-1 double-transgenic mouse model of Alphalzheimer's disease. Exp Ther Med. 2021;21(4):310.
118. Choi SH, et al. Wound healing effect of gintonin involves lysophosphatidic acid receptor/vascular endothelial growth factor signaling pathway in keratinocytes. Int J Mol Sci. 2021;22(18):10155.
119. Yin F, Watsky MA. LPA and S1P increase corneal epithelial and endothelial cell transcellular resistance. Invest Ophthalmol Vis Sci. 2005;46(6):1927–33.
120. van Nieuw Amerongen GP, Vermeer MA, van Hinsbergh VW. Role of RhoA and Rho kinase in lysophosphatidic acid-induced endothelial barrier dysfunction. Arterioscler Thromb Vasc Biol. 2000;20(12):E127–33.

121. Hisano Y, et al. Lysolipid receptor cross-talk regulates lymphatic endothelial junctions in lymph nodes. J Exp Med. 2019;216(7):1582–98.
122. Ridley AJ, Hall A. Signal transduction pathways regulating Rho-mediated stress fibre formation: requirement for a tyrosine kinase. EMBO J. 1994;13(11):2600–10.
123. Cross MJ, et al. Stimulation of actin stress fibre formation mediated by activation of phospholipase D. Curr Biol. 1996;6(5):588–97.
124. Wojciak-Stothard B, Ridley AJ. Rho GTPases and the regulation of endothelial permeability. Vasc Pharmacol. 2002;39(4–5):187–99.
125. Tran KC, Zhao J. Lysophosphatidic acid regulates Rho family of GTPases in lungs. Cell Biochem Biophys. 2021;79(3):493–6.
126. Hirakawa M, et al. Sequential activation of RhoA and FAK/paxillin leads to ATP release and actin reorganization in human endothelium. J Physiol. 2004;558(Pt 2):479–88.

Regulation of Vascular Endothelial Barrier Integrity and Function by Lipid-Derived Mediators

Panfeng Fu, Ramaswamy Ramchandran, Steven M. Dudek, Narasimham L. Parinandi, and Viswanathan Natarajan

Abstract Increased vascular permeability is a cardinal feature of acute lung injury and restoration of the disrupted endothelial barrier function is a requisite to stem the fluid and inflammatory cells in the alveolar space. Maintenance of endothelial cell (EC) integrity is a critical determinant of vascular permeability and inflammatory responses in a variety of pulmonary disorders including sepsis, ventilator-induced lung injury, and bacterial and viral infections. Although it is well established that disruption of EC tight and adherens junctions causes increased permeability, alveolar flooding, and pulmonary edema, there is compelling evidence to support that ECs have the inherent ability to anneal the junctions and restore the barrier function. This process of barrier restoration and resealing of the gaps are facilitated by several naturally occurring barrier-enhancing molecules such as hepatocyte growth factor, sphingosine-1-phosphate, prostaglandins, oxidized phospholipids, and hyperosmolarity. Many of the barrier-protective agents are generated and released either in circulation by the ECs or other cells and in proximity to the disrupted endothelium.

P. Fu
The Affiliated Hospital of Medical School, Medical School of Ningbo University, Ningbo, China

R. Ramchandran
Departments of Pharmacology & Regenerative Medicine, University of Illinois at Chicago, Chicago, IL, USA

S. M. Dudek
Department of Medicine, University of Illinois at Chicago, Chicago, IL, USA

N. L. Parinandi
Bob and Corinne Frick Center for Heart Failure and Arrhythmia, Division of Pulmonary, Critical Care, and Sleep Medicine, Davis Heart and Lung Research Institute, The Ohio State University Wexner Medical Center, The Ohio State University, Columbus, OH, USA

V. Natarajan (✉)
Departments of Pharmacology & Regenerative Medicine, University of Illinois at Chicago, Chicago, IL, USA

Department of Medicine, University of Illinois at Chicago, Chicago, IL, USA
e-mail: visnatar@uic.edu

N. L. Parinandi, T. J. Hund (eds.), *Cardiovascular Signaling in Health and Disease*, https://doi.org/10.1007/978-3-031-08309-9_17

While mechanisms of EC barrier disruption have been extensively investigated, the process of barrier restoration of the endothelium is inadequately understood. Several lipid-derived mediators have been identified to facilitate the healing process and this article addresses the role of these lipid mediators in EC barrier restoration and gap closure with an emphasis on the signaling pathways related to lamellipodia formation and barrier restoration. Since no specific therapy is available currently to target the EC barrier disruption, an in-depth understanding of the mechanisms underlying the regulation of barrier restoration and stabilization of the junctions will lead to development of novel therapies.

Keywords Endothelial cells · Lipid-derived mediators · Sphingosine-1-phosphate · Lysophosphatidic acid · Phosphatidic acid · Oxidized phospholipids · Prostanoids · Hepatocyte growth factor · Barrier restoration · Annealing of gaps

Introduction

The balance between opening and resealing of endothelial adherens junction (AJ) barrier is essential for the maintenance of lung vascular barrier integrity in health and disease. Many lung disorders, including acute respiratory distress syndrome (ARDS), bacterial infection, pulmonary fibrosis, and bronchopulmonary dysplasia, are characterized by the increased lung vascular permeability and pulmonary edema. Disassembly of the adherens junctions (AJs) of lung endothelial cell (EC) monolayers by edemagenic agents causes microvascular hyperpermeability and protein-rich pulmonary edema formation, and if uncorrected, it leads to deterioration of the lung gas exchange. These changes reflect failure of the lung's intrinsic homeostatic mechanisms, and as such, they are hallmarks of the inflammatory lung disorders. Despite our current understanding of the multifaceted mechanisms regulating the vascular permeability [1–3], little is known about the molecular regulation of endothelial barrier restoration following the lung injury and edema. Lipids constituting glycerophospholipids, sphingolipids, and cholesterol are essential components of all biological membranes of the eukaryotic cells. Activation of endothelial and inflammatory cells by the edemagenic agents stimulates catabolism of cellular lipids resulting in generation of the lipid-derived biomolecules. Many of these lipid-derived mediators act on receptors on the endothelial or epithelial cell surface and signal to mediate intracellular responses by modulation of the protein kinases and phosphatases, and transcriptional factors that regulate the genes involved in barrier integrity. The signaling pathways increasing endothelial permeability via the effector proteins such as VE-cadherin, Rho, and EC-specific myosin light chain kinase have been well described, and the focus of this review is to define the role of lipid-derived mediators such as prostanoids, phosphatidic acid (PA), and sphingosine-1-phosphate (S1P) by enzymes involved in the catabolism of glycerol- and sphingo-phospholipids on the disruption and formation of adhesive contacts between adjacent ECs via interaction of adherens junction and cytoskeletal proteins to restore lung endothelial barrier integrity.

Modulation of Endothelial Barrier Restoration by Barrier-enhancing Agents

A number of naturally occurring agents such as S1P [4], high molecular weight hyaluronic acid (HMW HA) [5, 6], adenosine-5′-triphosphate (5′-ATP) [7, 8], activated protein C (APC) [9], oxidized phosphatidylcholine [10, 11], prostaglandins [12, 13], hepatocyte growth factor (HGF) [14, 15], and simvastatin [15, 16] have been identified as barrier enhancers, and promoters of barrier integrity. In sepsis, elevated levels of the vascular endothelial growth factor (VEGF) and HGF have been reported in plasma, and elevated plasma VEGF levels in sepsis are associated with disease severity and mortality [17]. While elevated VEGF levels increase vascular endothelial permeability, HGF levels also have been shown to increase in early sepsis [18], suggesting initiation of the tissue protection and regeneration post sepsis. Similarly, S1P levels were decreased in plasma of sepsis patients and animal models of acute lung injury (ALI) [19, 20] and infusion of S1P offers protection against the lipopolysaccharide (LPS) or cecal ligation puncture (CLP)-mediated lung inflammation and injury [21]. In contrast to sepsis, S1P levels are elevated in lung tissues and plasma of patients with pulmonary disorders such as asthma, pulmonary hypertension, pulmonary fibrosis, and bronchopulmonary dysplasia in preclinical animal models [22–26]. Blocking S1P production by inhibiting sphingosine kinase (SPHK) 1 and/or 2 offers protection against the lung inflammation and injury in several lung disorders [27]. The barrier enhancement caused by HGF, S1P, simvastatin, 5′-ATP, and oxidized phospholipids (OxPLs) are associated with increased interaction between actin and cortactin, and α/β catenin and VE-cadherin at leading edge of cells [28–32]. HGF and S1P promote endothelial barrier enhancement by interacting with its cognate receptors c-Met [28, 33] and S1P 1–5 [34], respectively. Specific receptor(s) for OxPLs such as 1-palmitoyl-2-(5-oxovaleroyl)-*sn*-glycero-3-phosphocholine (POVPC) and 1-palmitoyl-2-glutaroyl-*sn*-glycero-3-phosphocholine (PGPC) in inducing barrier enhancement is yet to be identified; however, OxPLs bind to low-density lipoprotein receptor-related protein 6 (LRP6) [35]. Here, we will discuss mechanisms of endothelial barrier enhancement by prostaglandins, glycerophospholipids, sphingolipids, oxidized phospholipids, and HGF.

Prostaglandins and Leukotrienes in Endothelial Barrier Integrity

Arachidonic acid, released in response to stimuli from the *sn-2* position of cell membrane glycerophospholipids by phospholipase A_2 action, is oxygenated by cyclooxygenases (COX)1 or 2 to generate prostaglandins (PGs) of the type PGE_2, PGD_2, PGI_2, $PGF_{2\alpha}$, and thromboxane (TX)A_2 while 5-lipoxygenease (5LO) produces leukotrienes (LTs) of the type LTB_4 and cysteinyl LTs. PGs and LTs are released during acute and chronic inflammation by pro-inflammatory leukocytes and macrophages [36], and their major physiological effects include vasodilation, vascular leakage, and chemotaxis. In addition to the recruitment and activation of immune cells, PGs modulate vasculature during inflammation. Thus, the action of PGs and LTs depend on the target tissue, concentration, and activation of different G-protein coupled receptors (GPCRs).

Prostaglandin D_2: PGD_2 and agonism of its receptor DPI, reduce vascular leakage and enhance endothelial barrier via cAMP/PKA/Tiam1/Rac1 signaling axis in a murine model of acute lung injury (ALI) [37–39]. Surprisingly, activation of EP4, but not DPI, by PGD_2 strengthens EC barrier against thrombin-induced barrier disruption suggesting promiscuity of EP4 signaling in the endothelium [12], and transient knockdown of *EP4* abolishes PGD_2-and Bw245c-mediated barrier enhancement in human pulmonary microvascular ECs. The PGD_2-mediated barrier enhancement parallels a reduction in Ser 473AKT phosphorylation, but not the tyrosine phosphorylation of VE-cadherin, FAK, or paxillin. However, the role of PGD_2 in sepsis is controversial. In a low-dose LPS model of ALI, PGD_2 exacerbates lung inflammation [40]; however, a high dose of LPS offers limited protection [41] in a murine sepsis model. LPS induces PGD_2 production in bone marrow-derived macrophages and high level of PGD_2 may limit inflammation by strengthening endothelial barrier. Surprisingly, in the thrombin-induced barrier disruption of human pulmonary artery ECs, activation of EP4 but not DPI strengthens the barrier [12], suggesting PGD_2 crosstalk with EP4 in the absence of PGE_2.

Prostaglandin E_2: Like PGD_2, PGE_2 also promotes the lung microvascular integrity and inhibits the neutrophil trafficking in the vasculature via EP4 [42]. It has been shown that the alveolar epithelial-derived PGE_2 is a key regulator of endothelial barrier integrity as blocking the endothelial EP4 reduces the PGE_2-mediated barrier protection [43], suggesting the therapeutic potential of PGE_2 and EP4 in endothelial barrier restoration. PGE_2-based therapeutic strategies to accelerate the recovery of impaired lung endothelium would boost the endogenous PGE_2 levels by blocking 15-hydroxy prostaglandin dehydrogenase, and/or by activating EP4 with agonist. A potential mechanism of PGE_2 protection of EC barrier via EP4 might involve elevated production of 3′,5′-cyclic adenosine monophosphate (cAMP) catalyzed by adenylate cyclase in the endothelium.

Prostaglandin I_2: PGI_2, the predominant product of COX_2 in the endothelium, is generated from PGH_2 by PGI_2 synthase. The functional characteristics of PGI_2 include inhibition of platelet aggregation and adhesion to the neutrophils and ECs, dilation of bronchial and vascular smooth muscle cells, and promotion of endothelial barrier enhancement in response to inflammation [44, 45]. PGI_2 is highly unstable in vivo and exerts its physiological effects via the G-protein-coupled receptor, IP. PGI_2 stimulates intracellular pathways of adenylate cyclase and GTPases and modulates the cytoskeletal organization and stabilization of adherens and tight junctions in ECs [13, 46–48]. These physiological and morphological changes promote endothelial barrier enhancement and integrity during lung inflammation. PGI_2 or its synthetic analogs, iloprost and beraprost, exhibit protective action on the endothelium, and reduce the lung inflammatory injury induced by mechanical ventilation or bacterial infections [49, 50]. The barrier protective role of PGI_2 and its synthetic analogs is mediated via its receptor, IP, by inhibiting Rap1/Rac1, and PKA-dependent Rho, which regulate the assembly of cytoskeletal and junctional proteins of the endothelium. The low stability of PGI_2 in vivo makes it a less attractive therapeutic agent; however, the synthetic analogs of PGI_2, iloprost and beraprost, show greater therapeutic efficacy in preclinical animal models of lung injury and in EC culture models. Incorporation of iloprost in phospholipase-resistant phospholipid scaffold enhances its protective action on the LPS-induced ALI as compared to the treatment with free iloprost in mice [51]. The cyclopentenone prostaglandin, PGA_2 generated by elimination of a molecule of H_2O within the cyclopentane ring of PGE_2, also provides protection against the LPS-induced vascular leak and inflammation in vivo in a murine model and in vitro against the thrombin-mediated EC hyperpermeability [52]. Thus, the therapeutical potential of free and phospholipid-bound iloprost needs to be investigated in various lung inflammatory injury models with vascular leak.

Leukotrienes: In contrast to PGD_2, PGE_2 and PGI_2, LTB_4, and cysteinyl leukotrienes LTC_4 and LTD4 stimulate the EC inflammation and proliferation through CysLT2R/Rho Kinase and CysLT1R/ERK-dependent pathways [53]. Inhibition or deletion of CysLT2R stabilizes the tumor EC integrity and reduces metastasis in the mouse tumor model, suggesting CysLT2R as a possible target for the barrier restoration in tumors [54]. The barrier protective role of PGD_2, PGE_2, and PGI_2 is depicted in Fig. 1.

Phospholipase D/Phosphatidic Acid Signaling and Endothelial Barrier Integrity

Phosphatidic Acid: Phosphatidic acid (PA) is a glycerophospholipid and key intermediate in the biosynthesis of major glycerophospholipids and triglycerides in mammalian cells. PA is anionic and has a unique cone-shaped geometry that confers

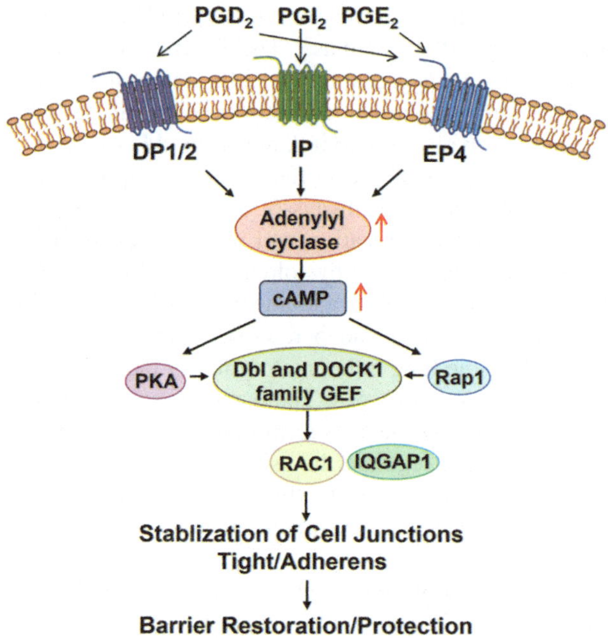

Fig. 1 Prostaglandins modulate endothelial barrier function. Schema depicts PGD_2, PGI_2, and PGE_2 signaling via its G-protein-coupled receptors in endothelial barrier restoration and barrier integrity. Ligation of PGD_2, PGI_2, and PGE_2 to its cell surface receptors DP1/2, IP and EP4, respectively, stimulates adenylate cyclase activity and enhances cAMP production. cAMP activates PKA or RAP1 and activates RAC1/IQGAP1 via DB1/DOCK1 GEFs leading to stabilization of tight and adherens junction (AJ) proteins and barrier integrity. PGD_2 can also bind to EP4, the PGE_2 receptor for signaling and activation of RAC1 via cAMP/PKA/RAP1. PGD_2, prostaglandin D_2; PGI_2, prostaglandin I2; PGE_2, prostaglandin E_2; DP1/2, PGD_2 receptor; IP, PGI_2 receptor; EP4, PGE_2 receptor; cAMP, cyclic adenosine monophosphate; PKA, protein kinase A; RAP1, Ras-related protein 1; DOCK1, Dedicator of cytokinesis 1; GEF, guanine nucleotide exchange factor; RAC1, Ras-related C3botulium toxin substrate; IQGAP1, IQ motif containing GTPase activating protein 1

fusiogenic and binding properties to proteins in biological systems. In cells, PA can be generated at least by four mechanisms (Fig. 2). In the first pathway, PA is generated de novo by acylation of dihydroxyacetone phosphate (DHAP) and/or glycerol-3-phosphate (G3P) catalyzed by DHAP or G3P acyltransferases, respectively. The second pathway involves hydrolysis of membrane glycerophospholipids such as phosphatidylcholine (PC), phosphatidylethanolamine (PE) or phosphatidylserine (PS) by phospholipase D (PLD) to PA and choline, ethanolamine, or serine, respectively. In the third pathway, PA is generated by phosphorylation of diacylglycerol (DAG) catalyzed by DAG kinase [55]. DAG can be derived from triglycerides through the action of lipases or from inositol phospholipids by phospholipase C (PLC) [56]. PA can also be dephosphorylated to DAG by PA phosphatases [57]. Finally, lysophosphatidic acid can be acylated by acyltransferases to PA [58, 59]. LPA can be generated by phosphorylation of acylglycerol mediated by acylglycerol

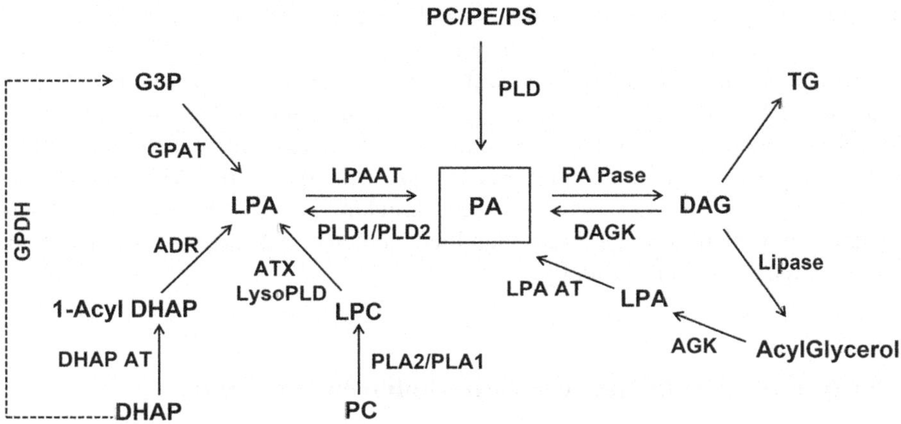

Fig. 2 Biochemical pathways of phosphatidic acid generation in mammalian cells. At least four biosynthetic pathways have been identified for the generation of phosphatidic acid (PA) in mammalian cells. For the de novo biosynthesis of PA, glycerol-3-phosphate (GP) or dihydroxyacetone phosphate (DHAP) is acylated by GP acyl transferase (GPAT) or DHAP acyltransferase (DHAPAT) to lysophosphatidic acid (LPA) or acyl DHAP, respectively. Acyl DHAP undergoes enzymatic reduction to LPA by acyldihydroreductase (ADR) to LPA. LPA is converted to PA by LPA acyltransferase (LPAAT). The second pathway involves hydrolysis of phosphatidylcholine (PC), phosphatidylethanolamine (PE), or phosphatidylserine (PS) by phospholipase D (PLD) 1 and/or 2 to PA. In the third pathway, diacylglycerol (DAG) generated from PC, phosphatidylinositol (PI), or polyphosphoinositides by phospholipase C is phosphorylated by DAG kinase to produce PA. DAG can be acylated by a DAG acyltransferase (DAGAT) to triglycerides (TG) or deacylated by lipase to acylglycerol, which is phosphorylated by AG kinase to LPA, and LPA to PA by LPAAT. In the fourth pathway, LPC in lipoproteins is hydrolyzed by lysophospholipase D (Lyso PLD) or autotaxin (ATX) to generate LPA that is acylated to PA by LPAAT. PA can be dephosphorylated to DAG by PA phosphatases or lipid phosphate phosphatases or hydrolyzed by P-specific phospholipase (PL) A_1 or A_2 to generate 1- or 2-acyl glycerol-3-phosphate (1- or 2-acyl LPA), respectively

kinase [60, 61] or hydrolysis of lysophosphatidylcholine (LPC) by autotaxin (ATX) or LysoPLD [62, 63]. Spatiotemporal localization of PA generated by these different mechanisms is different, thereby the generated PA is committed to different functions in the cell [64]. Among the four mechanisms, the PLD pathway has been widely investigated to understand its physiological and pathophysiological role in cell function. PLDs have been implicated in a variety of cell functions: vesicle trafficking, cytoskeletal dynamics, chemotaxis, cell motility, and reactive oxygen species (ROS) generation [65].

Phospholipase D: Phospholipase D (PLD), a ubiquitous glycerophospholipid-catabolizing enzyme in the mammalian cells, catalyzes hydrolysis of the membrane-associated phosphatidylcholine (PC) and other phospholipids to yield PA and free choline or a base, respectively [66]. There are six isoforms of PLD, PLD_{1-6} [67], of which PLD_1 and PLD_2 isoforms exhibit catalytic activity and have been widely recognized as key players in several human pathophysiologies, including cancer, hypertension, neurodisorders, diabetes, and acute lung injury [68, 69]. Although both PLD_1 and PLD_2 generate PA, they are localized in different cellular compartments and are differentially regulated [65, 70–72]; hence, the spatiotemporal generation of PA by PLD_1 and PLD_2 dictates specific function of PA [73–75]. PA is an important second messenger generated inside the cell that directly regulates several cellular processes, including apoptosis, cytoskeletal organization, cell morphogenesis, membrane biogenesis, and vesicular trafficking [76–78]. To date, no cell surface receptor(s) of PA has been identified, although PA exerts its action both extracellularly and intracellularly.

Phosphatidic Acid Induces Endothelial Permeability

Several agonists such as thrombin, HGF, LPS, ROS, oxidized low-density lipoprotein (ox-LDL), and S1P stimulate PLD_1 and PLD_2 in the ECs, and modulate permeability. Exogenous PA, but not DAG or LPA, significantly increases the EC monolayer permeability, and PA-induced permeability is attenuated by the tyrosine kinase inhibitor, herbimycin, and enhanced by vanadate, a tyrosine phosphatase inhibitor [79], suggesting a role for PA in the EC permeability. The ectopic PA-induced permeability of ECs is mimicked by PA confined to the neutrophil plasma membrane providing a potential link between the membrane-associated PA and EC permeability. Hydrogen peroxide in micromolar concentration activates PLD_1 and PLD_2 in bovine pulmonary artery ECs and blocking PLD_1 and PLD_2 activity with adenoviral dominant negative PLD_1 and PLD_2 mutants attenuates the hydrogen peroxide-induced EC permeability, cytoskeletal reorganization, and distribution of VE-cadherin and focal adhesion proteins [66]. The mechanism of ROS-mediated modulation of EC permeability by PLD_1/PLD_2-dependent PA signaling is unclear, but seems to involve reorganization of cytoskeletal actin and VE-cadherin at the junction. Further, treatment with hydrogen peroxide causes the EC barrier dysfunction and no significant barrier recovery in the time frame has been studied. However, PA generated by PLD_1/PLD_2 pathway has been shown to activate PKC ζ [80], alter actin cytoskeleton, and modify actomyosin contraction [81], and earlier studies have demonstrated an important role for RhoA family of GTPases in regulating the endothelial barrier function in response to agonists [82–84]. It is well recognized that actin polymerization leads to the formation of stress fibers is RhoA-dependent [85] process that is partly regulated by the PLD/PA signaling axis [86–91]. PA activates phosphatidylinositol-4-phosphate 5-kinase (PI4P5K) to generate phosphatidylinositol-4,5-bisphosphate (PIP_2) [92], an activator of actin cytoskeleton and modulator of interactions between actin and actin-binding proteins such as vinculin and filamin [93]. Thus, ROS-induced EC permeability mediated by PLD/PA signaling may involve modulation of Rho GTPases, PIP_2 formation, and actin cytoskeleton.

Phospholipase D_2 Facilitates Restoration of Endothelial Barrier Function In Vivo

Intratracheal administration of protease-activated receptor-1 activating peptide (PAR-1-AP) induced accumulation of inflammatory cells, mainly neutrophils, in lung alveolar spaces in the wild-type (WT) mice; however, *Pld2*$^{-/-}$ mice show greater neutrophil accumulation as compared to the WT and *Pld1*$^{-/-}$ mice. Further, *Pld2*$^{-/-}$ mice show significantly higher protein content in the bronchoalveolar lavage (BAL) fluid and an increase in IL-6 level in the BAL fluids compared to the WT mice and *Pld1*$^{-/-}$ mice after 3 h of PAR-1-AP challenge. Further, PAR-1-AP induces a much greater increase in the Evans blue dye uptake in *Pld2*$^{-/-}$ mice lungs as compared to the WT or *Pld1*$^{-/-}$ mice, indicating significantly greater lung vascular leakage in response to PAR-1-AP in the *Pld2*-deficient mice. Similar to PAR-1-AP, the *Pld2*$^{-/-}$ mice show elevated lung inflammatory injury and pro-inflammatory cytokine levels in the BAL fluids as compared to the WT mice. Collectively, the in vivo data show that deletion of PLD_2 in mice enhances lung vascular leakage and inflammatory injury in response to PAR-1-AP or LPS challenge [94], suggesting a protective role of PLD_2 in the thrombin- and LPS-induced lung injury.

Central Role of VE-Cadherin Trafficking to Nascent Adherens Junctions (AJs) in Restoring Endothelial Barrier Integrity

VE-cadherin, the EC-specific adhesion protein, through its extracellular domain, forms calcium-dependent homotypic interaction with VE-cadherin expressed on the adjacent ECs. Edemagenic agents induce disassembly of the AJs secondary to the internalization of VE-cadherin through VE-cadherin tyrosine phosphorylation by Src family kinases and c-Abl at Y658, whereas other agents referred to as the "permeability-decreasing mediators" (e.g., angiopoietin-1) inhibit VE-cadherin internalization to maintain barrier integrity [95]. Internalized VE-cadherin is trafficked in vesicles and some of which is tagged for the degradation by ubiquitination [96]. Therefore, vesicular VE-cadherin represents an important reservoir of VE-cadherin, which, upon redirection to the cell surface, serves as a critical source for the reassembly of the AJs during the post-injury restoration of the endothelial barrier [96, 97].

PLD_2/PA Signaling Facilitates Endothelial Barrier Restoration by Enhancing VE-Cadherin Dephosphorylation Via Tyrosine-Protein Phosphatase Non-Receptor Type 14 (PTPN14)

In the HLMVECs, inhibition or deletion of PLD_2, but not of PLD_1, delays the endothelial barrier recovery in response to thrombin. Further, thrombin stimulation of the human lung microvascular endothelial cells (HLMVECs) increases the co-localization of PLD_2-generated PA and VE-cadherin at cell–cell adhesion junctions, while inhibition of PLD_2 activity results in prolonged phosphorylation of Tyr-658 in VE-cadherin during the recovery phase after thrombin challenge. Immunoprecipitation experiments reveal strong association between PLD_2, VE-cadherin, and tyrosine-protein phosphatase non-receptor type 14 (PTPN14) following the thrombin stimulation. Depletion of PTPN14 delays the VE-cadherin dephosphorylation, reannealing of the AJs, and barrier function recovery (Fig. 3). PLD_2 inhibition attenuates the PTPN14 activity and reverses the PTPN14-dependent VE-cadherin dephosphorylation following thrombin stimulation [94]. These findings suggest that PLD_2 promotes the PTPN14-mediated dephosphorylation of VE-Cadherin and that redistribution of VE-cadherin at the AJs is essential for recovery of the endothelial barrier function resulting from an edemagenic insult.

PLD_2 Modulates Cortactin Phosphorylation in Formation of Lamellipodia, Resealing of AJ Barrier, and Restoration of Lung Vascular Barrier Integrity

Cortactin, an important regulator of cortical actin rearrangement, is an actin-binding protein that promotes lamellipodia protrusion [98, 99]. Recent studies suggest the importance of lamellipodia in maintenance and restoration of the endothelial barrier

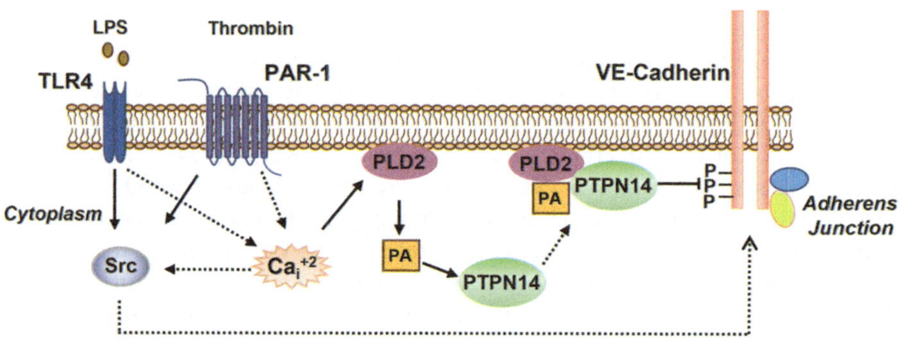

Fig. 3 PLD2/PA-regulation of VE-cadherin via PTPN14 and stabilization of endothelial adherens junctions (AJs) and barrier restoration. Stimulation of endothelial cells with lipopolysaccharide (LPS) via Toll-like receptor (TLR) 4 or thrombin via protease-activated receptor-1 (PAR-1) results in increase of intracellular calcium $[Ca^{2+}]_i$, and Src activation. Changes in intracellular calcium and/or Src activation stimulates phospholipase D_2 (PLD_2)-dependent phosphatidic acid (PA) generation during the recovery phase of lung endothelial barrier restoration. PA, a second messenger, activates tyrosine-protein phosphatase non-receptor type 14 (PTPN14), which maintains vascular endothelial (VE)-cadherin at AJs in a dephosphorylated state resulting in reannealing of adherens junctions (AJs) and barrier recovery. Blocking PLD_2 or PTPN14 results in retention of phospho-VE-cadherin for a prolonged time affecting annealing of the gaps formed by LPS or thrombin

integrity under basal and pathological conditions. Expression of the Y421F/Y466F/Y482F cortactin phospho-defective mutant blocks the thrombin-induced lamellipodia formation and co-localization of actin with cortactin in cell protrusions of the HLMVECs and prevents the recycling of VE-cadherin to AJs and restoration of endothelial barrier. Further, expression of catalytically inactive PLD_2 mutant attenuates the thrombin-induced tyrosine phosphorylation of cortactin at Y466 in the HLMVECs (Fig. 4). The PLD_2-mediated co-localization of cortactin with actin in cell periphery is dependent on IQGAP1 and downregulation of IQGAP1 attenuates the lamellipodia formation [100]. Further, PA generated from PLD_2 has been shown to promote IQGAP1 recruitment to the plasma membrane in smooth muscle cells [101, 102], which could bind directly to the actin filaments [102] and provide a direct molecular link between GTPases and the actin cytoskeleton. These data suggest a key role of PLD_2 and its activity in cortactin tyrosine phosphorylation and lamellipodia formation in the HLMVECs, which is a prerequisite for barrier annealing and restoration.

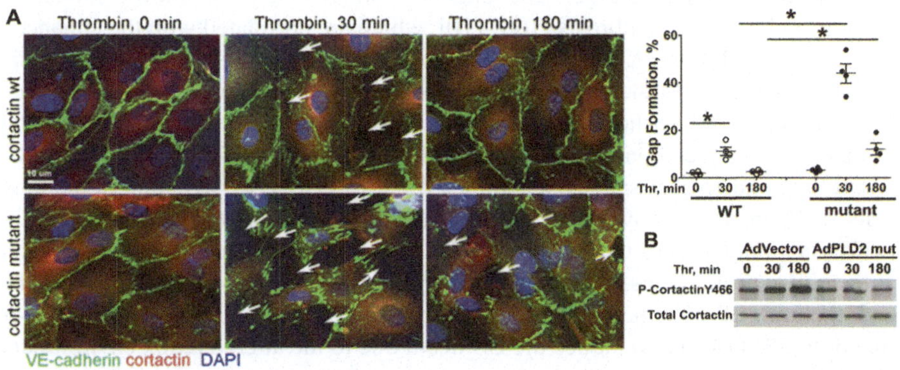

Fig. 4 PLD_2-dependent tyrosine phosphorylation of cortactin is essential for VE-cadherin localization at AJs. Wild type (WT) human umbilical vein endothelial cells (HUVECs) and HUVECs expressing green fluorescence protein (GFP)-phospho-defective cortactin mutants (Y421F/Y466F/Y482F) were challenged with thrombin for 30 min and 3 h. Cells were immunostained with anti-VE-cadherin antibody to assess formation of inter-endothelial gaps, reflecting loss of AJ integrity. Results of all experiments were quantified as shown on right with statistical significance set at $*p < 0.05$. Results show far greater disruption of AJs within 30 min post-thrombin challenge in cells transduced with phospho-defective cortactin as compared to controls with WT cortactin. Since cortactin is a key regulator of lamellipodia formation, these results suggest that formation of lamellipodia involved in the development of VE-cadherin homotypic interactions may require tyrosine phosphorylation of cortactin. Infection of WT human lung microvascular ECs with vector control or dominant negative PLD_2 catalytically inactive mutant (10 MOI, 24 h) resulted in inhibition of thrombin-induced phosphorylation of VE-cadherin at 30 min and loss of VE-cadherin at adherens junctions

Lysophosphatidic Acid/LPARs in Endothelial Barrier Regulation

Lysophosphatidic acid (LPA), the simplest phospholipid, is a natural constituent of all mammalian cells, plasma, and BAL fluid. Extracellular LPA is produced by autotaxin (ATX) or lysophospholipase D (LysoPLD) activity on LPC present in the lipoproteins, while intracellular LPA is generated from PA by PA-specific phospholipase A_1 (PLA_1) or phospholipase A_2 (PLA_2) or phosphorylation of acylglycerol (AG) by AG kinase (AGK) [61, 103]. LPA exhibits both pro- and anti-inflammatory properties and most of the biological effects of LPA are mediated via G-protein-coupled LPA1–6 receptors that are expressed on the cell surface [104]. Extracellular LPA stimulates expression of cytokines, chemokines, and cytokine receptors, and regulates the cytoskeletal rearrangement, and confers protection against lung injury by tightening of the epithelial barrier integrity [105]. The LPA-induced interleukin-8 (IL-8) secretion in human bronchial epithelial cells is mediated by the upstream activation of p38-mitogen-activated protein kinase (MAPK) and C-Jun N-terminal kinase (JNK) [106], and lipid phosphate phosphatase 1 [107], suggesting a complex regulation at cell surface and intracellular pathways. LPA levels are elevated in BAL fluid from segmental allergen challenged asthmatic patients, murine model of asthma, and pulmonary fibrosis [108–110], which are correlated with the enhanced expression of ATX and ATX activity. Inhibition of ATX activity with small molecule inhibitor(s) attenuates the house mite-induced airway hyperresponsiveness in mice [111]; however, it has shown no effect on the collagen deposition in murine model of bleomycin-induced pulmonary fibrosis [110]. The opposing roles of LPA in different lung pathologies suggest a complex role of LPA/LPAR signaling in the lung cells. LPA in the airway bronchial epithelial cells tightens the epithelial barrier [112] via modulation of the PKC δ- and PKC ζ-mediated E-cadherin accumulation at the cell–cell junctions, and cytoskeletal rearrangement of cortactin [113]. In the endothelium, LPA has been shown to reduce permeability and stabilize the barrier function [79, 114] but to induce the EC death due to the altered redox environment [115]. LPAR1 seems to play a key role in LPA-mediated lung inflammatory responses. Inhibition or downregulation of LPAR1 attenuates the LPS-induced lung inflammation [116], and bleomycin-mediated pulmonary fibrosis in mice [117]. In contrast to the in vivo studies, inhibition of LPAR1 with an antagonist AM996 in vitro causes the activation of Rho/Rho kinase, MLC-phosphorylation, and increased permeability. Some of these discrepancies could be due to the differential expression of LPARs, differences in the interactions of various LPA molecular species with their receptors [118, 119], and spatiotemporal generation of LPA in cells [120]. Further evaluation of LPA and LPARs in the endothelial barrier integrity is necessary to characterize the role of LPA in vascular permeability in human lung pathologies.

Sphingolipids Modulate Endothelial Barrier Restoration and Integrity

Sphingolipids are essential components of the eukaryotic cell membranes, which regulate the cellular functions such as apoptosis, senescence, proliferation, and motility by signaling through receptors and binding to target proteins. Metabolism of the sphingolipids is altered in several human pathologies and recent evidence suggests that sphingolipid metabolites such as sphingosine, ceramide, S1P, and Δ2-hexadecenal derived from the enzymatic action of sphingsine-1-phosphate lyase (S1P lyase) play a key role in the pathology of respiratory disorders [27]. Sphingomyelin (SM), the most abundant sphingolipid present in cells, is generated from serine and palmitoyl-CoA through the enzymatic action of serine palmitoyl-transferase (SPT) to form 3-keto-dihydrosphingosine, which is rapidly reduced to dihydrosphingosine (sphinganine) catalyzed by ketosphinganine reductase [121]. Dihydrosphinganine is converted to dihydroceramide and then to ceramide by six ceramide synthase isoenzymes with varying fatty acid chain lengths ranging from C18:0 to C24:1 [122]. Ceramide is then channeled to complex sphingolipids such as SM and glycosphingolipids or converted to ceramide-1-phosphate. Mammalian cells cannot convert dihydrosphinganine to sphingosine; however, sphingosine is derived from ceramide catalyzed by ceramidase(s) [123]. In addition, ceramide can also be directly generated from SM by the action of acid or neutral sphingomyelinase [124]. Sphingosine derived from ceramide is phosphorylated by sphingosine kinases (SPHK) 1 & 2 to S1P [125, 126]. S1P can be converted to sphingosine by S1P phosphatases or to Δ2-hexadecenal by S1P lyase, a pyridoxal phosphate-dependent enzyme localized in the endoplasmic reticulum (ER) membranes [127] and cell nucleus [128]. Ceramide metabolism is central to sphingolipid homeostasis and imbalance in ceramide metabolism could impact sphingolipidome and cell function [129].

Ceramide in Endothelial Apoptosis and Permeability

Ceramide has been implicated in the pathogenesis of human diseases including cancer, lung disorders, atherosclerosis, and diabetes [130–132]. Ceramide stimulates apoptosis [133] in cells and elevated ceramide levels have been observed in the lungs of patients with emphysema [131]; however, opposing findings have been reported about ceramide levels in cystic fibrosis (CF) patients and in mouse model of CF where reduced ceramide levels correlated with defects in fatty acyl chain molecular species [134]. These contradicting results in CF may be due to differences in animal models used and/or limitation in detection of only selected species of ceramides by mass spectrometry. Increased acid sphingomyelinase (ASMase) activity or expression directly corelates with ceramide levels in pulmonary disorders, including CF and COPD [135, 136]. Genetic knockdown of ASMase normalizes ceramide levels, decreases inflammation and vascular permeability in cigarette smoke-induced emphysema [131], pulmonary fibrosis [137] and CF in mice [138]. ASMase deficiency also protects against the LPS-mediated endothelial apoptosis in mice by modulating ceramide levels [139]; however, ASMase-dependent increase in ceramide decreases the platelet activating factor (PAF)-mediated vascular permeability [140] without affecting apoptosis. The disparate effect of ceramide in LPS and PAF models of lung injury may be due to the differences in the fatty acid composition of ceramides generated by the stimulation of ASMase. Further, one can infer that endothelial apoptosis is an important mechanism in inducing the increased vascular permeability in lung disorders. At present, there is no approved therapy to ameliorate the ceramide-induced endothelial permeability. Recent study suggests that modulation of elongase-mediated elongation of very long-chain fatty acids protein 4 (ELOVL4) prevents diabetes-induced retinal vascular permeability by stabilizing the tight junctions [141, 142]. Further investigation on the role of ELOVL4 is necessary to determine its efficacy in the endothelial barrier protection during lung injury.

S1P Metabolism and Cell Function

S1P, the simplest sphingophospholipid, is a natural constituent of plasma, biological fluids, and cells. The plasma levels of S1P are several folds higher (0.1–1.0 μM) compared to tissues (<0.1 μM) and most of the plasma S1P is bound to apoprotein M carrier of the high-density lipoprotein (HDL) (~60%) [142, 143]. Cellular S1P levels are regulated by its biosynthesis catalyzed by sphingosine kinase (SPHK)1 and 2 and its catabolism mediated by S1P phosphatases (SPP) 1 and 2, S1P lyase, and lipid phosphate phosphatases [2, 125]. Compared to endothelial and epithelial cells, erythrocytes and platelets have much higher levels of intracellular S1P due to lack of S1P lyase [144, 145]. Intracellularly generated S1P is also transported from inside to outside by ABC transporters [146–149], and spinster homolog 2 (SPNS2) transporter [150, 151]. S1P is a potent angiogenic factor and lipid mediator involved in diverse cellular processes such as cell growth, and survival [152], motility [153, 154], cytoskeletal organization [155], endothelial permeability [3], vascular tone [156], AJs [157] and tight junctions assembly [158], autophagy [159, 160], immune regulation [161–163], and morphogenesis [158]. These actions of S1P are due to its unique inside-out (extracellular), and intracellular signaling, highlighting its role as a signaling sphingolipid. Intracellularly, S1P is known to act as a second messenger and plays a role in calcium homeostasis, and more recently, S1P has been shown to bind to intracellular targets such as histone deacetylase 1/2 (HDAC1/2) [164–166] and human telomerase reverse transcriptase (hTERT) [167]. Further, release of S1P in the human lung ECs by photolysis of the caged S1P significantly enhances the barrier function, which is independent of S1P1, but is dependent on RAC1 [168]. Interestingly, S1P generated in the nucleus by the action of SPHK2 is shown to directly target HDAC1/2 and an integral component of the HDAC repressor complex [164–166]. S1P has been identified as a missing co-factor required for the E3 ligase activity of the tumor necrosis factor (TNF) receptor-associated factor 2 (TRAF2) [169].

S1P and Endothelial Barrier Enhancement

Initial studies with infusion of platelet-rich medium to preserve vascular integrity in organ preparations show a reduction in pulmonary edema, and this barrier enhancing effect may be due to S1P [170]. Further studies both in vivo and in vitro have demonstrated the ability of S1P to promote both barrier enhancement and restoration in the endothelium. In vivo, S1P infusion reduces the endotoxin-induced BAL fluid protein accumulation and alveolar edema in the murine and canine models of lung injury [171, 172]. Similarly, exogenously added S1P to EC monolayers increases the endothelial electrical resistance, which is rapid, and in a time-dependent fashion [155, 173]. S1P binds to high affinity G-protein-coupled receptors S1P1-5 on the EC plasma membrane. Compelling studies with genetically engineered and endothelial-specific S1PR knockout mice show a definitive role of S1P1 as a barrier protective receptor against vascular leak induced by LPS, TNF-α, radiation, and ventilator-induced lung injury [171, 174, 175]. Consistent with the barrier-protective role of S1P1, the WT mice treated with an S1P1 inverse agonist, SB649146, reduces the S1P/SEW2871-induced barrier protection after the LPS challenge [176]. However, downregulation of the S1P3 expression with a specific siRNA has conferred significant protection against the LPS-induced barrier disruption and leakage, as compared with the WT mice, suggesting a barrier disruptive function of S1P3. In contrast to LPS-induced lung injury, genetic deletion or knockdown of S1P1, S1P2, and S1P3 increases the susceptibility to lung injury in preclinical murine model of radiation-induced lung injury [177], suggesting differential role for S1PRs in these two models of lung injury. This might be due to the differential transduction of signals via multimeric G proteins (Fig. 5).

Although infusion of S1P has proven to be beneficial against lung injury in the preclinical animal models [35, 39], a tight regulation of circulating S1P level needs to be maintained as elevated plasma S1P may exhibit side effects that are likely to limit the usefulness as a drug against the pulmonary leak. While the IV administration of S1P decreases severity of the ALI, the intratracheal administration of S1P produces pulmonary edema through disruption of the epithelial/endothelial barrier via the ligation of S1P1 or S1P3. In the human lung ECs, high concentration of S1P (>10 μM) disrupts the EC permeability through the ligation of S1P3, suggesting a limited therapeutic efficacy for S1P in the barrier enhancement [31]. S1P also exhibits other side effects such as cardiac toxicity (bradycardia) through the activation of S1P3 in the heart [66], stimulates the human airway and bronchial smooth muscle cell contraction [67], and increases the airway hyperresponsiveness in the allergen-challenged mice [68]. These limitations have led to the development of synthetic S1P analogs such as fingolimod (FTY720), FTY720-phosphate,

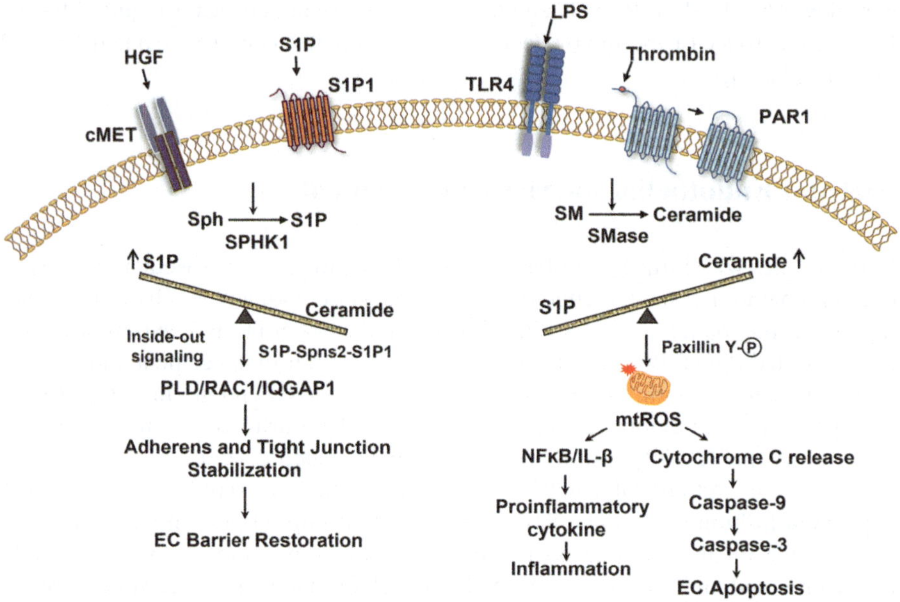

Fig. 5 Regulation of endothelial barrier function by HGF/S1P and LPS/Thrombin. Exposure of lung ECs to HGF or S1P enhances the barrier (left panel), while LPS or thrombin increases endothelial permeability (right panel). HGF or S1P stimulates SPHK1 in ECs and increases S1P levels. Intracellularly generated S1P is transported via SPNS2 transporter to extracellular milieu where S1P binding to G protein-coupled S1P1 activates PLD_2/RAC1/PI3K and induces a series of signaling cascades, including cytoskeletal reorganization, the assembly of AJ and tight junction proteins, and the formation of focal adhesions that act together to enhance endothelial barrier function. However, challenge of lung ECs with LPS or thrombin reduces intracellular S1P levels by modulation of SPHK1 activity. Decreased S1P levels induces actin stress fiber formation and disrupts the assembly of adherens junctions and tight junction and focal adhesion proteins, resulting in barrier disruption. Also, increase in ceramide levels modulates mitochondrial ROS production via paxillin tyrosine phosphorylation. Mito ROS can release cytochrome c from the mitochondria and induce apoptosis via activation of caspase 9/caspase 3 pathway. MitoROS can also activate NF-κB signaling and induce pro-inflammatory cytokines release and inflammation. HGF, hepatocyte growth factor; S1P, sphingosine-1-phosphate; Sph, sphingosine; SPHK, sphingosine kinase 1; SPNS2, spinster homolog 2; S1P1, S1P receptor 1; PLD, phospholipase D; RAC1, Ras-related C3 botulinum toxin 1; PI3K, phosphatidylinositol 3-kinase; EC, endothelial cell; LPS, lipopolysaccharide; SM, sphingomyelin; SMase, sphingomyelinase; ROS, reactive oxygen species; TLR4, toll-like receptor 4

FTY720-Phosphonate, SEW2871, ozanimod, and siponimod, which have shown beneficial effects in preclinical animal models, and in vitro systems [2, 178].

The S1P Analogs FTY720 and FTY-720 Phosphonate Modulate Endothelial Barrier Integrity

FTY720 (2-amino-2-(2-[4-octylphenyl]ethyl)-1,3-propanediol) is a synthetic derivative of the fungal metabolite myriocin and exhibits immunosuppressive and barrier-enhancing properties both in vitro and in vivo [171, 179]. FTY720 and other analogs such as ozanimod and siponimod have been approved by FDA for treatment of multiple sclerosis, and therefore, offers a potential future therapeutic opportunity for inflammatory lung diseases. Studies in vitro demonstrate that FTY720 enhances EC barrier function, in part, via a novel S1P1R-independent mechanism that involves an alternative Gi-coupled receptor [180]. A single IP injection of FTY720 significantly ameliorates the pulmonary leak and injury in mice 24 h after the LPS administration [171]. In vitro, lower doses of FTY720 enhance the endothelial barrier function in the human umbilical vein ECs (HUVECs), while higher concentrations of FTY720 induce irreversible barrier breakdown and apoptosis [181]. Similarly, low concentrations of FTY720 (0.1 mg/kg) reduce the lung permeability in mechanically ventilated mice. However, higher concentrations (2 mg/kg) increase the pulmonary leakage and apoptosis in ventilated mice, without affecting permeability in the non-ventilated mice [181]. Similar to S1P, FTY720 induces bradycardia via S1PR3 [182], which may limit its therapeutic utility in patients with ALI. FTY720, as an immunosuppressant, induces lymphopenia via downregulation of the lymphocyte S1P1R signaling [183, 184], which may be detrimental to patients with ALI. Further, FTY720 significantly increases rates of dyspnea and decreases lung function (lower FEV1) via mechanism(s) like the S1P-induced airway hyperresponsiveness. While at modest concentrations FTY720 is EC barrier-enhancing, the (S)- and (R)-FTY720 regioisomers potently disrupt the EC barrier integrity [185].

Due to limitations of FTY720 as a therapeutic agent for ALI, explorations of several FTY720 analogs have revealed FTY720-phosphonate (TyS) as a novel agent that regulates the endothelial barrier permeability among several screened novel and synthetic analogues of FTY720 for their barrier-regulatory capacities. In vivo data demonstrate that TyS significantly reduces the LPS-, bleomycin-, and MRSA-induced vascular leak in three different murine models of inflammatory lung injury [177, 185–187]. The superior efficacy of TyS with less side effects could be attributed to its stability in vivo as it is not hydrolyzed by phosphatases and retention of S1P1 on the surface of the endothelium. These studies advance our understanding of S1P analogs as improved therapeutic tools for prevention and restoration of the vascular leak in various pulmonary disorders.

SPHK1 as Potential Intracellular Target in Facilitating Endothelial Barrier Restoration

Cellular S1P levels are regulated by the synthesis and degradation catalyzed by SPHK1/SPHK2 [188, 189], S1P lyase [190, 191], SPP phosphatases [192], and transport from the cell to outside by SPNS2/ABC/MFSD2B transporters [147, 193, 194]. Modulation of these targets provides an opportunity to modulate intracellular S1P levels that could affect endothelial barrier function under normal and inflammatory conditions. The role of SPHK1- and SPHk2-derived S1P in lung inflammatory injury and recovery is dependent on the lung cell type, nature of insult, duration of insult, and animal model. Deletion of *Sphk1* or *Sphk2* in mice has no effect on endotoxin-mediated inflammatory responses and neutrophil function; however, knockdown of *Sphk2*, but not *Sphk1*, accelerates bacterial lung infection [195]. Inhibition of SPHK1 activity with *N,N*-dimethylsphingosine attenuates lung permeability, indicating an inflammatory role of SPHK1 [196]. However, total knockdown of *Sphk1* (*Sphk1*$^{-/-}$) in mice accentuates the LPS-induced lung injury and edema compared to the WT controls, whereas overexpression of h*Sphk1* in *Sphk1*$^{-/-}$ lung protects the mice from lung inflammation and injury [197]. Similarly, in RILI model, deletion of *Sphk1*exacerbates the radiation-induced lung inflammation [177], suggesting an anti-inflammatory role of SPHK1/S1P signaling axis in the endothelial barrier function. Interestingly, in other models of lung injury such as hyperoxia [25], pulmonary hypertension [198], and pulmonary fibrosis [24, 199], *Sphk1* deletion or inhibition of SPHK1 activity with SPHK1 inhibitor PF543 has been shown to be beneficial, indicating an inflammatory role of SPHK1/S1P signaling [27, 103, 200]. In many pulmonary disorders, SPHK2 seems to have no major role to play in the development of lung inflammation and injury; however, in the bacterial lung infection model, SPHK2-dependent S1P generation seems to regulate the lung inflammatory injury. Genetic deletion of *Sphk2*, but not *Sphk1*, attenuates the *Pseudomonas aeruginosa*-mediated lung inflammation and pulmonary edema, which involves SPHK2 activation and generation of S1P in the alveolar epithelial and EC nuclei, thus providing evidence for the pro-inflammatory function of SPHK2 [165]. Thus, targeting SPHK1 or SPHK2 may offer a therapeutic approach in maintaining or enhancing the endothelial barrier during various lung pathologies to reduce the pulmonary edema.

S1P Lyase Targeting Promotes Endothelial Barrier Integrity

S1P lyase-catalyzed hydrolysis of S1P to hexadecenal and ethanolamine phosphate represents an important metabolic step in regulating the intracellular S1P levels [154, 159, 190]. Inhibition or deletion of *SGPL1* enhances the intracellular S1P levels that may stabilize the endothelial barrier function. A role for S1P lyase and intracellularly generated S1P in the LPS-induced ALI has been demonstrated in vivo and in vitro. Intratracheal instillation of LPS to mice enhances the lung S1P lyase expression, decreases S1P concentrations in lung tissue, and induces lung injury [20]. Partial deletion of *Sgpl1* in mice (*Sgpl1*$^{+/-}$ mice) has increased S1P concentrations in the lung tissue and BAL fluid, with reduced lung injury and inflammation in response to the LPS challenge [20]. Furthermore, reducing S1P lyase activity by oral administration of THI shows a direct correlation between elevated S1P concentrations in the lung tissue and BAL fluid and reduced concentrations of neutrophils and IL-6 in mice receiving LPS intratracheally. Further, in vitro treatment of human lung microvascular ECs with LPS results in the reduced concentrations of intracellular S1P and increased mRNA and protein expression of S1P lyase and downregulation of *Sgpl1* expression by small interfering RNA (siRNA) increases S1P concentration in the cells and medium, which has attenuated the LPS-mediated phosphorylation of p38-MAPK and inhibitor of kB (I-kB) and has decreased IL-6 secretion. Also, S1P lyase siRNA treatment of lung ECs attenuates the LPS-induced endothelial barrier disruption by inducing the activation and redistribution of RAC1 to the cell periphery [20]. Similarly, knockdown of *Sgpl1* in human cerebral microvascular EC cell line HCMEC3 modulates barrier integrity in a dual manner. Under basal condition, *Sgpl1* knockdown destabilizes the EC barrier integrity but in an inflammatory setting, *Sgpl1* knockdown has conferred the protection from the pro-inflammatory cytokine-mediated permeability change [201]. Thus, targeting S1P lyase using a small molecule inhibitor might be a novel approach to enhance the cellular S1P levels and modulate the endothelial barrier restoration during lung injury. However, excess accumulation of S1P in cells and plasma causes lung inflammation and injury as seen in the *Sgpl1*$^{-/-}$ mice [202], and further evaluation of partial inhibition of S1P lyase with inhibitor(s) is necessary in preclinical animal models of lung injury.

S1P Transporter SPNS2 Regulates Lamellipodia Formation and Endothelial Barrier Function

The paradigm of inside-out signaling by S1P [203] requires export of intracellularly generated S1P out of cells to signal via its GPCRs either in an autocrine or paracrine manner. Three S1P transporters, namely ABC, SPNS2, and MFSD2B, have been described and SPNS2 is highly expressed in the lung ECs [193]. EC knockdown of SPNS2 reduces the circulating S1P to the same extent as the global knockout with similar lymphopenic effects [193, 204]. SPNS2 also regulates the inflammation and development of adaptive immune responses in various pathologies. There is evidence for a regulatory function of SPNS2 in the endothelial barrier function. The *Spns2* mRNA expression has been downregulated in the HUVECs exposed to a mixture of LPS, IL-1β, and TNF-α [205]. The *Spns2*-deficient mice have exhibited pulmonary edema and lung ECs isolated from *Spns2* knockout mice show increased permeability, which contributes to the EC barrier disruption. Studies carried out with human lung microvascular ECs further support a novel role for SPNS2 in annealing of AJs and barrier restoration. Downregulation of *SPNS2* with siRNA attenuates the thrombin-induced redistribution of VE-cadherin to the AJs of the HLMVECs, suggesting a requirement of the inside-out S1P signaling for barrier restoration. Loss of the *Spns2* expression also blocks the VE-cadherin dephosphorylation at Y658 in response to thrombin. VE-cadherin dephosphorylation by VE-PTPμ [206] or PTPN14 [94] is necessary for the redistribution of VE-cadherin back to the AJs as part of the recovery process. A role for SPNS2 in the HGF-mediated lamellipodia formation has been demonstrated. HGF signaling via its receptor, C-Met, stimulates lamellipodia formation and cell motility of lung ECs through the SPHK1/S1P/SPNS2 signaling axis [33]. Further studies are necessary to unravel mechanism(s) of SPNS2-mediated regulation of endothelial barrier.

Mechanisms of S1P- and HGF-Mediated Endothelial Barrier Restoration

Endothelial barrier integrity is a balance between two competing forces that include the intracellular contraction and adhesive cell–cell and cell matrix tethering. The actin microfilament is a critical determinant of EC adhesion and tight junctions; however, actin is also responsible for the generation of tensile intracellular forces via an actomyosin motor, which results in the EC barrier disruption. The actin rearrangement is driven by a coordinated activation of calcium/calmodulin dependent myosin light chain kinase (MLCK) and Rho Kinase, and together MLCK and Rho activity affect the MLC phosphorylation and actin polymerization that regulate the contractile or relaxed phenotype of the ECs [3, 120]. On the other hand, barrier enhancing agents such as S1P, FTY720, FTY720-phosphonate (Tys), and HGF enhance the endothelial barrier integrity by stimulating lamellipodia formation at cell periphery, which are critical for resealing of endothelial gaps and restoration of adherens and tight junctions [207–209]. However, barrier disrupting agents such as thrombin or LPS induce the actin stress fiber formation via calcium/calmodulin phosphorylation of non-muscle EC MLCK, activation of Rho, MLC phosphorylation, and VE-cadherin phosphorylation that disrupt the assembly of AJs and tight junctions and focal adhesion proteins, causing the barrier disruption.

Mechanisms of S1P-, and HGF-mediated regulation of endothelial barrier enhancement have been investigated in vitro in the lung ECs. Human lung ECs express S1P1 and S1P3 with the S1P1 signaling coupled to Gi and Rac1 activation whereas the S1P3 signaling is coupled to Gi, Gq/11, and G12/13 pathways that activate Rho to a greater extent as compared to Rac1 [2, 155, 176, 210, 211]. S1P ligation of S1P1 stimulates the Rac1-mediated translocation and co-localization of cortactin and non-muscle MLCK, MLC phosphorylation, and cortical actin formation to enhance the barrier function [29]. Recent evidence shows that S1P-induced activation and recruitment of phosphoinositide 3-kinase (PI3K) and TIAM1/RAC1 to the caveolin-enriched microdomains (CEM) is necessary for the α-actinin mediated cortical actin rearrangement and endothelial barrier enhancement [212]. Additionally, integrin β4 (ITGB4) recruited to the CEM forms a complex with S1P1 in lung ECs and downregulation of ITGB4 expression reduces the S1P-induced barrier enhancement [213]. These results support a role for S1P1 and ITGB4 complexes in the CEM for enhanced barrier function and vascular integrity. Coronin, an actin-binding protein, has been identified as critical co-factor for cofilin-dependent signaling pathways. S1P stimulated coronin 1B phosphorylation and enhanced co-localization with cortactin at lamellipodia and chemotaxis of lung ECs, which is PLD_2-, PKC-, and Rac-dependent [214]. Additionally, HGF stimulates the c-MET phosphorylation at Y1003, Y1313, Y1234, Y1235, Y1349, and Ser985 and enhances lamellipodia formation via the PI3K/AKT signal transduction [28]. Further, in addition to the PI3K/AKT pathway, other distinct mechanisms such as the microtubule-independent TIAM1 activation and microtubule- and activated protein C (APC)-dependent activation of Asef, a novel RAC activator, in the HGF-induced endothelial barrier enhancement has been reported [215–217]. Additionally, IQGAP1, an effector of RAC1 and CDC42, has been identified as a key regulator of actin-cytoskeleton dynamics, thereby regulating the lamellipodia formation, and barrier protection [218]. HGF- and S1P-mediated lamellipodia formation and barrier enhancement has been shown to be PI3K-dependent in the lung ECs. HGF-stimulated tyrosine phosphorylation of c-MET potentiates the lamellipodia formation via PI3K and AKT activation [28]. Inhibition of PI3K with LY294002 attenuates the Akt phosphorylation and suppresses the lamellipodia formation and EC migration, implicating the involvement of PI3K in the endothelial barrier enhancement. Mechanism(s) of HGF- and S1P-mediated activation of PI3K is unclear, but might involve PLD/PA signaling in the Chinese hamster ovary cells [219]. HGF stimulates S1P production in the HLMVECs by activation of SPHK1 [33] and enhances the co-localization of SPHK1/p-SPHK1 with actin/cortactin in the lamellipodia and downregulation of SPHK1 or inhibition of SPHK1 activity with a SPHK1-specific inhibitor, PF-543, attenuates the HGF-induced lamellipodia formation. In addition, downregulation of SPNS2 also suppresses the HGF-induced lamellipodia formation and EC migration, suggesting a key role for the "inside-out" S1P signaling [33]. The interdependence between the HGF-c-MET signaling and S1P has been further evidenced by knocking of S1P1, but not S1P2 or S1P3, which abolishes the lamellipodia formation in the HLMVECs. In addition to SPHK1, mammalian cells also express the second isoenzyme of SPHK, namely SPHK2; however, it appears that blocking SPHK2 with an inhibitor or downregulation of *Sphk2* with siRNA has no effect on the HGF-induced lamellipodia formation [33]. The crosstalk between the HGF/c-Met and SPHK1/S1P/SPNS2/S1P1 signaling axis in the HGF-mediated lamellipodia formation and endothelial barrier enhancement provides potential interaction between the receptor tyrosine kinase and G-protein-coupled receptors in regulating the EC barrier function (Fig. 6).

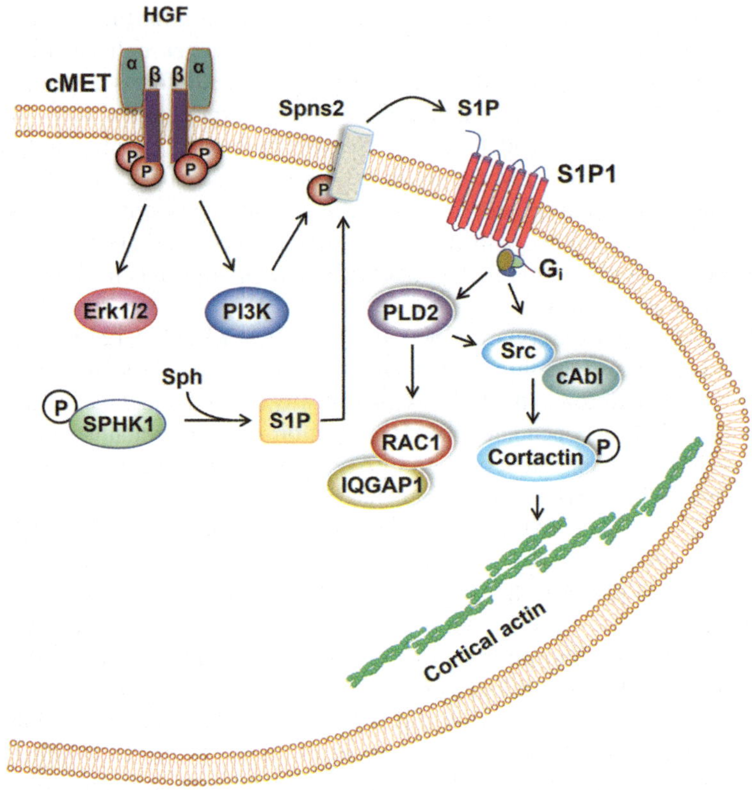

Fig. 6 Crosstalk between HGF/c-MET and SPHK1/S1P/SPNS2/S1P1 signaling axis in lamellipodia formation. HGF activation of its receptor c-Met initiates phosphorylation of ERK and Akt. Activation of ERK by HGF stimulates SPHK1 phosphorylation, which in turn converts sphingosine into S1P. HGF-mediated activation of PI3K/Akt pathway phosphorylates S1P transporter, SPNS2, which facilitates efflux of intracellular S1P to extracellular milieu. The secreted S1P binds to its G-protein-coupled receptor S1P1 to initiate downstream pathways of PLD, SRC, C-ABL, RAC1, and IQGAP1 mediating cortactin phosphorylation and stabilization. Mechanisms of OxPAPC-induced EC barrier protection: OxPAPC induces the activation of multiple signaling pathways that leads to the activation of Rap1 and Rac. The cytoskeletal remodeling facilitated by the cortical actin formation and assembly of tight junctions and adherens junctions proteins enhances lung endothelial barrier. In addition, EP4 receptor and lipoxin A4 also mediate the barrier protective effects of OxPAPC against LPS/TNFα-induced inflammation and lung injury in vitro and in vivo cortical actin and lamellipodia formation of lung ECs. This proposed model is a static model and does not convey the possibility that in a dynamic model, the interacting partners might function differently over time. MET, Mesenchymal-Epithelial transition factor; HGF, hepatocyte growth factor; ERK, extracellular signal regulated kinase; PI3K, phosphatidylinositol 3-kinase; AKT, protein kinase B; ABL, Abelson kinase; RAC, Ras-related C3 botulinum toxin substrate; IQGAP1, IQ motif containing GTPase activating protein 1

Mechanisms of FTY720- and FTY720-Phosphonate-Mediated Endothelial Barrier Enhancement

The mechanisms of FTY720-mediated EC barrier enhancement are unclear, but seem very different from the S1P signaling of EC barrier protection. FTY720 barrier enhancement is independent of S1P1, changes in intracellular $[Ca^{2+}]_i$, RAC1 activation, cortactin tyrosine phosphorylation, and cytoskeletal reorganization to cell periphery [180]. Further, phosphorylation of FTY720 to FTY720-P, not required for the EC barrier protection, is coupled to Gi, and requires signaling through the membrane lipid rafts. Also, FTY720-induced barrier enhancement is independent of adherens and tight junction proteins and inhibition of protein kinases has no effect on the FTY720-induced barrier enhancement [220]. Interestingly, FTY720 increases c-ABL phosphorylation and tyrosine kinase activity and downregulation of c-ABL partly attenuates the barrier enhancement by FTY720. The downstream targets of c-ABL in FTY720-mediated barrier enhancement are yet to be identified; however, the role of focal adhesion proteins such as FAK and paxillin needs to be evaluated. Earlier study has shown that S1P and HGF stimulate c-ABL and paxillin tyrosine phosphorylation, and inhibiting c-ABL activity attenuated S1P-induced tyrosine phosphorylation of paxillin, lamellipodia formation, and barrier enhancement, suggesting a plausible link between FTY720/C-ABL and FAK/paxillin in the EC barrier integrity [16]. FTY720 also reverses the vascular permeability increase mediated by *Plasmodium falciparum* [221].

Recent in vivo and in vitro studies show that FTY720-phophonate or Tysiponate (Tys) is a superior barrier enhancing agent as compared to FTY720. Tys administered IP to the bleomycin-injured mice decreases pulmonary inflammation and leak in mice [186]. In vitro, Tys enhances the lung EC barrier function by preserving S1P1 expression and blocking β-arrestin recruitment, and ubiquitination and proteosomal degradation of S1P1 [186]. The barrier protective role of Tys is also observed in the murine models of lung bacterial infection. In vivo intratracheal administration of live MRSA in mice causes significant vascular leakage and leukocyte infiltration into the alveolar space, and pre- or post-treatment with Tys attenuates the MRSA-induced lung permeability and levels of infiltrating neutrophils into lung alveolar space [187]. In vitro, Tys attenuates the heat-inactivated MRSA- or methicillin-resistant staphylococcal α-toxin-induced lung EC barrier disruption by reverting cytoskeletal rearrangement and VE-cadherin junctional disruption. Further, Tys inhibits the MRSA-induced Rho activation, MLC phosphorylation, IkB phosphorylation, and IL-6/IL-8 secretion [187]. These novel findings provide new insights into the Tys signaling and barrier protection in lung ECs (Fig. 7).

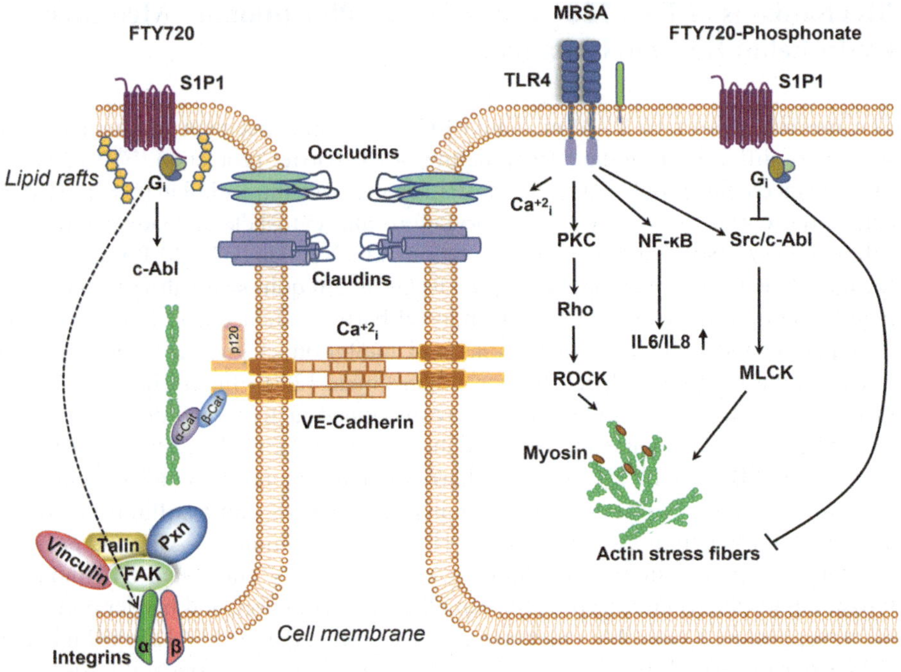

Fig. 7 Signaling pathways of FTY720 and FTY720-phosphonate (TyS) in enhancing endothelial barrier function. Both FTY720 (left panel) and FTY720-phosphonate (TyS) (right panel) increase endothelial cell (EC) barrier function in vitro. FTY720 and Tys bind to S1P1, but differ in signaling pathways involved in barrier regulation. FTY720-phosphonate or Tys, similar to S1P, rapidly induces cytoskeletal reorganization, cortical actin stabilization, RAC1 activation, tightening of junctional proteins, and barrier enhancement. FTY720 (left panel) induced barrier enhancement through Gi and lipid raft-coupled signaling. However, no significant $[Ca^{2+}]_i$ increase was observed in lung ECs, and no cytoskeletal rearrangement or cortical actin formation occurred during the time frame associated with maximal barrier effects. FTY720 stimulated c-ABL tyrosine kinase activity, but the downstream targets involved in barrier regulation are unclear. Recent studies indicate that c-ABL, paxillin, and FAK signaling are necessary for optimal barrier enhancement, and focal adhesion complexes also appear to participate in this process after FTY720. FTY720-phosphonate failed to induce the ubiquitination and subsequent proteosomal degradation of barrier-promoting S1P1. EC barrier enhancement by FTY720 is slower in onset and may involve an alternative, but not yet identified, G protein-coupled receptor (GPCR) in addition to S1P1. EC, endothelial cell; Ub, ubiquitination; PXN, paxillin; FAK, focal adhesion kinase; GIT, G protein-coupled receptor kinase interactor–1; Tyr, tyrosine; c-ABL, Abelson tyrosine kinase

Oxidized Phospholipids and Endothelial Barrier Protection

Oxidized phospholipids are primarily generated by the enzymatic and non-enzymatic peroxidation of the membrane phospholipids containing polyunsaturated fatty acids (PUFA) such as arachidonic, docosahexaenoic, and eicosapentanoic acids. Oxidized phospholipids (OxPLs) in OxLDL are a mixture of

1-palmitoyl-2-glutaroyl-*sn*-glycero-3-phosphocholine (PGPC) and 1-palmitoyl-2-oxovaleroyl-*sn*-glycero-3-phosphocholine (POVPC) [222], which are pro- or anti-inflammatory. Early studies show that lipids like 1-palmitoyl-2-arachidonoyl-*sn*-glycero-3-phosphorylcholine (PAPC) undergo oxidation to OxPAPC, which reverse the barrier disruption mediated by thrombin, LPS, mechanical ventilation, and heat-inactivated *Staphylococcus aureus* [10, 223–225]. Signaling pathways involved in the OxPAPC-mediated barrier protection have been described in the human lung ECs [226, 227]. Barrier protective effects of OxPAPC are attenuated by the inhibitors of small GTPases, PKA, PKC, and SRC, while MAPKs and PI3K have no role to play in the barrier protection [216, 223]. OxPAPC blocks the interaction of LPS with LPS-binding protein and CD14, but not the TNFα-induced activation of NF-κB [228]. OxPAPC also protects the endothelial barrier from thrombin-induced disruption and ventilator-induced vascular leakage [224]. Low dose OxPAPC (10 μg/ml) is barrier protective, whereas a high dose (100 μg/ml) is barrier disruptive and involves ROS and Src activation [229]. OxPAPC transactivation of S1P receptor plays a role in the barrier protective function. OxPAPC-recruited S1P1 into the caveolin-enriched membrane microdomains lead to the activation of Akt and Rac1, and cytoskeletal reorganization leading to the endothelial barrier enhancement [230]. These studies suggest a role for S1P1 in OxPAPC-mediated barrier protective responses; however, further investigations are necessary to determine if OxPAPC activation of S1P1 is mediated by the intracellular S1P generation followed by the "inside-out" signaling through S1P transporter Spns2. The contribution of S1P2 and S1P3 in the OxPAPC-mediated enhancement of vascular integrity needs further clarification. Direct binding of OxPAPC to the chaperone GRP78 associated with its cofactor HTJ-1 has been demonstrated to establish the beneficial effects [231]. Recently, a lipase, acyloxyacyl hydrolase (AOAH), which deacylates the two common OxPLs, 1-palmitoyl-2-glutaroyl-*sn*-glycero-3-phosphocholine (PGPC) and 1-palmitoyl-2-(5-oxovaleroyl)-*sn*-glycero-3-phosphocholine (POVPC) mitigates the acute lung injury [232, 233]. How AOAH regulation of the vascular ECs and barrier function is unclear. Understanding the prevalence of similar efficient endogenous clearance mechanisms, specific to the pro-inflammatory OxPL mediators, may offer clues towards their role in vascular leak and resolution of acute lung injury.

Conclusion and Perspective

Disassembly of adherens junctions (AJs) of lung endothelial cell (EC) monolayers by edemagenic agents causes microvascular hyperpermeability and protein-rich pulmonary edema formation, and if uncorrected, it leads to the deterioration of lung gas exchange. These changes reflect the failure of the lung's intrinsic homeostatic mechanisms, and as such, they are the hallmarks of acute respiratory distress syndrome (ARDS). Despite our current understanding of the multifaceted mechanisms regulating the vascular permeability, little is known about the molecular regulation of endothelial barrier restoration following lung injury and edema. A number of lipid-derived mediators and signaling pathways have been identified to regulate the stabilization of tight junctions and AJs and formation of lamellipodia at the leading edge of ECs for repair and closure of the wound. It will be critical to characterize how the spatiotemporal generation of some of these lipid mediators regulates the lamellipodia formation and facilitates assembly of the AJ and tight junction proteins in the protrusions for endothelial barrier restoration. The future focus has to define and identify intracellular signaling mechanisms that regulate the formation of adhesive contacts between the adjacent ECs via homotypic interaction of VE-cadherin in order to restore the lung endothelial barrier integrity. Recent studies also suggest the involvement of lamellipodin, a key protein component of lamellipodia in the stabilization cortical actin; however, not much is known on lamellipodin and its regulation of lamellipodia formation and the vascular endothelial barrier regulation.

Acknowledgements This work was supported by National Institutes of Health Grants P01 HL P01 HL060678 (to VN) and P01 HL126609 (to VN and SMD), and Zhejiang Province Medicinal Technology Grant (No. 2022482706), and Ningbo Nature Science Foundation Grant (No. 2021J244) (to PF).

References

1. Vandenbroucke E, Mehta D, Minshall R, Malik AB. Regulation of endothelial junctional permeability. Ann N Y Acad Sci. 2008;1123:134–45. https://doi.org/10.1196/annals.1420.016.
2. Natarajan V, Dudek SM, Jacobson JR, Moreno-Vinasco L, Huang LS, Abassi T, et al. Sphingosine-1-phosphate, FTY720, and sphingosine-1-phosphate receptors in the pathobiology of acute lung injury. Am J Respir Cell Mol Biol. 2013;49(1):6–17. https://doi.org/10.1165/rcmb.2012-0411TR.
3. Wang L, Dudek SM. Regulation of vascular permeability by sphingosine 1-phosphate. Microvasc Res. 2009;77(1):39–45. https://doi.org/10.1016/j.mvr.2008.09.005.
4. Abbasi T, Garcia JG. Sphingolipids in lung endothelial biology and regulation of vascular integrity. Handb Exp Pharmacol. 2013;216:201–26. https://doi.org/10.1007/978-3-7091-1511-4_10.
5. Singleton PA, Mirzapoiazova T, Guo Y, Sammani S, Mambetsariev N, Lennon FE, et al. High-molecular-weight hyaluronan is a novel inhibitor of pulmonary vascular leakiness. Am J Phys Lung Cell Mol Phys. 2010;299(5):L639–51. https://doi.org/10.1152/ajplung.00405.2009.
6. Behling-Kelly E, Vonderheid H, Kim KS, Corbeil LB, Czuprynski CJ. Roles of cellular activation and sulfated glycans in Haemophilus somnus adherence to bovine brain microvascular endothelial cells. Infect Immun. 2006;74(9):5311–8. https://doi.org/10.1128/IAI.00614-06.
7. Gunduz D, Hirche F, Hartel FV, Rodewald CW, Schafer M, Pfitzer G, et al. ATP antagonism of thrombin-induced endothelial barrier permeability. Cardiovasc Res. 2003;59(2):470–8.
8. Kolosova IA, Mirzapoiazova T, Adyshev D, Usatyuk P, Romer LH, Jacobson JR, et al. Signaling pathways involved in adenosine triphosphate-induced endothelial cell barrier enhancement. Circ Res. 2005;97(2):115–24. https://doi.org/10.1161/01.RES.0000175561.55761.69.
9. Okajima K. Prevention of endothelial cell injury by activated protein C: the molecular mechanism(s) and therapeutic implications. Curr Vasc Pharmacol. 2004;2(2):125–33.
10. Nonas S, Miller I, Kawkitinarong K, Chatchavalvanich S, Gorshkova I, Bochkov VN, et al. Oxidized phospholipids reduce vascular leak and inflammation in rat model of acute

lung injury. Am J Respir Crit Care Med. 2006;173(10):1130–8. https://doi.org/10.1164/rccm.200511-1737OC.
11. Birukova AA, Chatchavalvanich S, Oskolkova O, Bochkov VN, Birukov KG. Signaling pathways involved in OxPAPC-induced pulmonary endothelial barrier protection. Microvasc Res. 2007;73(3):173–81. https://doi.org/10.1016/j.mvr.2006.12.004.
12. Rittchen S, Rohrer K, Platzer W, Knuplez E, Barnthaler T, Marsh LM, et al. Prostaglandin D2 strengthens human endothelial barrier by activation of E-type receptor 4. Biochem Pharmacol. 2020;182:114277. https://doi.org/10.1016/j.bcp.2020.114277.
13. Birukova AA, Zagranichnaya T, Fu P, Alekseeva E, Chen W, Jacobson JR, et al. Prostaglandins PGE(2) and PGI(2) promote endothelial barrier enhancement via PKA- and Epac1/Rap1-dependent Rac activation. Exp Cell Res. 2007;313(11):2504–20. https://doi.org/10.1016/j.yexcr.2007.03.036.
14. Liu F, Schaphorst KL, Verin AD, Jacobs K, Birukova A, Day RM, et al. Hepatocyte growth factor enhances endothelial cell barrier function and cortical cytoskeletal rearrangement: potential role of glycogen synthase kinase-3beta. FASEB J. 2002;16(9):950–62. https://doi.org/10.1096/fj.01-0870com.
15. Yamada N, Nakagawa S, Horai S, Tanaka K, Deli MA, Yatsuhashi H, et al. Hepatocyte growth factor enhances the barrier function in primary cultures of rat brain microvascular endothelial cells. Microvasc Res. 2014;92:41–9. https://doi.org/10.1016/j.mvr.2013.12.004.
16. Fu P, Usatyuk PV, Jacobson J, Cress AE, Garcia JG, Salgia R, et al. Role played by paxillin and paxillin tyrosine phosphorylation in hepatocyte growth factor/sphingosine-1-phosphate-mediated reactive oxygen species generation, lamellipodia formation, and endothelial barrier function. Pulm Circ. 2015;5(4):619–30. https://doi.org/10.1086/683693.
17. van der Flier M, van Leeuwen HJ, van Kessel KP, Kimpen JL, Hoepelman AI, Geelen SP. Plasma vascular endothelial growth factor in severe sepsis. Shock. 2005;23(1):35–8.
18. Sekine K, Fujishima S, Aikawa N. Plasma hepatocyte growth factor is increased in early-phase sepsis. J Infect Chemother. 2004;10(2):110–4. https://doi.org/10.1007/s10156-004-0301-y.
19. Winkler MS, Nierhaus A, Holzmann M, Mudersbach E, Bauer A, Robbe L, et al. Decreased serum concentrations of sphingosine-1-phosphate in sepsis. Crit Care. 2015;19:372. https://doi.org/10.1186/s13054-015-1089-0.
20. Zhao Y, Gorshkova IA, Berdyshev E, He D, Fu P, Ma W, et al. Protection of LPS-induced murine acute lung injury by sphingosine-1-phosphate lyase suppression. Am J Respir Cell Mol Biol. 2011;45(2):426–35. https://doi.org/10.1165/rcmb.2010-0422OC.
21. Szczepaniak WS, Zhang Y, Hagerty S, Crow MT, Kesari P, Garcia JG, et al. Sphingosine 1-phosphate rescues canine LPS-induced acute lung injury and alters systemic inflammatory cytokine production in vivo. Transl Res. 2008;152(5):213–24. https://doi.org/10.1016/j.trsl.2008.09.002.
22. Jolly PS, Rosenfeldt HM, Milstien S, Spiegel S. The roles of sphingosine-1-phosphate in asthma. Mol Immunol. 2002;38(16–18):1239–45.
23. Gairhe S, Joshi SR, Bastola MM, McLendon JM, Oka M, Fagan KA, et al. Sphingosine-1-phosphate is involved in the occlusive arteriopathy of pulmonary arterial hypertension. Pulm Circ. 2016;6(3):369–80. https://doi.org/10.1086/687766.
24. Huang LS, Natarajan V. Sphingolipids in pulmonary fibrosis. Adv Biol Regul. 2015;57:55–63. https://doi.org/10.1016/j.jbior.2014.09.008.
25. Harijith A, Pendyala S, Reddy NM, Bai T, Usatyuk PV, Berdyshev E, et al. Sphingosine kinase 1 deficiency confers protection against hyperoxia-induced bronchopulmonary dysplasia in a murine model: role of S1P signaling and Nox proteins. Am J Pathol. 2013;183(4):1169–82. https://doi.org/10.1016/j.ajpath.2013.06.018.
26. Tabasinezhad M, Samadi N, Ghanbari P, Mohseni M, Saei AA, Sharifi S, et al. Sphingosin 1-phosphate contributes in tumor progression. J Cancer Res Ther. 2013;9(4):556–63. https://doi.org/10.4103/0973-1482.126446.
27. Ebenezer DL, Fu P, Natarajan V. Targeting sphingosine-1-phosphate signaling in lung diseases. Pharmacol Ther. 2016;168:143–57. https://doi.org/10.1016/j.pharmthera.2016.09.008.

28. Usatyuk PV, Fu P, Mohan V, Epshtein Y, Jacobson JR, Gomez-Cambronero J, et al. Role of c-Met/phosphatidylinositol 3-kinase (PI3k)/Akt signaling in hepatocyte growth factor (HGF)-mediated lamellipodia formation, reactive oxygen species (ROS) generation, and motility of lung endothelial cells. J Biol Chem. 2014;289(19):13476–91. https://doi.org/10.1074/jbc.M113.527556.
29. Dudek SM, Jacobson JR, Chiang ET, Birukov KG, Wang P, Zhan X, et al. Pulmonary endothelial cell barrier enhancement by sphingosine 1-phosphate: roles for cortactin and myosin light chain kinase. J Biol Chem. 2004;279(23):24692–700. https://doi.org/10.1074/jbc.M313969200.
30. Birukova AA, Malyukova I, Mikaelyan A, Fu P, Birukov KG. Tiam1 and betaPIX mediate Rac-dependent endothelial barrier protective response to oxidized phospholipids. J Cell Physiol. 2007;211(3):608–17. https://doi.org/10.1002/jcp.20966.
31. Jacobson JR, Dudek SM, Singleton PA, Kolosova IA, Verin AD, Garcia JG. Endothelial cell barrier enhancement by ATP is mediated by the small GTPase Rac and cortactin. Am J Phys Lung Cell Mol Phys. 2006;291(2):L289–95. https://doi.org/10.1152/ajplung.00343.2005.
32. Jacobson JR, Dudek SM, Birukov KG, Ye SQ, Grigoryev DN, Girgis RE, et al. Cytoskeletal activation and altered gene expression in endothelial barrier regulation by simvastatin. Am J Respir Cell Mol Biol. 2004;30(5):662–70. https://doi.org/10.1165/rcmb.2003-0267OC.
33. Fu P, Ebenezer DL, Berdyshev EV, Bronova IA, Shaaya M, Harijith A, et al. Role of sphingosine kinase 1 and S1P transporter Spns2 in HGF-mediated lamellipodia formation in lung endothelium. J Biol Chem. 2016;291(53):27187–203. https://doi.org/10.1074/jbc.M116.758946.
34. McVerry BJ, Garcia JG. In vitro and in vivo modulation of vascular barrier integrity by sphingosine 1-phosphate: mechanistic insights. Cell Signal. 2005;17(2):131–9. https://doi.org/10.1016/j.cellsig.2004.08.006.
35. Wang D. Image guidance technologies for interventional pain procedures: ultrasound, fluoroscopy, and CT. Curr Pain Headache Rep. 2018;22(1):6. https://doi.org/10.1007/s11916-018-0660-1.
36. Aoki T, Narumiya S. Prostaglandins and chronic inflammation. Trends Pharmacol Sci. 2012;33(6):304–11. https://doi.org/10.1016/j.tips.2012.02.004.
37. Murata T, Aritake K, Tsubosaka Y, Maruyama T, Nakagawa T, Hori M, et al. Anti-inflammatory role of PGD2 in acute lung inflammation and therapeutic application of its signal enhancement. Proc Natl Acad Sci U S A. 2013;110(13):5205–10. https://doi.org/10.1073/pnas.1218091110.
38. Kobayashi K, Tsubosaka Y, Hori M, Narumiya S, Ozaki H, Murata T. Prostaglandin D2-DP signaling promotes endothelial barrier function via the cAMP/PKA/Tiam1/Rac1 pathway. Arterioscler Thromb Vasc Biol. 2013;33(3):565–71. https://doi.org/10.1161/ATVBAHA.112.300993.
39. Horikami D, Toya N, Kobayashi K, Omori K, Nagata N, Murata T. L-PGDS-derived PGD2 attenuates acute lung injury by enhancing endothelial barrier formation. J Pathol. 2019;248(3):280–90. https://doi.org/10.1002/path.5253.
40. Xiao L, Ornatowska M, Zhao G, Cao H, Yu R, Deng J, et al. Lipopolysaccharide-induced expression of microsomal prostaglandin E synthase-1 mediates late-phase PGE2 production in bone marrow derived macrophages. PLoS One. 2012;7(11):e50244. https://doi.org/10.1371/journal.pone.0050244.
41. Jandl K, Stacher E, Balint Z, Sturm EM, Maric J, Peinhaupt M, et al. Activated prostaglandin D2 receptors on macrophages enhance neutrophil recruitment into the lung. J Allergy Clin Immunol. 2016;137(3):833–43. https://doi.org/10.1016/j.jaci.2015.11.012.
42. Konya V, Ullen A, Kampitsch N, Theiler A, Philipose S, Parzmair GP, et al. Endothelial E-type prostanoid 4 receptors promote barrier function and inhibit neutrophil trafficking. J Allergy Clin Immunol. 2013;131(2):532–40.e1–2. https://doi.org/10.1016/j.jaci.2012.05.008.
43. Barnthaler T, Maric J, Platzer W, Konya V, Theiler A, Hasenohrl C, et al. The role of PGE2 in alveolar epithelial and lung microvascular endothelial crosstalk. Sci Rep. 2017;7(1):7923. https://doi.org/10.1038/s41598-017-08228-y.

44. Bos CL, Richel DJ, Ritsema T, Peppelenbosch MP, Versteeg HH. Prostanoids and prostanoid receptors in signal transduction. Int J Biochem Cell Biol. 2004;36(7):1187–205. https://doi.org/10.1016/j.biocel.2003.08.006.
45. Melian EB, Goa KL. Beraprost: a review of its pharmacology and therapeutic efficacy in the treatment of peripheral arterial disease and pulmonary arterial hypertension. Drugs. 2002;62(1):107–33. https://doi.org/10.2165/00003495-200262010-00005.
46. Baumer Y, Drenckhahn D, Waschke J. cAMP induced Rac 1-mediated cytoskeletal reorganization in microvascular endothelium. Histochem Cell Biol. 2008;129(6):765–78. https://doi.org/10.1007/s00418-008-0422-y.
47. Landert M, Baumert BG, Bosch MM, Lutolf UM, Landau K. The visual impact of fractionated stereotactic conformal radiotherapy on seven eyes with optic nerve sheath meningiomas. J Neuroophthalmol. 2005;25(2):86–91. https://doi.org/10.1097/01.wno.0000165105.78365.22.
48. Birukova AA, Fu P, Xing J, Birukov KG. Rap1 mediates protective effects of iloprost against ventilator-induced lung injury. J Appl Physiol (1985). 2009;107(6):1900–10. https://doi.org/10.1152/japplphysiol.00462.2009.
49. Birukova AA, Fu P, Xing J, Cokic I, Birukov KG. Lung endothelial barrier protection by iloprost in the 2-hit models of ventilator-induced lung injury (VILI) involves inhibition of Rho signaling. Transl Res. 2010;155(1):44–54. https://doi.org/10.1016/j.trsl.2009.09.002.
50. Birukova AA, Wu T, Tian Y, Meliton A, Sarich N, Tian X, et al. Iloprost improves endothelial barrier function in lipopolysaccharide-induced lung injury. Eur Respir J. 2013;41(1):165–76. https://doi.org/10.1183/09031936.00148311.
51. Oskolkova O, Sarich N, Tian Y, Gawlak G, Meng F, Bochkov VN, et al. Incorporation of iloprost in phospholipase-resistant phospholipid scaffold enhances its barrier protective effects on pulmonary endothelium. Sci Rep. 2018;8(1):879. https://doi.org/10.1038/s41598-018-19197-1.
52. Ke Y, Oskolkova OV, Sarich N, Tian Y, Sitikov A, Tulapurkar ME, et al. Effects of prostaglandin lipid mediators on agonist-induced lung endothelial permeability and inflammation. Am J Phys Lung Cell Mol Phys. 2017;313(4):L710–L21. https://doi.org/10.1152/ajplung.00519.2016.
53. Duah E, Adapala RK, Al-Azzam N, Kondeti V, Gombedza F, Thodeti CK, et al. Cysteinyl leukotrienes regulate endothelial cell inflammatory and proliferative signals through CysLT(2) and CysLT(1) receptors. Sci Rep. 2013;3:3274. https://doi.org/10.1038/srep03274.
54. Duah E, Teegala LR, Kondeti V, Adapala RK, Keshamouni VG, Kanaoka Y, et al. Cysteinyl leukotriene 2 receptor promotes endothelial permeability, tumor angiogenesis, and metastasis. Proc Natl Acad Sci U S A. 2019;116(1):199–204. https://doi.org/10.1073/pnas.1817325115.
55. Sim JA, Kim J, Yang D. Beyond lipid signaling: pleiotropic effects of diacylglycerol kinases in cellular signaling. Int J Mol Sci. 2020;21(18):6861. https://doi.org/10.3390/ijms21186861.
56. Cocco L, Follo MY, Manzoli L, Suh PG. Phosphoinositide-specific phospholipase C in health and disease. J Lipid Res. 2015;56(10):1853–60. https://doi.org/10.1194/jlr.R057984.
57. Carman GM, Han GS. Phosphatidic acid phosphatase, a key enzyme in the regulation of lipid synthesis. J Biol Chem. 2009;284(5):2593–7. https://doi.org/10.1074/jbc.R800059200.
58. Koeberle A, Shindou H, Harayama T, Yuki K, Shimizu T. Polyunsaturated fatty acids are incorporated into maturating male mouse germ cells by lysophosphatidic acid acyltransferase 3. FASEB J. 2012;26(1):169–80. https://doi.org/10.1096/fj.11-184879.
59. Korbes AP, Kulcheski FR, Margis R, Margis-Pinheiro M, Turchetto-Zolet AC. Molecular evolution of the lysophosphatidic acid acyltransferase (LPAAT) gene family. Mol Phylogenet Evol. 2016;96:55–69. https://doi.org/10.1016/j.ympev.2015.12.001.
60. Bektas M, Payne SG, Liu H, Goparaju S, Milstien S, Spiegel S. A novel acylglycerol kinase that produces lysophosphatidic acid modulates cross talk with EGFR in prostate cancer cells. J Cell Biol. 2005;169(5):801–11. https://doi.org/10.1083/jcb.200407123.
61. Kalari S, Zhao Y, Spannhake EW, Berdyshev EV, Natarajan V. Role of acylglycerol kinase in LPA-induced IL-8 secretion and transactivation of epidermal growth factor-receptor in

human bronchial epithelial cells. Am J Phys Lung Cell Mol Phys. 2009;296(3):L328–36. https://doi.org/10.1152/ajplung.90431.2008.
62. Perrakis A, Moolenaar WH. Autotaxin: structure-function and signaling. J Lipid Res. 2014;55(6):1010–8. https://doi.org/10.1194/jlr.R046391.
63. Valdes-Rives SA, Gonzalez-Arenas A. Autotaxin-lysophosphatidic acid: from inflammation to cancer development. Mediat Inflamm. 2017;2017:9173090. https://doi.org/10.1155/2017/9173090.
64. Thakur R, Naik A, Panda A, Raghu P. Regulation of membrane turnover by phosphatidic acid: cellular functions and disease implications. Front Cell Dev Biol. 2019;7:83. https://doi.org/10.3389/fcell.2019.00083.
65. Cummings R, Parinandi N, Wang L, Usatyuk P, Natarajan V. Phospholipase D/phosphatidic acid signal transduction: role and physiological significance in lung. Mol Cell Biochem. 2002;234–235(1–2):99–109.
66. Usatyuk PV, Kotha SR, Parinandi NL, Natarajan V. Phospholipase D signaling mediates reactive oxygen species-induced lung endothelial barrier dysfunction. Pulm Circ. 2013;3(1):108–15. https://doi.org/10.4103/2045-8932.109925.
67. Frohman MA. The phospholipase D superfamily as therapeutic targets. Trends Pharmacol Sci. 2015;36(3):137–44. https://doi.org/10.1016/j.tips.2015.01.001.
68. Nelson RK, Frohman MA. Physiological and pathophysiological roles for phospholipase D. J Lipid Res. 2015;56(12):2229–37. https://doi.org/10.1194/jlr.R059220.
69. Abdulnour RE, Howrylak JA, Tavares AH, Douda DN, Henkels KM, Miller TE, et al. Phospholipase D isoforms differentially regulate leukocyte responses to acute lung injury. J Leukoc Biol. 2018;103(5):919–32. https://doi.org/10.1002/JLB.3A0617-252RR.
70. Colley WC, Sung TC, Roll R, Jenco J, Hammond SM, Altshuller Y, et al. Phospholipase D2, a distinct phospholipase D isoform with novel regulatory properties that provokes cytoskeletal reorganization. Curr Biol. 1997;7(3):191–201.
71. Iyer SS, Barton JA, Bourgoin S, Kusner DJ. Phospholipases D1 and D2 coordinately regulate macrophage phagocytosis. J Immunol. 2004;173(4):2615–23.
72. Sung TC, Altshuller YM, Morris AJ, Frohman MA. Molecular analysis of mammalian phospholipase D2. J Biol Chem. 1999;274(1):494–502.
73. Zhang F, Wang Z, Lu M, Yonekubo Y, Liang X, Zhang Y, et al. Temporal production of the signaling lipid phosphatidic acid by phospholipase D2 determines the output of extracellular signal-regulated kinase signaling in cancer cells. Mol Cell Biol. 2014;34(1):84–95. https://doi.org/10.1128/MCB.00987-13.
74. Liang D, Wu K, Tei R, Bumpus TW, Ye J, Baskin JM. A real-time, click chemistry imaging approach reveals stimulus-specific subcellular locations of phospholipase D activity. Proc Natl Acad Sci U S A. 2019;116(31):15453–62. https://doi.org/10.1073/pnas.1903949116.
75. Carneiro AF, Carneiro CN, Pires LN, Teixeira LSG, Azcarate SM, Dias FS. D-optimal mixture design for the optimization of extraction induced by emulsion breaking for multielemental determination in edible vegetable oils by microwave-induced plasma optical emission spectrometry. Talanta. 2020;219:121218. https://doi.org/10.1016/j.talanta.2020.121218.
76. Jenkins GM, Frohman MA. Phospholipase D: a lipid centric review. Cell Mol Life Sci. 2005;62(19–20):2305–16. https://doi.org/10.1007/s00018-005-5195-z.
77. Kooijman EE, Burger KN. Biophysics and function of phosphatidic acid: a molecular perspective. Biochim Biophys Acta. 2009;1791(9):881–8. https://doi.org/10.1016/j.bbalip.2009.04.001.
78. Tanguy E, Wang Q, Moine H, Vitale N. Phosphatidic acid: from pleiotropic functions to neuronal pathology. Front Cell Neurosci. 2019;13:2. https://doi.org/10.3389/fncel.2019.00002.
79. English D, Cui Y, Siddiqui R, Patterson C, Natarajan V, Brindley DN, et al. Induction of endothelial monolayer permeability by phosphatidate. J Cell Biochem. 1999;75(1):105–17.
80. Stasek JE Jr, Natarajan V, Garcia JG. Phosphatidic acid directly activates endothelial cell protein kinase C. Biochem Biophys Res Commun. 1993;191(1):134–41. https://doi.org/10.1006/bbrc.1993.1194.

81. Cross MJ, Roberts S, Ridley AJ, Hodgkin MN, Stewart A, Claesson-Welsh L, et al. Stimulation of actin stress fibre formation mediated by activation of phospholipase D. Curr Biol. 1996;6(5):588–97. https://doi.org/10.1016/s0960-9822(02)00545-6.
82. Monaghan-Benson E, Burridge K. The regulation of vascular endothelial growth factor-induced microvascular permeability requires Rac and reactive oxygen species. J Biol Chem. 2009;284(38):25602–11. https://doi.org/10.1074/jbc.M109.009894.
83. Nelson CM, Pirone DM, Tan JL, Chen CS. Vascular endothelial-cadherin regulates cytoskeletal tension, cell spreading, and focal adhesions by stimulating RhoA. Mol Biol Cell. 2004;15(6):2943–53. https://doi.org/10.1091/mbc.e03-10-0745.
84. Minshall RD, Vandenbroucke EE, Holinstat M, Place AT, Tiruppathi C, Vogel SM, et al. Role of protein kinase Czeta in thrombin-induced RhoA activation and inter-endothelial gap formation of human dermal microvessel endothelial cell monolayers. Microvasc Res. 2010;80(2):240–9. https://doi.org/10.1016/j.mvr.2010.04.007.
85. Pellegrin S, Mellor H. Actin stress fibres. J Cell Sci. 2007;120(Pt 20):3491–9. https://doi.org/10.1242/jcs.018473.
86. Kam Y, Exton JH. Phospholipase D activity is required for actin stress fiber formation in fibroblasts. Mol Cell Biol. 2001;21(12):4055–66. https://doi.org/10.1128/MCB.21.12.4055-4066.2001.
87. Zeiller C, Mebarek S, Jaafar R, Pirola L, Lagarde M, Prigent AF, et al. Phospholipase D2 regulates endothelial permeability through cytoskeleton reorganization and occludin downregulation. Biochim Biophys Acta. 2009;1793(7):1236–49. https://doi.org/10.1016/j.bbamcr.2009.04.001.
88. Komati H, Naro F, Mebarek S, De Arcangelis V, Adamo S, Lagarde M, et al. Phospholipase D is involved in myogenic differentiation through remodeling of actin cytoskeleton. Mol Biol Cell. 2005;16(3):1232–44. https://doi.org/10.1091/mbc.e04-06-0459.
89. Hastie LE, Patton WF, Hechtman HB, Shepro D. Metabolites of the phospholipase D pathway regulate H2O2-induced filamin redistribution in endothelial cells. J Cell Biochem. 1998;68(4):511–24.
90. Rudge SA, Wakelam MJ. Inter-regulatory dynamics of phospholipase D and the actin cytoskeleton. Biochim Biophys Acta. 2009;1791(9):856–61. https://doi.org/10.1016/j.bbalip.2009.04.008.
91. Jiang Y, Sverdlov MS, Toth PT, Huang LS, Du G, Liu Y, et al. Phosphatidic acid produced by RalA-activated PLD2 stimulates caveolae-mediated endocytosis and trafficking in endothelial cells. J Biol Chem. 2016;291(39):20729–38. https://doi.org/10.1074/jbc.M116.752485.
92. Jenkins GH, Fisette PL, Anderson RA. Type I phosphatidylinositol 4-phosphate 5-kinase isoforms are specifically stimulated by phosphatidic acid. J Biol Chem. 1994;269(15):11547–54.
93. Wakelam MJ, Hodgkin MN, Martin A, Saqib K. Phospholipase D. Semin Cell Dev Biol. 1997;8(3):305–10. https://doi.org/10.1006/scdb.1997.0152.
94. Fu P, Ramchandran R, Shaaya M, Huang L, Ebenezer DL, Jiang Y, et al. Phospholipase D2 restores endothelial barrier function by promoting PTPN14-mediated VE-cadherin dephosphorylation. J Biol Chem. 2020;295(22):7669–85. https://doi.org/10.1074/jbc.RA119.011801.
95. Gavard J, Patel V, Gutkind JS. Angiopoietin-1 prevents VEGF-induced endothelial permeability by sequestering Src through mDia. Dev Cell. 2008;14(1):25–36. https://doi.org/10.1016/j.devcel.2007.10.019.
96. Orsenigo F, Giampietro C, Ferrari A, Corada M, Galaup A, Sigismund S, et al. Phosphorylation of VE-cadherin is modulated by haemodynamic forces and contributes to the regulation of vascular permeability in vivo. Nat Commun. 2012;3:1208. https://doi.org/10.1038/ncomms2199.
97. Vestweber D, Winderlich M, Cagna G, Nottebaum AF. Cell adhesion dynamics at endothelial junctions: VE-cadherin as a major player. Trends Cell Biol. 2009;19(1):8–15. https://doi.org/10.1016/j.tcb.2008.10.001.

98. Usatyuk PV, Romer LH, He D, Parinandi NL, Kleinberg ME, Zhan S, et al. Regulation of hyperoxia-induced NADPH oxidase activation in human lung endothelial cells by the actin cytoskeleton and cortactin. J Biol Chem. 2007;282(32):23284–95. https://doi.org/10.1074/jbc.M700535200.
99. Kelley LC, Hayes KE, Ammer AG, Martin KH, Weed SA. Cortactin phosphorylated by ERK1/2 localizes to sites of dynamic actin regulation and is required for carcinoma lamellipodia persistence. PLoS One. 2010;5(11):e13847. https://doi.org/10.1371/journal.pone.0013847.
100. Usatyuk PV, Gorshkova IA, He D, Zhao Y, Kalari SK, Garcia JG, et al. Phospholipase D-mediated activation of IQGAP1 through Rac1 regulates hyperoxia-induced p47phox translocation and reactive oxygen species generation in lung endothelial cells. J Biol Chem. 2009;284(22):15339–52. https://doi.org/10.1074/jbc.M109.005439.
101. Wang Z, Cai M, Tay LWR, Zhang F, Wu P, Huynh A, et al. Phosphatidic acid generated by PLD2 promotes the plasma membrane recruitment of IQGAP1 and neointima formation. FASEB J. 2019;33(6):6713–25. https://doi.org/10.1096/fj.201800390RR.
102. Bashour AM, Fullerton AT, Hart MJ, Bloom GS. IQGAP1, a Rac- and Cdc42-binding protein, directly binds and cross-links microfilaments. J Cell Biol. 1997;137(7):1555–66. https://doi.org/10.1083/jcb.137.7.1555.
103. Suryadevara V, Ramchandran R, Kamp DW, Natarajan V. Lipid mediators regulate pulmonary fibrosis: potential mechanisms and signaling pathways. Int J Mol Sci. 2020;21(12):4257. https://doi.org/10.3390/ijms21124257.
104. Zhao Y, Natarajan V. Lysophosphatidic acid (LPA) and its receptors: role in airway inflammation and remodeling. Biochim Biophys Acta. 2013;1831(1):86–92. https://doi.org/10.1016/j.bbalip.2012.06.014.
105. Cummings R, Zhao Y, Jacoby D, Spannhake EW, Ohba M, Garcia JG, et al. Protein kinase Cdelta mediates lysophosphatidic acid-induced NF-kappaB activation and interleukin-8 secretion in human bronchial epithelial cells. J Biol Chem. 2004;279(39):41085–94. https://doi.org/10.1074/jbc.M404045200.
106. Saatian B, Zhao Y, He D, Georas SN, Watkins T, Spannhake EW, et al. Transcriptional regulation of lysophosphatidic acid-induced interleukin-8 expression and secretion by p38 MAPK and JNK in human bronchial epithelial cells. Biochem J. 2006;393(Pt 3):657–68. https://doi.org/10.1042/BJ20050791.
107. Zhao Y, Usatyuk PV, Cummings R, Saatian B, He D, Watkins T, et al. Lipid phosphate phosphatase-1 regulates lysophosphatidic acid-induced calcium release, NF-kappaB activation and interleukin-8 secretion in human bronchial epithelial cells. Biochem J. 2005;385(Pt 2):493–502. https://doi.org/10.1042/BJ20041160.
108. Zhao Y, Tong J, He D, Pendyala S, Evgeny B, Chun J, et al. Role of lysophosphatidic acid receptor LPA2 in the development of allergic airway inflammation in a murine model of asthma. Respir Res. 2009;10:114. https://doi.org/10.1186/1465-9921-10-114.
109. Sinclair KA, Yerkovich ST, Hopkins PM, Fieuw AM, Ford P, Powell JE, et al. The autotaxin-lysophosphatidic acid pathway mediates mesenchymal cell recruitment and fibrotic contraction in lung transplant fibrosis. J Heart Lung Transplant. 2021;40(1):12–23. https://doi.org/10.1016/j.healun.2020.10.005.
110. Black KE, Berdyshev E, Bain G, Castelino FV, Shea BS, Probst CK, et al. Autotaxin activity increases locally following lung injury, but is not required for pulmonary lysophosphatidic acid production or fibrosis. FASEB J. 2016;30(6):2435–50. https://doi.org/10.1096/fj.201500197R.
111. Collin HB, Carroll N. In vivo effects of thimerosal on the rabbit corneal endothelium: an ultrastructural study. Am J Optom Physiol Optic. 1987;64(2):123–30. https://doi.org/10.1097/00006324-198702000-00008.
112. He D, Su Y, Usatyuk PV, Spannhake EW, Kogut P, Solway J, et al. Lysophosphatidic acid enhances pulmonary epithelial barrier integrity and protects endotoxin-induced epithe-

lial barrier disruption and lung injury. J Biol Chem. 2009;284(36):24123–32. https://doi.org/10.1074/jbc.M109.007393.
113. Zhao J, He D, Berdyshev E, Zhong M, Salgia R, Morris AJ, et al. Autotaxin induces lung epithelial cell migration through lysoPLD activity-dependent and -independent pathways. Biochem J. 2011;439(1):45–55. https://doi.org/10.1042/BJ20110274.
114. Alexander JS, Patton WF, Christman BW, Cuiper LL, Haselton FR. Platelet-derived lysophosphatidic acid decreases endothelial permeability in vitro. Am J Phys. 1998;274(1):H115–22. https://doi.org/10.1152/ajpheart.1998.274.1.H115.
115. Brault S, Gobeil F Jr, Fortier A, Honore JC, Joyal JS, Sapieha PS, et al. Lysophosphatidic acid induces endothelial cell death by modulating the redox environment. Am J Phys Regul Integr Comp Phys. 2007;292(3):R1174–83. https://doi.org/10.1152/ajpregu.00619.2006.
116. Zhao J, He D, Su Y, Berdyshev E, Chun J, Natarajan V, et al. Lysophosphatidic acid receptor 1 modulates lipopolysaccharide-induced inflammation in alveolar epithelial cells and murine lungs. Am J Phys Lung Cell Mol Phys. 2011;301(4):L547–56. https://doi.org/10.1152/ajplung.00058.2011.
117. Tager AM, LaCamera P, Shea BS, Campanella GS, Selman M, Zhao Z, et al. The lysophosphatidic acid receptor LPA1 links pulmonary fibrosis to lung injury by mediating fibroblast recruitment and vascular leak. Nat Med. 2008;14(1):45–54. https://doi.org/10.1038/nm1685.
118. Ackerman SJ, Park GY, Christman JW, Nyenhuis S, Berdyshev E, Natarajan V. Polyunsaturated lysophosphatidic acid as a potential asthma biomarker. Biomark Med. 2016;10(2):123–35. https://doi.org/10.2217/bmm.15.93.
119. Bandoh K, Aoki J, Taira A, Tsujimoto M, Arai H, Inoue K. Lysophosphatidic acid (LPA) receptors of the EDG family are differentially activated by LPA species. Structure-activity relationship of cloned LPA receptors. FEBS Lett. 2000;478(1–2):159–65. https://doi.org/10.1016/s0014-5793(00)01827-5.
120. Simmons S, Erfinanda L, Bartz C, Kuebler WM. Novel mechanisms regulating endothelial barrier function in the pulmonary microcirculation. J Physiol. 2019;597(4):997–1021. https://doi.org/10.1113/JP276245.
121. Stoffel W. Studies on the biosynthesis and degradation of sphingosine bases. Chem Phys Lipids. 1970;5(1):139–58. https://doi.org/10.1016/0009-3084(70)90014-9.
122. Stiban J, Tidhar R, Futerman AH. Ceramide synthases: roles in cell physiology and signaling. Adv Exp Med Biol. 2010;688:60–71. https://doi.org/10.1007/978-1-4419-6741-1_4.
123. Chalfant CE, Spiegel S. Sphingosine 1-phosphate and ceramide 1-phosphate: expanding roles in cell signaling. J Cell Sci. 2005;118(Pt 20):4605–12. https://doi.org/10.1242/jcs.02637.
124. Perrotta C, Clementi E. Biological roles of acid and neutral sphingomyelinases and their regulation by nitric oxide. Physiology (Bethesda). 2010;25(2):64–71. https://doi.org/10.1152/physiol.00048.2009.
125. Pyne S, Pyne NJ. Sphingosine 1-phosphate signalling in mammalian cells. Biochem J. 2000;349(Pt 2):385–402. https://doi.org/10.1042/0264-6021:3490385.
126. Pitson SM. Regulation of sphingosine kinase and sphingolipid signaling. Trends Biochem Sci. 2011;36(2):97–107. https://doi.org/10.1016/j.tibs.2010.08.001.
127. Saba JD. Fifty years of lyase and a moment of truth: sphingosine phosphate lyase from discovery to disease. J Lipid Res. 2019;60(3):456–63. https://doi.org/10.1194/jlr.S091181.
128. Ebenezer DL, Ramchandran R, Fu P, Mangio LA, Suryadevara V, Ha AW, et al. Nuclear sphingosine-1-phosphate lyase generated 2-hexadecenal is a regulator of HDAC activity and chromatin remodeling in lung epithelial cells. Cell Biochem Biophys. 2021;79(3):575–92. https://doi.org/10.1007/s12013-021-01005-9.
129. Merrill AH Jr. De novo sphingolipid biosynthesis: a necessary, but dangerous, pathway. J Biol Chem. 2002;277(29):25843–6. https://doi.org/10.1074/jbc.R200009200.
130. Ogretmen B, Hannun YA. Biologically active sphingolipids in cancer pathogenesis and treatment. Nat Rev Cancer. 2004;4(8):604–16. https://doi.org/10.1038/nrc1411.

131. Petrache I, Natarajan V, Zhen L, Medler TR, Richter AT, Cho C, et al. Ceramide upregulation causes pulmonary cell apoptosis and emphysema-like disease in mice. Nat Med. 2005;11(5):491–8. https://doi.org/10.1038/nm1238.
132. Holland WL, Brozinick JT, Wang LP, Hawkins ED, Sargent KM, Liu Y, et al. Inhibition of ceramide synthesis ameliorates glucocorticoid-, saturated-fat-, and obesity-induced insulin resistance. Cell Metab. 2007;5(3):167–79. https://doi.org/10.1016/j.cmet.2007.01.002.
133. Mostegl MM, Richter B, Nedorost N, Maderner A, Dinhopl N, Kubber-Heiss A, et al. Identification of a putatively novel trichomonad species in the intestine of a common quail (Coturnix coturnix). Vet Parasitol. 2012;183(3–4):369–72. https://doi.org/10.1016/j.vetpar.2011.07.036.
134. Wojewodka G, De Sanctis JB, Radzioch D. Ceramide in cystic fibrosis: a potential new target for therapeutic intervention. J Lipids. 2011;2011:674968. https://doi.org/10.1155/2011/674968.
135. Li C, Guo S, Pang W, Zhao Z. Crosstalk between acid sphingomyelinase and inflammasome signaling and their emerging roles in tissue injury and fibrosis. Front Cell Dev Biol. 2019;7:378. https://doi.org/10.3389/fcell.2019.00378.
136. Becker KA, Riethmuller J, Luth A, Doring G, Kleuser B, Gulbins E. Acid sphingomyelinase inhibitors normalize pulmonary ceramide and inflammation in cystic fibrosis. Am J Respir Cell Mol Biol. 2010;42(6):716–24. https://doi.org/10.1165/rcmb.2009-0174OC.
137. Dhami R, He X, Schuchman EH. Acid sphingomyelinase deficiency attenuates bleomycin-induced lung inflammation and fibrosis in mice. Cell Physiol Biochem. 2010;26(4–5):749–60. https://doi.org/10.1159/000322342.
138. Luheshi NM, Giles JA, Lopez-Castejon G, Brough D. Sphingosine regulates the NLRP3-inflammasome and IL-1beta release from macrophages. Eur J Immunol. 2012;42(3):716–25. https://doi.org/10.1002/eji.201142079.
139. Tsuchiya K, Nakauchi M, Hondo I, Nihei H. [Progress on diagnosis and therapy of water-electrolyte imbalance--hypernatremia and hyponatremia]. Nippon Naika Gakkai Zasshi. 1997;86(10):1831–7.
140. Goggel R, Winoto-Morbach S, Vielhaber G, Imai Y, Lindner K, Brade L, et al. PAF-mediated pulmonary edema: a new role for acid sphingomyelinase and ceramide. Nat Med. 2004;10(2):155–60. https://doi.org/10.1038/nm977.
141. Kady NM, Liu X, Lydic TA, Syed MH, Navitskaya S, Wang Q, et al. ELOVL4-mediated production of very long-chain ceramides stabilizes tight junctions and prevents diabetes-induced retinal vascular permeability. Diabetes. 2018;67(4):769–81. https://doi.org/10.2337/db17-1034.
142. Christoffersen C, Nielsen LB, Axler O, Andersson A, Johnsen AH, Dahlback B. Isolation and characterization of human apolipoprotein M-containing lipoproteins. J Lipid Res. 2006;47(8):1833–43. https://doi.org/10.1194/jlr.M600055-JLR200.
143. Christoffersen C, Obinata H, Kumaraswamy SB, Galvani S, Ahnstrom J, Sevvana M, et al. Endothelium-protective sphingosine-1-phosphate provided by HDL-associated apolipoprotein M. Proc Natl Acad Sci U S A. 2011;108(23):9613–8. https://doi.org/10.1073/pnas.1103187108.
144. Ito K, Anada Y, Tani M, Ikeda M, Sano T, Kihara A, et al. Lack of sphingosine 1-phosphate-degrading enzymes in erythrocytes. Biochem Biophys Res Commun. 2007;357(1):212–7. https://doi.org/10.1016/j.bbrc.2007.03.123.
145. Hanel P, Andreani P, Graler MH. Erythrocytes store and release sphingosine 1-phosphate in blood. FASEB J. 2007;21(4):1202–9. https://doi.org/10.1096/fj.06-7433com.
146. Kim RH, Takabe K, Milstien S, Spiegel S. Export and functions of sphingosine-1-phosphate. Biochim Biophys Acta. 2009;1791(7):692–6. https://doi.org/10.1016/j.bbalip.2009.02.011.
147. Kobayashi N, Kobayashi N, Yamaguchi A, Nishi T. Characterization of the ATP-dependent sphingosine 1-phosphate transporter in rat erythrocytes. J Biol Chem. 2009;284(32):21192–200. https://doi.org/10.1074/jbc.M109.006163.

148. Mitra P, Oskeritzian CA, Payne SG, Beaven MA, Milstien S, Spiegel S. Role of ABCC1 in export of sphingosine-1-phosphate from mast cells. Proc Natl Acad Sci U S A. 2006;103(44):16394–9. https://doi.org/10.1073/pnas.0603734103.
149. Sato K, Malchinkhuu E, Horiuchi Y, Mogi C, Tomura H, Tosaka M, et al. Critical role of ABCA1 transporter in sphingosine 1-phosphate release from astrocytes. J Neurochem. 2007;103(6):2610–9. https://doi.org/10.1111/j.1471-4159.2007.04958.x.
150. Fukuhara S, Simmons S, Kawamura S, Inoue A, Orba Y, Tokudome T, et al. The sphingosine-1-phosphate transporter Spns2 expressed on endothelial cells regulates lymphocyte trafficking in mice. J Clin Invest. 2012;122(4):1416–26. https://doi.org/10.1172/JCI60746.
151. Kawahara A, Nishi T, Hisano Y, Fukui H, Yamaguchi A, Mochizuki N. The sphingolipid transporter spns2 functions in migration of zebrafish myocardial precursors. Science. 2009;323(5913):524–7. https://doi.org/10.1126/science.1167449.
152. Olivera A, Spiegel S. Sphingosine-1-phosphate as second messenger in cell proliferation induced by PDGF and FCS mitogens. Nature. 1993;365(6446):557–60. https://doi.org/10.1038/365557a0.
153. Xu SZ, Muraki K, Zeng F, Li J, Sukumar P, Shah S, et al. A sphingosine-1-phosphate-activated calcium channel controlling vascular smooth muscle cell motility. Circ Res. 2006;98(11):1381–9. https://doi.org/10.1161/01.RES.0000225284.36490.a2.
154. Berdyshev EV, Gorshkova I, Usatyuk P, Kalari S, Zhao Y, Pyne NJ, et al. Intracellular S1P generation is essential for S1P-induced motility of human lung endothelial cells: role of sphingosine kinase 1 and S1P lyase. PLoS One. 2011;6(1):e16571. https://doi.org/10.1371/journal.pone.0016571.
155. Garcia JG, Liu F, Verin AD, Birukova A, Dechert MA, Gerthoffer WT, et al. Sphingosine 1-phosphate promotes endothelial cell barrier integrity by Edg-dependent cytoskeletal rearrangement. J Clin Invest. 2001;108(5):689–701. https://doi.org/10.1172/JCI12450.
156. Levkau B. Sphingosine-1-phosphate in the regulation of vascular tone: a finely tuned integration system of S1P sources, receptors, and vascular responsiveness. Circ Res. 2008;103(3):231–3. https://doi.org/10.1161/CIRCRESAHA.108.181610.
157. Mehta D, Konstantoulaki M, Ahmmed GU, Malik AB. Sphingosine 1-phosphate-induced mobilization of intracellular Ca2+ mediates rac activation and adherens junction assembly in endothelial cells. J Biol Chem. 2005;280(17):17320–8. https://doi.org/10.1074/jbc.M411674200.
158. Lee MJ, Thangada S, Claffey KP, Ancellin N, Liu CH, Kluk M, et al. Vascular endothelial cell adherens junction assembly and morphogenesis induced by sphingosine-1-phosphate. Cell. 1999;99(3):301–12. https://doi.org/10.1016/s0092-8674(00)81661-x.
159. Huang LS, Berdyshev EV, Tran JT, Xie L, Chen J, Ebenezer DL, et al. Sphingosine-1-phosphate lyase is an endogenous suppressor of pulmonary fibrosis: role of S1P signalling and autophagy. Thorax. 2015;70(12):1138–48. https://doi.org/10.1136/thoraxjnl-2014-206684.
160. Lavieu G, Scarlatti F, Sala G, Carpentier S, Levade T, Ghidoni R, et al. Regulation of autophagy by sphingosine kinase 1 and its role in cell survival during nutrient starvation. J Biol Chem. 2006;281(13):8518–27. https://doi.org/10.1074/jbc.M506182200.
161. Chi H. Sphingosine-1-phosphate and immune regulation: trafficking and beyond. Trends Pharmacol Sci. 2011;32(1):16–24. https://doi.org/10.1016/j.tips.2010.11.002.
162. Spiegel S, Milstien S. The outs and the ins of sphingosine-1-phosphate in immunity. Nat Rev Immunol. 2011;11(6):403–15. https://doi.org/10.1038/nri2974.
163. Walzer T, Chiossone L, Chaix J, Calver A, Carozzo C, Garrigue-Antar L, et al. Natural killer cell trafficking in vivo requires a dedicated sphingosine 1-phosphate receptor. Nat Immunol. 2007;8(12):1337–44. https://doi.org/10.1038/ni1523.
164. Hait NC, Allegood J, Maceyka M, Strub GM, Harikumar KB, Singh SK, et al. Regulation of histone acetylation in the nucleus by sphingosine-1-phosphate. Science. 2009;325(5945):1254–7. https://doi.org/10.1126/science.1176709.
165. Ebenezer DL, Berdyshev EV, Bronova IA, Liu Y, Tiruppathi C, Komarova Y, et al. Pseudomonas aeruginosa stimulates nuclear sphingosine-1-phosphate generation and epi-

genetic regulation of lung inflammatory injury. Thorax. 2019;74(6):579–91. https://doi.org/10.1136/thoraxjnl-2018-212378.
166. Fu P, Ebenezer DL, Ha AW, Suryadevara V, Harijith A, Natarajan V. Nuclear lipid mediators: role of nuclear sphingolipids and sphingosine-1-phosphate signaling in epigenetic regulation of inflammation and gene expression. J Cell Biochem. 2018;119(8):6337–53. https://doi.org/10.1002/jcb.26707.
167. Panneer Selvam S, De Palma RM, Oaks JJ, Oleinik N, Peterson YK, Stahelin RV, et al. Binding of the sphingolipid S1P to hTERT stabilizes telomerase at the nuclear periphery by allosterically mimicking protein phosphorylation. Sci Signal. 2015;8(381):ra58. https://doi.org/10.1126/scisignal.aaa4998.
168. Usatyuk PV, He D, Bindokas V, Gorshkova IA, Berdyshev EV, Garcia JG, et al. Photolysis of caged sphingosine-1-phosphate induces barrier enhancement and intracellular activation of lung endothelial cell signaling pathways. Am J Phys Lung Cell Mol Phys. 2011;300(6):L840–50. https://doi.org/10.1152/ajplung.00404.2010.
169. Alvarez SE, Harikumar KB, Hait NC, Allegood J, Strub GM, Kim EY, et al. Sphingosine-1-phosphate is a missing cofactor for the E3 ubiquitin ligase TRAF2. Nature. 2010;465(7301):1084–8. https://doi.org/10.1038/nature09128.
170. Gimbrone MA Jr, Aster RH, Cotran RS, Corkery J, Jandl JH, Folkman J. Preservation of vascular integrity in organs perfused in vitro with a platelet-rich medium. Nature. 1969;222(5188):33–6. https://doi.org/10.1038/222033a0.
171. Peng X, Hassoun PM, Sammani S, McVerry BJ, Burne MJ, Rabb H, et al. Protective effects of sphingosine 1-phosphate in murine endotoxin-induced inflammatory lung injury. Am J Respir Crit Care Med. 2004;169(11):1245–51. https://doi.org/10.1164/rccm.200309-1258OC.
172. McVerry BJ, Peng X, Hassoun PM, Sammani S, Simon BA, Garcia JG. Sphingosine 1-phosphate reduces vascular leak in murine and canine models of acute lung injury. Am J Respir Crit Care Med. 2004;170(9):987–93. https://doi.org/10.1164/rccm.200405-684OC.
173. Schaphorst KL, Chiang E, Jacobs KN, Zaiman A, Natarajan V, Wigley F, et al. Role of sphingosine-1 phosphate in the enhancement of endothelial barrier integrity by platelet-released products. Am J Phys Lung Cell Mol Phys. 2003;285(1):L258–67. https://doi.org/10.1152/ajplung.00311.2002.
174. Oo ML, Chang SH, Thangada S, Wu MT, Rezaul K, Blaho V, et al. Engagement of S1P(1)-degradative mechanisms leads to vascular leak in mice. J Clin Invest. 2011;121(6):2290–300. https://doi.org/10.1172/JCI45403.
175. Sanchez T, Skoura A, Wu MT, Casserly B, Harrington EO, Hla T. Induction of vascular permeability by the sphingosine-1-phosphate receptor-2 (S1P2R) and its downstream effectors ROCK and PTEN. Arterioscler Thromb Vasc Biol. 2007;27(6):1312–8. https://doi.org/10.1161/ATVBAHA.107.143735.
176. Sammani S, Moreno-Vinasco L, Mirzapoiazova T, Singleton PA, Chiang ET, Evenoski CL, et al. Differential effects of sphingosine 1-phosphate receptors on airway and vascular barrier function in the murine lung. Am J Respir Cell Mol Biol. 2010;43(4):394–402. https://doi.org/10.1165/rcmb.2009-0223OC.
177. Mathew B, Jacobson JR, Berdyshev E, Huang Y, Sun X, Zhao Y, et al. Role of sphingolipids in murine radiation-induced lung injury: protection by sphingosine 1-phosphate analogs. FASEB J. 2011;25(10):3388–400. https://doi.org/10.1096/fj.11-183970.
178. Fronza M, Lorefice L, Frau J, Cocco E. An overview of the efficacy and safety of ozanimod for the treatment of relapsing multiple sclerosis. Drug Des Devel Ther. 2021;15:1993–2004. https://doi.org/10.2147/DDDT.S240861.
179. Camp SM, Chiang ET, Sun C, Usatyuk PV, Bittman R, Natarajan V, et al. Pulmonary endothelial cell barrier enhancement by novel FTY720 analogs: methoxy-FTY720, fluoro-FTY720, and beta-glucuronide-FTY720. Chem Phys Lipids. 2015;191:16–24. https://doi.org/10.1016/j.chemphyslip.2015.08.004.

180. Dudek SM, Camp SM, Chiang ET, Singleton PA, Usatyuk PV, Zhao Y, et al. Pulmonary endothelial cell barrier enhancement by FTY720 does not require the S1P1 receptor. Cell Signal. 2007;19(8):1754–64. https://doi.org/10.1016/j.cellsig.2007.03.011.
181. Muller HC, Hocke AC, Hellwig K, Gutbier B, Peters H, Schonrock SM, et al. The sphingosine-1 phosphate receptor agonist FTY720 dose dependently affected endothelial integrity in vitro and aggravated ventilator-induced lung injury in mice. Pulm Pharmacol Ther. 2011;24(4):377–85. https://doi.org/10.1016/j.pupt.2011.01.017.
182. Brown BA, Kantesaria PP, McDevitt LM. Fingolimod: a novel immunosuppressant for multiple sclerosis. Ann Pharmacother. 2007;41(10):1660–8. https://doi.org/10.1345/aph.1G424.
183. Matloubian M, Lo CG, Cinamon G, Lesneski MJ, Xu Y, Brinkmann V, et al. Lymphocyte egress from thymus and peripheral lymphoid organs is dependent on S1P receptor 1. Nature. 2004;427(6972):355–60. https://doi.org/10.1038/nature02284.
184. Kovarik JM, Schmouder R, Barilla D, Wang Y, Kraus G. Single-dose FTY720 pharmacokinetics, food effect, and pharmacological responses in healthy subjects. Br J Clin Pharmacol. 2004;57(5):586–91. https://doi.org/10.1111/j.1365-2125.2003.02065.x.
185. Camp SM, Bittman R, Chiang ET, Moreno-Vinasco L, Mirzapoiazova T, Sammani S, et al. Synthetic analogs of FTY720 [2-amino-2-(2-[4-octylphenyl]ethyl)-1,3-propanediol] differentially regulate pulmonary vascular permeability in vivo and in vitro. J Pharmacol Exp Ther. 2009;331(1):54–64. https://doi.org/10.1124/jpet.109.153544.
186. Wang L, Sammani S, Moreno-Vinasco L, Letsiou E, Wang T, Camp SM, et al. FTY720 (s)-phosphonate preserves sphingosine 1-phosphate receptor 1 expression and exhibits superior barrier protection to FTY720 in acute lung injury. Crit Care Med. 2014;42(3):e189–99. https://doi.org/10.1097/CCM.0000000000000097.
187. Wang L, Letsiou E, Wang H, Belvitch P, Meliton LN, Brown ME, et al. MRSA-induced endothelial permeability and acute lung injury are attenuated by FTY720 S-phosphonate. Am J Phys Lung Cell Mol Phys. 2022;322(1):L149–L61. https://doi.org/10.1152/ajplung.00100.2021.
188. Pyne S, Lee SC, Long J, Pyne NJ. Role of sphingosine kinases and lipid phosphate phosphatases in regulating spatial sphingosine 1-phosphate signalling in health and disease. Cell Signal. 2009;21(1):14–21. https://doi.org/10.1016/j.cellsig.2008.08.008.
189. Hatoum D, Haddadi N, Lin Y, Nassif NT, McGowan EM. Mammalian sphingosine kinase (SphK) isoenzymes and isoform expression: challenges for SphK as an oncotarget. Oncotarget. 2017;8(22):36898–929. https://doi.org/10.18632/oncotarget.16370.
190. Aguilar A, Saba JD. Truth and consequences of sphingosine-1-phosphate lyase. Adv Biol Regul. 2012;52(1):17–30. https://doi.org/10.1016/j.advenzreg.2011.09.015.
191. Saba JD, Hla T. Point-counterpoint of sphingosine 1-phosphate metabolism. Circ Res. 2004;94(6):724–34. https://doi.org/10.1161/01.RES.0000122383.60368.24.
192. Mandala SM. Sphingosine-1-phosphate phosphatases. Prostaglandins Other Lipid Mediat. 2001;64(1–4):143–56. https://doi.org/10.1016/s0090-6980(01)00111-3.
193. Nagahashi M, Kim EY, Yamada A, Ramachandran S, Allegood JC, Hait NC, et al. Spns2, a transporter of phosphorylated sphingoid bases, regulates their blood and lymph levels, and the lymphatic network. FASEB J. 2013;27(3):1001–11. https://doi.org/10.1096/fj.12-219618.
194. Kobayashi N, Kawasaki-Nishi S, Otsuka M, Hisano Y, Yamaguchi A, Nishi T. MFSD2B is a sphingosine 1-phosphate transporter in erythroid cells. Sci Rep. 2018;8(1):4969. https://doi.org/10.1038/s41598-018-23300-x.
195. Zemann B, Urtz N, Reuschel R, Mechtcheriakova D, Bornancin F, Badegruber R, et al. Normal neutrophil functions in sphingosine kinase type 1 and 2 knockout mice. Immunol Lett. 2007;109(1):56–63. https://doi.org/10.1016/j.imlet.2007.01.001.
196. Vlasenko LP, Melendez AJ. A critical role for sphingosine kinase in anaphylatoxin-induced neutropenia, peritonitis, and cytokine production in vivo. J Immunol. 2005;174(10):6456–61. https://doi.org/10.4049/jimmunol.174.10.6456.
197. Wadgaonkar R, Patel V, Grinkina N, Romano C, Liu J, Zhao Y, et al. Differential regulation of sphingosine kinases 1 and 2 in lung injury. Am J Phys Lung Cell Mol Phys. 2009;296(4):L603–13. https://doi.org/10.1152/ajplung.90357.2008.

198. Chen J, Tang H, Sysol JR, Moreno-Vinasco L, Shioura KM, Chen T, et al. The sphingosine kinase 1/sphingosine-1-phosphate pathway in pulmonary arterial hypertension. Am J Respir Crit Care Med. 2014;190(9):1032–43. https://doi.org/10.1164/rccm.201401-0121OC.
199. Huang LS, Berdyshev E, Mathew B, Fu P, Gorshkova IA, He D, et al. Targeting sphingosine kinase 1 attenuates bleomycin-induced pulmonary fibrosis. FASEB J. 2013;27(4):1749–60. https://doi.org/10.1096/fj.12-219634.
200. Cheresh P, Kim SJ, Huang LS, Watanabe S, Joshi N, Williams KJN, et al. The sphingosine kinase 1 inhibitor, PF543, mitigates pulmonary fibrosis by reducing lung epithelial cell mtDNA damage and recruitment of fibrogenic monocytes. Int J Mol Sci. 2020;21(16):5595. https://doi.org/10.3390/ijms21165595.
201. Stepanovska B, Lange AI, Schwalm S, Pfeilschifter J, Coldewey SM, Huwiler A. Downregulation of S1P lyase improves barrier function in human cerebral microvascular endothelial cells following an inflammatory challenge. Int J Mol Sci. 2020;21(4):1240. https://doi.org/10.3390/ijms21041240.
202. Allende ML, Bektas M, Lee BG, Bonifacino E, Kang J, Tuymetova G, et al. Sphingosine-1-phosphate lyase deficiency produces a pro-inflammatory response while impairing neutrophil trafficking. J Biol Chem. 2011;286(9):7348–58. https://doi.org/10.1074/jbc.M110.171819.
203. Spiegel S, Maczis MA, Maceyka M, Milstien S. New insights into functions of the sphingosine-1-phosphate transporter SPNS2. J Lipid Res. 2019;60(3):484–9. https://doi.org/10.1194/jlr.S091959.
204. Donoviel MS, Hait NC, Ramachandran S, Maceyka M, Takabe K, Milstien S, et al. Spinster 2, a sphingosine-1-phosphate transporter, plays a critical role in inflammatory and autoimmune diseases. FASEB J. 2015;29(12):5018–28. https://doi.org/10.1096/fj.15-274936.
205. Jeya Paul J, Weigel C, Muller T, Heller R, Spiegel S, Graler MH. Inflammatory conditions disrupt constitutive endothelial cell barrier stabilization by alleviating autonomous secretion of sphingosine 1-phosphate. Cell. 2020;9(4):928. https://doi.org/10.3390/cells9040928.
206. Nawroth R, Poell G, Ranft A, Kloep S, Samulowitz U, Fachinger G, et al. VE-PTP and VE-cadherin ectodomains interact to facilitate regulation of phosphorylation and cell contacts. EMBO J. 2002;21(18):4885–95. https://doi.org/10.1093/emboj/cdf497.
207. Adams CL, Nelson WJ. Cytomechanics of cadherin-mediated cell-cell adhesion. Curr Opin Cell Biol. 1998;10(5):572–7. https://doi.org/10.1016/s0955-0674(98)80031-8.
208. McNeill H, Ryan TA, Smith SJ, Nelson WJ. Spatial and temporal dissection of immediate and early events following cadherin-mediated epithelial cell adhesion. J Cell Biol. 1993;120(5):1217–26. https://doi.org/10.1083/jcb.120.5.1217.
209. Breslin JW, Zhang XE, Worthylake RA, Souza-Smith FM. Involvement of local lamellipodia in endothelial barrier function. PLoS One. 2015;10(2):e0117970. https://doi.org/10.1371/journal.pone.0117970.
210. Waeber C, Blondeau N, Salomone S. Vascular sphingosine-1-phosphate S1P1 and S1P3 receptors. Drug News Perspect. 2004;17(6):365–82.
211. Lee JF, Gordon S, Estrada R, Wang L, Siow DL, Wattenberg BW, et al. Balance of S1P1 and S1P2 signaling regulates peripheral microvascular permeability in rat cremaster muscle vasculature. Am J Physiol Heart Circ Physiol. 2009;296(1):H33–42. https://doi.org/10.1152/ajpheart.00097.2008.
212. Singleton PA, Dudek SM, Chiang ET, Garcia JG. Regulation of sphingosine 1-phosphate-induced endothelial cytoskeletal rearrangement and barrier enhancement by S1P1 receptor, PI3 kinase, Tiam1/Rac1, and alpha-actinin. FASEB J. 2005;19(12):1646–56. https://doi.org/10.1096/fj.05-3928com.
213. Ephstein Y, Singleton PA, Chen W, Wang L, Salgia R, Kanteti P, et al. Critical role of S1PR1 and integrin beta4 in HGF/c-Met-mediated increases in vascular integrity. J Biol Chem. 2013;288(4):2191–200. https://doi.org/10.1074/jbc.M112.404780.
214. Usatyuk PV, Burns M, Mohan V, Pendyala S, He D, Ebenezer DL, et al. Coronin 1B regulates S1P-induced human lung endothelial cell chemotaxis: role of PLD2, protein kinase C

and Rac1 signal transduction. PLoS One. 2013;8(5):e63007. https://doi.org/10.1371/journal.pone.0063007.
215. Singleton PA, Salgia R, Moreno-Vinasco L, Moitra J, Sammani S, Mirzapoiazova T, et al. CD44 regulates hepatocyte growth factor-mediated vascular integrity. Role of c-Met, Tiam1/Rac1, dynamin 2, and cortactin. J Biol Chem. 2007;282(42):30643–57. https://doi.org/10.1074/jbc.M702573200.
216. Birukova AA, Alekseeva E, Mikaelyan A, Birukov KG. HGF attenuates thrombin-induced endothelial permeability by Tiam1-mediated activation of the Rac pathway and by Tiam1/Rac-dependent inhibition of the Rho pathway. FASEB J. 2007;21(11):2776–86. https://doi.org/10.1096/fj.06-7660com.
217. Higginbotham K, Tian Y, Gawlak G, Moldobaeva N, Shah A, Birukova AA. Hepatocyte growth factor triggers distinct mechanisms of Asef and Tiam1 activation to induce endothelial barrier enhancement. Cell Signal. 2014;26(11):2306–16. https://doi.org/10.1016/j.cellsig.2014.07.032.
218. Watanabe T, Wang S, Kaibuchi K. IQGAPs as key regulators of actin-cytoskeleton dynamics. Cell Struct Funct. 2015;40(2):69–77. https://doi.org/10.1247/csf.15003.
219. Banno Y, Takuwa Y, Akao Y, Okamoto H, Osawa Y, Naganawa T, et al. Involvement of phospholipase D in sphingosine 1-phosphate-induced activation of phosphatidylinositol 3-kinase and Akt in Chinese hamster ovary cells overexpressing EDG3. J Biol Chem. 2001;276(38):35622–8. https://doi.org/10.1074/jbc.M105673200.
220. Wang L, Chiang ET, Simmons JT, Garcia JG, Dudek SM. FTY720-induced human pulmonary endothelial barrier enhancement is mediated by c-Abl. Eur Respir J. 2011;38(1):78–88. https://doi.org/10.1183/09031936.00047810.
221. Oggungwan K, Glaharn S, Ampawong S, Krudsood S, Viriyavejakul P. FTY720 restores endothelial cell permeability induced by malaria sera. Sci Rep. 2018;8(1):10959. https://doi.org/10.1038/s41598-018-28536-1.
222. Fu P, Birukov KG. Oxidized phospholipids in control of inflammation and endothelial barrier. Transl Res. 2009;153(4):166–76. https://doi.org/10.1016/j.trsl.2008.12.005.
223. Birukov KG, Bochkov VN, Birukova AA, Kawkitinarong K, Rios A, Leitner A, et al. Epoxycyclopentenone-containing oxidized phospholipids restore endothelial barrier function via Cdc42 and Rac. Circ Res. 2004;95(9):892–901. https://doi.org/10.1161/01.RES.0000147310.18962.06.
224. Nonas S, Birukova AA, Fu P, Xing J, Chatchavalvanich S, Bochkov VN, et al. Oxidized phospholipids reduce ventilator-induced vascular leak and inflammation in vivo. Crit Care. 2008;12(1):R27. https://doi.org/10.1186/cc6805.
225. Meliton AY, Meng F, Tian Y, Sarich N, Mutlu GM, Birukova AA, et al. Oxidized phospholipids protect against lung injury and endothelial barrier dysfunction caused by heat-inactivated Staphylococcus aureus. Am J Phys Lung Cell Mol Phys. 2015;308(6):L550–62. https://doi.org/10.1152/ajplung.00248.2014.
226. Birukov KG, Karki P. Injured lung endothelium: mechanisms of self-repair and agonist-assisted recovery (2017 Grover Conference Series). Pulm Circ. 2018;8(1):2045893217752660. https://doi.org/10.1177/2045893217752660.
227. Karki P, Birukov KG. Oxidized phospholipids in control of endothelial barrier function: mechanisms and implication in lung injury. Front Endocrinol (Lausanne). 2021;12:794437. https://doi.org/10.3389/fendo.2021.794437.
228. Bochkov VN, Kadl A, Huber J, Gruber F, Binder BR, Leitinger N. Protective role of phospholipid oxidation products in endotoxin-induced tissue damage. Nature. 2002;419(6902):77–81. https://doi.org/10.1038/nature01023.
229. Starosta V, Wu T, Zimman A, Pham D, Tian X, Oskolkova O, et al. Differential regulation of endothelial cell permeability by high and low doses of oxidized 1-palmitoyl-2-arachidonyl-sn-glycero-3-phosphocholine. Am J Respir Cell Mol Biol. 2012;46(3):331–41. https://doi.org/10.1165/rcmb.2011-0153OC.

230. Singleton PA, Chatchavalvanich S, Fu P, Xing J, Birukova AA, Fortune JA, et al. Akt-mediated transactivation of the S1P1 receptor in caveolin-enriched microdomains regulates endothelial barrier enhancement by oxidized phospholipids. Circ Res. 2009;104(8):978–86. https://doi.org/10.1161/CIRCRESAHA.108.193367.
231. Birukova AA, Singleton PA, Gawlak G, Tian X, Mirzapoiazova T, Mambetsariev B, et al. GRP78 is a novel receptor initiating a vascular barrier protective response to oxidized phospholipids. Mol Biol Cell. 2014;25(13):2006–16. https://doi.org/10.1091/mbc.E13-12-0743.
232. Zou B, Jiang W, Han H, Li J, Mao W, Tang Z, et al. Acyloxyacyl hydrolase promotes the resolution of lipopolysaccharide-induced acute lung injury. PLoS Pathog. 2017;13(6):e1006436. https://doi.org/10.1371/journal.ppat.1006436.
233. Munford RS, Weiss JP, Lu M. Biochemical transformation of bacterial lipopolysaccharides by acyloxyacyl hydrolase reduces host injury and promotes recovery. J Biol Chem. 2020;295(51):17842–51. https://doi.org/10.1074/jbc.REV120.015254.

Hyperglycemic Oxoaldehyde (Glyoxal)-Induced Vascular Endothelial Cell Damage Through Oxidative Stress Is Protected by Thiol Iron Chelator, Dimercaptosuccinic Acid – Role of Iron in Diabetic Vascular Endothelial Dysfunction

Travis O. Gurney, Patrick J. Oliver, Sean M. Sliman, Anita Yenigalla, Timothy D. Eubank, Drew M. Nassal, Jiaxing Miao, Jing Zhao, Thomas J. Hund, and Narasimham L. Parinandi

Abstract In diabetic patients, the vascular endothelium is susceptible to damage through carbonyl stress caused in part by elevated levels of the glucose-derived oxoaldehyde glyoxal. Here, we investigated our hypothesis that glyoxal mediates the cytotoxicity, cytoskeletal alterations, and barrier dysfunction through the reactive oxygen species (ROS)-induced oxidative stress involving intracellular iron (Fe) and determined protection of the thiol heavy metal chelator, dimercaptosuccinic acid (DMSA) to establish role of iron in the glyoxal-induced damage of the bovine pulmonary ECs (BPAECs). Our results revealed that glyoxal induced cytotoxicity, mitochondrial dysfunction, loss of angiogenic potential, ROS generation, and loss of reduced glutathione (GSH) in a dose- (0.1, 0.5, and 1.0 mM) and time-dependent

T. O. Gurney · P. J. Oliver · S. M. Sliman · A. Yenigalla · N. L. Parinandi (✉)
Division of Pulmonary, Critical Care, and Sleep Medicine, Department of Medicine, Dorothy M. Davis Heart & Lung Research Institute, The Ohio State University Wexner Medical Center, Columbus, OH, USA
e-mail: narasimham.parinandi@osumc.edu

T. D. Eubank
Department of Microbiology, Immunology, and Cell Biology, Robert C. Byrd University Health Sciences Center, West Virginia University, Morgantown, WV, USA

D. M. Nassal · T. J. Hund
Frick Center for Heart Failure and Arrhythmia and Division of Cardiology, Dorothy M. Davis Heart & Lung Research Institute, The Ohio State University Wexner Medical Center, Columbus, OH, USA

J. Miao · J. Zhao
Department of Physiology and Cell Biology, The Ohio State University Wexner Medical Center, Columbus, OH, USA

N. L. Parinandi, T. J. Hund (eds.), *Cardiovascular Signaling in Health and Disease*, https://doi.org/10.1007/978-3-031-08309-9_18

(4, 6, 12, and 24 h) fashion in the BPAECs, which were all significantly attenuated by DMSA (5.0 mM) treatment. Glyoxal also induced barrier dysfunction (loss of trans-endothelial electrical resistance), enhanced transcellular permeability of dextran, and caused cytoskeletal reorganization (actin stress fiber formation and rearrangement of cortactin, occludin, and ZO-1) in the BPAEC monolayers, which were all significantly attenuated by DMSA treatment. Exogenous iron (50 μM $FeSO_4$) significantly enhanced the glyoxal-induced cytotoxicity and the ROS production in the BPAECs, which were all significantly attenuated by DMSA treatment. To support this, we visualized the intracellular chelatable iron in the ECs by fluorescence microscopy. Immunofluorescence microscopy revealed the glyoxal-induced formation of advanced glycation end products (AGEs) in the BPAECs, which was attenuated by DMSA treatment. For the first time, this study revealed that the thiol heavy metal chelator, DMSA, protected against the vascular EC damage caused by the hyperglycemic oxoaldehyde AGE precursor through the action of iron and oxidative stress that culminates into diabetic vascular endothelial dysfunction.

Keywords Glyoxal · Hyperglycemic oxoaldehyde · AGEs · Advanced glycation end products · Diabetic vascular endothelial dysfunction · Iron · Reactive oxygen species · DMSA · Cytoskeletal reorganization

Introduction

Hyperglycemia (elevated blood glucose levels) is becoming increasingly prevalent as a major public health concern worldwide. This can be attributed to the rampant increase in diabetes mellitus, a metabolic endocrine disorder characterized by the chronic state hyperglycemia. Diabetes mellitus is classified as either type 1, where the body is unable to produce insulin, or type 2, where the body loses its sensitivity to insulin [1]. While the severity of type 1 diabetes is not to be downplayed, recent epidemiological studies have shown a considerable elevation in the development of specifically type 2 diabetes due to increasingly poor diets and sedentary lifestyles. Diabetes mellitus is often associated with, and occasionally directly responsible for, several systemic complications, including the cardiovascular diseases (CVDs), neurodegeneration, neuropathy, and nephropathy [1]. At the biochemical level, oxidatively modified sugar products (derived from endogenous glucose oxidation), known as the advanced glycation end products (AGEs), have been identified as possible culprits in the damaging effects of chronic state hyperglycemia. AGEs are ultimately produced as a result of the oxidative environment and glycoaldehydes, whose presence is increased drastically under hyperglycemic conditions [2].

It has been previously reported that high circulating levels of AGEs in diabetic patients are a direct result of the endogenous formation of AGE precursors, such as glyoxal, methylglyoxal, and 3-deoxyglucosone, through the Schiff-base and Amadori rearrangement [2]. However, it has been observed that exogenous AGEs, obtained from diet upon regular consumption, can also contribute directly to the increase in oxidative stress responsible for the harmful inflammation encountered in diabetes mellitus [3]. Regardless, high levels of both AGEs and their precursors are highly damaging to the vasculature (both microvasculature and macro vessels) of the diabetic patients and are particularly toxic to the blood vessel endothelial cell (EC) layer [2]. AGEs can damage the vascular endothelium by cross-linking with proteins, which increase vascular rigidity and produce high levels of reactive oxygen species (ROS), leading to cytotoxicity, barrier dysfunction, and cytoskeletal rearrangement [2]. The most common classes of drugs prescribed for management of diabetes are the biguanides (metformin), the thiazolidinediones (pioglitazone), and the sulfonylureas [4]. The primary action of these frequently prescribed drugs is either stimulating insulin release or decreasing insulin sensitivity. Few drugs have been designed to directly combat the adverse AGE-induced vascular damage in diabetes mellitus [4].

The role of iron in a myriad of human disease states is being reinvestigated as more is learned about its biochemical properties at the cellular level. Considering the highly oxidative environment presented by hyperglycemic conditions, studies are now being performed exploring the relevance of the role of iron in diabetes. Through the Fenton and Haber-Weiss reactions, iron can cycle rapidly through its oxidation states, resulting in the generation of hydroxyl radical (·OH) and superoxide (O_2^{-}), eventually depleting the endogenous ROS scavengers (antioxidants), such as reduced glutathione (GSH), the non-enzymatic cellular antioxidant and the enzymatic antioxidants catalase and superoxide dismutase (SOD) [5]. ROS buildup then causes damage through oxidative stress, hindering insulin production, gluconeogenesis, and the ability to use insulin, and subsequently aiding in the development of diabetes. Pancreatic islet β-cells accrue more iron than other tissues and are therefore more susceptible to the effects of excessive iron storage, or iron overload [6]. This association between iron and diabetes is well documented, and it is believed that reducing iron overload can significantly improve diabetic conditions by increasing insulin production and sensitivity [7]. Despite this, there have been very few studies on decreasing the levels or overload of iron in the diabetic patients.

Therefore, we focus on investigating the protective effects of established, widely used, and nontoxic, metal chelator drugs as we consider them to be candidates for combination therapy in future treatment protocols for the hyperglycemic (diabetic) vascular disorders. Our earlier study has demonstrated that the sugar-derived aldehyde, glyoxal, is formed during diabetes and causes endothelial dysfunction through cytotoxicity, barrier dysfunction, cytoskeletal alterations, leading ultimately to the inhibition of angiogenesis [2]. However, since iron overload has been associated with diabetic complications involving oxidative stress, in the current study, we hypothesized that glyoxal (Fig. 1) causes vascular endothelial damage through endogenous cellular iron dysregulation and subsequent redox alterations and oxidative stress through elevated ROS generation. Additionally, we also hypothesized that the well-established, widely used, and nontoxic thiol (-SH)-containing heavy metal chelator, dimercaptosuccinic acid (DMSA) (Fig. 2), would be able to attenuate the vascular EC damage caused by the hyperglycemic oxoaldehyde, glyoxal (that would lead to elevated levels of AGEs) through complexing with the endogenous cellular iron, reducing oxidative stress, and normalizing the redox alterations.

Fig. 1 Chemical structure of hyperglycemic sugar-derived oxoaldehyde, Glyoxal

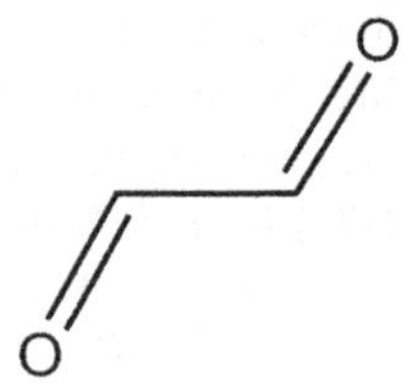

Fig. 2 Chemical structure of thiol-containing heavy metal chelator, *meso*-2,3-Dimercaptosuccinic acid (DMSA)

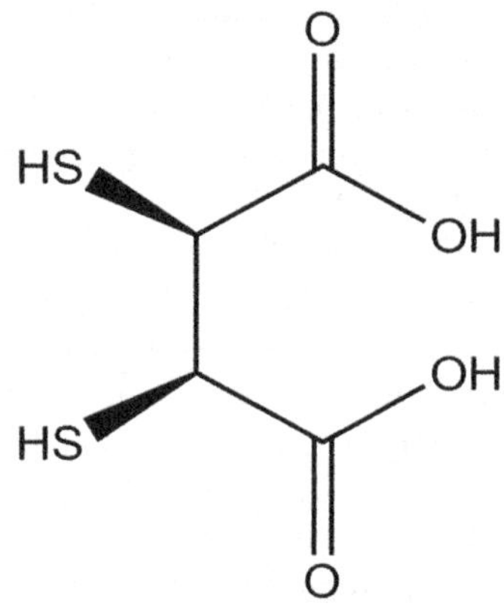

In order to test our hypothesis, in the current study, we utilized our well-established bovine pulmonary artery EC (BPAEC) system and investigated the glyoxal-induced cytotoxicity, ROS formation, oxidative stress, thiol-redox alteration, formation of AGEs, cytoskeletal rearrangement, and barrier dysfunction. We also utilized the widely used EC Matrigel tube formation assay with the vascular ECs and investigated the adverse actions of glyoxal on the in vitro angiogenesis. In addition, we also investigated the protective actions of DMSA on the glyoxal-induced EC dysfunctions and inhibition of angiogenesis. For the first time, results of the current study revealed that the hyperglycemic oxoaldehyde, glyoxal caused cytotoxicity, cytoskeletal reorganization, barrier dysfunction, formation of AGEs, and inhibition of angiogenesis in vascular ECs in vitro through ROS production, thiol-redox dysregulation, and oxidative stress involving endogenous cellular iron, which were all effectively attenuated by the thiol-containing heavy metal chelator, DMSA.

Materials and Methods

Materials

Bovine pulmonary artery endothelial cells (BPAECs) (passage 2) were obtained from the Cell Applications Inc. (San Diego, CA). Fetal bovine serum (FBS), trypsin, minimum essential medium (MEM), and non-essential amino acids were obtained from the Gibco Invitrogen Corp. (Grand Island, NY). Glyoxal, meso-2,3-dimercaptosuccinic acid (DMSA), ferrous sulfate, 2,2′-bipyridyl, N-acetyl-L-cysteine (NAC), t-octylphenoxypolyethoxyethanol (Triton X-100), bovine serum albumin (BSA), 36.5% formaldehyde solution, 3-[4,5-dimethylthiazol-2-yl]-2, 5-diphenyl tetrazolium bromide (MTT) assay kit and lactic acid dehydrogenase cytotoxicity assay (LDH) kit were obtained from the Sigma Chemical Co. (St. Louis, MO). [^{3}H]-Thymidine was obtained from the American Radiolabeled Chemicals, Inc. (St. Louis, MO). Electrical cell impedance system (ECIS) electrode arrays were obtained from the Applied Biophysics (Troy, NY). AlexaFluor 488, AlexaFluor 568, 4′,6-diamidino-2-phenylindole dihydrochloride (DAPI), and rhodamine–phalloidin were purchased from the Molecular Probes Invitrogen Co. (Carlsbad, CA). Mouse anti-ZO-1, anti-occludin, anti-cortactin antibodies were obtained from the Zymed Laboratories (San Francisco, CA). Polyoxyethylene sorbitan monolaurate (Tween-20) and 10x Tris-buffered saline (TBS) were obtained from the Bio-Rad Laboratories (Hercules, CA). Endothelial cell growth supplement was obtained from the Upstate Biotechnology (Lake Placid, NY, U.S.A.). Phen green SK (PGSK) diacetate and 6-carboxy-2,7-dichlorodihydroxyfluorescein diacetate dicarboxy methyl ester (DCFDA) and dihydroethidium (DHE) were purchased from the Molecular Probes (Eugene, OR, U.S.A.).

Cell Culture

BPAECs were cultured in MEM supplemented with 10% FBS, non-essential amino acids, antibiotics, and endothelial cell growth supplement (factor) according to our published method as described previously [2]. Cells in culture were maintained at 37 °C in a humidified environment of 5% CO_2 – 95% air and grown to contact-inhibited monolayers with typical cobblestone morphology. When confluence was reached, cells were trypsinized and subcultured in T-75-cm^2 flasks or 15.5-mm or 35-mm tissue culture dishes. Confluent cells showed cobblestone morphology under light microscope and stained positive for Factor VIII. All experiments were conducted between 5 and 20 passages (70–95% confluence).

Cell Treatment

Glyoxal was prepared by dissolving in PBS at 70 °C and then diluted to desired concentrations in MEM according to our earlier published method [2]. Other pharmacological agents were prepared directly in MEM to desired concentrations. Cells were then treated with MEM or solutions prepared in MEM as desired under sterile conditions and incubated for the chosen lengths of time at 37 °C in a humidified environment of 5% CO_2 – 95% air.

Cell Morphology Assay

Morphological changes in BPAECs grown in 35-mm sterile dishes up to 90% confluence, following their exposure to different concentrations of glyoxal (0–1 mM) in absence or presence of chosen pharmacological agents (0–5.0 mM) in MEM for 0–24 h at 37 °C in a humidified environment of 5% CO_2 – 95% air, were examined according to our published method as an index of cytotoxicity [2]. Images of cell morphology were digitally captured with the Olympus 1 × 50 microscope at 20 X magnification.

LDH Assay of Cytotoxicity

BPAECs were grown up to 90% confluence in sterile 15.5-mm dishes (24-well culture plates) and exposed to different concentrations of glyoxal (0–1.0 mM) in absence or presence of DMSA (0–5.0 mM) for 4–24 h. At the end of incubation, supernatant was removed, the experiment was terminated with 1 N HCl, and LDH release was determined spectrophotometrically according to the manufacturer's recommendations (Sigma Chemical Co., St. Louis, MO) as described in our published method [2]. LDH activity was expressed as the difference in absorbance (Δ) between 490 and 690 nm.

MTT Assay of Cytotoxicity

BPAECs were grown up to 90% confluence in 15.5-mm dishes (24-well culture plate) and exposed to different concentrations of glyoxal (0–1.0 mM) in absence or presence of DMSA (0–5.0 mM) for 4–24 h. At the end of incubation, MTT solution (10% of culture volume) was added and incubated for 3–4 h, the solution was removed, and MTT solvent was added in an amount equal to original culture volume. Absorbance of the reduced MTT was measured according to the manufacturer's recommendations (Sigma Chemical Co., St. Louis, MO) as described in our published method [2]. Extent of MTT reduction was expressed as the difference in absorbance (Δ) between 570 and 690 nm.

[^{3}H]-Thymidine Incorporation Assay for Cell Proliferation

BPAECs were grown to 60% confluence in 35-mm dishes. Complete medium was removed from the culture dishes and the treatment solutions were added to the dishes. The treatment medium was then removed and 1 mL of [^{3}H]-thymidine

(1.0 μCi/mL) in MEM was added to each well and incubated for 24 h. After incubations, [^{3}H]-thymidine was removed, and cells were washed twice with ice-cold PBS. Cells were then washed twice with 5% trichloroacetic acid (TCA) in distilled water. Washings of 5% TCA were then removed and the cells were treated with 10.25 M NaOH (500 μL) for 30 min to solubilize the cells. The solubilized cell solution (400 μL) was transferred to the scintillation vials, followed by the addition of 10 mL of scintillation cocktail, and then the [^{3}H] radioactivity was counted in the Packard Tri-carb 2900TR Liquid Scintillation Counter, as described in our published method [2]. Cell replication was expressed as DPM of [^{3}H]-thymidine incorporation into cells/dish.

Fluorescence Microscopy of Actin Stress Fibers

Formation of actin stress fibers as an index of endothelial cytoskeletal reorganization was analyzed by confocal laser scanning microscopy (CLSM), as described in our published method [2]. BPAECs grown on sterile glass coverslips (90% confluence) were treated with different concentrations of glyoxal, DMSA, or glyoxal + DMSA following which they were rinsed twice with PBS and then fixed with 3.7% formaldehyde in PBS for 10 min at room temperature. The cells were then permeabilized with 0.5% Triton X-100 prepared in tris-buffered saline (TBS) containing 0.01% Tween-20 (TBS-T) for 2 min. The cells were then washed four times with PBS and treated with 1% BSA in TBS-T for 1 h. Actin stress fibers were visualized by staining the cells with rhodamine–phalloidin (1:50 dilution) in 1% BSA in TBS-T for 1 h. The cells were then rinsed four times with PBS to remove excess stain, stained with 1% DAPI in PBS for 5 min to visualize the nuclei following which they were washed four times with PBS, mounted, and then examined under Zeiss LSM 510 Confocal/Multiphoton Microscope at 543-nm excitation and 565-nm emission at 63 X magnification. DAPI image acquisition was performed by 2-photon Verdi V-10 Laser in IR spectrum with excitation at 750 nm. These images were captured digitally.

Immunofluorescence Microscopy of ZO-1, Occludin, and AGEs

Tight junction protein reorganization and formation of AGEs were analyzed by confocal laser scanning microscopy (CLSM) according to our earlier published methods [2]. BPAECs grown on sterile glass coverslips (90% confluence) were treated with different concentrations of glyoxal, DMSA, or glyoxal + DMSA, then rinsed three times with PBS, and fixed with 3.7% formaldehyde in PBS for 10 min at room temperature. The cells were then rinsed three times with PBS and permeabilized with 0.25% Triton X-100 prepared in TBS containing 0.01% Tween-20 (TBS-T) for

5 min. The cells were then washed three times with TBS-T and treated with TBS-T containing 1% BSA blocking buffer for 30 min at room temperature. The cells were then incubated overnight at room temperature with the primary antibody [anti-ZO-1, anti-Occludin, anti-Amadori antibodies (1:200 dilution)] in 1% BSA solution in TBS-T. After rinsing three times with TBS-T, the cells were treated with AlexaFluor 488 (1:100 dilution) in 1% BSA in TBS-T for 1 h. For the visualization of nuclei, the cells were then stained with 1% DAPI in PBS for 5 min. Finally, the cells were washed three times with TBS-T, mounted, and examined under Zeiss LSM 510 Confocal/Multiphoton Microscope at 543 nm excitation and 600 nm emission at 63 X magnification. DAPI image acquisition performed by 2-photon Verdi-V10 laser in IR spectrum with excitation at 750 nm. The images were captured digitally.

Immunofluorescence Confocal Microscopy of Cortactin

Rearrangement of cortical actin was analyzed by confocal laser scanning microscopy (CLSM). BPAECs grown on sterile glass coverslips (90% confluence) were treated with different concentrations of glyoxal, DMSA, or glyoxal + DMSA, then rinsed three times with PBS, and fixed with 3.7% formaldehyde in PBS for 10 min at room temperature. The cells were then rinsed three times with PBS and permeabilized with 0.25% Triton X-100 prepared in TBS containing 0.01% Tween-20 (TBS-T) for 5 min. The cells were then washed three times with TBS-T and treated with TBS-T containing 1% BSA blocking buffer for 30 min at room temperature. The cells were then incubated overnight at room temperature with the primary antibody [anti-cortactin (1:100 dilution)] in 1% BSA solution in TBS-T. After rinsing three times with TBS-T, the cells were treated with AlexaFluor 568 (1:100 dilution) in 1% BSA in TBS-T for 1 h. Finally, the cells were washed three times with TBS-T, mounted, and examined under Zeiss LSM 510 Confocal/Multiphoton Microscope with 560 nm long pass filter at 63 X magnification. The images were captured digitally.

Measurement of Transendothelial Cell Electrical Resistance (TER)

BPAECs were cultured in complete MEM on gold electrodes (Applied Biophysics Inc., Troy, NY) at 37 °C in a humidified atmosphere of 5% CO_2 – 95% air and grown to contact-inhibited monolayer with typical cobblestone morphology. TER of the BPAEC monolayer cultured on the gold electrodes was measured on an electric cell substrate impedance-sensing system (ECIS; Applied Biophysics Inc., Troy,

N.Y.) according to our earlier published procedure following treatment of the cells with MEM containing the chosen concentrations of glyoxal, pharmacological agent (e.g., DMSA), or glyoxal + pharmacological agent in a humidified atmosphere of 5% CO_2 – 95% air at 37 °C [2]. The total endothelial electrical resistance, as measured across the EC monolayer, was determined by the combined resistance between the basal and/or cell matrix adhesion. TER measurements were done in duplicate and expressed as normalized resistance for each of the treatments.

In Vitro Endothelial Cell Tube Formation Assay for Angiogenesis

In vitro angiogenesis assay was performed using our established Matrigel tube formation assay [2]. BPAECs were grown to 80% confluence in complete medium (10% FBS, nonessential amino acids, antibiotics, and growth factor) in two separate flasks under a humidified 5% CO_2 – 95% air environment at 37 °C. Both flasks of cells were starved in minimal media (MEM) at 37 °C under a humidified 5% CO_2 – 95% air environment at 37 °C for 30 min following which DMSA in minimal medium was added at a final concentration of 5.0 mM to one of the flasks and continued to incubate for an additional 45 min under a humidified 5% CO_2 – 95% air environment at 37 °C. During this time, 120 μL of growth factor-reduced Matrigel (BD Discovery Labware, Bedford, MA) was coated in the wells of a 48-well culture plate and placed at 37 °C for 1 h to polymerize. The cells from both flasks were then trypsinized, centrifuged, washed, resuspended in minimal medium and counted. BPAECs (2×10^4) that were not pre-treated with DMSA were added to 1 mL of rich medium containing glyoxal at a final concentration of 1.0 mM and then immediately transferred to the Matrigel-coated plate (rich medium + glyoxal [1.0 mM]). BPAECs (2×10^4) that were pretreated with DMSA were added to either 1 mL of minimal medium, or 1 mL of rich medium, or to 1 mL of rich medium supplemented with glyoxal (0–1.0 mM) or DMSA (5.0 mM) + glyoxal (0–1.0 mM). The cells were then allowed to incubate for 12 h at 37 °C under a humidified 5% CO_2 – 95% air environment at 37 °C and subsequently analyzed for the formation of tubes connecting one colony to the next. At the termination of the experiment, the medium was removed, and the cells were fixed with 4% paraformaldehyde. Four digital images, under a Nikon light microscope at 40 X magnification, were captured per well at the same location in each well and the tubes were counted in a blinded fashion. The data represent $n = 6$ for each condition. The statistical analysis was performed using an ANOVA with SYSTAT12 software (San Jose, CA).

Reactive Oxygen Species (ROS) Determination by DCFDA Fluorescence

Formation of ROS in BPAECs in 35-mm dishes (5 × 10^5 cells/dish) was determined by 2′,7′-dichlorofluorescein (DCF) fluorescence in cells preloaded with 10 μM DCFDA for 30 min in complete MEM at 37 °C in a 5% CO_2 – 95% air environment prior to exposure to glyoxal (0–1.0 mM) in absence or presence of DMSA (0–5.0 mM) for 2 h according to our earlier published method [8, 9]. After treatment, the dishes containing the cells were either photographed using Olympus 1 × 50 fluorescent microscope with FITC filter set, or the cells were detached with a Teflon cell scraper where medium containing cells was transferred to 1.5 mL microcentrifuge tubes and centrifuged at 8000 × *g* for 10 min. The supernatant medium was aspirated, and the cell pellet was washed twice with ice-cold phosphate-buffered saline (PBS). To prepare cell lysates, pellets were sonicated on ice with a probe sonicator at a setting of 2 for 15 s in 150 μL of ice-cold PBS. Fluorescence of oxidized DCF in cell lysates, an index of formation of ROS, was measured on a Bio-Tek Synergy HT fluorescent plate reader set at 490 nm excitation and 530 nm emission, respectively, using appropriate blanks. The extent of ROS formation was expressed as arbitrary fluorescence units.

Superoxide (O^{2-}) Determination by DHE Fluorescence

Generation of superoxide ($O_2^{-\cdot}$) in BPAECs was determined by DHE (dihydroethidium) fluorescence according to our earlier published procedure [9]. BPAECs grown in 35-mm dishes were treated with different concentrations of glyoxal, DMSA, or glyoxal + DMSA for 2 h under a humidified 5% CO_2 – 95% air environment at 37 °C. After treatment, the cells were washed once with warm PBS and incubated with DHE (5.0 μM) in warm PBS for 30 min. After incubation, cells were washed once with warm PBS and examined under Zeiss LSM 510 Confocal/Multiphoton Microscope at 543 nm excitation with a 560 nm long pass filter at 10 X magnification. These images were captured digitally. Fluorescent intensity was determined using the NIH sponsored program ImageJ.

Reduced Glutathione (GSH) Determination

Intracellular soluble thiol-reduced glutathione (GSH) levels were determined using the GSH-Glo GSH assay kit. BPAECs grown to 100% confluence in 96 well plates under a humidified 5% CO_2 – 95% air environment at 37 °C were treated with MEM alone or MEM containing desired concentrations of glyoxal for 12 h under a humidified 5% CO_2 – 95% air environment at 37 °C. Following incubation, intracellular

GSH levels were determined according to the manufacturer's recommendations (Promega Corp. Madison, WI), as described in our earlier published method [10].

FITC Paracellular Transport (Leak) Through the Endothelial Cell Monolayer

BPAECs were grown to 90–100% confluence in 22-mm dishes (12-well culture plate) on sterile inserts (0.4 μm) under a humidified 5% CO_2 – 95% air environment at 37 °C and treated with phenol red-free MEM alone or MEM containing the chosen concentrations of glyoxal and/or DMSA for 12–24 h. Following treatment, cells were washed with MEM (phenol red-free) and incubated for 1 h with 1 mg/mL fluorescein isothiocyanate-dextran (FITC-dextran, 70 kDa) solution under a humidified 5% CO_2 – 95% air environment at 37 °C. At the end of the incubation period, the supernatant from the insert chamber was removed, and fluorescence was measured at 480 nm excitation and 530 nm emission which served as an index of paracellular transport of large molecular weight polymer of 70 kDa across the BPAEC monolayer that signifies the vascular endothelial leak.

Visualization of Free (Labile) Chelatable Iron in BPAECs by Fluorescence Microscopy

The free, chelatable iron pool was analyzed by confocal laser scanning microscopy (CLSM) according to Petrat et al. [11, 12]. BPAECs were incubated with PG SK diacetate (20 μM) for 10 min. at room temperature and washed twice with PBS. Selected wells were then incubated with 2,2′-bipyridyl (5 mM) for 10 min at room temperature, washed twice with PBS, samples were mounted, and examined under Zeiss LSM 510 Confocal/Multiphoton Microscope with 500 nm excitation and 530 nm emission BP filter at 63 X magnification. The images were captured digitally. Fluorescent intensity was determined using the NIH sponsored program ImageJ.

Statistical Analysis

Standard deviation (S.D.) for each data point was calculated from triplicate samples. Data were subjected to one-way analysis of variance, and pair wise multiple comparisons was done by Dunnett's method with $p < 0.05$, indicating significance using SigmaStat (Jandel Scientific, San Rafael, California). The statistical analysis of angiogenesis (endothelial tube formation assay) was performed using an ANOVA with SYSTAT12 software (San Jose, CA).

Results

Glyoxal Induces Cytotoxicity to BPAECs as Evidenced by LDH Release by Cells

We first investigated the effects of glyoxal at physiologically relevant concentrations on the healthy endothelium. We measured glyoxal-induced LDH release from the BPAECs as an assay for overall cell health and cytotoxicity. Increased LDH release was observed as early as 4–6 h post-treatment with high dose (1.0 mM) glyoxal (Fig. 3a and b). By 12–24 h of treatment, increased LDH release was observed with both moderate (0.5 mM) and high-dose (1.0 mM) glyoxal (Fig. 3c and d). Low-dose (0.1 mM) glyoxal did not have a significant effect on LDH release from BPAECs at any time point. These results show a dose- and time-dependent cytotoxicity in BPAECs induced by glyoxal treatment.

DMSA Protects Against Glyoxal-Induced LDH Release by BPAECs

We hypothesized that iron plays a major role in the complications caused by AGEs and their precursors. Therefore, we investigated whether the thiol-containing heavy metal chelator, 2,3-dimercaptosuccinic acid (DMSA) would attenuate the glyoxal-induced LDH release by the BPAECs. The BPAECs were treated with MEM alone, MEM containing DMSA (1.0, 2.0, and 5.0 mM), MEM containing glyoxal (0.5 mM), and MEM-containing DMSA (1.0, 2.0, and 5.0 mM) + glyoxal (0.5 mM). Glyoxal (0.5 mM) caused a significant increase in the LDH release by cells at 12 h of treatment as compared to untreated control cells or cells co-incubated with glyoxal and DMSA (Fig. 4). These results clearly indicated that DMSA effectively protected the glyoxal-induced cytotoxicity as determined by the LDH release by the ECs.

Glyoxal Inhibits MTT Reduction by BPAECs

Based on the observation that glyoxal induced significant cytotoxicity in the BPAECs, we investigated mitochondrial integrity as another index of cell health by determining the mitochondrial dehydrogenase-catalyzed MTT reduction in the BPAECs treated with glyoxal [2]. Glyoxal at the tested doses of 0.1, 0.5, and 1.0 mM, caused a significant time- and dose-dependent inhibition of MTT reduction in the BPAECs at 4–24 h as compared to control cells (Fig. 5a–d). Also, the extent of glyoxal-induced inhibition of reduction of MTT by the BPAECs treated with glyoxal at 0.5 and 1.0 mM for longer periods of time (12 and 24 h) was significantly

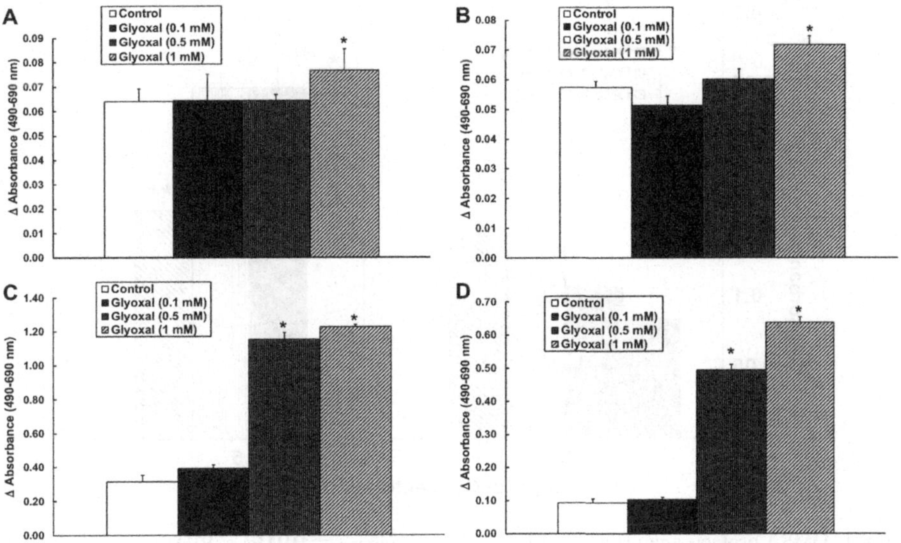

Fig. 3 Glyoxal induces cytotoxicity to BPAECs as evidenced by LDH release by cells. BPAECs grown to 95% confluence in 15.5–mm dishes (24-well culture plate) were treated with MEM alone or MEM-containing glyoxal (0.1, 0.5, and 1.0 mM) for (**a**) 4 h, (**b**) 6 h, (**c**) 12 h, and (**d**) 24 h at 37 °C in a humidified environment of 5% CO_2 – 95% air. At the end of treatment, release of LDH (as an index of cytotoxicity) from the cells into medium was determined spectrophotometrically as described in the *Materials and Methods*. Data represent means ±S.D. of triplicate experiments conducted under identical conditions. *Significantly different at $p < 0.05$ as compared with the control cells treated with MEM alone

at the maximum (Fig. 5c and d). More strikingly, at the doses of 0.5 and 1.0 mM, glyoxal caused more than 90% of inhibition of MTT reduction in the BPAECs at 24 h of treatment (Fig. 5d). These results clearly revealed that glyoxal caused drastic inhibition of MTT reduction in BPAECs suggesting the cytotoxicity of glyoxal in the ECs through interfering with the mitochondrial function.

DMSA Protects Against Glyoxal-Induced Inhibition of MTT Reduction by BPAECs

We next investigated whether DMSA was able to protect mitochondrial function by preserving MTT reduction in the presence of glyoxal. BPAECs were treated with MEM alone, MEM containing DMSA (1.0, 2.0, and 5.0 mM), MEM containing glyoxal (0.5 mM), and MEM-containing DMSA (1.0, 2.0, and 5.0 mM) + glyoxal (0.5 mM). Glyoxal at a dose of 0.5 mM caused a significant and drastic decrease in MTT reduction by the BPAECs at 12 h of treatment as compared to control (Fig. 6). Cells co-incubated with glyoxal and DMSA showed a significant attenuation of the glyoxal-induced decline of MTT reduction in the BPAECs, suggesting the

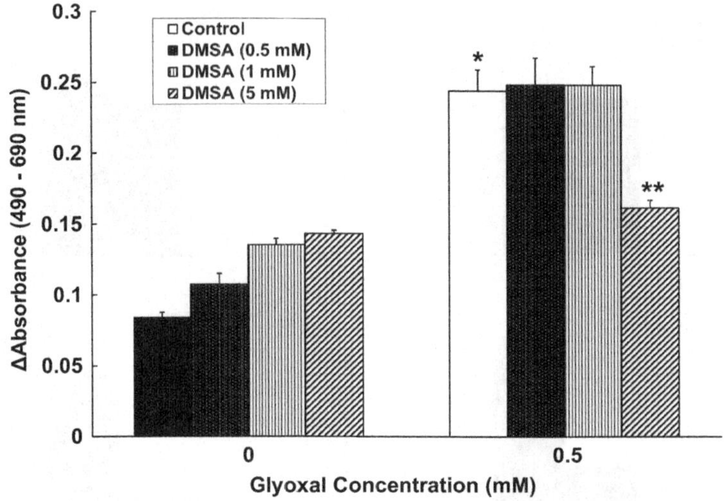

Fig. 4 DMSA protects against glyoxal-induced LDH release by BPAECs. BPAECs grown to 95% confluence in 15.5-mm dishes (24-well culture plate) were treated with MEM alone, MEM containing DMSA (1.0, 2.0, and 5.0 mM), MEM-containing glyoxal (0.5 mM) alone, and MEM containing DMSA (1.0, 2.0, and 5.0 mM) + glyoxal (0.5 mM) for 12 h at 37 °C in a humidified environment of 5% CO_2 – 95% air. At the end of treatment, release of LDH (as an index of cytotoxicity) from the cells was determined spectrophotometrically as described in *Materials and Methods*. Data represent means ±S.D. of triplicate experiments conducted under identical conditions. *Significantly different at $p < 0.05$ as compared with the control cells treated with MEM alone. **Significantly different at $p < 0.05$ as compared with the cells treated with MEM-containing glyoxal

protection of the inhibition of the mitochondrial dehydrogenase activity by DMSA in the BPAECs (Fig. 6).

Glyoxal-Induced Cell Morphological Alterations Are Attenuated by DMSA in BPAECs

We next investigated the effects of glyoxal on cell morphology as another index of cytotoxicity. Consistent with our previous measures of cytotoxicity, glyoxal induced intense cell morphological alterations in the BPAECs in a dose- and time-dependent fashion at doses of 0.5 and 1.0 mM at 6, 12, and 24 h of treatment (Fig. 7). As evidenced from light microscopy, glyoxal-treated cells exhibited loss of normal elongated cell morphology of ECs and appeared circular or rounded in a dose- and time-dependent fashion (Fig. 7). Importantly, DMSA at 5.0 mM dose protected the glyoxal (0.5 mM)-induced cell morphology alterations in BPAECs in a time-dependent manner (6, 12, and 24 h) (Fig. 8a). Our study also revealed that the well-established thiol-antioxidant, N-acetyl-L-cysteine (NAC) at 0.5, 1.0, and 5.0 mM

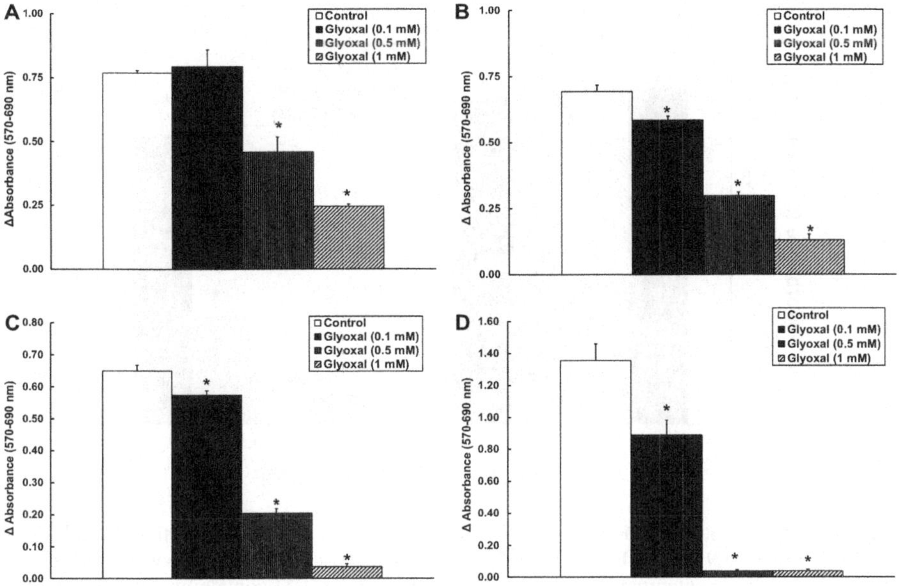

Fig. 5 Glyoxal inhibits MTT reduction by BPAECs. BPAECs grown to 95% confluence in 15.5-mm dishes (24-well culture plate) were treated with MEM alone or MEM-containing glyoxal (0.1, 0.5, and 1.0 mM) for (**a**) 4 h, (**b**) 6 h, (**c**) 12 h, and (**d**) 24 h at 37 °C in a humidified environment of 5% CO_2 – 95% air. At the end of treatment, MTT reduction (as an index of cell health and mitochondrial integrity) was determined spectrophotometrically as described in *Materials and Methods*. Data represent means ±S.D. of triplicate experiments conducted under identical conditions. *Significantly different at $p < 0.05$ as compared with the control cells treated with MEM alone

doses, effectively protected against the glyoxal (0.5 mM)-induced cell morphology alterations of ECs at 12 h of treatment (Fig. 8b). These results clearly demonstrated that the glucose-derived oxoaldehyde, glyoxal caused intense cell morphology alterations that were protected by the thiol-containing heavy metal chelator, DMSA in the ECs.

DMSA Protects Against Glyoxal-Induced Inhibition of Cell Proliferation as Evidenced by Decrease in [^{3}H]-Thymidine Incorporation in BPAECs

As a final measure of BPAEC function, we assessed the effects of glyoxal on BPAEC proliferation utilizing the well-established [^{3}H]-thymidine incorporation assay. The results showed that BPAECs treated with glyoxal (0.1, 0.5, and 1.0 mM) for 12 h exhibited a significant decrease in cell proliferation (DNA synthesis), demonstrating that glyoxal suppressed the EC division in a dose-dependent fashion. Co-treatment with DMSA (5.0 mM), however, resulted in significant and effective

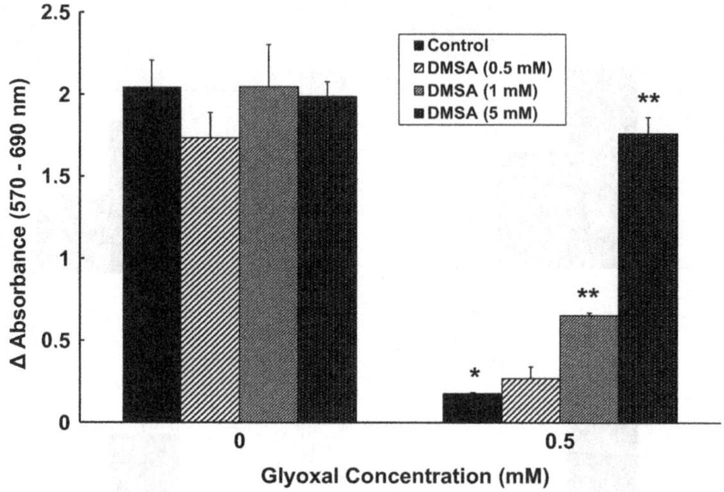

Fig. 6 DMSA protects against glyoxal-induced inhibition of MTT reduction by BPAECs. BPAECs grown to 95% confluence in 15.5-mm dishes (24-well culture plate) were treated with MEM alone, MEM containing DMSA (1.0, 2.0, and 5.0 mM), MEM-containing glyoxal (0.5 mM), and MEM containing DMSA (1.0, 2.0, and 5.0 mM) + glyoxal (0.5 mM) for 12 h at 37 °C in a humidified environment of 5% CO_2 – 95% air. At the end of treatment, MTT reduction (as an index of cell health and mitochondrial integrity) was determined spectrophotometrically as described in *Materials and Methods*. Data represent means ±S.D. of triplicate experiments conducted under identical conditions. *Significantly different at $p < 0.05$ as compared with the control cells treated with MEM alone. **Significantly different at $p < 0.05$ as compared with the control cells treated with MEM-containing glyoxal

cell proliferation (DNA synthesis) as compared to the same in the glyoxal-treated cells alone, demonstrating that the thiol-containing heavy metal chelator, DMSA offered protection against glyoxal-induced inhibition of EC proliferation (Fig. 9a).

DMSA Protects Against Glyoxal-Induced Inhibition of In Vitro Angiogenesis in BPAECs

Here, we investigated the action of glyoxal on the angiogenic potential (in vitro tube formation on Matrigel) of the BPAECs without or with the treatment of DMSA. As shown in Fig. 9b and c, glyoxal at doses of 0.5 and 1.0 mM drastically and significantly inhibited the in vitro tube formation of the ECs, wherein the higher tested dose of glyoxal (1.0 mM) almost completely abolished the angiogenic potential of the BPAECs. DMSA at 5.0 mM dose effectively and significantly protected against the glyoxal-induced arrest of the in vitro tube formation (angiogenesis) of the BPAECs (Fig. 9b and c).

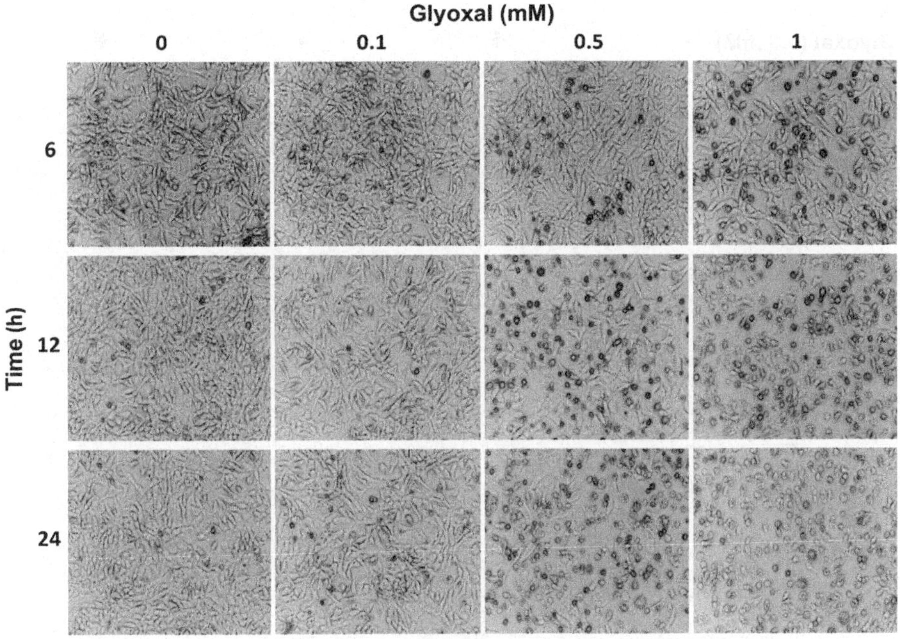

Fig. 7 Glyoxal induces cell morphological alterations in dose- and time-dependent fashion in BPAECs. BPAECs grown to 95% confluence in 35-mm dishes (6-well culture plate) were treated with MEM alone or MEM-containing glyoxal (0.1, 0.5, and 1.0 mM) for 6 h, 12 h, and 24 h at 37 °C in a humidified environment of 5% CO_2 – 95% air. At the end of treatment, cells were examined under light microscope at 20 X magnification as described in *Materials and Methods*. Each image is a representative picture obtained from three independent experiments conducted under identical conditions

Glyoxal Induces ROS Formation in BPAECs Which Is Attenuated by DMSA

Our results show that glyoxal induced a wide range of functional deficits in BPAECs, which were at least partially mitigated by the thiol-containing heavy metal chelator. Based on these results, we hypothesized that the cytotoxic effects of glyoxal are mediated by iron-mediated redox cycling and ROS formation, which are attenuated by the redox heavy metal chelator, DMSA in the ECs. Therefore, we measured (i) total ROS production and (ii) superoxide formation in glyoxal-treated BPAECs +/– DMSA using fluorescent reporters DCFDA (ROS reporter) and DHE (superoxide reporter), respectively (REFs instead of explaining in next sentences). The results showed that glyoxal (0.1, 0.5, and 1.0 mM) caused a significant increase in ROS generation in the BPAECs in a dose-response manner at 2 h of treatment that was attenuated by DMSA (5.0 mM) (Fig. 10a and b). Our experiments also revealed that glyoxal (0.5 mM) induced the formation of superoxide ($O_2^{\cdot-}$) that was effectively attenuated by DMSA (5.0 mM) in the BPAECs at 2 h of treatment (Fig. 10c). These results

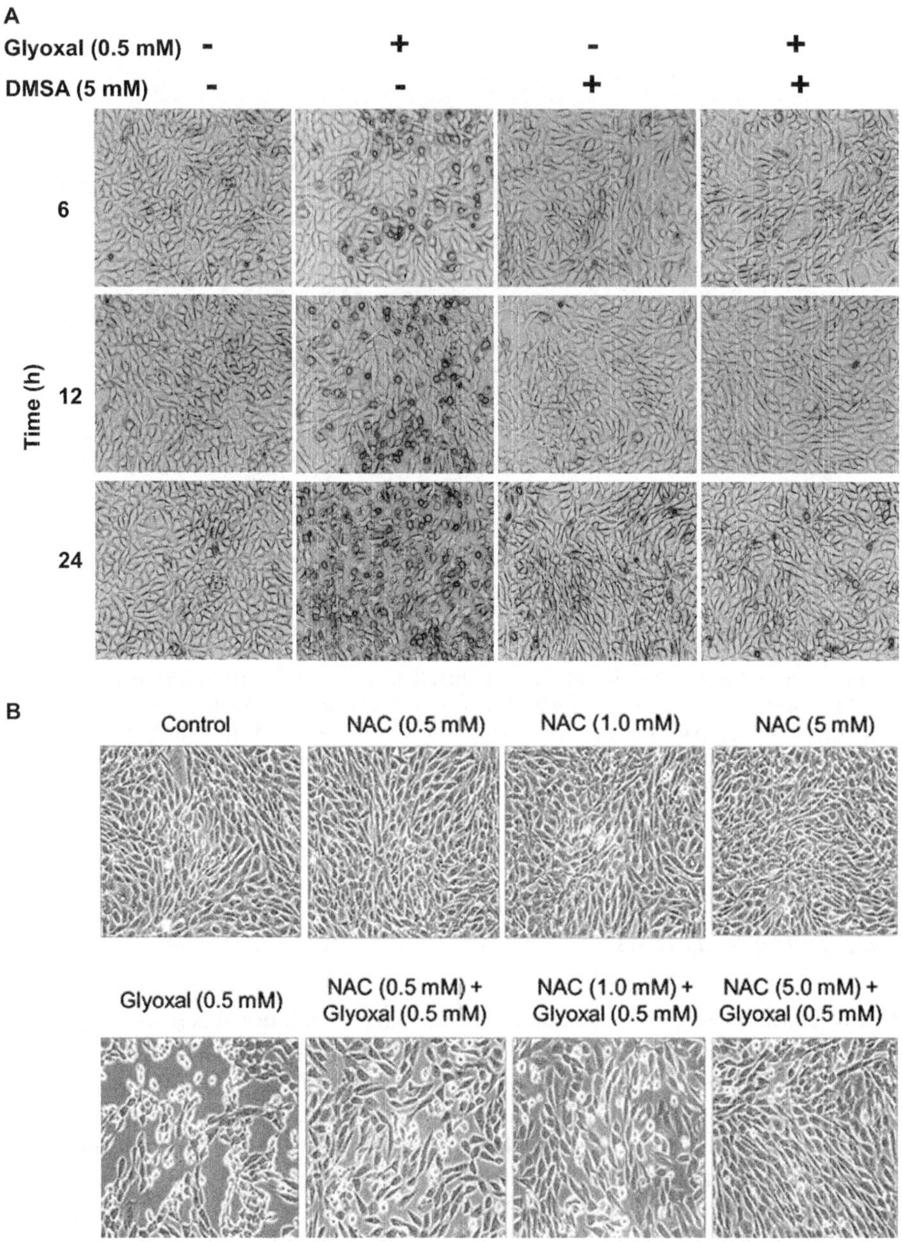

Fig. 8 (**a**) DMSA protects against glyoxal-induced alterations of cell morphology in BPAECs. BPAECs grown to 95% confluence in 35-mm dishes (6-well culture plate) were treated with MEM alone, MEM containing DMSA (5.0 mM), MEM containing glyoxal (0.5 mM), and MEM containing DMSA (5.0 mM) + glyoxal (0.5 mM) for 6 h, 12 h, and 24 h at 37 °C in a humidified environment of 5% CO_2 – 95% air. At the end of treatment, cells were examined under light microscope at 20 X magnification as described in *Materials and Methods*. Each image is a representative picture

clearly indicated that glyoxal induced ROS and superoxide ($O_2^{-\cdot}$) formation in the BPAECs at an early time of insult that was attenuated by the thiol-containing heavy metal chelator, DMSA, implicating the role of iron (Fe) in redox cycling in the ECs.

Glyoxal Causes Thiol-Redox Alteration Through GSH Depletion Which Is Attenuated by DMSA in BPAECs

GSH recruitment is the natural cellular response to scavenge intracellular ROS in a living cell system. Without this defense mechanism, ECs are susceptible to oxidant attack. Our earlier experiments of this study established that glyoxal induced a significant and robust increase in the generation of cellular oxidant species (ROS and superoxide) in the BPAECs. We sought to determine whether glyoxal also altered the levels of reduced glutathione (GSH) to further compromise the cell's resistance to oxidative stress. Glyoxal (0.1, 0.5, and 1.0 mM) caused significant and almost total loss of GSH at 8 h of treatment in a dose-dependent fashion which was significantly and effectively attenuated by DMSA (5.0 mM), the thiol-containing heavy metal chelator (Fig. 11). This experiment clearly revealed that the hyperglycemic oxoaldehyde, glyoxal caused the cellular small molecule thiol-redox, GSH depletion which was almost completely restored by DMSA.

DMSA Protects Against Glyoxal-Induced Loss of Transendothelial Electrical Resistance (TER) in BPAEC Monolayer

We previously reported that glyoxal alters cell-to-cell adhesion and cell-to-substratum attachment in the EC monolayer to modulate EC barrier function and vascular leak [2]. We therefore investigated whether DMSA attenuates glyoxal-induced EC monolayer perturbation (barrier dysfunction) using the Electric Cell Impedance Sensing (ECIS) to determine transendothelial resistance (TER) in a monolayer of BPAECs. Upon treatment with glyoxal (0.1, 0.5, and 1.0 mM) for 20 h, the TER drastically decreased as compared to untreated control BPAECvmonolayers (Fig. 12a). DMSA (5.0 mM) effectively and completely

Fig. 8 (continued) obtained from three independent experiments conducted under identical conditions. (**b**) NAC protects against glyoxal-induced alterations of cell morphology in BPAECs. BPAECs grown to 95% confluence in 35-mm dishes (6-well culture plate) were treated with MEM alone, MEM containing NAC (0.5, 1.0, and 5.0 mM) alone, MEM containing glyoxal (0.5 mM) alone, and MEM containing NAC (0.5, 1.0, and 5.0 mM) + glyoxal (0.5 mM) for 12 at 37 °C in a humidified environment of 5% CO_2 – 95% air. At the end of treatment, cells were examined under light microscope at 20 X magnification as described in *Materials and Methods*. Each image is a representative picture obtained from three independent experiments conducted under identical conditions

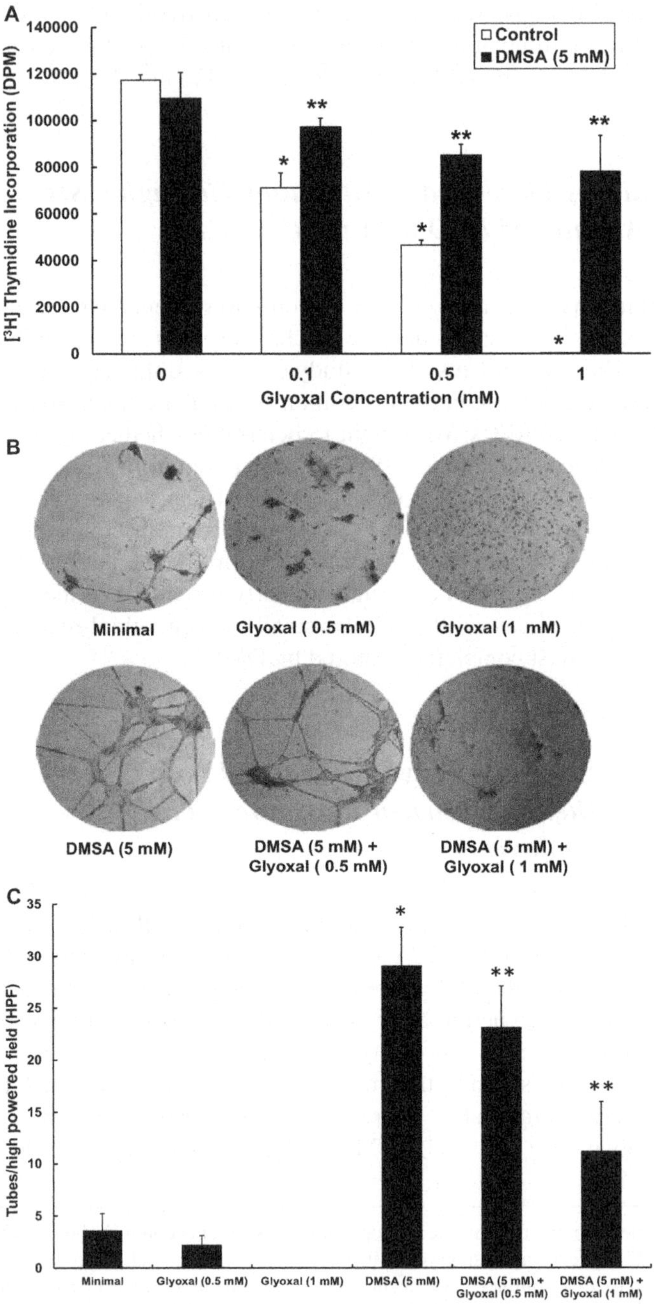

Fig. 9 (**a**) DMSA protects against glyoxal-induced inhibition of cell proliferation as evidenced by decrease in [^{3}H]-thymidine incorporation in BPAECs. BPAECs grown to 95% confluence in 35-mm dishes (6-well culture plate) were treated with MEM alone, MEM containing DMSA (5.0 mM), MEM containing glyoxal (0.1, 0.5, and 1.0 mM), and MEM containing DMSA (5.0 mM)

restored the glyoxal-induced loss of TER to the extent exhibited by the control untreated BPAEC monolayer at 20 h of treatment (Fig. 12b). These results showed that the thiol-containing heavy metal chelator, DMSA effectively protected against the loss of TER and barrier dysfunction caused by the hyperglycemic oxoaldehyde, glyoxal implicating the role of redox and transition metal such as iron.

DMSA Protects Against Glyoxal-Induced Paracellular Permeability (Leak) in BPAEC Monolayer

Barrier function (tightness of EC monolayer) is crucial for the paracellular gaps and leakiness of the EC monolayer. As our current ECIS experiments revealed that the glyoxal-induced loss of TER (as an index of EC barrier dysfunction) and its protection by DMSA, here we investigated whether the oxoaldehyde would cause paracellular permeability (leak) of high molecular weight polymer (FITC-dextran of 70 kDa) across the BPAEC. Glyoxal (0.5 mM) was observed to cause a significant increase in the paracellular permeability (leak) of FITC-dextran across the BPAEC monolayer at 12 and 24 h of treatment that was significantly and effectively attenuated by DMSA (5.0 mM) (Fig. 13a and b). These results clearly revealed that the thiol-containing heavy metal chelator, DMSA effectively protected the hyperglycemic oxoaldehyde-induced EC monolayer paracellular permeability, thus reverting the EC barrier dysfunction.

Fig. 9 (continued) + glyoxal (0.1, 0.5, and 1.0 mM) for 12 h at 37 °C in a humidified environment of 5% CO_2 – 95% air. At the end of treatment, cells were labeled with [^{3}H]-thymidine (1 μCi/mL) in complete BPAEC medium for 24 h. Cell replication at the end of the treatment was assayed by determining the [^{3}H]-thymidine incorporated into the cells as described in *Materials and Methods*. Data represent means ±S.D. of triplicate experiments conducted under identical conditions. *Significantly different at $p < 0.05$ as compared with the control cells treated with MEM alone. **Significantly different at $p < 0.05$ as compared with the cells treated with MEM-containing glyoxal. (**b, c**) DMSA protects against glyoxal-induced inhibition of in vitro angiogenesis in BPAECs. BPAECs were grown up to 80% confluence in complete medium in two separate flasks. Both flasks of cells were starved in minimal medium (MEM) at 37 °C for 30 min and following that DMSA (5 mM) in MEM was added to one of the flasks and continued to incubate for an additional 45 min. The cells from both flasks were then added to the Matrigel-coated plates and then treated with 1.0 mL of minimal medium alone, 1.0 mL of rich medium supplemented with glyoxal (0.5 and 1 mM) alone, 1.0 mL of rich medium supplemented with DMSA (5.0 mM) alone, and 1 mL of rich medium supplemented with DMSA (5.0 mM) and glyoxal (0.5 and 1.0 mM). Following that, the cells were incubated for 12 h at 37 °C and then examined for the formation of tubes connecting in adjacent colonies. The medium was then aspirated, and cells were fixed with 4% paraformaldehyde. Four digital images (**b**) were captured per well at the same location in each well and the tubes were counted in a blinded fashion (**c**) under a Nikon light microscope at 40 X magnification. The data represent mean ±S.D. of samples for each condition. *Significantly different at $p < 0.05$ as compared with the control cells treated with minimal medium alone. **Significantly different at $p < 0.05$ as compared with the cells treated with rich medium containing glyoxal alone. The statistical analysis was performed using an ANOVA with SYSTAT12 software (San Jose, CA)

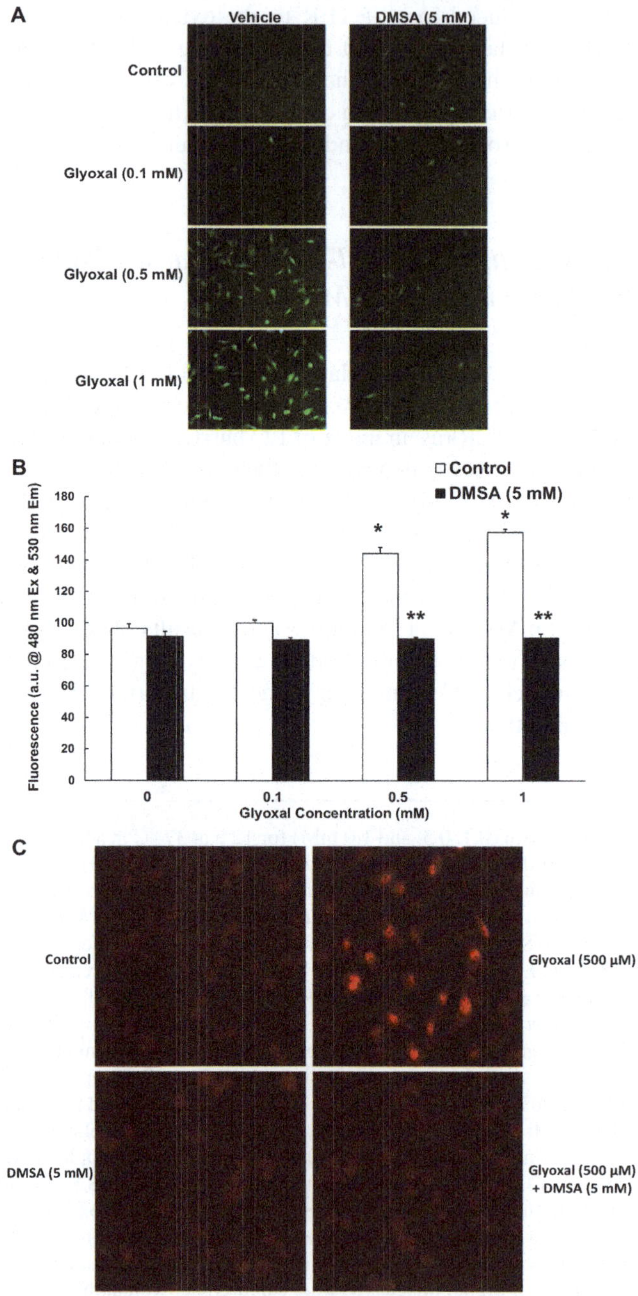

Fig. 10 Glyoxal induces ROS formation in BPAECs which is attenuated by DMSA. BPAECs grown to 95% confluence in 35-mm dishes (6-well culture plate) were preloaded with either DCFDA (10.0 µM/dish for 30 min in complete medium) or DHE (5.0 µM/dish for 30 min in complete medium), following which the fluorophores were removed and the cells were treated with

DMSA Protects Glyoxal-Induced Cytoskeletal Rearrangement in BPAECs

Oxidants are known to cause cytoskeletal rearrangement in vascular ECs upstream of the EC monolayer barrier dysfunction and vascular endothelial leak/paracellular permeability. Also, earlier we have reported that the glucose-derived oxoaldehyde, glyoxal induces cytoskeletal rearrangement in the vascular ECs [2]. The intense barrier dysfunction (both decline in the TER and increase in paracellular leak)

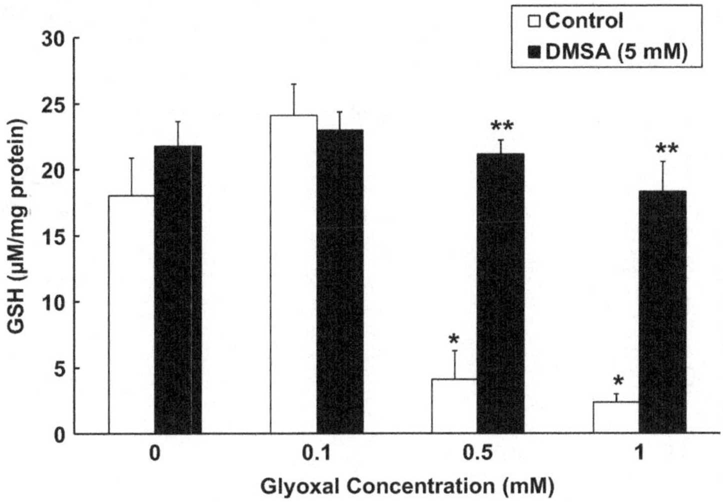

Fig. 11 Glyoxal causes thiol-redox alteration through GSH depletion, which is attenuated by DMSA in BPAECs. BPAECs grown to 100% confluent in 96-well sterile plates were treated with MEM alone, MEM containing DMSA (5.0 mM), MEM containing glyoxal (0.1, 0.5, and 1.0 mM), and MEM containing DMSA (5.0 mM) + glyoxal (0.1, 0.5, and 1.0 mM) for 8 h at 37 °C in a humidified environment of 5% CO_2 – 95% air. At the end of treatment, the intracellular GSH concentrations were determined by the GSH-Glo chemiluminescence assay as described in *Materials and Methods*. Data represent mean ±S.D. of triplicate experiments conducted under identical conditions. *Significantly different at $p < 0.05$ as compared with the control cells treated with MEM alone. **Significantly different at $p < 0.05$ as compared with the cells treated with MEM-containing glyoxal

Fig. 10 (continued) MEM alone, MEM containing DMSA (5.0 mM), MEM containing glyoxal (0.1, 0.5, and 1.0 mM), and MEM containing DMSA (5.0 mM) + glyoxal (0.1, 0.5, and 1.0 mM) for 2 h at 37 °C in a humidified environment of 5% CO_2 – 95% air. At the end of treatment, (**a**) images were captured digitally with Olympus 1 × 50 fluorescence microscope to visualize the DCF fluorescence as an index of ROS formation, and (**b**) quantification of intracellular ROS production by determining the fluorescence of oxidized DCF spectrofluorimetrically as described in *Materials and Methods*. (**c**) Intracellular DHE fluorescence, as an index of superoxide formation, was examined on Olympus 1 × 50 fluorescence microscope and images were captured digitally as described in *Materials and Methods*. Data represent mean ±S.D. of triplicate experiments conducted under identical conditions. *Significantly different at $p < 0.05$ as compared with the control cells treated with MEM alone. **Significantly different at $p < 0.05$ as compared with the cells treated with MEM containing glyoxal. Each image is a representative of three different images captured from triplicate samples for each experimental condition

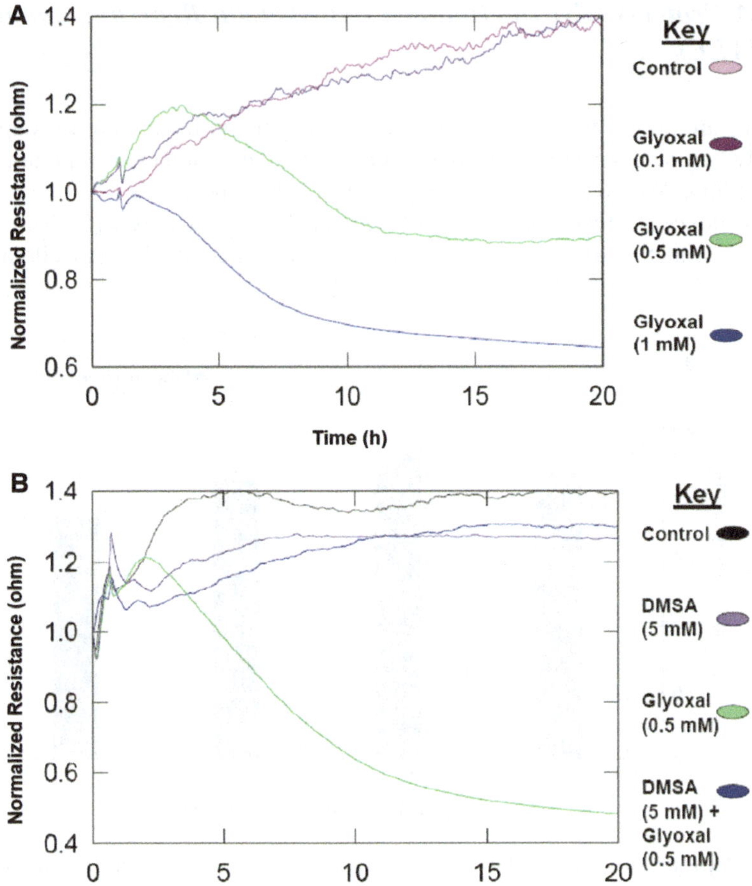

Fig. 12 DMSA protects against glyoxal-induced loss of transendothelial electrical resistance (TER) in BPAEC monolayer. BPAEC monolayers were cultured in complete MEM on sterile ECIS gold electrodes and treated with (**a**) MEM alone and MEM containing glyoxal (0.1, 0.5, and 1.0 mM), and (**b**) MEM alone, MEM containing DMSA (5.0 mM) alone, MEM containing glyoxal (0.5 mM) alone, and MEM containing DMSA (5.0 mM) + glyoxal (0.5 mM) for 20 h in a humidified atmosphere of 5% CO_2 – 95% air at 37 °C following which the TER was measured continuously on electrical cell impedance system (ECIS) as described in *Materials and Methods*. Each tracing is a representative of the normalized resistance values of two (duplicate) independent experiments conducted under identical conditions

induced by glyoxal in the BPAEC monolayer and its protection by DMSA as observed in our earlier experiments initiated these experiments to determine whether DMSA would protect the glyoxal-induced cytoskeletal rearrangements in BPAECs, including the actin stress fiber formation and rearrangement/redistribution of the cytoskeletal proteins such as cortactin, ZO-1, and occludin. Results of these experiments revealed that glyoxal (0.1, 0.5, and 1.0 mM), in a dose-dependent manner at 4 h of treatment, caused actin stress fiber formation indicative of the actin cytoskeletal rearrangement (Fig. 14a and b), intense cortactin reorganization/redistribution

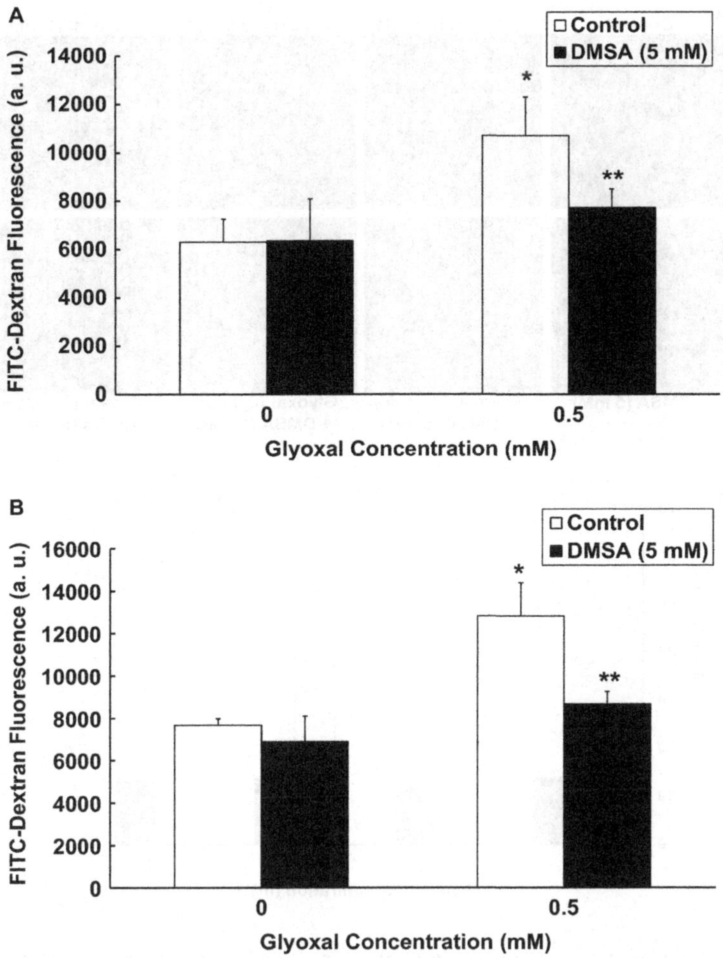

Fig. 13 DMSA protects against glyoxal-induced paracellular permeability (leak) in BPAEC monolayer. BPAECs grown to 100% confluence in 22-mm dishes (12-well culture plate) were treated with MEM alone, MEM containing DMSA (5.0 mM), MEM containing glyoxal (0.5 mM), and MEM containing DMSA (5.0 mM) + glyoxal (0.5 mM) for (**a**) 12 h and (**b**) 24 h at 37 °C in a humidified environment of 5% CO_2 – 95% air. At the end of treatment, FITC fluorescence (as an index of EC leak of FITC-tagged dextran of 70 kDa mol. wt.) at 480 nm excitation and 530 nm emission was determined spectrofluorimetrically as described in *Materials and Methods*. Data represent means ±S.D. of triplicate experiments conducted under identical conditions. *Significantly different at $p < 0.05$ as compared with the control cells treated with MEM alone. **Significantly different at $p < 0.05$ as compared with the cells treated with MEM-containing glyoxal

(Fig. 14c and d), loss of ZO-1 (Fig. 14e and f), and loss of occludin (Fig. 14g and h), all of which were significantly and affectively attenuated by DMSA (5.0 mM). These results clearly showed that the hyperglycemic oxoaldehyde, glyoxal caused the EC cytoskeletal reorganization that was effectively protected by the thiol-containing heavy metal chelator, DMSA.

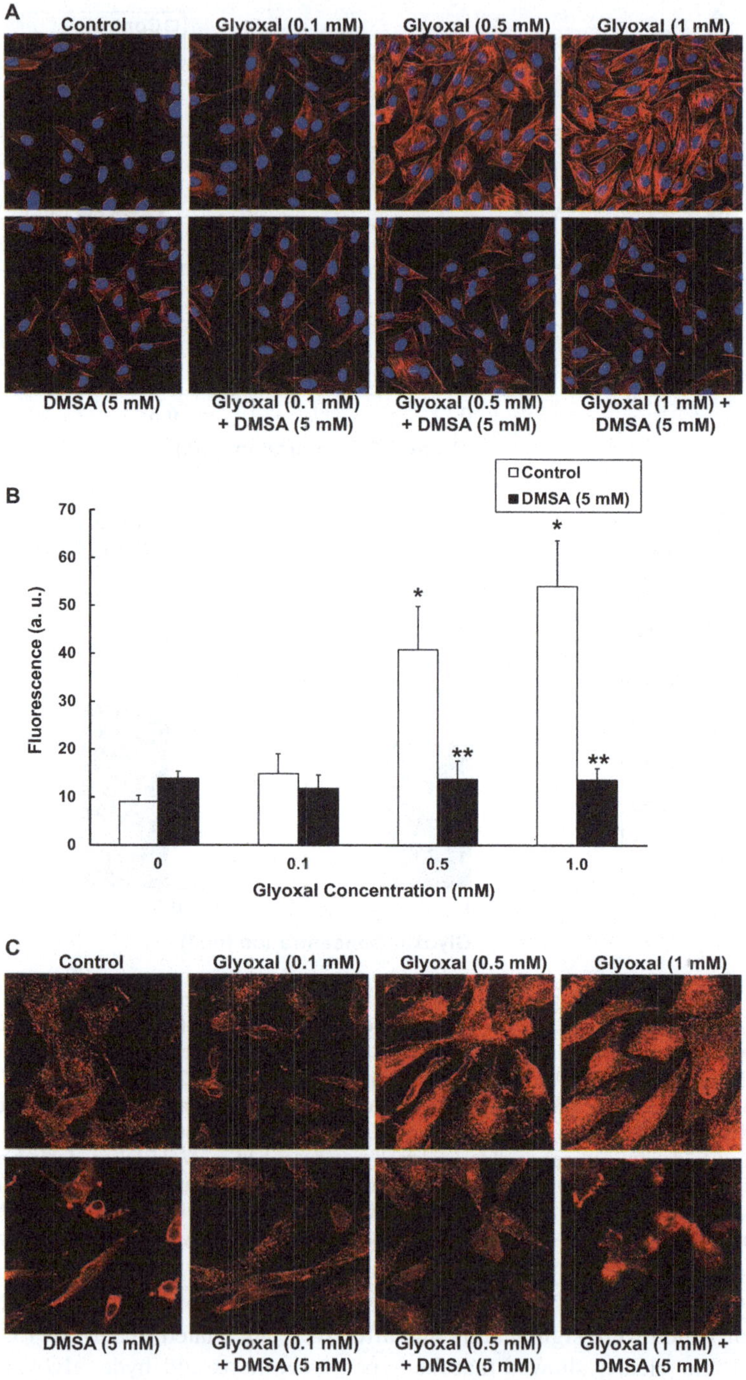

Fig. 14 DMSA protects glyoxal-induced cytoskeletal rearrangement in BPAECs. BPAECs grown (continued)

DMSA Protects Against Glyoxal-Mediated Formation of AGEs in BPAECs

Sugar-derived hyperglycemic oxoaldehydes such as glyoxal and methylglyoxal are known to cause formation of the advanced glycation end products (AGEs), and we have reported earlier that glyoxal causes AGE formation in the vascular ECs [2]. Having this as the premise, here we investigated whether DMSA would attenuate the glyoxal-mediated formation of AGEs in BPAECs. As shown in Fig. 15a, glyoxal (0.5 and 1.0 mM) significantly caused an intense formation of AGEs in BPAECs at 4 h of treatment that apparently was completely attenuated by DMSA (5.0 mM) treatment (Fig. 15a and b). This experiment clearly established that the oxoaldehyde, glyoxal mediated the formation of AGEs which was blocked by the thiol-containing heavy metal chelator, DMSA.

Iron (Fe^{2+}) Exacerbates Glyoxal-Induced Cytotoxicity and ROS Formation in BPAECs

Earlier experiments of the current study clearly revealed that the thiol-containing heavy metal chelator, DMSA effectively attenuated the glyoxal-induced cytotoxicity, redox dysregulation, ROS formation, oxidative stress, and barrier dysfunction in the BPAECs, suggesting role of the redox-active transition metal such as iron (Fe). Therefore, here we investigated whether low concentrations of exogenously added iron as iron (II) sulfate ($FeSO_4$) would affect the glyoxal-induced cytotoxicity and ROS formation in the BPAECs. The results revealed that $FeSO_4$ (50 μM) significantly exacerbated the glyoxal (0.1 and 0.5 mM)-induced LDH release (Fig. 16a) and decrease in MTT reduction (Fig. 16b) at 12 h of treatment. Furthermore, the results showed that $FeSO_4$ (50 μM) significantly and intensely enhanced the

Fig. 14 (continued) to 95% confluence grown on sterile glass coverslips in sterile 35-mm dishes (6-well culture plates) were treated with MEM alone, MEM containing DMSA (5.0 mM), MEM containing glyoxal (0.1, 0.5, and 1.0 mM), and MEM containing DMSA (5.0 mM) + glyoxal (0.1, 0.5, and 1.0 mM) for 4 h at 37 °C in a humidified environment of 5% CO_2 – 95% air. At the end of treatment, cells were fixed and stained for (**a**, **b**): actin stress fibers/DAPI; (**c**, **d**): cortactin; (**e**, **f**) ZO-1; and (**g**, **h**) occludin as described in *Materials and Methods*. Each image is a representative picture obtained from three independent experiments conducted under identical conditions. The fluorescence images of the cytoskeletal elements were captured on the Zeiss LSM 510 Confocal/Multiphoton fluorescence microscope at 63 X magnification as described in *Materials and Methods*. The fluorescence intensities were obtained from the captured images at the respective excitation and emission wavelengths for each of the cytoskeletal elements examined for each experimental condition and plotted graphically. Data represent means ±S.D. of triplicate experiments conducted under identical conditions. *Significantly different at $p < 0.05$ as compared with the control cells treated with MEM alone. **Significantly different at $p < 0.05$ as compared with the cells treated with MEM containing glyoxal

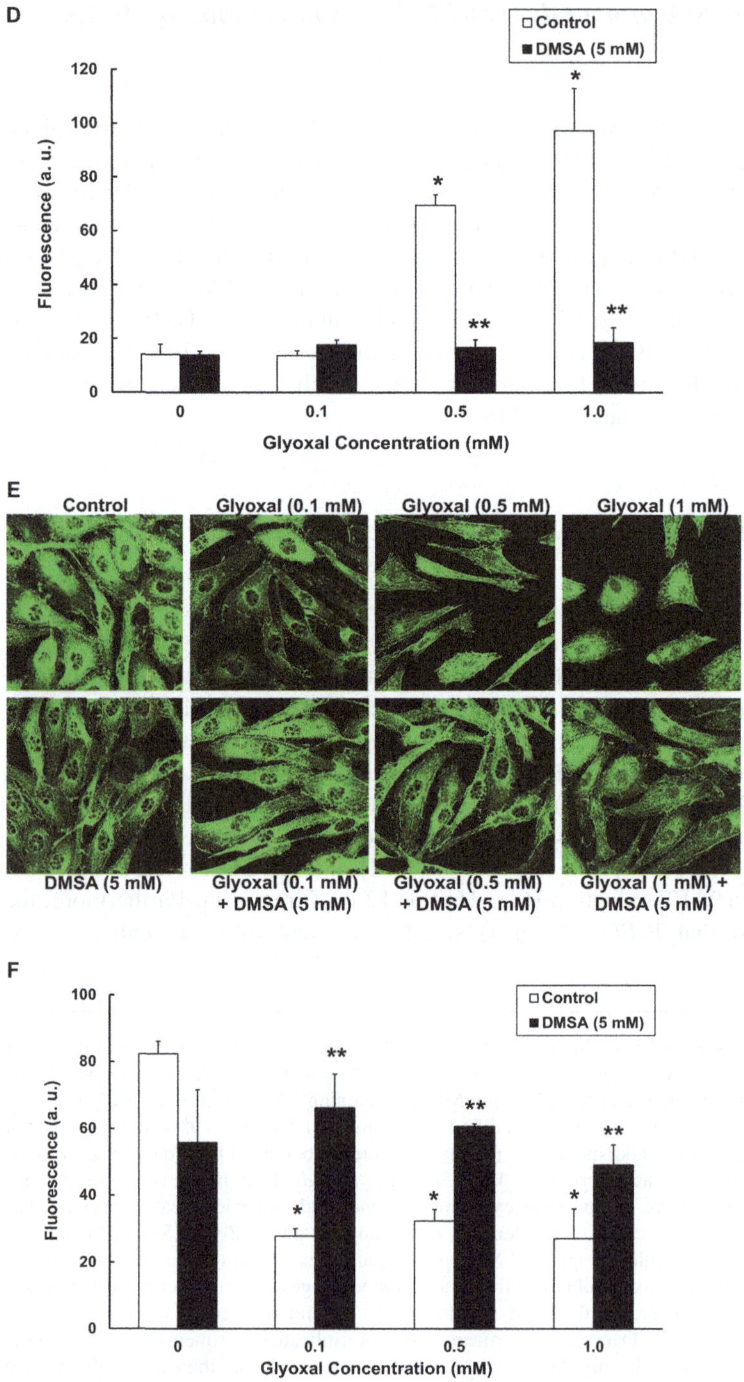

Fig. 14 (continued)

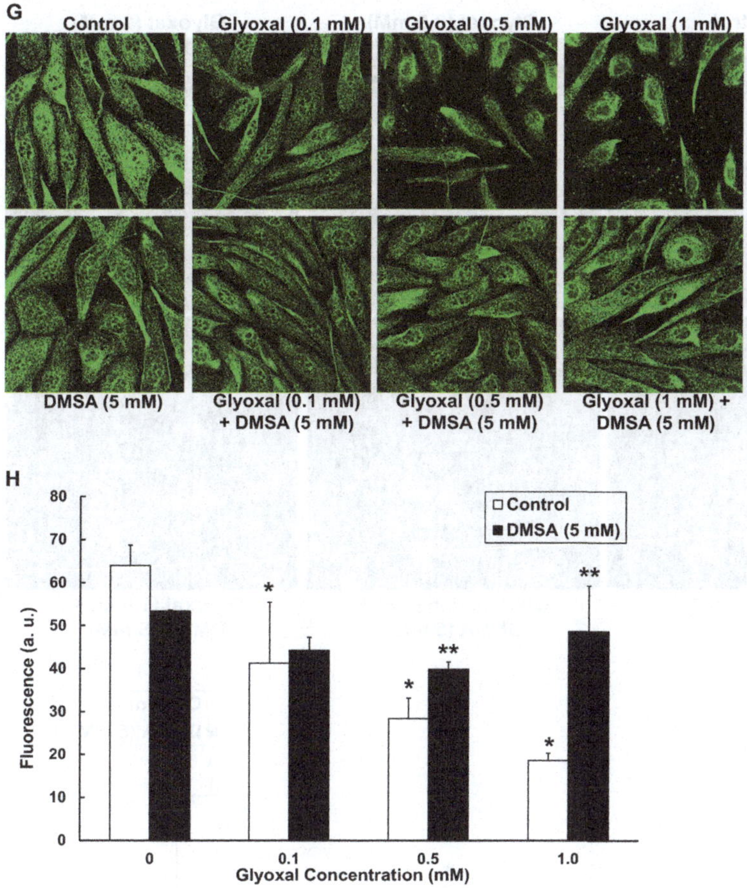

Fig. 14 (continued)

glyoxal-induced ROS formation (intracellular DCF fluorescence) in a dose-dependent fashion (0.1, 0.5, and 1.0 mM) at 1 h of treatment as compared to same in the untreated control BPAECs and BPAECs treated with glyoxal alone (Fig. 16c). These results strikingly established that a low concentration of iron (Fe^{2+}) at 50 μM exacerbated the glyoxal-induced cytotoxicity and ROS formation (oxidative stress) synergistically, suggesting the redox-active role of iron (II) therein.

Determination of Chelatable Iron (Fe) Availability in BLMVECs

Since all the preceding experiments clearly revealed that the thiol-containing heavy metal chelator, DMSA effectively attenuated the glyoxal-induced cytotoxicity, ROS production, thiol-redox dysregulation, cytoskeletal reorganization,

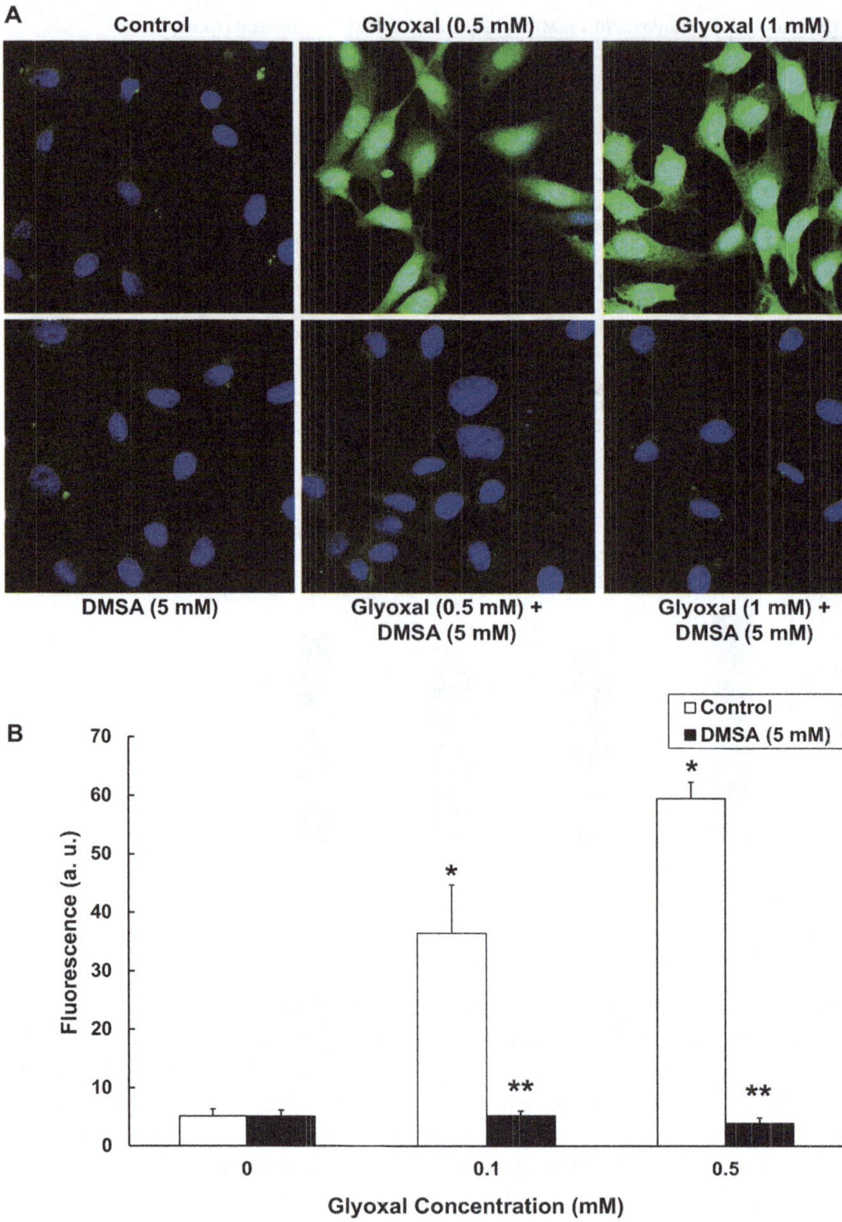

Fig. 15 DMSA protects against glyoxal-mediated formation of AGEs in BPAECs. BPAECs grown to 95% confluence grown on sterile glass coverslips in sterile 35-mm dishes (6-well culture plates) were treated with MEM alone, MEM containing DMSA (5.0 mM), MEM containing glyoxal (0.1, 0.5, and 1.0 mM), and MEM containing DMSA (5.0 mM) + glyoxal (0.1, 0.5, and 1.0 mM) for 4 h at 37 °C in a humidified environment of 5% CO_2 – 95% air. At the end of treatment, cells were fixed and stained for AGE antiserum (Anti-Amadori) as described in *Materials and Methods*. Fluorescence images were captured with intensities on Zeiss LSM 510

and paracellular permeability in the BPAECs indicating the role of iron (Fe) therein, here we determined the intracellular chelatable iron availability in the BPAECs by the fluorescence microscopy, utilizing the iron-specific fluorophore PG SK in conjunction with 2,2′-dipyridyl (dequencher) as described in the materials and methods. The results clearly revealed intense fluorescence of the intracellular chelatable iron (Fe) pool in the untreated ECs at basal level indicating the availability of intracellular iron for the redox reactions in BPAECs (Fig. 17a and b).

Discussion

Diabetes cases are on the rise globally, and subsequently, the rates of hyperglycemia are increasing rapidly. Hyperglycemia is the primary cause of major diabetic complications and is connected to both micro- and macrovascular diseases, including retinopathy, nephropathy, coronary heart disease, and stroke [2]. Since insulin is essential by cells for the absorption of glucose, insulin deficiencies found in both type 1 and type 2 diabetic patients lead to hyperglycemia. The hyperglycemic conditions lead to an increased production of both AGE precursors and ultimately AGEs, which have recently been linked primarily to vascular complications. Glyoxal is the precursor to an estimated 40–50% of AGEs, and it can form adducts with DNA- and RNA-causing mutations and has been proven to cause cytotoxicity and oxidative stress [2]. The current study is focused on the effect of the AGE precursor, glyoxal, which is formed in excess under hyperglycemic conditions, on vascular ECs.

The current study showed that the AGE precursor glyoxal has a wide range of adverse effects on BPAEC function, including morphological changes, ROS production, membrane permeability, and cytoskeleton rearrangement, while also decreasing mitochondrial activity, DNA synthesis, ROS scavenger levels, and cell barrier function. Importantly, we found that the heavy metal chelator dimercaptosuccinic acid (DMSA) was able to successfully attenuate all the adverse effects brought on by glyoxal, while exogenous Fe exacerbated glyoxal-induced damage.

Fig. 15 (continued) Confocal/Multiphoton Microscope at 543 nm excitation and 600 nm emission under 63 X magnification (**a**). Each image is a representative picture selected from three independent experiments conducted under identical conditions. The fluorescence intensities were obtained from the captured images at the respective excitation and emission wavelengths for each of the cytoskeletal elements examined for each experimental condition and plotted graphically (**b**). Data represent means ±S.D. of triplicate experiments conducted under identical conditions. *Significantly different at $p < 0.05$ as compared with the control cells treated with MEM alone. **Significantly different at $p < 0.05$ as compared with the cells treated with MEM containing glyoxal

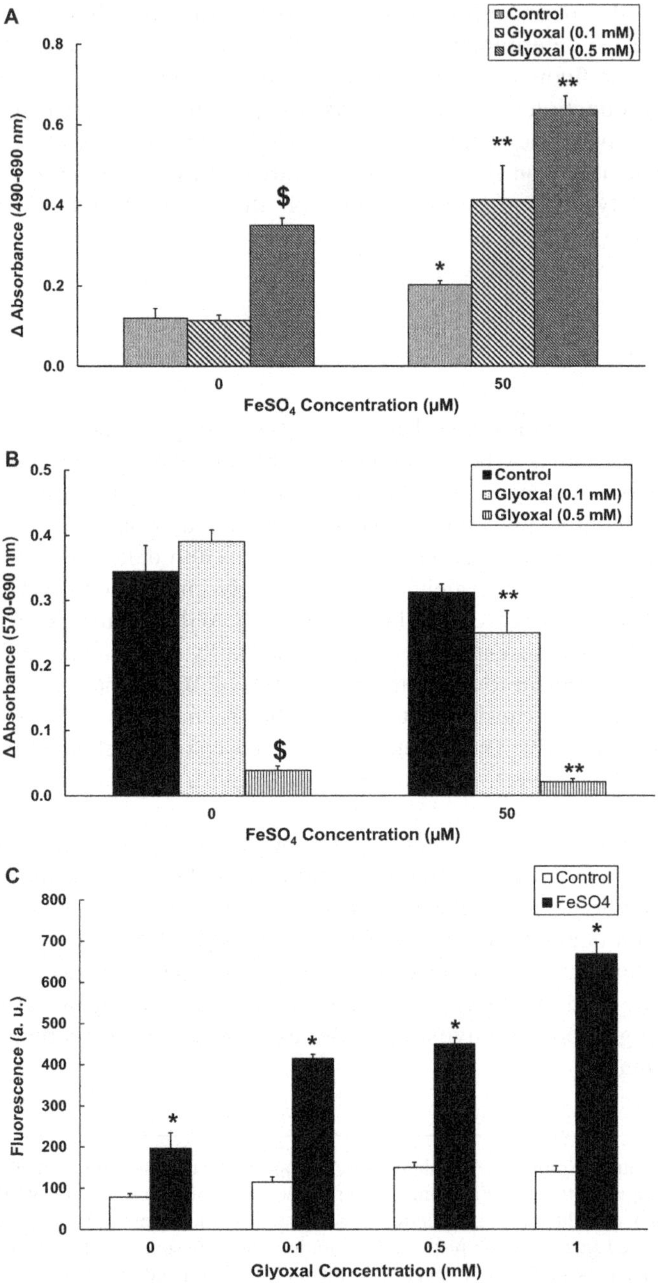

Fig. 16 Iron (Fe^{2+}) exacerbates glyoxal-induced cytotoxicity and ROS formation in BPAECs. BPAECs grown to 95% confluence in sterile 15.5-mm dishes (24-well culture plates) were treated with MEM alone, MEM containing glyoxal (0.1 and 0.5 mM), MEM containing $FeSO_4$ (50 μM), and MEM-containing glyoxal (0.1 and 0.5 mM) + $FeSO_4$ (50 μM) for 12 h at 37 °C in a humidified

There is natural ROS production and scavenging that accompanies many biochemical processes, particularly those involved in metabolism. Oxidative stress, however, occurs when either too much ROS are produced or not enough are scavenged. Oxidative stress has been shown to cause β cell dysfunction and insulin resistance, further promoting hyperglycemia [13, 14]. Therefore, AGEs can be one of the primary sources of excessive ROS production and oxidative stress prediabetes, and since AGEs are produced through various oxidation and glycation reactions, their production is further increased in diabetics from the higher serum glucose concentrations and oxidative stress. Additionally, it has recently been shown that AGEs can be obtained from poor diet, and about one-tenth of AGEs in foods are absorbed and added to circulating levels [13]. It has also been shown that diet high in AGEs had greater damaging impact on vascular functionality compared to diet low in AGEs [15].

AGE formation occurs when a saccharide (for example, glucose, fructose, or pentose) makes nonenzymatic posttranslational modifications to a macromolecule (including lipids, proteins, or nucleic acid). This process, under normal physiological conditions, is relatively slow and occurs in three stages via the Maillard reaction. First, Schiff bases are formed slowly. Then, unstable AGE precursors are generated. Last, irreversible AGEs are produced. Additionally, AGEs can bond with metals like Cu^{2+} or Fe^{2+} to catalyze the rapid generation of ROS [15]. AGEs can damage protein structure and, subsequently, function, generally in the extracellular matrix, causing cell differentiation, impairing migration and adhesion, and potentially cell death. One of the most detrimental effects of diabetic hyperglycemia is vascular damage, which may occur as a result of AGE formation [13]. Diabetic vasculopathy is associated with multiple cardiovascular diseases. This can be a result of AGEs cross-linking with extracellular proteins as well as AGEs binding to cellular AGE receptors (RAGE) [2, 13]. Increases in oxidative stress, inflammation, and endothelial dysfunction can occur when AGEs bind to cellular RAGEs [15]. Elevated AGE levels in diabetic patients have been shown to cause endothelial (the lining that surrounds bodily cavities, organs, and

Fig. 16 (continued) environment of 5% CO_2 – 95% air. At the end of treatment, (**a**) release of LDH from cells into medium (as an index of cytotoxicity) and (**b**) MTT reduction (as an index of cell health and mitochondrial activity) were determined spectrophotometrically as described in *Materials and Methods*. BPAECs grown to 95% confluence in 35-mm dishes (6-well culture plates) were preloaded with 10 μM DCFDA and then treated with MEM alone, MEM containing glyoxal (0.1, 0.5, and 1 mM), MEM containing $FeSO_4$ (50 μM), and MEM-containing glyoxal (0.1, 0.5, and 1 mM) + $FeSO_4$ (50 μM) for 1 h at 37 °C in a humidified environment of 5% CO_2 – 95% air. At the end of treatment, (**c**) intracellular ROS production was determined by measuring the fluorescence of oxidized DCF spectrophotometrically as described in *Materials and methods*. Data represent mean ±S.D. of triplicate experiments conducted under identical conditions. $Significantly different at $p < 0.05$ as compared with the control cells treated with MEM alone. *Significantly different at $p < 0.05$ as compared with the control cells treated with MEM alone. **Significantly different at $p < 0.05$ as compared to the cells treated with MEM containing glyoxal alone

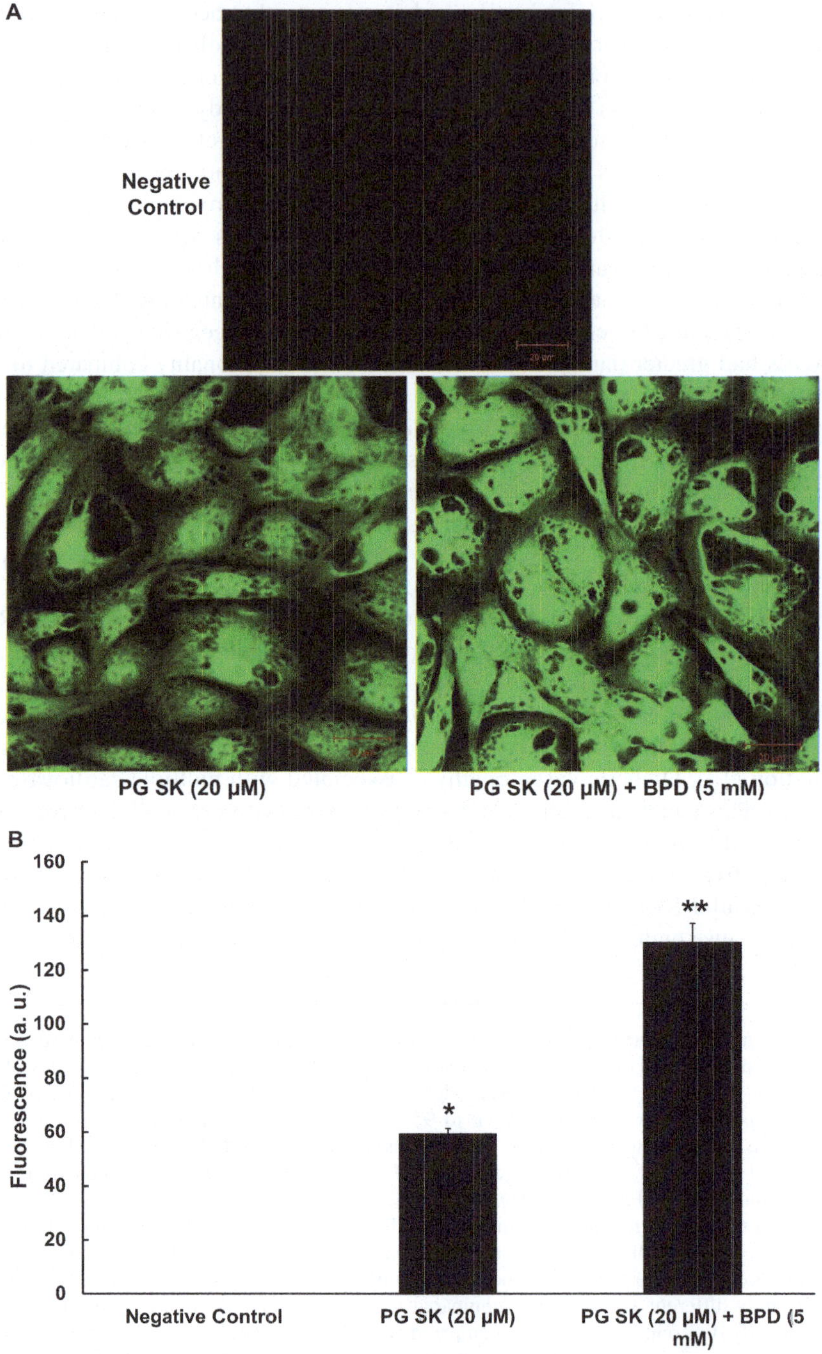

Fig. 17 Determination of chelatable iron (Fe) availability in BLMVECs. BPAECs grown to 95% confluence grown in 35-mm dishes (6-well culture plate) on glass coverslips were analyzed to

structures, including blood vessels) dysfunction [16]. Vascular endothelium is vital to maintaining the vascular integrity and is responsible for barrier function, clotting, angiogenesis, leukocyte function, and vasoconstriction and dilation, which regulate the properties and function of blood and blood vessel, including the regulation of blood pressure [2]. Since AGEs have greater affinity to bind to protein macromolecules, circulating AGEs can damage the vascular endothelial cells (ECs) upon interaction with the endothelial barrier by increasing oxidative stress [2]. The RAGE receptor is involved in immunity pathophysiology; activation of a RAGE by an AGE can induce several immune dysfunction responses, including increasing proinflammatory cytokines production and oxidative stress [15].

Methylglyoxal, another AGE precursor, has been shown to contribute to the development of diabetic cardiomyopathy. Under hyperglycemic conditions, methylglyoxal is overproduced as a byproduct of glycolysis. The subsequent buildup of methylglyoxal leads to vascular damage, cytotoxicity, inflammation, and EC dysfunction [17]. A study by Sotokawauchi et al. showed that increased fructose levels directly increased AGE production and likely induce damage to ECs, as fructose reacts ten times faster with proteins to form AGEs compared to glucose [18]. There have been studies that indicate protecting against AGEs can mitigate EC damage [2, 19, 20]. Sliman et al. showed that aminoguanidine can protect against glyoxal-induced EC dysfunction and damage [2]. Additionally, Sun et al. demonstrated that curcumin can inhibit the formation of AGEs and decrease oxidative stress from methylglyoxal [19]. Additionally, there is evidence to support low levels of the antioxidant vitamin B6 can increase AGE production and that supplementation may mitigate those effects [20].

Iron is a crucial component of many bodily processes, including cell growth, oxygen transport, DNA synthesis, electron transport, metabolism, etc. [21] Recent studies have indicated that iron overload can play a role in the development of diabetes and diabetic complications [22–24]. Iron levels are mainly controlled by absorption from dietary sources. The liver stores iron as ferritin and regulates that storage via the hormone hepcidin; hepcidin, when secreted, decreases intestinal absorption of iron [21, 25]. Iron has two stable oxidation states, Fe^{2+} and Fe^{3+}, and

Fig. 17 (continued) visualize and quantify chelatable intracellular iron pool by fluorescence microscopy. At the end of incubation, cells were fixed and stained for determination of free iron as described in *Materials and Methods*. BPAECs were incubated with PG SK diacetate (20 μM) for 10 min. at room temperature and washed twice with PBS. Selected wells were then incubated with 2,2′-bipyridyl (5 mM) for 10 min at room temperature, washed twice with PBS, samples were mounted, and examined under Zeiss LSM 510 Confocal/Multiphoton Microscope with 500 nm excitation and 530 nm emission at 63 X magnification. The images were captured digitally (**a**). Fluorescence intensities were determined using the NIH sponsored program ImageJ and plotted (**b**). Each image is a representative picture chosen from the independent experiment conducted under identical conditions. *Significantly different at $p < 0.05$ as compared with the negative control cells treated. **Significantly different at $p < 0.05$ as compared to the cells treated with 2,2′-bipyridyl alone

acts as an electron donor and electron acceptor, respectively, during biochemical reactions. However, this cycling can generate ROS byproducts [5]. Under normal physiological conditions, natural scavengers may remove these byproducts without damage; however, under hyperglycemia, iron overload may cause excessive ROS production and increased oxidative stress. There is a plethora of studies that have linked iron overload to the development of type 2 diabetes [5]. Iron overload can result in inhibition of glucose functionality by hindering insulin secretion and signaling and increasing insulin resistance [21]. Conversely, iron deficiency can also impact glucose levels, especially in diabetic patients, supporting the concept that the body requires a specific range of iron to optimally function [26]. This is supported by several studies [5], including one by Mirlohi et al. that demonstrated that patients with iron overload were positively correlated with significantly increased blood glucose levels, serum AGE levels, total ROS, and oxidative stress [27]. Additionally, studies have shown that excessive dietary intake of iron, such as through red meat consumption, especially in diabetic patients, has a direct correlation with increased iron levels and insulin dysfunction, elevated glucose levels, and oxidative stress [5, 28]. Iron overload in hyperglycemic conditions can potentially increase the formation of AGEs through Fenton reactions and ROS formation. Under hyperglycemia, iron-binding proteins may be glycated during the formation of AGEs, which releases additional free iron, ultimately creating a positive feedback loop [5].

Dimercaptosuccinic acid is a heavy metal chelator, typically used to treat metal poisoning. Chelation occurs when an organic compound bonds with a metal (forming a chelate). Heavy metals can have harmful effects, including toxicity. Therefore, a chelator is used so the metal will bond with it instead, and the metal loses its biochemical properties and is excreted [29]. In the current study, we observed that treatment with glyoxal significantly damaged BPAECs on several fronts. Glyoxal increased cytotoxicity, ROS production, cell morphology changes, decreased cellular integrity and mitochondrial activity, depleted GSH, and induced cytoskeleton rearrangement. Since studies support our belief that iron plays a significant role in the oxidative stress and other mechanisms that lead to the BPAEC damage, we hypothesized that using DMSA to scavenge free iron would mitigate the destructive capability of glyoxal (Fig. 18). The present study demonstrated that treatment with DMSA successfully attenuated all the adverse cellular effects. Treating with DMSA alone did not produce any significant impact compared to the control. DMSA, or a similar heavy metal chelator, with its demonstrated ability to protect the integrity and function of the endothelial cells against the iron and AGE attack, has the potential for use within treatment for diabetic vascular endothelial disorders.

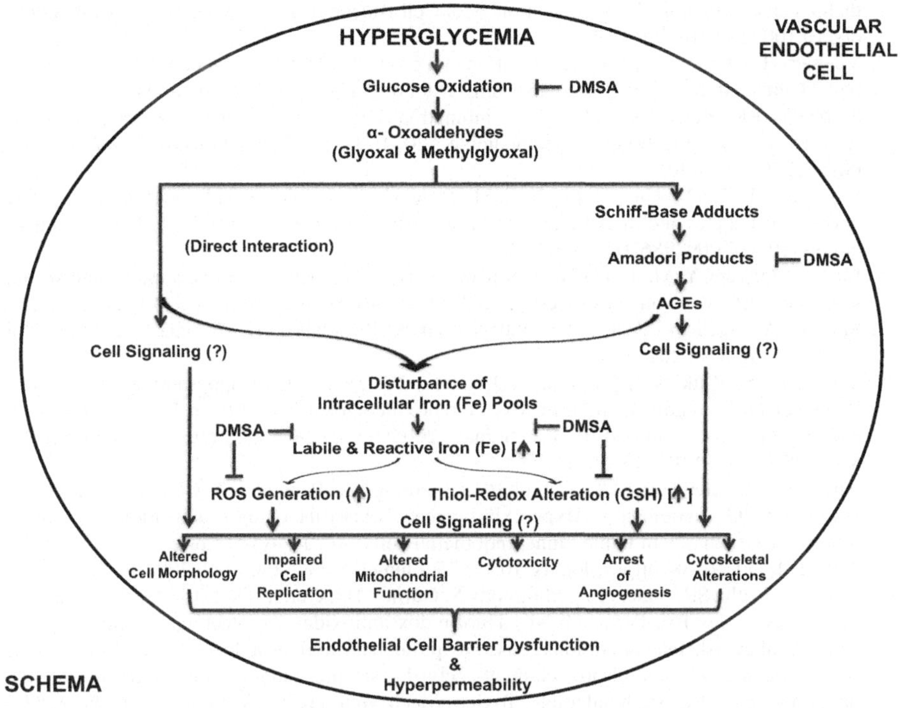

Fig. 18 Proposed mechanism of DMSA protection against glyoxal-induced cytotoxicity, cytoskeletal rearrangement, and barrier dysfunction through oxidative stress involving iron in vascular ECs. Hyperglycemia causes glucose oxidation leading to the formation of oxoaldehydes (e.g., glyoxal and methylglyoxal) which either (1) directly and/or (2) indirectly (through formation of Schiff-base adducts and AGEs) cause altered cell signaling and intracellular iron (Fe) pools. The altered intracellular iron (Fe) is redox-active and causes the generation of ROS and altered thiol-redox pools leading to elevated oxidative stress, which causes cytotoxicity, impaired cell division, loss of angiogenic potential, cytoskeletal alterations, EC barrier dysfunction, and paracellular hyperpermeability of the EC monolayer. The thiol-containing heavy metal chelating drug, DMSA, effectively protects against the hyperglycemic oxoaldehyde-induced vascular EC damage, cytoskeletal alterations, and barrier dysfunction suggesting the role of intracellular iron in mediation of cellular oxidant injury which could culminate into diabetic vascular dysfunction/damage

Acknowledgements Support from the Division of the Pulmonary, Critical Care, and Sleep Medicine, Department of Medicine, and the Heart & Lung Research Institute of the Ohio State University Wexner Medical Center is appreciated. Funding to Jing Zhao (1R01HL151513 and R01GM115389-01) and Thomas J. Hund (R01HL156652 and R01HL135096) are acknowledged.

References

1. Blake R, Trounce IA. Mitochondrial dysfunction and complications associated with diabetes. Biochim Biophys Acta. 2014;1840(4):1404–12.
2. Sliman SM, Eubank TD, Kotha SR, Kuppusamy ML, Sherwani SI, Butler ES, Kuppusamy P, Roy S, Marsh CB, Stern DM, Parinandi NL. Hyperglycemic oxoaldehyde, glyoxal, causes barrier dysfunction, cytoskeletal alterations, and inhibition of angiogenesis in vascular endo-

thelial cells: aminoguanidine protection. Mol Cell Biochem. 2010;333(1–2):9–26. https://doi.org/10.1007/s11010-009-0199-x.

3. Vlassara H, Uribarri J. Advanced glycation end products (AGE) and diabetes: cause, effect, or both? Curr Diab Rep. 2014;14(1):453. https://doi.org/10.1007/s11892-013-0453-1.
4. Prabhakar PK, Kumar A, Doble M. Combination therapy: a new strategy to manage diabetes and its complications. Phytomedicine. 2014;21(2):123–30. https://doi.org/10.1016/j.phymed.2013.08.020.
5. Liu Q, Sun L, Tan Y, Wang G, Lin X, Cai L. Role of iron deficiency and overload in the pathogenesis of diabetes and diabetic complications. Curr Med Chem. 2009;16(1):113–29. https://doi.org/10.2174/092986709787002862.
6. Liu J, Li Q, Yang Y, Ma L. Iron metabolism and type 2 diabetes mellitus: a meta-analysis and systematic review. J Diabetes Investig. 2020;11(4):946–55. https://doi.org/10.1111/jdi.13216.
7. Simcox JA, McClain DA. Iron and diabetes risk. Cell Metab. 2013;17(3):329–41. https://doi.org/10.1016/j.cmet.2013.02.007.
8. Varadharaj S, Watkins T, Cardounel AJ, Garcia JG, Zweier JL, Kuppusamy P, Natarajan V, Parinandi NL. Vitamin C-induced loss of redox-dependent viability in lung microvascular endothelial cells. Antioxid Redox Signal. 2005;7(1–2):287–300. https://doi.org/10.1089/ars.2005.7.287. PMID: 15650416.
9. Parinandi NL, Kleinberg MA, Usatyuk PV, Cummings RJ, Pennathur A, Cardounel AJ, Zweier JL, Garcia JG, Natarajan V. Hyperoxia-induced NAD(P)H oxidase activation and regulation by MAP kinases in human lung endothelial cells. Am J Physiol Lung Cell Mol Physiol. 2003;284(1):L26–38. https://doi.org/10.1152/ajplung.00123.2002. Epub 2002 Jul 26.
10. Patel RB, Kotha SR, Sauers LA, Malireddy S, Gurney TO, Gupta NN, Elton TS, Magalang UJ, Marsh CB, Haley BE, Parinandi NL. Thiol-redox antioxidants protect against lung vascular endothelial cytoskeletal alterations caused by pulmonary fibrosis inducer, bleomycin: comparison between classical thiol-protectant, N-acetyl-L-cysteine, and novel thiol antioxidant, *N,N'*-bis-2-mercaptoethyl isophthalamide. Toxicol Mech Methods. 2012;22(5):383–96. https://doi.org/10.3109/15376516.2012.673089.
11. Petrat F, Rauen U, de Groot H. Determination of the chelatable iron pool of isolated rat hepatocytes by digital fluorescence microscopy using the fluorescent probe, phen green SK. Hepatology. 1999;29(4):1171–9. https://doi.org/10.1002/hep.510290435.
12. Petrat F, de Groot H, Rauen U. Determination of the chelatable iron pool of single intact cells by laser scanning microscopy. Arch Biochem Biophys. 2000;376(1):74–81. https://doi.org/10.1006/abbi.2000.1711.
13. Nowotny K, Jung T, Höhn A, Weber D, Grune T. Advanced glycation end products and oxidative stress in type 2 diabetes mellitus. Biomol Ther. 2015;5(1):194–222. https://doi.org/10.3390/biom5010194.
14. Papachristoforou E, Lambadiari V, Maratou E, Makrilakis K. Association of glycemic indices (hyperglycemia, glucose variability, and hypoglycemia) with oxidative stress and diabetic complications. J Diabetes Res. 2020;2020:7489795. https://doi.org/10.1155/2020/7489795.
15. Shen CY, Lu CH, Wu CH, Li KJ, Kuo YM, Hsieh SC, Yu CL. The development of Maillard reaction, and advanced glycation end product (AGE)-receptor for AGE (RAGE) signaling inhibitors as novel therapeutic strategies for patients with AGE-related diseases. Molecules (Basel, Switzerland). 2020;25(23):5591. https://doi.org/10.3390/molecules25235591.
16. Tan KC, Chow WS, Ai VH, Metz C, Bucala R, Lam KS. Advanced glycation end products and endothelial dysfunction in type 2 diabetes. Diabetes Care. 2002;25(6):1055–9. https://doi.org/10.2337/diacare.25.6.1055.
17. Vulesevic B, McNeill B, Giacco F, Maeda K, Blackburn NJ, Brownlee M, Milne RW, Suuronen EJ. Methylglyoxal-induced endothelial cell loss and inflammation contribute to the development of diabetic cardiomyopathy. Diabetes. 2016;65(6):1699–713. https://doi.org/10.2337/db15-0568.

18. Sotokawauchi A, Matsui T, Higashimoto Y, Yamagishi SI. Fructose causes endothelial cell damage via activation of advanced glycation end products-receptor system. Diab Vasc Dis Res. 2019;16(6):556–61. https://doi.org/10.1177/1479164119866390.
19. Sun YP, Gu JF, Tan XB, Wang CF, Jia XB, Feng L, Liu JP. Curcumin inhibits advanced glycation end product-induced oxidative stress and inflammatory responses in endothelial cell damage via trapping methylglyoxal. Mol Med Rep. 2016;13(2):1475–86. https://doi.org/10.3892/mmr.2015.4725.
20. Mascolo E, Vernì F. Vitamin B6 and diabetes: relationship and molecular mechanisms. Int J Mol Sci. 2020;21(10):3669. https://doi.org/10.3390/ijms21103669.
21. Fernández-Real JM, McClain D, Manco M. Mechanisms linking glucose homeostasis and iron metabolism toward the onset and progression of type 2 diabetes. Diabetes Care. 2015;38(11):2169–76. https://doi.org/10.2337/dc14-3082.
22. Rajpathak SN, Crandall JP, Wylie-Rosett J, Kabat GC, Rohan TE, Hu FB. The role of iron in type 2 diabetes in humans. Biochim Biophys Acta. 2009;1790(7):671–81. https://doi.org/10.1016/j.bbagen.2008.04.005.
23. Swaminathan S, Fonseca VA, Alam MG, Shah SV. The role of iron in diabetes and its complications. Diabetes Care. 2007;30(7):1926–33. https://doi.org/10.2337/dc06-2625.
24. Beutler E, Hoffbrand AV, Cook JD. Iron deficiency and overload. Hematology Am Soc Hematol Educ Program. 2003;2003(1):40–61. https://doi.org/10.1182/asheducation-2003.1.40.
25. Zhang C, Rawal S. Dietary iron intake, iron status, and gestational diabetes. Am J Clin Nutr. 2017;106(Suppl 6):1672S–80S. https://doi.org/10.3945/ajcn.117.156034.
26. Soliman AT, De Sanctis V, Yassin M, Soliman N. Iron deficiency anemia and glucose metabolism. Acta Biomed. 2017;88(1):112–8. https://doi.org/10.23750/abm.v88i1.6049.
27. Mirlohi MS, Yaghooti H, Shirali S, Aminasnafi A, Olapour S. Increased levels of advanced glycation end products positively correlate with iron overload and oxidative stress markers in patients with β-thalassemia major. Ann Hematol. 2018;97(4):679–84. https://doi.org/10.1007/s00277-017-3223-3.
28. Misra R, Balagopal P, Raj S, Patel TG. Red meat consumption (heme iron intake) and risk for diabetes and comorbidities? Curr Diab Rep. 2018;18(11):100. https://doi.org/10.1007/s11892-018-1071-8.
29. Sears ME. Chelation: harnessing and enhancing heavy metal detoxification – a review. Sci World J. 2013;2013:219840. https://doi.org/10.1155/2013/219840.

Index

N. L. Parinandi, T. J. Hund (eds.), *Cardiovascular Signaling in Health and Disease*, https://doi.org/10.1007/978-3-031-08309-9

The manufacturer's authorised representative in the EU is Springer Nature Customer Service Centre GmbH, Europaplatz 3, 69115 Heidelberg, Germany. If you have any concerns regarding our products, please contact ProductSafety@springernature.com

Printed and bound by CPI Group (UK) Ltd, Croydon, CR0 4YY
28/09/2026
02224067-0007